ASTRONOMICAL DATA ANALYSIS SOFTWARE AND SYSTEMS XX

COVER ILLUSTRATION:

An X-ray image constructed from 3 stacked observations of M17, obtained by the *Chandra* X-ray Observatory. The yellow ellipses depict the size of the 90% encircled counts fraction of the *Chandra* point spread function at the locations of X-ray sources detected using the CIAO wavdetect tool. The sources are color-coded by source likelihood, computed using the maximum likelihood estimator developed using the Sherpa modeling and fitting engine.

Image created by J. Miller using ChIPS, and provided courtesy of D. Nguyen et al., this volume (page 517).

ASTRONOMICAL SOCIETY OF THE PACIFIC
CONFERENCE SERIES

A SERIES OF BOOKS ON RECENT DEVELOPMENTS IN ASTRONOMY AND ASTROPHYSICS

Volume 442

EDITORIAL STAFF

Managing Editor: Joseph Jensen
Associate Managing Editor: Jonathan Barnes
Publication Manager: Pepita Ridgeway
Editorial Assistant: Cindy Moody
LATEX Consultant: T. J. Mahoney

MS 179, Utah Valley University, 800 W. University Parkway, Orem, Utah 84058-5999
Phone: 801-863-8804   E-mail: aspcs@aspbooks.org
E-book site: http://www.aspbooks.org

PUBLICATION COMMITTEE

ASPCS volumes may be found online with color images at http://www.aspbooks.org.
ASP monographs may be found online at http://www.aspmonographs.org.

For a complete list of ASPCS volumes, ASP monographs, and
other ASP publications see http://www.astrosociety.org/pubs.html.

All book order and subscription inquiries should be directed to the ASP at
800-335-2626 (toll-free within the USA) or 415-337-2126,
or email service@astrosociety.org

ASTRONOMICAL SOCIETY OF THE PACIFIC
CONFERENCE SERIES

Volume 442

ASTRONOMICAL DATA ANALYSIS
SOFTWARE AND SYSTEMS XX

Proceedings of a Conference held at
Seaport World Trade Center, Boston, Massachusetts, USA
7–11 November 2010

Edited by

Ian N. Evans
Smithsonian Astrophysical Observatory, Cambridge, Massachusetts, USA

Alberto Accomazzi
Smithsonian Astrophysical Observatory, Cambridge, Massachusetts, USA

Douglas J. Mink
Smithsonian Astrophysical Observatory, Cambridge, Massachusetts, USA

Arnold H. Rots
Smithsonian Astrophysical Observatory, Cambridge, Massachusetts, USA

SAN FRANCISCO

ASTRONOMICAL SOCIETY OF THE PACIFIC
390 Ashton Avenue
San Francisco, California, 94112-1722, USA

Phone: 415-337-1100
Fax: 415-337-5205
E-mail: service@astrosociety.org
Web site: www.astrosociety.org
E-books: www.aspbooks.org

First Edition
© 2011 by Astronomical Society of the Pacific
ASP Conference Series
All rights reserved.

ISBN: 978-1-58381-764-3
e-book ISBN: 978-1-58381-765-0
ISSN: 1080-7926

Library of Congress (LOC) Cataloging in Publication (CIP) Data:
Main entry under title
Library of Congress Control Number (LCCN): 2011904255

Printed in the United States of America by Sheridan Books, Ann Arbor, Michigan.
This book is printed on acid-free paper.

Contents

Part I. Archives

Part VIII. Pipelines (Space-Based Instruments)

Part IX. Scientific Computing

Part XI. Solar Astronomy

Part XII. Virtual Observatory

Part XIII. Visualization

Preface

This volume of the Astronomical Society of the Pacific (ASP) Conference Series contains papers that were presented (and submitted) to the 20th annual conference on Astronomical Data Analysis Software and Systems (ADASS XX), which was held at the Seaport World Trade Center, Boston, Massachusetts, USA, on 7–11 November 2010. The ADASS XX conference was hosted by the Harvard-Smithsonian Center for Astrophysics (CfA), which is a collaboration between the Harvard College Observatory (HCO) of Harvard University and the Smithsonian Astrophysical Observatory (SAO) of the Smithsonian Institution.

1.1. Conference Overview

Boston is the capital of the US state of Massachusetts. Puritan colonists from England founded the Massachusetts Bay Colony on the site of the current city of Boston in the year 1630, just ten years after the arrival of the first Pilgrims in the new world. The city played a key role in the American Revolution, with tumultuous events such as the Boston Tea Party, the Battle of Bunker Hill, and the Siege of Boston taking place in the city and surrounding areas. Boston is home to the oldest university in the United States (Harvard) and also hosts the running of the world's oldest annual marathon (the Boston Marathon).

The conference commenced with a pair of well-attended tutorials that explored methods for Cross-Identifying Data Resources in the the Virtual Observatory (VO) Era, and Techniques for Time Domain Astronomy. The key topics for the ADASS XX conference were Education and Public Outreach, Scientific Computing, Large Observatory Challenges, Cross Catalog Matching, Solar Astronomy, and Grid and Grid Virtualization. These and a wide range of other topics were addressed by a complete program of 14 invited, 32 oral, and 126 poster presentations. The conference keynote speaker was Stephen Wolfram of Wolfram Research, who delivered a fascinating live demonstration and discourse on Making Scientific Data Computable.

Many attendees participated in one or more of five "Birds of a Feather" (BoF) sessions. The topics of this year's BoFs were IRAF and Beyond, FITS, Towards HDF5: Encapsulation of Large and/or Complex Astronomical Data, Astronomy Visualization for Education and Outreach, and an ADASS Publications BoF. The latter raised the question of whether there should be a journal for the astronomical computing community, as well as a lively discussion regarding the future of the ADASS conference proceedings book. Five floor demonstrations and an equal number of focus demonstrations rounded out the conference and provided attendees with the opportunity to learn about the latest tools and technologies in a live setting.

To encourage excellence in the broad areas of focus of the conference, Wolfram Research, developers of Mathematica and Wolfram Alpha, sponsored a competition for the best conference contribution by a student, the best general oral contribution, and the best poster contribution. The winners were Benjamin Barsdell (student), Andrew Connolly (oral), and Séverin Gaudet (poster). Each winner received a Mathematica license with 1 year of support.

1.2. Proceedings

As these are the proceedings for ADASS's banner 20th year, as there was a spirited discussion of options for the future of the proceedings at the meeting, and as there is a very visible change in their appearance, we thought it appropriate to devote some attention to the proceedings themselves in the preface.

For the first time, the proceedings are printed in full color. This is possible because of recent changes made at ASP. Also because of these changes, we expect to get this volume published and out the door roughly six months after the conference.

The previous volume presented a paper on the State of ADASS (Rots 2010) that touched on the relevance of the proceedings, and we can update its two tables with current information. Table 1 lists venues, attendance, and registration fees since 2000, while Table 2 provides statistics on presentations and proceedings papers since 2003.

Table 1.   ADASS Venues, Attendance, and Cost

ADASS	Year	Location	Attendance	Registration[a]
X	2000	Boston, MA, USA	287	$190
XI	2001	Victoria, BC, Canada	211	$325
XII	2002	Baltimore, MD, USA	291	$325
XIII	2003	Strasbourg, France[b]	363	$369[c]
XIV	2004	Pasadena, CA, USA	311	$325
XV	2005	El Escorial, Spain[b]	358	$363[c]
XVI	2006	Tucson, AZ, USA	293	$325[c]
XVII	2007	London, UK[b]	314	$350
XVIII	2008	Québec, Canada	283	$340
XIX	2009	Sapporo, Japan	211	$456
XX	2010	Boston, MA, USA	302	$550[d]
XXI	2011	Paris, France[e]		

[a]In US dollars
[b]Paired with IVOA fall Interop meeting
[c]Included lunches
[d]Included some lunches and banquet
[e]Future venue

As noted, there was discussion during the conference about the future and fate of the proceedings (see Lewis 2011), and of the publication of software-related papers in general (see Gray & Mann 2011). Drawing on those papers we would like to make a few observations, some of which are covered in the papers while others are not.

It was generally acknowledged that the proceedings provide a valuable opportunity for the community to publish software-related articles. We should emphasize that the discussion isn't so much about whether or not the conference proceedings should be published, nor about whether they should be published on-line, but is instead about whether they should also be printed and distributed in hardcopy form.

There is also the question about refereed publication of papers. The ADASS proceedings are not refereed, and there certainly is a legitimate need for some papers from the ADASS community to obtain that level of certification. But it is possible to have software papers published in refereed journals. Perhaps one may draw the distinction between publishing in ADASS proceedings on the one hand, and publishing in PASP

Table 2. ADASS Presentations and Papers

Year	Attend	FlD^a	FcD^b	BoF^c	Inv^d	Oral^e	Poster^f	Pap^g	%Pap^h	Cit^i	Cit/P^j
2003	363	12		4	15	23	185	194	81%	303	1.56
2004	311	12	10	5	15	22	157	138	62%	315	2.28
2005	358	9	4	5	14	31	167	178	77%	361	2.03
2006	293	7	3	7	13	35	131	162	83%	172	1.06
2007	314	14	3	4	13	39	157	166	72%	115	0.69
2008	283	10	7	6	12	33	131	123	62%	238	1.87
2009	211	6	2	2	12	32	90	117	81%	8	0.07
2010	302	5	5	5	15	32	126	152	81%		

^{a}Number of floor demos
^{b}Number of focus demos
^{c}Number of Birds of a Feather sessions
^{d}Number of invited presentations
^{e}Number of contributed oral presentations
^{f}Number of poster presentations
^{g}Number of papers in proceedings; also includes prefaces
  and some description of BoFs, Floor and Focus Demos
^{h}Percentage of presentations in proceedings
^{i}Total number of citations to proceeding papers, as of 2011 March 07
^{j}Average number of citations per paper, as of 2011 March 07

or A&A on the other, as being that "I have something *useful* to say" is appropriate for ADASS proceedings, while "I have something *new* and *important* to say" qualifies for the refereed journals.

Lewis (2011) reports that a majority, though not overwhelming, is in favor of dropping the hardcopies. The question is "what does this buy us?" It is perfectly clear that the task of the proceedings editors is not going to become easier — most of the work is in the correction of spelling, language, lay-out, references, and the construction of the various indices. That is where the overwhelming majority of the time to publish is spent — and we hope to prove with the publication of this volume that hardcover proceedings can be published in about six months. Eliminating the hardcopies would enable us to relax the page limits. These are currently set by the maximum number of pages that can physically be bound into a single volume. However, one should not expect dramatic changes. There is considerable merit in conciseness, and the editors would have to impose a limit on the length of each contribution. It might be a little longer than the equivalent of the current four pages for contributed papers, but probably not much. The inclusion of other media has raised as another advantage: they cannot be printed, admittedly, but there is nothing that prevents the authors from linking them to the on-line version.

Finally, there is the matter of cost. Nobody volunteered after the vote to say that (s)he would be happy to take on the responsibility for setting up the framework for online-only publication, curation, and preservation. It is easy to say that the proceedings will be published only on-line, and there is an unspoken suggestion that that will be much more cost-effective. However, if no one can commit to making this work for, say, the next 30 years, we do not that that it is a realistic option. In reality, most of the cost (just like the effort) is not specific to the production of the hardcopy volumes. Instead, the cost is in publishing, curation, and preservation.

So we suspect that, even if we drop the hardcopy proceedings, we will still want to contract with a publisher to publish, curate, and preserve the on-line proceedings. The difference in cost was at one point calculated to be of the order of $5,000. A large sum? That amounts to ~ $17 per attendee. More to the point, one should try to do a full-cost calculation for an ADASS conference, including venue costs, food, LOC and POC labor, lodging, travel, and time of all participants, preparation of manuscripts, editors' labor, on-line publishing cost, and, yes, printing and mailing the hardcopies. When all of the costs are tallied, the last item amounts to no more than 0.25% of the total.

1.3.   Organizing Committees and Sponsors

The Program Organizing Committee (POC) responsible for the content of the ADASS XX conference consisted of Carlos Gabriel (ESA-ESAC, Spain; Chair), Pascal Ballester (ESO, Germany), David Barnes (Swinburne University of Technology, Australia), Daniel Durand (NRC, Canada), Daniel Egret (Observatoire de Paris, France), Tony Krueger (STScI, USA), Deborah Levine (IPAC, USA), Jim Lewis (University of Cambridge, UK), François Ochsenbein (CDS, France), Ray Plante (NCSA, USA), Nuria Lorente (NRAO, USA), Arnold Rots (SAO, USA), Betty Stobie (NOAO, USA), Tadafumi Takata (NAO, Japan), and Christian Veillet (CFHT, USA).

The Local Organizing Committee (LOC) consisted of Arnold Rots (SAO; Chair), Alberto Accomazzi (SAO), Ian Evans (SAO), Douglas Mink (SAO), Pavlos Protopapas (Harvard University), and Sherry Winkelman (SAO).

The LOC members want to thank especially the local administrative and technical support staff who worked so diligently to ensure that the ADASS XX conference was well organized and ran smoothly. In particular, we wish to acknowledge the efforts of Patricia Buckley, Paul Grant, Raymond Hemond, Michelle Henson, Debra Nickerson, Frank Nigro, Lisa Paton, and Susan Tuttle. Finally, we want to thank the many volunteers who facilitated the conference operations by providing support for the tutorial presentations, demos, BoFs, and poster room setup, manning the registration and copyright desks and the speaker ready room, and performing microphone runner duties.

The organizers wish to express their deep appreciation to the conference host institution, sponsors, and supporting organizations. The ADASS XX conference was sponsored by Wolfram Research, NetApp, and the *Hinode* X-Ray Telescope (XRT) science team. The ADASS XX conference was supported by the *Chandra* X-ray Center (CXC), Canada-France-Hawaii Telescope (CFHT), Centre de Données astronomiques de Strasbourg (CDS), European Space Agency (ESA), European Southern Observatory (ESO), Infrared Processing and Analysis Center (IPAC), National Astronomical Observatory of Japan (NAOJ), National Center for Supercomputing Applications (NCSA), National Optical Astronomy Observatory (NOAO), National Radio Astronomy Observatory (NRAO), and Space Telescope Science Institure (STScI). The generous contributions of the host institution, sponsors, and supporting organizations helped to keep costs manageable, which in turn kept conference fees modest, and maximized the value of the conference.

1.4. Further Information

ADASS XXI will be help in Paris, France from 6–10 November 2011. Details concerning this and all of the ADASS meetings, as well as electronic versions of many of the proceedings, may be found on the ADASS web site: http://www.adass.org.

References

Gray, N., & Mann, R. G. 2011, in Astronomical Data Analysis Software and Systems XX, edited by I. N. Evans, A. Accomazzi, D. J. Mink, & A. H. Rots (San Francisco, CA: ASP), vol. 442 of ASP Conf. Ser., 655
Lewis, J. R. 2011, in Astronomical Data Analysis Software and Systems XX, edited by I. N. Evans, A. Accomazzi, D. J. Mink, & A. H. Rots (San Francisco, CA: ASP), vol. 442 of ASP Conf. Ser., xxiii
Rots, A. H. 2010, in Astronomical Data Analysis Software and Systems XIX, edited by Y. Mizumoto, K.-I. Morita, & M. Ohishi (San Francisco, CA: ASP), vol. 434 of ASP Conf. Ser., xiv

Ian N. Evans
Alberto Accomazzi
Douglas J. Mink
Arnold H. Rots
Smithsonian Astrophysical Observatory
Cambridge, MA, USA
The ADASS XX Proceedings editors, March 2011

Astronomical Data Analysis Software and Systems XX
ASP Conference Series, Vol. 442
Ian N. Evans, Alberto Accomazzi, Douglas J. Mink, and Arnold H. Rots, eds.
© *2011 Astronomical Society of the Pacific*

The Future of ADASS Proceedings

James R. Lewis

Cambridge Astronomy Survey Unit, University of Cambridge, UK

Abstract. The 20 years that ADASS has been running has seen amazing leaps in our ability to disseminate information around the globe. In 2009 a sub-committee of the POC was set up to investigate how we publish the proceedings of this conference. This paper primarily is a summary of the general introduction to the problem that was given on the Monday of ADASS XX. A BoF on Monday night was used to discuss the problem in more depth. Finally on Wednesday a vote was taken during one of the plenary sessions to gauge the feelings of the ADASS community at large.

1. Background

The ADASS conference is twenty years old this year. Every year we publish a record of what was discussed at ADASS, the so-called 'conference proceedings'. This is not only meant as a historical record of the progress of our various disciplines. It also serves as a memory-aid for all of the attendees as well as a means to catch-up for those who have missed the conference. This is a model adopted by most science conferences, and in the case of ADASS, we publish our proceedings in the form of a book in the Astronomical Society of the Pacific Conference Series. A copy of this book goes to each participant as well as to any subscribing organization. Additional copies are also put on sale to any who would like one.

We at ADASS are particularly fortunate to be living in a very interesting time in terms of the advances in technology used to disseminate information. Some of the major changes that have taken place in the past twenty years include:

The Internet Our whole lives have been transformed by the rise of the internet.

Fast Search Engines By providing simple and fast access to the vast quantities of information available on the internet, search engines such as Google are probably the single thing that has contributed most the internet's success.

Inexpensive Publishing Tools When I say 'inexpensive' I sometimes mean 'free'. It is now possible for everyone to produce professional looking publications.

Electronic Readers By this I mean items such as the Amazon Kindle and the iPad from Apple. These are not that prevalent yet, but they soon will be. Their simple user interfaces are making reading electronic documents a much more pleasant experience than one gets from staring at a computer screen.

These innovations have helped lead to an explosion of information which is now available at the touch of a button. As a result, more and more people are opting to get

---

their information electronically, rather than through printed media. This is evidenced by the current downturn in business in traditionally printed news media such as newspapers and magazines.

The format of a traditional hardback book does have some limitations in respect to the ADASS proceedings. Many of these problems have been around for a long time and it's now with our experience of electronic media that some are becoming truely annoying.

- There is a limit to the number of pages that can be accommodated in a given book in the ASP format. This is simply a limitation to the physical size of the binding. In order that as many articles can be printed as possible, it is necessary to restrict the length of articles to 4 pages for contributed talks or posters and 10 pages for invited talks. This includes any and all graphics. With such small page limits a great deal of what was originally presented has to be cut.

- The graphical content is restricted to that which can be accomodated on a single page of a book. Any detailed graphics can only be included if, when reduced, the salient information is still legible. It is, of course, impossible to include any animation or movie graphics that may have been presented during the talk. Until this year, no color graphics was allowed either.

- We are as readers becoming more used to having the functionality of electronic readers and web browsers available. These include the ability to pan, zoom and change fonts. Especially important for reading scientific articles is the ability to link articles together as one does on a well designed web page.

- Editors have a huge task working with publishers and authors. The main task they have with respect to the latter is to ensure that page limits are adhered to and that any graphics that are included look right when they are reduced. No matter how the proceedings are published, the job of editor will never be easy, though.

- The proceedings are expensive, usually costing approximately $20,000. This covers the cost to the publishers for printing and distribution.

- Probably the most worrying aspect of producing proceedings for ADASS is that the process is very slow. It is usual these days that the proceedings will take a year or more to appear. We work in a discipline that is moving very quickly and by the time a proceedings volume reaches the public much of the information in it is badly out of date. To my mind this seriously devalues the proceedings as a research tool.

2. Solutions?

In response to these concerns the Program Organizing Committee of ADASS during its meeting at ADASS 2009 created a subcommittee to look into the question of whether the way that the proceedings are published needs to be changed. The committee consists of Daniel Durand, Carlos Gabriel, Francois Ochsenbein, Ray Plante, and the chair, James Lewis.

The following options were considered by this subcommittee:

- The 'status quo' — that is continue publishing with ASP, exactly as we always have. This has the advantage that we have a good and long standing relationship with ASP and they are quite accomodating.

- Continue to produce a book, but with a publisher who is faster and cheaper. We made some enquiries in this area, but the savings in time and money usually turned out to be minimal.

- Continue to produce a book, but try to negotiate an open access policy with ASP. Such a policy would allow the individual proceedings articles to be made available electronically before the book was published, hence allowing faster access to the results of the conference. During the course of the committee's negotiations with ASP, the latter was very reluctant to offer open access as it was felt that it would discourage people from buying the book. This might mean that ASP would be unable to recover the cost of producing it. This year, after long discussions with the LOC, ASP has been convinced to provide open access to the proceedings so long as ADASS can guarantee that it will buy at least 250 volumes.

- Stop producing a book altogether and publish the proceedings with an electronic publisher.

- Stop producing a book altogether and publish the proceedings electronically ourselves.

The concept of abandoning the book format and going with electronic-only publication is a little radical, but there are some obvious and immediate advantages. I feel that the content of the proceedings would be enhanced greatly in an electronic format. Page limits could be relaxed significantly (although not scrapped altogether as we don't necessarily want to encourage bloated articles), and there would be a much greater scope for including a wide variety of graphics in the contributions. The ability to link articles and references would aid the proceedings' use as a research tool. Publishing electronically would also undoubtedly be much faster. Apart from the time spent on physically printing, binding and distributing the finished book, there is the issue of the great deal of the time spent by the authors and editors alike which is tied up in trying to make the content of articles conform to what can be accomodated in a book. Finally an electronic publication is more environmentally friendly than printing.

All these advantages notwithstanding, the idea of electronic proceedings is more than a little scary. This really is the 'devil we don't know' and whether we go with an established publisher or do it ourselves, there will be a learning curve to be conquered. It is very true that for a large percentage of ADASS attendees, the proceedings is one of the only places where their work can be published. We are all only too aware that anyone can publish the most incredible nonsense on the Internet and that has led to something of an image problem for the Internet in some quarters. This begs the question of whether this kind of predjudice would lower the value of the ADASS proceedings in the eyes of the scientific public or indeed in the eyes of potential employers. The biggest worry for me with electronic-only publication is whether we can be assured of the continued long term existence and curation of the ADASS proceedings even if the ADASS conference itself eventually halts. We all know that books last centuries, if not millenia — can any of us say that that will be true of a website or its content? Allowing the proceedings to go into a parlous state would be a great disservice to the many people who rely on their contributions to enhance their future job prospects.

3. BoF

The publication subcommittee decided that it could not make such an important decision with such far-reaching consequences without consulting the ADASS community at large. In an attempt to do this we decided to to hold a 'Birds of a Feather' session at ADASS 2010 where this could be discussed. This was held on the Monday night of the conference and was combined with a second discussion on the question of setting up a new journal.

To start the BoF, Ray Plante spoke about the software tools he developed to edit the proceedings. He also reviewed a website for the International Journal of Data Curation,[1] which is a fully electronic journal made using free software from the Public Knowledge Project.[2] He felt that the quality of the publicly available software was high and that the amount of technical effort required to move to an electronic-only model would be small.

Arnold Rots then gave a brief presentation outlining his opposition to abandoning paper publication. He felt that although there are certainly advantages to electronic publication, the ADASS proceedings provide an historical record of our conference and a unique publishing opportunity for some of our attendees. We are currently unable to say how moving to an electronic model will affect the long term availability of the proceedings. This is could be a real problem if either an electronic publisher goes out of business or the ADASS conference itself ceases.

Most people in the room felt frustration at the limitations imposed by the books as outlined earlier and a majority felt that an electronic proceedings would be highly desireable. However, the enthusiasm of many was seriously dampened by concern for the curation issue.

4. A Vote

During a plenary session on Wednesday I reviewed what had been discussed in the BoF. In the end I offered two possibilities for how ADASS should proceed with this issue. First was to continue with ASP and see if the promised improvements materialise but, at the same time, continue to monitor developments in the electronic publishing world. The second was to go ahead and move to electronic publication now. A vote by all present showed a majority in favour of the latter choice, but not with a significant enough majority to be a real mandate for change. So for the time being, paper proceedings will continue to be produced, but the publications subcommittee will continue to research the curation issues highlighted here.

---

[1]http://www.ijdc.net/index.php/ijdc

[2]http://pkp.sfu.ca/

Participants

A. Accomazzi, Smithsonian Astrophysical Observatory, 60 Garden Street, MS-83, Cambridge, MA 02138, USA

C. Alcock, Smithsonian Astrophysical Observatory, 60 Garden Street, MS-45, Cambridge, MA 02138, USA

T. Aldcroft, Smithsonian Astrophysical Observatory, 60 Garden Street, MS-70, Cambridge, MA 02138, USA

A. Alexov, University of Amsterdam, Astronomical Institute 'Anton Pannekoek', Postbus 94249, Amsterdam, 1090 GE, Netherlands

C. Anderson, Smithsonian Astrophysical Observatory, 60 Garden Street, MS-34, Cambridge, MA 02138, USA

K. Anderson, Instituut Anton Pannekoek, Universiteit van Amsterdam, Nieuwezijds Voorburgwal 258, Amsterdam, 1012RT, Netherlands

M. Andrecut, ISIS, University of Calgary, 2500 University Drive, Calgary, Alberta, T2N 1N4, Canada

T. Aoki, Waseda University, 1-6-1 Nishiwaseda, Shinjuku-ku, Tokyo, 169-8050, Japan

K. Arcand, Smithsonian Astrophysical Observatory, 60 Garden Street, MS-70, Cambridge, MA 02138, USA

C. Arviset, ESA-ESAC, POBOX 78, Villanueva Canada, 28691 Madrid, Spain

C. Baffa, INAF Osservatorio di Arcetri, Largo E. Fermi 5, I-50125 Firenze, Italy

C. Bailer-Jones, Max Planck Institute for Astronomy, Heidelberg, Koenigstuhl 17, 69117 Heidelberg, Germany

P. Ballester, European Southern Observatory, Karl-Schwarzshild-Str. 2, 85748 Garching bei München, Germany

S. Bamford, University of Nottingham, School of Physics and Astronomy, University Park, Nottingham, NG7 2RD, United Kingdom

K. Banse, European Southern Observatory, Karl-Schwarzshild-Str. 2, 85748 Garching bei München, Germany

S. Bardeau, IRAM, 300 rue de la piscine, 38400 Saint Martin d'Heres, France

I. Barg, National Optical Astronomy Observatory, 950 North Cherry Ave., Tucson, AZ 85719, USA

B. Barsdell, Swinburne University of Technology, 91 Jolimont Rd., Forest Hill, VIC, 3131, Australia

U. Becciani, INAF - Catania, Via S. Sofia 78, I95125 Catania, Italy

G. Becker, Smithsonian Astrophysical Observatory, 60 Garden Street, MS-67, Cambridge, MA 02138, USA

J. Becla, SLAC National Accelerator Laboratory, 2575 Sand Hill Road, M/S 97, Menlo Park, CA 94025, USA

I. Bernst, 1.Physikalisches Institut, University of Cologne, Zuelpicher Strasse 77, 50937 Cologne, Germany

E. Bertin, IAP, 98bis, bd Arago, F-75014 Paris, France

T. Blanchard, NetApp, 168 Highland St., Portsmouth, NH 03801, USA

S. Block, Smithsonian Astrophysical Observatory, 60 Garden Street, MS-42, Cambridge, MA 02138, USA

T. Boch, CDS, Observatoire de Strasbourg, 11 rue de l'Université, Strasbourg, France

J. Bogart, SLAC National Accelerator Laboratory, 2575 Sand Hill Road, Mail Stop 71, Menlo Park, CA 94025, USA

N. Bonaventura, Smithsonian Astrophysical Observatory, 60 Garden Street, MS-21, Cambridge, MA 02138, USA

M. Borkin, Harvard University, 60 Garden Street, MS-42, Cambridge, MA 02138, USA

D. Broguière, IRAM, 300 rue de la piscine, 38400 Saint Martin d'Heres, France

T. Budavári, Johns Hopkins University, 3400 N. Charles Street, Baltimore, MD 21218, USA

R. Burgon, The Open University, UK, PSSRI, The Open University, Walton Hall, Milton Keynes, Buckinghamshire, MK7 6AA, United Kingdom

D. Burke, Smithsonian Astrophysical Observatory, 60 Garden Street, MS-02, Cambridge, MA 02138, USA

H. Bushouse, Space Telescope Science Institute, 3700 San Martin Drive, Baltimore, MD 21218, USA

N. Caon, IAC, C/ Via Lactea s/n, La Laguna, 38200 Tenerife, Spain

N. Cardiel, Universidad Complutense de Madrid, Dept. Astrofisica, Fac. Fisicas, Avenida Complutense s/n, 28040 Madrid, Spain

M. T. Ceballos, Instituto de Física de Cantabria (CSIC-UC), Avda Los Castros s/n, Santander, 39005 Cantabria, Spain

P.-Y. Chabaud, CNRS/LAM CeSAM, Technopole Chateau Gombert, 38 rue Frédéric Joliot-Curie, 13013 Marseille, France

D. Chance, Space Telescope Science Institute, 3700 San Martin Drive, Baltimore, MD 21218, USA

M. Charcos-LLorens, Universities Space Research Association, NASA Ames Research Center, Mail Stop N211-3, Room 223, Moffet Field, CA 95035, USA

J. Chen, Smithsonian Astrophysical Observatory, 60 Garden Street, MS-81, Cambridge, MA 02138, USA

I. Chilingarian, SAI MSU / Observatoire de Strasbourg, 13 Universitetsky prospect, Moscow, 119992, Russia

M. Clark, National Radio Astronomy Observatory, PO Box 2, Green Bank, WV 24944, USA

M. Clarke, Gemini North Observatory, 670 N. A'ohoku Place, Hilo, HI 96720, USA

A. Connolly, University of Washington, Box 351580, Seattle, WA 98195, USA

A. Connors, Eureka Scientific, 46 Park Street, Arlington, MA 02474, USA

M. Conroy, Smithsonian Astrophysical Observatory, 60 Garden Street, MS-19, Cambridge, MA 02138, USA

S. Conseil, CNRS/LAM CeSAM, Technopole Chateau Gombert, 38 rue Frédéric Joliot-Curie, 13013 Marseille, France

M. Cornell, McDonald Observatory, The University of Texas at Austin, 1 University Station C1402, Austin, TX 78712-0259, USA

A. Costa, INAF Catania Astrophysical Observatory, Via S. Sofia 78, 95123 Catania, Italy

B. Cowan, National Solar Observatory, 950 N. Cherry Ave, Tucson, AZ 85726, USA

M. Cresitello-Dittmar, Smithsonian Astrophysical Observatory, 60 Garden Street, MS-81, Cambridge, MA 02138, USA

N. Cross, Institute for Astronomy, Edinburgh, Blackford Hill, Edinburgh, EH9 3HJ, United Kingdom

A. Csillaghy, University od Applied Sciences i4Ds, Steinackerstrasse 5, Windisch, 5210, Switzerland

L. De Bilbao, IDOM, Avda. Lehendakari Aguirre 3, 48014 Bilbao Spain

S. De Castro, Rhea System SA, Avenue Einstein 2a, Louvain-La-Lneuve, 1348, Belgium

J.-P. De Cuyper, Royal Observatory of Belgium, Ringlaan 3, Ukkel, B1180, Belgium

A. Deep, Leiden Observatory, PO Box 9513, Leiden, ZH, 2300 RA, Netherlands

J. De Jong, Max Planck Institute for extraterrestial Physics (MPE), Giessenbachstrasse 1, 85748 Garching bei München, Germany

A. Delgado, European Southern Observatory, Karl-Schwarzschild-Str. 2, 85748 Garching bei München, Germany

N. Dencheva, Space Telescope Science Institute, 3700 San Martin Drive, Baltimore, MD 21218, USA

J. DePasquale, Smithsonian Astrophysical Observatory, 60 Garden Street, MS-67, Cambridge, MA 02138, USA

S. Derriere, CDS, Observatoire de Strasbourg, CNRS - Observatoire astronomique, 11 rue de l'Université, 67000 Strasbourg, France

E. Deul, Leiden Observatory, P.O. Box 9513, Leiden, 2300 RA, Netherlands

S. Doe, Smithsonian Astrophysical Observatory, 60 Garden Street, MS-81, Cambridge, MA 02138, USA

T. Donaldson, Space Telescope Science Institute, 3700 San Martin Drive, Baltimore, MD 21218, USA

B. Dorner, CRAL - Observatoire de Lyon, 9, avenue Charles Andre, 69230 Saint Genis Laval, France

T. Dower, Space Telescope Science Institute, 3700 San Martin Drive, Baltimore, MD 21218, USA

P. Dowler, National Research Council Canada, 5071 West Saanich Road, Victoria, BC, V9E 2M7, Canada

M. Droettboom, Space Telescope Science Institute, 3700 San Martin Drive, Baltimore, MD 21218, USA

R. Dubois, SLAC National Accelerator Laboratory, 2575 Sand Hill Rd, Menlo Park, CA 94025, USA

G. Dubois-Felsmann, SLAC / LSST, SLAC, MS 29, 2575 Sand Hill Road, Menlo Park, CA 94025, USA

K. DuPrie, CGSJ, Ester Minami Azabu 501, 2-9-12 Minami Azabu, Minato-ku, Tokyo, 106-0047, Japan

D. Durand, National Research Council Canada, 5071 W. Saanich Rd., Victoria, BC, V9A2Z5, Canada

G. Duvert, JMMC/LAOG/Obs. Grenoble, 414 rue de la Piscine, Domaine Universitaire, F-38410 Saint-Martin d'Hères, France

F. Economou, JAC, 660 N Aohoku Pl, Hilo, HI 96720, USA

C. Erdmann, Smithsonian Astrophysical Observatory, 60 Garden Street, MS-56, Cambridge, MA 02138, USA

I. Evans, Smithsonian Astrophysical Observatory, 60 Garden Street, MS-81, Cambridge, MA 02138, USA

J. Evans, Smithsonian Astrophysical Observatory, 60 Garden Street, MS-81, Cambridge, MA 02138, USA

G. Fabbiano, Smithsonian Astrophysical Observatory, 60 Garden Street, MS-06, Cambridge, MA 02138, USA

N. Fajersztejn, ESAC (European Space Astronomy Center), Urb. Villafranca del Castillo, Madrid, Spain

J. Fay, Microsoft Research, 19007 NE 132nd Street, Woodinville, WA 98077, USA

P. Federl, University of Calgary, 2500 University Dr NW, Calgary, Alberta, T2N1N4, Canada

T. Fenouillet, CNRS/LAM CeSAM, Technopole Chateau Gombert, 38 rue Frédéric Joliot-Curie, 13013 Marseille, France

M. Fitzpatrick, National Optical Astronomy Observatory, 950 N. Cherry Ave, Tucson, AZ 85719, USA

M. Folk, The HDF Group, 1901 So. First St., Suite C-2, Champaign, IL 61820, USA

V. Forchi, European Southern Observatory, Karl-Schwarzshild-Str. 2, 85748 Garching bei München, Germany

J. Freixenet, University of Girona VAT:ESQ6750002E, Campus Montilivi, P4, E17071 Girona, Spain

C. Gabriel, ESA / ESAC, Puerto Rico 8A, Madrid, 28016 Madrid, Spain

E. Galle, Smithsonian Astrophysical Observatory, 60 Garden Street, MS-21, Cambridge, MA 02138, USA

P. García-Lario, Herschel Science Center - ESAC/ESA, P.O. Box 78, Villafranca del Castillo, 28691 Madrid, Spain

R. Gastaud, CEA DAPNIA IRFU SEDI, CEA Saclay, IRFU Sédi, 91191 Gif-sur-Yvette, France

S. GAUDET, NRC/HIA/CADC, 2225 Allenby St., 5071 West Saanich Rd., Victoria, BC, V8R 3C4, Canada

W. GENTZSCH, DEISA, An der Pirkacher Breite 11, 93073 Neutraubling, Germany

C. GHELLER, CINECA, Via Magnanelli 6/3, Casalecchio di Reno, I-40033 Bologna, Italy

E. GIANI, INAF Osservatorio di Arcetri, Largo E. Fermi 5, I-50125 Firenze, Italy

D. GIBBS, Smithsonian Astrophysical Observatory, 60 Garden Street, MS-81, Cambridge, MA 02138, USA

R. GIBSON, University of Washington, Box 351580 UW, Seattle, WA 98195, USA

S. GIMENEZ, CNRS/LAM CeSAM, Technopole Chateau Gombert, 38 rue Frédéric Joliot-Curie, 13013 Marseille, France

K. GLOTFELTY, Smithsonian Astrophysical Observatory, 60 Garden Street, MS-81, Cambridge, MA 02138, USA

P. GÓMEZ-ALVAREZ, INSA, ESA, European Space Astronomy Center (ESAC), P.O. Box 78, V. de la Cañada, 28691 Madrid, Spain

A. GOODMAN, Harvard-Smithsonian Center for Astrophysics, 60 Garden Street, MS-42, Cambridge, MA 02138, USA

A. GOPU, Indiana University Pervasive Technology Institute, 2719 E 10th Street, Bloomington, IN 47401, USA

E. GOTTSCHALK, Fermilab, Fermilab, MS-122, Batavia, IL 60510-0500, USA

D. GRAESSLE, Smithsonian Astrophysical Observatory, 60 Garden Street, MS-70, Cambridge, MA 02138, USA

Y. GRANET, CNRS/LAM CeSAM, Technopole Chateau Gombert, 38 rue Frédéric Joliot-Curie, 13013 Marseille, France

N. GRAY, University of Glasgow, UK, Physics and Astronomy, University of Glasgow, Glasgow, G12 8QQ, United Kingdom

G. GREENE, Space Telescope Science Institute, 3700 San Martin Drive, Baltimore, MD 21218, USA

P. GREENFIELD, Space Telescope Science Institute, 3700 San Martin Drive, Baltimore, MD 21218, USA

J. GRIER, Smithsonian Astrophysical Observatory, 60 Garden Street, MS-81, Cambridge, MA 02138, USA

S. GROOM, IPAC / Caltech, 1200 E. California Blvd, MC100-22, Pasadena, CA 91125, USA

D. GRUMM, Space Telescope Science Institute, 3700 San Martin Drive, Baltimore, MD 21218, USA

S. GURAM, University of Calgary, Dept of Physics and Astronomy, 2500 University Dr NW, Calgary, Alberta, T2N1N4, Canada

J. HAASE, ST-ECF, Karl-Schwarzschild-Str. 2, 85748 Garching bei München, Germany

B. HAEUSSLER, University of Nottingham, School of Physics & Astronomy, Univ. of Nottingham, University Park, Nottingham, NG7 2RD, United Kingdom

R. HAIN, Smithsonian Astrophysical Observatory, 60 Garden Street, MS-81, Cambridge, MA 02138, USA

D. HALL, Smithsonian Astrophysical Observatory, 60 Garden Street, MS-81, Cambridge, MA 02138, USA

T. HAMADA, Nagasaki Advanced Computing Center, Nagasaki University, 1-14 Bunkyo-machi, Nagasaki, 852-8521, Japan

R. HANISCH, Virtual Astronomical Observatory, Suite 730, 1400 16th St. NW, Washington, DC 20036, USA

R. HANUSCHIK, European Southern Observatory, Karl-Schwarzschild-Str. 2, 85748 Garching bei München, Germany

P. HARBO, Smithsonian Astrophysical Observatory, 60 Garden Street, MS-81, Cambridge, MA 02138, USA

A. HASSAN, Swinburne University of Technology, Center for Astrophysics and Supercomputing, PO Box 218, Hawthorne, VIC, 3221, Australia

H. HE, Smithsonian Astrophysical Observatory, 60 Garden Street, MS-81, Cambridge, MA 02138, USA

J. L. HERNANDEZ MUNOZ, ESAC, European Space Agency, European Space Astronomy Center, Urbanizacion Villafranca del Castillo, 28692 Madrid, Spain

P. HIRST, Gemini Observatory, 670 N. Aohoku Pl, Hilo, HI 96720, USA

P. HODGE, Space Telescope Science Institute, 3700 San Martin Drive, Baltimore, MD 21218, USA

M. HOLLIMAN, Institute for Astronomy, University of Edinburgh, 1c Maxwell St, Flat 2f2, Edinburgh, EH10 5HT, United Kingdom

R. HOOK, ESO / ST-ECF, Karl-Schwarzschild-Str. 2, 85748 Garching bei München, Germany

W. HOVEST, Max Planck Institut f. Astrophysik, Karl-Schwarzschild-Str. 1, 85741 Garching bei München, Germany

A. L. IBARRA IBAIBARRIAGA, XMM-Newton SOC (ESAC/ESA), Camino bajo del Castillo, s/n., Villanueva de la Cañada, 28692 Madrid, Spain

M. IRWIN, Institute of Astronomy, Madingley Road, Cambridge, United Kingdom

T. JENNESS, Joint Astronomy Center, 660 N. A'ohoku Place, Hilo, HI 96720, USA

C. JOHNSON, Space Telescope Science Institute, 3700 San Martin Drive, Baltimore, MD 21218, USA

W. JOYE, Smithsonian Astrophysical Observatory, 60 Garden Street, MS-67, Cambridge, MA 02138, USA

Y. JUNG, European Southern Observatory, Karl-Schwarzshild-Str. 2, 85748 Garching bei München, Germany

M. KAROVSKA, Smithsonian Astrophysical Observatory, 60 Garden Street, MS-70, Cambridge, MA 02138, USA

I. KATKOV, Sternberg Astronomical Institute (SAI MSU), Universitetsky pr., 13, Kaluga, 248032, L.Tolstoy, 3, 90, Moscow, 119992, Russia

H. KELLY, SLAC National Accelerator Laboratory, 2 Mead Drive, Brookline, NH 03033, USA

B. KENT, National Radio Astronomy Observatory, 520 Edgemont Road, Charlottesville, VA 22903, USA

A. KHALATYAN, AIP, An der Sternwarte 16, Potsdam, 14482 Brandenburg, Germany

C. KIDDLE, University of Calgary, Department of Computer Science, 2500 University Drive NW, Calgary, Alberta, T2N1N4, Canada

D.-W. (DAE-WON) KIM, Harvard-Smithsonian Center for Astrophysics, 60 Garden Street, MS-67, Cambridge, MA 02138, USA

D.-W. (DONG-WOO) KIM, Smithsonian Astrophysical Observatory, 60 Garden Street, MS-06, Cambridge, MA 02138, USA

C. KNAPIC, INAF OATs, via G.B. Tiepolo 11, 34143 Trieste, Italy

Y. KOMIYA, NAOJ, 2-21-1, Osawa, Mitaka, Tokyo, Mitaka, Tokyo, 181-8588, Japan

S. KOUZUMA, Chukyo University, 101 Tokodate, Kaizu-cho, Toyota, Aichi, 470-0348, Japan

D. KOZIKOWSKI, , 3 Ridgeway Terrace, Newton, MA 02461-1331, USA

M. KÜMMEL, ST-ECF, Karl-Schwarzschild-Str. 2, 85748 Garching bei München, Germany

M. KURTZ, Smithsonian Astrophysical Observatory, 60 Garden Street, MS-20, Cambridge, MA 02138, USA

M. KYPRIANOU, Space Telescope Science Institute, 3700 San Martin Drive, Baltimore, MD 21218, USA

K. LABRIE, Gemini Observatory - Hilo, 670 N. A'ohoku Pl., Hilo, HI 96720, USA

M. LACY, NAASC/NRAO, 520 Edgemont Road, Charlottesville, VA 22903, USA

U. LAMMERS, European Space Agency, P.O. Box 78, Villanueva de Canada, 28691 Madrid, Spain

V. LATTANZI, Harvard-Smithsonian Center for Astrophysics, 60 Garden Street, MS-72, Cambridge, MA 02138, USA

W. LATTER, NHSC/Caltech, MS100-22, Caltech, Pasadena, CA 91125, USA

O. LAURINO, Smithsonian Astrophysical Observatory, 60 Garden Street, MS-81, Cambridge, MA 02138, USA

M. LEE, Wentworth Institute of Technology, 15 Shepherd Ave, Apt 1, Boston, MA 02115, USA

N. LEE, Smithsonian Astrophysical Observatory, 60 Garden Street, MS-21, Cambridge, MA 02138, USA

J.-P. LE FEVRE, CEA Irfu, Saclay, 91191 Gif-sur-Yvette, France

S. LEON TANNE, ALMA (JAO), Alonso de Cordova, 3107, Vitacura, Santiago, Chile

K. LEVAY, Space Telescope Science Institute, 3700 San Martin Drive, Baltimore, MD 21218, USA

Z. LEVAY, Space Telescope Science Institute, Office of Public Outreach, 3700 San Martin Drive, Baltimore, MD 21218, USA

D. Levine, IPAC, Caltech, 9539 Creemore Place, Tujunga, CA 91042, USA

T. Levoir, CNES, 18 Avenue Edouard Belin, 31401 Toulouse, France

J. Lewis, Institute of Astronomy, Cambridge, Madingley Road, Cambridge, CB3 0HA, United Kingdom

X. Lladó, University of Girona VAT:ESQ6750002E, Campus Montilivi, P4, E17071 Girona, Spain

A. López García, Valencia Astronomical Observatory, Edificio de Investigacion, Campus de Burjassot, 46010 Valencia, Spain

N. Lorente, NRAO / ALMA, PO BOX O, 1003 Lopezville Road, Socorro, NM 87801, USA

E. Los, Harvard College Observatory, 7 Cheyenne Dr., Nashua, NH 03063, USA

S. Lubow, Space Telescope Science Institute, 3700 San Martin Drive, Baltimore, MD 21218, USA

J. Lyn, Smithsonian Astrophysical Observatory, 60 Garden Street, MS-81, Cambridge, MA 02138, USA

D. Magee, UC Santa Cruz, 15 Vista Verde Cir., Watsonville, CA 95076, USA

V. Mahadevan, University of British Columbia Okanagan, Barber School Unit 5, 3333 University Way, Kelowna, BC, V1V 1V7, Canada

J.-C. Malapert, CNES, 18 Avenue Edouard Belin, 31401 Toulouse, France

O. Malkov, Institute of Astronomy, 48 Pyatnitskaya St., Moscow, 119017, Russia

R. Mann, University of Edinburgh, Royal Observatory, Blackford Hill, Edinburgh, EH9 3HJ, United Kingdom

D. Marcos, European Southern Observatory, Juan de Urbieta 44, Madrid, 28007 Madrid, Spain

P. Martens, Montana State University, Physics Department, EPS 247, Bozeman, MT 59717, USA

J. Masters, National Radio Astronomy Observatory (NRAO), 520 Edgemont Rd, Charlottesville, VA 22903, USA

A. Matthews, Smithsonian Astrophysical Observatory, 60 Garden Street, MS-20, Cambridge, MA 02138, USA

S. McConnell, Trent University, 1600 West Bank Drive, Peterborough, On, K9J 7B8, Canada

J. McDowell, Smithsonian Astrophysical Observatory, 60 Garden Street, MS-06, Cambridge, MA 02138, USA

K. McGarvey, University of Houston, 4800 Calhoun Rd., Houston, TX 77204-3010, USA

W. McLaughlin, Smithsonian Astrophysical Observatory, 60 Garden Street, MS-81, Cambridge, MA 02138, USA

B. McLean, Space Telescope Science Institute, 3700 San Martin Drive, Baltimore, MD 21218, USA

S. Mele, CERN, Scientific Information Service, Geneve 23, CH 1211, Switzerland

L. MICHEL, Observatory of Strasbourg, 11 Rue de l'Université, 67000 Strasbourg, France

J. MILLER, Smithsonian Astrophysical Observatory, 60 Garden Street, MS-81, Cambridge, MA 02138, USA

D. MINK, Smithsonian Astrophysical Observatory, 60 Garden Street, MS-20, Cambridge, MA 02138, USA

Y. MIZUMOTO, NAOJ, Ohsawa, Mitaka, Tokyo, 181-8588, Japan

C. MOREAU, CNRS/LAM CeSAM, Technopole Chateau Gombert, 13013 Marseille, France

D. MORGAN, Smithsonian Astrophysical Observatory, 60 Garden Street, MS-34, Cambridge, MA 02138, USA

J. MORRISON, University of AZ, Steward Observatory, 933 N Cherry Ave, Tucson, AZ 85721, USA

C. MOTCH, Observatoire de Strasbourg, 11 rue de l'Université, 67000 Strasbourg, France

A. MUENCH-NASRALLAH, Smithsonian Astrophysical Observatory, 60 Garden Street, MS-42, Cambridge, MA 02138, USA

R. MUSHOTZKY, University of Maryland, Dept of Astronomy, College Park, MD 20742, USA

D. NGUYEN, Smithsonian Astrophysical Observatory, 60 Garden Street, MS-81, Cambridge, MA 02138, USA

J. NOMARU, Subaru Telescope, 650 N.A'ohoku Place, Hilo, HI 96720, USA

A. OBERTO, CDS (Strasbourg observatory, France), 11 rue de l'université, 67000 Strasbourg, France

F. OCHSENBEIN, CDS, Strasbourg, 11, rue de l'Université, 67000 Strasbourg, France

M. OHISHI, NAOJ, 2-21-1, Osawa, Mitaka, Tokyo, 181-8588, Japan

E. OJERO-PASCUAL, ESA/European Space Astronomy Center, Apdo. Correos 78, Villanueva de la Cañada, Madrid, 28691 Madrid, Spain

C. OLSEN, NetApp, 372 Lang Road, Portsmouth, NH 03801, USA

W. O'MULLANE, European Space Astronomy Center, P.O. Box 78,, 28691 Villanueva de la Cañada,, Madrid, 28691 Madrid, Spain

S. OTT, ESA, Keplerlaan 1, Noordwijk, 2200AG, Netherlands

J. OVERLY, Smithsonian Astrophysical Observatory, 60 Garden Street, MS-81, Cambridge, MA 02138, USA

M. PAEGERT, Vanderbilt University, Department of Physics and Astronomy, 6301 Stevenson Center, VU Station B #351807, Nashville, TN 37235, USA

A. PANASYUK, Smithsonian Astrophysical Observatory, 60 Garden Street, MS-50, Cambridge, MA 02138, USA

S. PASCUAL, Universidad Complutense de Madrid, Dpto Astrofisica, Facultad CC Fisicas, 28040 Madrid, Spain

W. PENCE, NASA/GSFC, NASA, Greenbelt Road, Greenbelt, MD 20771, USA

N. P<small>ENG</small>, National Astronomical Observatories of CAS, 20A Datun Road, Beijing, 100012, China

A. P<small>EPE</small>, Harvard-Smithsonian Center for Astrophysics, 60 Garden Street, MS-99, Cambridge, MA 02138, USA

M. P<small>ERACAULA</small>, University of Girona VAT:ESQ6750002E, Campus Montilivi, P4, E17071 Girona, Spain

M. P<small>ETREMAND</small>, LSIIT - University of Strasbourg, Pôle API, Bd Sébastien Brant, BP 10413, 67412 Illkirch, France

F. P<small>IERFEDERICI</small>, Space Telescope Science Institute, 3700 San Martin Drive, Baltimore, MD 21218, USA

F.-X. P<small>INEAU</small>, CNRS, Observatoire de Strasbourg, 11, rue de l'Université, 67000 Strasbourg, France

R. P<small>LANTE</small>, UIUC/NCSA, 1205 W. Clark Street, Urbaba, IL 61801, USA

D. P<small>LUMMER</small>, Smithsonian Astrophysical Observatory, 60 Garden Street, MS-81, Cambridge, MA 02138, USA

D. P<small>ONZ</small>, INSA, Pintor Rosales, 34, 28008 Madrid, Spain

M. P<small>OUND</small>, University of Maryland, Astronomy Department, College Park, MD 20742, USA

P. P<small>ROTOPAPAS</small>, Smithsonian Astrophysical Observatory, 60 Garden Street, MS-67, Cambridge, MA 02138, USA

A. P<small>TAK</small>, NASA, 8800 Greenbelt Road, Greenbelt, MD 20771, USA

L. Q<small>UICK</small>, Space Telescope Science Institute, 3700 San Martin Drive, Baltimore, MD 21218, USA

T. R<small>AFFERTY</small>, McDonald Observatory, 1 University Station C1400, Austin, TX 78712, USA

T. R<small>AUCH</small>, Eberhard Karls University, Sand 1, 72076 Tübingen, Germany

J. R<small>ECTOR</small>, Caltech, 221 Vista Circle Drive, Sierra Madre, CA 91024, USA

R. R<small>EDMAN</small>, NRC of Canada, 5071 West Saanich Road, Victoria, BC, V9E 2E7, Canada

B. R<small>EFSDAL</small>, Smithsonian Astrophysical Observatory, 60 Garden Street, MS-81, Cambridge, MA 02138, USA

G. A. R<small>ENTING</small>, ASTRON, Aquamarijnstraat 527, Groningen, Groningen, 9743PM, Netherlands

L. R<small>IZZI</small>, Joint Astronomy Center, 660 N. A'ohoku Place, University Park, Hilo, HI 96720, USA

J.-C. R<small>OCHE</small>, IRAM, 300 rue de la piscine, 38400 Saint Martin d'Heres, France

Y. R<small>OEHLLY</small>, CNRS/LAM CeSAM, Technopole Chateau Gombert, 38 rue Frédéric Joliot-Curie, 13013 Marseille, France

A. R<small>OGERS</small>, Space Telescope Science Institute, 3700 San Martin Drive, Baltimore, MD 21218, USA

A. R<small>OJAS</small>, Carnegie Mellon University, Luqta Street, Doha, Qatar, 24866, Qatar

S. Rostopchin, McDonald Observatory, 17 Solar Dr, McDonald Observatory, TX 79734, USA

A. Rots, Smithsonian Astrophysical Observatory, 60 Garden Street, MS-67, Cambridge, MA 02138, USA

R. D. Saxton, ESAC / VEGA, Satellite tracking station, P.O. box (Apartado) 78, Villanueva de la Cañada, 28691 Madrid, Spain

B. Schulz, NASA Herschel Science Center/IPAC/Caltech, Mail Code 100-22, Pasadena, CA 91125, USA

J. Schwarz, European Southern Observatory, Karl-Schwarzschild-Str. 2, 85748 Garching bei München, Germany

R. Seaman, National Optical Astronomy Observatory, 950 N. Cherry Ave., Tucson, AZ 85719, USA

M. Servillat, Harvard-Smithsonian Center for Astrophysics, 60 Garden Street, MS-67, Cambridge, MA 02138, USA

Y. Shirasaki, NAOJ, 2-21-1 Osawa, Mitaka, Tokyo, Japan

R. Shuping, SOFIA, 1716 NE 52nd Ave., Portland, OR 97213, USA

A. Siemiginowska, Smithsonian Astrophysical Observatory, 60 Garden Street, MS-04, Cambridge, MA 02138, USA

L. Sitongia, NCAR/High Altitude Observatory, 3090 Center Green Drive, Boulder, CO 80301, USA

K. Smalley, Smithsonian Astrophysical Observatory, 60 Garden Street, MS-18, Cambridge, MA 02138, USA

R. Smareglia, INAF-OATs, Via G.B. Tiepolo, 11, 34143 Trieste, Italy

I. Smith, University of Washington, Box 351580, U.W., Seattle, WA 98195, USA

C. Sontag, Space Telescope Science Institute, 3700 San Martin Drive, Baltimore, MD 21218, USA

M. Sosey, Space Telescope Science Institute, 3700 San Martin Drive, Baltimore, MD 21218, USA

T. Stephens, NASA GFSC/Wyle IS, 977S. 1830E., Spanish Fork, UT 84660, USA

E. Stobie, National Optical Astronomy Observatory, 950 N. Cherry Ave., Tucson, AZ 85719, USA

F. Stoehr, European Southern Observatory, Karl-Schwarzschild-Str 2, 85748 Garching bei München, Germany

L. Storrie-Lombardi, Spitzer Science Center, Caltech, MS 314-6, Pasadena, CA 91125, USA

O. Streicher, Astrophysikalisches Institut Potsdam, An der Sternwarte 16, 14482 Potsdam, Germany

C. Surace, CNRS/LAM CeSAM, Technopole Chateau Gombert, 38 rue Frédéric Joliot-Curie, 13388 Marseille, France

D. Swade, Space Telescope Science Institute, 3700 San Martin Drive, Baltimore, MD 21218, USA

M. Swam, Space Telescope Science Institute, 3700 San Martin Drive, Baltimore, MD 21218, USA

J. Swinbank, University of Amsterdam, Sterrenkundig Instituut Anton Pannekoek, Postbus 94249, Amsterdam, 1090 GE, Netherlands

A. Szalay, Johns Hopkins University, Department of Physics and Astronomy, 3701 San Martin Drive, Baltimore, MD 21218, USA

T. Takata, National Astronomical Observatory of Japan, Osawa 2-21-1, Mitaka, Tokyo, 181-8588, Japan

H. Teplitz, IRSA/Caltech, MS 100-22, Caltech, Pasadena, CA 91125, USA

P. Teuben, University of Maryland, Astronomy Department, University of Maryland, College Park, MD 20769, USA

B. Thomas, National Optical Astronomy Observatory, 18 8th Street NE, Washington, DC 20002, USA

M. Tibbetts, Smithsonian Astrophysical Observatory, 60 Garden Street, MS-81, Cambridge, MA 02138, USA

D. Tody, NRAO, VAO, 908 Paisano Dr., Socorro, NM 87801, USA

S. Tokarz, Smithsonian Astrophysical Observatory, 60 Garden Street, MS-20, Cambridge, MA 02138, USA

A. Torrent, University of Girona VAT:ESQ675002E, Campus Montilivi, P4, E17071 Girona, Spain

P. Udomprasert, Smithsonian Astrophysical Observatory, 60 Garden Street, MS-42, Cambridge, MA 02138, USA

F. Valdes, National Optical Astronomy Observatory, P.O. Box 26732, Tucson, AZ 85726, USA

D. Van Stone, Smithsonian Astrophysical Observatory, 60 Garden Street, MS-81, Cambridge, MA 02138, USA

C. Veillet, Canada-France-Hawaii Telescope, 65-1238 Mamalahoa Hwy, Kamuela, HI 96743, USA

I. Vera, European Southern Observatory, Karl-Schwarzschild-Str. 2, 85748 Garching bei München, Germany

D. Vibert, LAM - CNRS, 38 rue Frédéric Joliot-Curie, cedex 13, 13388 Marseille, France

M. Vuong, European Southern Observatory, Karl-Schwarzschild-Str. 2, 85748 Garching bei München, Germany

N. Walton, IoA, University of Cambridge, Madingley Road, Cambridge, Cambs, CB3 0HA, United Kingdom

R. Warmels, European Southern Observatory, Karl-Schwarzschild-Str. 2, 85748 Garching bei München, Germany

C. Warner, University of Florida, 211 Bryant Space Science Center, Department of Astronomy, University of Florida, Gainesville, FL 32611, USA

A. Watry, Smithsonian Astrophysical Observatory, 60 Garden Street, MS-67, Cambridge, MA 02138, USA

D. Westman, University of Washington, 4606 Palatine Ave. N., Seattle, WA 98103, USA

R. White, Space Telescope Science Institute, 3700 San Martin Drive, Baltimore, MD 21218, USA

K. Wiley, Department of Astronomy, University of Washington, 3910 15th ave, rm C319, Seattle, WA 98195-0002, USA

M. Wilson, Smithsonian Astrophysical Observatory, 60 Garden Street, MS-81, Cambridge, MA 02138, USA

T. Winegar, Subaru Telescope - NAOJ, 650 N. Aohoku Place, Hilo, HI 96720, USA

S. Winkelman, Smithsonian Astrophysical Observatory, 60 Garden Street, MS-67, Cambridge, MA 02138, USA

M. Wise, ASTRON (Netherlands Institute for Radio Astronomy), P.O. Box 2, Dwingeloo, 7990 AA, Netherlands

S. Wolfram, Wolfram Research, 1166 Massachusetts Ave, Cambridge, MA 02138, USA

H. Woodruff, CFHT, 65-1238 Mamalahoa Hwy., Kamuela, HI 96743, USA

Y. Yamada, Kyoto University, Oiwake-cho Kita-Shirakawa, Kyoto, 606-8502, Japan

C. Yamauchi, Japan Aerospace Exploration Agency, 3-1-1, Yoshinodai, Sagamihara, Kanagawa, 229-8510, Japan

N. Yasuda, IPMU, UTokyo, 5-1-5, Kashiwanoha, Kashiwa, Chiba, Japan

Y. Zhang, National Astronomical Observatory,CAS, Datun Road 20A, Chaoyang District, Beijing, 100012, China

P. Zografou, Smithsonian Astrophysical Observatory, 60 Garden Street, MS-81, Cambridge, MA 02138, USA

I. Zolotukhin, Observatoire de Paris, 61 avenue de l'Observatoire, 75014 Paris, France

Part I

Archives

Astronomical Data Analysis Software and Systems XX
ASP Conference Series, Vol. 442
Ian N. Evans, Alberto Accomazzi, Douglas J. Mink, and Arnold H. Rots, eds.
© *2011 Astronomical Society of the Pacific*

Optimizing Architectures for Multi Mission Archives

Gretchen Greene

Space Telescope Science Institute Baltimore, MD

Abstract. Data management systems for new missions are often the end of the chain for design and an area that was in earlier days underestimated. Thanks to the visionaries of our times and measurable increases in science user bandwidth, the value of building a robust archive system is now seen to provide a critical capability for producing newer and greater science. Throughout the astronomical community, teams of scientists and engineers are focusing on how we can build optimized architectures to support multiple missions, both space-based and ground, local and distributed. At Space Telescope Science Institute, such a team is using the successful foundation of the Hubble Space Telescope archive, incorporating lessons learned into the design and development of the James Webb Space Telescope data management systems, and unifying the MAST public science archive with the operational mission archives. The process of optimizing the architecture components combine the resource efficiency of an internal storage cloud while increasingly leveraging collaborative efforts for shared community development of archive and data processing technology, such as the standard protocols and data models developed by the international Virtual Observatory.

1. Introduction

Throughout the astronomical community, archive and data centers are evolving advanced data management subsystems for the support of both space and ground based mission data search and discovery. At Space Telescope Science Institute (STScI) there are currently several big active missions including HST, Kepler, and GALEX. STScI will also host the primary science data processing and archive operations for the James Webb Space Telescope (JWST) which is currently in the early design phase of the data management system. JWST ground test data is currently being archived in the STScI Science instrument Integration and test Data (SID) archive. The Multi-mission Archive (MAST) at Space Telescope was established in 1997 as NASA's optical/UV archive to support both active and legacy missions. As part of it's holdings MAST serves images, including the all-sky DSS, spectra, catalogs (GSC), and time series. For specialized science the community may generate derived data sets from the holdings of the STScI MAST archive which result in High Level Science Products (HLSP). These community contributed products may be archived and retrieved from MAST. MAST is currently in the process of integrating the Hubble Legacy Archive (HLA) web portal access and services for images, source lists, and spectra data products created with the latest calibrations of the HST observations into a more modern framework. Figure 1 illustrates the various strata of the STScI Archive and Data System which characterize the architectural elements of the data management systems.

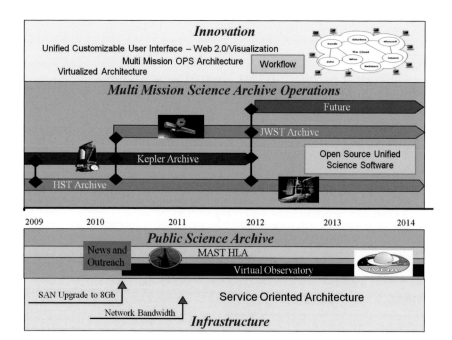

Figure 1.   STScI Data Management System Strata.

1.1.   Balancing Architecture Goals

When designing architecture to support multiple missions, one critical key to planning is finding the balance between the goals required for individual missions and organizational infrastructure. Identifying common mission oriented requirements helps to address this issue. For example, mission operations will require High Reliability, Availability, and Serviceability (HRAS) system design while at the same time providing the capability to adjust the DMS operational priorities. For multiple missions this means identifying areas where there are shared costs for architecture support, such as common server hardware frameworks, enterprise licensing and maintenance and virtualized development environments. Planning optimal hardware life cycles through the life of a mission will require N refreshes depending on the complexity of technology. For planning purposes at STScI, a nominal five year cycle has been demonstrated across several periods for test and operational server systems, safe store offline media, and networking switches. Performance management, both continuous and benchmarked, is necessary for monitoring end-to-end mission systems functions and finding similar methodologies that address memory management, CPU loads, and network utilization at all junctions throughout the DMS. This is especially needed for acceptance testing following deployment of new software builds, system patches, and hardware upgrades. Organizational architecture requirements add complexity to the mission goals in the features which work toward optimization of the usefulness of the full archive content. With the rapid pace of technology change, organizations require more now than before building systems which have flexibility for innovation. The tradeoffs and costs can be minimized by utilizing the design phases, and hardware life cycle refreshes to upgrade and improve architectural elements in DMS systems. Another aspect to the growth in organizational archives is the increasing co-dependency with external archive resources. Identifying mission based science contributions from external collaborations can foster the architectures needed for archive mirroring.

1.2. Data Management System Work Areas

The DMS work areas at STScI which require architectural robustness to support multi-mission archive operations are fairly comprehensive. The data receipt and processing of science and engineering telemetry data utilizes pipeline workflow frameworks for multiple instrument specific pipelines, including the processing for calibration enhancements and generation of higher level science products. STScI is currently in the process of designing the future workflow framework for JWST. The OPUS framework developed for HST is designed to be fairly mission specific and has technology components that require extensive engineering to meet the throughput, operational, and functional requirements for JWST. The data processing architectural design which requires larger throughput includes moving to a Linux cluster environment to support distributed and parallel pipelines. The archive systems for storage and distribution of raw and processed science and engineering data include servers for operational monitoring, file brokering, secondary archive safe-store appliances, and database systems. The functional architecture which supports these work areas can be viewed in Figure 2. Web servers are required for the science user interfaces, web applications and retrieval services which interface to the file brokering. Science calibration software, the heart of the processing, requires careful design to meet the CPU load, memory, and IO requirements for each mission. There are supporting data analysis tools that are quite varied in nature and require production accessibility via web servers or development and testing systems for community distributed software. All these systems include interface and service support for other mission subsystems such as the planning and scheduling of operations, NASA centers, and for JWST, the flight operations systems.

2. Architecture Tiers

The elements of architecture can be viewed as an interleaved system with views and models of the various tiers for the development, test, and operational environments. There are various types of software applications which support the DMS work areas which require maintenance, enhancements, and modernized functionality to efficiently utilize the operational architecture. The compute, database, and application servers require technology which can scale for increased performance demand and fail tolerance. Multi mission architectures can address this need by designing clustered systems with failover and front end load balancing servers. To meet the increasing bandwidth and IOPS DMS systems are experiencing, the network infrastructure for both internal usage and connectivity to the external community is highly critical. At STScI, the storage tiers rely on the network infrastructure for access to an internal private cloud, yet we also continue work to identify areas of the DMS systems which may benefit from external solutions. Large scale archive capacity reaching in the PBs can offset supporting architectural safe-store and mirroring with external archive and data center mirror sites. We are also factoring into the architectural design considerations regarding the evolving interoperability standards of the Virtual Observatory, which will increase the use of both internal and external networked systems.

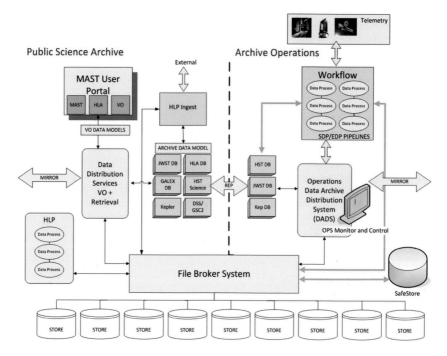

Figure 2.   Functional Architecture Conceptual Design.

The big issues to address with architectural tiers are migration of legacy systems (full or partial system replacements), how mission schedules interface with technology life cycles, and choosing the right path forward to allow scaling and flexibility for growth. The decisions for solutions will largely be influenced by who the primary customer is. However, balancing architectural goals may best be achieved by both building system reusable and replaceable tier components that evolve during design and refresh opportunity, and by utilizing flexibility inherent to service oriented architecture design to minimize interfaces and dependencies.

2.1.   Development Strategies

Building an archive and data center to support multi missions requires the capabilities of optimizing, innovating, consolidating and balancing requirements. At STScI we have built a roadmap which includes strategies for achieving these essential goals. We have established working teams across organizational groups to combine the science, software engineering and IT expertise to bridge communications. Hardware systems share, where possible, baseline configurations for consistent deployment and maintenance. We build dedicated systems to address the unique specifications for development, test, and operations environments. Identifying common functional systems allows architectural solutions to be combined and shared where possible with physical and system level configurations. Designing a scalable infrastructure for networked and distributed science operations with end-to-end performance level monitoring has proven time and time again to be highly beneficial for optimizing functional designs and for troubleshooting. Engaging in resource sharing through collaborative partnerships will reduce the need to reinvent, re-engineer, and maintain over the long term.

2.2. Storage Architecture

The storage architecture that has evolved at STScI is based on a private cloud storage area network (SAN). This network is the kernel to the infrastructure and has been recently upgraded to 8Gb fiber channel switch network. The network has commodity integration support across platforms with EMC storage frameworks. The connectivity of host servers is through Host Bus Adapters to dual enterprise class switches with multiple fabrics. Within this network, there is support for various tiered storage, depending on performance and recovery/failover requirements, using flash drives for high IO, SATA for the large file brokered archive, and fiber channel drives for databases. The storage file brokering systems are currently custom, yet we are working toward integration of a community shared solution, IRODS, along with internal solutions to manage both batch and direct file access. To better understand the cost effectiveness of this private cloud solution, we are developing a Total Cost of Ownership (TCO) model, commonly termed in the commercial sector, for establishing optimal architectural designs for increasing capacity solutions. Figure 3a shows the multi-mission deployment across the STScI DMS SAN with our current data holdings reaching the PB regime.

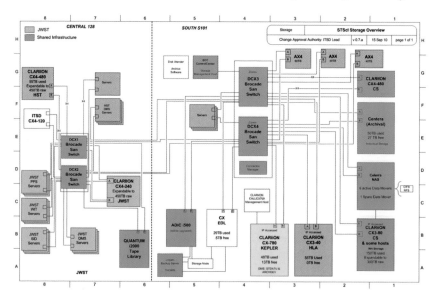

Figure 3.   SAN Storage Architecture for multiple active missions.

2.3.   Application Architecture Considerations

The focus at STScI is evolving from a legacy Sunfire 15K systems with reconfigurable domains to a more highly distributed cluster model. This allows component upgrades and application deployment and modification to have node isolation with lower associated costs. Our current archive operations have maxed out the 72 CPU Sparc technology in a high maintenance cost system. In benchmark comparisons, migrating to newer Nehalem Intel quad six-core processors, we can achieve a 5× performance increase on a single Linux server box up with up to 24 simultaneous processes. With a singular large scale server model we have experienced difficulty with separation of loads between the data ingest and processing, user request and data retrieval from calibration pipelines. Moving toward a cluster distributed model will provide the functional decoupling and monitoring in addition to flexibility in prioritizing application deployments. One are we are exploring is shared file systems for the clustered models. Another area of optimizing application architectures includes the use of modern web techniques for both archive user portal services and operational control systems applications. The web allows development and deployment to decouple from user installations and version control. Database architectures which support applications via service layer interfaces also greatly reduce the management costs and architectural dependencies for large numbers of meta files. The database services can be reusable through the implementation of a common archive data model.

2.4.   Virtualized Development Infrastructure and Experience

One of the most active areas of common multi-mission architecture designs is the use of a virtualized development infrastructure (Figure 4). STScI has adopted the VMWare enterprise solution as an optimal technology. In both the public science and missions operations development architectures, we are building ESX clusters to support the various DMS software development activity in the public science and operations archive. We have realized pros and cons to the virtualized environments. On the side of the pros, we have demonstrated with success the ability to map multiple server OS environments to a single physical node for multi-function purposes. The migration and sharing of VM server deployments can be 'motioned' across physical nodes for load management. Storage can be presented across ESX servers to facilitate migration of servers without data copy and the commercial tools work well for automated physical to virtual system configuration migrations and vice versa. Fully configured VM servers have the added feature that they can be distributed as a file to an external location. This alleviates new installation of OS, databases and applications. For regression testing across multiple platforms, a suite of VMs are ideal. On the con side we discovered there are CPU allocation limits which have slow technology evolution. Our testing demonstrated that the VM does not scale for hyper threaded CPUs beyond 8 processors, which means that for systems with more than 4 cores we have system under-utilization. VMWare ESX operating systems limit storage LUN allocations to 2TB. For large scale configuration this builds increased complexity for abstraction layers. The raw-device mapped LUNs may appear on servers not hosting VMs, which offers the potential for data corruption. VMs have no awareness of SAN HBS and therefore you can not use SAN backup. It may be possible to bridge the gap using VEEAM, which we are currently testing, yet additional licensing is required. The licensing and costs for implementation are based on the number of cores, thus for large physical nodes this can become quite expensive.

2.5.   Evolving Operations Architectures

The current active mission science operations for data management at STScI archive and data center is evolving architecturally following the models and concepts addressed in this paper. Figure 5 illustrates the idea for adopting common architectural solutions in both the public science archive and mission data management system operations for

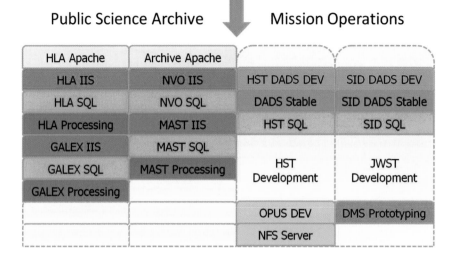

VMware ESX cluster – Expand as needed

Figure 4.   Virtualized development environments.

Linux Clusters and database server environments. The Test and Operations systems will share base configurations, maintenance and service elements which allow resource sharing between Test and Operations as well as cross mission for application frameworks (note the diagonal lines designating shared architectural solutions). While the public science archive and mission operations use independent environments for testing and operations, the systems consist of common technological solutions for infrastructure, e.g. the SAN, as well as shared virtual development environments. The overall Science Operations Center network backbone provides the interfaces between the DMS systems and the other mission subsystems.

2.6.   External Architecture Collaborations

Mirroring STScI archive holdings requires an exchange of the Public Science Archive calibrated data products. The metadata exchange may be simplified by adopting VO data models and standard service protocols. These standards allow external sites to optimize their architectures independently while at the same time minimizing the hardware and staffing resources needed to support the synchronization of hardware and software. We are currently considering external cloud solutions to mitigate bandwidth requirements (see Figure 6). One specific science initiative project for which we adopt architectural collaborations is catalog cross-matching. This is particularly beneficial for spatial indexing of large scale catalogs which are data intensive. While in the process of building these capabilities, we are working to adopt common data models, services, and deployments across servers to merge local catalogs with those externally produced to provide more complete wavelength coverage for science operations analysis of fields of view. STScI is diligently pursuing archive architectures which support interoper-

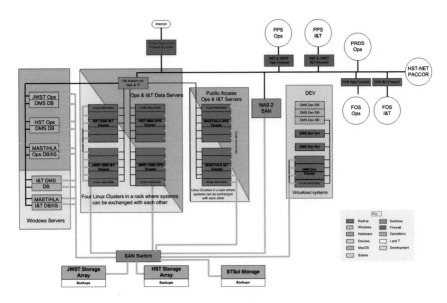

Figure 5.   Multi Mission development, test, and operations notional architecture.

ability, innovation, and technology to enable scientific data mining across astronomical archives and data centers. One area previously mentioned, the Virtual Observatory, synergistically provides standards which will serve as external networking infrastructure for the multi mission architectural development.

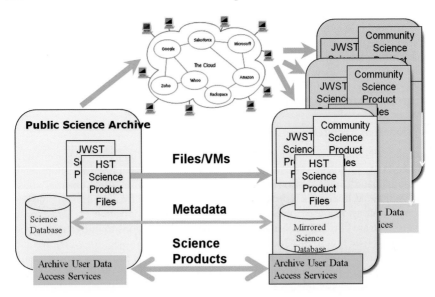

Figure 6.   Mirrored Solutions which utilize external cloud for single push and multiple pulls.

3. Conclusions

Architectures for multi missions require shared solutions across a wide range of data management system elements. Building a common yet flexible infrastructure is key for scalability, performance and maintenance. Long term planning for technology life cycle peaks in a few years, and requires a refresh of architectural components. Building systems which can adapt in units rather than full replacements will be more cost effective. Adoption of the virtualized server solutions works efficiently for development and distribution of preconfigured systems. The synergy both with internal and external architectural designs and solutions provides the opportunity for advancing science returns from the expanding network of archives and data centers.

Acknowledgments. The author would like to the thank the STScI Archive Management team along with the engineering and information technology technical staff that are working toward the multimission architectural development efforts at STScI.

Astronomical Data Analysis Software and Systems XX
ASP Conference Series, Vol. 442
Ian N. Evans, Alberto Accomazzi, Douglas J. Mink, and Arnold H. Rots, eds.
©2011 Astronomical Society of the Pacific

Digital Preservation and Astronomy: Lessons for Funders and the Funded

Norman Gray and Graham Woan

School of Physics and Astronomy, University of Glasgow, UK

Abstract. Astronomy looks after its data better than most disciplines, and it is no coincidence that the consensus standard for the archival preservation of all types of digital assets — the OAIS Reference Model — emerged originally from the space science community. It is useful to highlight both what is different about astronomy (and indeed about Big Science in general), what could be improved, and what is exemplary, and in the process I will give a brief introduction to the framework of the OAIS model, and its useful conceptual vocabulary. I will illustrate this with a discussion of the spectrum of big-science data management practices from astronomy, through gravitational wave (GW) data, to particle physics.

1. Introduction

In this paper we will briefly discuss some of the issues surrounding the 'long-term' management of data for 'big science' projects. It is therefore intended to illuminate the ways in which scientists in these areas have distinctive data management requirements, and a distinctive data culture, which contrasts informatively with other disciplines.

Below, we give a brief account of what we mean by 'big science', illustrate some of the features of astronomy, particle-physics and gravitational wave data, present an overview of the OAIS model, and summarise the work we have done in this area.

2. Big Data in Big Science

What is 'big science'?

Big science projects tend to share many features which distinguish them from the way that experimental science has worked in the past. Such projects share (non-independent) features such as:

Big money These are decades-long projects, supported by country-scale funders and billion-currency-unit budgets;

Big author lists The (gravitational wave) LSC author list runs at around 0.8 kAuthor, and the LHC's ATLAS detector is 3 kAuthor;

Big data rates Projects produce petabytes of data per year (ATLAS will flatten out at about 10 PB/yr, and Advanced LIGO at around 1 PB/yr);

Big administration These include MOUs, councils, and more.

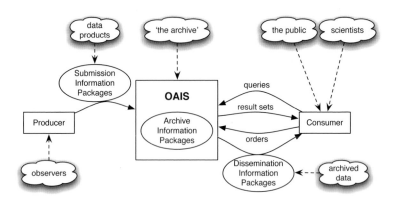

Figure 1.   A high-level view of the OAIS model, with annotations indicating the non-OAIS names for objects (redrawn from CCSDS 650.0)

There is an interesting historical discussion of the features of 'big science', and LIGO's progress towards that style of working, in Collins (2003), with an extended history of the sub-discipline in Collins (2004). The study we are reporting on is restricted to the case of such large-scale projects.

What is the 'long term'?

Astronomy has a long tradition of preserving data, of exploiting old data, and also, in the teeth of centuries of technological change, of finding old data to be readable. The 'Venus of Ammisaduqa' cuneiform tablet (a 7th century BCE media refresh of 17th century observations) is intelligible only to antiquaries, but the 12th century Toledan tables are intelligible, with help, to a contemporary astronomer, and Kepler's 17th century Rudolphine tables are readable without difficulty, suggesting that the intelligibility timescale for astronomy is of order one millennium (note that each of these examples are disseminated publications, not raw data). Particle physics is more mayfly-like, with HEP data becoming unreadable and uninteresting on a timescale of a few decades. Gravitational Wave astronomy is somewhere in the middle, with an experimental practice reminiscent of particle physics, but (eventual) data products which match astronomy.

Data preservation requires a consideration of both media preservation and data intelligibility on timescales potentially ranging from months (in the case of researchers wanting to re-find their own data) to multiple millennia (in the case of very long term nuclear waste disposal (US Department of Energy 2004)). The Open Archival Information System (OAIS) standard, however, pragmatically defines the 'long-term' to mean, essentially, 'long enough for storage technology to change'.

The OAIS standard (CCSDS 650.0) is a high-level model for archives (see Fig. 1). Since the OAIS model ultimately emerged from the space-data community, it is a natural fit to astronomical data. Although the OAIS model has been criticised for being excessively general — it is hard to think of any half-way respectable archive that is not at some level 'conformant' with OAIS (cf., Rosenthal et al. 2005) — it is nonetheless very useful as a conceptual toolbox, and as the basis for costings and other deliberations.

Once data sets have been deposited in an archive, and thus reprocessed from Submission Information Packages (SIP) to Archive ones (AIP), they become the exclusive responsibility of the archive, which must plan, and be funded, accordingly. One of the key questions to be decided by the archive is the nature of the Designated Communities who are expected to retrieve information from the archive, and the description of these communities — who they are, what they know, how hard they can be expected to work in order to retrieve archived data — is a key part of the archive's documentation. That documentation is written with enough detail that anyone in the designated communities can use it to make the data intelligible, *without* recourse to any of the data producers, who are presumed to have died, retired, or forgotten all they know, and the creation of this 'Representation Information' is a key part of the archive's initial negotiation with the data producers.

3. Sharing and Preserving Astronomy Data

In science, we preserve data so that we can make it available later. This is on the grounds that scientific data should generally be universally available, partly because it is usually publicly paid for, but also because the public display of corroborating evidence has been part of science ever since the modern notion of science began to emerge in the 17th century (CE) — witness the Royal Society's motto, which loosely translates as 'take nobody's word for it'. Of course, the practice is not quite as simple as the principle, and a host of issues, ranging across the technical, political, social and personal, make this more complicated than it might be, but it is worth noting that the physical sciences generally perform better here than other disciplines, both in the technical maturity of the existing archives and in the community's willingness to allocate the time and money to see this done effectively.

In 2009–2010 the authors were commissioned by JISC, which is responsible for the exploitation of digital technology in the UK HE system, to examine the way in which the gravitational wave community, as a proxy for big science in general, managed its data, and to make recommendations as appropriate to JISC, the relevant funding councils, and to the GW community. The results of this work will appear online[1] in due course. The brief summary of the conclusions is that this community, accurately representing the broader astronomical community, is functioning very well in this respect.

In particular, the commendable features of the community's approach to data management are:

- There is an expectation of explicit and costed data management planning (though this could arguably benefit from being more systematic).

- There is implicit identification of 'designated communities'.

- There has always been a recognition that clearly available and described science-data products (AIPs in OAIS-speak) are a vital part of the communication between observers and their colleagues.

- The community is honest about both its need for proprietary periods, and the acknowledgement that these restrictions must only be temporary and short.

---

[1]http://purl.org/nxg/projects/mrd-gw

All of these features are naturally re-expressible in terms of the OAIS model, which conveniently couples them to the work being done elsewhere on the costing of OAIS models, and on developing tests of OAIS conformance.

Our project has not yet come to specific conclusions about funding. However, it becomes clear that funders are, at some level, as concerned with predictability and auditability as with economy, so that if conventional practice can be demonstrated to be additionally good practice, or within easy reach of it, then this facet of funders' goals can be marked as achieved.

Finally, we believe that our recommendations will resemble the following.

1. Big-science funders should require projects to develop plans based on the OAIS model, or profiles of it;

2. they should additionally develop or support expertise in criticising the result;

3. and use that modelled plan as a framework for validation of the project's efforts, both during the lifetime of the project and at its end.

Glossary *ATLAS*: one of the detectors at the LHC; *LIGO*: the US-based gravitational wave experiment; *LSC*: the multinational LIGO Scientific Collaboration;

Acknowledgments. This talk describes work at the University of Glasgow, funded by JISC, UK. We are grateful for helpful discussions with the LSC; LIGO document P1000179-v1.

References

CCSDS 650.0 2002, Reference Model for an Open Archival Information System (OAIS) – CCSDS 650.0-B-1, CCSDS Recommendation. Identical to ISO 14721:2003, URL http://public.ccsds.org/publications/archive/650x0b1.pdf
Collins, H. 2004, Gravity's shadow: the search for gravitational waves (University of Chicago Press)
Collins, H. M. 2003, Hist. Stud. Phys. Biol. Sci., 33, 261
Rosenthal, D. S. H., Robertson, T., Lipkis, T., Reich, V., & Morabito, S. 2005, D-Lib Magazine, 11. URL http://www.dlib.org/dlib/november05/rosenthal/11rosenthal.html
US Department of Energy 2004, Permanent Markers Implementation Plan, Tech. Rep. DOE/WIPP 04-3302, United States Department of Energy. URL http://www.wipp.energy.gov/picsprog/test1/Permanent_Markers_Implementation_Plan_rev1.pdf

Astronomical Data Analysis Software and Systems XX
ASP Conference Series, Vol. 442
Ian N. Evans, Alberto Accomazzi, Douglas J. Mink, and Arnold H. Rots, eds.
© 2011 Astronomical Society of the Pacific

CeSAM: The Astrophysical Data Center of Marseille

T. Fenouillet, C. Surace, and the CeSAM Team

Laboratoire d'Astrophysique de Marseille (LAM),
UMR 6110, 38 Rue Frédéric Joliot-Curie, 13013 Marseille, France

Abstract. The "Centre de donnéeS Astrophysiques de Marseille" (CeSAM) from "Laboratoire Astrophysique de Marseille" (LAM) has been set up to provide access to quality controlled data via web based applications, tools, pipeline developments and VO compliant applications to the astrophysical community. This paper describes the organization, the infrastructure, tools as well as resources available for downloading. It includes DAL, SOS, ETC, CoRoT tools and the project's data interfaces which are described elsewhere in these proceedings.

1. CeSAM

New astrophysical projects provide large amounts of data and need high quality checks. Based on its five years of experience and its participation in international projects, the former DIS (Département d'Informatique Scientifique) from LAM has become a data center: The CeSAM (Centre de donnéeS Astrophysique de Marseille).

Its activities broaden in scope from Information System design and concepts to astrophysical software. Its competences deal with images, spectra, data, signal processing, database and application development as well as applications and database systems administration. Eighteen engineers belong to this data center and participate in more than 20 projects.

Missions of CeSAM are:

- To design and provide systems to projects,

- To design, develop and maintain quality control systems for data provided by the LAM,

- To design, develop and maintain tools used by projects in the astrophysics community,

- To design, develop and maintain pipelines packages,

- To provide technical support (on applications, infrastructures, databases and pipelines),

- To provide expertise on R&D for data processing to science teams.

2. Team Activities

The CeSAM team performs multiple activities as shown in Figure 1. With a wide range of required skills, CeSAM is involved at many levels in data processing projects.

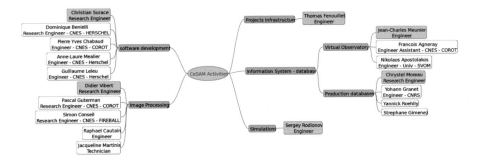

Figure 1.   The CeSAM Team and activities.

3. Tools

CeSAM has already developed different tools to help the researcher, from pipeline tools to visualization interfaces.

- CIGALE: Code Investigating GALaxy Emission has been developed to study the evolution of galaxies by comparing modelled galaxy spectral energy distributions (SEDs) to observed ones from the far UV to far IR. CIGALE is software that extends the SED fitting up to the far infrared (development: JAVA, ICE).

- EMPhot-pprior: Photometry of astrophysical sources, galaxies, and stars, in crowded field images to estimate the flux in a low resolution band using prior information and using a Bayesian approach under the Poisson noise assumption (development: IDL,C). Described in Conseil et al. (2011).

- VO DAL: The VO DAL is an interface layer based on Struts to give access to data using VO-compliant queries including SSA, SIA, and ConeSearch. It is based on access to a metadata database and makes use of the pgsphere package capabilities. This tool is available on requests from the authors (development: JAVA, TOMCAT — postgreSQL).

- ETC-42: This tool is a VO compliant exposure time calculator. The goal is to provide a generic Exposure time calculator flexible enough to be customized with one's own instrument, source, site and characteristics (JAVA development — XML — embedded database). Described in Surace et al. (2011).

- SOS: This tool is designed to perform analysis of light curves data (development: IDL).

- PAD: The Detections Analysis program is an exoplanet detection validator software (using folding and multi visualisation techniques) (development: IDL). Described in Chabaud et al. (2011).

- AITAS: This tool is an archiving system for AIT data. It is automatically built from the data models themselves and provides access to multi-criteria forms to select data (development: JAVA, Servlet).

- CRT: CoRoTool is a step by step interactive analysis detection pipeline of CoRoT exoplanets search (development: IDL). Described in Chabaud et al. (2011).

4. Projects

CeSAM works on many projects for LAM on international collaborations as well as internal ones as shown in Figure 2.

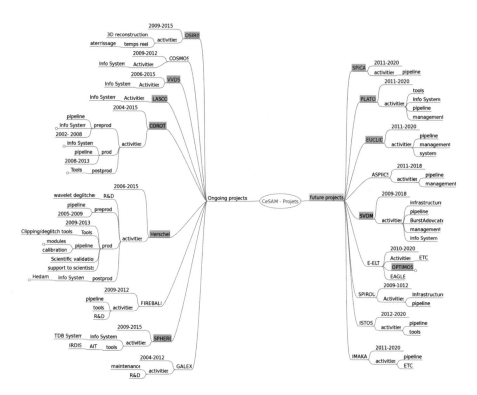

Figure 2.   List of Project involving CeSAM.

The CENCOS Project and Hershel I. S. are highlighted this year in ADASS XX. Moreau et al. (2011) and Roehlly et al. (2011)

5. Infrastructures

The CeSAM infrastructure is based on a three-tier architecture (see Figure 3). Backup systems and test servers are configured to allow automatic recoveries. The servers are connected to the Internet with a 10 Gb link to the Renater network (French telecommunication infrastructure). The next step is about to start: It's a cluster of virtual servers that will smooth the servers usage, as each real server will host 3-tiers servers (data, database, web). The database systems will be hot replicated in a transparent manner as new hosts are required.

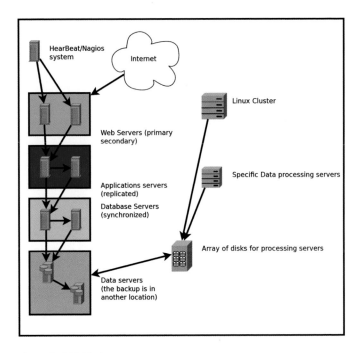

Figure 3.   CeSAM Infrastructure.

6. Links

- Website: http://lamwws.oamp.fr/cesam
- Head Department: Christian Surace mailto:christian.surace@oamp.fr
- System Administrator: Thomas Fenouillet mailto:thomas.fenouillet@oamp.fr

References

Chabaud, P.-Y., Agneray, F., Meunier, J.-C., Guterman, P., Cautain, R., Surace, C., & Deleuil, M. 2011, in Astronomical Data Analysis Software and Systems XX, edited by I. N. Evans, A. Accomazzi, D. J. Mink, & A. H. Rots (San Francisco, CA: ASP), vol. 442 of ASP Conf. Ser., 339

Conseil, S., Vibert, D., Arnouts, S., Milliard, B., Zamojski, M., Llebaria, A., & Guillaume, M. 2011, in Astronomical Data Analysis Software and Systems XX, edited by I. N. Evans, A. Accomazzi, D. J. Mink, & A. H. Rots (San Francisco, CA: ASP), vol. 442 of ASP Conf. Ser., 107

Moreau, C., Gimenez, S., Kneib, J.-P., & Tasca, L. 2011, in Astronomical Data Analysis Software and Systems XX, edited by I. N. Evans, A. Accomazzi, D. J. Mink, & A. H. Rots (San Francisco, CA: ASP), vol. 442 of ASP Conf. Ser., 21

Roehlly, Y., Buat, V., Heinis, S., Moreau, C., & Gimenez, S. 2011, in Astronomical Data Analysis Software and Systems XX, edited by I. N. Evans, A. Accomazzi, D. J. Mink, & A. H. Rots (San Francisco, CA: ASP), vol. 442 of ASP Conf. Ser., 25

Surace, C., et al. 2011, in Astronomical Data Analysis Software and Systems XX, edited by I. N. Evans, A. Accomazzi, D. J. Mink, & A. H. Rots (San Francisco, CA: ASP), vol. 442 of ASP Conf. Ser., 559

Astronomical Data Analysis Software and Systems XX
ASP Conference Series, Vol. 442
Ian N. Evans, Alberto Accomazzi, Douglas J. Mink, and Arnold H. Rots, eds.
©2011 *Astronomical Society of the Pacific*

The COSMOS Information Systems at LAM

Chrystel Moreau, Stéphane Gimenez, Jean-Paul Kneib, and Lidia Tasca

*Laboratoire d'Astrophysique de Marseille / C.N.R.S.—Université de Provence,
38, rue Frédéric Joliot-Curie, 13388 Marseille, France*

Abstract. The last decades have witnessed a strong increase in the amount of data coming from astronomical surveys. These data are exploited by large international collaborations of scientists working around the same scientific goals. The regular request of high data-quality control, fast data access via easy-to-use graphic interfaces, as well as the possibility to cross correlate information coming from different observations motivate the use of scientific information systems. The CeSAM (Centre de donneéS Astronomiques de Marseille) data center answers to the aforementioned needs offering a specialized database service to the contemporary large astrophysical surveys (VVDS, GALEX, HST-COSMOS, Hershel, CoRoT, etc.). We here focus our attention on the HST-COSMOS and zCOSMOS information systems, recently opened to the scientific community. We are dealing with data coming from the largest ever-undertaken cosmological survey. The associated database has the specificity to archive, visualize and correlate multi-wavelength and spectroscopic datasets.

1. Introduction

For the HST-COSMOS and zCOSMOS projects, COSMOS Information Systems were created and are managed by the CeSAM. They have been developed to offer:

- A unique data set, with data quality control,
- Tracking of previous releases,
- Easy data access and query interface,
- Public release, large visibility.

2. HST-COSMOS

The Cosmological Evolution Survey (Scoville et al. 2007) is an astronomical survey designed to probe the formation and evolution of galaxies as a function of cosmic time (redshift) and large scale structure environmement. The project PI is Nicolas Scoville (California of technology, USA), while the local Managing Scientist is J.-P. Kneib (CNRS, LAM, France).

The survey covers a 2 square degree equatorial field with imaging from most of the major space based telescope (Hubble, Spitzer, GALEX, XMM, Chandra) and a number of large ground based telescope (Subaru, VLA, ESO-VLT, CFHT, and others). Over 2 million galaxies have been detected, covering 75% of the age of the universe. The COSMOS survey involves almost 100 scientists in a dozen countries.

The Project Scientist for the HST-Cosmos information system is J.-P. Kneib (CNRS, LAM, France), the Scientist is L. Tasca (CNRS, LAM, France), the Information System Manager is C. Moreau (CNRS, LAM, France), and the developer is S. Gimenez (CNRS, LAM, France).

The HST-COSMOS Information System offers access to:

- Multi-λ COSMOS catalogs,
- BRIGHT & DEEP zCOSMOS catalogs,
- Ancillary data (mass, magnitude, morphology, spectroscopic features),
- Postage-stamp images for all bands.

Data can be searched by several criteria, by position or by SQL query. The service can be accessed online at http://cencos.oamp.fr/hstcosmos/.

3. zCOSMOS

The zCOSMOS (Lilly et al. 2007) project is an approved Large Program on the ESO VLT. The PI is Simon Lilly (ETH Zurich), and the local Managing Scientist is L. Tasca (CNRS, LAM, France). 600 hours of observation are used to carry out a major redshift survey with the VIMOS spectrograph on the COSMOS field to yield spectra for:

- Approximately 28,000 galaxies at $0.2 < z < 1.2$ selected to have $I\_AB < 22.5$ at a sampling rate of 70%,
- Approximately 12,000 galaxies at $1.2 < z < 3$ with $B\_AB < 25$ and chosen by two color-selection criteria (B-Z) vs. (Z-K) and (U-B) vs. (V-R) at a sampling rate of 70%.

The Project Scientist for the zCOSMOS information system is L. Tasca (CNRS, LAM, France), the Information System Manager is C. Moreau (CNRS, LAM, France). The zCosmos Information System provides access to:

- BRIGHT & DEEP parent catalog,
- 10k, 20k BRIGHT + 8k DEEP spectroscopy,
- Ancillary data (masses, magnitudes, morphology, spectroscopic features),
- HST/ACS i-band postage-stamps images,
- 1D spectra.

Data can be searched by several criteria, by position or by SQL query. The service can be accessed online at http://cencos.oamp.fr/zCosmos/.

Acknowledgments. The COSMOS Information Systems team of CeSAM (see Fenouillet et al. 2011) would like to thank the surveys used in the creation of the databases.

References

Fenouillet, T., et al. 2011, in Astronomical Data Analysis Software and Systems XX, edited by I. N. Evans, A. Accomazzi, D. J. Mink, & A. H. Rots (San Francisco, CA: ASP), vol. 442 of ASP Conf. Ser., 17
Lilly, S. J., et al. 2007, ApJS, 172, 70
Scoville, N., et al. 2007, ApJS, 172, 1

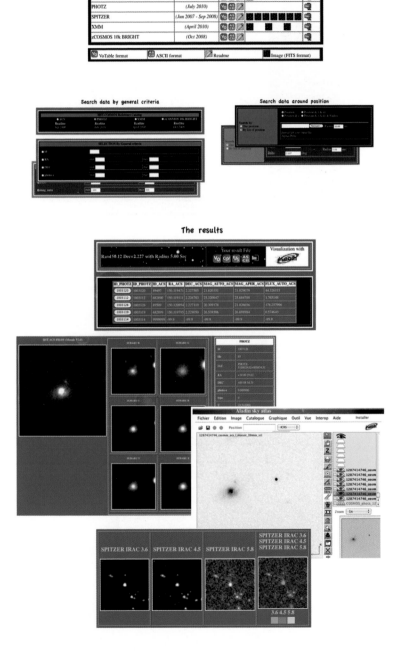

Figure 1.   The HST-COSMOS Information System

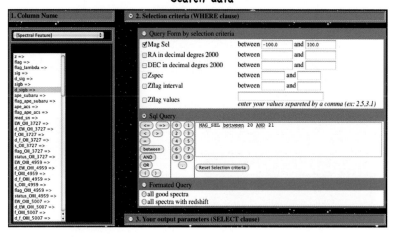

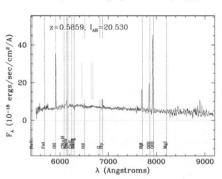

Figure 2.   The zCosmos Information System

Astronomical Data Analysis Software and Systems XX
ASP Conference Series, Vol. 442
Ian N. Evans, Alberto Accomazzi, Douglas J. Mink, and Arnold H. Rots, eds.
© *2011 Astronomical Society of the Pacific*

HeDaM: The Herschel Database in Marseille

Yannick Roehlly, Véronique Buat, Sébastien Heinis, Chrystel Moreau, and Stéphane Gimenez

Laboratoire d'Astrophysique de Marseille / C.N.R.S.—Université de Provence, 38, rue Frédéric Joliot-Curie, 13388 Marseille, France

Abstract.   The Herschel Database in Marseille (HeDaM) is an information system operated by the Laboratoire d'Astrophysique de Marseille (LAM) datacenter — CeSAM. Gathering data from various Herschel Space Observatory surveys, HeDaM serves both as a collaboration tool between scientists and as a way to distribute public data to a wider community. With HeDaM, users have access to the direct download of catalogs and maps, but also to value-adding services such as: multi-λ FITS cut-outs, integration with Aladin applet and Topcat, and list-based cone searches.

1. Introduction

Since 2009, ESA's Herschel Space Observatory has been providing various scientific *consortia* with a lot of data in the far-infrared domain. Scientists from the LAM are deeply involved in some Herschel cosmological and extra-galactic surveys. These projects required a system to centralize the archival of Herschel science and ancillary data during the proprietary period and, eventually, distribute these data to the scientific community. To fill this need, the *Centre de donnéeS Astrophysiques de Marseille* — CeSAM (see Fenouillet et al. 2011) — created the Herschel Database in Marseille (http://hedam.oamp.fr).

2. Functionality Provided by HeDaM

HeDaM is based on CNES SITools (Malapert 2011) Tomcat application.[1] Datasets can be sorted according to a number of criteria and users can browse, perform searches, and display data tables (Figure 1). HeDaM offers the convenience to access both Herschel and ancillary data using the same interface. SITools was modified to support:

- Direct download of catalogs in FITS and VOTable formats as well as images;
- Creation of multi-λ FITS postage stamps around each source;
- List-based multi-position cone-searches on selected datasets;
- Dataset display using the Aladin (Bonnarel et al. 2000) applet or through a *java-webstart* instance of Topcat (Taylor 2005);
- The creation of various illustrations such as composite images.

---

[1]Note: this paper describes version 2 of SITools while we are still using version 1.

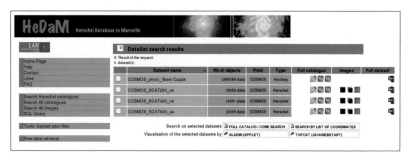

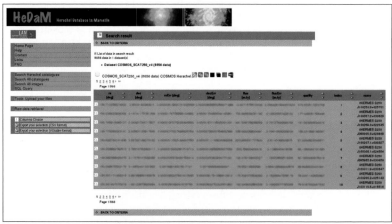

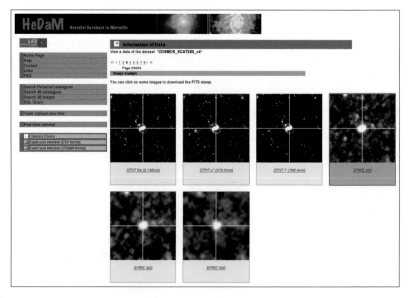

Figure 1.   Various screenshots of the HeDaM site. Top: the dataset selection screens; middle: a catalog table display; bottom: the detailed view of a source with multi-λ postage stamps around it.

3. Advanced Feature Implementations

List-Based Cone Search

HeDaM users can perform list-based cone searches on several catalogs simultaneously and get the results very quickly thanks to the PostgreSQL / pgSphere[2] combination, and to the possibility for VOTable files to be generated *on the fly*.[3]

Another advantage of the VOTable format is that, being an XML format, it can be transformed into XHTML directly by the browser using an XSLT stylesheet. Thus, the same "search by list" result can be either downloaded to be opened by a software application, or displayed as shown in Figure 2.

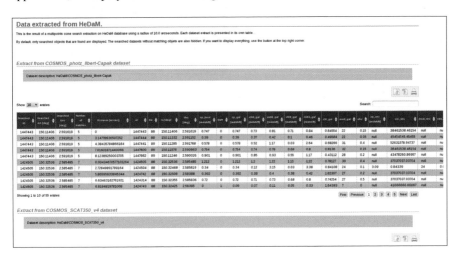

Figure 2.   Displaying a VOTable file directly in the browser.

Multi-λ FITS Postage Stamps

We adapted the source detail page from SITools to display multi-λ cut-outs around each source. Each FITS and PNG image URL corresponds to a meaningful scheme and missing ones are generated on demand — using the HTTP *404 error redirection* mechanism — and cached by the server. A screenshot of the detail page can be found in Figure 1.

4. Evolution of the Service

We are currently working to provide our users with a way to visualise the footprints of the data available in HeDaM — using the Google Sky API — and to get pretty prints

---

[2]http://pgsphere.projects.postgresql.org

[3]Because VOTable is in XML format and is thus text, as opposed to a binary format.

of them (Figure 3). For this, we plan to use Python with *footprintfinder*[4] (Stoehr 2008) for footprint creation and *matplotlib* with *pywcsgrid2*[5] for printing.

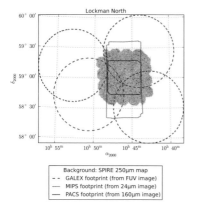

Figure 3.   A planned enhancement: displaying footprints of data available in HeDaM.

Acknowledgments.   The HeDaM team would like to thank the surveys used in the creation the database, namely: the *Herschel Multi-tiered Extragalactic Survey*[6] (Oliver et al. 2010), the *Herschel Reference Survey* (Boselli et al. 2010), the *Very Nearby Galaxy Survey* (Bendo et al. 2010; Roussel et al. 2010), the *Dwarf Galaxy Survey* (Galametz et al. 2010; O'Halloran et al. 2010) and the *Great Observatories Origins Deep Survey / Herschel* (Elbaz et al. 2011, in preparation).

References

Bendo, G. J., et al. 2010, A&A, 518, L65
Bonnarel, F., et al. 2000, A&AS, 143, 33
Boselli, A., et al. 2010, PASP, 122, 261
Fenouillet, T., et al. 2011, in Astronomical Data Analysis Software and Systems XX, edited by I. N. Evans, A. Accomazzi, D. J. Mink, & A. H. Rots (San Francisco, CA: ASP), vol. 442 of ASP Conf. Ser., 17
Galametz, M., et al. 2010, A&A, 518, L55
Malapert, J.-C. 2011, in Astronomical Data Analysis Software and Systems XX, edited by I. N. Evans, A. Accomazzi, D. J. Mink, & A. H. Rots (San Francisco, CA: ASP), vol. 442 of ASP Conf. Ser., 103
O'Halloran, B., et al. 2010, A&A, 518, L58
Oliver, S. J., et al. 2010, A&A, 518, L21
Roussel, H., et al. 2010, A&A, 518, L66
Stoehr, F. 2008, ST-ECF Newsletter, 7
Taylor, M. B. 2005, in Astronomical Data Analysis Software and Systems XIV, edited by P. Shopbell, M. Britton, & R. Ebert (San Francisco, CA: ASP), vol. 347 of ASP Conf. Ser., 29

---

[4]http://www.stecf.org/software/ASTROsoft/Footprintfinder/

[5]From Lee J. Joon, see http://leejjoon.github.com/pywcsgrid2/

[6]http://hermes.sussex.ac.uk

Astronomical Data Analysis Software and Systems XX
ASP Conference Series, Vol. 442
Ian N. Evans, Alberto Accomazzi, Douglas J. Mink, and Arnold H. Rots, eds.
© *2011 Astronomical Society of the Pacific*

Storing Data in Science Archives: Striving for a Common Architecture

N. Fajersztejn, C. Arviset, D. Baines, I. Barbarisi, J. Castellanos, N. Cheek, H. Costa, M. Fernandez, J. Gonzalez, A. Laruelo, I. Leon, B. Martinez, I. Ortiz, P. Osuna, C. Rios, J. Salgado, M. H. Sarmiento, and D. Tapiador

ESA-ESAC, Science Operations Department, PO Box 78, 28691 Villanueva de la Canada, Madrid, Spain

Abstract.   The Science Archive Team at ESAC (European Space Astronomy Center) is responsible for developing, maintaining and operating the Science Archives for all the Astronomy and Planetary missions. Due to the different nature of the missions, each one with its peculiarities, the data produced has a great variety of formats and is always delivered in very project-specific way. Taking this restriction into account, managing the storage and ingestion of data into the Science Archives in a homogeneous way to make a better use of resources has never been easy. The continuous evolution of technologies involved has made this an even more challenging process since the release of the first archive a decade ago.

1.   Introduction

The European Space Astronomy Center (ESAC) hosts ESA's space-based Astronomy and Planetary mission scientific archives. This currently includes the ISO Data Archive,[1] the XMM-Newton Science Archive,[2] the Integral SOC Science Data Archive,[3] all ESA's Planetary mission archives (Rosetta, Mars Express, Venus Express, Smart-1, Huygens and Giotto),[4] the Herschel Science Archive,[5] the SOHO Science Archive[6] and the EXOSAT Science Archive.[7] More archives are currently under development including Planck Legacy Archive, Lisa Pathfinder and possibly others. A dedicated Science Archives and Virtual Observatory Team (SAT henceforth) is in charge of the design, development, operations and maintenance of the aforementioned archives.

---

[1]http://iso.esac.esa.int/ida

[2]http://xmm.esac.esa.int/xsa

[3]http://integral.esac.esa.int/isda

[4]http://www.rssd.esa.int/psa

[5]http://herschel.esac.esa.int/Science_Archive.shtml

[6]http://soho.esac.esa.int/data/archive/index_ssa.html

[7]http://www.rssd.esa.int/index.php?project=EXOSAT&page=archive

2.   First Generation of Science Archives

The SAT started building Astronomical Data Archives back in 1996. IT standards, tools, languages, etc. have had an evolution which could hardly be foreseen at the time. Mission-specific requirements and limited available technology at the time led to very heterogeneous ways of extracting and ingesting data from the various Science Products into many different relational databases. In spite of these limitations, the team succeeded in building the first generation of Science Archives.

Figure 1 gives an overview of the different technologies, RDBMS and operating systems that have been used in the ingestion of data for the 1st generation of archives.

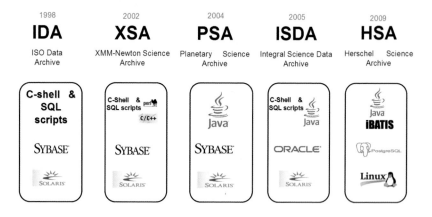

Figure 1.   Technologies used in the 1st generation of Science Archives.

3.   Second Generation of Science Archives

Based on the experience accumulated over the past decade building several archives, it is now possible for us to build a second generation of archives with a common infrastructure to support the storage and ingest of data in a more flexible way.

Since the first public version of the Infrared Space Observatory (ISO) Archive, the SAT has undertaken the effort to build a state of the art **Archives Building System Infrastructure (ABSI)** that provides the building blocks for the creation of ESA Space Based Missions archives with renewed technologies and standards. This new infrastructure consists of a common scientific archives architecture, divided into three tiers. Inside every tier, a set of tools and libraries allows a quick building and deployment of individual and independent modules which together build an archive system. Advantages of this approach are:

- Easier and faster building of new archives,
- Re-use of knowledge and tools to maximize resources,
- Generation of multiple generic libraries and components shared across projects,
- Scalability along with an easy maintenance.

Figure 2 illustrates the evolution undergone by the archives when the ABSI approach was adopted. The same OS, same RDBMS (with the same modules required to run geometrical searches), and same programming language with its associated frameworks were used throughout.

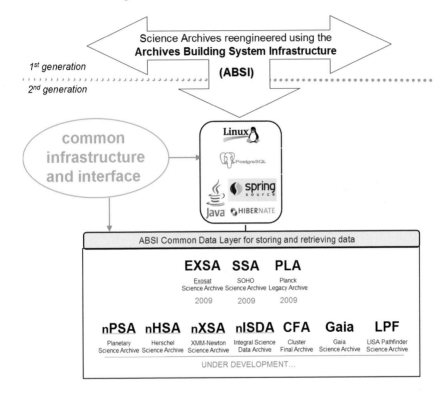

Figure 2.   Evolution of Science Archives into the 2nd generation framework.

4.   Conclusion

The Archives Building Systems Infrastructure has allowed the SAT to create three new archives from scratch. Archives already available to the community are performing well, efficient and scalable, even under demanding requirements.

As an example, the SOHO Science Archive currently has almost two million observations in the database and nearly four million files (and growing) in the data repository while providing fast and reliable service.

As a result of this success, several archives are under development following this infrastructure, and some existing ones are being re-engineered in order to benefit from all the advantages provided by the ESA 2nd Generation Science Archives. Getting to the point where all the archives are sharing the same infrastructure is something one could hardly imagine some time ago, but experience and results are proving the success of our approach and indicate that this is the best way to go forward.

Astronomical Data Analysis Software and Systems XX
ASP Conference Series, Vol. 442
Ian N. Evans, Alberto Accomazzi, Douglas J. Mink, and Arnold H. Rots, eds.
© *2011 Astronomical Society of the Pacific*

Centralized FITS Metadata Handling Framework at ESO

I. Vera, A. Dobrzycki, M.-H. Vuong, and A. Brion

ESO, European Organization for Astronomical Research in the Southern Hemisphere, Karl-Schwarzschild-Str. 2, 85748 Garching bei München, Germany

Abstract. The European Organization for Astronomical Research in the Southern Hemisphere (ESO) headquartered in Garching, Germany, operates three state-of-the-art observatories in Chile: at La Silla, Paranal and Chajnantor. The observatories produce huge amounts of data, which are transferred to the ESO headquarters and stored in the ESO Archive. The archive also stores the products generated by pipeline processing of raw data by the ESO Quality Control (QC) group. In addition, the Archive also stores science products delivered by the principal investigators. Historically, these three processes were independent from one another which resulted in different metadata handling frameworks for each kind of product. This architecture led to duplicate efforts for building metadata-based services as well as creating challenging scenarios for integrating metadata from different processes. We present the metadata handling framework recently implemented at the European Southern Observatory, in which all metadata from FITS compliant sources are stored in a centralized repository database. The framework allows building metadata-based services in a descriptive way, transparently propagating any modifications/updates performed in the central repository. In addition, through a set of common functions, the framework allows for enrichment of FITS-based metadata with derived values and/or with metadata from other ESO operational databases.

1. Introduction

The ESO archive contains every type of astronomical data, from raw data observed at the telescopes to highly processed data products submitted by the Principal Investigators (PI). Practically all this astronomical data are stored in Flexible Image Transport System (FITS) files. FITS is the standard astronomical data format, which consists of an unlimited number of Header-Data Units (HDUs). Each HDU contains a data part and a metadata header, which describes the structure and contents of that HDU.

Query services at ESO are built on top of a relational database and the contents are currently extracted directly from the metadata headers of FITS files. Every query service has been built around its own set of tools and their own architecture, which in the long term has made it very difficult to integrate the different services. We can distinguish three main data services:

- Raw data: Services that provide the user with data obtained at the observatories.

- Internal data products: Services that provide users with data products generated by ESO internal pipeline processing.

- External data products: Services that provide users with data products submitted by the PIs.

In this paper, we first describe how the services enumerated above have been built and we will discuss their current limitations. We then present the new architecture designed to handle FITS based query services.

2.   Legacy Metadata Handling Frameworks

Different data services provided by ESO have been built mostly independently of each other. This was due to the different groups of people working on each framework and the needs of the organization at that point of time.

2.1.   Data Generated by the Observatories: Raw Data

ESO operates three state-of-the-art observatories in the Atacama dessert, Chile. Data generated by those observatories is archived at ESO headquarters in Garching, Germany (Dobrzycki et al. 2006). Historically, due to the limitation of bandwidth, data was transferred to ESO headquarters via Diplobag, which takes between one or two weeks to arrive to the archive.

Due to the necessity of having almost real-time information of what had been observed, some metadata from the headers was extracted on the mountain and replicated to the headquarters using database replication. This metadata was used to build the query services to access data generated at the observatories.

The architecture is shown in Fig. 1. The main problem with this data flow is the difficulty to perform any changes in the contents of the metadata service. Changes are required in the databases at the observatory and at the headquarters, in addition to changes in the replication definitions.

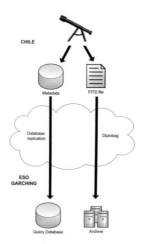

Figure 1.   Archival of raw data.

2.2. Data Generated by ESO Pipeline Processing: Internal Data Products

As part of the back-end data flow, data from the VLT/ VLTI are processed and their quality is checked and monitored (Hanuschik et al. 2002). This process is done daily, providing feedback to the observatory about the health of the instruments. This process is done through ESO internal pipelines.

Apart from quality control of the instrument, science raw files belonging to the VLT/VLTI are processed using the internal ESO pipelines and they are made available to the PI of the program. The architecture is shown in Fig. 2. Currently it is not possible to access those processed products for anyone but the PI even when data have been made public.

2.3. Data Generated by Principal Investigators: External Data Products

ESO requests from PIs of large programs and public surveys the delivery of final data products at the time of publication of their results. This data can later be retrieved using specialized query forms or using VO tools (Rité et al. 2008). The architecture is shown in Fig. 3, where highly specialized tools were developed to check the validity of the products as well as to develop a mapping tool for keywords recognition. Currently ESO is developing a new set of tools were external data products are easy to manipulate inside ESO data flow.

Figure 2.   Archival of ESO internal data products.

Figure 3.   Archival of ESO external data products.

3.   Centralized Metadata Handling Framework

The new ESO metadata framework is a centralized architecture to process metadata coming from FITS headers. New developments in the ESO data flow have allowed an improvement in the data architecture. First, the introduction of a dedicated line to transfer information from the observatories to the headquarters (Zampieri et al. 2009) has allowed the direct transfer of FITS files in minutes after the exposure. Secondly, the development of the keywords repository (Vuong et al. 2008), has provided permanent storage for all FITS headers at ESO.

Fig. 4 shows the schematics for the new architecture. During the archival process, FITS metadata headers are extracted and inserted in the keywords repository. Query services are fed from the keywords repository in an incremental fashion, and any change in FITS metadata is automatically propagated to the query services. The mapping between keywords and database fields is done using a configuration file which allows the use of a set of common functions. These functions are easily created and can be used to enrich the metadata with derived values and/or with metadata from other ESO operational databases.

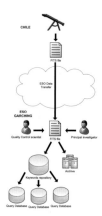

Figure 4.   Centralized metadata handling framework.

4. Conclusions

Several advantages have been achieved with the new architecture. The most important ones are listed here:

- Well defined information flow, with no parallel paths. Data modifications occur in one place and are propagated to all dependant services.
- Introduction of derived metadata items which can be shared by different metadata services.
- It is possible to efficiently combine information from all parts of ESO data flow.
- Homogenization of information for metadata-based services.

Currently it has been implemented for raw data, but it is planned to soon support different types of processed data.

References

Dobrzycki, A., Brandt, D., Giot, D., Lockhart, J., Rodriguez, J., Rossat, N., & Vuong, M. H. 2006, in Observatory Operations: Strategies, Processes, and Systems, edited by D. R. Silva, & R. E. Doxsey (Bellingham, WA: SPIE), vol. 6270 of Proc. SPIE, 627027

Hanuschik, R. W., Hummel, W., Sartoretti, P., & Silva, D. R. 2002, in Observatory Operations to Optimize Scientific Return III, edited by P. J. Quinn (Bellingham, WA: SPIE), vol. 4844 of Proc. SPIE, 139

Rité, C., Slijkhuis, R., Rosati, P., Delmotte, N., Rino, B., Chéreau, F., & Malapert, J. 2008, in Astronomical Data Analysis Software and Systems XVII, edited by R. W. Argyle, P. S. Bunclark, & J. R. Lewis (San Francisco, CA: ASP), vol. 394 of ASP Conf. Ser., 605

Vuong, M., Brion, A., Dobrzycki, A., Malapert, J., & Moins, C. 2008, in Observatory Operations: Strategies, Processes, and Systems II, edited by R. J. Brissenden, & D. R. Silva (Bellingham, WA: SPIE), vol. 7016 of Proc. SPIE, 70161M

Zampieri, S., Forchi, V., Gebbinck, M. K., Moins, C., & Padovan, M. 2009, in Astronomical Data Analysis Software and Systems XVIII, edited by D. A. Bohlender, D. Durand, & P. Dowler (San Francisco, CA: ASP), vol. 411 of ASP Conf. Ser., 540

Astronomical Data Analysis Software and Systems XX
ASP Conference Series, Vol. 442
Ian N. Evans, Alberto Accomazzi, Douglas J. Mink, and Arnold H. Rots, eds.
© *2011 Astronomical Society of the Pacific*

Feedback from the ESO Archive: Towards Observatory Use Optimization

M. Vuong, A. Delgado, N. Delmotte, J. Santander-Vela, R. Stevenson, and I. Vera

ESO, European Organization for Astronomical Research in the Southern Hemisphere, Karl-Schwarzschild-Str. 2, 85748 Garching bei München, Germany

Abstract. The ESO Archive contains data collected at the three observatory sites in Chile, on Cerro Paranal, Cerro La Silla and the Chajnantor plateau. The data originates from 10 telescopes ranging from 2.2 to 12 meters equipped with 35 state-of-the-art instruments. Together, they cover a large observational domain from optical to submillimeter. In the near future, ESO will also host a complete synchronised copy of the Atacama Large Millimeter/submillimeter Array (ALMA) archive through the European ALMA Regional Center (ARC). In this paper we present the large and varied collection of archival holdings stored within the ESO data center, and services for users to access the Archive. We illustrate a few examples of archive feedback: *(a)* APEX observation planning optimization based on meteorological archival data; *(b)* historical PWV record over Paranal based on analysis of archival spectra taken during 2001–2008; *(c)* finding Solar System Bodies in ESO archive observations; *(d)* feeding back the Archive with pipelined processed spectra public through the ESO Archive services. The provision of such value-added services is a vital step in fully exploiting all the precious resources that comprise a state-of-the-art astronomical observatory.

1. ESO Data and Services

The European Southern Observatory constructs, operates and continues to upgrade optical/infrared telescopes on Cerro La Silla and Cerro Paranal. In partnership with the Max-Planck Gesellschaft (MPG) and Sweden, ESO operates the submillimiter antenna APEX on Chajnantor plateau, where it is constructing ALMA in partnership with North America and East Asia. Collectively, the telescopes produce huge amounts of data (e.g. 18 Terabytes for 2009), which are stored at the ESO Science Archive Facility[1] in Garching.

1.1. Data Holdings

As of October 2010, the ESO Archive contains more than 12 million observations collected by 35 state-of-the art instruments in the optical, infra-red and submillimeter wavebands. Those observations include: 17900 and 26500 hours of imaging and spectroscopy exposures, respectively, using optical and infra-red instruments on the NTT, MPG-ESO 2.2m and ESO-3.6m telescopes at La Silla since 1991; 14700 and 41100 hours of imaging and spectroscopy exposures, respectively, using optical and

---

[1]http://archive.eso.org

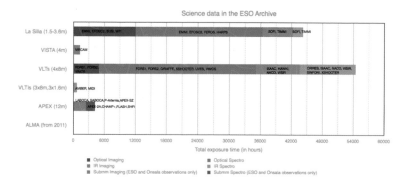

Figure 1.   Science observations performed at ESO telescopes and observatory sites. The total exposure time has been calculated using science frames having a target pointing, which are available in the Archive.

infra-red instruments on the VLTs, VLTIs and VISTA telescopes at Paranal since 1999; and at least 4200 hours of exposures using bolometer and heterodyne instruments on the APEX telescope using ESO and Swedish observing time (50% of the total) since 2005.

1.2.   Services Sccessing the ESO Archive

Most of the ESO data are available world-wide. After a proprietary period (usually 1 year), science data can be requested by the archive users. To access the ESO Archive contents, several services are offered[2]: *(a)* ESO archive query form[3]: a unified and observatory-oriented access to the ESO collection of raw data for astronomers with no previous experience with ESO instruments; *(b)* Reduced data query form[4]: a unified and science-oriented access to the ESO collection of both imaging and spectroscopic advanced data products; *(c)* VirGO[5]: the next generation visual archive browser; *(d)* Instrument specific query forms for raw data[6]: instrument specific and technically-oriented access points to the ESO archive for astronomers already familiar with ESO instrument setups and observing strategies.

The ESO Archive database also stores measurements of environmental conditions collected at three observatory sites: *(a)* on Cerro La Silla and Cerro Paranal[7] since Jan. 1991 and Aug. 1998 respectively; *(b)* on Chajnantor Plateau[8] since Jan. 2006.

---

[2]VO services delivering images and spectra are also available to VO clients

[3]http://archive.eso.org/eso/eso_archive_main.html

[4]http://archive.eso.org/eso/eso_archive_adp.html

[5]http://archive.eso.org/cms/virgo

[6]http://archive.eso.org/cms/eso-data/instrument-specific-query-forms

[7]http://archive.eso.org/asm/ambient-server

[8]http://archive.eso.org/wdb/eso/meteo_apex/form

2. Feedback from the ESO Archive

2.1. Climatology at Chajnantor Plateau Based on Meterological Data Stored in the ESO Archive

We illustrate an example of feedback from archive use towards a more optimal planning of observations. Fig. 2 shows the amount of precipitable water vapor (PWV) in the atmosphere, as measured by the radiometer of the APEX telescope, located at 5100m at the Chajnantor plateau. The radiometer data starting from 2006 are transferred in real time to Garching and are stored in the ESO archive. These data have been extracted to calculate the 25, 50 and 75 percentile levels of each week of the year. This plot clearly shows the impact of the altiplanic winter, leading to higher PWV between late December and March. Based on the analysis, the APEX team has optimised their technical and science operations activities, scheduling major maintenance tasks during January and March. The new experience from the 5 years of data shown here has led to a 2 week shift in the period during which the science operations are concentrated. The figure also shows that the best weather months are July and August, which is when the shortest wavelength instruments will be operated. While this information is intuitive from experience in observing, it is only after quantifying it by Archival study that major changes in the operations schedule can be approved (Lundgren et al. 2010).

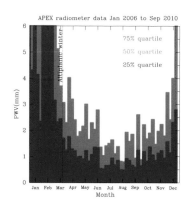

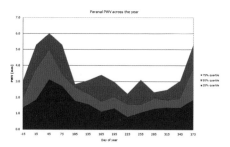

Figure 2. Annual PWV variation at Chajnantor Plateau from meteorological data in the ESO Archive.

Figure 3. Annual PWV variation at Paranal based on UVES flux standard observations from 2001 to 2008 (Courtesy of R. Hanuschik). Day=-15 corresponds to December 16th

2.2. Precipitable Water Vapor over Cerro Paranal Based on Archival UVES Data

In supporting the characterization of potential sites for the European Extremely Large Telescope (E-ELT), a campaign has been established to improve the understanding of atmospheric precipitable water vapor (PWV) at ESO's La Silla Paranal Observatory (Kerber et al. 2010). Fig. 3 shows the annual PWV variation derived from the reprocessing of UVES archival data for the period 2001 to 2008. About 1500 UVES flux standard observations extracted from the Archive were analyzed to reconstruct the PWV history at Paranal.

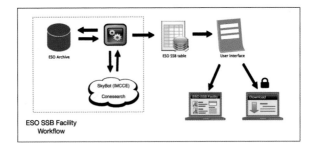

Figure 4.   Solar System Body workflow at ESO

2.3.   Finding Solar System Bodies in ESO Archival Observations

As shown in Fig. 1 above, thousands of images reside in the Archive awaiting further exploitation. There is a very good chance that many of those had their field of view crossed by solar system bodies (SSB), whether planned or by chance. Nevertheless, their identification in astronomical archives proves to be very difficult given that the traditional name resolvers like NED/SIMBAD do not account for moving targets. Therefore, investigation for a Solar System Body service (Delgado et al. 2011) has been initiated at the ESO Archive. Fed by images taken with the Wide Field Imager (WFI) on the MPG/ESO-2.2m telescope at La Silla, the Skybot service[9] is being used to identify the list of SSB present in matching WFI frames that could be stored later on in a database and exposed to the users via a search interface (Fig. 4).

2.4.   A Step Towards a Science-Ready Archive: Re-Ingest of Data Products into the ESO Archive

The ESO Science Archive Facility aims at optimising the scientific return from its large science observation collection. As part of this objective, services for offering the user community access to science ready data are planned. For example, all HARPS observations (about 200,000 frames) taken during the period 2003–2010 have been extracted from the Archive and analyzed through a pipeline. Processed data products will be then massively re-ingested into the Archive to be served to the astronomical community. A dedicated query form to access those data is in preparation.

Acknowledgments.   We thank R. Hanuschik for the record of PWV diagram over Paranal, F. Mac Auliffe, I. Percheron, S. Moehler and J. Pritchard for computing the total exposure time on LABOCA/APEX and VLT/I and P. Eglitis, N. Fourniol for helpful comments.

References

Delgado, A., Delmotte, N., & Vuong, M. 2011, in Astronomical Data Analysis Software and Systems XX, edited by I. N. Evans, A. Accomazzi, D. J. Mink, & A. H. Rots (San Francisco, CA: ASP), vol. 442 of ASP Conf. Ser., 111
Kerber, F., Querel, R., Hanuschik, R., Chacón, A., Sarazin, M., & Project Team 2010, The Messenger, 141, 9
Lundgren, A., et al. 2010, in Observatory Operations: Strategies, Processes, and Systems III, edited by D. R. Silva, A. B. Peck, & B. T. Soifer (Bellingham, WA: SPIE), vol. 7737 of Proc. SPIE, 773708

---

[9]http://vo.imcce.fr/webservices/skybot

Astronomical Data Analysis Software and Systems XX
ASP Conference Series, Vol. 442
Ian N. Evans, Alberto Accomazzi, Douglas J. Mink, and Arnold H. Rots, eds.
© *2011 Astronomical Society of the Pacific*

LBT Distributed Archive: Status and Features

C. Knapic,[1] R. Smareglia,[1] D. Thompson,[2] and G. Gredel[3]

[1]*INAF / Astronomical Observatory of Trieste, Italy*

[2]*Large Binocular Telescope Observatory, AZ, USA*

[3]*Max-Planck-Institute fuer Astronomie, Heidelberg, Germany*

Abstract.   After the first release of the LBT Distributed Archive, this successful collaboration is continuing within the LBT corporation. The IA2 (Italian Center for Astronomical Archive) team had updated the LBT DA with new features in order to facilitate user data retrieval while abiding by VO standards. To facilitate the integration of data from any new instruments, we have migrated to a new database, developed new data distribution software, and enhanced features in the LBT User Interface. The DBMS engine has been changed to MySQL. Consequently, the data handling software now uses java thread technology to update and synchronize the main storage archives on Mt. Graham and in Tucson, as well as archives in Trieste and Heidelberg, with all metadata and proprietary data. The LBT UI has been updated with additional features allowing users to search by instrument and some of the more important characteristics of the images. Finally, instead of a simple cone search service over all LBT image data, new instrument specific SIAP and cone search services have been developed. They will be published in the IVOA framework later this fall.

1. Requirements and General Choices

The LBT Data Archive (LBT/DA) must provide the storage, maintenance, efficient transmission and release of all LBT scientific data to the LBTO and Astronomical Communities with an appropriate policy and by VO standards (Smareglia et al. 2006). After the first release of the LBT Distributed Archive, this successful collaboration is continuing within the LBT corporation (Ragazzoni et al. 2006; Speziali et al. 2008). LBT/DA has been developed by the IA2 (Italian Center for Astronomical Archives) team.[1]

The structure of the whole database and archive is designed for easy integration of FITS format data from new instruments, allowing insertion of new FITS header keywords if they respect FITS standards.[2] The technologies used are open source and VO compliant.

Multi-threading Java technology is used to send up to ten files at a time in order to maximize the efficiency of data transmission. The software is fast and has robust error tracking in order to monitor all operations.

The main archives on Mt. Graham and in Tucson are identical and store all LBT data. They provide an efficient distribution of proprietary data to the archives in Trieste and Heidelberg. All metadata and calibration files are present in each archive, while proprietary data is sent only where appropriate. At the moment, only INAF raw data (Trieste) have a one year proprietary policy after which files are considered freely distributable.

---

[1]http://ia2.oats.inaf.it/

[2]http://fits.gsfc.nasa.gov

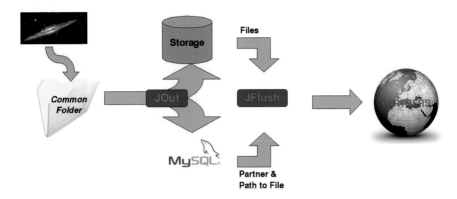

Figure 1.   LBT-DA Data Flow.

At all archive sites a User Interface (UI) for data retrieval is present, allowing authorized users to download single FITS files, tar archives with more than one FITS file or a simple VOTable describing the current results of a query. The UI and the consequent query results are formed depending on the user account type (public, administrator, partner or PI). SIAP and Cone Search VO web services are published; they release LBT INAF public raw data to VO Alliance clients (Euro-VO registry).

2. Data Handling

Data handling applications are distributed throughout the four archive sites. They are used they are used in different modes depending on location but the code is the same in order to be consistent everywhere (Smareglia et al. 2008).

2.1. Data Management

Newly acquired data at the LBT are temporarily stored in a common folder. This folder is monitored by a Java system utility (JNotify) that starts the file management process (JOut).

A MySQL RDBMS is used to store metadata. Once a file is detected, JOut inserts a subset of the FITS keywords in an instrument specific schema table and in a general table (LBT) of all the LBT data. The general LBT table contains also technical information like the file location, file status, transfer file status and so on.

Some operations are performed on metadata using MySQL procedures to adjust occasionally forgotten keywords or null values and to synchronize instrument tables with LBT table. Then a new record is inserted in the database, the corresponding file is moved/stored in the Archive and, in the case of the Main Archives, also in a well-defined Repository directory directly accessible from the local computers.

In the Archives files are stored in a compressed form while in the Repository files are stored as they are. Supported compression modes are gzip (.gz) or fpack (.fz). The selection between the two can be done when JOut is started. Gzip is the default mode.

If an error occurs, different alternatives are possible depending on the error type. In the case of fatal errors (Jnotify error), JOut terminates. Errors related to FITS file content cause the file to be moved to a warning directory and an error to be reported to a database. In any case, all errors are reported in a dedicated log file.

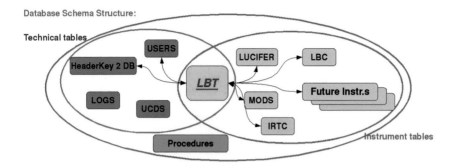

Figure 2.   LBT-DA Database Schema Structure.

2.2.   Data Forwarding

Once data are correctly inserted into the database, stored in the archive and all updates are performed, the files are ready to be sent to recipients (Partners/Owners). Depending on which recipient is selected, a java multi-threading program (JFlush) finds the associated URL for the recipients and searches for files with that name in the partner column of the general table. It uses a blocking queue to create a list of commands to be executed and a "WorkerThread" object that executes the operations as independent threads. The constructor takes as input an array of strings (list of commands), the number of threads to run simultaneously, and the owner of the data. It works by instantiating N WorkerThread objects that execute the command strings that are the first N elements available of the blocking queue list. Each thread executes independently and when one thread ends, another starts the next command until the end of the list. The procedure terminates when the list is empty and a special string is sent. The command is basically an invocation of a shell script that uses Rsync. The Rsync software application minimizes data transfer and serves files in a remote shell SSH mode. If Rsync ends successfully, a flag with information about destination and current time is set, otherwise a "communication Interrupted" message is issued. Every interrupted transfer is reentered in the blocking queue for a later run of Jflush. The program provides a detailed error log file. All metadata in all databases are synchronized.

3.   Data Retrieval: User Interface and Web Services

An IVOA compliant and user friendly web application (the User Interface) is available which could also work in a clustered web server. In it, all data from all LBT instruments are presented. The UI can perform a general query on all LBT data or a more specific query about a single instrument. Release of public data is dependent on Partner policies, but in general all metadata are accessible and they are issued in the form of simple VOTables. Currently all files, VOTables and tar archives are served directly from the UI. In the future we plan to use a FileServer, a web service that works as an application-independent validator for the incoming file request: if the query fully meets the file access policy, the requested file is served, otherwise an exception is thrown.

The LBT application is dedicated to system administration, public data access or user-authenticated use. In the first case all data (files and metadata) present in archive are available to the administrators. In the second case, all metadata, calibration and INAF public files are available to the public. In the third case, all metadata, calibration and data files are available to download either to research partners or to the individual PIs.

Another application we developed is the LBT web service, a SIAP and Cone Search compliant service that is published in the Euro-VO registry. It serves a VOTable with all available public data, with the delivery of the files is delegated to the FileServer service. Currently only INAF has an LBT data policy: all INAF raw data older than one year are public.

4. Conclusions

Java applications, MySQL procedures and shell scripts handle files automatically from the receipt of data from the instruments through the end-delivery to the science user. All metadata and non-proprietary data are available to the community. Proprietary data are secured by robust logic and dedicated database queries. VO-compliant services have been developed and are working.

New features will be developed to make data retrieval more efficient and utilities more user friendly. A tool to easily build a UI web application is planed to be developed to allow users to build their own applications for the retrieval of all the FITS format data, requiring only a database and a web server. Data managed from IA2 staff will be shifted to this new technology first.

Acknowledgments. We are very grateful to all people involved in the LBT consortium for many helpful discussions on data handling, security, and data policy, all of which have helped us improve the LBT Distributed Archive.

References

Ragazzoni, R., et al. 2006, in Ground-based and Airborne Telescopes, edited by L. M. Stepp (Bellingham, WA: SPIE), vol. 6267 of Proc. SPIE, 626710

Smareglia, R., Gasparo, F., Manzato, P., Fontana, A., Thompson, D., & Gredel, R. 2008, in Astronomical Data Analysis Software and Systems XVII, edited by R. W. Argyle, P. S. Bunclark, & J. R. Lewis (San Francisco, CA: ASP), vol. 394 of ASP Conf. Ser., 442

Smareglia, R., Pasian, F., Becciani, U., Longo, G., Prete Martinez, A., & Volpicelli, A. 2006, Memorie della Societa Astronomica Italiana Supplementi, 9, 423

Speziali, R., et al. 2008, in Ground-based and Airborne Instrumentation for Astronomy II, edited by I. S. McLean, & M. M. Casali (Bellingham, WA: SPIE), vol. 7014 of Proc. SPIE, 70144T

Astronomical Data Analysis Software and Systems XX
ASP Conference Series, Vol. 442
Ian N. Evans, Alberto Accomazzi, Douglas J. Mink, and Arnold H. Rots, eds.
© *2011 Astronomical Society of the Pacific*

Automated Curation of Infra-Red Imaging Data in the WFCAM and VISTA Science Archives

Nicholas Cross, Ross Collins, Eckhard Sutorius, Nigel Hambly, Rob Blake, and Mike Read

Scottish Universities' Physics Alliance (SUPA), Institute for Astronomy, School of Physics, University of Edinburgh, Royal Observatory, Blackford Hill, Edinburgh EH9 3HJ, UK

Abstract.   The two fastest near infrared survey telescopes are UKIRT-WFCAM and VISTA. The data from both these instruments are being archived by Wide Field Astronomy Unit (WFAU) at the IfA, Edinburgh, using the same curation pipeline, with some instrument-specific processing. The final catalogs from these surveys will contain many tens of billions of detections. Data are taken for a range of large surveys and smaller PI programs. The surveys vary from shallow hemisphere surveys to ultra-deep single pointings with hundreds of individual epochs, each with a wide range of scientific goals, leading to a wide range of products and database tables being created. Processing of the main surveys must allow for the inclusion of specific high-level requirements from the survey teams, but automation reduces the amount of work by archive operators allowing a higher curation efficiency. The decision making processes which drive the curation pipeline are a crucial element for efficient archiving. This paper describes the main issues involved in automating the pipeline.

1.   The WFCAM and VISTA Science Archives

The WFCAM and VISTA Science Archives (VSA, Hambly et al. (2008)) are the main access for data from WFCAM (Casali et al. 2007) and VISTA (Emerson & Sutherland 2010). The majority of time on both instruments is spent on large surveys: UKIDSS (Lawrence et al. 2007), and the VISTA Public Surveys (Arnaboldi et al. 2007). There are also a range of smaller Principal Investigator (PI) programs allocated by the Telescope Allocation Committees each semester that require curating. We run the same set of tasks on all the surveys and programs, with the amount of processing in each task dependent on the type of program. For instance a wide survey will spend more time on band-merging and neighbor tables, but a deep survey will spend more time on deep stack creation and multi-epoch tables. PI programs are set up completely automatically[1] because of the large number of programs, but surveys are set up in a semi-automatic way receiving special instructions, quality control and sometimes additional products from the science teams. PI programs usually obtain all their data in one observing semester and are processed completely at the end of the semester, so new releases will be due to software or calibration improvements necessitating a complete reprocessing. Surveys build up data over many semesters and will be appended to, as well as occasionally reprocessed. These different scenarios have to be factored into the pipeline control.

---

[1] Occasionally we have manually grouped together several related PI programs from different semesters before automatic processing.

2. Overview of Data Pipeline

Data consisting of images and catalogs are first transferred to WFAU by the Cambridge Astronomy Survey Unit (CASU) who process each observing block and calibrate the data. We ingest these into the science archives along with any external data or quality control provided by the science teams for public surveys. The automated curation pipeline is then run, executing the following tasks:

- Quality Control,
- Program set-up (using **ProgrammeBuilder** class),
- Creation and ingestion of deep products and catalogs,
- Creation of band-merged Source table from deepest products,
- Recalibration of each epoch,
- Creation of band-merged catalogs for each epoch,
- Creation of neighbor tables,
- Creation of synoptic tables for light-curves and variability analysis.

The dataflow plan for automated curation of a single program is shown in Fig 1a. The automated pipeline has changed in a couple of ways since Collins et al. (2009). We have removed the distinction between deep and shallow programs in a way that also allows us to do an easier comparison between the curation and data tables and removes the need to copy data. Appending either in width (more pointings) or depth (more epochs) is much easier. We have also designed a much more sophisticated SQL schema template that can be used by both surveys and PI programs so that only one template is needed per instrument. These changes have significantly increased the amount of automation of public surveys so that many of the curation tables are filled in the same way as PI programs.

3. Setting up a Program

The curation tables that control the pipeline are filled by the **ProgrammeBuilder** class, see Fig 1b. The image metadata is used to group pawprint frames by position, filter, microstepping and in the case of VISTA position angle and offset position within a standard tile. Unique pointings are found by grouping by position alone and then products are found for each filter in these pointings. The information for each pawprint product is put into the `RequiredStack` table. For VISTA, the pawprint stacks are grouped into tiles in `RequiredTile` and the two tables are linked via `ProductLinks`. For each filter used, the number of epochs at each pointing is found, and this is used to determine whether multi-epoch tables are required. This information goes into `RequiredFilters`, `Programme` and `RequiredNeighbours`.

`RequiredNeighbours` contains the list of all the neighbor and cross-match tables which need to be created. In the case of PI programs, a common set of internal neighbor tables and cross-matches to all-sky surveys is produced and, in addition, surveys are cross-matched with wide range of specified external surveys.

Once these curation tables are set-up, a program-specific schema is created using them and a template SQL schema. The template schema is composed of SQL definitions for different attributes, substitution strings and control structures. We give an example below from the template for VISTA `Source` tables:

```
++c:a
**s*&a&m&b&Pnt      real not null,      --/D Point source colour
&As&-&Bs& (using aperMag3)  --/U mag  --/C PHOT_COLOR  --/Q
&a&AperMag3,&b&AperMag3  --/N -0.9999995e9  --/G
allSource::colours
==c:a
```

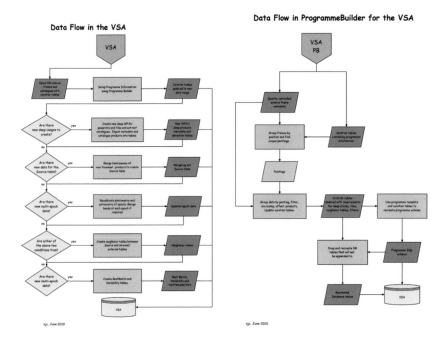

Figure 1.   Left (a): Overall dataflow for WSA or VSA automated processing.
Right (b): Dataflow for the ProgrammeBuilder class. In both figures, the rectangular yellow boxes represent tasks, the parallelograms represent data (light blue for temporary and purple for permanent) and the light yellow diamonds represent control structures. The cylinder represents the database that the permanent data are ingested into and the pentagon represents the pipeline that the dataflow is embedded in. In Fig a), processing of each task to create a permanent data object is determined by a control structure. New data feeds into the control for the next task. Each control structure compares expected products, based on the sorting and grouping of data in Fig b), to actual created products and before determining what still needs to be created.

The + + c : a line is a control structure which repeats each subsequent line for all the program filters until the == c : a line for all color combinations in the survey. The ∗ ∗ s∗ structure controls which table(s) each line goes into when several narrow tables are created for curation purposes and subsequently joined at release into one table. Lines will only go into the Source table, but not the MergeSource. &a&, &A&, &b& and &B& are substitution strings, where a and b refer to the first and second filter in a color respectively and a and A refer to lower and upper case respectively. When the template is processed for the VISTA-VMC (Cioni, M.-R. et al. 2011, in preparation), which contains Y, J and Ks band data, the following piece of schema is produced:

```
ymjPnt       real not null,       --/D Point source colour Y-J
(using aperMag3)  --/U mag  --/C PHOT_COLOR   --/Q yAperMag3,
jAperMag3  --/N -0.9999995e9  --/G allSource::colours
jmksPnt       real not null,       --/D Point source colour J-Ks
(using aperMag3)  --/U mag  --/C PHOT_COLOR   --/Q jAperMag3,
ksAperMag3  --/N -0.9999995e9  --/G allSource::colours
```

The SQL schema is used to create the database with the correct tables, control the code that produces the table data and create a schema browser and glossary for scientists. Having a single template and control structures reduces the need to repeat the SQL, which makes it much easier to update and maintain.

4.   Triggering the Automated Pipeline

Each task is triggered by comparing the curation tables with the data tables. For instance, the deep product curation task will compare the products specified in `RequiredStack` and `RequiredTile` with data in the tables `Multiframe` and `ProgrammeFrame` matching on the program, the product identifier and the release number. If all required products have been created then the pipeline moves onto the next task. If not, the remaining products will be created. Thus the pipeline can be restarted easily if there is a network error or software bug. Other curation tasks are similarly triggered and the data tables are updated at the end of each task. Log files are produced and curation history tables are kept up to date, so any failures can easily be identified.

References

Arnaboldi, M., Neeser, M. J., Parker, L. C., Rosati, P., Lombardi, M., Dietrich, J. P., & Hummel, W. 2007, The Messenger, 127, 28
Casali, M., et al. 2007, A&A, 467, 777
Collins, R., Cross, N. J., Sutorius, E., Read, M., & Hambly, N. 2009, in Astronomical Data Analysis Software and Systems XVIII, edited by D. A. Bohlender, D. Durand, & P. Dowler (San Francisco, CA: ASP), vol. 411 of ASP Conf. Ser., 226
Emerson, J. P., & Sutherland, W. J. 2010, in Ground-based and Airborne Telescopes III, edited by L. M. Stepp, R. Gilmozzi, & H. J. Hall (Bellingham, WA: SPIE), vol. 7733 of Proc. SPIE, 773306
Hambly, N. C., et al. 2008, MNRAS, 384, 637
Lawrence, A., et al. 2007, MNRAS, 379, 1599

Astronomical Data Analysis Software and Systems XX
ASP Conference Series, Vol. 442
Ian N. Evans, Alberto Accomazzi, Douglas J. Mink, and Arnold H. Rots, eds.
©2011 Astronomical Society of the Pacific

LOFAR Long Term Archive

G. A. Renting and H. A. Holties

Netherlands Institute for Radio Astronomy (ASTRON),
P.O. Box 2, 7990 AA Dwingeloo, The Netherlands

Abstract. An overview of the LOFAR Long Term Archive (LTA) is given. The role of the LTA within the LOFAR project is described, as well as the procedure to ingest new data products into the archive and the data model used to store them.

1. LOFAR

LOFAR is a large new radio telescope being constructed in Europe with its center in the Netherlands. It has a total of almost 50 stations with its core in north-eastern Netherlands.

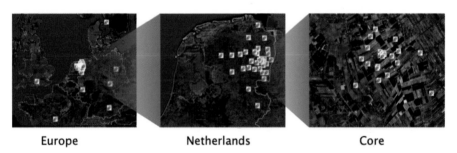

Europe   Netherlands   Core

LOFAR stations consist of two types of antennas, the Low Band Antenna (LBA) for observing 10–90 MHz and the High Band Antenna for 110–250MHz.

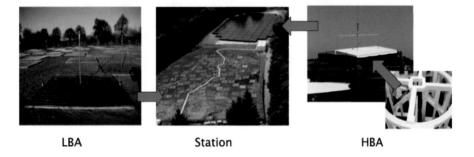

LBA   Station   HBA

Each station has 48 to 192 of both antenna types. These are used as a phased array to form one or more digital beams on the sky. The number of beams that can be formed simultaneously is only limited by the electronics and software. The LOFAR core with 3 beams observing different objects is conceptually shown in the left image of Figure 1.

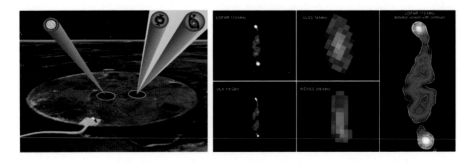

Figure 1.   Left: LOFAR core; right: early result image from LOFAR.

LOFAR was officially opened on the 12th of June by Queen Beatrix and just over 20 stations and the central processing facility are now operational. An early result is shown in Figure 1.

2.   Long Term Archive

The LOFAR data processing (Fig. 2) consists of three main stages:

- A realtime step in which data from the stations is streamed to a central IBM BlueGene supercomputer,
- An offline step in which the data is flagged, calibrated, compressed and initial scientific results are generated,
- The LOFAR Long Term Archive (LTA) for storage of scientific results and further reduction of intermediate results from offline processing.

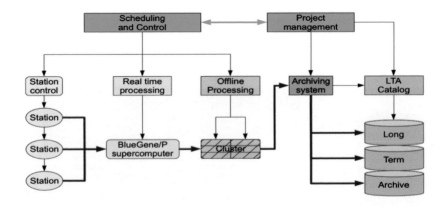

Figure 2.   LOFAR data processing system.

The LTA consists of three main parts:

- A staging area and systems at the central processing facility for managing the ingest into the archive,

- A catalog of the contents for querying and user access,
- Petabyte storage and compute facilities at several large data centers (Target, RuG, SARA, Jülich).

The LTA is expected to store about 5 Petabyte a year from 2011 onward. In addition to this storage it has access to several large GRID clusters for further processing of data. The storage is also connected to dedicated computing systems of some of the LOFAR Key Science Projects.

Finally there is a web based access portal where any registered users can query the catalog and retrieve data. This will also be used for future public access.

3. Ingest Procedure

The ingest procedure to transfer data into the LTA consists of six steps. These are detailed in Figure 3. A unique identifier called StorageTicket is used to uniquely identify a dataproduct between project management and LTA. An XML-based Submission Information Package (SIP) is used to transfer the metadata to the LTA catalog, separately from the actual data transfer.

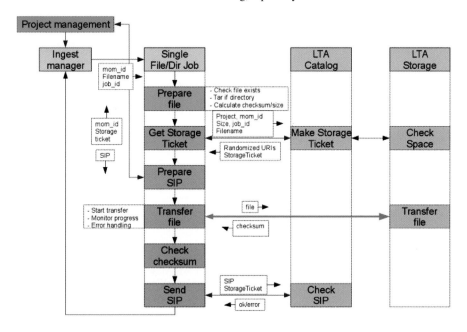

Figure 3.   LOFAR Long Term Archive data ingest procedure.

4. Data Model

The LOFAR data in the archive is described using a data lineage model, where the stored data product and the process that created it are described. Also any data procucts that were used as inputs for the process in a similar fashion are described even if they are not actually in the archive. This creates a tree of processes and data products that chain back to the original raw data and measurements. In addition to this there are detailed interface documents of the different data products themselves, see also Wise et al. (2011) and Anderson et al. (2011).

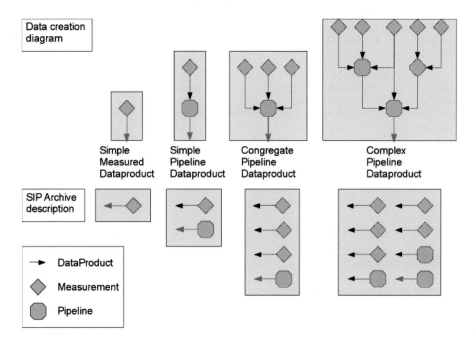

Figure 4.   LOFAR data lineage model.

Acknowledgments.   The author would like to thank the colleagues in the Target Project, BigGrid project, ASTRON, the ICD development group and the Netherlands Organization for Scientific Research (NWO) and the ADASS XX organizers for what proved to be an excellent conference!

References

Anderson, K. R., Alexov, A., Bähren, L., Grießmeier, J.-M., Wise, M., & Renting, G. A. 2011, in Astronomical Data Analysis Software and Systems XX, edited by I. N. Evans, A. Accomazzi, D. J. Mink, & A. H. Rots (San Francisco, CA: ASP), vol. 442 of ASP Conf. Ser., 53

Wise, M., Alexov, A., Folk, M., Pierfederici, F., Anderson, K. R., & Bähren, L. 2011, in Astronomical Data Analysis Software and Systems XX, edited by I. N. Evans, A. Accomazzi, D. J. Mink, & A. H. Rots (San Francisco, CA: ASP), vol. 442 of ASP Conf. Ser., 663

Astronomical Data Analysis Software and Systems XX
ASP Conference Series, Vol. 442
Ian N. Evans, Alberto Accomazzi, Douglas J. Mink, and Arnold H. Rots, eds.
©*2011 Astronomical Society of the Pacific*

LOFAR and HDF5: Toward a New Radio Data Standard

Kenneth Anderson,[1] A. Alexov,[1] L. Bähren,[1] J.-M. Grießmeier,[2] M. Wise,[3] and
G. A. Renting[3]

[1]*Astronomical Institute "Anton Pannekoek", University of Amsterdam,
Postbus 94249, 1090 GE Amsterdam, The Netherlands*

[2]*Laboratoire de Physique et Chimie de l'Environnement et de l'Espace (LPC2E),
3A Avenue de la Recherche, 45071 Orléans Cedex 2, France*

[3]*Netherlands Institute for Radio Astronomy (ASTRON),
P.O. Box 2, 7990 AA Dwingeloo, The Netherlands*

Abstract. For decades now, scientific data volumes have experienced relentless, exponential growth. As a result, legacy astronomical data formats are straining under a burden not conceived of when these formats were first introduced. With future astronomical projects ensuring this trend, ASTRON and the LOFAR project are exploring the use of the Hierarchical Data Format version 5 (HDF5) for LOFAR radio data encapsulation. Most of LOFAR's standard data products will be stored natively using the HDF5 format. In addition, HDF5 analogues for traditional radio data structures such as visibility data and spectral image cubes are also being developed. The HDF5 libraries allow for the construction of potentially distributed, entirely unbound files. The nature of the HDF5 format further provides the ability to custom design a data encapsulation format. The LOFAR project has designed several data formats that will accommodate all LOFAR data products, examples of which are presented in this paper. With proper development and support, it is hoped that these data formats will be adopted by other astronomical projects as they, too, attempt to grapple with a future filled with mountains of data.

1. Introduction

The commencement of the operational phase of The LOw Frequency ARray (LOFAR) telescope holds forth both great scientific potential and challenges to current and legacy information technologies: volume and complexity of the data will continue to push the envelope of commonly used data protocols.

Recognizing that this envelope is already strained, the LOFAR project has embarked on an ambitious project to design and define a set of radio data standard formats that are capable of encapsulating the full spectrum of not just LOFAR data products, but astronomical radio data in general.

It is with this ambition in mind that the LOFAR data formats group has been developing these format specifications and associated software infrastructure, an ongoing, two year effort to date. It was determined that HDF5 would be a robust, viable data framework capable of handling the size, scope, diversity, and parallel processing requirements of LOFAR data (Wise et al. 2011). This work certainly has potential use beyond the radio community. New large scale optical telescopes, such as the LSST, are also investigating the viability of using HDF5. Furthermore, the 20 year history of HDF and its continuing use by NASA, the NOAA, and other agencies, ensure broad use and and long term support.

In addition to the format descriptions themselves, the LOFAR project is currently developing a set of software tools for creating and working with these formats. The Data Access

Library (DAL) in C++, along with an associated Python interface (pyDAL), are designed to allow for the easy construction and manipulation of these data formats. There are also a number of tools already available to read and visualize HDF5 files, such as: HDFView,[1] VisIt[2] + plugin, PyTables,[3] h5py,[4] MATLAB[5] and IDL.[6] Additionally, the LOFAR project has started work on an HDF5 plugin for VisIt, a visualization tool specializing in handling large datasets. Initial communication has been opened with William Joye regarding implementation of DS9 compatibility with LOFAR Sky Image Cubes.

2. The LOFAR Radio Telescope

LOFAR is nearing the end of construction in The Netherlands and throughout Europe. There are currently 30 Dutch stations and 5 other European stations actively observing commissioning proposals. Once LOFAR is completed, there will be 40 stations in The Netherlands and at least 8 additional European/International stations. LOFAR's Low Band Antenna (LBA) functions in the range of 30-80 MHz and High Band Antenna (HBA) is in the 120-240 MHz range; the telescope bandwidth is 48MHz. LOFAR can create 8 (possibly more) simultaneous beams, has a spectral resolution of 0.76 kHz (1 sec) and a time resolution of 5.1 nano-seconds.

The features of LOFAR that point to massive data volumes: 30,000 networked, passive phased-array antennæ in The Netherlands; international baselines to 1500 km; 6 Gb/s Data Rate, correlated by Blue Gene/P supercomputer, Groningen, NL; Ability to form 8 (and possibly more) concurrent digital beams.

3. LOFAR Data: Variety, Complexity, Volume

Datasets produced by LOFAR observations will vary tremendously in size and complexity. Images, Beam-formed (BF) data, Transient Buffer board time-series data are expected to produce large files, with some observations potentially creating files of several tens of terabytes (see Table 1).

LOFAR's observational modes are capable of producing highly complex, large volume datasets. Legacy protocols fall short of being able to describe or store these data. This looming predicament is especially germane to the SKA pathfinder LOFAR project, wherein certain operational modes will be capable of generating datasets comprising hundreds of gigabytes to tens of terabytes. Therefore, the LOFAR project has been driven to consider viable alternatives to "standard" astronomical data formats, such as FITS and CASA Tables. For more detailed discussion of LOFAR data rates, volume, complexity, and HDF5, see Wise et al. (2011).

4. LOFAR Data Format Specifications

A viable solution was needed for potentially massive LOFAR data products. HDF5 provides a framework allowing users to essentially design their own files to appropriately accommodate a known variety of datatypes (Wise et al. 2011). The LOFAR project has been engaged in

---

[1]HDFView: http://www.hdfgroup.org/hdf-java-html/hdfview/

[2]VisIt: https://wci.llnl.gov/codes/visit/

[3]PyTables: http://www.pytables.org/moin

[4]h5py: http://code.google.com/p/h5py/

[5]MATLAB: http://www.mathworks.com/products/matlab/

[6]IDL: http://www.ittvis.com/ProductServices/IDL.aspx

Exposure Time	Number of Subbands	Number of Stations	File Size Known Mode	File Size Search Mode
1 min	248	20	11.2GB	244GB
10 min	248	30	112GB	3.3TB
1 hr	248	10	672GB	6.7TB
1 hr	248	20	672GB	13.4TB
1 hr	248	30	672GB	26.8TB
2 hr	248	5	1.3TB	6.7TB
12 hr	248	5	8.0TB	40.3TB
12 hr	248	15	24.0TB	120.1TB

Table 1.   Sample, LOFAR BF dataset sizes for different observation times and number of stations used. The difference between Known and Search Mode is that data are saved per station for searching, as opposed to combined when observing a known object (to increase signal-to-noise and save disk space).

developing and designing a complete set of specifications for all LOFAR observational data. This has necessarily required differing file designs for differing data, with a certain structural parallelism maintained across all file designs.

These Interface Control Documents (ICD)[7] provide detailed descriptions of a range of expected LOFAR Data Products, including Radio Sky Images, Transient Time Series data, Beam-Formed data, Dynamic Spectra, UV Visibility, Rotation Measure Synthesis, Near-field Imaging. By way of example, structures of a LOFAR Radio Sky Image and a Dynamic Spectra file are outlined in Figures 1 and 2.

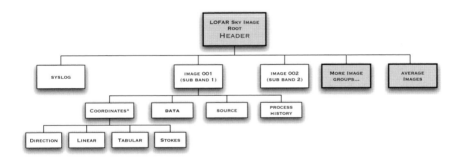

Figure 1.   LOFAR Radio Sky Image Data Structure.

The LOFAR ICD team has also produced a document describing the coordinate system for the file structure; these coordinates are closely related to the WCS for FITS. It is not the intention of the project to redo decades of WCS work which is already in place, but to map the LOFAR coordinates to standard WCS.

5.   Summary and Future Considerations

In order for the adoption of LOFAR data formats to prove useful in the real world, the LOFAR project is committing resources to help develop the next generation of astronomical tools for LOFAR data. A major effort of the LOFAR project has been the development of the Data Access Library (DAL), which ultimately will provide interfaces through FITS, the CASA/AIPS++

---

[7]http://usg.lofar.org/wiki/doku.php?id=documents:lofar_data_products

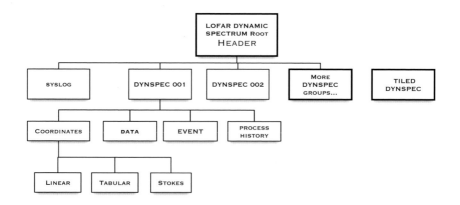

Figure 2.   LOFAR Dynamic Spectra Data Structure.

Measurement Sets and HDF5. All LOFAR products will be accessible through DAL tools, which are part of the LOFAR User Software (LUS)[8] repository.

The LOFAR project has set up a moderated majordomo email list, called **nextgen-astro-data@astron.nl**. Interested parties are encouraged to sign up via **majordomo@astron.nl** with "**subscribe nextgen-astrodata**" as the only text in the body of the message. Ultimately we would like to see these formats grow into a true set of standards for radio data that can meet the demands of the next generation of radio observatories. Such standards are sorely lacking in the radio community at present and are clearly needed as radio astronomy moves into the SKA era.

Acknowledgments.   The authors would like to thank colleagues of the ICD development group, ASTRON, the *Sterrunkundig Instituut Anton Pannekoek*, the HDF Group, and the ADASS XX organizers for what proved to be an excellent conference!

References

Wise, M., Alexov, A., Folk, M., Pierfederici, F., Anderson, K. R., & Bähren, L. 2011, in Astronomical Data Analysis Software and Systems XX, edited by I. N. Evans, A. Accomazzi, D. J. Mink, & A. H. Rots (San Francisco, CA: ASP), vol. 442 of ASP Conf. Ser., 663

---

[8]LOFAR User Software Repository: `http://usg.lofar.org/wiki/doku.php?id=development:getting_started`

Astronomical Data Analysis Software and Systems XX
ASP Conference Series, Vol. 442
Ian N. Evans, Alberto Accomazzi, Douglas J. Mink, and Arnold H. Rots, eds.
© *2011 Astronomical Society of the Pacific*

Data Processing and Archiving at the North American ALMA Science Center

Mark Lacy, David Halstead, and Mike Hatz

North American ALMA Science Center, National Radio Astronomy Observatory, 520, Edgemont Road, Charlottesville, VA 22903

Abstract. The Atacama Large mm/submm Array (ALMA) is expected to begin Early Science operations in approximately one year. The data rate is expected to ramp up from ~20TB/yr (using the first 16 antennas) in 2011–2012 to the fully operational rate (with all 64 antennas) of ~ 200TB/yr in 2013. During this time, our data processing capabilities will also evolve, as we move from being able to process most datasets on desktop machines, to needing to use computing clusters to process a typical dataset. In this poster we present the data processing and archiving plans for the NAASC, and how these relate to the observatory pipeline processing and archiving activities at the Joint ALMA Observatory in Santiago.

1. Introduction

Advances in correlator and receiver technologies have placed radio telescopes at the forefront of the challenge to deal with the data volume and complexity obtainable from modern astronomical instrumentation. ALMA is typical in having a correlator that is capable of outputting data at very high rates (~ 1GB/s), though post-correlator hardware currently restricts the maximum data rate to a more manageable maximum rate of 64MB/s, and a rate of 6MB/s is the specification for the mean rate during science operations (Lucas et al. 2004). As the ALMA array increases in size from its Early Science configuration of 16 antennas (expected to begin operations in September 2011) to its final size of 66 antennas (50 12m antennas, 12 7m antennas in the Compact Array, and 4 Total Power antennas) in 2013, the data rate will increase very steeply (roughly as the square of the number of antennas for the interferometric arrays). Thus the data transfer, archiving and processing capabilities will need to ramp up over a two year timescale.

A particular challenge with ALMA is that it is an observer-driven telescope, with a wide variety of observing modes, producing data with a varying set of processing requirements. A simple pipeline processing model that might be appropriate for large surveys is thus not appropriate for ALMA, where users may well wish to rerun the pipeline scripts with different parameters to tune the results to their specific science needs.

2. Data Transfer and Archiving

The ALMA telescope and correlator is situated at the high Array Operations Site (AOS), at an altitude of 5000m. From here, it will be transferred via an optical fiber link to the Operations Support Facility (OSF) at 2900m, 30km away. The OSF contains a small data archive, capable of holding a couple of months of data. Telescope monitoring and logging data are also kept at the OSF for a similar timescale. Data is transferred from the OSF to the main archive in the Santiago Central Office (SCO) via a dedicated internet link (up to 100Mb/s), supplemented by shipping of storage media if required. Here, the data processing pipeline is run, and the results archived. (For a detailed overview of ALMA science operations see Nyman et al. (2010).)

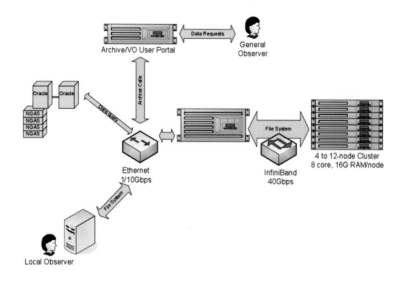

Figure 1.   The NAASC cluster configuration during Early Science.

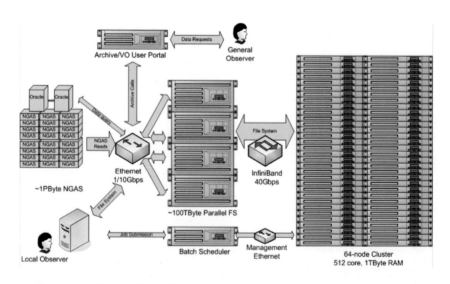

Figure 2.   The final NAASC cluster configuration during full ALMA operations.

The main archive is mirrored to the three ALMA Regional Centers (ARCs) (Europe [the European Southern Observatory (ESO) in Garching], East Asia [the National Astronomical Observatory of Japan in Tokyo] and North America [the North American ALMA Science Center (NAASC) in Charlottesville, VA]). Transfer to the ARCs is performed via a combination of data network and shipping of storage media, with metadata always transferred via internet. The NAASC has negotiated an agreement for a share (currently 100Mb/s) of the dedicated link between Santiago and Internet 2 at Florida International University that is being set up for the National Optical Astronomical Observatory to support large surveys, particularly the Large Synoptic Survey Telescope. We expect this link will be sufficient to transfer all data, except possibly some of the largest datasets.

The role of the ARC archives is to provide local mirrors for faster downloading of data, redundancy in the case of data loss or corruption, and a local copy for pipeline reruns. The archive at all five sites (OSF, SCO and the three ARCs) uses the Next Generation Archive System (NGAS) developed at ESO (Wicenec et al. 2002).

A mean data rate of 6MB/s corresponds to a cumulative volume of 200TB/year of raw data. Processed data is expected to add to this, but in most cases will be a relatively small fraction of the raw data. Nevertheless, the actual data rate from the array will depend on the science needs that can only be assessed after the approved science projects are known. We will therefore rely on the scalability of the NGAS system to increase our storage capacity should the need arise.

3. Data Processing

The data processing plan for ALMA is to run the ALMA Science Pipeline (Davis et al. 2004) at the SCO, but also to have it available at the ARCs for reprocessing of ALMA data, both in a bulk sense (if, for example, a processing or calibration algorithm is changed), and, (to a degree still to be determined, and likely to be ARC-dependent) for user-requested reprocessing using different parameters in the pipeline. The ALMA Science Pipeline is built using tools from the CASA software package (Jaeger 2008). At the NAASC, we are also planning to have visitors come to run data processing in a more interactive manner using CASA directly.

Our processing will be based on a cluster, which we will build in close collaboration with our colleagues in Socorro, NM, who are building a similar cluster for analysis of data from the Extended Very Large Array, and with the ALMA Computing IPT, who will be responsible for building the ALMA pipeline cluster at the SCO. The key feature of our design is a Lustre fast file system that allows data from the archive to be staged and accessed with high bandwidth (via an Infiniband switch) by the cluster nodes. By doing this we hope to alleviate the I/O bound nature of much of our processing. The cluster will be built up slowly over the next two years, with the first nodes being purchased in time to analyze ALMA Early Science data, expected starting in September 2011 (Figure 1). The cluster will then grow with the size of the array, reaching a final size of about 64 nodes in full operations (Figure 2). The details of the node specifications (numbers of cores, memory, etc.) are still being discussed, as development of parallelization capability in CASA is ongoing.

Visitors to the NAASC will be able to run jobs on the cluster directly. Resources permitting, we also hope to be able to offer a service whereby users will be able to fill out a webform which allows the adjustment of several pipeline parameters, such as *uv*-weighting, clean algorithm parameters, and output image size and channelization. They would then submit a job to the cluster and be able to retrieve their results after the pipeline has run.

References

Davis, L. E., Glendenning, B. E., & Tody, D. 2004, in Astronomical Data Analysis Software and Systems XIII, edited by F. Ochsenbein, M. G. Allen, & D. Egret (San Francisco, CA: ASP), vol. 314 of ASP Conf. Ser., 89

Jaeger, S. 2008, in Astronomical Data Analysis Software and Systems XVII, edited by R. W. Argyle, P. S. Bunclark, & J. R. Lewis (San Francisco, CA: ASP), vol. 394 of ASP Conf. Ser., 623

Lucas, R., Richer, J., Shepherd, D., Testi, L., Wright, M., & Wilson, C. 2004, Alma memo 51

Nyman, L., Andreani, P., Hibbard, J., & Okumura, S. K. 2010, in Observatory Operations: Strategies, Processes, and Systems III, edited by D. R. Silva, A. B. Peck, & B. T. Soifer (Bellingham, WA: SPIE), vol. 7737 of Proc. SPIE, 77370G

Wicenec, A., Knudstrup, J., & Johnston, S. 2002, in Astronomical Data Analysis Software and Systems XI, edited by D. A. Bohlender, D. Durand, & T. H. Handley (San Francisco, CA: ASP), vol. 281 of ASP Conf. Ser., 95

Astronomical Data Analysis Software and Systems XX
ASP Conference Series, Vol. 442
Ian N. Evans, Alberto Accomazzi, Douglas J. Mink, and Arnold H. Rots, eds.
©*2011 Astronomical Society of the Pacific*

Virtualization and Grid Utilization within the CANFAR Project

Séverin Gaudet,[1] Patrick Armstrong,[2] Nick Ball,[1] Ed Chapin,[3] Pat Dowler,[1] Ian Gable,[2] Sharon Goliath,[1] Sébastien Fabbro,[2] Laura Ferrarese,[1] Stephen Gwyn,[1] Norman Hill,[1] Dustin Jenkins,[1] J. J. Kavelaars,[1] Brian Major,[2] John Ouellette,[1] Mike Paterson,[2] Michael Peddle,[1] Chris Pritchet,[2] David Schade,[1] Randall Sobie,[2] David Woods,[3] Kristen Woodley,[3] and Alinga Yeung[2]

[1]*Herzberg Institute of Astrophysics, National Research Council Canada, 5071 West Saanich Road, Victoria, BC V9E 2E7, Canada*

[2]*Department of Physics and Astronomy, University of Victoria, 3800 Finnerty Rd, Victoria, BC V8P 5C2, Canada*

[3]*Department of Physics and Astronomy, University of British Columbia, 6224 Agricultural Road, Vancouver, BC V6T 1Z1, Canada*

Abstract. The Canadian Advanced Network For Astronomical Research (CANFAR) is an operational system for the delivery, processing, storage, analysis, and distribution of very large astronomical datasets. CANFAR combines the Canadian national research network (CANARIE), grid processing and storage resources (Compute Canada) and a data center (CADC) into a unified Platform-as-a-service (PaaS) cyberinfrastructure supporting Canadian astronomy projects. The CANFAR processing service is based on virtualization and combines features of the grid and cloud processing models to provide a self-configuring virtual cluster deployed on multiple cloud clusters. The service makes use of many technologies from the grid, cloud and Virtual Observatory communities.

1. Introduction

The Canadian Advanced Network For Astronomical Research[1] (CANFAR) is an operational system for the delivery, processing, storage, analysis, and distribution of very large astronomical datasets. CANFAR combines the Canadian national research network (CANARIE), a geographically distributed collection of grid processing and storage resources (Compute Canada) and a data center (Canadian Astronomy Data Center) into a Platform-as-a-service (PaaS) cyberinfrastructure supporting Canadian astronomy projects. An overview of the whole CANFAR architecture can be found in Gaudet et al. (2010, 2009). The CANFAR processing service is based on virtualization and combines features of the grid and cloud processing models by providing a self-configuring virtual cluster deployed on multiple cloud clusters. The service makes use of many technologies from the grid, cloud and Virtual Observatory communities such as Condor, Nimbus (or OpenNebula, Eucalyptus or Amazon EC2), Xen, Cloud Scheduler, VOSpace, UWS, SSO, CDP and GMS. See Dowler et al. (2011) for a description of how VO technologies are incorporated into CADC services including CANFAR.

---

[1]`http://astrowww.phys.uvic.ca/~canfar/`

2. The Processing Context

There are existing models for providing CPU cycles to users, including well-established Grid systems and nascent cloud services. From anecdotal evidence, data intensive users have tried to use existing Grid infrastructure with unsatisfactory results. The problems reported by astronomers are:

- Environment customization and maintenance issues — Grid operators install and maintain the environment for their users. Astronomers tend to run complex software with many dependencies. Making their software run correctly in the environment provided by a cluster operator is very difficult. Cluster operators modifying their system to incorporate updates or support other users make the software maintenance issue an ongoing problem.

- Poor responsiveness — Astronomers complained that it would often take days for a job to start running. One common response to the poor responsiveness was to seek out other clusters that may be less busy. This exacerbated the environment configuration issues described above.

We are not aware of serious attempts by astronomers to use cloud infrastructure, however we anticipate several problems with the cloud:

- Until recently, only commercial clouds have been available. Paying the usage fees for significant CPU cycles will lead to funding and administrative issues.

- Clouds do not inherently provide job scheduling. A user can incorporate a job scheduling system into their virtual environments, however this is a non-trivial process.

- Clouds do not inherently share resources between multiple users. Existing clouds tend to allocate resources on a first-come-first-served basis.

To address the limitations of the grid and cloud processing models, the CANFAR project has produced a hybrid processing system that makes use of scheduling while abstracting the grid processing resources as a cloud.

3. The Processing System

The CANFAR processing system presents a grid-like interface to users and creates a virtual cluster built from resources supplied by multiple cloud providers. The technologies used or supported by the CANFAR processing system are:

- **Virtual Image (VI) Management:** A service to allow users to boot and configure, save and share Virtual Images. The Virtual Images are made available to the Cloud Scheduler, where they are booted into the virtual cluster. The IVOA VOSpace[2] standard was implemented to provide the functionality for the saving of and, by setting access permissions, the sharing of Virtual Images.

- **Virtualization**: Xen virtualization is being used. Both Xen and KVM were considered as potential virtualization technologies. Xen was selected because it was the most popular virtualization technology at the time. In addition, it was the only one utilized experimentally by facility operators.

- **Job Scheduler**: The CANFAR virtual cluster requires a batch job processing system to provide the functionality of a Grid cluster. Although both Condor and Grid Engine were considered, Condor was selected because it allows Virtual Machines to join the virtual

---

[2]http://www.ivoa.net/Documents/VOSpace/

cluster without modifying the Condor configuration. Grid Engine would have required the cluster configuration to be modified each time a Virtual Machine joins or leaves the virtual cluster.

- **Cloud Scheduler**: The scheduler is the glue between the cloud clusters provisioned by Nimbus, and the batch interface provided by Condor. The Cloud Scheduler (Armstrong et al. 2010; Fransham et al. 2010) examines the workload in the Condor queue, and uses the resources from multiple cloud clusters to create a virtual cluster suitable for the current workload. A schematic of the Cloud Scheduler illustrating multi-cluster use is shown in Figure 1.

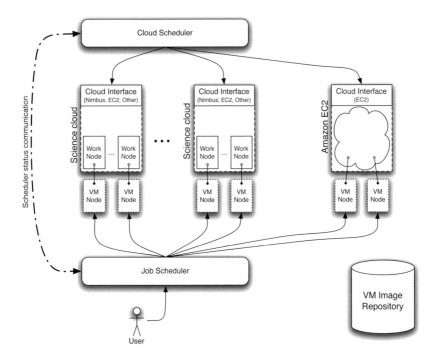

Figure 1.   Cloud Scheduler schematic showing multi-cluster use.

- **Cloud functionality**: The primary cloud technology supported by the Cloud Scheduler is the Nimbus toolkit. Partial support was also developed for openNebula and Eucalyptus and Amazon EC2. Nimbus was selected as the primary development target because it is open source and allows the cloud workload to be intermixed with conventional batch jobs unlike the other systems. It is believed that this flexibility makes the deployment more attractive to facility operators.

- **Operating systems**: Both host and guest operating systems are assumed to be some flavor of Linux. The guest operating system currently in use is Scientific Linux 5.5.

4.   Status

Since beta release in mid-June, the system has been dynamically using 3 clusters and executed 40 core-years of processing in 21 weeks. Members of eight different science projects are involved as early adopters and their feedback has been valuable to improving the usability of the system. Scientists are tackling the learning curve on virtual machines and once familiar with the environment they have successfully used the system to their advantage. Once the system is judged to be sufficiently robust (or at least the problems well understood and avoidable), and additional Compute Canada grid facilities are added to the processing pool, CANFAR services will be made available the whole community. CANFAR development is scheduled to finish in 2011 at which point the Canadian Astronomy Data Center will assume responsibility for operations.

Acknowledgments.   The support from CANARIE, Compute Canada, the Canadian Space Agency and the National Research Council are acknowledged.

References

Armstrong, P., et al. 2010, ArXiv e-prints. 1007.0050
Dowler, P., Gaudet, S., & Schade, D. 2011, in Astronomical Data Analysis Software and Systems XX, edited by I. N. Evans, A. Accomazzi, D. J. Mink, & A. H. Rots (San Francisco, CA: ASP), vol. 442 of ASP Conf. Ser., 603
Fransham, K., et al. 2010, in Proceedings of the High Performance Computing Symposium
Gaudet, S., Dowler, P., Goliath, S., Hill, N., Kavelaars, J. J., Peddle, M., Pritchet, C., & Schade, D. 2009, in Astronomical Data Analysis Software and Systems XVIII, edited by D. A. Bohlender, D. Durand, & P. Dowler (San Francisco, CA: ASP), vol. 411 of ASP Conf. Ser., 185
Gaudet, S., et al. 2010, in Software and Cyberinfrastructure for Astronomy, edited by N. M. Radziwill, & A. Bridger (Bellingham, WA: SPIE), vol. 7740 of Proc. SPIE, 774011

Astronomical Data Analysis Software and Systems XX
ASP Conference Series, Vol. 442
Ian N. Evans, Alberto Accomazzi, Douglas J. Mink, and Arnold H. Rots, eds.
©2011 Astronomical Society of the Pacific

Preparing Data Files for Long-Term Archiving

Russell O. Redman

National Research Council of Canada,
5071 West Saanich Road, Victoria, BC, Canada V9E 2E7

Abstract.   With proper preparation, data files can be made easy to archive, informative for users who download the files from the archive, and ultimately rewarding for those who prepared the files. This paper sketches the Common Archive Object Model in use at the Canadian Astronomy Data Center, highlights key decisions that should be made as early as possible in the design of the files, and summarizes the metadata that should be included in each file.

1.   Common Archive Object Model

Storing data files in an archive is a means to publish the data. Like papers in journals, the ideal is for data files to be easy to find, easy to understand, and easy to relate to each other. From the metadata stored in the file headers it should be possible to recover the physical properties of the observation, the organization arrangements of the people who took the data, and any relationships amongst the archived files introduced by the data reduction system.

Within an archive the metadata should be organized following a well thought out data model. One such model is the Common Archive Object Model (CAOM), which has been in active use for about four years at the Canadian Astronomy Data Center (CADC), part of the Herzberg Institute of Astrophysics within the National Research Council of Canada. CAOM successfully describes the data files in nine different collections at the CADC containing data at wavelengths from the radio through optical, including images, spectra, polarimetry, data cubes and hypercubes.

CAOM is largely stable, but a new version is under development that will extend the model to even more file structures, to better represent the contents of raw data files, non-FITS data formats like HDF5, and common but nonstandard FITS formats. The changes anticipated in the new version should have little or no impact on the metadata needed to fill the model. For more details, see (Dowler 2011, in preparation).

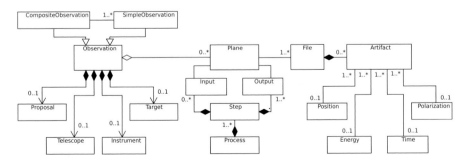

Figure 1.   CAOM Overview

Figure 1 shows the relationships amongst the major classes within CAOM, simplified for pedagogical reasons. This document can only summarize the significance of each of the classes; a more complete discussion is in preparation for publication along with the CAOM version 2 specification and can be found in Redman (2010).

2. Observation

Observations are the highest level of organization explicitly built into CAOM. An observation can be summarized as "the set of files derived from a batch of observed photons." Observation is a virtual base class; actual instances can be created for one of the two specialized subclasses, SimpleObservation or CompositeObservation.

Simple observations contain files derived from a single observatory-defined observation, at least for observatories that archive raw data. Some observatories do not archive raw data, in which case all of the basic data products will be recorded in simple observations. Metadata to fill the proposal, telescope, instrument and target classes should be supplied by the data providers for every simple observation.

A composite observation does not contain its own raw data but records files derived from the data in a set of simple observations that are referred to as its members. A composite cannot be a member of another composite. A composite can be predictive, generated before any files are produced by the data reduction system, in which case it will not contain any planes. Members can be added or removed from a composite. The metadata to fill the proposal, telescope, instrument and target objects are usually derived from its simple members, filling each field for which the members all contain the same value and leaving undefined any field that is ambiguous.

Observations are uniquely identified by the collection (usually the telescope or project acronym) and collectionID, a string supplied by the data provider.

A proposal instance records the proposal ID assigned by the observatory, as well as the PI name, proposal title and an optional set of keywords. If the proposal was part of a larger project like a survey, the project name can also be recorded.

A telescope instance records the telescope name, its location as an (X, Y, Z)-tuple in meters, and an optional set of keywords. The location can be left undefined for moving observatories such as those mounted on satellites or balloons.

An instrument instance records the instrument name and an optional set of keywords. The keywords normally summarize the instrument configuration with a set of key=value pairs, a scheme that works well for discrete settings but is usually not appropriate for environmental conditions or floating-point numerical settings.

The target records the object name, classification and redshift.

3. Plane and Provenance

Planes group together files that would normally be downloaded together. For example, some data reduction systems store images and weight maps, or even data and headers, in separate files. Also, spectral-imaging systems often produce datacubes that are too large to handle and are broken into "tiles." In all these cases we would normally want the entire plane of files if we get any part of it. Planes are the targets of search operations under CAOM. For example, a cone search around a position will return a set of planes whose positional metadata intersect the cone.

Planes are uniquely identified by the attribute "productID" and data providers will need to specify how to generate or extract the productID value from each file in the plane. Also defined for each plane is a pair of date-time values that set when the data and metadata in the plane become "public." Before such time the data or metadata are "proprietary," accessible only to the project team.

The provenance classes, Process, Step, Input and Output, record a simplified model of the data reduction system, recording relations amongst whole planes of data. For example, a two-step data reduction

```
raw -> image -> source catalog
```

would create one process instance with two steps, the first of which inputs data from a single raw data plane and outputs a single image plane, while the second inputs the image plane generated in the first step and outputs a single plane recording the source catalog. Each input has a name that defines the role the input plane played in the data reduction. Each output has a name that identifies what kind of product was produced, and a version number. The process itself records the process name, process version, a URL giving a reference describing the data and data reduction, and the name of the producer. Each instance of the process should also record a runID string that identifies the processing instance, and a lastExecuted date-time telling when the processing ran. Automated pipelines often supply the value of the runID string, but less formal systems may wish to record a data reduction script/configuration file name as the runID.

4. File and Artifact

The file is the basic unit of astronomical packaging and whole files are normally downloaded by default. Files often have complicated internal structures, so CAOM decomposes a file into a set of "artifacts," each of which can be described in some World Coordinate System (WCS), perhaps as an axis in an array, or as an interval (or list of intervals) along a virtual axis. Beware that this part of the model has been drastically simplified in Figure 1; the model as shown is adequate for arrays of data stored in FITS files, but the new version of CAOM will also handle a much larger set of file structures. Regardless, the required metadata in the files should be the same.

Each artifact optionally contains structures describing the position (e.g. RA/Dec), photon energy (wavelength or frequency), time and polarization. A discussion of these structures is well outside the scope of this article, but the metadata should follow conventions like those established for FITS WCS in Hanisch et al. (2001); Greisen & Calabretta (2002a,b); Greisen et al. (2006).

5. Other Issues

Data providers need to define several key policies in an interface control document (ICD) with the archive operators:

- Input/Output: For each step in data processing, define the sets of input roles and output products names, thus establishing input/output relations and set of sibling planes that will affect how data can be grouped for downloading.

- Version Numbers: Establish the form of versioning to be used, under what circumstances to create new version numbers, and whether to allow/forbid the replacement of existing files for particular version numbers.

- Generation of productID: The productID must be unique for each plane and can be read from a header, parsed from a header by extracting a common value from related groups of values, or constructed from sets of headers.

- File Names: Ensure that file names are unique in the archive, preferably constructed from the collection, collectionID, productID and a fileID that distinguishes files within a plane.

- Order of Ingestion: Specify the order of ingestion for files if it cannot be determined from the input/output relations amongst planes.

- Validity Checking: Specify the level of security required, the tools used to verify file integrity, and constraints that must be satisfied before files are accepted into the archive. Tools like fitsverify[1] are useful, recognizing that warnings from fitsverify should be treated as errors to be corrected before a file is acceptable for long-term archiving. Checksums are important. It may be useful to specify lists of mandatory headers and their allowed values, as well as a list of permitted headers, so that invalid files can be recognized and rejected.

- Previews: Specify whether they will be generated during data reduction or with archive-supplied software, what naming conventions will be used, and how to find metadata for the files if they do not carry their own headers.

- Raw Data: Specify whether raw data will be archived, whether some other form of minimally processed data will be provided, and how to find the required metadata if the files do not carry their own headers in a convenient form.

- Ancillary Data: Specify whether the archive will hold ancillary data files, the file naming conventions, and dependencies involving the ancillary files, and how to find metadata if the files will be recorded in CAOM.

References

Greisen, E. W., & Calabretta, M. R. 2002a, A&A, 395, 1077
— 2002b, A&A, 395, 1061
Greisen, E. W., Calabretta, M. R., Valdes, F. G., & Allen, S. L. 2006, A&A, 446, 747
Hanisch, R. J., Farris, A., Greisen, E. W., Pence, W. D., Schlesinger, B. M., Teuben, P. J., Thompson, R. W., & Warnock III, A. 2001, A&A, 376, 359
Redman, R. O. 2010, Preparing Data Files for Long-Term Archiving. In preparation, URL ftp://ftp.hia.nrc.ca/pub/users/ror/ADASS_2010/File_Preparation.pdf

---

[1]http://heasarc.gsfc.nasa.gov/docs/software/ftools/fitsverify

Astronomical Data Analysis Software and Systems XX
ASP Conference Series, Vol. 442
Ian N. Evans, Alberto Accomazzi, Douglas J. Mink, and Arnold H. Rots, eds.
© *2011 Astronomical Society of the Pacific*

The XID Results Database of the XMM-Newton Survey Science Center

L. Michel and C. Motch, on behalf of the Survey Science Center of the XMM-Newton satellite

CNRS, Université de Strasbourg, Observatoire Astronomique,
11 rue de l'Université, 67000 Strasbourg, France

Abstract. The Survey Science Center (SSC) of the XMM-Newton satellite has carried out several large optical campaigns aiming at the spectroscopic identification of samples of about a thousand X-ray sources at various X-ray flux levels and towards different Galactic directions. In addition, the SSC has obtained multi-color wide-field imaging for hundreds of XMM-Newton fields. Building learning samples for the statistical identification of all 2XMM sources was one of the main drivers for undertaking these observing campaigns. However, as demonstrated by the amount of papers published, these collections of data also constitute a very valuable resource which can be used for addressing a wide range of astrophysical issues. We describe the content and architecture of the XID results database recently opened by the SSC and containing a first installment of these data. The interface provides easy selection and browsing through catalogs and access to all optical images and spectral data associated with any given X-ray source as well as all relevant XMM-Newton data. The database was created using the database generator Saada and, together with the XCat-DB already deployed at the Observatoire de Strasbourg, provides another example of the flexibility, ease of use and scalability offered by Saada.

1. The XID Program

The XID results database archives the scientific outcome of the X-ray follow-up and Identification program (XID) carried out by the Survey Science Center of the XMM-Newton satellite. The XID project is designed to ensure that the potential of the XMM-Newton serendipitous survey can be exploited by the community in the context of a wide range of scientific programs. One of the principal objectives of the program is to obtain well-defined completely identified groups of X-ray sources using dedicated optical and infrared spectroscopic observations. These identified samples can allow us to characterize the parameter space occupied by the different X-ray source populations encountered in the XMM-Newton survey. The basic X-ray and optical source parameters can eventually be used to assign a statistical identification for a large fraction of all the sources in the XMM-Newton serendipitous source catalogs.

The XID program[1] began in mid-2000, shortly after the start of operations. Most of the program is now complete. The XID results database collects the main results from the XID program in a uniform way, providing access to the individual source identifications and to the key optical and X-ray data for each object.

Its main elements are a number of complementary sub-programs designed to cover a wide range of limiting X-ray fluxes and Galactic latitudes. Currently the database contains three key XID program sub-samples all of which are published:

---

[1]`http://xmmssc-www.star.le.ac.uk/XID/`

1. XBS: Bright flux sample, 400 sources, 238 spectra (Caccianiga et al. 2008).
2. XMS: Medium flux sample, 319 sources, 280 spectra (Barcons et al. 2007).
3. GPS: Galactic Plane sample, 43 sources, 27 spectra (Motch et al. 2010).

Optical identifications from other sub-programs will enter the XID results database at the time of their main publication:

4. XWAS: XMM-2dF Wide Angle Survey (Tedds et al. 2007).
5. SXDS: Subaru/XMM-Newton Deep Survey (Ueda et al. 2008).
6. GPSE: Galactic Plane Survey Extended (Motch 2006).

The XID program is also supported by a substantial imaging program covering several hundreds of XMM-Newton fields usually observed in more than one band. The imaging program, needed for the selection of the spectroscopic targets, is also an important resource in its own right given the large sky area and hence X-ray source sample it covers. Imaging data will be made available in the forthcoming months.

2.   The XID Results Database[2]

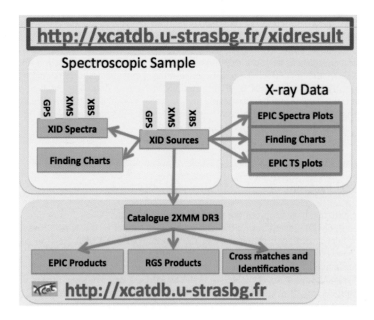

Figure 1.   Content of the XID Results DB and connection to the XCat-DB

2.1.   Database Content

Identified X-ray sources are made available together with their optical spectra, finding charts, parameters (object class, position, magnitude, redshift, etc.), and combined multi-detection X-ray quantities extracted from the 2XMM catalog (see Fig. 1). Some of the X-ray data files

---

[2]http://xcatdb.u-strasbg.fr/xidresult

created for the 2XMM catalog processing, image thumbnails, EPIC spectra and EPIC time series are also shown. These flat images add useful qualitative information to Web pages (e.g. spectral shapes, see Fig. 2). In addition, links to the XCat-DB[3] (Motch et al. 2009) provide a direct access to most XMM-Newton data products (EPIC, RGS, cross-matches) related to the X-ray sources.

2.2. Search Interface and Output Format

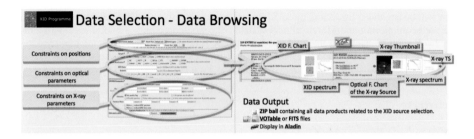

Figure 2.   Search interface and data layout

Queries can be constrained by position, by optical parameters (magnitude, object class) or by X-rays parameters (flux, hardness ratio) (see Fig. 2). Search requests are first setup by a classical HTML form, but users can also edit by hand the query string in order to refine any search criterion. Selected data can either be displayed on Web pages or can be downloaded in flat tables (FITS table, VO table). A last option is to extract the selected sources together with their associated optical and X-ray data (spectra, time series, plots, etc.) in composite directories, all packed in a ZIP file.

3.   Identifications with the XCat-DB

The XCat-DB proposes possible identifications for the 262,902 XMM-Newton serendipitous sources contained in the 2XMM DR3. These identifications are based on cross-correlations with archival catalogs mostly done at pipeline processing time. For a number of large catalogs (GSC2.2, USNO A-2, USNO B-1, 2MASS and SDSS DR7), a specific likelihood ratio algorithm (Pineau et al. 2009) was designed to compute the identification probability of the X-ray source with the corresponding optical or infra-red entry. The database also provides a source classification in terms of stellar or extragalactic nature. The method is based on a Kernel Density Classification and relies on three different parameter spaces. In addition, the database offers full access to all 2XMM-DR3 associated data (images, spectra and time series) and allows for complex queries.

4.   Database Built with SAADA

Both XCat-DB and XID results database have been developed with SAADA[4] (Michel et al. 2005). SAADA makes the creation of data collections merging heterogeneous data-sets (e.g. XID samples) quite easy. Once the data have been loaded, SAADA offers the possibility to

---

[3]http://xcatdb.u-strasbg.fr

[4]http://saada.u-strasbg.fr

compute and store links between them. These links are used either to browse or to select data. This feature has been applied to associate each XID source with various products such as X-ray time series, spectra, images or plots. SAADA also provides a generic template for the Web interface which provided many of the components that have been used here.

5.   Conclusion and Prospects

The XID results database is publicly available at http://xcatdb.u-strasbg.fr/xidresult. Its content will be expanded by adding more XID spectroscopic samples (XWAS, SXDS and GPS) in the forthcoming months. Moreover, a new data collection consisting of wide-field optical images will soon be loaded in the XID results database following the same scheme (data linking and connection to the XCat-DB). Both cone search[5] and spectrum based access[6] services have already been published into the VO. A Table Access Protocol[7] service will be declared soon.

References

Barcons, X., et al. 2007, A&A, 476, 1191
Caccianiga, A., et al. 2008, A&A, 477, 735
Michel, L., Nguyen, H. N., & Motch, C. 2005, in Astronomical Data Analysis Software and Systems XIV, edited by P. Shopbell, M. Britton, & R. Ebert (San Francisco, CA: ASP), vol. 347 of ASP Conf. Ser., 71
Motch, C. 2006, in The X-ray Universe 2005, edited by A. Wilson, vol. 604 of ESA Special Publication, 383
Motch, C., Michel, L., & Pineau, F. 2009, in Astronomical Data Analysis Software and Systems XVIII, edited by D. A. Bohlender, D. Durand, & P. Dowler (San Francisco, CA: ASP), vol. 411 of ASP Conf. Ser., 466
Motch, C., et al. 2010, A&A, 523, A92
Pineau, F., Derriere, S., Michel, L., & Motch, C. 2009, in Astronomical Data Analysis Software and Systems XVIII, edited by D. A. Bohlender, D. Durand, & P. Dowler (San Francisco, CA: ASP), vol. 411 of ASP Conf. Ser., 259
Tedds, J., et al. 2007, in XMM-Newton: The Next Decade, 34P
Ueda, Y., et al. 2008, ApJS, 179, 124

---

[5]http://www.ivoa.net/Documents/latest/ConeSearch.html

[6]http://www.ivoa.net/Documents/latest/SSA.html

[7]http://www.ivoa.net/Documents/TAP

Astronomical Data Analysis Software and Systems XX
ASP Conference Series, Vol. 442
Ian N. Evans, Alberto Accomazzi, Douglas J. Mink, and Arnold H. Rots, eds.
©*2011 Astronomical Society of the Pacific*

Improving Position Accuracy in Archived HST Images

N. Dencheva, W. Hack, M. Droetboom, A. Fruchter, and P. Greenfield

Space Telescope Science Institute, 3700 San Martin Dr., Baltimore, MD 21210

Abstract. We present changes to the HST pipeline, which aim at increasing the astrometric accuracy of archived HST images through successive World Coordinate System (WCS) corrections. In addition, all distortion information and astrometric corrections are stored in the science files, decreasing the size of a typical HST archive request. These changes allow the development of a WCS based version of Multidrizzle and image alignment software. We have developed two software packages to support these changes: PyWCS and STWCS. PyWCS is a general purpose WCS library. STWCS extends PyWCS and defines an HST specific WCS object. This paper provides details on how various WCS conventions have been merged to create a unified comprehensive description of the WCS of HST imaging observations.

1. Introduction

HST images can exhibit significant distortion. The total distortion solution for ACS/WFC is described as consisting of a polynomial part, filter dependent non-polynomial fine scale residuals and a detector defect correction (Anderson 2002). In addition, the distortion can vary with time (Anderson 2007). Currently, distortion solutions are stored in reference files which users have to download from the archive in order to reprocess the data. The IDCTAB files contain the polynomial distortion and the DGEO files contain the combined solution for the detector defect and the filter dependent fine scale residuals. Storing the distortion solution in the science files will decrease the overall size of the requested data.

When considering the available conventions for distortion representation, namely the SIP convention (Shupe et al. 2005) and WCS Paper IV (Calabretta et al. 2004), we had the following requirements in mind:

- The size of the science files should not be increased considerably,
- The distortion corrections should be kept independent.

None of the available distortion conventions fit these requirements. Two of the major deficiencies are:

- Inability to chain corrections: This type of operation is needed because the polynomial and non-polynomial corrections must be applied to coordinates which were corrected for the detector defect existing on some CCDs.
- Lack of a way to represent more than one type of distortion correction: In order to avoid a high order polynomial the distortion for some HST instruments was split into two parts — a polynomial component and a residual lookup table. Paper IV defines a Lookup and a Polynomial distortion representation but only allows one or the other.

These requirements on the application of the calibrations to HST data leave us with no alternative within current FITS standards. As a result, we developed a set of rules which allow us to take advantage of the most appropriate conventions for each separate component of the distortion model and combine them in an efficient manner eliminating the need for external reference data.

2. Distortion Representation

Figure 1 represents the transformation from detector to world coordinates for a typical ACS/WFC observation. Mathematically this transformation can be approximated by:

$$\begin{pmatrix} \alpha \\ \delta \end{pmatrix} = \begin{pmatrix} CRVAL1 \\ CRVAL2 \end{pmatrix} + \begin{pmatrix} CD11 & CD12 \\ CD21 & CD22 \end{pmatrix} \begin{pmatrix} u' + f(u',v') + LT_x(x',y') \\ v' + g(u',v') + LT_y(x',y') \end{pmatrix}$$

where $(x',y') = DET2IM(x,y)$ is a small correction for a detector defect expressed as a periodic change in the pixel width. It is represented by an extension with EXTNAME = D2IMARR following WCS Paper IV Lookup distortion description. The (x',y') coordinates are the input to all distortion corrections.

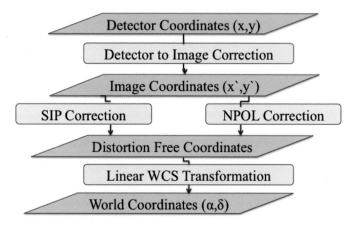

Figure 1.   Coordinate transformation pipeline.

The polynomial distortion $f(u',v')$ and $g(u',v')$ is stored in the header using the Simple Imaging Polynomial (SIP) convention:

$$f(u',v') = \sum_{p+q=2}^{AORDER} A_{pq}u'^{p}v'^{q} \qquad g(u',v') = \sum_{p+q=2}^{BORDER} B_{pq}u'^{p}v'^{q}$$

where $(u',v') = (x' - CRPIX1, y' - CRPIX2)$.

The non-polynomial distortion LTx, LTy, is stored in the science files using Paper IV Lookup convention. For each X and Y axis a separate extension with an EXTNAME value of WCSDVARR is attached to the science file.

The linear WCS, SIP coefficients and non-polynomial distortion extensions include corrections for the time dependent component of the distortion and the velocity aberration of the observation. Table 1 shows the new structure of the science files.

3. Work in Progress — Distributing WCS Solutions

We are developing an optional method which allows for tracking of all updates through a binary table with EXTNAME=WCSCORR.

Aligning an image with a catalog typically only requires small corrections to the WCS, which can be stored as alternate WCS's in the headers. By definition these share the same distortion model.

Table 1.   Science file structure

EXT	FITSNAME	FILENAME	EXTVE	DIMENS	BITPI	OBJECT
0	j94f05bgq_flt	j94f05bgq_flt.fits			16	
1	IMAGE	SCI	1	4096x2048	-32	
2	IMAGE	ERR	1	4096x2048	-32	
3	IMAGE	DQ	1	4096x2048	16	
4	IMAGE	SCI	2	4096x2048	-32	
5	IMAGE	ERR	2	4096x2048	-32	
6	IMAGE	DQ	2	4096x2048	16	
7	IMAGE	D2IMARR	1	4096	-32	
8	IMAGE	WCSDVARR	1	65x33	-32	
9	IMAGE	WCSDVARR	2	65x33	-32	
10	IMAGE	WCSDVARR	3	65x33	-32	
11	IMAGE	WCSDVARR	4	65x33	-32	

For cases in which a WCS registration solution was generated for a specific astrometric solution a mechanism for passing all components of the WCS is needed. We are considering using FITS files, tentatively called 'headerlets', based on the distortion and WCS representation described above for this purpose. Table 2 shows the structure of a headerlet file.

Table 2.   Headerlet file structure

EXT	FITSNAME	FILENAME	EXTVE	DIMENS	BITPI	OBJECT
0	j94f05bgq_flt	j94f05bgq_flt.fits			16	
1	IMAGE	SIPWCS	1		16	
2	IMAGE	WCSDVARR	1	65x33	-32	
3	IMAGE	WCSDVARR	2	65x33	-32	
4	IMAGE	SIPWCS	2		16	
5	IMAGE	WCSDVARR	3	65x33	-32	
6	IMAGE	WCSDVARR	4	65x33	-32	
7	IMAGE	D2IMARR	1	4096	-32	

4.   Software Implementation

4.1.   PyWCS — General WCS Library

PyWCS provides Python bindings to Mark Calabretta's WCSLIB library. It also provides an implementation of the SIP convention and the WCS Paper IV Lookup table convention.

4.2.   STWCS — HST Specific WCS Package

STWCS defines an HST specific WCS object with methods for detector to sky coordinate transformations. It is used in the HST pipeline to compute the WCS of an image based on its distortion model and the position angle of the observation. It provides a function which given a list of HSTWCS objects computes a common frame on the sky and its WCS based on their distortion corrected footprints. In addition, it provides support for alternate WCS descriptions in headers as described in WCS Paper I (Greisen & Calabretta 2002).

4.3. Betadrizzle — WCS-Based Implementation of Multidrizzle

Betadrizzle is a new Python and C only implementation of Multidrizzle with its functionality split in two — coordinate transformation part based on PyWCS and STWCS and a resampling part based on the drizzle algorithm.

4.4. TweakReg — Image Registration Software

TweakReg is an extended implementation in Python and C of the current tweakshifts script. It updates the headers directly with the determined WCS correction necessary to align the image with the reference image used for the fit without the need for image resampling (drizzling). It is independent of IRAF.

5. Conclusion

The changes presented here aim at increasing the astrometric accuracy of HST images and facilitating reprocessing. Once implemented in the HST pipeline and archive, users will be able to retrieve calibrated images and align them to astrometric references without the need to download additional reference files. Distributing WCS solutions should dramatically simplify the process of generating photometrically uniform, distortion-free mosaics from HST data regardless of when the data was taken.

References

Anderson, J. 2002, in The 2002 HST Calibration Workshop : Hubble after the Installation of the ACS and the NICMOS Cooling System, edited by S. Arribas, A. Koekemoer, & B. Whitmore (Baltimore, MD: STScI), 13

Anderson, J. 2007, Variation of the Distortion Solution of the WFC, Tech. Rep. ISR ACS 2007-08, STScI

Calabretta, M. R., Valdes, F., Greisen, E. W., & Allen, S. L. 2004, in Astronomical Data Analysis Software and Systems XIII, edited by F. Ochsenbein, M. G. Allen, & D. Egret (San Francisco, CA: ASP), vol. 314 of ASP Conf. Ser., 551

Greisen, E. W., & Calabretta, M. R. 2002, A&A, 395, 1061

Shupe, D. L., Moshir, M., Li, J., Makovoz, D., Narron, R., & Hook, R. N. 2005, in Astronomical Data Analysis Software and Systems XIV, edited by P. Shopbell, M. Britton, & R. Ebert (San Francisco, CA: ASP), vol. 347 of ASP Conf. Ser., 491

Part II

Cross Catalog Matching

Astronomical Data Analysis Software and Systems XX
ASP Conference Series, Vol. 442
Ian N. Evans, Alberto Accomazzi, Douglas J. Mink, and Arnold H. Rots, eds.
©2011 Astronomical Society of the Pacific

Cross-Identification of Astronomical Objects: Playing with Dice

Tamás Budavári

Department of Physics and Astronomy, The Johns Hopkins University
3400 North Charles Street, Baltimore, MD 21218, USA

Abstract. The cross-identification of objects in separate observations is one of the most fundamental problems in astronomy. Scientific analyses typically build on combined, multicolor and/or multi-epoch datasets, and heavily rely on the quality of their associations. Cross-matching, however, is a hard problem both statistically and computationally. We will discuss a probabilistic approach through an unusual example of dice. This analog problem is simpler in terms of the math and provides valuable insight into the conceptual problems of the matching. The results are directly applicable to astronomy. On the sky the method yields simple, intuitive formulas in the usual limits that are easily calculable, but also generalizes to more complicated situations. It naturally accommodates more sophisticated physical models, such as that of the spectral energy distribution of galaxies or the proper motion of stars. Building on this new mathematical framework, new tools are being developed to enable automated associations.

1. Introduction

Instead of being a review of the topic, this proceedings aims to provide a gentle introduction to the latest statistical developments that are at the core of the cross-identification problems. Following closely the presented material at the Conference, we will not detail any of the analytical calculations but offer a simple analog that provides great insights and simplifies the algebra. Here we will solve the cross-identification of dice and show how that problem maps exactly on the astronomical case. Through this example we can appreciate the simplicity of the probabilistic method. On a personal and historical note, this is exactly how the author originally recognized the importance and applicability of the following Bayesian approach.

Figure 1.   From a bag of dice we draw twice with replacement. First we roll a ⊡ followed by a ⊡. Is it the same die? This problem is essentially identical to crossmatching astronomical detections.

2.   Same or Not?

Part of the reason why crossmatching is a hard problem is that we often ask the wrong question. How far are the detections on the sky? What is their common position? Using Bayesian hypothesis testing we can sidestep these misleading meta-questions and try to decide directly whether the observations belong to the same object. The Bayes factor is the likelihood ratio of two hypotheses. One that says that the observations are of the same object, the other is its complement,

$$B = \frac{L_{\text{same}}}{L_{\text{not}}}. \tag{1}$$

How do we calculate the likelihood of a hypothesis?

First we do it with dice. The analogy is that a die is an object. By rolling once, we make an observation of its position. If the dice are fair, the observations provide no constraint on their identity. However, if the dice are loaded, they prefer one side over the others. If a small piece of lead is placed under the dot of ⚀, the die will have a higher probability of landing on that side and hence come out to show the opposite side ⚅. Let us assume that all our dice are loaded in various directions. If we know how the dice are loaded, we know the probability of the possible outcomes for an die. For example, a die with *loadedness* of $l = 1$ will prefer that side as described by some known probabilities, e.g.,

$$P_1(⚀) = \frac{3}{12}, \quad P_1(⚁) = \frac{2}{12}, \quad \dots, \quad P_1(⚅) = \frac{1}{12}.$$

Similarly we can write the rest of the possible combinations as

$$P_2(⚀) = \frac{2}{12}, \quad P_2(⚁) = \frac{3}{12}, \quad \dots, \quad P_2(⚅) = \frac{2}{12}$$

$$\vdots \qquad\qquad \vdots \qquad\qquad \vdots$$

$$P_6(⚀) = \frac{1}{12}, \quad P_6(⚁) = \frac{2}{12}, \quad \dots, \quad P_6(⚅) = \frac{3}{12}.$$

This matrix of probabilities is the analog of the known astrometric accuracy: the probability (density) of the possible outcomes for a given true location.

If the dice drawn with replacement are indeed the same, their loadedness has to be identical. It is the same die after all. The likelihood of a given loadedness l is the product of the $P_l(⚀)$ and $P_l(⚃)$ probabilities. But we do not know what l is. We could use maximum likelihood estimation to figure out the best guess value(s), but now we are not interested in that. Instead we have to consider all the l values to account for all possibilities in our hypothesis. The uniform prior on l is 1/6, as it can take 6 possible values. The result is the likelihood of the dice being the same

$$L_{\text{same}} = \frac{1}{6} \sum_l P_l(⚀) P_l(⚃). \tag{2}$$

The sum is calculated directly from our data, which are the faces we rolled. For more dice we can still use this same calculation. Only the product in the likelihood will contain more terms, one for each observations.

The complement hypothesis says that the two dice are different, hence their loadedness could differ. We have 2 independent variables and the sum conveniently falls apart

$$L_{\text{not}} = \left[\frac{1}{6} \sum_{l_1} P_{l_1}(⚀) \right] \left[\frac{1}{6} \sum_{l_2} P_{l_2}(⚃) \right]. \tag{3}$$

This works similarly with multiple observations, not just for two.

On the sky the calculation involves integrals of continuous probability density functions, but conceptually everything is the same. Our data consist of the measured positions $D = \{x_i\}$ unit vectors. The spherical normal distribution, called the Fisher distribution, has a precision parameter w, which is $1/\sigma^2$ for high accuracies. The $w = 0$ value means no spatial constraint (like a fair die.) The Bayes factor is calculated analytically and takes the following simple form in the general case,

$$B = \frac{\sinh w}{w} \prod_{i=1}^{n} \frac{w_i}{\sinh w_i} \quad \text{with} \quad w = \left| \sum_{i=1}^{n} w_i x_i \right|. \tag{4}$$

If all positional measurements are highly accurate ($w_i \gg 1$), we get back the a more familiar exponential formula

$$B = 2^{n-1} \frac{\prod w_i}{\sum w_i} \exp \left\{ -\frac{\sum_{i<j} w_i w_j \psi_{ij}^2}{2 \sum w_i} \right\}, \tag{5}$$

where ψ_{ij} represents the angle between x_i and x_j. In the 2-way case, the dimensionless Bayes factor simplifies to

$$B = \frac{2}{\sigma_1^2 + \sigma_2^2} \exp \left\{ -\frac{\psi^2}{2(\sigma_1^2 + \sigma_2^2)} \right\}, \tag{6}$$

where all quantities are in radians, and $\sigma_i^2 = 1/w_i$ as before. The top panel of Figure 2 illustrates equation (6) on a logarithmic scale for fixed $0\rlap{.}''1$ and $0\rlap{.}''8$ uncertainties, that roughly correspond to the precision of the Sloan Digital Sky Survey (SDSS) and the Two Micron All Sky Survey (2MASS). Note that for constant accuracies, a cut on the Bayes factor $B = B(\psi; \sigma_1, \sigma_2)$ is equivalent to a thresholding of the angular separation.

When the value of B is much larger than 1, the data suggest a good match, and when B is close to 0, the evidence points to separate objects. While in practice these extrema certainly occur, the interesting regime is the intermediate. What the measured values really correspond to is difficult to see at first.

3. Probability

The Bayes factor is a the fundamental quantity we rely on. Its interpretation, however, may not be obvious at first. Our goal is the assign probabilities that we can relate to. It surely must be possible, right? The interesting observation to make is that deriving the probability for a single association is, in fact, impossible without considering the entire datasets. The Bayes factor merely tells us how much the data prefer one hypothesis over the other. If our bag contains a single die, no matter what we roll, we will know that we pick the same die every time we draw one. This comes in as a prior on the hypotheses. By definition the Bayes factor is the missing link that connects the prior and the posterior probabilities. For two complementary hypotheses (same or not), B tells us how the prior probability P_0 is updated based on the data to yield the posterior,

$$P = \left[1 + \frac{1 - P_0}{B P_0} \right]^{-1}. \tag{7}$$

This means that we can only compute the probability if we have a prior. What is it? If we have N dice in our bag, the probability of drawing the same die for the second time is $P_0 = 1/N$. When we draw k times, the prior is $P_0 = N^{k-1}$. Astronomy is just a little bit different from this. The added complication comes from the fact that separate observations have different selection functions. In this general case the prior becomes

$$P_0 = \frac{N_\star}{\prod N_i}, \tag{8}$$

where N_i are the number of sources in the ith dataset and $N_\star$ is the size of the crossmatch set. The latter is unknown but an educated guess usually works reasonably well. The bottom panel of Figure 2 shows the SDSS-2MASS matching scenarios with the same styles as the Bayes factors in the top panel. Here we use N values that correspond to 27,000 and 11,500 detections per square degree for SDSS and 2MASS, respectively. We see that the exquisite SDSS astrometry provides great constraints. When we match SDSS against SDSS (solid line), probabilities peak at around 1. The larger uncertainty of 2MASS means a lower maximum posterior for the 2MASS-2MASS matching and a slower drop as a function of the angular separation. The SDSS-2MASS crossmatch shown in dotted lines can only be calculated with an estimate of the overlap. Here we plot the results 100%, 75%, and 50% of the 2MASS density.

We also note that the prior can be determined accurately from the ensemble statistics of the input datasets (Budavári & Szalay 2008). Iteratively solving a simple set of equations takes the guesswork completely out of the problem and provides a self-consistent result. Its properties can be simulated in details to evaluate its accuracy (Heinis et al. 2009).

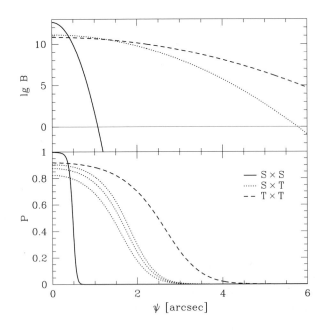

Figure 2.   The Bayes factor is shown in the *top* panel as a function of angular separation for three different matching scenarios. The *solid* line corresponds to detections with $\sigma = 0\rlap{.}''1$ accuracy and the *dashed* line is for $0\rlap{.}''8$. These correspond roughly to the precision of the SDSS and 2MASS, respectively. The *dotted* line is the analog of the SDSS-2MASS matching. The *bottom* panel illustrates the probability using the same line styles. We use 27,000 SDSS and 11,500 2MASS sources per square degree. The 3 dotted lines are based on estimates of the selection functions, whose overlap is 100%, 75%, and 50% of the 2MASS. As the intersection decreases, so does the probability.

4. Implementation in SQL and on GPUs

Using the proper probabilistic method is just as hard or easy as traditional methods. We have to find the candidates and evaluate their quality. The computational work is prohibitively expensive in the naive way. The trick is to narrow down the possible candidates before looking at the combinatorial list of candidates. One can do this by sorting the detections into buckets on the sky, whose neighborhood relations are known. One especially successful approach has been the so called *zone algorithm* (Gray et al. 2007) that groups the sources into constant declination zones, while sorting them by R.A. within. A zone has only 2 neighbors topologically, unlike other pixel based buckets, e.g., HEALPix or HTM. Also inside a zone, the computers can stream the objects and perform the matching in sliding window style.

The first implementations were in SQL, utilizing B-tree indexing on the composite (Zone-ID,RA) key. A short query can describe the matching that can run in about 15 minutes on a single machine for 2 of the largest catalogs SDSS vs. GALEX. The problem is limited by computation but can be parallelized very well. Using multiple cores and more machines can speed things up linearly.

Another approach is to leverage new architectures that were designed for parallel computing in the first place. The latest generation of Graphics Processing Units (GPUs) can not only render millions of pixels on our screen 60 times a second, but are also capable of general purpose computation. These video cards can run 25,000 threads at the same time on 512 cores, while still consuming comparable power to normal CPUs. Using NVIDIA's C for CUDA (Compute Unified Device Architecture), a prototype has been implemented that performs over 10 times faster than a 16-core Xeon processor. Limited by the 1.5 GB of global memory on a GTX 480, we can at the same time store roughly 2×30 million detections with their IDs and positions. Preprocessing of these datasets takes 2 seconds to sort into zones and by R.A.; performing a $5''$ search using $5''$ zones takes 11 seconds. The prototype can be optimized further, but the most important bottlenecks have been eliminated from the parallel implementation. To run fast one has to make sure that all memory writes are committed at the same time. When the threads write neighboring memory slots, the hardware can speed up the execution tremendously. Once the matching candidates have been identified, calculating the Bayes factor is negligibly fast even if the astrometric uncertainty changes from object to object.

Multiple datasets can be matched in a recursive algorithm, where every step adds a new set of detections to previous sub-matches (Budavári & Szalay 2008). This is possible because if two detections are far away they can never be part of any association, and similarly other sub-matches can be rejected early on. This helps to keep the list of candidates short and enable processing in less than the naive combinatorial time.

5. Summary

The success of the Bayesian approach has been demonstrated on several datasets, e.g., SDSS and GALEX in a paper by Heinis et al. (2009), or another on the Chandra catalog versus the SDSS by Rots & Budavári (2011). A recent project is applying this method to the entire Hubble Legacy Archive (Lubow et al. 2011). The calculation is efficient and the results have proven to be reliable in simulations (Heinis et al. 2009). But the beauty of the method is that it is generic enough to make use of other kinds of measurements, such as photometry. The Bayes factors are similarly computed for the observed magnitudes. Bayes factors based on different measurements multiply to yield the combined result

$$B_{\text{total}} = B_{\text{position}} \cdot B_{\text{photometry}} \cdots B_{\text{other}}. \qquad (9)$$

By listing various B values along with the list of candidates, the crossmatch datasets can be tailored for specific analyses in a straightforward way. The method also extends beyond that. Kerekes et al. (2010) have shown that it is possible to cross-identify stars even if they move at an unknown speed. The proper motion of the objects becomes another parameter of the model that we have to integrate over. The calculation will have to be done numerically but only for a smaller set of objects.

There are several directions to move forward and new models to include, but perhaps it is more important today to create a facility that can efficiently perform various cross-identifications. The early prototype of SkyQuery and later Open SkyQuery quickly became very popular, but suffered from serious limitations of the technology. At the Johns Hopkins University we are working on the next reincarnation that will make use of the aforementioned theoretical developments and the latest implementation tricks on a cluster of machines to deliver the matches as fast as possible. Within the Virtual Astronomical Observatory (VAO), a set of crossmatch engines are being planed that together aim to house the largest databases. The new system will also be able to use smaller legacy and the latest datasets that are published via standard VO interfaces.

References

Budavári, T., & Szalay, A. S. 2008, ApJ, 679, 301
Gray, J., Nieto-Santisteban, M. A., & Szalay, A. S. 2007, ArXiv Computer Science e-prints. arXiv:cs/0701171
Heinis, S., Budavári, T., & Szalay, A. S. 2009, ApJ, 705, 739
Kerekes, G., Budavári, T., Csabai, I., Connolly, A. J., & Szalay, A. S. 2010, ApJ, 719, 59
Lubow, S., Budavári, T., & Cole, N. 2011, in Astronomical Data Analysis Software and Systems XX, edited by I. N. Evans, A. Accomazzi, D. J. Mink, & A. H. Rots (San Francisco, CA: ASP), vol. 442 of ASP Conf. Ser., 97
Rots, A. H., & Budavári, T. 2011, ApJS, 192, 8

Astronomical Data Analysis Software and Systems XX
ASP Conference Series, Vol. 442
Ian N. Evans, Alberto Accomazzi, Douglas J. Mink, and Arnold H. Rots, eds.
©*2011 Astronomical Society of the Pacific*

Efficient and Scalable Cross-Matching of (Very) Large Catalogs

François-Xavier Pineau, Thomas Boch, and Sébastien Derriere

CDS, Observatoire Astronomique de Strasbourg, Université de Strasbourg, CNRS, 11 rue de l'Université, 67000 Strasbourg, France

Abstract.   Whether it be for building multi-wavelength datasets from independent surveys, studying changes in objects luminosities, or detecting moving objects (stellar proper motions, asteroids), cross-catalog matching is a technique widely used in astronomy. The need for efficient, reliable and scalable cross-catalog matching is becoming even more pressing with forthcoming projects which will produce huge catalogs in which astronomers will dig for rare objects, perform statistical analysis and classification, or real-time transients detection. We have developed a formalism and the corresponding technical framework to address the challenge of fast cross-catalog matching. Our formalism supports more than simple nearest-neighbor search, and handles elliptical positional errors. Scalability is improved by partitioning the sky using the HEALPix scheme, and processing independently each sky cell. The use of multi-threaded two-dimensional kd-trees adapted to managing equatorial coordinates enables efficient neighbor search. The whole process can run on a single computer, but could also use clusters of machines to cross-match future very large surveys such as GAIA or LSST in reasonable times. We already achieve performances where the 2MASS ($\sim$ 470M sources) and SDSS DR7 ($\sim$ 350M sources) can be matched on a single machine in less than 10 minutes. We aim at providing astronomers with a catalog cross-matching service, available on-line and leveraging on the catalogs present in the VizieR database. This service will allow users both to access pre-computed cross-matches across some very large catalogs, and to run customized cross-matching operations. It will also support VO protocols for synchronous or asynchronous queries.

Introduction

The largest catalogs of astronomical sources built so far, e.g., the USNOB1, contain about one billion sources. Projections for the LSST anticipate a number of unique sources about three times greater after 5 years of exploitation. With a minimum of 6 parameters — identifier (integer), positions (doubles) and associated errors (floats) — it will represent about 100 GB of data.

The method used to cross-correlate such catalogs has to take into account the current trend in computer hardware improvements: increasing and faster memory, more cores but stable clock frequency, cheaper machines grouped in clusters. Therefore it has to be scalable with its performance depending on both available machines and individual process efficiency. A catalog cross-match task must thus be split into pieces of various sizes, which can be processed independently on different threads (multi-threading) distributed on different machines (parallel processing).

This article is organized as following: § 1 deals with the partitioning and multi-threading of a cross-match task, § 2 with modified kd-tree for counterparts searches, § 3 with data loading, and the § 4 presents some results.

1.   HEALPix Partitioning and Multi-Threaded Pixel Processing

HEALPix (Górski et al. 2005) is a hierarchical sky partitioning developed at NASA. At level 0, the sky is divided into 12 pixels. Then, at each successive level, the pixels are divided into four new pixels so that for a given level l, the number of pixels is $N_l = 12 \times 4^l$.

We use HEALPix to divide the sky into pixels that can be processed independently, one by one on a single machine and simultaneously on a cluster of machines. The chosen HEALPix-pixels level depends on several parameters, such as the density of sources in both catalogs and the available memory. To cross-match the sources contained in a pixel of a catalog A with a catalog B, we first load the sources of the catalog B which are in the pixel. In order not to miss some correlations, we also have to load catalog B sources which are in an extra border around the pixel (see Fig. 1). We then build a modified 2d-tree (see § 2) containing the sources

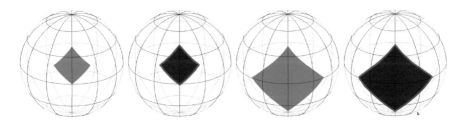

Figure 1.   HEALPix pixels of level 1 and 0 for a two catalogs A and B. No to miss some correlations, an extra border is required around the catalog B pixels.

retrieved from the catalog B. Then, for each catalog A source in the pixel, we perform a cone-search query in the 2d-tree. The process is multi-threaded: we create a pool of threads and, until all sources have been processed, we 1) take a worker from the pool, 2) fill the worker queue with catalog A sources, 3) wake-up the worker thread for correlation and result writing and 4) put back the worker in the pool, waiting for new sources to correlate.

2.   Multi-Threaded Modified kd-tree

A kd-tree is an Euclidean space-partitioning data structure especially adapted for fast k-nearest-neighbor (kNN) queries. We want the lightest possible data structure storing spherical coordinates and allowing us to perform fast cone-search or kNN queries relying on angular distances. The solution we have developed and adopted is a 2 dimensional kd-tree (2d-tree) stored in an array for which we have modified the query algorithm. The two dimensions are the spherical coordinates, α and δ. The creation of the tree is standard (see Fig. 2): the algorithm is a simple *quicksort* with alternating sorted coordinate.

In a standard fixed radius 2d-tree query, the algorithm first goes down the tree to the leaf node containing the target. It then backs up, and at each parent node it decides to go down the other sub-tree if the disk — defined by the target and the radius of the query — overlaps the rectangular area covered by the sub-tree (see Moore (1991) for more details). With spherical coordinates, the area covered by a sub-tree is no longer rectangular, and the distance between the target and a node is not Euclidean. We thus had to change the standard algorithm by computing target–node angular distances — resorting to the Haversine formula — and implementing a boolean function testing if a cone overlaps a range in α and δ.

The generation of a kd-tree is quite straightforward to multi-thread since each sub-tree is built independently (see Fig. 2). Nevertheless, resorting to a multi-threaded sort algorithm for the first nodes would even accelerate the process.

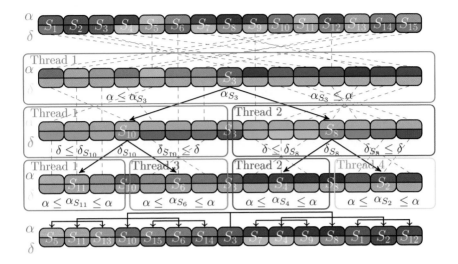

Figure 2.   Multi-threaded creation of a 15 sources array 2d-tree.

3. HEALPix Indexed Binary File

An efficient cross-match requires an efficient way to retrieve the needed data: positions, positional errors if necessary, and possibly identifiers. Disk accesses are expensive operations and to avoid too much head movement overhead it is better to access unfragmented data. The basic ideas to optimize data loading are to read only the necessary data, to read them from contiguous blocks, and in a binary format to avoid conversions. For fast access to the different sky cells' contents, data have to be grouped by HEALPix pixels and the file must be indexed.

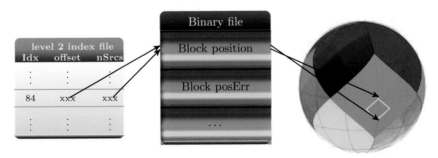

Figure 3.   Indexed binary catalog file.

We have implemented an indexed binary file format which stores a catalog by blocks (Fig. 3). Each catalog file contains one block for identifiers, one block for positions, one block for positional errors, and some other blocks. In each block, rows are sorted by the HEALPix pixel indices of the sources they belong to, so that the pixels of different HEALPix levels are stored contiguously in each block.

For each HEALPix level, for all pixels, an index file stores the index of the first row and the number of rows the pixel contains. So for each block, a pixel on the sky maps to a contiguous portion of the file, which is known perfectly thanks to the index files.

4. Performances Tests

We have performed some tests on several large catalogs: SDSS DR7 ($\sim$350M sources), 2MASS ($\sim$470M sources), and USNOB1 ($\sim$1G sources).

Our code is full Java and tests have been performed on a unique machine running Ubuntu 10.04 with Java 1.6.0_20 and a sun 64-Bit JVM (Java Virtual Machine). The machine is a Dell server with 24 GB of 1333 MHz RAM, two hyper-threaded Intel Xeon quad-cores at 2.27 GHz (16 threads available) and a 10 000 rpm HDD. Results are presented Table 1.

Table 1. Results of large catalogs cross-correlation tests. d is the cone-search aperture and d_σ the distance in sigma when taking into account individual elliptical errors on positions.

Catalogs	d	d_σ	nMatch	exec. time
SDSS7 vs 2MASS	5″		49.2 M	$\sim$9 min
SDSS7 vs 2MASS	5″	3.44	37.5 M	$\sim$10 min
2MASS vs USNOB1	5″		583.3 M	$\sim$30 min

All tests have been performed for a HEALPix level 3 (see Górski et al. 2005, Table 1) with border pixels of level 9. The execution time includes: data loading, tree creation, cross-correlation with (d_σ not empty) or without individual elliptical errors on positions, writing of a join file containing for each association the two sources identifiers and the cross-match distance. The candidate selection criteria when using positional errors is described in Pineau et al. (2010). On a similar machine with 2 hyper-threaded six-cores (24 threads available) the SDSS7 versus 2MASS cross-correlation execution time is under 7 minutes.

References

Górski, K. M., Hivon, E., Banday, A. J., Wandelt, B. D., Hansen, F. K., Reinecke, M., & Bartelmann, M. 2005, ApJ, 622, 759

Moore, A. W. 1991, An introductory tutorial on kd-trees, Tech. Rep. Technical Report No. 209, Computer Laboratory, University of Cambridge, Carnegie Mellon University, Pittsburgh, PA

Pineau, F.-X., Motch, C., Carrera, F., Della Ceca, R., Derriere, S., Michel, L., Schwope, A., & Watson, M. G. 2010, ArXiv e-prints. 1012.1727

Astronomical Data Analysis Software and Systems XX
ASP Conference Series, Vol. 442
Ian N. Evans, Alberto Accomazzi, Douglas J. Mink, and Arnold H. Rots, eds.

VAMDC: The Virtual Atomic and Molecular Data Center

Nicholas A. Walton,[1] Marie Lise Dubernet,[2,3] Nigel J. Mason,[4] Nikolai Piskunov,[5] Guy T. Rixon,[1] and the VAMDC Consortium[6]

[1]*Institute of Astronomy, University of Cambridge, Cambridge, CB3 0HA, UK.*
Email: `naw@ast.cam.ac.uk`

[2]*Laboratoire de Physique Moléculaire pour l'Atmosphère et l'Astrophysique,*
UMR7092 CNRS/INP, Université Pierre et Marie Curie, Case 76, 4 Place Jussieu,
75252 Paris Cedex 05, France

[3]*Laboratoire Univers et Théories, UMR8102 CNRS/INSU, Observatoire de Paris,*
Section Meudon, 5 Place Janssen, 92195 Meudon Cedex, France

[4]*Open University, Faculty of Science, Walton Hall, Milton Keynes, MK7 6AA, UK*

[5]*Uppsala University, Department of Physics and Astronomy, Lägerhyddsvägen 1,*
Uppsala 75120, Sweden

[6]`http://www.vamdc.eu`

Abstract. The Virtual Atomic and Molecular Data Center (VAMDC) is a European Union funded collaboration between groups involved in the generation, evaluation, and use of atomic and molecular data. VAMDC aims to build a secure, documented, flexible and interoperable e-science environment-based interface to existing atomic and molecular data. The project will cover establishing the core consortium, the development and deployment of the infrastructure and the development of interfaces to the existing atomic and molecular databases. This paper describes the organization, its objectives and introduces the VAMDC level one service release.

1. Introduction

Atomic and molecular (A+M) data are of critical importance across a wide range of applications such as astrophysics, atmospheric physics, fusion, environmental sciences, combustion chemistry, health and clinical science including radiotherapy, and underpin a range of industries ranging from technological plasmas to lighting. However, currently A+M data resources are highly fragmented and only available through a variety of highly specialised and often poorly documented interfaces, thus severely limiting their access and exploitation. This, in turn, provides a severe handicap to the development of research across a wide range of topics including space exploration (the characterization of extrasolar planets, understanding the chemistry of our local solar system and of the wider universe); the study of the terrestrial atmosphere and quantification of climate change; the development of the international fusion program for energy; our understanding of radiation damage within biological systems; and the development of plasma technology in materials processing (leading to nanoscale architectures) and as a tool for clinical and environmental studies.

Accordingly in the past decade the wider research community has appreciated the need to collate and make available the A+M data that describes fundamental atomic and molecular processes recognising how access to such data is central to achieving scientific breakthroughs across a range of disciplines.

However such increasing demands of the research community for large amounts of A+M data present major challenges to the expert research teams in Europe, the USA, Asia and elsewhere that measure, derive and collate such data as demand outstrips supply. The interface between the producers of A+M data and the wide body of users of that data is a major bottleneck, slowing scientific discovery.

Recognising this, the Virtual Atomic and Molecular Data Center (VAMDC,[1] Dubernet et al. 2010), funded through the EU FP7 Research Infrastructure program, has begun the task of providing a universal and pervasive data infrastructure providing a unified system to integrate the major sources of A+M data and, through standard interfaces, provide direct and simple access to such data for wider user community. In the following sections we provide a brief overview of the project.

2. The VAMDC Initiative and Objectives

The VAMDC project commenced in July 2009 and will run until the end of 2012. The consortium is composed of fifteen institutes spread over the European Union, Russia, Serbia and Venezuela. In addition there are a number of external partners located in the USA. These groups are all involved in the generation, evaluation and use of A+M data. In addition there is expertise within the consortium of development of baseline e-infrastructures including the Euro-VO.[2]

VAMDC is creating a secure, flexible, interoperable and fully documented e-science research infrastructure for A+M data. This data underpins many areas of research and potential users come from industry as well as academia. The amount of A+M data is complex and increasingly large with the handling of such data often requiring the use of application tools. However there remain many problems in ensuring data completeness and quality. The key VAMDC objectives are therefore: to implement a VAMDC interface for accessing major existing databases containing heterogeneous data and aimed at different users; to enable data queries across multiple databases each focused on specific research topics; to facilitate the data publishing and quality control process for major A+M data producers; and to involve wide user and producer communities in the development and use of VAMDC tools. The key end user communities are from astrophysical, atmospheric, plasma and combustion science, and from the industrial applications fields. They perform simulations, observations, and diagnostic interpretation. There is also a teaching component to VAMDC in that it is necessary to acclimatise users in the VAMDC methodology and architecture.

3. VAMDC Provision of Atomic and Molecular Data

The VAMDC infrastructure provides access to a comprehensive number of major A+M resources provided by the VAMDC partners. These resources are published by the expert teams generating the A+M data, and are located across Europe and elsewhere, e.g., France, UK, Sweden, Austria, Germany, Italy, Serbia, Russia, Venezuela, USA, and include for example:

1. The VALD database of atomic and molecular data provides a robust and consistent analysis of radiation from astrophysical objects.

2. CHIANTI a well-established atomic database for ions of astrophysical importance. Combined with IDL spectroscopic diagnostic programs, it is used in the analysis of optically thin collisionally-ionised plasmas, relevant in solar physics.

---

[1]http://www.vamdc.eu

[2]http://www.euro-vo.org

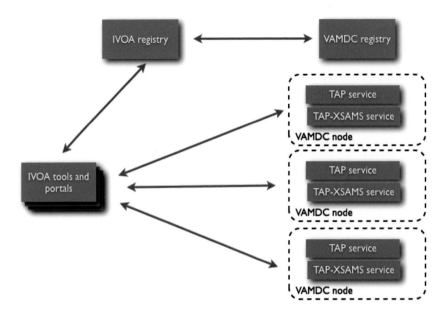

Figure 1.   This shows the base infrastructure, with the key components, portal, registry, interfaces to data and applications. TAP is a access protocol to tabular data, XSAMS is the schema for A+M data developed actively by the VAMDC project.

3. The Cologne Database for Molecular Spectroscopy (CDMS) provides recommendations for of spectroscopic transition frequencies and intensities for atoms and molecules of interest to the astronomical community, or for studying the Earth atmospheres.

4. The BASECOL database provides excitation rate coefficients for ro-vibrational excitation of molecules by electrons, He and H_2 including error estimates. This database has been mainly used for the study of interstellar, circum stellar and cometary atmospheres.

5. NIST Atomic Databases, the largest of these, the Atomic Spectra Database, contains evaluated data on about 77,000 energy levels and 144,000 spectral lines from atoms and ions of 99 elements.

4.   The VAMDC Organization

The project is organized around three core activities: Networking, Service and Research. The networking Activities foster cooperation between A+M scientists, database providers and data users throughout Europe. Thus, this activity both manages and directs the internal activities of the project, but also ensures a full engagement program with the wider community, including the organization of workshops and training events. The service activities focus on the deployment of VAMDC services, and technical support of the functioning infrastructure. To date the key service activities have included the development of standard services, a basic monitoring service, and support of a simple portal access to the VAMDC infrastructure. The research activities have focussed on standards and interoperability, together with future planned data mining functionality. A key year one output has been the development of the XSAMS schema[3] for describing molecular and atomic data.

5.   The VAMDC Level One Service Release

In the VAMDC infrastructure it is foreseen that the user accesses a database through some application that is also part of VAMDC. The application connects with the relevant databases and code services and communication takes place according to standardised formats agreed by VAMDC. Following this methodology, the VAMDC project has recently released its baseline level one service release, where the initial basic data framework has been established, with the connections made to the major VAMDC data resources. Figure 1 shows diagrammatically the service structure. In this, there is reuse of standard interfaces to tabular data developed in the context of the International Virtual Observatory Alliance[4] together with newly developed interfaces, for instance the rich XSAMS interface to A+M data held in relational databases.

Acknowledgments.   VAMDC is funded under the "Combination of Collaborative Projects and Coordination and Support Actions" Funding Scheme of The Seventh Framework Program. Call topic: INFRA-2008-1.2.2 Scientific Data Infrastructure. Grant Agreement number: 239108.

References

Dubernet, M.-L., et al. 2010, JQSRT, 111, 2151

---

[3]http://www.xsams.org

[4]http://www.ivoa.net

Astronomical Data Analysis Software and Systems XX
ASP Conference Series, Vol. 442
Ian N. Evans, Alberto Accomazzi, Douglas J. Mink, and Arnold H. Rots, eds.
© *2011 Astronomical Society of the Pacific*

Astronomical Image Processing with Hadoop

Keith Wiley,[1] Andrew Connolly,[1] Simon Krughoff,[1] Jeff Gardner,[2]
Magdalena Balazinska,[3] Bill Howe,[3] YongChul Kwon,[3] and Yingyi Bu[3]

[1] *University of Washington Department of Astronomy*

[2] *University of Washington Department of Physics*

[3] *University of Washington Department of Computer Science*

Abstract. In the coming decade astronomical surveys of the sky will generate tens of terabytes of images and detect hundreds of millions of sources every night. With a requirement that these images be analyzed in real time to identify moving sources such as potentially hazardous asteroids or transient objects such as supernovae, these data streams present many computational challenges. In the commercial world, new techniques that utilize cloud computing have been developed to handle massive data streams. In this paper we describe how cloud computing, and in particular the map-reduce paradigm, can be used in astronomical data processing. We will focus on our experience implementing a scalable image-processing pipeline for the SDSS database using Hadoop (`http://hadoop.apache.org/`). This multi-terabyte imaging dataset approximates future surveys such as those which will be conducted with the LSST. Our pipeline performs image coaddition in which multiple partially overlapping images are registered, integrated and stitched into a single overarching image. We will first present our initial implementation, then describe several critical optimizations that have enabled us to achieve high performance, and finally describe how we are incorporating a large in-house existing image processing library into our Hadoop system. The optimizations involve prefiltering of the input to remove irrelevant images from consideration, grouping individual FITS files into larger, more efficient indexed files, and a hybrid system in which a relational database is used to determine the input images relevant to the task. The incorporation of an existing image processing library, written in C++, presented difficult challenges since Hadoop is programmed primarily in Java. We will describe how we achieved this integration and the sophisticated image processing routines that were made feasible as a result. We will end by briefly describing the longer term goals of our work, namely detection and classification of transient objects and automated object classification.

1. Introduction

Future astronomical surveys will generate data in quantities which cannot be processed by single computers. One potential solution to this problem is to harness large clusters of computers by using *cloud computing*. In this paper we describe the development of a cloud computing based image coaddition system using the Hadoop MapReduce cluster framework. We describe our system, then show an example of a coadded mosaic and analyze its improved detection threshold.

2.   Experimental Setup

This research was performed on the *CluE* cluster (see Acknowledgements). At the time, the cluster had 700 nodes, each with 4×2.8GHz cores, 8GB ram, and 800GB storage for a total cluster storage capacity of 560TBs.

We chose as our dataset *Sloan Digital Sky Survey* (SDSS) Stripe 82 (Abazajian et al. 2009; SDSS 2007). The SDSS camera has 30 CCDs (2048×1489 pixels, 6MB FITS) in 5 bandpass filters which capture 6 parallel strips of sky at a time. Stripe 82 is a 30TB, 4 million image dataset gathered near the equatorial plane ($\pm 1.25°$ declination) with an average coverage of ~ 75. In theory, coaddition at such coverage should yield a SNR improvement of $\sim 8.7\times$ or an improved limiting magnitude of ~ 2.3 mags.

Our research has focused on the development of a massively parallel *image coaddition* system. Given the variety of uses of the term, we define image coaddition as the process of background-subtracting, warping, PSF-matching, registering, and per-pixel averaging a set of partially overlapping images into a final image called a *mosaic*. Much of this process can be trivially parallelized since many of the steps are performed on the input images prior to their incorporation into the mosaic.

3.   Massively Parallel Data Processing

In recent years, a new approach to massively parallel data processing called *cloud computing* has gained popularity. A cloud consists of a large network (1000s) of relatively cheap commodity computers which is then made accessible over the internet. This economical construction and internet-based access permit clouds to be offered as a generic service wherein users program and submit their own jobs remotely and as third party customers. One popular example of such a general-purpose cloud is Amazon's EC2.

MapReduce is a framework for designing cloud-computing programs (Dean & Ghemawat 2004) which encapsulates the cluster-related aspects of parallel computing, namely intra-network communication, resiliency to task/node failure, etc. This design alleviates much of the complexity that parallel programing would otherwise impose. A MapReduce program is performed in two sequential stages. The *mapper* stage performs a parallel computation on the input data. The results are distributed to the *reducer* stage which conglomerates the mapper outputs into the final job output. *Hadoop* is an open-source implementation of MapReduce (Apache 2007; White 2009) which has quickly grown in popularity in large part due to is relatively easy learning curve and the large and active online community of support.

Hadoop is programmed in Java. However, our research group has already developed a sophisticated C++ image-processing library. In order to access this library from Hadoop, we use the *Java Native Interface* (JNI). Using JNI, our Java-based mapper and reducer serve primarily to interface with the Hadoop framework and distributed file system, but delegate most of the computational demands (image coaddition) to C++.

While many forms of data-processing require processing the entire input dataset, image coaddition does not. A *query* (a bounds on the sky within which to generate a mosaic) only covers the small subset of the input images. Therefore, we use a front-end relational database containing metadata about the input images, including their sky bounds. Our Hadoop job first performs a SQL query to retrieve the filenames of the images which are relevant to the coaddition process. Those filenames then represent the input to our MapReduce image coaddition system.

4.   Image Coaddition in Hadoop

In order to adapt image coaddition to Hadoop, we perform the initial processing on each input image in a highly parallelized mapper stage. This processing includes background-subtraction, warping to the final coordinate system, and PSF-matching. The results are then sent to a single

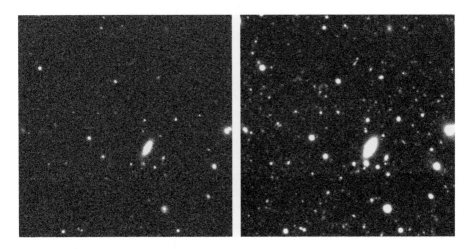

Figure 1.   This figure shows a single r-band frame on the left and a mosaic of 96 frames on the right (with a max coverage of ~ 75). The mosaic reveals more faint sources as a result of coaddition.

reducer which performs the per-pixel average and generates the final mosaic. The serialized nature of the reducer is acceptable since the overall computational demands are dominated by the steps performed in the mappers.

5.   Results

Fig. 1 shows an example of image coaddition. A single r-band frame is shown on the left and a mosaic of 96 frames is shown on the right. We would expect the point source detection threshold for such a mosaic to be improved by ~ 2 mags over the single frame and in Fig. 2 we observe that the expected improvement was achieved. On the CluE cluster, our system was able to generate this mosaic in ~ 34 minutes. However, many factors can influence this result: Hadoop restarts failed tasks, we did not enable compiler optimizations, and our image-processing routines are still under development. We estimate that when properly configured, this job time may drop well below 13 minutes, which corresponds to a per-image (mapper) processing time of < 8 minutes.

6.   Future Work

In the near future we hope to improve our coaddition system in many ways. We would like to improve the overall algorithm by parallelizing the reducer, implementing better memory management, and continuing to improve our image-processing routines. We intend to extend the query description to include time-bounded queries and to ultimately perform automated object detection and classification on the mosaics. Finally, we intend to wrap our system in more accessible scripting languages and perhaps to offer it through a web-based graphic user interface (GUI).

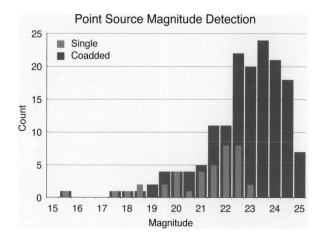

Figure 2.   This plot shows the point source magnitude detections achieved by the single image and the mosaic shown in Fig. 1. We observe that the mosaic's point source detection threshold is improved by ~ 2 mags, as expected.

7. Conclusions

This research demonstrates a massively-parallel cloud-computing based image coaddition system. We described Hadoop and our image coaddition system within Hadoop. We then showed an example mosaic generated from SDSS Stripe 82 and demonstrated that it achieved the expected improvement in point source detection threshold.

Acknowledgments.   This work is funded by the NSF Cluster Exploratory (CluE) grant (IIS-0844580) and NASA grant 08-AISR08-0081. The cluster is maintained by IBM and Google. We thank them for their continued support. We further wish to thank both the LSST group in the astronomy department and the database research group in the computer science department at the University of Washington.

References

Abazajian, K. N., et al. 2009, ApJS, 182, 543
Apache 2007, Apache hadoop. URL http://hadoop.apache.org/
Dean, J., & Ghemawat, S. 2004, in Sixth Symposium on Operating System Design and Implementation (San Francisco, CA, USA), OSDI'04
SDSS 2007, SDSS Stripe 82, http://www.sdss.org/legacy/stripe82.html, http://www.sdss.org/dr7/coverage/sndr7.html
White, T. 2009, Hadoop The Definitive Guide (Sebastopol, CA: O'Reilly), 1st ed.

Astronomical Data Analysis Software and Systems XX
ASP Conference Series, Vol. 442
Ian N. Evans, Alberto Accomazzi, Douglas J. Mink, and Arnold H. Rots, eds.
© *2011 Astronomical Society of the Pacific*

Cross Matching Sources In the Hubble Legacy Archive (HLA)

Stephen Lubow,[1] Tamás Budavári,[2] and Nathan Cole[2]

[1] *Space Telescope Science Institute, 3700 San Martin Drive, Baltimore MD 21218*

[2] *Department of Physics and Astronomy, Johns Hopkins University, Baltimore MD 21218*

Abstract. We have begun a project to cross identify sources observed in different HST exposures. We first determine the set of overlapping exposures which would have objects in common. We apply the object source lists for the exposures currently available in the HLA. For each set of overlapping exposures, we determine trimmed source list of possible cross matches using a Bayesian method. We apply a novel algorithm for shifting the exposures into common alignment. We then update the positions of the objects and repeat the computation of the trimmed source list and exposure shifts iteratively until no further improvement is obtained. In preliminary tests, we have been able to obtain mean relative astrometric shifts between exposures of order of a few milliarcseconds.

1. Source Lists

The Hubble Legacy Archive[1] (HLA) provides astronomers with new capabilities for exploring HST data in an interactive manner over the internet by means of a web browser. The HLA has undertaken a large reprocessing effort for images obtained by the NICMOS, WFPC2, and ACS instruments. This effort has provided high quality drizzled science images with an astrometric accuracy of a few tenths of an arc-sec for nearly all the science images obtained by these instruments. In addition, the HLA provides information about other HST instruments. One of the main categories of products available from HLA are the source lists (Whitmore et al. 2008). Sources are detected from multi-band WFPC2 and ACS images that are combined within a visit of HST (pointing of the telescope over a limited time). These "white light" images produce the deepest image possible within a visit of HST from which sources can be detected. The source detections are carried with DAOPHOT and Source Extractor. The source lists contain positional information about each source, together with photometric and other source properties. To date over 30,000 white light images have been analyzed to produce over 3×10^7 sources.

These source lists fall short of providing a "catalog" of sources. The main reason is that different white light images may contain the same source. One major goal of our project is to cross match source across white light images. Overlaps of white light images occur in many cases. As seen in Fig. 1, the number of cases of overlapping images decreases as a power law with the number of overlapping images in each case. This power law behavior continues to cases involving about 20 overlapping white light images. Beyond this regime, there is a long tail that extends to more than 50 overlapping white light images. The long tail typically involves astronomical objects that are popular or were of interest for particular purposes, such as the globular cluster 47 Tuc. This tail poses challenges for algorithms in the cross matching process.

HST observations are scattered across the sky and often overlap to form irregular geometric patterns. Consequently, the exposure time coverage of any area in the sky is quite variable.

---

[1] http://hla.stsci.edu

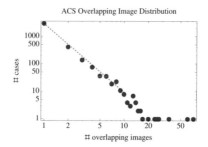

Figure 1.   Distribution of the number of overlapping ACS images plotted on a log-log scale plotted as blue points. On the vertical axis is the number of cases of overlapping images. On the horizontal axis is the number of overlapping images in each case (for a value of 1 the images are single, without overlap). The red dotted line follows a power law with exponent -2.5 that fits the points well for less than 10 overlapping images.

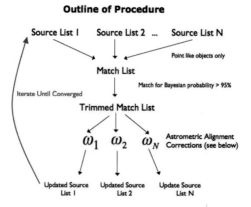

Figure 2.   The process flow for cross matching.

These data characteristics also pose challenges for cross matching. As seen in Fig. 2, these source lists provide the first step in the cross matching process.

2. Matching Step

To perform the cross matching, it is necessary to determine a set of sources that nearly overlap (approximately match) across white light images to a certain level of confidence. To determine this set of sources, we apply a Bayesian technique developed by Budavári & Szalay (2008). This technique takes into account the expected errors associated with the astrometry of the sources in the source lists to determine a probability of overlap among the sources across white light images. The sources that meet a certain threshold for overlap probability constitute the "trimmed souce lists." This step is the second one in Fig. 2.

3. Alignment Step

Once we identify a set of high probability matches, we align the overlapping white light images to reduce the shifts among the matched sources in the trimmed source lists. To carry out the alignment, we assume that the images can differ by translation and rotation. Both are described as a rotation on a sphere that represents the sky. The shifts are then described by a 3D rotation vector ω_k for the kth image. The rotations are adjusted to minimize the RMS shifts between the matched sources in the different white light images. More precisely, we determine $\omega_1 \ldots \omega_N$ for the N overlapping white light images such that

$$F(\omega_1, \omega_2, .., \omega_N) = \sum_{k,m} \frac{1}{\sigma_{k,m}^2} \left| \mathbf{c}^{(k,m)} - (\mathbf{r}^{(k,m)} + \omega^{(k)} \times \mathbf{r}^{(k,m)}) \right|^2 \qquad (1)$$

is minimized. In this sum, k ranges over images $1 \ldots N$ and m ranges over the trimmed sources in the kth image. $\mathbf{r}$ is the initial position of the source on the unit sphere, with associated RMS positional error σ. Vector $\mathbf{c}$ is the is the (suitably weighted) mean position on the unit sphere among the corresponding members of a set of matching objects in the trimmed source lists. That is, assuming the mth source on each image matches across white light images, $\mathbf{c}$ is given by

$$\mathbf{c}^{(k,m)} = \sum_k w_{k,m} \mathbf{r}^{(k,m)}, \qquad (2)$$

for weighting factors $w_{k,m}$. The minimization in equation (1) can be done analytically by assuming that the rotations are small. The rotation vectors can then be determined by inverting a matrix. This alignment step is the third step in Fig. 2.

4. Update Step

Given the optimal rotations that minimize relative shifts, we can determine improved positions for each source by applying the rotations determined by the previous step for the appropriate image. That is, we determine a new set of source positions for each source in the trimmed list as

$$\mathbf{r}^{(k,m)} \to \mathbf{r}^{(k,m)} + \omega^{(k)} \times \mathbf{r}^{(k,m)}. \qquad (3)$$

This step the the fourth step listed in Fig. 2.

5. Iteration

Of course, these updated positions do not bring the matching sources into perfect alignment, since function F in equation (1) is nonzero (although minimized). To make further improvements, we apply this updated source lists to the first step in the process outlined in Fig. 2. The

Lubow, Budavári, and Cole

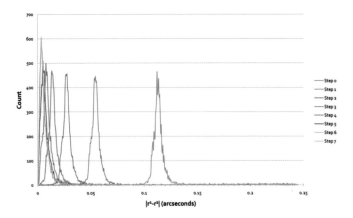

Figure 3.   The distribution of shifts between sources in two images. The rightmost curve is the initial distribution of shifts from the HLA source lists. The other curves represent subsequent iteration steps in which the images are aligned through the procedure described by Fig. 2. Each step involves a loop in the Fig. 2. After 6 iterations, the shifts are reduced by almost a factor of 100.

four steps are repeated. The iteration of the steps continues until there appears to be no further improvement to F. At this stage, we regard the process as converged. We typically obtain convergence after about 6–12 iterations.

A typical result is shown in Fig. 3. The plot shows the distribution of postional shifts among the matched sources in a case involving 2 overlapping ACS white light images. In the first iteration, the shifts are of order 0.1 arcseconds, which is consistent with the typical astrometric accuracy of the HLA. Subsequent iterations reduce the errors to a few milliarcseconds — almost 2 orders of magnitude improvement.

Acknowledgments.   We acknowledge support from NASA AISRP grant NNX09AK62.

References

Budavári, T., & Szalay, A. S. 2008, ApJ, 679, 301
Whitmore, B., Lindsay, K., & Stankiewicz, M. 2008, in Astronomical Data Analysis Software and Systems XVII, edited by R. W. Argyle, P. S. Bunclark, & J. R. Lewis (San Francisco, CA: ASP), vol. 394 of ASP Conf. Ser., 481

Part III

Data Analysis

Astronomical Data Analysis Software and Systems XX
ASP Conference Series, Vol. 442
Ian N. Evans, Alberto Accomazzi, Douglas J. Mink, and Arnold H. Rots, eds.
© 2011 Astronomical Society of the Pacific

SITools2: Framework for Data Access Layer

Jean-Christophe Malapert

CNES (Center National d'Etudes Spatiales), 18 avenue Edouard Belin,
31 401 TOULOUSE CEDEX 9, France

Abstract. SITools2 is an open source framework developed by CNES and AKKA technology. The purpose of this framework is two-fold. The first aim is to setup easily an archival system based on the Open Archival Information System standard (OAIS).[1] The OAIS is composed of several functions: access, data management, storage, ingestion, data preservation and administration. SITools2 is currently implementing the access and the administration parts. The second aim is to federate the development in scientific laboratories. In this perspective, the framework has been designed to be extensible so that developers in scientific laboratories register their own services in the SITools2's API. A first version of SITools2 will be released by June 2011.

1. The Server

1.1. The REST Architecture

SITools2 is a client-server architecture based on a REST (Representational State Transfer; Fielding 2000) architecture. This style of architecture has been choosen for its features:

- Providing an API that is independent of any specific technologies,

- Allowing URIs to be used for both the access preservation and the cache system,

- Providing a set of representations that allow the interoperability of the same URI with different kind of clients,

- Offering a simple way to provide security to the resources.

RESTlet has been choosen as REST framework because it offers a lot of possibilities in terms of features and extensions.

1.2. The Functionalities

The server side is built in JAVA and it is composed of two main APIs:

- One for the system administration: user management and system configuration,

- A second one for search capabilities.

From the APIs, several functionalities are available: application registration, user management, access rights, handling datasource, dataset, application and project. Other features are also available such as open search, RSS flux, dictionary and query form definition. Each functionality is split into two subfunctions: configuration and request.

---

[1]See (CCSDS 650.0).

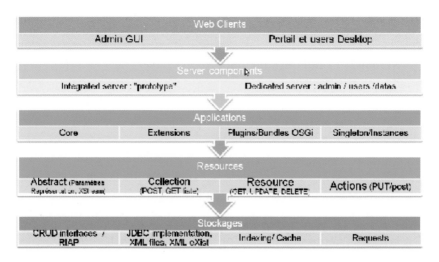

Figure 1.   Model of software layers.

1.3.   The Extension Point

We designed SITools2 to be able to add easily some new functionalities to its API by a set of extension features. The first way is to add an extension as an OSGI module to share this module with a community. The second way is to register an external application by the use of a proxy. The registration of applications lets the administrator configure the application's access rights and to insert the access information in an unique log file, making easier the web mining of the access log.

2.   The Clients

In addition to the server, two user interfaces have been built in top of the REST web services:

- An administration panel for setting up the access services,
- A portal and a user panel for discovering, retrieving the data.

2.1.   The Rich Internet Application (RIA)

The client is based on a RIA. This technology uses AJAX (Asynchronous Javascript and XML) and offers a better interaction with the user. Among the different RIA framework existing on the market, we have choosen EXT-JS because of its graphical components based on the inheritance. This EXT-JS feature makes possible the customization of graphical components by developers and fits with our need to build a framework.

2.2.   The Web User Interface

Figure 2(a) shows a view of the portal that is composed of different windows: a list of archived projects, a RSS flux and an open search query on the whole list of projects. The functionalities of the portal are easily extensible in the sens that the graphical components can be inserted dynamically by the use of portlets. Figure 2(b) shows a user desktop for a specific project. This interface contains some graphical componants to discover/query datasets defined by the administrator. New functionalities can be added to this web interface by dropping a package in a specific directory.

Figure 2.   The user interface (from left to right): (a)portal and (b)user desktop.

2.3.   The Web Administrator Interface

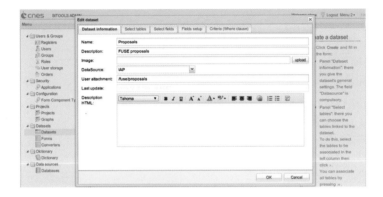

Figure 3.   The administrator interface.

Figure 3 shows a view of the current administration interface. From this interface, the administrator can create datasets that he wishes to put online. He can also configure the access rights on the metadata/data, manage the users/groups/roles and define some query capabilities such as open search query or query forms.

3.   Messages Transferred Between Server and Client

The message exchanged between the server and the clients has two features:

- The message contains only data and no presentation information. In this way, others clients could interact with the server. The web technology evolving quickly, we cannot predict if EXT-JS will be present in the future. Having a weak coupling between the client and the server should let us change the clients without breaking the server part.

- The message format between web interfaces and the server is in JSON (JavaScript Object Notation) to improve the data transfer. Nevertheless, each URI supports XML format to interact with other clients.

4.   Sofware Components Deployment

The deployment is realized on Jetty as an embedded application server. Moreover, the configuration of the system is located at one place on the disk in "/data" directory. All these characteristics bring some advantages:

Figure 4.   Software components deployment.

- Easier deployment,
- Easier configuration,
- Better security management (e.g., Unix access rights on the configuration directory).

In addition, we anticipate packaging the sofware using of Izpack to bring nice features during the installation.

References

CCSDS 650.0 2002, Reference Model for an Open Archival Information System (OAIS) – CCSDS 650.0-B-1, CCSDS Recommendation. Identical to ISO 14721:2003, URL `http://public.ccsds.org/publications/archive/650x0b1.pdf`
Fielding, R. T. 2000, Ph.D. thesis, University of California, Irvine

Astronomical Data Analysis Software and Systems XX
ASP Conference Series, Vol. 442
Ian N. Evans, Alberto Accomazzi, Douglas J. Mink, and Arnold H. Rots, eds.
©*2011 Astronomical Society of the Pacific*

EMphot — Photometric Software with Bayesian Priors: Application to GALEX

Simon Conseil,[1] Didier Vibert,[1] Stéphane Arnouts,[2] Bruno Milliard,[1]
Michel Zamojski,[3] Antoine Llebaria,[1] and Mireille Guillaume[4]

[1]*Laboratoire d'Astrophysique de Marseille, OAMP, Université Aix-Marseille & CNRS, Marseille, France*

[2]*Canada France Hawaii Telescope, Kamuela, Hawaii, United States*

[3]*California Institute of Technology, Pasadena, United States*

[4]*Institut Fresnel, Université Aix-Marseille, Marseille, France*

Abstract.   EMphot is a software tool for the photometry of astrophysical sources, galaxies and stars, in crowded field images. Its goal is to estimate the flux in a low resolution band using prior information (position and shape) from a better resolved band, in a Bayesian approach under the Poisson noise assumption. The solution is reached with an Expectation-Maximization (EM) algorithm for solving the photometry and includes several steps: prior shapes deblending in high resolution images, astrometry correction, PSF optimization, background correction from the residual.

1.   Introduction

Photometry of astrophysical sources, galaxies and stars, in crowded field images, if an old problem, is still a challenging goal, with current and future survey missions releasing new data with increased sensitivity, resolution and field of view.

This paper gives an overview of EMphot, details on the different components can be found in several papers from the same team: Guillaume et al. (2006) present the original procedure used to face on this challenge, using the Poisson statistics to define the Bayes assumption and solve for the maximum posterior likelihood with the constraint for all fluxes to be positive. Llebaria et al. (2008) discuss the photometric performance and behavior of the method when dealing with imperfect knowledge of background, PSF and object positions. Vibert et al. (2009) describe the improvement of using the extended shape inferred by deblending the high resolution optical images and not only the position of the optical sources.

2.   Maximum Likelihood Parametric Estimation with Priors: Expectation-Maximization (EM)

The specificity of the proposed photometric procedure is the Bayesian approach under the Poisson noise assumption. The solution is reached with an EM algorithm.

Let be x_i for $i \in \{1,...M\}$ the observed value on pixel i of the UV image considered as a sample of the random variable X_i following a Poisson statistics. Let be $\mu_i = E\{X_i\}$ for $i \in \{1,...M\}$ the expected values for this image. Let be $h_{k,i}$ the known relative value of object $k \in \{1,...K\}$ on pixel i deduced from the visible catalog. Let be $\alpha^T = (\alpha_1,...\alpha_k)$ the vector of unknown fluxes of these objects. Let be r_i the relative instrument response taking into account exposure time and efficacy of the system. Last, let be b_i be the known background level value on pixel i of the UV image. We define the model for the UV image as follows:

$$\forall i \in \{1, \ldots M\} \begin{cases} \mu_i = r_i \sum_{k=1}^{K} \alpha_k h_{k,i} + r_i b_i \\ P\{X_i = x_i\} = \exp(-\mu_i)\frac{\mu_i^{x_i}}{x_i!} \end{cases} \quad \text{where} \quad h_{k,i} = \sum_j o_{k,j} f_{i-j} \quad (1)$$

here $h_{k,i}$ results from the convolution between each object known profile $o_{k,i}$ with the point spread function f_i of the imaging system. By normalizing the function $h_{k,i}$ to unity, this model allows to estimate directly α, the flux vector of the set of sources in the image. As was shown in more detail in (Guillaume et al. 2006), introducing the expectation maximization scheme, we get the iterative algorithm:

$$\alpha_k^{(n+1)} = \alpha_k^{(n)} \frac{\sum_{i=1}^{M} \frac{x_i}{\mu_i^{(n)}} r_i h_{k,i}}{\sum_{i=1}^{M} r_i h_{k,i}} \quad \text{where} \quad \mu_i^{(n)} = r_i \sum_{j=1}^{K} \alpha_j^{(n)} h_{j,i} + r_i b_i \quad (2)$$

The E step compares the data image x_i to the projection $\mu_i^{(n)}$ of the $\alpha_i^{(n)}$ estimates. The result is introduced in the M step as the corrective ratio needed for the new set of $\alpha_k^{(n+1)}$ estimates. The background level b is considered known, we found it in a previous procedure.

3. Features

- **Prior shapes computation:** Firstly, a deblending is done using SExtractor ellipses to define objects contour. Central symmetry is used to determine the flux assigned to each object blended in one pixel. Secondly, the image is degraded to the resolution of the GALEX image.

- **Astrometry correction:** cross-correlate the positions of the brightest objects (detected with SExtractor) with the brightest objects of the prior catalog and warp with a 2nd order polynomial fitting.

- **Image processed by tiles**.

- **Initial fluxes** $\alpha_i^{(0)}$: use U-band value or estimate from the image using a PSF weighted sum.

- **Prior flux constraint:** $\text{mag}_{UV} > \text{mag}_U - 1$

- **PSF rescaling:** we optimally rescale the PSF maximizing the likelihood with respect to the PSF scale for known fluxes.

- **Error estimation** from the residual: $\widehat{\sigma}_k^2 = \sum_{i=1}^{M} h_{k,i}(x_i - \widehat{\mu_i})^2 / \sum_{i=1}^{M} h_{k,i}^2$

- **Background correction:** mask and do the inpainting around objects artifacts in the residual, filter high frequencies and redo EM iterations.

- **Post-processing** of the output catalog: flag objects inside GALEX and CFHTLS masks, compute statistics on nearest neighbors, compare with GALEX catalog.

4. Simulations

The resulting photometric accuracy is quantified with both completely simulated crowded fields and simulated objects added on top of the real images. Error estimation is done using Monte-Carlo simulations, adding simulated objects to the real image or simulating all the objects, using the number counts from GALEX. Simulations use astrometry corrections, stamps or an optimal PSF scale value to be consistent with the processing algorithm.

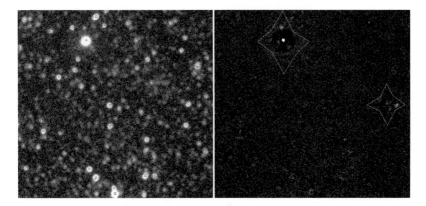

Figure 1.   XMMLSS_00 NUV field with 200 iterations, left: GALEX image (priors in red), right: EM residual (masked region in green).

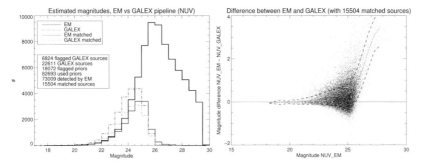

Figure 2.   Comparison SExtractor / EMphot.

5.   Results

We apply this software to the Deep Imaging Survey (DIS) of the GALEX mission, which observes in two UV bands with long exposure times ($\sim 70,000$s), and produces deep sky images of 1 square degree, with hundreds of thousands of galaxies or stars. Priors are computed from CFHTLS data. These UV observations are of lower resolution than same field observed in visible bands, and with a very faint signal dominated by the photon shot noise, with background level around 100 (resp. 10) counts in the near (resp. far) UV band. Figures 1, 2 and 3 show the results on XMMLSS_00 NUV field.

6.   Conclusion

Finally, compared to blind photometry estimation, the method leads to small and flat residuals, increases the faint source detection threshold, and provides a better accuracy for bright contaminated objects. On the processed DIS fields, EM provides good photometry and completeness down to magnitude 25.5, which is 1 magnitude deeper than the GALEX pipeline. As an important by-product, the method automatically solves the problem of determining the optical counterparts to UV sources, and shares the UV flux between partly resolved or unresolved nearby objects. The method is also an optimal approach for measuring drop-outs in FUV and NUV.

110   *Conseil et al.*

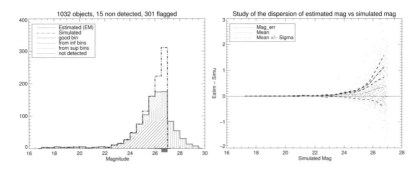

Figure 3.   Monte-Carlo simulations on XMMLSS_00 NUV with 3 × 500 objects added to the image.

Future developments involve PSF parametrization, Gaussian noise (BVLS), application to HERSCHEL, Model selection method for reducing prior number and Astrometry improvement with a maximum-likelihood recentering.

References

Guillaume, M., Llebaria, A., Aymeric, D., Arnouts, S., & Milliard, B. 2006, in Image Processing: Algorithms and Systems, Neural Networks, and Machine Learning, edited by E. R. Dougherty, J. T. Astola, K. O. Egiazarian, N. M. Nasrabadi, & S. A. Rizvi (Bellingham, WA: SPIE), vol. 6064 of Proc. SPIE, 332
Llebaria, A., Magnelli, B., Arnouts, S., Pollo, A., Milliard, B., & Guillaume, M. 2008, in Image Processing: Algorithms and Systems VI, edited by J. T. Astola, K. O. Egiazarian, & E. R. Dougherty (Bellingham, WA: SPIE), vol. 6812 of Proc. SPIE, 68121F
Vibert, D., Zamojski, M., Conseil, S., Llebaria, A., Arnouts, S., Milliard, B., & Guillaume, M. 2009, in Computational Imaging VII, edited by C. A. Bouman, E. L. Miller, & I. Pollak (Bellingham, WA: SPIE), vol. 7246 of Proc. SPIE, 72460U

Astronomical Data Analysis Software and Systems XX
ASP Conference Series, Vol. 442
Ian N. Evans, Alberto Accomazzi, Douglas J. Mink, and Arnold H. Rots, eds.
© *2011 Astronomical Society of the Pacific*

Solar System Body Observations Discovery on ESO Archival Data

A. Delgado, N. Delmotte, and M. Vuong

European Southern Observatory, Karl-Schwarzschild-Straße 2,
85748 Garching bei München, Germany

Abstract.   The ESO Science Archive Facility (SAF) aims at optimising the scientific return from its wide collection of astronomical observations and as part of this evolution, the creation of a Solar System Bodies (SSB) Facility is being investigated. Starting with the data coming from the instruments with wider field of view, the mining of the ESO Archive is in progress to discover which observations have their frames crossed by solar system objects, intentionally or not. Skybot is the service chosen to identify such observations. Two scenarios are considered: (1) The identification of any kind of SSB having crossed ESO observation frames and (2) The identification of ESO observation frames having been crossed by a particular SSB using the computed ephemerides of that specific SSB. Based on this framework, prospective results are presented.

1.   Scope and Goal

Aiming at the creation of a Solar System Bodies (SSB) Facility, the ESO Archive[1] is being harvested to discover frames containing potential SSB exposures using the IMCCE-Sybot service (Berthier et al. 2006). The data mining has started with the observations obtained from instruments with wider field of view and could be progressively extended to all imagers.

A SSB facility would provide an added value to the SAF allowing further scientific exploitation of the data. During the ESO archive user community survey carried out in 2006 (Delmotte et al. 2006), 16% of the users expressed interest in having search capabilities by Solar System objects/moving targets in the ESO science archive interface. The user benefits from a simple access to SSB observations will be the following: Better characterization of archive holdings (especially, but not only, regarding planets/satellite observations), deriving and refining asteroid orbital parameters and calculating physical properties of asteroids.
Within this framework, two scenarios have been considered:

1. The identification of ESO observation frames having been crossed by a particular SSB using the computed ephemerides of that specific SSB as a faster method to identify observations of SSB in the ESO archive.

2. The identification of any kind of SSB having crossed ESO observation frames for the creation of a general database of ESO SSB observations.

2.   Identification of Frames Crossed by a Particular SSB

A use case for the discovery of SSB observations is the identification of ESO frames having been crossed by a specific solar system object using the computed ephemerides for a given epoch interval. The service used for obtaining the computed ephemerides of a selected object

---

[1]`http://archive.eso.org`

is JPL-Horizons.[2] It is straightforward to cross-match these positions/epochs with the ESO Archive for a given instrument (= field of view).

The scenario to identify ESO frames can be divided in two phases:

1. An extensive search is performed using imaging instruments and the widest field of view. The results (a list of ESO frames) are saved in an ASCII file. The subsequent refining step, via SkyBot[3] conesearch service, will be done in a small dataset instead of the complete archive.

2. The previous search is refined by using the preceding output file. Every right ascension, declination, epoch and instrument name of the frames obtained in the previous phase are now the input for SkyBot conesearch requests. This allows us to confirm if the object was in fact in the field of view of the specific instrument. If the position returned by Skybot is not in the field of view of the instrument used for taking the observation, the frame is discarded.

3.   Identification of Frames Crossed by any Kind of SSB

3.1.   Building of the Solar System Bodies Facility

Building of a SSBF can be broken down into the following steps:

- *Data selection and preparation.* The archive is queried constraining on instrument name and observation technique. A set of parameters containing position/epoch is obtained for every frame. Unit conversion is carried out to meet SkyBot — IMCCE input request requirements.

- *Querying the ephemerides service (Skybot — IMCCE).* The frames set of parameters selected in the previous step is used for sending conesearch requests to Skybot. Apart from fixed parameters like observatory location and type of output, the right ascension, declination and epoch are used as input of the conesearch requests made on a per frame basis to SkyBot. Also the field of view of the instrument used for taking these observations is part of the input.

- *Persisting the results.* The results from the previous step are stored in ASCII files and could be dumped afterwards to a database for further exploitation and access by users via query forms and applications. This feedback, as described in Vuong et al. (2011), will provide an added value to the ESO archive.

- *User interface.* A user interface shall be developed as part of the Archive services. The interface should contain a form with the list of fields for querying, filtering and sorting of the query results.

  The aim of the interface shall be to allow archive users to mine the archive by SSB scientific parameters like: right ascension, declination, modified julian date, uncertainty on the position, SSB type and class, SSB number/name, visual magnitude, solar elongation, geocentric and heliocentric distance, phase angle and proper motion.

Seeking for asteroids in an astronomical archive can present difficulties in interpretation[4] and the actual detection of an object on an image depends not only on the magnitude but also on the position uncertainties and the precision of the epoch.

---

[2]ftp://ssd.jpl.nasa.gov/pub/ssd/Horizons_doc.pdf

[3]http://vo.imcce.fr/webservices/skybot/

[4]http://vo.imcce.fr/webservices/skybot

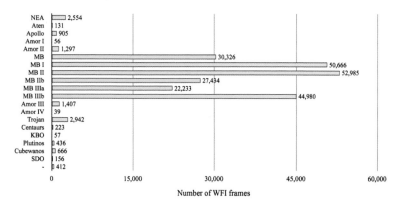

Figure 1.   Number of WFI observations per asteroid class.

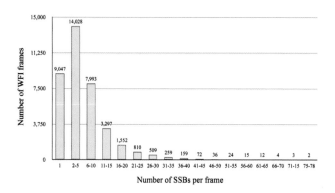

Figure 2.   Distribution of the number of SSBs per frame.

3.2.   Description of the Sample

This prototype has been developed and tested using only WFI[5] frames from 2001 to 2008 (inclusive), but any imagers are suitable for detecting SSBs and the mining should be extended to medium size field imagers[6] like EMMI, VIMOS, FORS1/2.

Subsequently, the SSB detections database can be expanded to all imaging instruments, and in the near future VIRCAM/VISTA could be an extraordinary asset given its wide field of view.

For this harvest, only asteroids, planets and satellites have been considered, since SkyBot is very slow when the search is extended to comets, but faster ways of checking comet serendipitous detections on images should be investigated. The large majority of SSBs are located in the ecliptic plane but constraining the search to that area, all the intriguing objects with rare orbits would be dismissed.

---

[5]http://www.eso.org/sci/facilities/lasilla/instruments/wfi

[6]http://www.eso.org/sci/facilities/develop/

Only public scientific frames have been examined for this project and no constraints were applied on exposure time. In the future it would be interesting to consider the possibility of refining the results based on the observation filter since SSBs are expected to be detected more efficiently in the B and V bands. U and R are less effective. Another important parameter to consider is integration time since longer exposure times would let the SSBs show their motions on the frames.

3.3. Prospective Results on WFI Frames

In this context, detection means the discovery of an observation that meets the requirements in terms of epoch and position (right ascension, declination) and the SSB is in the instrument field of view but could or could not be exposed by the detector because of its low magnitude, too short exposure time or because of instrument astrometric position deviation. The total number of WFI observations taken from 2001 to 2008 is 85,176. 240,080 SSB detections were obtained on 37,822 different frames. The number of distinct asteroids detected were 35,504, 3 planets and 16 satellites.

The amount of WFI frames containing detections with exposure times greater than 1,000 seconds was 1,333 out of 37,822. 5,789 observations have exposures in the range of 500 to 1000 seconds and 27,667 from 50 to 500 seconds.

The number of WFI observations per asteroid class is shown in Figure 1. Note that the majority of the asteroids detected are part of the Main Belt.

Figure 2 shows the distribution of the number of SSBs per frame. 25% of the frames with potential detections contain one possible SSB exposure.

Acknowledgments.   This work is part of the Spanish in-kind contribution to ESO. This development has been supported from the Spanish MICINN through grant CAC-2006-47.

References

Berthier, J., Vachier, F., Thuillot, W., Fernique, P., Ochsenbein, F., Genova, F., Lainey, V., & Arlot, J. 2006, in Astronomical Data Analysis Software and Systems XV, edited by C. Gabriel, C. Arviset, D. Ponz, & S. Enrique (San Francisco, CA: ASP), vol. 351 of ASP Conf. Ser., 367
Delmotte, N., et al. 2006, The Messenger, 125, 41
Vuong, M., Delgado, A., Delmotte, N., Santander-Vela, J., Stevenson, R., & Vera, I. 2011, in Astronomical Data Analysis Software and Systems XX, edited by I. N. Evans, A. Accomazzi, D. J. Mink, & A. H. Rots (San Francisco, CA: ASP), vol. 442 of ASP Conf. Ser., 37

Astronomical Data Analysis Software and Systems XX
ASP Conference Series, Vol. 442
Ian N. Evans, Alberto Accomazzi, Douglas J. Mink, and Arnold H. Rots, eds.
©2011 Astronomical Society of the Pacific

GALFACTS RFI Excision Methods

M. Andrecut, S. S. Guram, S. J. George, and A. R. Taylor

Institute for Space Imaging Science,
University of Calgary, Alberta, T2N 1N4, Canada

Abstract.   Radio astronomical observations are susceptible to Radio Frequency Interference (RFI) contamination. In this case, the signal from astrophysical sources is distorted due to close, and relatively strong, radio emissions from other sources operating on the same frequency spectrum (communication services, for example). As a consequence, the observed data needs to be cleaned, by removing the undesired RFI components, while preserving as much of the underlying useful information as possible. As the data acquisition rates of radio telescopes increases and observations bandwidth extend beyond protected spectral allocations, software systems to mitigate RFI signals are becoming critical. Here, we discuss some of the RFI excision methods implemented in the data processing pipeline of the Galactic ALFA Continuum Survey (GALFACTS), which is a large-area spectro-polarimetric survey being carried out with the Arecibo Radio telescope in Puerto Rico.

1.   Introduction

The GALFACTS observations with the Arecibo Radio Telescope will map about 40% of the sky, over the period 2008-2011. The objective of GALFACTS is to produce high sensitivity Stokes I, Q, U and V multi-frequency images of the polarised emission over the full range of Galactic latitudes from the mid-plane to the pole. The technical specifications of the project are as follows: $\Delta \nu$ = 300 MHz (1225 - 1525 MHz); $1\sigma \sim 80~\mu Jy$ per Stokes (over full band), $\Delta T_{rms} \sim 0.8$ mK; 8192 frequency channels; 3.4′ resolution (maximum baseline 220m); Sky coverage 12,734 degrees; total observing time 1600 hrs. Since the GALFACTS survey produces terabyte sized data sets, the data processing pipeline raises considerable challenges. A detailed description of the GALFACTS pipeline is given elsewhere (Guram & Taylor 2009; Guram et al. 2011). Here, we limit our discussion to the RFI excision methods implemented in the data processing pipeline. The GALFACTS experiment is relatively narrow band, which reduces the exposure to RFI problems. However, some RFI contamination is still present, and its removal plays an important role in producing well-calibrated data. Unfortunately, there is no universal method of RFI excision in radio astronomy. Different types of radio telescopes, different types of observations and different types of RFI determine different methods for RFI excision (Fridman & Baan 2001; Gilloire & Sizun 2009; Kesteven 2009). In this paper, we describe a simple adaptive method, which assumes that the time-frequency signals of radio sources and system noise are Gaussian distributed, while the RFI component is not Gaussian distributed. Based on this assumption, the RFI component is separated from the radio source signal, using an adaptive thresholding approach, applied both in frequency and time domains.

2.   RFI Excision Problem

The purpose of RFI excision in radio astronomy is to reduce (ideally to eliminate) the effects of RFI on the observations. The RFI contamination potentially affects all measured signals, corresponding to the auto-correlation (xx, yy) and respectively the cross-correlation (xy, yx) components. Since the implemented RFI excision algorithms are identical for all signal components, in the following description we drop the xx, yy, xy, yx arguments, and we consider a generic signal $s_{n,t}$, where $n = 0, 1, ..., N - 1$ is the frequency channel index (N is the total number of channels), and $t = 0, 1, ..., T - 1$, is the observation time index (T is the total number of observations). Also, we assume that the radio telescope receiver output signal can be described as a superposition of three independent components: $s_{n,t} = s_{n,t}^{rs} + s_{n,t}^{sys} + s_{n,t}^{RFI}$, where $s_{n,t}^{rs}$ is the radio source noise signal, $s_{n,t}^{sys}$ is the system noise signal (receiver, feed, antenna), and $s_{n,t}^{RFI}$ is the RFI contamination signal. The system and radio source signals are characterized by a Gaussian distribution function $G(\langle s \rangle, \sigma)$, with mean $\langle s \rangle$ and standard deviation σ. The typical RFI signals are strong impulse-like bursts in both temporal and spectral domains, and they behave as outliers in the data with Gaussian distribution. Therefore, the observation data is characterized by a contaminated Gaussian distribution, $F(s) = G(\langle s \rangle, \sigma) + \varepsilon F_{RFI}(s)$, where ε is the fraction associated with the unknown RFI contribution $F_{RFI}(s)$. As a result, the distribution has a normal shape in the central region but has tails that are heavier than those of a normal distribution. In such a model, one establishes a threshold to distinguish the RFI-contaminated state from the RFI-free state, and the data exceeding some threshold is rejected. The implemented algorithms are run in both frequency and time domains, and they use an adaptive thresholding strategy. Detection of an RFI signal will result on a flag bit being set to identify bad data points. The search is done first in the frequency domain, which flags bad channels for each time step, then it is done in the time domain to flag bad time samples.

3. RFI Excision in Frequency Domain

The RFI contamination in frequency domain is characterized by the presence of strong impulsive signals, which must be detected and removed from the spectrum. Since the goal is to detect channels with large power values, comparing to their neighbors, an obvious strategy is to use first order differences (or equivalently the first derivative) in the frequency domain: $d_{n,t} = s_{n+1,t} - s_{n,t}$. Since $s_{n,t}$ is assumed to be a random Gaussian process, the new variable $d_{n,t}$ also has a Gaussian distribution with mean zero. Thus, the detection method is based on a Gaussian model for the amplitude of the new variable $d_{n,t}$. Assuming that τ_t is the threshold for the statistical significance test, the algorithm rejects all the values $d_{n,t}$ with $|d_{n,t} - \mu_t| > \tau_t$, where $\mu_t = (N-1)^{-1} \sum_{n=0}^{N-2} d_{n,t}$ is the band averaged value of $d_{n,t}$ at time t. While in principle the thresholding approach is sound and simple, it still raises the question on how to choose the optimal threshold? This question can be addressed by using an adaptive method to set the threshold. We use an iterative approach, where at each iteration step the threshold is defined as $\tau_t = c\sigma_t$, where c is a constant, and $\sigma_t = \sqrt{(N-1)^{-1} \sum_{n=0}^{n=N-2}(d_{n,t} - \mu_t)^2}$ is the standard deviation of the spectral values of $d_{n,t}$, at time t. Obviously, the standard deviation depends on the current data considered free of RFI contamination, and therefore the threshold is adapting at each new iteration. The iteration process stops, when there is no bad data satisfying the rejection criterion. Also, in order to keep track of the flagged data values, we consider that $f_{n,t}$ is a binary variable, such that $f_{n,t} = true$ if the data in channel n at time t is bad (RFI-contaminated), and respectively $f_{n,t} = false$ if the data is good (RFI-free). The frequency domain algorithm can be formulated as following:

```
c ← 5; //Define a statistical significance constant
//Assume that all data is good
for(n = 0...N − 1) for(t = 0...T − 1) f_{n,t} ← false;
for t = 0...T − 1{//Adaptive thresholding iteration for each time step t
    do{//Compute the mean and standard deviation
        μ ← 0; σ ← 0; M ← 0;
        for(n = 0...N − 2){
            if(f_{n,t} = false and f_{n+1,t} = false) then {
                M ← M + 1; d ← s_{n+1,t} − s_{n,t};
```

$$\delta \leftarrow d - \mu; \mu \leftarrow \mu + \delta/M;$$
$$\sigma \leftarrow \sigma + \delta \cdot (d - \mu);\}\}$$
$$\sigma \leftarrow \sqrt{\sigma/M};$$

$BadData \leftarrow false;$ //Assume that there is no bad data
//Apply detection/rejection test
$for(n = 0...N - 1)\{$
    $d \leftarrow s_{n+1,t} - s_{n,t};$
    $if(|d - \mu| > c\sigma)$ then$\{$
        $f_{n,t} \leftarrow true; f_{n+1,t} \leftarrow true; BadData \leftarrow true;\}\}$
$\}while(BadData = true)\}$
return f;

4. RFI Excision in Time Domain

In this case, we assume that the RFI contamination affects the whole spectral band at time t, and therefore we can use the average band signal: $\mu_t = \frac{1}{N}\sum_{n=0}^{N-1} s_{n,t}$. Also, because in the time domain the RFI effect may be wider (over several time steps), we use the second order differences (or the second derivative), instead of the first order differences: $D_t = \mu_{t+1} + \mu_{t-1} - 2\mu_t$. The method assumes a Gaussian model for the second difference variable, and the data exceeding the adaptive threshold is rejected. We consider that g_t is a binary rejection flag, such that $g_t = true$ if the data at time t is bad (RFI contaminated), and respectively $g_t = false$ if the data is good (RFI free). The time domain algorithm, based on standard deviation, can be formulated as following:

$c \leftarrow 5;$ //Define a statistical significance constant
//Compute the mean signal for each time t
$for(t = 0...T - 1)\{$
    $\mu_t \leftarrow 0; M \leftarrow 0;$
    $for(n = 0...N - 1)\{$
        $M \leftarrow M + 1; \mu_t \leftarrow \mu_t + s_{n,t};\}$
    $\mu_t \leftarrow \mu_t/M;\}$
//Assume that all data is good
$for(t = 0...T - 1)\ g_t \leftarrow false;$
//Adaptive thresholding iteration
$do\{$//Compute the mean and standard deviation
    $\theta \leftarrow 0; \sigma \leftarrow 0; M \leftarrow 0;$
    $for(t = 1...T - 2)\{$
        $if\ (g_t = false$ and $g_{t+1} = false$ and $g_{t-1} = false)$ then $\{$
            $M \leftarrow M + 1; D \leftarrow \mu_{t+1} + \mu_{t-1} - 2\mu_t; \delta \leftarrow D - \theta;$
            $\theta \leftarrow \theta + \delta/M; \sigma \leftarrow \sigma + \delta \cdot (D - \theta);\}\}$
    $\sigma \leftarrow \sqrt{\sigma/M};$
    $BadData \leftarrow false;$ //Assume that there is no bad data
    /Apply detection/rejection test
    $for(t = 0...T - 1)\{$
        $D \leftarrow \mu_{t+1} + \mu_{t-1} - 2\mu_t;$
        $if(|D - \theta| > c\sigma)$ then$\{$
            $g_t \leftarrow true; g_{t+1} \leftarrow true; g_{t-1} \leftarrow true;$
            $BadData \leftarrow true;\}\}$
$\}while(BadData = true)$
return g;

5. Discussion and Conclusion

One can see that the algorithms requires the setting of a threshold constant c, which latter is multiplied with the standard deviation, in order to define the adaptive threshold. This threshold constant c is related to the statistical significance of the rejection test. Assuming a Gaussian model, about 99.730% lie within three standard deviations, $c = 3$; about 99.993% are within four standard deviations, $c = 4$; and about 99.999% are within five standard deviations, $c = 5$. Thus, at each iteration the algorithm rejects all the data which lies above c standard deviations. For example, for $c = 5$ the probability that such a data comes from a Gaussian distribution is smaller than 0.0001% and therefore one may assume that it is RFI contaminated. By setting c to smaller or higher values one can simply set the aggressivity of the detection/rejection test. Also, we should note that in order to minimize memory allocation, we use the Knuth on-line algorithm for calculating the mean and standard deviation.

In conclusion, we have successfully implemented several RFI detection and rejection algorithms, that work well in both frequency and time domains. The algorithms are fast and suitable for large data volumes produced by the GALFACTS survey. Improvements are ongoing to detect RFI at weaker levels, and to implement different algorithms based on higher order statistics.

References

Fridman, P. A., & Baan, W. A. 2001, A&A, 378, 327

Gilloire, A., & Sizun, H. 2009, Ann. Telecom., 64, 625

Guram, S. S., Andrecut, M., George, S. J., & Taylor, A. R. 2011, in Astronomical Data Analysis Software and Systems XX, edited by I. N. Evans, A. Accomazzi, D. J. Mink, & A. H. Rots (San Francisco, CA: ASP), vol. 442 of ASP Conf. Ser., 317

Guram, S. S., & Taylor, A. R. 2009, in The Low-Frequency Radio Universe, edited by D. J. Saikia, D. A. Green, Y. Gupta, & T. Venturi (San Francisco, CA: ASP), vol. 407 of ASP Conf. Ser., 282

Kesteven, M. 2009, The Current Status of RFI Mitigation in Radioastronomy, Tech. rep., Australia Telescope National Facility

Astronomical Data Analysis Software and Systems XX
ASP Conference Series, Vol. 442
Ian N. Evans, Alberto Accomazzi, Douglas J. Mink, and Arnold H. Rots, eds.
© 2011 Astronomical Society of the Pacific

CUDA-Accelerated SVM for Celestial Object Classification

Nanbo Peng, Yanxia Zhang, and Yongheng Zhao

*Key Laboratory of Optical Astronomy, National Astronomical Observatories,
Chinese Academy of Sciences, 100012 Beijing, China*

Abstract. Recently, the development in highly parallel Graphics Processing Units (GPUs) provides us a new method to solve advanced computation problems. We introduce an automated method called Support Vector Machine (SVM) based on Nvidia's Compute Unified Device Architecture (CUDA) platform for classifying celestial objects. SVM has been proved a good algorithm for separating quasars from stars, but it takes a lot of time for training and predicting with large samples. Using the data adopted from the Sloan Digital Sky Survey (SDSS) Data Release Seven (DR7), CUDA-accelerated SVM shows achieving greatly improved speedups over commonly used SVM software running on a CPU. It achieves speedups of 1.25–9.96× in training and 9.29–364.4× in predicting. This approach is effective and applicable for quasar selection in order to compile an input catalog for the Large Sky Area Multi-Object Fiber Spectroscopic Telescope (LAMOST).

1. Introduction

Quasars are the most luminous, powerful, and energetic celestial objects known in the universe. They show a very high redshift, which is an effect of the expansion of the universe. So, many astronomical survey projects make a schedule for finding new quasars as many as possible. Until now, the largest number of known quasars over 100,000 have been founded by the Sloan Digital Sky Survey, which is a major multi-filter imaging and spectroscopic redshift survey using a dedicated 2.5-m wide-angle optical telescope.

The key to discovering new quasars is how to separate quasar candidates from stars in the photometric data, which often is several orders of magnitude more than spectroscopic data. In fact, this problem actually belongs to the general problem of classification in Data Mining (DM). In astronomy, a lot of DM algorithms have been successfully applied on celestial object classification (See a comprehensive review by Ball & Brunner 2010)). Support Vector Machine (SVM) is one of the most promising tool and we have successfully used it to get a more than 95% performance both on the precision and the recall of selecting quasar candidates in the article of Peng, Zhang, & Zhao (2010).

Although SVM has a stable high performance, training a SVM model and using this model to predict class labels is time-consuming, especially with massive data sets. A lot of research has been done to accelerate the training speed, such as Osuna's decomposition approach (Osuna, Freund, & Girosi 1997), Platt's Sequential Minimal Optimization (Platt 1998), and Joachims's shrinking and caching techniques (Joachims 2002). These methods reduce solution time of SVM through improving the algorithm by itself and usually do not consider the hardware environment. However, in the last few years, many researchers and scientists have parallelized SVM on clusters or graphic cards. Graf et al. (2005) proposed a method for parallelizing SVM on clusters and Catanzaro et al. (2008) implemented a SVM solver on GPUs using a SMO algorithm.

Table 1.   Computation time of functions in SVM^{light}

Function name	Computation time [s]						
	q=2	q=4	q=8	q=16	q=32	q=64	q=128
select_working_set	29.58	24.11	15.75	12.32	8.20	5.61	2.44
*cache_multiple_kernel_row**	64.21	43.12	56.92	79.46	113.6	159.0	119.5
*optimize_svm**	7.45	5.28	2.68	3.88	6.32	19.61	73.44
*update_linear_component**	15.77	16.98	21.70	27.40	35.41	43.66	38.06
calculate_svm_mode	0.03	0.02	0.09	0.07	0.06	0.04	0.07
check_optimality	19.95	9.42	7.46	6.69	3.77	3.10	1.19
reactivate_inactive_examples	23.70	22.24	21.42	22.76	20.74	21.79	19.71
shrink_problem	0.50	0.23	0.22	0.14	0.12	0.06	0.06
kernel_cache_shrink	152.6	124.0	65.23	42.30	22.76	13.91	9.77

In this paper, we implement a parallelized SVM using CUDA which is similar to the work of Liao et al. (2009). This version of SVM is based on SVM^{light}[1] software created by Joachims and Radial Basis Function (RBF) kernel is adopted. The performance of our program shows that CUDA-Accelerated SVM for celestial object classification is fast and reliable.

2.   Method

Support Vector Machine (SVM) which was originally created by Vapnik (1995) is derived from the theory of structural risk minimization which belongs to statistical learning theory. The core idea of SVM is to map input vectors into a high-dimensional feature space and construct the optimal separating hyperplane in this space. This hyperplane will attempt to separate positive examples from negative examples with the largest distance from the hyperplane to the nearest of the positive and negative examples.

The main idea of parallelizing SVM is to find the most time-consuming routines and let them computed by GPU. Predicting class labels is intuitive to be parallelized, but training a SVM model belongs to a quadratic optimization problem, which is the most challenge. SVM^{light} employs the decomposition idea of Osuna et al. (1997) to turn the optimization problem into a series of smaller tasks. We can control the size of the small task by setting the parameter q in SVM^{light}. In Sect. 3, we will give a detailed description about how the size of working set affects the computation time of each function associated with optimization.

In this work, we develop CUDA-Accelerated SVM based on the latest version 6.02 of SVM^{light} and CUDA SDK 3.1 which can read the NVIDIA CUDA reference manual for more details (http://nvidia.com/cuda). In terms of hardware, we use an Intel Core 2 Quad 2.66 GHz processor and newly developed NVIDIA GeForce GTX 470, which supports vectorized single and double precision floating-point operations.

Data sets used in this article are extracted from the **SpecPhotoAll** table of SDSS DR7 (Abazajian et al. 2009) following the criterion mentioned in Sect. 2 of Peng et al. (2010). These samples consist of four SDSS colors ($u - g, g - r, r - i, i - z$) and one magnitude ($r$). According to the conclusion of Peng et al. (2010), the optimal parameter configuration of a SVM model for training applies ($C_{-+} = 2$, $C_{+-} = 2$, kernel = RBF, $\gamma = 3.2$) in the following experiments.

---

[1]http://svmlight.joachims.org/

Table 2.   Comparison of the SVM Models Derived from CPU and GPU

Data size	Iterations		Support vectors		Threshold b	
	CPU	GPU	CPU	GPU	CPU	GPU
500	13	25	202	202	0.6698	0.6698
5000	186	203	859	858	0.7582	0.7650
25000	1325	1323	2790	2788	0.7340	0.7400
50000	727	731	4736	4727	0.7481	0.7574

Table 3.   Comparison of computation times on CPU and GPU

Data size	Training			Classifying		
	CPU [s]	GPU [s]	Speedup [x]	CPU [s]	GPU [s]	Speedup [x]
500	0.49	0.99	0.50	0.65	0.07	9.29
5000	9.61	7.66	1.25	6.55	0.11	59.54
25000	236.6	59.13	4.00	32.79	0.13	252.2
50000	398.6	40.01	9.96	65.64	0.18	364.6

3.   Results and Discussion

We tested the computation time of each function which repeats many times in the optimization procedure of SVM. As shown in Table 1, all of them are affected by parameter q which
means maximum size of working set in SVM^{light}. The running time of the three functions
*cache_multiple_kernel_row**, *optimize_svm** and *update_linear_component** marked with asterisk increase as q becomes larger, while the other functions decrease except *calculate_svm_mode*
which is stable. We also found that when q increases, the three functions marked with asterisks
handle most of the computing tasks and they are suited to be accelerated by CUDA. Therefore,
we parallelized every routine of *cache_multiple_kernel_row** and *update_linear_component**,
and some part of *optimize_svm**. Since a very large q will make the quadratic programming
problem become more complex, we decided to use $q = 128$ (default $q = 10$) to test our program.

Table 2 gives the slight difference of training models produced by SVM^{light} and CUDA-
accelerated SVM. Although our implementation of SVM is completely equivalent to SVM^{light}
in logic, there is still a very small difference between the calculated results of CPU and GPU.
This problem is caused by the structure of GPU itself and we do not need to pay much attention
to it. Table 3 shows the performance of CUDA-accelerated SVM in training and predicting.
Because the predicting module is computing intensive and easily be parallelized, the increase
in prediction speed is much higher than that of training and can even reach 364.6×. Since the
training strategy of SVM is rather complicated, the best speedup of training only reaches 9.96×
in our data sets. However the exception for the small sample happens, for example, when the
size of data set is just 500. In that case the run time becomes even longer due to the failure of
hiding the I/O latency of GPU. Conversely, for a large sample the acceleration based on GPU
shows its superiority.

4. Conclusion

This work has demonstrated the utility of CUDA-Accelerated SVM for celestial object classification. Training run time can be reduced by a factor 1.25–9.96 and predicting time can be reduced by a factor 9.29–364.4 compared to the original SVM^{light} program. GPU provides astronomers a powerful tool to solve the large-scale computation problems faced in astronomy, i.e., rapidly labeling billions of sources in the photometric data is possible in short time. In addition, GPU is a very low cost way to achieve high performance compared with clusters. Our future work will focus on improving the performance of our algorithm in training session and applying GPU to more astronomical problems.

Acknowledgments.   This paper is funded by National Natural Science Foundation of China under grant No.10778724 and No.11033001, 863 project under Grant No. 2006AA01A120, and by the Young Researcher Grant of National Astronomical Observatories, Chinese Academy of Sciences. We acknowledgment the SDSS database.

References

Abazajian, K. N., et al. 2009, ApJS, 182, 543
Ball, N. M., & Brunner, R. J. 2010, IJMPD, 19, 1049
Catanzaro, B., Sundaram, N., & Keutzer, K. 2008, in Proceedings of the 25th International Conference on Machine Learning (ICML 2008) (Helsinki, Finland), 104
Graf, H. P., Cosatto, E., Bottou, L., Dourdanovic, I., & Vapnik, V. 2005, in In Advances in Neural Information Processing Systems (MIT Press), 521
Joachims, T. 2002, Learning to Classify Text using Support Vector Machines (Norwell, MA, USA: Kluwer Academic Publishers)
Liao, Q., Wang, J., Webster, Y., & Watson, I. A. 2009, J. Chemical Information and Modeling, 49, 2718
Osuna, E., Freund, R., & Girosi, F. 1997, in Neural Networks for Signal Processing [1997] VII (IEEE), vol. 7740, 276
Peng, N., Zhang, Y., & Zhao, Y. 2010, in Software and Cyberinfrastructure for Astronomy, edited by N. M. Radziwill, & A. Bridger (Bellingham, WA: SPIE), vol. 7740 of Proc. SPIE, 77402T
Platt, J. C. 1998, in Advances in Kernel Methods: Support Vector Learning (Cambridge, MA: MIT Press), 185
Vapnik, V. 1995, The nature of statistical learning theory (New York.: Springer)

Astronomical Data Analysis Software and Systems XX
ASP Conference Series, Vol. 442
Ian N. Evans, Alberto Accomazzi, Douglas J. Mink, and Arnold H. Rots, eds.
©*2011 Astronomical Society of the Pacific*

LS-SVM Applied for Photometric Classification of Quasars and Stars

Yanxia Zhang, Yongheng Zhao, and Nanbo Peng

Key Laboratory of Optical Astronomy, National Astronomical Observatories, Chinese Academy of Sciences, 100012 Beijing, China

Abstract. The major drawback of Support Vector Machines (SVM) is their higher computational cost for a quadratic programming (QP) problem. In order to overcome this problem, we propose using Least Squares Support Vector Machines (LS-SVM). LS-SVM's solution is given by a linear system, which makes SVM method more generally simple and applicable. In this paper, LS-SVM is used for classification of quasars and stars from SDSS and UKIDSS photometric databases. The result shows that LS-SVM is highly efficient and powerful especially for large scale problem and has comparable performance with that of SVM.

1. Introduction

With the construction of the space- and ground-based telescopes and the advancement of detection and reception technologies, astronomy enters an era abundant in data and information. Astronomical data increase at a breathtaking speed, up to Petabyte. How to collect, save, analyze and mine so huge data is a challenge for astronomers. Astronomers have to rely on the emerging and flourishing data mining technology. Most of astronomical problems belong to the different tasks of data mining, and can be solved by data mining methods. The detailed review about this issue can refer to the review papers (Zhang et al. 2002, 2008a; Ball & Brunner 2010).

Up to now, there have been successful applications of various algorithms in astronomy (Ball & Brunner 2010, and references therein). The most common data mining problems in astronomy are classification and function estimation. Taking classification for example, Zhang & Zhao (2003) employed Support Vector Machines (SVM), Learning Vector Quantization (LVQ) for multi-wavelength data classification. Zhao & Zhang (2008) applied various decision tree methods for the same sample. Zhang et al. (2008b) used decision table for astronomical object classification. Gao et al. (2008) implemented Support Vector Machines (SVM) and K-Dimensional Tree (KD-tree) for separating quasars from stars based on large survey databases. Gao et al. (2009) put forward random forest algorithm for classification of multiwavelength data. Pei et al. (2010) explored a fast and effective CUDA-based k-nearest neighbor algorithm for classification. Linear Discriminant Analysis (LDA), K-Dimensional Tree (KD-tree), Support Vector Machines (SVM) were used for classification of pointed sources from SDSS photometric database (Peng et al. 2010b). K-nearest neighbor method was raised for solving classification based on optical and infrared databases (Peng et al. 2010a). A kd-tree based k-nearest neighbor method (KD-KNN) was employed to discriminate quasars from stars (Zhang et al. 2010a). SVM for quasar candidate selection was used on the data from SDSS DR7 (Peng et al. 2010). Based on random sample approach for feature selection and weighting, SVM was investigated to classify quasars from stars (Zhang et al. 2010b). Obviously, only in terms of accuracy, SVM is an effective and reliable approach for classification, but its training and predicting speed is rather slow. In order to overcome this disadvantage, many variant SVMs have been put forward, such as Least-Squared Support Vector Machine (LS-SVM) and CUDA-based Support Vector Machine.

Quasars and stars all belong to pointed sources from images, so successful selection of quasars from stars is of great importance. Quasar candidate preselection methods applied by previous surveys include radio selection, color selection, slitless spectroscopy (SS) selection, X-ray selection, selection by infrared sources, by variability, or by zero proper motion, and automated methods. All of these methods have merits and demerits. In order to improve the effectiveness and efficiency of large sky survey, careful preparation of input catalog is necessary. With large photometric data, the high-efficient rapid classification algorithms are in need.

This paper focuses on studying the capabilities of LS-SVM to derive accurate and robust classification models for the type prediction of the unclassified photometric data from large sky survey data. Specifically, we discriminate quasars from stars with optical data (SDSS) and infrared data (UKIDSS) with LS-SVM. The performance of LS-SVM classification will be comparable to SVM in terms of accuracy and be superior to SVM on running time.

2. Data

We adopted the same sample as that is studied in Wu & Jia (2010). The quasar sample is obtained by cross-identification of all quasars in SDSS Data Release 7 (DR7) with the UKIDSS Data Release 3 (DR3) within 3 arcsecond radius. The star sample is similarly obtained by cross-match of the two survey databases. The final sample includes 8498 quasars and 8996 stars with both SDSS *ugriz* and UKIDSS *YJHK*. The Vega magnitudes are used in this paper. All SDSS AB magnitudes are converted to Vega magnitudes by the following definition (Wu & Jia 2010; Hewett et al. 2006): $u = u(AB) - 0.927, g = g(AB) + 0.103, r = r(AB) - 0.146, i = i(AB) - 0.366$ and $z = z(AB) - 0.533$. The SDSS magnitudes are corrected by the Galactic extinction map of Schlegel et al. (1998).

3. Method

Least Squares Support Vector Machine (LS-SVM), a semi-parametric modeling technique, is the least squares version of Support Vector Machine (SVM). Its main advantage is that it is computationally more efficient than the standard SVM method. SVM constructs an optimal separating hyperplane between the positive and negative with the maximal margin, which can be formulated as a quadratic programming problem involving inequality constraints. While LS-SVM involves the equality constraints only, training requires the solution of a linear equation set instead of the long and computationally hard quadratic programming problem involved by the standard SVM. Although this method effectively reduces the algorithmic complexity, for really large problems including a very large quantity of training samples, this least-squares solution can still become highly memory and time consuming. LS-SVM, just like SVM, has shown to be effective for many classification, regression, time-series prediction problems. The main reference and overview on LS-SVM is detailed in Suykens et al. (2002). We apply the toolbox LS-SVM in Matlab implementation. The software can be downloaded from the website: http://www.esat.kuleuven.be/sista/lssvmlab/.

4. Results and Discussion

We tried many experiments with the above sample by LS-SVM and compared the performance based on different input patterns and different model parameters. In order to determine the useful input pattern, we randomly separated the sample into three parts: two thirds for training and one third for testing. The result is shown in Table 1. The model parameters γ and σ^2 may be determined by the program itself. Considering the run time, we randomly tested some γ and σ^2 values, and kept the better result. When the model parameters are set, the time to construct a predicting model is very short, only costing no more than a few minutes. The computer we used is Intel(R) Core(TM)2 Quad CPU Q9550@2.83 GHz with 6 GB of memory.

As Table 1 shows, the performance based on $(u, g, r, i, z, Y, J, H, K)$ is superior to that on (u, g, r, i, z) or (Y, J, H, K); that of (Y, J, H, K) is better than that of (u, g, r, i, z). Moreover the accuracy of $(u - g, g - r, r - i, i - z, z - Y, Y - J, J - H, H - K)$ is higher than that of $(u - g, g - r, r - i, i - z)$ or $(Y - J, J - H, H - K)$; that of $(Y - J, J - H, H - K)$ outperforms that of $(u - g, g - r, r - i, i - z)$. The best performance adds up to 97.98% with input pattern of $(u - g, g - r, r - i, i - z, z - Y, Y - J, J - H, H - K)$ and model parameters ($\gamma = 10$ and $\sigma^2 = 4$); the better one is 97.38% with $(u, g, r, i, z, Y, J, H, K)$ and the same model parameters.

Since the result with colors showed superiority to that with magnitudes, we further tested the sample with colors by three-fold cross-validation. The result is indicated in Table 2. The accuracy with 8 colors amounts to $(98.81 \pm 0.60)\%$; that with 3 infrared colors is $(95.36 \pm 1.97)\%$; that with 4 optical colors is only $(82.37 \pm 7.21)\%$. Here we only discuss the result by three-fold cross-validation, not by ten-fold cross-validation because the latter needs much larger computer memory which surpasses the present memory resources of our computer. For a large enough sample, the three-fold cross-validation is reliable and applicable.

Table 1.   The performance of different input patterns

input pattern/model parameter	accuracy	
	$\gamma = 10,$ $\sigma^2 = 0.4$	$\gamma = 10,$ $\sigma^2 = 4$
$u, g, r, i, z, Y, J, H, K$	96.69%	97.38%
u, g, r, i, z	91.96%	90.67%
Y, J, H, K	92.23%	92.03%
$u - g, g - r, r - i, i - z, z - Y, Y - J, J - H, H - K$	95.94%	97.98%
$u - g, g - r, r - i, i - z$	92.99%	92.39%
$Y - J, J - H, H - K$	94.48%	94.41%

Table 2.   The classification accuracy of three-fold cross-validation

	average accuracy
$u - g, g - r, r - i, i - z, z - Y, Y - J, J - H, H - K$	$(98.81 \pm 0.60)\%$
$u - g, g - r, r - i, i - z$	$(82.37 \pm 7.21)\%$
$Y - J, J - H, H - K$	$(95.36 \pm 1.97)\%$

5.   Conclusion

We explored LS-SVM to classify quasars from stars with SDSS DR7 and UKIDSS DR3 databases. Efficient methods to preselect quasar candidates are of great value for large sky survey projects, such as the Chinese Large Sky Area Multi-Object Fibre Spectroscopic Telescope (LAMOST, now renamied Guoshoujing Telescope). In our case, LS-SVM shows its advantage of short running time and good performance. Nevertheless this approach is expensive in memory resources. Therefore, LS-SVM reveals its weakness for the much larger sample. In future work, we will study further how to accelerate the speed of data mining techniques while at the same time maintaining good performance.

Acknowledgments.   This paper is funded by National Natural Science Foundation of China under grant No.10778724 and No.11033001, 863 project under Grant No. 2006AA01A120, and by the Young Researcher Grant of National Astronomical Observatories, Chinese Academy of Sciences. We acknowledgment SDSS and UKIDSS databases.

References

Ball, N. M., & Brunner, R. J. 2010, IJMPD, 19, 1049
Gao, D., Zhang, Y., & Zhao, Y. 2008, MNRAS, 386, 1417
— 2009, Res. Astron. Astrophys., 9, 220
Hewett, P. C., Warren, S. J., Leggett, S. K., & Hodgkin, S. T. 2006, MNRAS, 367, 454
Pei, T., Zhang, Y., & Zhao, Y. 2010, in Software and Cyberinfrastructure for Astronomy, edited by N. M. Radziwill, & A. Bridger (Bellingham, WA: SPIE), vol. 7740 of Proc. SPIE, 77402G
Peng, N., Zhang, Y., Pei, T., & Zhao, Y. 2010a, in Software and Cyberinfrastructure for Astronomy, edited by N. M. Radziwill, & A. Bridger (Bellingham, WA: SPIE), vol. 7740 of Proc. SPIE, 77402X
Peng, N., Zhang, Y., & Zhao, Y. 2010b, in Software and Cyberinfrastructure for Astronomy, edited by N. M. Radziwill, & A. Bridger (Bellingham, WA: SPIE), vol. 7740 of Proc. SPIE, 77402M
Peng, N., Zhang, Y., & Zhao, Y. 2010, in Software and Cyberinfrastructure for Astronomy, edited by N. M. Radziwill, & A. Bridger (Bellingham, WA: SPIE), vol. 7740 of Proc. SPIE, 77402T
Schlegel, D. J., Finkbeiner, D. P., & Davis, M. 1998, ApJ, 500, 525
Suykens, J. A. K., Van Gestel, T., De Brabanter, J., De Moor, B., & Vandewalle, J. 2002, Least Squares Support Vector Machines (Singapore: World Scientific)
Wu, X., & Jia, Z. 2010, MNRAS, 406, 1583
Zhang, Y., & Zhao, Y. 2003, PASP, 115, 1006
Zhang, Y., Zhao, Y., & Cui, C. 2002, Progress Astron., 20, 312
Zhang, Y., Zhao, Y., & Gao, D. 2008a, Adv. Space Res., 41, 1949
Zhang, Y., Zhao, Y., & Zheng, H. 2010a, in Software and Cyberinfrastructure for Astronomy, edited by N. M. Radziwill, & A. Bridger (Bellingham, WA: SPIE), vol. 7740 of Proc. SPIE, 77402S
Zhang, Y., Zheng, H., & Zhao, Y. 2008b, in Advanced Software and Control for Astronomy II, edited by A. Bridger, & N. M. Radziwill (Bellingham, WA: SPIE), vol. 7019 of Proc. SPIE, 701938
— 2010b, in Software and Cyberinfrastructure for Astronomy, edited by N. M. Radziwill, & A. Bridger (Bellingham, WA: SPIE), vol. 7740 of Proc. SPIE, 77402Z
Zhao, Y., & Zhang, Y. 2008, Adv. Space Res., 41, 1955

Astronomical Data Analysis Software and Systems XX
ASP Conference Series, Vol. 442
Ian N. Evans, Alberto Accomazzi, Douglas J. Mink, and Arnold H. Rots, eds.
©*2011 Astronomical Society of the Pacific*

A Calibration and Mapping Pipeline for the Green Bank Telescope

Joe Masters,[1] Bob Garwood,[1] Glen Langston,[2] and Amy Shelton[2]

[1]*NRAO Headquarters, 520 Edgemont Rd, Charlottesville, VA 22903-2475*

[2]*NRAO Green Bank, P.O. Box 2, Rt. 28/92, Green Bank, WV 24944-0002*

Abstract. We present the first data calibration and spectral line mapping pipeline for the Green Bank Telescope (GBT), a 100 meter single-dish radio telescope operated by the National Radio Astronomy Observatory in West Virginia. The pipeline was developed for a prototype 7-pixel, K-band Focal Plane Array receiver (commissioned in the Fall of 2010) but is generalized for use with other receivers, operating at different wavelengths. One of the key design goals of the pipeline was to solve, on a smaller scale, issues that will arise with the entrance of the next generation of GBT detection equipment. New focal plane arrays, spectrometers and other instruments will produce much larger datasets, thus reinforcing the need for automated calibration and mapping. We discuss the implementation of the pipeline, near-term improvements and the issues we foresee as GBT hardware evolves over coming decade.

1. Introduction

The Robert C. Byrd Green Bank Telescope (GBT), at 100 meters diameter, is the world's largest fully steerable radio telescope, operated by the National Radio Astronomy Observatory in Green Bank, West Virginia. We present a prototype software calibration and mapping pipeline designed for a new receiver on the GBT: the K-band Focal Plane Array (KFPA). Despite its origins in the KFPA project, the pipeline is suitable for general GBT mapping data reduction.

2. KFPA Hardware

The KFPA is a seven pixel receiver operating at 18–27.5 GHz and is primarily intended for imaging ammonia (NH_3) and other molecules found within this frequency range. The instrument is a prototype for future Focal Plane Array hardware on the GBT. Proposals include a 61-pixel K-band receiver and a 100+ pixel W-band receiver, which would greatly increase the volume of data coming from the GBT during a single observation.

3. Mapping Techniques

The On the Fly (OTF) mapping technique is assumed for use of the GBT mapping pipeline. OTF mapping has the advantage of covering a given area in less time than traditional mapping using many single pointings. Two types of OTF mapping are generally available for GBT observers: position-switched and frequency switched. Each technique provides a different method for establishing an off-target reference observation for calibration. Position-switched references are generated by pointing at an off-source region of the sky. Frequency-switched references are generated by shifting the observed frequency band a small amount from one switching phase to another within an integration. In that way, a line-free region of the spectrum can be subtracted from an observed line in the same spectral bin. The current pipeline supports position-switched observations. The next release will support frequency-switched observations as well.

4.   Pipeline Implementation

The pipeline is primarily written in Python, taking advantage of many features from the standard library including the multiprocessing, subprocess, argparse and logging modules. Dependencies are limited to:

- Numpy (for large array manipulation),
- Pyfits (for I/O),
- AIPS (for imaging, with Obit and parseltongue for Python support),
- A tool for converting NRAO SD-FITS to AIPS-readable data (idlToSdfits), and
- Weather prediction tools for the GBT (in Tcl).

4.1.   Parallelization

Parallel calibration of feeds and polarizations is a key aspect in pipeline processing efficiency. While parallelization gives a modest performance boost today, it will be essential as the data volume increases with more feeds simultaneously receiving data. One aspect of the mapping pipeline that is not yet parallelizable is the gridding of calibrated spectra. We currently need to combine all calibrated spectra of a map to create a single spectral image cube. In the future, we may be able to create image cubes from each feed and polarization, then combine them in the final step. There are also opportunities to split the spectral cubes into separate frequency chunks to add futher parallelization.

4.2.   Map Discovery

Another key feature of the pipeline is its ability to determine the reference and mapping scans of an observation without explicit user instruction. This is achieved with the support of column keyword values being set by user-controlled software, which are used by the pipeline for the identification of each scan. For example,

```
scan 12:    OFF \
scan 14:    MAP  |   First map with two off-source reference scans.
scan 15:    MAP  |
scan 16:    OFF /
scan 17:    OFF \
scan 19:    MAP  |   Second map with one reference scan.
scan 20:    MAP /
```

indicates two maps with off-source reference scans.

5.   Other Aspects of the Pipeline

5.1.   User Control

The user is able to control the inputs of the pipeline via the command line and/or a parameter file. The parameter file allows commonly set options to be stored for future use. For example:

```
gbtpipeline @options.par --verbose 4
```

is a command to invoke the pipeline and the contents of the file 'options.par' are:

```
-i /home/sdfits/AGBT10C_045_01/AGBT10C_045_01.raw.acs.fits
--allmaps
--clobber
```

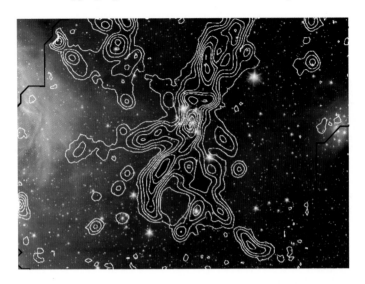

Figure 1.   NH3 (1,1) integrated intensity contours over a Spitzer 3.6, 4.5 and 8 micron RGB image of the Serpens South cluster. Credit: Rachel Friesen, NRAO.

5.2.   Reprocessing

The pipeline is available for reprocessing of archived data and the imaging step may be run separately from the calibration. The pipeline has been useful for commissioning the instrument in that various calibration parameters can be modified and the data can be recalibrated. Thus far, the sizes of the datasets and the processing times have not been prohibitively large for multiple iterations of calibration and mapping.

5.3.   Documentation

Developer documentation (as HTML and PDF) is provided and made with the Sphinx documentation package. A pipeline user's guide is also provided as a pdf (Langston et al. 2010). Documentation is available on the NRAO wiki as well as on github, where the source code is hosted (Masters & Garwood 2010).

6.   Future Plans

The KFPA pipeline is the first pipeline implementation at NRAO, Green Bank, and the KFPA is a prototype instrument. Consequently, future automated processing plans are numerous. Future planned developments include:

- Support for frequency-switched observations,
- Retooling the pipeline for use with a higher data rate FPGA spectrometer,
- Operating in near real-time to optimize the pipeline as an automated imaging tool,
- Moving away from FITS in favor of a HDF5-based container to provide for a more scalable framework (Anderson et al. 2010),
- Improved calibration with more frequency-dependent corrections,
- Eliminating the dependency on AIPS in favor of a lightweight gridding code, and
- Providing a full-featured, portable pipeline for off-site use.

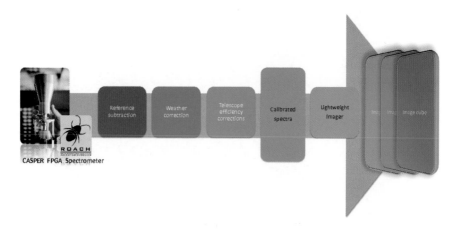

Figure 2.   The pipeline is moving towards a model that minimizes the use of intermediate files. A new spectrometer with higher time and freqency resolution recording is driving this change, as is the anticipated need to support future instrumentation
with high data rates.

Acknowledgments.   We would like to thank the National Science Foudation for funding
this work.

References

Anderson, K. R., Alexov, A., Båhren, L., Brießmeier, J., Wise, M., & Adriaan Renting, G.
    2010, in ISKAF2010 Science Meeting
Langston, G., Garwood, B., & Masters, J. 2010, KFPA Observer's Guide. Observing instruc-
    tions which include use of the pipeline.
Masters, J., & Garwood, B. 2010, KFPA Code and Documentation Repository. URL `http:
    //github.com/jmasters/KFPA-Pipeline`

Astronomical Data Analysis Software and Systems XX
ASP Conference Series, Vol. 442
Ian N. Evans, Alberto Accomazzi, Douglas J. Mink, and Arnold H. Rots, eds.
© *2011 Astronomical Society of the Pacific*

Python Scripting for CIAO Data Analysis

Elizabeth C. Galle, Craig S. Anderson, Nina R. Bonaventura, D. J. Burke,
Antonella Fruscione, Nicholas P. Lee, and Jonathan C. McDowell

Smithsonian Astrophysical Observatory,
60 Garden Street, Cambridge, MA 02138, USA

Abstract.   The Chandra X-ray Center has adopted Python as the primary scripting language in the Chandra Interactive Analysis of Observations software package
(CIAO). Python is a dynamic object-oriented programming language that offers strong
support for integration with other languages and tools and comes with extensive standard libraries. Integrating Python into CIAO allows us to develop powerful new scripts
for data analysis, as well as rewrite and improve upon popular CIAO contributed scripts.
We discuss the coding guidelines that we have developed during this process, using specific CIAO contributed scripts — available for download online — as examples.

1. The CIAO Contributed Package

The CIAO contributed tarfile[1] contains analysis scripts and modules that automate repetitive
tasks and extend the functionality of the CIAO[2] software package by filling specific analysis
needs. Many of the scripts were conceived and written by CXC scientists in the early years of
the mission, before CIAO became a robust and mature software package. Over the course of ten
years, the contributed package evolved into a collection of shell, Perl, S-Lang, and slsh scripts.
     Since Python was adopted as the primary scripting language in CIAO, a targeted effort
has been made to review the code of these scripts and to rewrite them in Python, bringing them
up-to-date and making them easier to maintain over the future of the mission.

2. Style Guide

The contributed tarfile had been developed in a multiple-author environment without coding
standards in place. The scripts spanned a number of languages with varying levels of parameter
file support, consistency of warning and error messages, and cross-platform compatibility.
     In addition to being written in Python, a baseline requirement was that every script should
have a parameter file and an XML help file for used in the CIAO ahelp system.
     The next step was to establish style guidelines for the script authors. The standard Python
style guide, known as PEP 8,[3] was used as a primary resource. In addition, a scripting style
guide was developed to ensure that the scripts were CIAO-like in their appearance and operation.

---

[1]`http://cxc.harvard.edu/ciao/download/scripts/`

[2]`http://cxc.harvard.edu/ciao/`

[3]`http://www.python.org/dev/peps/pep-0008/`

2.1.   An Excerpt from the Style Guide

The following is an excerpt from the scripts style guide section on verbosity levels and error messages.

```
Verbose levels
The script should default to verbose=0 or 1. If the script doesn't
really create any screen output that the user really needs to see then
use 0 otherwise 1.

A verbose level of 2 should be considered to be useful for a user
tracking what the script has done - so at some level this replaces an
explicit log file. Things like listing the parameters used, and
what steps are being taken would happen at this level. The
ciao_contrib.runtool module will print out the command line for each
tool (i.e. so the user can see what is actually run).

A verbose level of 3, 4 and 5 are for debugging the tool/script (some
of the contributed modules will print copious amounts of verbiage at
level 5).

How to write the messages
The ciao_contrib.logger_wrapper module, which internally uses the
Python logging module, provides a slightly simpler interface for the
script writer. By using this module, you can get information from some
of the other contributed modules too (e.g. ciao_contrib.runtool).

Once the module has been loaded, the initialize_logger() routine is
used to make sure the messages will get displayed on screen, and the
set_verbosity() routine sets the verbose level of the tool and any
libraries that also use this setup. As shown below, the
make_verbose_level() routine is used to create a routine that will
display a message if the verbosity is a given level or higher. We
assume that the variable toolname has been defined previously:
```

```
    import ciao_contrib.logger_wrapper as lw

    lw.initialize_logger(toolname)
    v1 = make_verbose_level(toolname, 1)
    v2 = make_verbose_level(toolname, 2)
```

```
(you only need to call make_verbose_level() for the verbose levels
your script uses, and the choice of v1, v2, etc. is up to you).
```

3.   A Case Study: Rewriting acisspec

The acisspec script is a textbook example of a contributed script. It was written by a CXC scientist for specific analysis needs, but was found to be useful to many users doing imaging spectroscopy. The shell script was added to the contributed package in 2001.

Originally, acisspec was designed to:

- Extract ACIS PI spectra and associated WMAPs for both extended sources (and background)

- Coadd or average two ACIS PI spectra and build weighted responses

The extended source extraction was replaced by the specextract tool in CIAO 3.3 (November 2005), but acisspec was still required for coadding imaging spectra.

There were two initial goals for the acisspec rewrite:

- Replicate the coadding and weighting functionality but remove the spectral extraction steps

- Extend the script to be capable of combining N spectra and responses (acisspec is restricted to two inputs)

The script, renamed to "combine_spectra," was released in August 2010 for use in CIAO 4.2. combine_spectra sums multiple imaging source PHA spectra and (optionally) the ARFs, background spectra, and background ARFs.

The Python source code is easier to maintain and update than the shell syntax used for acisspec, creating a lighter and faster development cycle. As a Python script, combine_spectra can also be imported as a module into other scripts. The script itself, as well as any functions it contains, can be invoked by other scripts to extend their functionality. The scripts team is evaluating using this modularity to incorporate combine_spectra into the spectral extraction tool, specextract.

3.1. Example of Code Improvements

The original acisspec tool applied a combination of the UNIX echo and awk commands to the CIAO output, from which the BACKSCALE header values were calculated, e.g.

```
f1=`echo "$backscbgd $expbgd $expsou $expsou1 $expbgd1 ... " |
awk '{printf "%.6e", $1*($2/$3)*($4/$5)*($6/$7); }' -`
f1=`printf '%7.5f \n' $f1`
```

In combine_spectra, we used the mathematical functions from the Python module Num-Py[4] to simplify this operation. (Notice also the excellent code commenting — another point for the style guide.)

```
#+++++++++++++++++++++++++++++++++++++++++++++++++++++++++++++++++++++
# Define the total source and background exposure values, the HEASARC
# BACKSCAL value for the combined background spectrum, and the
# coefficients used to scale the background spectra before combining.
# Reference: http://heasarc.gsfc.nasa.gov/docs/asca/abc_backscal.html
#+++++++++++++++++++++++++++++++++++++++++++++++++++++++++++++++++++++

    srcpha = numpy.array(spha_strarray)
    expsrc = expsrc_array
    totexpsrc = sum(expsrc_array)

    backscalsrc = backsrc_array
    cbackscalsrc = 1.0  # backscal value for combined source spectrum

    if bkg != "NULL":

        bkgpha = numpy.array(bpha_strarray)
        expbkg = expbkg_array
        totexpbkg =  sum(expbkg_array)
```

---

[4]http://numpy.scipy.org/

```
backscalbkg = backbkg_array
bkg_coadd_factor =  expsrc*(backscalsrc/backscalbkg)

cbackscalbkg = totexpsrc/sum(bkg_coadd_factor)   # backscal value
                                                 # for combined
                                                 # background

f = range(fcount)  # initialize the array of background

  for i in fc:
    ...
    if bkg != "NULL":
        f[i] = cbackscalbkg*(totexpbkg/totexpsrc)*(expsrc[i]/expbkg[i])
*(backscalsrc[i]/backscalbkg[i])
```

4. The `ciao_contrib.runtool` Module

The `ciao_contrib.runtool` module allows CIAO tools to be run as if they were Python functions and supports a pset-like parameter mode. The easy access to and handling of CIAO tools, accessing header and table data from FITS files, and writing data to headers and tables of output FITS files are used extensively in `combine_spectra`.

More information on `ciao_contrib.runtool` is available from the "How to run CIAO tools from within Python" webpage[5] and in "Charming Users into Scripting CIAO with Python — the `ciao_contrib.runtool` Module" (Burke, *et al.*) in this proceedings.

Acknowledgments. This work was supported by the Chandra X-ray Center under NASA contract NAS8-03060.

---

[5]`http://cxc.harvard.edu/ciao/scripting/runtool.html`

Astronomical Data Analysis Software and Systems XX
ASP Conference Series, Vol. 442
Ian N. Evans, Alberto Accomazzi, Douglas J. Mink, and Arnold H. Rots, eds.
©*2011 Astronomical Society of the Pacific*

Chandra X-ray Center Science Data Systems Regression Testing of CIAO

Nicholas P. Lee, Margarita Karovska, Elizabeth C. Galle, and Nina R. Bonaventura

Smithsonian Astrophysical Observatory,
60 Garden Street, Cambridge, MA 02138, USA

Abstract. The *Chandra Interactive Analysis of Observations* (CIAO) is a software system developed for the analysis of *Chandra X-ray Observatory* observations. An important component of a successful CIAO release is the repeated testing of the tools across various platforms to ensure consistent and scientifically valid results. We describe the procedures of the scientific regression testing of CIAO and the enhancements made to the testing system to increase the efficiency of run time and result validation.

1. Introduction

The *Chandra Interactive Analysis of Observations* (CIAO)[1] system is a suite of tools and applications developed by the *Chandra* X-ray Center Science Data Systems (SDS) and Data Systems (DS) groups to aid users in the reduction and analysis of *Chandra X-ray Observatory* observations and as a mission independent system, for the analysis of other X-ray and non-X-ray observations. CIAO is supported on several platforms, including: Linux (32- and 64-bit), Intel Mac OSX (32- and 64-bit), and Solaris.

The commitment to support a platform means that a binary installation is provided for the end-user, compilation of the source code is confirmed, and that fixes are developed for bugs on the supported platforms. The SDS commitment to the user includes providing "helpdesk" support where the supported platforms provide a level of commonality that aids in the replication of user problems and finding work arounds.

A key aspect of a successful CIAO release is regression testing to ensure the cross-platform performance and consistent results of the tools. SDS repeatedly tests every CIAO tool, supporting libraries, and contributed scripts on each supported platform to demonstrate new functionality and ensure that bugs are not introduced during the development process.

Regression testing goes through a set of phases (Figure 1), following the initial scientific unit testing. The SDS regression test suite currently contains over a thousand individual tests.

With each regression test phase, results are checked against platform-specific baselines of past results and verified for expected changes or emergence of problems with each beta code freeze.

2. Unit Testing

Unit tests are performed by SDS scientists on individual tools that have been created or updated by the developers to meet specified scientific requirements. The focus of this testing is the science performance of a tool, judged in conjunction with other tools as would be done in analysis by an end-user. That is, input files for the tool under testing have been previously processed by another tool and the output files of the tool under test may be used as input for another tool.

---

[1]`http://cxc.harvard.edu/ciao`

SDS Regression Test Steps

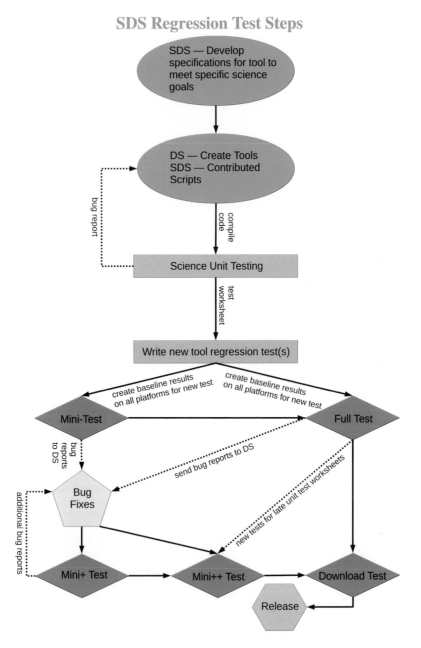

Figure 1.   Flow diagram of the steps of a typical regression test cycle.

3. Regression Testing

3.1. The Testing Environment

The "ciaotest" environment used for regression testing is a set of tools written in ksh used to run CIAO using the provided test scripts as individual tests or as a set using a list of tests. The environment also provides tools to compare the test results with baselines and generates a report of whether differences are detected within specified tolerance values.

3.2. Writing Regression Tests

The regression test team is provided with test worksheets, test input files, and the validated results from the science unit tester. A worksheet is a standard form providing information on the status of the tool; pointers to the necessary files; and expected results based on the unit test.

The regression test writer takes the worksheet and writes a test script compatible with our "ciaotest" regression test environment, using the provided input files. Steps include:

- Create test script based on worksheet outline, and if necessary, additional test scripts when there are multiple test cases.
- If necessary, to enhance test, write an external script (in Python or a shell language) callable by the test script to provide additional results to verify against.
- Run test script on all supported platforms; compare and validate test results with the results provided by unit tester.
- Observed cross-platform differences are reported to DS and SDS as necessary.
- Test run results are added to platform baseline data sets.

3.3. The Mini-Test, Mini+ Tests, and the Full Test

Following a code freeze, DS will verify that the compiled builds on supported platforms are ready to be tested. Upon verification, testing is handed off to SDS.

The "mini-test" contains a sub-sample (150-200 tests) of the entire test suite, composed of previously released CIAO tools, the new or modified tools, and tools that use libraries that had been updated. The mini-test is used to inspect the stability of the key tools in the current development code. Since it is a small suite, ~10–15% of the total number of tests, the mini-test can be executed and analyzed fairly quickly and frequently in between code freezes to examine the progress as changes are made.

Once problems are addressed and bug fixes applied, the failed tests — and any additional new tests that come from unit testers since the start of regression testing — are added to the existing mini-test list, which are referred to as a "mini+" test.

The "mini+" test is performed to verify that the changes made to CIAO have fixed previously reported problems, and new tests perform as expected. Iteratively, subsequent problems are reported to scientists and developers, and tested as a "mini++" test. Common to all the regression tests:

- Execution and analysis of tests performed on all supported platforms.
- Results compared and analyzed with previously saved regression tests results and evaluated as to whether they are physically sensible.
- Cross-platform comparison of results of the current regression test to ensure consistent behavior across all supported platforms.
- Identified differences are confirmed with scientists and developers as software enhancements and/or expected changes.
- Differences in results must be addressed by scientists and/or developers prior to CIAO release.

The multiple iterations of "mini" and "mini+" tests, and code fixes and changes, are followed by an execution of the full set of tests. The results are then compared against baseline result sets produced by previous CIAO releases and development drops.

4.   Recent Regression Test Enhancements

Although the basic regression testing system has been in place since shortly after the *Chandra* launch, over the past few years, significant improvements have been made to the testing infrastructure. Enhancements include:

- Upgrade of hardware and operating systems.
- Restructured testing procedures and increased automation.
- Introduction of tolerance files for automatic comparison.
- Ability to ignore path differences added.
- Input test files, test results, baseline data sets, and comparison results stored on NetApp servers.
- ASCII dumps of FITS tables during automated comparison.
- Elimination of outdated tests and updating existing tests by removing deprecated functions and replacing with newer functions.

These changes have significantly decreased the run and analysis time of tests, in spite of the significant increase of the number of tests over time. In particular, the migration to a NetApp server from a local disk remotely accessed by multiple machines has increased the read and write efficiency of the regression, cutting the full test run time by up to ~75%.

Including tolerances, loosening the constraints — within safe limits — and ignoring file path differences between baselines and results have decreased the number of flagged test "failure" during the automated result analysis process which subsequently require manual checks. The automated ASCII dumps of FITS tables for comparison allow for quicker manual checking of test failures without having to open large tables, which is machine intensive and very slow.

5.   Download Testing

In conjunction with DS, SDS conducts "download testing" of the CIAO system, as packaged for public release. CIAO is downloaded and installed on various platforms with various module configurations to verify the stability of the build.

Acknowledgments.   This work was supported by the Chandra X-ray Center under NASA contract NAS8-03060.

Astronomical Data Analysis Software and Systems XX
ASP Conference Series, Vol. 442
Ian N. Evans, Alberto Accomazzi, Douglas J. Mink, and Arnold H. Rots, eds.
©*2011 Astronomical Society of the Pacific*

ACIS Sub-Pixel Resolution: Improvement in Point Source Detection

Craig S. Anderson,[1] Amy E. Mossman,[1] Dong-Woo Kim,[1] Glenn E. Allen,[2] Kenny J. Glotfelty,[1] and Giuseppina Fabbiano[1]

[1]*Smithsonian Astrophysical Observatory, 60 Garden Street, Cambridge, MA 02138, USA*

[2]*MIT Kavli Institute for Astrophysics and Space Research, 77 Massachusetts Avenue, 37-287, Cambridge, MA 02139*

Abstract. We investigate how to achieve the best possible ACIS spatial resolution by binning in ACIS sub-pixel and applying an event repositioning algorithm after removing pixel-randomization from the pipeline data. We quantitatively assess the improvement in spatial resolution by (1) measuring point source sizes and (2) detecting faint point sources. The size of a bright (but no pile-up), on-axis point source can be reduced by 20–30%. With the improved resolution, we detect ∼ 20% more faint sources when embedded in the extended, diffuse emission in a crowded field. We further discuss the false source rate of ∼ 10% among the newly detected sources, using a few ultra-deep observations.

1. How to Obtain the Best Possible ACIS Resolution

In order to achieve the best possible ACIS resolution, we apply the following steps to the Chandra X-ray Center (CXC) pipeline products. Then we quantitatively assess the improvement by comparing the point source sizes (§ 2) and detections (§ 3) before and after the procedure.

Sub-pixel binning: Chandra coordinates contain positional accuracy finer than one ACIS pixel (0.492 arcsec) through dither and aspect correction. Imaging data binned by ACIS sub-pixel can already provide an improved resolution.

Remove pixel randomization: The current pipeline default is to apply pixel randomization by 1/2 ACIS pixel on the chip coordinate to remove the instrumental "gridded" appearance of the data and to avoid any possible aliasing affects associated with this spatial grid. This pixel randomization has to be removed before applying a sub-pixel algorithm.

ACIS sub-pixel algorithm: Positional accuracy can be improved by utilizing the positional information in 3×3 event islands. Several sub-pixel event repositioning algorithms have been developed during the first years of the Chandra mission. The first implementation by Tsunemi et al. (2001) applied the knowledge of charge cloud size in 3×3 event islands to corner events with ASCA grade 6 (4–16% of on-axis events). Mori et al. (2001) extended the algorithm to all split pixel events: event grades 2, 3, 4 are shifted 1/2 pixel in one direction, grade 6 events are shifted 1/2 pixel in two directions, while grade 0 events remain centered on the event island. Later Li et al. (2003, 2004) improved the algorithm to SER (subpixel event repositioning) and further to EDSER (energy dependent SER). CXC also implemented EDSER[1] in CIAO 4.3.[2]

PSF deconvolution: We do not discuss deconvolution here. We refer the reader to CXC Announcement #64.[3]

---

[1]http://space.mit.edu/CXC/docs/docs.html#subpix

[2]http://cxc.harvard.edu/ciao/releasenotes/ciao_4.3_release.html

[3]http://cxc.harvard.edu/announcements/announce_64.html

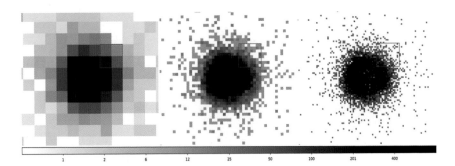

Figure 1.   SN1987A (Energy 300–5000eV)
The improved resolution is shown for a bright extended source with binning of 1, 1/4, and 1/8 pixel (left to right). Images have pixel randomization removed and a sub-pixel event repositioning algorithm applied. Red squares are 1 arcsecond per side. See also Park et al. (2002).

2.   Testing Improvement of Point Source Sizes

We selected seven bright sources (5 on-axis, 2 off-axis) with pile-up fraction less than 5%. Source sizes presented in Table 1 were measured by the CIAO tool srcextent[4] which calculates the size (sigma) and associated uncertainty of a photon-count source image using the Mexican Hat Optimization algorithm (Houck 2007). The uncertainty (at 90% confidence) is derived from Monte Carlo trials. Applying sub-pixel binning of 1/4 pixel reduces the on-axis source size by 19–24%. Additionally removing pixel randomizaton and applying the sub-pixel event repositioning algorithm further reduces the on-axis source size by another 6–8%. Off-axis sources show no statistical improvement.

3.   Testing Improvement of Faint Source Detections Embedded in Diffuse Emission

With the improved resolution, we present in Table 2 ~20% more faint source detections embedded in extended, diffuse emission in a crowded field. To check whether the new sources are real or suprious, we compare sources detected in shallow and deep images. We assume that the real sources which are newly found in the shallow image will be detected in the deep image. We also find an increase of ~10% in false source detections defined as those newly found in the shallow image, but not in the deep image. Because some point sources (LMXBs) in elliptical galaxies are variable, the false source rate is actually an upper limit. To lessen the effect of variable sources, we cut a deep observation (90–110 ks) into smaller pieces (10ks and 20ks), instead of merging multiple observations with months or years between exposures.

After sub-pixel binning by 1/2 pixel the fraction of newly detected sources is ~9% (21/240, 38/306) of which 12–14% (3/21, 3/28) may be false. Binning by 1/2 pixel and applying the SER algorithm the fraction of new sources is 15–28% (36/240, 86/306) of which 3–8% (3/36, 3/86) may be false. However, 8–10% (15/240, 31/306) of the original sources (6–20% of them may be false) detected in the pipeline product are lost, indicating that the original detections are still needed. We find no improvement when the background and diffuse emissions are low, e.g., Chandra Deep Field.

---

[4]http://cxc.harvard.edu/ciao/ahelp/srcextent.html

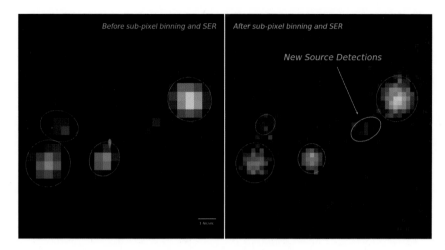

Figure 2.   Example of additional faint source detections.
Source regions output by wavdetect tool on ACIS image data after removal of pixel randomization, sub-pixel binning, and SER algorithm applied.

Table 1.   Point Source Size Improvement

ObsID	pipeline products		binning by 1 pixel pixel randomization off		sub-pixel algorithm applied			
02228	0.51	(0.47-0.54)	0.48	(0.44-0.53)	0.47	(0.43-0.52)		
02254	0.73	(0.69-0.76)	0.72	(0.68-0.75)	0.71	(0.68-0.75)		
00927	0.63	(0.60-0.66)	0.62	(0.59-0.64)	0.62	(0.59-0.64)		
01602	0.76	(0.71-0.81)	0.74	(0.69-0.79)	0.72	(0.67-0.77)		
03140	0.57	(0.51-0.62)	0.55	(0.50-0.60)	0.56	(0.51-0.61)		
04964	1.75	(1.66-1.83)	1.75	(1.67-1.84)	1.75	(1.66-1.84)		
04936	1.02	(0.98-1.05)	1.00	(0.97-1.04)	1.00	(0.96-1.03)		
			binning by 1/2 pixel				improvement(%)	
02228	0.44	(0.40-0.47)	0.41	(0.38-0.44)	0.40	(0.37-0.43)	14	22
02254	0.62	(0.59-0.65)	0.60	(0.57-0.63)	0.59	(0.56-0.63)	15	19
00927	0.50	(0.48-0.52)	0.49	(0.46-0.51)	0.48	(0.46-0.51)	21	24
01602	0.61	(0.57-0.65)	0.59	(0.55-0.64)	0.58	(0.54-0.62)	20	24
03140	0.44	(0.40-0.47)	0.41	(0.38-0.45)	0.38	(0.35-0.41)	23	33
04964	1.74	(1.65-1.82)	1.73	(1.65-1.82)	1.73	(1.65-1.82)	1	1
04936	1.00	(0.97-1.04)	0.98	(0.95-1.02)	0.98	(0.94-1.02)	2	4
			binning by 1/4 pixel				improvement(%)	
02228	0.43	(0.40-0.46)	0.40	(0.37-0.43)	0.39	(0.37-0.42)	16	24
02254	0.57	(0.54-0.60)	0.56	(0.54-0.59)	0.56	(0.54-0.59)	22	23
00927	0.45	(0.43-0.47)	0.43	(0.41-0.45)	0.42	(0.40-0.44)	29	33
01602	0.59	(0.56-0.63)	0.54	(0.51-0.57)	0.51	(0.48-0.55)	22	33
03140	0.39	(0.36-0.43)	0.40	(0.37-0.43)	0.30	(0.28-0.33)	32	47
04964	1.74	(1.66-1.83)	1.73	(1.65-1.82)	1.74	(1.65-1.82)	1	1
04936	1.03	(0.99-1.06)	0.99	(0.95-1.02)	0.98	(0.95-1.02)	0	4

Sources for ObsID's 04964,04936 are off-axis

Table 2.   Source Detection Comparison

		O	sub-pixel binning				binning + sub-pixel algorithm			
			A	B	C	D	A	B	C	D
			NGC 3379 obsid=7073 (87 ks)							
1	10ks	32	2-0	4-0	2-0	4-0	2-0	11-2	2-0	9-0
2	10ks	34	2-0	4-1	2-0	2-0	2-1	3-0	2-1	3-0
3	10ks	30	0-0	2-0	0-0	2-0	0-0	2-1	0-0	2-0
4	20ks	52	2-0	4-3	2-0	3-2	4-1	9-6	4-0	7-2
5	20ks	48	1-0	3-4	1-0	2-0	1-1	6-1	1-1	5-0
6	20ks	44	2-2	6-4	2-2	5-1	3-1	7-2	3-1	7-1
	Subtotal	240	9-2	23-12	9-2	18-3	12-4	38-12	12-3	33-3
			NGC 4278 obsid=7081 (114 ks)							
1	10ks	40	2-0	8-1	1-0	3-1	5-0	16-1	5-0	9-1
2	10ks	40	5-0	7-0	5-0	6-0	6-1	17-1	6-1	13-1
3	10ks	38	3-0	3-2	2-0	1-1	3-0	15-0	3-0	8-0
4	20ks	62	1-0	14-5	1-0	10-0	3-0	20-11	3-0	14-1
5	20ks	61	6-0	9-5	6-0	4-1	6-0	26-3	6-0	22-0
6	20ks	65	2-1	3-2	2-1	1-1	6-1	19-7	6-1	17-0
	Subtotal	306	19-1	44-15	17-1	25-3	29-2	113-23	29-2	83-3
	Total	546	28-3	67-27	26-3	43-6	41-6	151-35	41-5	126-6

For n-m; n,(m) = number of new sources (not) confirmed in deeper image
O. number of sources from the pipeline image binned by ACIS pixel (0.492")
A. number of lost sources (i.e., detected in the raw image, but not in the sub-pix image)
B. number of new sources (i.e., detected in the sub-pix image, but not in the raw image)
C, D. same as A, B but excludes those sources with 0 size by wavdetect (r_major=r_minor=0)

References

Houck, J. C. 2007, Measuring Detected Source Extent Using Mexican-Hat Optimization, Tech. rep., MIT Kavli Institute for Astrophysics and Space Science. URL http://cxc.cfa.harvard.edu/csc/memos/files/Houck_source_extent.pdf
Li, J., Kastner, J. H., Prigozhin, G. Y., & Schulz, N. S. 2003, ApJ, 590, 586
Li, J., Kastner, J. H., Prigozhin, G. Y., Schulz, N. S., Feigelson, E. D., & Getman, K. V. 2004, ApJ, 610, 1204
Mori, K., Tsunemi, H., Miyata, E., Baluta, C. J., Burrows, D. N., Garmire, G. P., & Chartas, G. 2001, in New Century of X-ray Astronomy, edited by H. Inoue & H. Kunieda (San Francisco, CA: ASP), vol. 251 of ASP Conf. Ser., 576
Park, S., Burrows, D. N., Garmire, G. P., Nousek, J. A., McCray, R., Michael, E., & Zhekov, S. 2002, ApJ, 567, 314
Tsunemi, H., Mori, K., Miyata, E., Baluta, C., Burrows, D. N., Garmire, G. P., & Chartas, G. 2001, ApJ, 554, 496

Astronomical Data Analysis Software and Systems XX
ASP Conference Series, Vol. 442
Ian N. Evans, Alberto Accomazzi, Douglas J. Mink, and Arnold H. Rots, eds.
©*2011 Astronomical Society of the Pacific*

A New Sky Subtraction Technique for Low Surface Brightness Data

Ivan Yu. Katkov[1] and Igor V. Chilingarian[2,1]

[1]*Sternberg Astronomical Institute, Moscow State University,*
13 Universitetski prospect, 119992, Moscow, Russia

[2]*Centre de Données Astronomiques de Strasbourg, Observatoire de Strasbourg,*
CNRS UMR 7550, Université de Strasbourg, 11 Rue de l'Université,
67000 Strasbourg, France

Abstract. We present a new approach to the sky subtraction for long-slit spectra that is suitable for low-surface brightness objects based on the controlled reconstruction of the night sky spectrum in the Fourier space using twilight or arc-line frames as references. It can be easily adopted for FLAMINGOS-type multi-slit data. Compared to existing sky subtraction algorithms, our technique is taking into account variations of the spectral line spread along the slit thus qualitatively improving the sky subtraction quality for extended targets. As an example, we show how the stellar metallicity and stellar velocity dispersion profiles in the outer disc of the spiral galaxy NGC5440 are affected by the sky subtraction quality. Our technique is used in the survey of early-type galaxies carried out at the Russian 6-m telescope, and it strongly increases the scientific potential of large amounts of long-slit data for nearby galaxies available in major data archives.

1. Introduction

Low-surface brightness ($\mu_B > 23$ mag/arcsec^2) outer regions of galaxies contain crucially important information for understanding the properties of their extended discs and dark matter haloes. Brightness profiles of dwarf early-type galaxies whose mean surface brightness is correlated with the luminosity, can be situated entirely in the low-surface brightness regime. Analysis of absorption line spectra at such surface brightness levels is often hampered by systematic errors of the sky subtraction that sometimes may lead to wrong astrophysical conclusions. Therefore, in order to analyze deep spectral data, it is important to improve the sky subtraction technique.

Here we present a new approach to the sky subtraction for long-slit spectra based on the controlled reconstruction of the night sky spectrum in the Fourier space using twilight or arc line frames as references.

Due to optical distortions, the shape of the spectral line spread function (LSF) in a long-slit spectrograph varies along the wavelength range as well as along the slit. In Fig. 1, we provide an example of the LSF shape of the SCORPIO (Afanasiev & Moiseev 2005) universal spectrograph at the Russian 6-m telescope reconstructed from the twilight frame (i.e. the Solar spectrum). The LSF is slightly asymmetrical and cannot be described by the Gaussian function, a usual parametrization in most data reduction packages. Here we use the Gauss-Hermite representation (van der Marel & Franx 1993) up-to the 4th order moment that allows one to describe first-order differences of the line profile from the Gaussian shape. These LSF variations affect the night sky spectrum which is subtracted from science frames during the data reduction. On the Fig. 2 we show a reduced long-slit spectrum of the spiral galaxy NGC 5440 before the sky subtraction step.

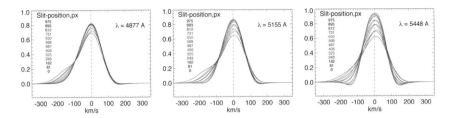

Figure 1.   An example of the LSF shape of the SCORPIO reconstructed from the twilight frame at different wavelength and slit positions. We used the Gauss-Hermite LSF representation. One can see that the profile asymmetry increases towards the outer slit regions. There is also a notable change of the overall spectral resolution from blue to red.

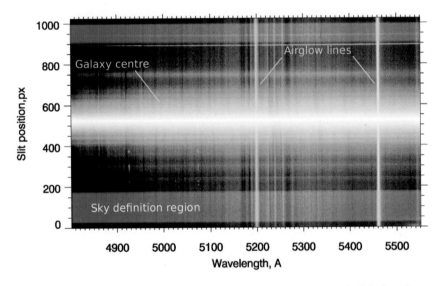

Figure 2.   A reduced long-slit spectrum of the spiral galaxy NGC 5440 before the sky subtraction step. Yellow areas denote a region of the frame used to construct the night sky spectrum used for the sky subtraction. Taking into account the profile variation shown in Fig. 1, it is clear that the intrinsic LSF shape in these regions will differ from that in regions of the galaxy close to the slit center.

2.   The Sky Subtraction Algorithm

In the traditional sky subtraction technique implemented in most standard data reduction packages (IRAF, MIDAS), the night sky spectrum is constructed from the outer regions of the slit which are (supposedly) free of galaxy light as a κ-σ clipped average. Then it is subtracted at every slit position.

The new algorithm presented here is an improvement of a method proposed in Chilingarian et al. (2009) aimed at increasing its stability with certain features taken from the technique by Kelson (2003). However, compared to the latter method, our approach allows one to take into account empirically the variations of the LSF along the slit.

The new technique includes several steps:

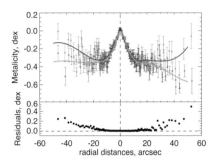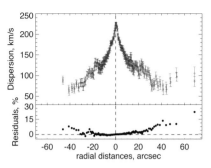

Figure 3.   Comparison between traditional technique and our deconvolution. The two panels on the left display the internal velocity dispersion profiles of NGC 5440. Blue data points are for the new technique, while red ones are for the "classical approach". Differences between the two approaches are displayed in the bottom panel. The two panels on the right hand side display the stellar metallicity profiles of NGC 5440 using the same symbols and colors as on the left hand side panels.

1. An oversampled sky spectrum is created from the non-linearized spectra using the wavelength solutions in order to perform the pixel-to-wavelength coordinate mapping. Then it is approximated using a b-spline. This approach was proposed by Kelson (2003) to improve the sky subtraction in undersampled datasets.

2. At every position along the slit, we change the LSF shape inside this night sky spectrum using a Fourier-based technique into the LSF at that slit position. The observed sky spectrum is a convolution of a true spectrum with the LSF:

$$R(\lambda, y) = R_0(\lambda) * LSF(\lambda, y); \quad S(\lambda, y) = S_0(\lambda) * LSF(\lambda, y), \tag{1}$$

where $R(\lambda, y)$ is a template spectrum (high signal-to-noise twilight frame), $S(\lambda)$ — the night sky spectrum. Then according to the convolution theorem the ratio between the Fourier transforms of the template spectrum and the object spectrum is a constant function on position along slit y:

$$\frac{FFT(S(\lambda, y))}{FFT(R(\lambda, y))} = \frac{FFT(S_0(\lambda))}{FFT(R_0(\lambda))} = \frac{FFT(S(\widetilde{y}, \lambda))}{FFT(R(\widetilde{y}, \lambda))} = F(\lambda), \tag{2}$$

where $\widetilde{y}$ — position at the sky definition region. The night sky spectrum at current position along slit can be expressed as follows:

$$S(y, \lambda) = FFT^{-1} \left(\frac{FFT(S(\widetilde{y}, \lambda))}{FFT(R(\widetilde{y}, \lambda))} \cdot FFT(R(y, \lambda)) \right). \tag{3}$$

3. The b-spline parameterization provides the necessary regularisation for the numerical stability of this procedure.

3.   Usage Example and Perspectives of the Method

In Fig. 3 we present the result of the data analysis of a long-slit spectrum of NGC 5440 for the two sky subtraction techniques. We fitted the reduced sky subtracted spectra with high resolution stellar population models with the NBURSTS full spectral fitting technique (Chilingarian et al. 2007b,a) and extracted kinematical (radial velocity and velocity dispersion) and stellar population (age and metallicity) parameters along the slit. The radial profiles of velocity dispersion and metallicity are shown in Fig. 3. While the measurements are very similar near the galaxy center, they differ notably in the peripheral regions. With the new sky subtraction technique the uncertainties are lower, and the general trend of the galaxy metallicity gradient corresponds to the physical expectations. The traditional sky subtraction technique possesses systematic errors in the low surface brightness regime, which propagate through the data analysis and may result in misleading astrophysical conclusions.

Our sky subtraction technique is adopted in the survey of nearby lenticular galaxies (P.I.: Prof. Zasov, Moscow State University) carried out at the Russian 6-m telescope using the SCO-PRIO spectrograph. Our approach can be easily modified for multi-slit spectroscopic data with parallel slits ("FLAMINGOS"-type spectra).

Since out technique improves the quality of data analysis at low signal-to-noise ratios, it can also be used to re-reduce and re-analyze long-slit spectroscopic datasets for hundreds of galaxies obtained with different telescopes and publicly available in data archives.

Acknowledgments.   IK thanks the ADASS organizing committee for the provided financial aid and RFBR for covering the remaining travel expenses.

References

Afanasiev, V. L., & Moiseev, A. V. 2005, Astron. Lett., 31, 194
Chilingarian, I., Prugniel, P., Sil'Chenko, O., & Koleva, M. 2007a, in IAU Symposium, edited by A. Vazdekis & R. F. Peletier, vol. 241 of IAU Symp., 175
Chilingarian, I. V., Novikova, A. P., Cayatte, V., Combes, F., Di Matteo, P., & Zasov, A. V. 2009, A&A, 504, 389
Chilingarian, I. V., Prugniel, P., Sil'Chenko, O. K., & Afanasiev, V. L. 2007b, MNRAS, 376, 1033
Kelson, D. D. 2003, PASP, 115, 688
van der Marel, R. P., & Franx, M. 1993, ApJ, 407, 525

Astronomical Data Analysis Software and Systems XX
ASP Conference Series, Vol. 442
Ian N. Evans, Alberto Accomazzi, Douglas J. Mink, and Arnold H. Rots, eds.
©*2011 Astronomical Society of the Pacific*

Using Boundary Fits to Determine Spectra Pseudo-Continua

N. Cardiel

Departamento de Astrofísica y Ciencias de la Atmósfera,
Facultad de Ciencias Físicas,
Universidad Complutense de Madrid, E28040–Madrid, Spain

Abstract. The appropriate determination of pseudo-continuum spectra is one of the critical steps when trying to retrieve scientifically relevant information from spectroscopic data. Typically one derives the pseudo-continua by fitting a low-order polynomial or splines to the data, excluding regions with conspicuous spectroscopic features. The problem with this approach is that, no matter how carefully one chooses the data to be fitted, those fits are always passing through the spectral data points, whereas in principle the goal is to obtain a representation of a pseudo-continuum that must be placed on top of the spectrum. The use of boundary fits constitutes an excellent approach for this particular problem.

1. Introduction

The work presented in this poster is a demonstration of the usefulness of using boundary fits to determine pseudo-continuum spectra. The basic idea behind the concept of boundary fitting was initially presented in a previous ADASS Conference (Cardiel 2009b). Full details of the method, showing that the technique is suitable for the determination of pseudo-continuum spectra, was presented in Cardiel (2009a).

The determination of an accurate pseudo-continuum constitutes a quite common step when correcting spectroscopic data from telluric absorptions using featureless (or almost featureless) calibration spectra, when subtracting the large scale variation of spectra when estimating velocity dispersions, or when measuring spectroscopic features, just to mention a few examples. This last case is extremely important because most of the time the data are analyzed by determining equivalent widths computed through an interactive (but hardly repeatable) determination of the local pseudo-continuum around the considered absorption or emission features.

2. Boundary Fits

Boundary fits can easily be computed by introducing in the fitting procedure an asymmetric weight for the data at both sides of a given fit, so the points located on one side exert a stronger effect on the fit than the points at the opposite side. In order to use the data asymmetrically, it is necessary to start with some initial estimate, that in practice can be obtained by employing the traditional least-squares method (with a symmetric data treatment). Once this initial estimate is available, it is straightforward to continue using the data asymmetrically and, in an iterative process, determine the sought boundary. Interested readers can find a full description of this method in Cardiel (2009a).

It is important to emphasize that the impact of flux uncertainties must be considered when computing a boundary fit, since the extra scatter introduced in the data by the errors will tend to bias the fit. This circumstance can be properly handled by allowing some data points to be placed outside of the boundary fit.

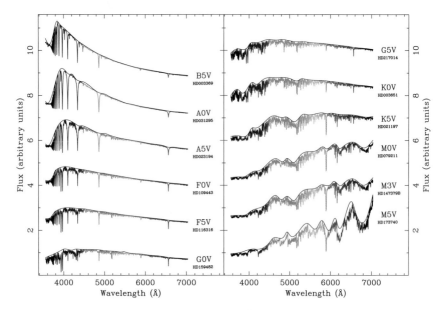

Figure 1.   Examples of pseudo-continuum fits obtained using adaptive splines. Several stars from the stellar library MILES (Sánchez-Blázquez et al. 2006), spanning different spectral types, have been selected. The fitted pseudo-continua (black lines) have been automatically determined employing 19 knots.

3.   Using Boundary Fits to Determine Pseudo-Continuum Spectra

3.1.   Low-Frequency Pseudo-Continuum

When trying to obtain a reasonable representation of the pseudo-continuum of a given spectrum, it is obvious that the tasks is more difficult as one increases the considered wavelength range. For this purpose, the use of adaptive splines provides a very good solution. For illustration, Fig. 1 shows the result of using adaptive splines to estimate the pseudo-continuum of 12 different spectra corresponding to stars exhibiting a wide range of spectral types (from B5V to M5V), selected from the empirical stellar library MILES (Sánchez-Blázquez et al. 2006).

Although in all the cases the fits have been computed blindly without considering the use of an initial knot arrangement appropriate for the particularities of each spectral type, it is clear from the figure that adaptive splines are flexible enough to give reasonable fits independently of the considered star.

More refined fits can be obtained using an initial knot pattern more adjusted to the curvature of the pseudo-continuum exhibit by the stellar spectra.

3.2.   High-Frequency Pseudo-Continuum

A different situation arises when trying to estimate the strength of spectroscopic features. Although with slight differences among them, most authors have employed line-strength indices with definitions close to the classical expression for an equivalent width

$$\mathrm{EW}(\text{Å}) = \int_{\text{line}} [1 - S(\lambda)/C(\lambda)] \, d\lambda, \tag{1}$$

where $S(\lambda)$ is the observed spectrum and $C(\lambda)$ is the local continuum, usually obtained by interpolation of $S(\lambda)$ between two adjacent spectral regions (e.g. Faber 1973; Faber et al. 1977;

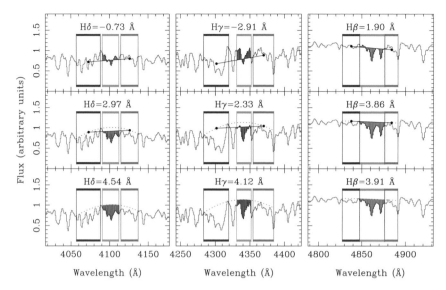

Figure 2.   Comparison of different strategies in the computation of the pseudo-continuum for the measurement of line-strength indices. In this example, three Balmer features are analyzed, namely Hδ, Hγ and Hβ (from left to right), showing the commonly employed blue, central and red sidebands. Top panels display the typical approach, whereas the middle and bottom panels make use of the boundary fits.

Whitford & Rich 1983). In practice, as pointed out by Geisler (1984) (see also Rich 1988), at low and intermediate spectral resolution the local continuum is unavoidably lost, and a pseudo-continuum is measured instead of a true continuum. The upper boundary fitting, by using either simple polynomials or adaptive splines, constitutes an excellent option for the estimation of that pseudo-continuum.

In order to investigate this possibility in more detail, Fig. 2 compares the actual line-strength indices derived for three Balmer lines (Hβ, Hγ and Hδ, from right to left) using three different strategies. Overplotted on each spectrum are the bandpasses typically used for the measurement of these spectroscopic features. In particular, the bandpass limits for Hβ are the revised values given by Trager (1997), whereas for Hγ and Hδ the limits correspond to $H\gamma_F$ and $H\delta_F$, as defined by Worthey & Ottaviani (1997).

For each feature, the corresponding line-strength has been computed by determining the pseudo-continuum using three different methods:

1. Computing the straight line joining the mean fluxes in the blue and red bandpasses (top panels), which is the traditional method.

2. Determining the straight line joining the values of the upper boundary fits evaluated at the centers of the same bandpasses (central panels).

3. Obtaining the upper boundary fits themselves (bottom panels).

For the cases 2 and 3 the upper boundary fits have been derived using a second order polynomial fitted to the three bandpasses.

The resulting line-strength indices, numerically displayed above each spectrum, have been computed as the area comprised between the adopted pseudo-continuum fit and the stellar spectrum within the central bandpass. For the three Balmer lines it is clear that the use of the boundary fit provides larger indices. The traditional method provides very bad values for Hγ and Hδ (which are even negative!), given that the pseudo-continuum is very seriously affected by the absorption features in the continuum bandpasses.

The previous example clearly illustrates that line-strength indices can be strongly biased depending on the method adopted for evaluating the local pseudo-continuum. This is a very important issue that deserves very careful analysis.

Acknowledgments. This work was supported by the Spanish Programa Nacional de Astronomía y Astrofísica under grants AYA2006–15698–C02–02 and AYA2009–10368.

References

Cardiel, N. 2009a, MNRAS, 396, 680
— 2009b, in Astronomical Data Analysis Software and Systems XVIII, edited by D. A. Bohlender, D. Durand, & P. Dowler (San Francisco, CA: ASP), vol. 411 of ASP Conf. Ser., 216
Faber, S. M. 1973, ApJ, 179, 731
Faber, S. M., Burstein, D., & Dressler, A. 1977, AJ, 82, 941
Geisler, D. 1984, PASP, 96, 723
Rich, R. M. 1988, AJ, 95, 828
Sánchez-Blázquez, P., et al. 2006, MNRAS, 371, 703
Trager, S. C. 1997, Ph.D. thesis, University of California, Santa Cruz
Whitford, A. E., & Rich, R. M. 1983, ApJ, 274, 723
Worthey, G., & Ottaviani, D. L. 1997, ApJS, 111, 377

Astronomical Data Analysis Software and Systems XX
ASP Conference Series, Vol. 442
Ian N. Evans, Alberto Accomazzi, Douglas J. Mink, and Arnold H. Rots, eds.
© *2011 Astronomical Society of the Pacific*

Segmentation and Detection of Extended Structures in Low Frequency Astronomical Surveys using Hybrid Wavelet Decomposition

Marta Peracaula,[1] Xavier Lladó,[1] Jordi Freixenet,[1] Arnau Oliver,[1] Albert Torrent,[1] Josep M. Paredes,[2] and Josep Martí[3]

[1]*Institut d'Informàtica i Aplicacions, Universitat de Girona, Spain*

[2]*Departament d'Astronomia i Meteorologia, Facultat de Física, Universitat de Barcelona, Spain*

[3]*Departamento de Física, Escuela Politécnica Superior, Universidad de Jaén, Spain*

Abstract. The morphological complexity of extended real structures (such as SNRs, HII regions, bow shocks, etc.), and their wide variety in scale and surface brightness make their automatic detection and segmentation in large surveys a difficult task. We propose in this paper a segmentation method based on applying wavelet decomposition in the residual thresholded images. This strategy avoids the artifacts produced by strong sources in a straight wavelet decomposition. Our method successfully segments extended structures at different scales and therefore is suitable for further morphological analysis and object recognition processes. Results using images from radio and infrared wavelengths surveys show the validity of our approach.

1. Motivation and Objectives

Large surveys reveal thousands of low spatial frequency objects, shown at different intensity scales. When imaging rich areas in the interstellar medium, many of the compact sources overlap with objects associated to extended, morphologically complex real structures, such as SNRs, HII regions, bow shocks, etc. The wide variety in spacial scale and surface brightness of these objects make their automatic detection and segmentation a difficult task.

To illustrate these facts, we use in this paper the image corresponding to the high galactic longitude end of the Phase I Canadian Galactic Plane Survey (CGPS hereafter, see Taylor et al. 2003). Figure 1 shows, on the top, the image composition corresponding to mosaics V1, V2, W1, W2, X1, X2, Y1 and Y2 of the CGPS 1420 MHz, continuum. We have eliminated 0.1% of the intensity outliers in order to visualize some dozens of sources. Nevertheless, the data corresponding to these mosaics contain thousands of objects, as it is illustrated when we display the sub-image contained in the red square at different intensity scales. On the bottom left of the figure we show the sub-image with 2% of the outliers eliminated, whereas on the right 5% of them have been removed. In addition to a great number of compact sources, several extended structures and their surrounding emission have appeared (the zoomed area contains Lynds Bright Nebula 679).

In previous work we focused our attention on the performance of techniques suitable for the detection of faint compact objects (see for example Peracaula et al. 2008). In the present work we focus on the automatic detection of extended and irregular structures for further cataloguing and morphological analysis. In this context, wavelet image decomposition has been proven as a tool that can detect and separate objects represented at different spatial frequencies. However, the high dynamic range in intensity of this kind of image diminishes the performance

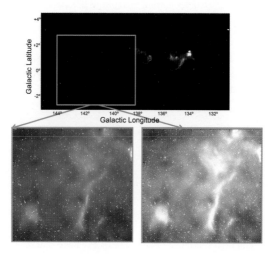

Figure 1.   Top: Contrast stretched (0.1% outliers eliminated) image corresponding to mosaics V1, V2, W1, W2, X1, X2, Y1 and Y2 of Phase I CGPS at 1420 MHz, continuum. Bottom: Zoomed area around Lynds Bright Nebula 679, displayed with 2% of outliers eliminated (left) and 5% of outliers eliminated (right).

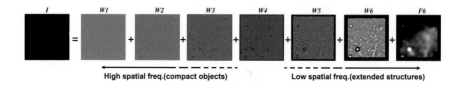

Figure 2.   Wavelet decomposition of the sample image.

of the decomposition and a strategy to avoid this problem has to be applied, as we propose in the next sections.

2.   Wavelet Decomposition Using the "à trous" Algorithm

Multiscale Vision Models (Bijaoui & Rué 1995) decompose an image in J scales or wavelet planes and segment independently each of the images representing a scale. In Figure 2 we use the image containing mosaics V1, V2, W1 and W2 of the CGPS at 1420 MHz (continuum), and decompose it in 6 scales plus the smoothed array using the "à trous" algorithm with a B_3 filtering function (see for example Starck & Murtagh 1994, and references therein). As it can bee seen, low index scales emphasize high spatial frequencies (which translates to compact objects and semi-compact in case of true signal). High index scales emphasize low spatial frequencies (in this case extended source structures).

Using this approach we encounter two major problems:

1. As shown in Figure 3, due to their high brightness, strong compact sources are not filtered out and show up in low spatial frequency planes,

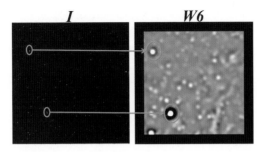

Figure 3.   Strong compact sources (as the ones circled in the original image on the left), show up in low spatial frequency planes (right).

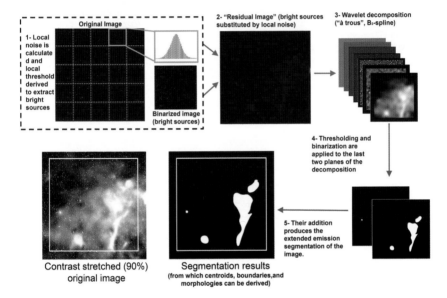

Figure 4.   Graphical representation of our algorithm and visual comparison of the segmentation result with the contrast stretched image.

2. To keep the wavelet coefficient mean at zero, negative artifacts around these strong structures appear and pollute the entire image.

3.   Our Approach

We propose to create an image where bright sources are substituted by local noise. Wavelet decomposition will then be applied to this new image in order to avoid the polluting effects of strong easily detectable sources. The algorithm we propose consists of the following steps:

1. We calculate local noise in the original image and derive a local threshold.

2. Pixels with intensity levels over the threshold are labeled and connected zones dilated in order to extract bright sources.

3. Two images are created: a "residual image" where bright sources have been substituted by local noise, and a binary image mask with the bright sources.

4. We apply a 6-scale Wavelet decomposition using the "à trous" algorithm and a B_3-spline filtering function to the "residual image".

5. Local thresholding and binary masking is applied to the last scale and the smoothed array.

6. The addition of the last two binary planes of the decomposition produces the extended emission segmentation of the image.

In Figure 4 we show a graphical representation of the algorithm using the sample image. On the bottom left we show the contrast stretched sample image in order to compare it with the segmentation result.

Acknowledgments.   The authors acknowledge support by DGI of the Spanish Ministerio de Educación y Ciencia (MEC) under grants AYA2007-68034-C03-01/02/03 and DPI2007-66796-C03-02, as well as partial support by the European Regional Development Fund (ERDF/FEDER). This research used the facilities of the Canadian Astronomy Data Center operated by the National Research Council of Canada with the support of the Canadian Space Agency. The research presented in this paper has used data from the Canadian Galactic Plane Survey a Canadian project with international partners supported by the Natural Sciences and Engineering Research Council.

References

Bijaoui, A., & Rué, F. 1995, Signal Proc., 46, 345
Peracaula, M., Freixenet, J., Martí, J., Martí, J., & Paredes, J. M. 2008, in Astronomical Data Analysis Software and Systems XVII, edited by R. W. Argyle, P. S. Bunclark, & J. R. Lewis (San Francisco, CA: ASP), vol. 394 of ASP Conf. Ser., 547
Starck, J., & Murtagh, F. 1994, A&A, 288, 342
Taylor, A. R., et al. 2003, AJ, 125, 3145

Astronomical Data Analysis Software and Systems XX
ASP Conference Series, Vol. 442
Ian N. Evans, Alberto Accomazzi, Douglas J. Mink, and Arnold H. Rots, eds.
©*2011 Astronomical Society of the Pacific*

Galapagos: A Semi-Automated Tool for Galaxy Profile Fitting

Boris Häußler,[1] Marco Barden,[2] Steven P. Bamford,[1] and Alex Rojas[3]

[1]*School of Physics & Astronomy, University of Nottingham, University Park, Nottingham NG7 2RD, UK*

[2]*Institute of Astro- and Particle Physics, University of Innsbruck, Technikerstraße 25, A-6020 Innsbruck, Austria*

[3]*Carnegie Mellon University in Qatar, P.O. Box 24866, Doha, Qatar*

Abstract.   When it comes to measuring galaxy parameters, e.g. sizes (half-light radii), shapes (axial ratios) or profiles (Sérsic indices), on a large sample of tens of thousands of galaxies, e.g. COSMOS, STAGES or GEMS, or even millions of galaxies, e.g. SDSS, an automated fitting routine is strongly required. In this work, we introduce GALAPAGOS (Galaxy Analysis over Large Areas: Parameter Assessment by GALFITting Objects from SEXTRACTOR), a code that enables users to carry out profile fitting on large surveys in a mostly automated manner. After initial setup and specification of a set of parameters, the code manages the whole process: source extraction (with SEXTRACTOR) on the individual survey frames, masking and deblending, setup of the galaxy fitting process (using GALFIT), the fitting itself, and finally the compilation of all the resulting parameters into an object catalog; without the need for any user interaction. We present these steps and highlight the strengths of GALAPAGOS compared to other codes used for similar purposes. We have carried out thorough tests of the current version of the code on both real and simulated data. We show results and discuss the statistical and systematic biases of fitting codes in general and GALAPAGOS in particular. We find that GALAPAGOS returns very accurate measurements of the galaxy parameters without systematic bias and with only small statistical uncertainties, at least on 1-orbit HST data and assuming an optimal setup. We also briefly present MEGAMORPH, which will open up Galapagos for various survey strategies (including ground-based with variable PSF), multi-wavelength data, Bulge/Disk decompositions and high-performance computing facilities.

1.   Galapagos

The aim of the GALAPAGOS (also see Häussler et al. 2007) code is to automate galaxy profile fitting as far as possible to allow the fitting of large surveys while keeping vital parameters regarding detection, deblending, masking and fitting under user control. The code, once set up and started, runs the entire process automatically without user interaction, down to the output of a fits table that includes all galaxy parameters, derived from both SEXTRACTOR (Bertin & Arnouts 1996) and GALFIT (Peng et al. 2002). In the following, we briefly explain the individual steps in this process. For more details and plots, please see Barden et al., (in preparation), Häussler et al. (2007) and visit the webpage[1] where the code, manual and more information can be obtained.

---

[1]`http://astro-staff.uibk.ac.at/~m.barden/GALAPAGOS/`

1.1.   SExtractor

GALAPAGOS applies SEXTRACTOR to detect objects in the individual survey images. Due to the high dynamic range of galaxies, in most big surveys it becomes largely impossible to find one single setup that suitably detects both bright and faint galaxies. GALAPAGOS therefore uses SEXTRACTOR in a (user specified) HDR mode, using a "cold" setup to detect bright sources without splitting them up into sub-components and a "hot" setup to detect faint sources. These catalogs are combined on each survey frame and then combined to form one big survey catalog. During the combination of the individual frames, it takes care of image overlap and double detections. User specified "bad detections" can also be removed in this process.

1.2.   Postage Stamp Cutting

For every object in the detection catalog, a 'postage stamp' image is cut out of the survey frames and stored. It is centered around the primary object and is big enough to allow for fitting and masking of neighboring objects and the sky background. The user can specify a size ratio to increase the postage stamp size. These images are used as input images to GALFIT. Working on smaller files, instead of the full survey frames, is — with the exception of over-crowded fields — much quicker and makes identification of interesting objects afterwards much easier as one image exists for every object.

1.3.   Sky Determination

The estimation of a good sky background level is absolutely essential to a successful fit of the galaxy profile. The output values are very sensitive to the sky (e.g. Häussler et al. 2007). However, deriving a good measure for the sky is very difficult in an automated fashion. Based on the assumption of an ideally flat-fielded image, GALAPAGOS determines the sky background value in a series of annuli (with user specified width and step size) masking neighboring objects to ensure a robust sky estimate. GALAPAGOS uses the original survey frames at this point as the postage stamps might be too small to do this properly. Using both real and simulated data, in Häussler et al. (2007), we have shown that this estimate is more accurate than other measurements (e.g. SEXTRACTOR) and that GALAPAGOS performs best when this sky estimate is used (in contrast to other fitting codes, which, due to their setup, might prefer other sky estimates).

1.4.   Deblending, Masking, Fitting

Before running GALFIT to fit the galaxy profile, any code needs to identify the objects which are to be fit and create mask images that blank out neighboring objects that are not to be fit simultaneously. Using all the available information from SEXTRACTOR (e.g. modified SEXTRACTOR ellipses), GALAPAGOS decides on an automated basis which neighboring galaxies need to be fit simultaneously (the so-called secondary object's ellipse overlaps with that of the primary source) and which neighbors can be masked out (ellipses don't overlap; a so-called tertiary object). GALAPAGOS creates a mask image that blanks out this entire ellipse. In case of multiple overlaps of secondaries and tertiaries, it employs a clever method to decide which pixels to mask and which not.

Every object in the survey will become the primary object at some point. Fitting results of secondary objects are not used further while fitting results from primary objects may be used as a "fixed" secondary source, i.e. all the secondary's parameters are kept at their starting values from the previous primary fit. Secondaries generally are only used to improve the fitting result of the primary. While this seems to increase the overall CPU time, as several secondaries might be fit as such on several occasions, it actually keeps the number of sources that have to be fit simultaneously small on an object-by-object basis. It furthermore ensures ideal deblending of neighbors in all cases while avoiding to fit all objects in a survey at the same time.

GALAPAGOS fits galaxies in decreasing order of brightness by:

- Writing out a GALFIT start file that includes all the information about the image, the previously determined sky, the primary and all secondary objects. In case one of the neighboring objects has previously been fit already, GALAPAGOS does not use free parameters during the fit but merely subtracts the object from the image using the already existing fitting result.

- Running GALFIT from the system command line. As each GALFIT only uses one of the CPU cores, GALAPAGOS is able to start a user specified number of fits in parallel to improve overall fitting speed, e.g. on a dual quad-core CPU machine, one could run 8 parallel fits, assuming it has enough memory.

- Reading in the fitting results of the primary source and writing them into the fitting catalog. These values are then used for further deblending of the sources where a fit already exists.

- Writing out an objects catalog (in form of a fits table that contains all information gathered by both SEXTRACTOR and GALFIT over the course of the process.

On large surveys, GALAPAGOS can be run in a 2-step mode to take full advantage of several multi-CPU machines. On a dual quad-core server machine (á 2.4GHz) on STAGES (Gray et al. 2009) data, we managed to run this whole process in around 9–10 days on all ~89,000 galaxies in the field. As GALFIT uses an effective downhill-gradient fitting method, it is fast compared to other codes and using a different minimisation algorithm. Without need for user interaction during the fitting process the total time needed simply depends on the survey size and data quality.

2. Testing Galapagos

In GEMS (Rix et al. 2004) and STAGES, we carried out extensive tests of GALAPAGOS using both real and simulated data (see both Häussler et al. 2007; Gray et al. 2009). We find that GALAPAGOS in general returns very good results. Only for faint galaxies with very high Sérsic indices, which are most sensitive to uncertainties in the sky determination, can a small systematic offset can be seen. While GALFIT generally underestimates the true errorbar of the fit, they can be derived statistically by using simulated images. We have also demonstrated the independence of galaxy parameters from both distance and magnitude of neighboring objects as measured and known from simulated data. Whereas other fitting methods are sensitive to neighbors, GALAPAGOS is not. This is a direct result of the methods employed in GALAPAGOS for masking and deblending of neighboring sources.

3. MegaMorph

It is becoming clear that the striking difference between elliptical and spiral galaxies is actually a result of variation in the relative prominence of their more fundamental spheroid and disk components. Our understanding of galaxies would therefore be greatly improved by disentangling these components. However, measuring the properties of the individual components within a galaxy is considerably more difficult than measuring its overall properties as done by GALAPAGOS in its current version. Galaxies are complex structures and, beyond the general distinction between spheroids and disks, they display a range of other features that make it difficult for computational methods to extract meaningful information. The MEGAMORPH (see proceedings by S. Bamford) project is tackling this problem through a combination of statistical techniques and by utilizing the full set of multi-color information available for each galaxy in most surveys. Our starting point is GALAPAGOS and GALFIT, two pieces of existing, tried and tested software, which we are adapting to perform robust, physically meaningful galaxy bulge-disk decompositions using data from many wavelength bands simultaneously. To control the computational intensity of the task, we need to use efficient algorithms and tools to optimally use high performance computing facilities. The MEGAMORPH project is in its early stages, and we are currently concerned with constructing test samples with which to evaluate the performance of our developments with respect to current methods.

Acknowledgments.   BH is grateful for support from the Science and Technology Facilities Council (STFC). MB was in part supported by the Austrian Science Foundation FWF under grant P18416. SPB is supported by an STFC Advanced Fellowship.

References

Bertin, E., & Arnouts, S. 1996, A&AS, 117, 393
Gray, M. E., et al. 2009, MNRAS, 393, 1275
Häussler, B., et al. 2007, ApJS, 172, 615
Peng, C. Y., Ho, L. C., Impey, C. D., & Rix, H. 2002, AJ, 124, 266
Rix, H., et al. 2004, ApJS, 152, 163

Astronomical Data Analysis Software and Systems XX
ASP Conference Series, Vol. 442
Ian N. Evans, Alberto Accomazzi, Douglas J. Mink, and Arnold H. Rots, eds.
© 2011 Astronomical Society of the Pacific

Detection of Periodic Variability in Simulated QSO Light Curves

David B. Westman,[1] Chelsea L. MacLeod,[1] and Željko Ivezić[1,2]

[1]*Department of Astronomy, Box 351580, University of Washington, Seattle, WA 98195-1580 USA*
[2] *University of Zagreb, Department of Physics, Bijenička c. 32, P.P. 331, Zagreb, Croatia*

Abstract. Periodic light curve behavior predicted for some binary black hole systems might be detected in large samples, such as the multi-million quasar sample expected from the Large Synoptic Survey Telescope (LSST). We investigate the false-alarm probability for the discovery of a periodic signal in light curves simulated using damped random walk (DRW) model. This model provides a good description of observed light curves, and does not include periodic behavior. We used the Lomb-Scargle periodogram to search for a periodic signal in a million simulated light curves that properly sample the DRW parameter space, and the LSST cadence space. We find that even a very conservative threshold for false-alarm probability still yields thousands of "good" binary black hole candidates. We conclude that the future claims for binary black holes based on Lomb-Scargle analysis of LSST light curves will have to be interpreted with caution.

1. Introduction

Modern surveys of the sky, such as the Sloan Digital Sky Survey (SDSS, York et al. 2000), have collected huge amounts of data (20 TB for SDSS), requiring the development of automated analysis methods. The Large Synoptic Survey Telescope (LSST, Ivezić et al. 2008) will gather even more data than the SDSS did (one SDSS equivalent per night over ten years of operations). Among other populations, LSST will identify several million quasars (QSO) and obtain their light curves. In this contribution, we discuss an automated analysis of a million simulated light curves to search for periodic variability.

The optical variability of QSOs has been recognized since they were first identified (Matthews & Sandage 1963) and is aperiodic and on the order of 20% on timescales of months to years (for recent results see, e.g., MacLeod et al. 2010, hereafter M10). Periodic variability has been suggested as one of the observational characteristics of a binary black hole system (Komossa 2003), (also see Shen & Loeb 2010, for an investigation of broad line emission spectra in binary black holes), but there is no convincing observational evidence for such systems yet.

2. Goal

The large QSO sample expected from LSST might enable a detection of periodic signal in observed light curves. Recently we employed the Lomb-Scargle periodogram (Lomb 1976; Scargle 1982; Horne & Baliunas 1986) to test ~9000 spectroscopically confirmed QSOs from SDSS Stripe 82 (S82) for periodic variability (see Appendix of M10). We reused the tools developed for the analysis of SDSS data and studied light curves simulated using a mathematical variability model trained on SDSS data, and sampled using simulated LSST cadences.

3.   Creation of Light Curves

Approximately 1×10^6 QSO light curves were generated using the damped random walk model (Kelly et al. 2009; Kozlowski et al. 2010, M10). The difference between this model and the well-known random walk is that an additional self-correcting term pushes any deviations back towards the mean flux on a time scale τ. The above studies have established that DRW can statistically explain the observed light curves of quasars at an impressive fidelity level (0.01–0.02 mag).

The input parameters to the model are the characteristic time scale, τ, and the root-mean-square (rms) variability on long time scales, or structure function, *sf*. The input parameters were determined using the scalings with black hole mass (M_{BH}), absolute magnitude (M_i), and redshift found by M10. These physical parameters were drawn from the distribution shown in Figure 1. After generating ~83,500 well-sampled light curves, each light curve was resampled to the 12 different simulated *r*-band LSST cadences from Delgado et al. (2006, ~200 observations spread over 10 years) to obtain $\sim 1 \times 10^6$ total light curves.

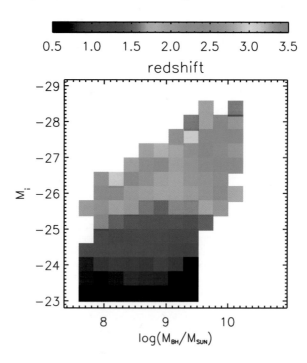

Figure 1.   QSO model parameter ranges used to derive inputs to light-curve generation

All light curves were analyzed with Lomb-Scargle periodogram software. If the maximum power spectral density (PSD) value was above the level set according to the false alarm probability, *fap*, then the curve became part of a set used for further examination.

4.   Results

We found that for high *fap* values, the actual number of light curves which exceeded the *fap* level was less than the theoretical value by an appreciable amount. For a *fap* value of 5%, there were 13,294 (1.3%) light curves that exceeded that level, for a *fap* value of 1%, there were 4696 (0.47%), and for a fap value of 0.1%, there were 1035 (0.1%).

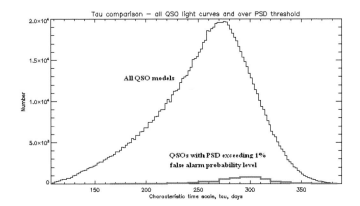

Figure 2.   Distribution of characteristic time, τ, for all QSO light curves compared with τ distribution for light curves with a peak SED value exceeding the 1% *fap* level

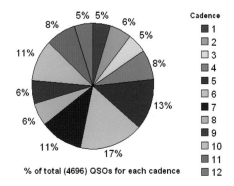

Figure 3.   Proportion of QSOs with peak PSD values over 1% *fap* level sorted by LSST cadence

Figure 2 shows that values of τ for QSO models with PSD exceeding the 1% *fap* level (red histogram) are distributed differently from those for all the QSO models (black histogram). This bias is due to the fact that when τ is long, only a few "oscillations" are observed over the duration of the light curve, causing the periodogram to mistake the damped random walk behavior for a periodic behavior.

Figure 3 compares the proportions of QSO models found to exceed the 1% *fap* level for each of the cadences. There is a marked difference between the various cadences here, showing that some of the cadences allow many more periodogram results for which the 1% *fap* level is exceeded. This may be because the variations in the sampling pattern and the "windowing" effect of the Lomb-Scargle periodogram can cause a resonance effect at some test frequencies. Therefore, the results obtained by using the Lomb-Scargle periodogram method can be greatly influenced by the pattern of the observations used.

5. Conclusion

This work shows that the Lomb-Scargle periodogram method may be useful for detecting potentially periodic behavior in QSO light curves in a large-scale surveys, such as the one to be carried out by the LSST. However, even with *fap* as small as 0.1%, the large LSST sample would yield ~1,000 **false** candidates. Therefore, future claims of periodic behavior based on Lomb-Scargle analysis of LSST light curves will have to be interpreted with caution. In particular, black hole binary candidates identified by this method would have to be examined individually and with supplemental observations.

Acknowledgments. We acknowledge support by NSF grant AST-0807500 to the University of Washington, and NSF grant AST-0551161 to the LSST for design and development activity. Ž. Ivezić thanks the University of Zagreb, where a portion of this work was completed, for its hospitality, and acknowledges support by the Croatian National Science Foundation grant O-1548-2009.

References

Delgado, F., Cook, K., Miller, M., Allsman, R., & Pierfederici, F. 2006, in Observatory Operations: Strategies, Processes, and Systems, edited by D. R. Silva, & R. E. Doxsey (Bellingham, WA: SPIE), vol. 6270 of Proc. SPIE, 62701D

Horne, J. H., & Baliunas, S. L. 1986, ApJ, 302, 757

Ivezić, Ž., Tyson, J. A., Allsman, R., Andrew, J., Angel, R., & for the LSST Collaboration 2008, ArXiv e-prints. `arXiv:0805.2366`

Kelly, B. C., Bechtold, J., & Siemiginowska, A. 2009, ApJ, 698, 895

Komossa, S. 2003, in The Astrophysics of Gravitational Wave Sources, edited by J. Centrella (Mellville, NY, USA: AIP), vol. 686 of AIP Conf. Proc., 161

Kozlowski, S., et al. 2010, ApJ, 708, 927

Lomb, N. R. 1976, Ap&SS, 39, 447

MacLeod, C. L., et al. 2010, ApJ, 721, 1014

Matthews, T. A., & Sandage, A. R. 1963, ApJ, 138, 30

Scargle, J. D. 1982, ApJ, 263, 835

Shen, Y., & Loeb, A. 2010, ApJ, 725, 249

York, D. G., et al. 2000, AJ, 120, 1579

Astronomical Data Analysis Software and Systems XX
ASP Conference Series, Vol. 442
Ian N. Evans, Alberto Accomazzi, Douglas J. Mink, and Arnold H. Rots, eds.
© *2011 Astronomical Society of the Pacific*

XMM-Newton Remote Interface to Science Analysis Software: First Public Version

A. Ibarra and C. Gabriel

XMM-Newton SOC, European Space Astronomy Center (ESAC) / ESA, Madrid, Spain.

Abstract. We present the first public beta release of the XMM-Newton Remote Interface to Science Analysis (RISA) software, available through the official XMM-Newton web pages. In a nutshell, RISA is a web based application that encapsulates the XMM-Newton data analysis software. The client identifies observations and creates XMM-Newton workflows. The server processes the client request, creates job templates and sends the jobs to a computer. RISA has been designed to help, at the same time, non-expert and professional XMM-Newton users. Thanks to the predefined threads, non-expert users can easily produce light curves and spectra. And on the other hand, expert user can use the full parameter interface to tune their own analysis. In both cases, the VO compliant client/server design frees the users from having to install any specific software to analyze XMM-Newton data.

1. Introduction

The astronomical data reduction software is definitely changing from local stand-alone application, with no internet access and needing lots of third party software installed locally in the desktop, to web based applications making use of modern internet technologies, that until now, were only used in business projects.

As examples of modern applications we can mention the Gaia data processing architecture (O'Mullane et al. 2009; Lammers et al. 2009), a complex system written in pure Java, using Interprocess Communication with RMI and rendered graphically by a Web server via Java Server Pages (JSP). Another modern application is the HERSCHEL common software system (Ott 2010), designed using the latest web based software architectures.

An example of an old-style application is XMM-Newton Science Analysis Software (SAS) (Gabriel et al. 2004), which is likely to remain heavily used for at least the next 10 years. SAS was designed and developed in the mid 90's using the latest technologies at the time, such as, C++, FORTRAN 90/95 and Perl. Since the very beginning, SAS was thought to be a "stand-alone" application, compiled and distributed for multiple platforms (including SunOS, DEC, many Linux flavors and MacOS). It is clear that applications like SAS, have to evolve together with the new technologies (Gabriel et al. 2008; Ibarra et al. 2010; Ibarra et al. 2009).

Recently, the Swift X-ray telescope is offering through a web page the possibility to analyze point source data.[1] This web form allows the creation of X-ray light curves, spectra and positions of any object in the Swift XRT field of view. Images of the field of view can also be created. Also HEASARC has released *Web Hera*.[2] *Web Hera* introduces to the general user community a new way to use Ftools, no installation is required for the majority of the software. No longer required is the process of installing the HEASOFT suite on the user's local machine.

---

[1] http://www.swift.ac.uk/user_objects/

[2] http://heasarc.nasa.gov/webHera/

Also astronomy archives are changing from basic data bases with almost no metadata inside to complex data bases offering information through VO protocols (Fajersztejn et al. year).

Not only is the software paradigm changing, but also astronomer skills and needs are evolving. The scientific community has access to ever larger amounts of data in different wavelength regimes and with different characteristics. The analysis of these data requires, as a rule, complex installation procedures of various analysis systems, some expertise on the systems, and large computing resources, all of which can be highly demanding for most users.

In this paper we present the first public release of RISA, the XMM-Newton SAS software evolution towards new technologies and new astronomer profiles.

2.   RISA: the XMM-Newton Design

RISA is a client/server application, using web services technology, that wrap, in an easy and flexible way, the functionalities of the XMM-Newton SAS. RISA embedded SAS tasks that are executed the SAS tasks in a GRID environment, using GridWay[3] as a Grid meta-scheduler.

RISA software allows scientists to discover, download, and reduce on-the-fly, XMM-Newton data without having to install any project specific software and it uses all XMM-Newton SAS capabilities (parameter interface and image selection expressions). It has been coded in Java, using AJAX and SOAP technologies, and taking into account the Virtual Observatory paradigm, using VO protocols such as SIAP[4] or SAMP.[5]

In Figure 1 we show the RISA software design, which is based on a client/server application, using tomcat as the server application. Currently RISA is using Grid environments to run the SAS workflows, but it can be easily adapted to any other system architecture such as cluster or cloud computing. The application has been designed as a mission independent analysis tool, capable of implementing tasks and workflows from different missions.

As it can be seen in Figure 1, the system allows the user to search for any XMM-Newton data (pointing or slew observation) using SIAP protocols and name resolver services. After processing SIAP response, RISA client creates the main window with all the XMM-Newton observation available from the original request. Then the user creates the SAS workflow and when finished the client serializes the information and sends it to the RISA server. RISA server sends the job(s) to the Grid through DRMAA OGF standard (GridWay implementation). Each node in the Grid makes a request to the XMM-Newton Science Archive (XSA) to retrieve the XMM-Newton ODF (Observation Data File) data set, corresponding to a given observation ID, using the AIOClient (ESAC archive team tool to automatically download XMM-Newton data) application.

Once the data processing has finished, the results are taken to the Storage Element automatically by GridWay. The RISA service then knows that the observations have been processed and informs the client (job DONE). The results can be displayed with VOSpec, DS9 or Aladin opened through the SAMP protocol, i.e., without having to download the data.

The RISA client is able to work starting from raw XMM-Newton data or from pipeline pre-processed files. The user can create tailored workflows fully configurable or can also select pre-defined workflows that automatically produce XMM-Newton spectra or light curves. The results can be sent through SAMP messages to viewer applications such as VOSpec or DS9. Finally, the user can retrieve the data when the jobs have finished.

---

[3]http://www.gridway.org

[4]http://www.ivoa.net/Documents/SIA/

[5]http://www.ivoa.net/Documents/latest/SAMP.html

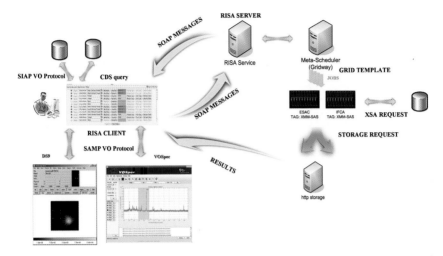

Figure 1. RISA architectural design. The Client application is in charge of the workflow creation and when finished, the information is sent through SOAP messages to the server. The server processes the request, send the jobs to the GRID and send back information to the client about the job status.

3. Summary

RISA is a true VO compliant application able to perform full data reduction of XMM-Newton observations in a GRID architecture. Our goal, using this approach, is to move beyond the paradigm of simply delivering products to providing a complete solution for the non-expert astronomer, offering a complete suite of programs to reduce and analyze XMM-Newton data or any other data set, without having to install any dedicated software and providing all functionalities in an easier way.

More and more astronomers are gathering astronomical data to cross-match results with catalogs in different energy ranges. This implies handling data from possible unfamiliar fields of astronomy and their associated software.

This new web interface to old software has been created to help non-expert astronomers to reduce and access astronomical data using flexible and intuitive applications. Providing a common and standarized framework that allows the user to reduce data from different energy ranges in a transparent way.

In the near future we will study the possibility of running these applications in a Cloud environment using middleware such as OpenNebula.[6]

References

Fajersztejn, N., et al. cpryear, in Astronomical Data Analysis Software and Systems XX, edited by I. N. Evans, A. Accomazzi, D. J. Mink, & A. H. Rots (San Francisco, CA: ASP), vol. 442 of ASP Conf. Ser., 29

Gabriel, C., Ibarra, A., de La Calle, I., Salgado, J., Osuna, P., & Tapiador, D. 2008, in Astronomical Data Analysis Software and Systems XVII, edited by R. W. Argyle and P. S. Bunclark and J. R. Lewis (San Francisco, CA: ASP), vol. 394 of ASP Conf. Ser., 183

---

[6]http://www.opennebula.org/

Gabriel, C., et al. 2004, in Astronomical Data Analysis Software and Systems XIII, edited by F. Ochsenbein and M. G. Allen and D. Egret (San Francisco, CA: ASP), vol. 314 of ASP Conf. Ser., 759

Ibarra, A., Calle, I., Gabriel, C., Salgado, J., & Osuna, P. 2009, in Astronomical Data Analysis Software and Systems XVIII, edited by D. A. Bohlender, D. Durand, & P. Dowler (San Francisco, CA: ASP), vol. 411 of ASP Conf. Ser., 322

Ibarra, A., Saxton, R., Ojero, E., & Gabriel, C. 2010, in Astronomical Data Analysis Software and Systems XIX, edited by Y. Mizumoto, K.-I. Morita, & M. Ohishi (San Francisco, CA: ASP), vol. 434 of ASP Conf. Ser., 293

Lammers, U., Lindegren, L., O'Mullane, W., & Hobbs, D. 2009, in Astronomical Data Analysis Software and Systems XVIII, edited by D. A. Bohlender, D. Durand, & P. Dowler (San Francisco, CA: ASP), vol. 411 of ASP Conf. Ser., 55

O'Mullane, W., Hernández, J., Hoar, J., & Lammers, U. 2009, in Astronomical Data Analysis Software and Systems XVIII, edited by D. A. Bohlender, D. Durand, & P. Dowler (San Francisco, CA: ASP), vol. 411 of ASP Conf. Ser., 470

Ott, S. 2010, in Astronomical Data Analysis Software and Systems XIX, edited by Y. Mizumoto, K.-I. Morita, & M. Ohishi (San Francisco, CA: ASP), vol. 434 of ASP Conf. Ser., 139

Part IV

Education and Public Outreach

Astronomical Data Analysis Software and Systems XX
ASP Conference Series, Vol. 442
Ian N. Evans, Alberto Accomazzi, Douglas J. Mink, and Arnold H. Rots, eds.
© *2011 Astronomical Society of the Pacific*

Data to Pictures to Data: Outreach Imaging Software and Metadata

Zoltan G. Levay

Space Telescope Science Institute,
3700 San Martin Drive
Baltimore, MD 21218, USA

Abstract.   A convergence between astronomy science and digital photography has enabled a steady stream of visually rich imagery from state-of-the-art data. The accessibility of hardware and software has facilitated an explosion of astronomical images for outreach, from space-based observatories, ground-based professional facilities and among the vibrant amateur astrophotography community. Producing imagery from science data involves a combination of custom software to understand FITS data (FITS Liberator), off-the-shelf, industry-standard software to composite multiwavelength data and edit digital photographs (Adobe Photoshop), and application of photo/image-processing techniques. Some additional effort is needed to close the loop and enable this imagery to be conveniently available for various purposes beyond web and print publication. The metadata paradigms in digital photography are now complying with FITS and science software to carry information such as keyword tags and world coordinates, enabling these images to be usable in more sophisticated, imaginative ways exemplified by Sky in Google Earth and World Wide Telescope.

1. Introduction

Over the last several years, a convergence between astronomy science and digital photography has enabled a steady stream of visually rich imagery from state-of-the-art data. Such imagery is popular not only among attentive and interested enthusiasts, but also with the general public (tax-payers), and even decision-makers (politicians). Because of the deep penetration of science imagery into popular culture, the most attractive, accessible presentation of science discoveries results in wider acceptance and acknowledgement of the importance of those results (and therefore can translate into more enthusiastic funding support).

The accessibility of hardware and software has facilitated an explosion of astronomical images for outreach, from space-based observatories, and ground-based facilities. But it is not limited to professional facilities. The relatively inexpensive accessibility of excellent optics, detectors, off-the-shelf image processing software and powerful computers means that the vibrant amateur astrophotography community is also routinely producing amazing images in increasing volume.

2. Producing Outreach Images

Producing color images intended for outreach starts with science data: separate black and white (grayscale) images made through different filters. Professional astronomical imaging instruments use grayscale detectors rather than detectors used in consumer digital cameras, which directly read out a multi-channel color image. Usually, multiple exposures have been used and combined to produce each filter exposure. Exposures may have been made at multiple pointings stitched together to build wider fields of view. Standard reduction and calibration is applied, as for science analysis. The main steps in producing color images are:

- Reduce data

- Register frames, stitch mosaics

- Scale data to optimize tonality

- Apply color and composite

- Adjust brightness, contrast, tonality color to maximize visible detail (information).

- Retouch artifacts

Initial processing depends on the nature of the data and images. Generally, routine pipeline processing of raw data suffices to produce calibrated, geometrically corrected images as a starting point for assembling an outreach image. In most cases, three images are required, from different spectral bands to reconstruct a color image. A dataset may represent an image mosaic to produce a wider-field panorama, in which case the individual frames would be stitched into the mosaic filter by filter. In any case the separate grayscale frames are registered before compositing. Rector et al. (2007) provides a more detailed and complete description of the philosophy and practical application of the techniques to produce presentation/outreach images from astronomy data.

The first step is to scale the data from the original (usually floating-point) in the FITS transforming into something usable for further processing. For this step we use FITS Liberator,[1] special-purpose software developed by ESA and NASA, primarily at ESA/Hubble. Originally developed as a plug-in to Adobe Photoshop, in its latest version (3.0) it is stand-alone software that outputs a standard TIFF image file. FITS Liberator allows significant flexibility in the transformation, including several transform functions (linear, log, asinh, etc.), in addition to black and white clipping and other scaling parameters. A GUI and visual display previews the effect of adjusting the parameters before saving a TIFF and optionally opening in the image editor.

Subsequent processing is accomplished using image editing software such as Adobe Photoshop.[2] The three images are scaled separately to maximize the tonal range in each image. This means setting white and black clipping points, and scaling to optimize visible structure throughout the gray levels in the image.

While much additional processing can recover structure and detail in the images, it is important to achieve a good tonal balance in the images initially, in order to reduce the noise levels, and certainly to avoid clipping the important bright areas in the image. For example, it is important to avoid clipping to white the brighter regions of nebulae or the cores of galaxies, but to achieve the brightest values in these areas while preserving "highlight" detail, and at the same time to maximize the brightness and contrast in the lowest values ("shadow" details) without overly enhancing the noise. By the same token, there is no reason to avoid clipping star images quite deeply as there is generally no important inherent structure in the stellar PSF for a presentation image.

To reconstruct color, different hues are applied to each image and the images composited together using the "Screen" layer blend mode. Using the three additive primary colors red, green and blue maximizes the color palette in the reconstructed image. In general, the lowest energy (reddest) image is assigned red, the highest energy (bluest) image is assigned blue, and the intermediate energy is assigned green. Any hue may be applied in principle, some variation may be used for aesthetic reasons to shift colors or to enhance particular features.

---

[1]http://www.spacetelescope.org/projects/fits_liberator/

[2]Photoshop is part of the industry-standard Creative Suite software developed by Adobe Systems, Inc. Photoshop includes extensive, powerful tools for editing digital images. Descriptions of using the software here should not be construed as an endorsement of a specific commercial product. Much of this sort of image processing can be accomplished in other software, including open-source software such as The GIMP. The terminology and paradigms embodied in Photoshop are useful for discussing these concepts.

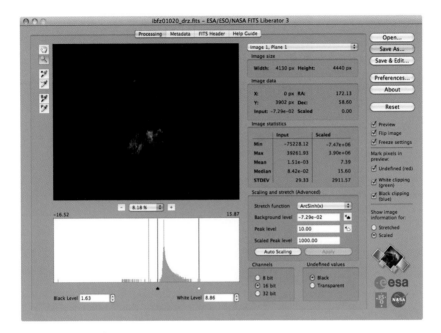

Figure 1.   Screen image of the FITS Liberator GUI.

This technique permits combining multiple images in different layers, with more flexibility than using the three primary color channels, including the use of more layers, and the possibility of using varying, subtle hues. One advantage of this technique is the opportunity to apply hues to the filter images that vary from the additive primaries, and which may not necessarily have any relationship to the color response of the filters and detectors used to make the observations. It is also possible to easily apply non-destructive adjustments to the separate images. Adjustments can be applied selectively to different spatial locations in the image or to selected ranges of tonal values, using selections, masks, and other techniques.

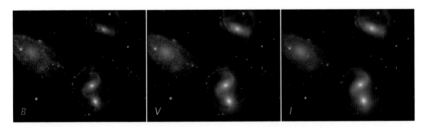

Figure 2.   Hubble Space Telescope Wide Field Camera 3 images of the galaxy group Stephan's Quintet in three broad-band visible-light filters; left: F439W (B), center: F555W (V) and right: F814W (I).

Beyond basic image compositing, established digital darkroom techniques make features in the data visible and art principles make science images more visually appealing. The kinds of adjustments we make are generally to improve the overall contrast and tonal range, with a sky background close to black, brightest areas such as cores of stars and galaxies a pure white. Color adjustments are intended to remove color casts and render features in "appropriate" colors,

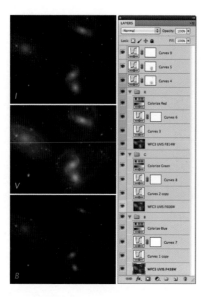

Figure 3.   Left panel: The same images as in Figure 2, with a different additive primary hue applied to each. Right: Photoshop layers palette representing separate image layers for each filter dataset and adjustment layers to change the brightness profile (curves adjustments) for each layer and for the composited image and apply hue to each filter layer (Hue/Saturation adjustments).

reddish HII regions, blue young clusters, red/yellow dust, etc. The intent of these adjustments is to avoid misrepresenting the data so as not to mislead the viewer. We will also retouch the images to remove telescope and instrument artifacts such as reflection ghosts, residual cosmic rays, hot pixels, etc., and judiciously apply some noise reduction and sharpening. In addition, we crop and rotate the images to present the visually most appealing composition, regardless of the orientation on the sky.

Figure 4.   Left: Initial color composite from HST WFC3 images of Stephan's Quintet rendered in hues assigned to datasets from separate filters. Right: The same image adjusted to improve the contrast, tonal range, and color.

The details of scaling, hue selection, etc., vary depending on the nature of the data. For example, broad-band filters respond differently from narrow band (nebular) filter images, or a combination of the two. Nevertheless, the overall visualization paradigm applies, not only to visible-light images, but any datasets that represent different physical conditions in the same spatial area. Effective images may be constructed spanning the entire available range of the spectrum, including X-rays, UV, visible, IR, and radio.

In some cases, the resulting images may be relatively close to a visually accurate rendering of the scene, provided the filter/detector response matches the human visual system model, given the enhancement provided by optics and detectors. In most cases, the rendering is a visualization of phenomena outside the limits of human perception. Nevertheless, the goal in producing presentation images remains to stay honest to the data, to avoid misleading the viewer, but also to produce an aesthetically pleasing image. The resulting images can be spectacularly beautiful as well as deeply informative. An attractive, aesthetically pleasing image is more likely to be viewed and appreciated by the general viewer, even though the subject matter may be unfamiliar, indeed quite abstract. We expect someone seeing a beautiful image would be more likely to want to learn more. Upon understanding the nature of the subject, the image takes on a deeper meaning, much like experiencing another dimension. Arcand et al. (2011) expand on viewers' reactions to presentation images, describing a study of public perceptions of astronomy visualization.

3. Beyond the Image

Beyond producing pretty pictures, what can we do with these outreach images? Each facility producing images for outreach has maintained a repository of its products, and serves these to the public with web services. In addition to the images, this includes captions, supplementary and explanatory graphics, animations, movies, podcasts, and much other content to provide a rich experience for the visitor. For example, NASA's three Great Observatories maintain extensive web sites devoted to serving information to the public, news media and educators.

- Chandra X-ray Observatory: `http://chandra.harvard.edu/`

- Hubble Space Telescope: `http://hubblesite.org/`

- Spitzer Space Telescope: `http://www.spitzer.caltech.edu/`

More recently there have been efforts to make these resources available more generally. There are now several desktop virtual telescope/planetarium software packages that make use of outreach images: Sky in Google Earth[3] and Microsoft WorldWide Telescope,[4] seamlessly incorporate images on the virtual sky, positioned and scaled correctly, and with supplemental, supporting information and links to the source sites. So far a somewhat ad-hoc process has been used to incorporate the imagery. A more robust, general solution is being developed. To accomplish this, the most important aspect is to develop metadata to describe each image, and make the content available.

3.1. Metadata

Metadata is vital to describe any image to make it useable in a variety of ways. To be most useful, the metadata must take into account how the image might be used. Because it has become important to mainstream digital photography, standard image formats such as TIFF and JPEG, and proprietary formats for "raw" digital photographs now make use of standardized paradigms for incorporating image metadata useful for digital photography. At the same time, there is a

---

[3]`http://www.google.com/earth/`

[4]`http://www.worldwidetelescope.org/`

recognition among providers of outreach content that only by expanding the associated information about the images through metadata can images and other content be made more useful to the consumers. Conceptually, this overlaps the functionality of the Virtual Observatory, although the needs of the outreach community and the professional astronomy community diverge in the details.

The Virtual Astronomy Multimedia Project[5] (VAMP) aims to define the metadata paradigm for outreach imaging as distinct from the uses of archived observatory data and mainstream digital photographs. In addition, VAMP aims to generalize the paradigm to include other outreach-oriented content such as informational graphics, animations, 3D visualizations, etc. The paradigm includes Astronomy Visualization Metadata (AVM) schema defining tags, including an outreach-oriented astronomical subject taxonomy and world coordinate system (WCS) tags, interactive tools, web and scripting resources to enable incorporating metadata tags in images, and tools to associate WCS with outreach images. searches. AVM metadata tags comprise several categories:

- Creator Content: information about the producer of the item
- Observation: information about the observation dataset(s)
- Coordinates: world coordinate system
- Publisher: information about the distributor of the content
- FITS Liberator: parameters used to scale the original data

3.2. Tools

The sort of information needed to associate with the images comes from various, diverse sources, not limited to the headers of the source data. Some of the metadata are generic and can be compiled once and inserted via "boilerplate" templates. Composite images usually comprise multiple, diverse datasets, even different observatories, so it may not be so simple to assemble the necessary pieces that are unique to each image. For outreach content other than images, there may not even be an original data source from which to draw metadata. Therefore, one aspect of the VAMP initiative is to develop and pool resources to help providers tag their content with metadata keywords. This includes capabilities of FITS Liberator to copy metadata values from the input FITS image header into metadata tags in the output (TIFF) image, a web-based form to manually fill in fields and produce a file containing metadata tags, and scripts to convert text-based (CSV) tables into metadata formats.

Extensible Metadata Platform (XMP) is a variant of XML developed by Adobe to facilitate interaction with image metadata for digital photography and graphic arts. Adobe software includes facilities for populating metadata including File Info panels in Photoshop and other Creative Suite applications, and metadata panels in Bridge and Photoshop Lightroom. Extensions are available that implement the VAMP metadata tags in the XMP paradigm.

3.3. WCS Resolution

World Coordinate System (WCS) metadata are the key element to enable positioning images properly within any representations of the sky, registering images of the same location from disparate sources, and provides a means to find images by coordinate searches. However, it is not entirely straightforward to establish and embed this information since the processed outreach images rarely conform to the spatial scope of the original image data. For aesthetic reasons the images are cropped, rotated, resampled with respect to the originals (flying in the face of traditional astronomy, which prefers to orient celestial images with north up).

While FITS Liberator can copy WCS metadata from the input FITS to the output TIFF, this has limited usefulness because Photoshop and other image editing tools do not recognize

---

[5]http://www.virtualastronomy.org//

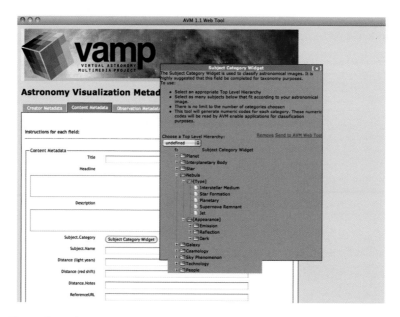

Figure 5.   The web-base AVM tagging tool to manually populate metadata tag values. The inset shows the subject category widget which generates the proper subject category code from user-selectable entries.

the WCS and so will not modify it to take into account any geometric changes to the image such as cropping or rotation. In addition, in most cases, more than one dataset was used to construct the composite image, and the datasets are scaled and imported one by one.

Astrometry.net[6] is a simple, powerful utility for resolving world coordinates of arbitrary astrophotographs with no user input other than the image. It does depend on catalogs to identify sources and resolve coordinates, which make it limited to relatively large fields of view and requires numerous cataloged features to be in the image. Many outreach images, in particular those from HST, comprise a very narrow field of view and often include few cataloged sources, so the WCS cannot be resolved this way. (As an interesting aside, a connection between astrometry.net and flickr.com has been established. *Flickr* is one of the largest online photo-sharing and social-networking sites. Photographers around the world upload photos, discuss them, and make social connections. For any photograph posted to the *flickr* group "Astrometry" the proper WCS will be submitted to astrometry.net via the *flickr* API. If an astrometric solution is accomplished, the coordinates and identified features will be posted in the notes and comments features on the photo's *flickr* page.)

One useful tool for resolving astrometric solutions for presentation images is Pinpoint WCS[7] (original algorithm and GUI development at STScI, and further development with a new multi-platform GUI at CfA). Pinpoint displays a FITS image and associated TIFF or JPEG, allows the user to identify the same features in both, derives a WCS and outputs into a FITS header and/or AVM keywords, importable into image metadata.

---

[6]http://www.astrometry.net/

[7]https://www.cfa.harvard.edu/~akapadia/pinpointwcs/

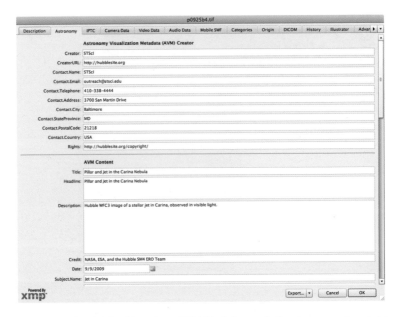

Figure 6.   The Adobe Photoshop CS5 File Info panel allowing manual metadata entry. The tabs at the top select different pre-defined metadata panels. The Astronomy panel is defined in XMP extensions provided by VAMP. The other tabs are default panels.

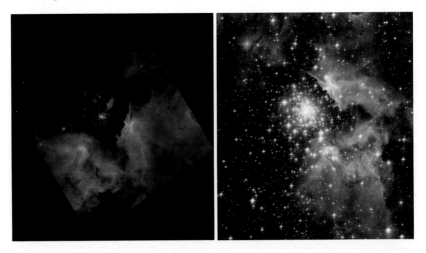

Figure 7.   Left: HST/ACS image of star-forming region NGC 3603, oriented north-up. Right: A cropped, rotated, and resampled presentation image derived from the same HST data. The WCS of these two images is not the same.

3.4.   Delivery

The ultimate goal for outreach, and part of the larger VAMP effort is to make available as much as possible of the store of outreach content to as wide an audience as possible. One approach

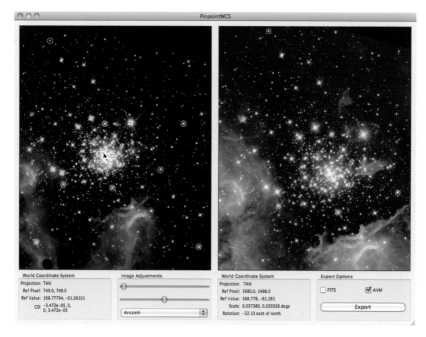

Figure 8.   Pinpoint WCS GUI; top left: FITS image display with WCS from the header and image adjustment controls below. Top right: color composite TIFF display with derived WCS and export options below. Yellow circles in each image correspond to matching user-selected reference features in the two images used to compute the new WCS.

is the potential development of a repository of education and public outreach content. Whether this is a single repository/archive to which producers will provide content, or a service linking to distributed content archives is still under development. Individual providers will certainly continue to serve their content via the web using their separate facilities. Including enhanced metadata will enable the content to be used in many ways, not necessarily even imagined yet. In addition, it is possible to provide alternate access to the archives via existing and developing facilities through VO services. Finally, a common repository would enable "one-stop shopping" by consumers of E/PO content.

References

Arcand, K. K., Watzke, M., Smith, L. F., & Smith, J. K. 2011, in Astronomical Data Analysis Software and Systems XX, edited by I. N. Evans, A. Accomazzi, D. J. Mink, & A. H. Rots (San Francisco, CA: ASP), vol. 442 of ASP Conf. Ser., 179

Rector, T., Levay, Z. G., Frattare, L. M., English, J., & Pu'uohau-Pummill, K. 2007, AJ, 133, 598

Astronomical Data Analysis Software and Systems XX
ASP Conference Series, Vol. 442
Ian N. Evans, Alberto Accomazzi, Douglas J. Mink, and Arnold H. Rots, eds.
© 2011 Astronomical Society of the Pacific

The Aesthetics of Astronomy:
Exploring the Public's Perception of Astronomy Images and the Science Within

Kimberly Kowal Arcand,[1] Megan Watzke,[1] Lisa F. Smith,[2] and Jeffrey K. Smith[2]

[1]*Chandra X-ray Center, Smithsonian Astrophysical Observatory, 60 Garden Street, Cambridge, MA 02138*

[2]*University of Otago*

Abstract. Every year, hundreds of astronomical images are released to the public via the efforts of professional education and public outreach (EPO) specialists, as well as from scientists themselves and amateur astronomers. Each of these images represents a variety of decisions from the individual or team that assembled it. These choices include cropping, color, contextualization, and many more. But how effective are these choices in both engaging the public's interest and communicating scientific information to them? In 2008, the Smithsonian Astrophysical Observatory began a unique research study — dubbed the Aesthetics & Astronomy (A&A) project — to examine these very questions. The study included images from across the electromagnetic spectrum and probed the effects of the scientific and artistic choices in processing astronomical data. This paper provides an overview of the results of the preliminary 2008 A&A study and includes a synopsis of the direction of the ongoing 2010 study.

1. Background

More than 400 years after Galileo Galilei deployed his first telescope in Italy, society enjoys what many consider to be a "golden age" of astronomy. Between the telescopes in space and those on the ground, scientists have access to information on the Universe that truly spans the electromagnetic spectrum using sophisticated software, hardware, and other technologies.

This abundance of observatories has also coincided with the dawn and now domination of the Internet. Data of all types are transmitted electronically with abundant ease and virtually omnipresent access. Astronomy has benefited from this development. No longer does the public have to travel to a science center or museum to see the latest imagery. Rather, they simply look at their computer or smart phone. New and freely downloadable tools now also exist for connected members of the public to explore astronomical data and even create their own images.

Therefore, it is plausible to say that a new era of an accessible Universe has been entered, in which people can participate and explore like never before. But there is a lack both of robust studies to understand how people — particularly non-experts — perceive these images and the information they attempt to convey. Virtually all astronomical images for public consumption have been processed to varying degrees through color mapping, artifact removal, smoothing, and/or cropping. These techniques — applied by scientists, or astronomical imaging experts, or a combination — are used to attempt to strike a balance between the science being highlighted and the aesthetics designed to engage the public. The extent, however, to which these choices affect the non-expert's perception, comprehension, and also trust, has never been rigorously studied. (Some informal surveys have been conducted, but a search of the literature produces a scant amount of data.) To address this dearth of proper research and analysis on these topics, a group at the Smithsonian Astrophysical Observatory, along with colleagues at the University of Otago in New Zealand, initiated the Aesthetics and Astronomy (A&A) project.

Figure 1.   Beautiful art or not? Left: "Untitled" serigraph on paper by Gene Davis, 1974. Credit: Smithsonian American Art Museum, Bequest of Florence Coulson Davis; Right: Multiwavelength NGC4696 in X-ray, Radio and Infrared, Credit: X-ray: NASA/CXC/KIPAC/S. Allen et al.; Radio: NRAO/VLA/G. Taylor; Infrared: NASA/ESA/McMaster Univ./W. Harris (used in 2008 online survey).

1.1.   Introduction to Aesthetics in Astronomy

> *def., Aesthetics* is the study of how human beings react in sensory and emotional fashion to the things we encounter in life, especially as being appealing or not appealing.
> — Smith, L.F., & Smith, J.K. (in press). Aesthetics. In I. Weiner & E. Craighead (Eds.). *Corsini's Encyclopedia of Psychology (4th Ed.).* John Wiley and Sons.

Aesthetics from a psychological perspective is the study of all things beautiful whether art or not, and all things art whether beautiful or not.

The Aesthetics & Astronomy group (A&A) combines the perspectives of professional astronomy communicators, astrophysicists, and experts in psychology and aesthetics. In 2008, A&A began when it conducted online studies and a series of in-person focus groups. The images came from a variety of space & ground-based observatories across wavelengths (Chandra, Hubble, Spitzer, Very Large Array, Hinode, and others).

The following research questions were asked in the development of the A&A project:

- How much do variations in presentation of color, explanation, and scale affect comprehension of astronomical images?

- What are the differences between various populations (experts, novices, students) in terms of what they learn from the images?

- What misconceptions do the non-experts have about astronomy and the images they are exposed to?

- Does presentation have an effect on the participant, whether aesthetic or in terms of learning?

2.   Highlights of the 2008 Study

In Autumn 2008, an invitation to participate in the web-based surveys and experiment was advertised on well-trafficked, public-friendly web sites such as NASA's Astronomy Picture of

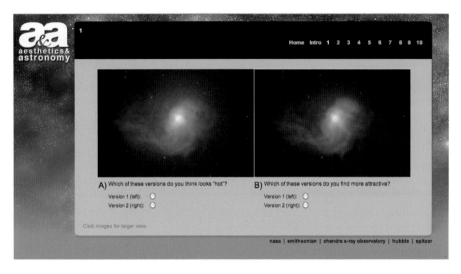

Figure 2.   First question exploring the red-blue/hot-cold correlation with NGC 4696 X-ray/Infrared/Optical.

the Day (APOD)[1] and the Chandra X-ray Observatory[2] web site. These web sites featured a box with an explanation of this project and an invitation to participate in an online survey. Clicking on the box led to the first page of the web-based survey/experiment.

The web-based survey/experiment began with an introductory page describing the study and containing an informed consent protocol. A control was included to prevent multiple responses from the same computer address. Following the introductory page, participants were given a short list of demographic items: age, gender, highest level of education, self-rating of expertise in astronomy on a 1 (complete novice) to 10 (expert) scale, residence, and familiarity with the APOD web site (which was expected to, and did, generate the largest number of responses).

Each participant then accessed the first of a series of items, each depicting an astronomical image. The images were presented with text or no text, variations in color, background imagery or no background imagery, and/or the presence or absence of scales. Participants were re-randomized for assignment to conditions for each item in the survey. After viewing an image with or without accompanying text, participants responded to various questions. The deep space images were taken from the Chandra, Hubble, Spitzer, GALEX, Very Large Array (VLA), and Hinode satellites and telescopes, among others, and included G292.0+1.8, M33 X-7, Whirlpool Galaxy (M51), and NGC 4696, and the Sun's poles. Latencies were collected to see how the experimental conditions influenced viewing times.

In the focus group portion of the study, visitors were solicited through advertisements in local papers, online events calendars, and personal contacts. None of the focus group participants had participated in the web-based survey. There was a total of n = 31 volunteers for the focus groups: local high school students (n = 8), lay public (n = 8), local high school science teachers (n = 6), and astrophysicists and astronomers who work on image development (n = 9). Semi-structured protocols were written to guide the discussion for the focus groups. The discussion questions centered on two images: Messier 101, and an image of the spicules on

---

[1] http://apod.nasa.gov

[2] http://chandra.si.edu

the pole of the Sun with superimposed scales intended to aid comprehension (see Appendix A). Participants also completed a paper-and-pencil activity with five images (see full published report for details and appendices[3]).

The focus group participants completed several activities in one 2-hour session per group at the Harvard-Smithsonian Center for Astrophysics in Cambridge, Mass. Using the two images, they responded to semi-structured items designed to explore how they viewed the image, their comprehension and misconceptions of the image and the underlying science associated with it, and their desires to learn more about the image and how it was made. The focus group of astrophysicists was also asked to discuss how they believed they communicate the science behind the images to non-experts.

The following sections present some highlights of the 2008 study. A complete report of the entire 2008 study, along with additional information, can be found online.[4]

2.1.   NGC 4696 and Color Mapping

NGC 4696 is a large elliptical galaxy in the Centaurus Galaxy Cluster. A vast cloud of hot gas (X-ray) surrounds high-energy bubbles (radio) on either side of the bright white area around the supermassive black hole. The point sources show infrared radiation from star clusters on the outer edges of the galaxy.

Participants were asked if they found one image more attractive than the other and if they thought one image looked hotter in temperature than the other. For the entire sample, the following results were analyzed:

- 71.4% thought the red was hotter than the blue.

- 53.2% found the blue more attractive than the red.

In terms of responses for the total sample, there was a clear indication of which looks hotter (the red). Almost 3 out of 4 chose red. If just the experts are segmented out, it was shown that they tended to like the red image better, and they reported that the blue image represented a hotter temperature as compared to the red.

In terms of color mapping astronomical images, most astronomical experts typically assign blue as a hot color and red as a colder color. However, when the public typically thinks hot, they think red, and not blue. They might understand that the hottest part of a flame is the blue part, but red still means fire.

Is the use of the red version here however, perpetuating a misconception? Is it the responsibility of the scientist to "correct" or disabuse notions of the public (that red is not as hot as blue). The primarily red image might actually convey the heat of this object better even though its color mapping does not follow the more "correct" version of chromatic order (red, green, and blue for low, medium, and high respectively). Does it make more sense to use the version that aligns with a cultural norm (red as hot), than to try to disabuse that notion?

2.2.   G292 and Text Comprehension

The study also investigated the effect of text on comprehension. First, we used a 2 × 2 design, where we randomly assigned participants to having text or not having text, and either seeing an image with stars or seeing that image without stars. Here are the four conditions: This is a young supernova remnant located in our galaxy. We showed the image with text, stars, no text and no stars.

A series of questions was then asked:

- How does it look, and how attractive is it to you?

---

[3]http://arxiv.org/abs/1009.0772

[4]http://arxiv.org/abs/1009.0772

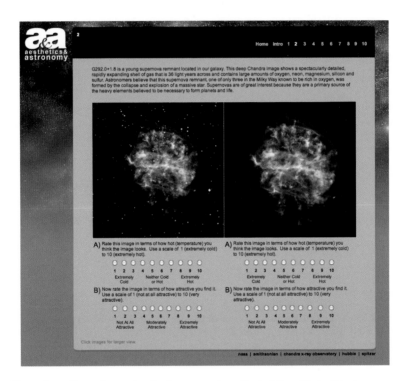

Figure 3.   Second question in the survey investigated the effect of text on comprehension using the young galactic supernova remnant G292 in X-ray (right) and X-ray/Optical (left) light.

- Then, in the text condition, the participant was asked how carefully he/she read the text.
- The participant was asked what it was, how big they thought it was and how sure they were of their answers

The results showed that people generally had a good idea what the object was, and that as expected, text matters, especially to non-experts.

An unexpected finding was the following: When asked how attractive the image was, there was a significantly higher rating for the image with the text, no matter which image was shown. This held up across all levels of expertise. And, experts found both images less attractive than novices, whether with or without text.

2.3. Overall Outcomes

- Providing context for the image is critical to comprehension.

- Experts prefer text that is shorter/to the point; novices prefer narrative expository style to accompany image.

- A sense of scale with the images is helpful for comprehension at all levels of expertise.

- Experts and novices view the images differently. Novices begin with a sense of awe/wonder, and focus first on the aesthetic qualities. Experts wonder how the image was produced, what information is being presented in the image, and what the creators of the image wanted to convey.

- Experts are much more likely to view blue as hot than are novices; about 80% of novices see red as hot compared to 60% of experts.

More details from 2008 study are available in the SAGE Journal of Science Communication[5] as well as an upcoming issue of Communicating Science with the Public.[6]

3. Direct Implementation

Within a few months of participating in the focus groups, the Education and Public Outreach (EPO) group at the Chandra X-ray Observatory began work on implementing some of the specific shared comments, particularly those expressed from the younger student group. Some of these applications on the Chandra web site were relatively straight-forward, such as including bulleted text for overview and quick comprehension in the body of new images releases and providing more "Wikipedia-style" links for the body of the text of new image releases. The second step was developing a simple interactive multiwavelength image feature which would allow the user to do right on the web page as had been expressed in the focus group — move from one color or wavelength to the next to "build" the complete image composite so they could "see" how it was made. In addition, an interactive, question-based text script was built into the Chandra "photo" pages with click-tracking methods to count the user clicks per question and per image, and to compare totals.

Each of these changes came out of the feedback that was received during the A&A online survey and focus groups. The feedback from the public on these relatively simple changes to the website have been overwhelmingly positive, through the comment and rating sections. The next step is to implement a questionnaire on the Chandra website to ask users specifically how these new features affect their enjoyment and comprehension of the image and the science behind it.

The Chandra EPO also approached a new series of print products with a similar implementation featuring multiwavelength astronomical images. Here, a familiar series of questions

---

[5]http://tiny.cc/t2mhx

[6]http://www.capjournal.org/

Figure 4.   Screenshot of the Chandra web site with the simple "interactive" multiwavelength image feature implemented, also show is the bulleted text design and additional links in the body of the text.

— who, what, when, where, why, and how — were used to engage the viewer in an approachable manner. The text highlights some of the content that was commonly asked for during the focus groups including how the images were made, the historical importance of the object, the location in the night sky, etc. Data collection and a brief summative evaluation of these six posters are being conducted to analyze the impact of the improved features on the public's understanding.

4.   Current and Future Studies

The A&A group is currently conducting a series of studies, funded in part by a grant from the Smithsonian Institution, that ask viewers to evaluate astronomical images with their corresponding descriptions across different media platforms: web, mobile, traditional print, and large format print. The images being used include those from the Chandra X-ray Observatory, Hubble Space Telescope, Spitzer Space Telescope, Solar Dynamics Observatory and others. Working with museum professionals and science center partners we have produced a traveling exhibit of the material. Touring through six locations in 2010, this exhibit allows participants to access the astronomical imagery and text through traditionally sized and large scale prints.[7] An online study of the same material was launched[8] to test user's perceptions on mobile devices in comparison with traditional online platforms.

Also, the A&A project employed in-person focus groups in December 2010 to explore the aesthetics-context correlation further, across all four of the platforms. Questions on interpretation of scientific principles (perception of temperature, for example), aesthetic appeal, and interpretation of unfamiliar (meaning non-terrestrial) objects are being included in all forms of the study.

While the 2008 and 2010 studies have brought forth many intriguing results thus far, this area of research is still considered to be rather unexplored. Therefore, there are many potential paths for future exploration. Some of the ideas for future studies being considered by the A&A group include:

- Eye tracking study with experts and non-experts comparing to existing baselines for art images. How does information influence tracking behavior?

- Study the intersection of misconceptions and images — what tends to increase misconceptions, and what decreases them?

5. Conclusion

Members of the professional astronomical community are operating in an unusual age. At the moment, an amazing amount of data from a multitude of advanced observatories is available to both professional astronomers and also the general public. At the same time, however, issues around "false color" and what is "real" in this age of Photoshop and other digital manipulation are widely spread, sometimes in the context of the very astronomical images that have been so carefully processed and preserved. It is the goal of the A&A project to examine the most effective ways of communicating these exciting discoveries through the aesthetic appeal astronomical images can offer. With so much data and so many tools at astronomy's disposal, are the best possible practices being employed? Potentially, studies such as A&A can uncover ways to help dispel some of the misinformation that exists about the veracity and legitimacy of what is distributed to the public. Colleagues in any of the related disciplines to the A&A project that are interested are encouraged to contact the authors of this paper for further discussion on these topics.

Acknowledgments.   The 2008 study was developed with funding from the Hinode X-ray Telescope, performed under NASA contract NNM07AB07C, and the Education and Outreach group for NASA's Chandra X-ray Observatory, operated by SAO under NASA Contract NAS8-03060. Additional A&A team members include Randall K. Smith and Jay Bookbinder of SAO as well as Kelly Keach of University of Otago. Special thanks to Jerry Bonnell and Robert J. Nemiroff, the authors of the NASA Astronomy Picture of the Day web site. Findings from this research were presented at the 2009 and 2010 Annual Meetings of the American Psychological Association (Smith et al. 2009; Smith et al. 2011), the XXI Congress of the International Association of Empirical Aesthetics (Smith et al. 2011), in a public inaugural professorial lecture at the University of Otago (Smith 2009), at an astronomy visualization symposium at the California Academy of Sciences (Arcand 2009) and at the Astronomical Society of the Pacific conference (Arcand et al. 2010).

---

[7]Visit http://astroart.cfa.harvard.edu/ for the schedule of locations.

[8]http://chandra.si.edu/mobile/aa.html

References

Arcand, K. K. 2009, Aesthetics and Astronomy. Presented at the Astronomy Visualization Conference, California Academy of Sciences, San Francisco, CA

Arcand, K. K., Smith, L. F., Smith, J., Watzke, M., Hove, K. H. T., & Smith, R. 2010, in Science Education and Outreach: Forging a Path to the Future, edited by J. Barnes, D. A. Smith, M. G. Gibbs, & J. G. Manning (San Francisco, CA: ASP), vol. 431 of ASP Conf. Ser., 139

Smith, L. F. 2009, Aesthetics and Astronomy: What do we see? Inaugural Professorial Lecture, University of Otago, Dunedin, NZ

Smith, L. F., Smith, J. K., Arcand, K. K., Smith, R. K., Bookbinder, J., & Keach, K. 2011, Science Comm., in press

Smith, L. F., Smith, J. K., Arcand, K. K., Smith, R. K., & Holterman ten Hove, K. 2009, Aesthetics and Astronomy: How experts and novices perceive astronomical images. Presented at the annual meeting of The American Psychological Association, Toronto, Canada

Astronomical Data Analysis Software and Systems XX
ASP Conference Series, Vol. 442
Ian N. Evans, Alberto Accomazzi, Douglas J. Mink, and Arnold H. Rots, eds.
©*2011 Astronomical Society of the Pacific*

The COSPAR Capacity Building Initiative

Carlos Gabriel,[1] Peter Willmore,[2] Mariano Méndez,[3] Pierre-Philippe Mathieu,[4] Ondrej Santolik,[5] and Randall Smith[6]

[1]*European Space Astronomy Center, ESA, E-28691 Villanueva de la Cañada, Madrid, Spain*

[2]*University of Birmingham, UK*

[3]*University of Groningen, the Netherlands*

[4]*ESRIN / ESA*

[5]*IAP, Academy of Sciences, Czech Republic*

[6]*SAO, USA*

Abstract. The COSPAR Capacity Building Workshops have been conceived to meet the following objectives: (1) To increase knowledge and use of public archives of space data in order both to broaden the scope of research programs in developing countries and to ensure that scientists in those countries are aware of the full range of facilities that are available to them; (2) To provide highly-practical instruction in the use of these archives and the associated publicly-available software; and (3) To foster personal links between participants and experienced scientists attending the workshops to contribute to reducing the isolation often experienced by scientists in developing countries. Since 2001 a total of twelve workshops have been successfully held in different scientific areas (X-ray, Gamma-ray, Space Optical and UV Astronomy, Magnetospheric Physics, Space Oceanography and Planetary Science) in nine developing countries (Brazil, India, China, South Africa, Morocco, Romania, Uruguay, Egypt and Malaysia). In this contribution we discuss the modalities of the workshops, the experience so-far gained, and the future including collaborations with other institutions sharing the aim of increasing the scientific activities in developing countries.

1. Introduction

For a couple of decades we have now been in what could be called the era of information dissemination. In the scientific world it has become normal practice that data, as well as specific data analysis tools, are often public and accessible, e.g., through web services.

A very collaborative attitude has also developed throughout the years, in the specific case of astronomy, from the public observatory concept in astrophysics to a high development of open archives, with the culmination in the Virtual Observatory paradigms.

All this is motivated by, and contributing to, a maximization of the scientific exploitation of the enormous amounts of data we are collecting in the different observatories, experiments, and/or simulations.

This situation offers a unique opportunity to developing countries to increase local scientific research in a way which is not linked to a large expenditure, otherwise necessary for carrying out basic science activities, but especially basic space science activities. It is well recognized that a certain level of basic science research is fundamental for a country in order to have productive research in applied sciences. Beyond that, the idea of universality of scientific

189

enquiry calls for dedicated efforts to achieve a broad access to scientific knowledge by everyone in the world.

The situation is therefore a typical win-win opportunity, with data and analysis tools providers interested in broadening their "customers", while there is a need of data and tools in countries with no participation so far in several basic (space) science fields, traditionally reserved to the "developed world".

Looking for the obstacles to achieving a larger exploitation of scientific data and tools in the developing countries, we find that one of the main reasons is simply the lack of awareness of the possibilities available for scientific research based on those data and tools. This is especially true for space sciences since several of the related fields of research are not among traditional fields of study in developing countries.

This situation can be improved through an educational program. This was recognized and proposed as a program to COSPAR ten years ago. The COSPAR Program of Capacity Building Workshops started in 2001, with the main aim of encouraging scientists from developing countries to use scientific data from space missions for active research and to build lasting bridges between scientists.

2.   The CBP Workshops — Selection

The Program is not directed but relies on proposals from recognized scientists within the international space science community. Proposals for workshops are evaluated according to a series of criteria in different categories (none of them representing an absolute go/no-go criterion, but with a certain weight in the final decision), such as:

- **General** Is the science involved within COSPAR's scope?

- **Data** Are space data involved? Are the missions involved current? Are the missions producing exciting results? Do the missions have open Guest Observer programs?

- **Archives and analysis**What is the quality for the archives involved (size, free, accessible)? How is the availability and quality of specific free software analysis? Is there an associated Help Desk?

- **General location** Has the workshop a regional character? Are the data suitable for research in the region? To what extent will the workshop be leading to active research? Will the workshop strengthen cross-country links within the region? What is the size of the relevant community in the region and/or host country?

- **Local characteristics** What is the standing of proposer/team, the quality of venue (internet bandwidth, number of computers, audio-visual facilities)?

- **Funding** Does funding come from both local and international sources?

3.   The CBP Workshops — Practicalities and Learning Needs

The two-week duration of the workshops is a compromise between need and financial affordability, taking into account that the aim is to organize an average of three workshops every two years, covering several disciplines.

The learning needs cover a broad range, starting in many cases with the type of science involved, when the specific area is not at all or not well known to the targeted audience or not exploited in other ranges. Instruments and data characteristics are very important and often completely new to the participants. The use of archives is fundamental for a pre-investigation and as the source of data. Key in the learning process for the future work are the specific analysis software packages, the main tools used during the workshop, together with certain analysis techniques, which can be specific to the research area (e.g. forward-folding fitting in X-ray astronomy).

While the whole program of the workshops includes lectures on missions and scientific instruments, software, science, etc., the central point is the individual research project every student is going to work on. This is where all the newly acquired knowledge has to be applied under the direct supervision of an expert in the field, who usually also acts as a lecturer in the workshop. The project is performed on a dedicated workstation with full access to programs, data and bibliography. Around 50% of the (official) time of the workshop[1] is spent on the project, which has to be discussed at the end of the workshop in form of a slide presentation or a poster.

4. The Program so Far and What We Have Learned

During the first ten years of the program a total of twelve workshops have been succesfully held in nine different countries, in very different disciplines (see Fig. 1).

Year	Topic	Missions	Local proj.	Where?
2001	X-ray Astronomy	Chandra, XMM-Newton		INPE, Brazil
2003	X-ray Astronomy	Chandra, XMM-Newton	Astrosat	Udaipur, India
2004	Magnetospheric Physics	Cluster	Double-Star	Beijing, China
2004	X-ray Astronomy	Chandra, XMM-Newton	SALT	Durban, South Africa
2005	Space Oceanography	GEOS		CRTS, Rabat, Morocco
2007	Solar-Terrestrial Interactions	Cluster	PECS Programme	Sinaia, Romania
2007	Planetary Science	PDS, PSA	SAPC	Montevideo, Uruguay
2008	X-ray Astronomy	Chandra, XMM & Swift		Alexandria, Egypt
2008	Optical and UV astronomy	Hubble, Fuse, Galex	Astronomy strategy	Kuala Lumpur, Malaysia
2009	Lunar & Planetary Surface Science	Rosetta & diverse Mars missions	Chang'e I, Chandrayaan-1	Harbin, China
2010	Gamma-ray Astronomy	Fermi	Astrosat	Bangalore, India
2010	Earth observation: water cycle	SMOS		Fortaleza, Brazil
2011	Earth observation: atmospheric aerosols	several, eg. MODIS, MISR, TOMIS, ENVISAT		Greater Noida, India
2011	X-ray Astronomy	Chandra, XMM & Suzaku		San Juan, Argentina

Figure 1. The workshops held so far (in purple) and the ones accepted for 2011 (in blue).

On the positive side we have observed in all workshops that most participants can handle an extremely steep learning curve, with a very efficient knowledge transfer. The format of the workshop contributes to fostering good contacts between students and lecturers. In most cases we feel that the participant at the end of a workshop is provided with basic tools for getting involved in international (space) research projects and collaborations.

A few problems that have been encountered have been, for instance, the limitations in bandwidth available at the sites of the workshops, although in our experience this has become less of a problem with time. Also in some of the workshops the competence in English of some of the students and/or the lack of experience with Linux can be limiting.

Overall, the program as a whole represents a win-win situation, both for participants accessing new science, and for missions reaching a wider community. Moreover, it is an extremely

---

[1]Most students tend to spend long hours working on their projects, outside the "official" hours, if conditions allow.

valuable experience for lecturers and providers of data and tools, who get confronted with needs and limitations that are quite different from the ones they encounter in their "traditional" communities.

5.   The CBP Associated Fellowship

In order to make the program yet more effective through "follow-up" measures, we have started in 2008 with a fellowship initiative. The fellowships should enable participants of a CBP workshop to build further on skills gained there, through a two to four weeks visit to carry joint research in a collaborating lab.

These visits should not be seen as a training exercise, but as intended to foster research collaborations. Therefore it is the quality of the proposed research that is the fundamental criterion for successful selection. The program has to be presented jointly by the candidate and the receiving scientist for approval. Around 20 institutions in several European countries and in the USA, but also in India and China, are collaborating with this initiative.

6.   Summary

The COSPAR Capacity Building Program, born around 10 years ago, is successful and expanding. It represents a rewarding experience for participants, countries/regions involved, and also lecturers. We are trying to maintain the commitment level of COSPAR, the supporting institutions, the developing countries, and the individuals involved in organization and execution. At the same time we are working to further broaden the scope of the workshops, while keeping the essential elements intact.

Part V

Grid and Grid Virtualization

Astronomical Data Analysis Software and Systems XX
ASP Conference Series, Vol. 442
Ian N. Evans, Alberto Accomazzi, Douglas J. Mink, and Arnold H. Rots, eds.
© *2011 Astronomical Society of the Pacific*

DEISA: A Distributed European HPC Ecosystem for Astrophysics Research

Wolfgang Gentzsch

The DEISA Project and Open Grid Forum, 93073 Neutraubling, Germany

Abstract. The paper presents a brief overview of the EU project DEISA, the Distributed European Infrastructure for Supercomputing Applications. We describe its vision, mission, objectives, infrastructure and services offered to the e-science community. The DEISA Extreme Computing Initiative for supercomputing applications is highlighted, and four grand-challenge projects of astrosciences simulations are presented: modeling of turbulent astrophysical flows, simulating the Local Universe, turbulent, active, and rotating stars, and simulating the formation of the Milky Way.

1. The DEISA Project

Fifteen years ago, grid infrastructures for pooling and sharing of compute resources and for fostering collaboration among scientists started to emerge world-wide. Powerful grids of continental scope evolved, such as TeraGrid and Open Science Grid (OSG) in the US; EGEE, EGI and DEISA in the EU; NAREGI in Japan; and APAC in Australia. Today, scientific grids have matured, their scope was extended, and "e-Infrastructures" or "Cyberinfrastructures" have been built up.

The DEISA Consortium has deployed and operated the "Distributed European Infrastructure for Supercomputing Application" (DEISA 2010b), co-funded through the EU FP6 DEISA project from 2004 to 2008. Since May 2008, the consortium has continued to support and further develop the distributed HPC infrastructure and its services through the EU FP7 DEISA2 project with funds until 2011. Eleven DEISA members are from seven countries: BSC (Barcelona, Spain), CINECA (Bologna, Italy), CSC (Espoo, Finland), ECMWF (Reading, UK), EPCC (Edinburgh, UK), FZJ (Juelich, Germany), HLRS (Stuttgart, Germany), IDRIS (Orsay, France), LRZ and RZG (Garching, Germany) and SARA (Amsterdam, The Netherlands). Further centers were integrated as associate partners: CEA-CCRT (France), CSCS (Manno, Switzerland), and KTH (Stockholm, Sweden). Activities and services relevant for applications enabling, operation, and technologies have been continued and further enhanced, as these are indispensable for the effective support of computational sciences in the HPC area. For details see Gentzsch et al. (2010, 2011), and Lederer (2008). The resulting infrastructure is unmatched world-wide in its heterogeneity and complexity, enabling the operation of a powerful Supercomputing Grid built on top of national services, and facilitating Europe's ability to undertake world-leading computational science research.

Launched in 2005, the DEISA Extreme Computing Initiative (DEISA 2010c) regularly selects leading grand challenge HPC projects, based on a peer review system and approved by the DEISA Executive Committee (Execom), to enhance DEISA's impact on the advancement of computational sciences. By selecting the most appropriate supercomputer architectures for each project, DEISA has opened up the most powerful HPC architectures available in Europe for the most challenging projects. This service provisioning model has been extended from single project support to supporting Virtual European Communities. Collaborative activities have been carried out with European and other international initiatives. Of essential importance is the cooperation with the Partnership for Advanced Computing in Europe (PRACE) which has started with the installation of leadership-class Tier-0 supercomputers in Europe (currently in the range of PetaFLOPS), and with the preparation of an integrated Tier-0 and Tier-1 HPC ecosystem (Tier-1, currently in the range of hundreds of TeraFLOPS).

2.   DEISA's Vision, Mission, and Objectives

Vision: DEISA2 aims at delivering a turnkey operational solution for a future persistent European HPC ecosystem, as suggested by ESFRI (ESFRI 2006). The ecosystem integrates national Tier-1 centers and the new Tier-0 centers.

Mission: In DEISA2, the following two-fold strategy is applied:

- Consolidation of the existing infrastructure developed in DEISA1 by guaranteeing the continuity of those activities and services that currently contribute to the effective support of world-leading computational science in Europe.

- Evolution of this infrastructure towards a robust and persistent European HPC ecosystem, by enhancing the existing services, by deploying new services including support for European Virtual Communities, and by cooperating and collaborating with new European initiatives, especially PRACE that has started to enable shared European PetaFLOPS supercomputer systems.

Objectives of the DEISA1 project running from 2004 to 2008:

- Enabling terascale science by integrating Europe's most powerful HPC systems. DEISA is a European supercomputing service built on top of existing national HPC services. This service is based on the deployment and operation of a persistent, production quality, distributed supercomputing environment with continental scope.

- The criterion for success: Enabling scientific discovery across a broad spectrum of science and technology. The integration of national facilities and services, together with innovative operational models, is expected to add substantial value to existing infrastructures.

Objectives of the DEISA2 project running from the years 2008 to 2011:

- Enhancing the existing distributed European HPC environment (built in DEISA1) towards a turnkey operational infrastructure.

- Enhancing services by offering a variety of options of interaction with computational resources. Preparation for the integration of European Tier-1 and Tier-0 centers.

3.   The DEISA Infrastructure Services

DEISA is operated on top of national HPC services. It includes the most powerful supercomputers in Europe with an aggregated peak performance of more than 2 PetaFLOPS which are interconnected with a trusted, dedicated 10 Gbit/s network, based on GEANT2 and the National Research and Education Networks (NRENs). The essential services to operate the infrastructure and support its efficient usage are organized in three Service Activities:

Operations Services refer to operating the infrastructure including all existing services, adopting approved new services from the Technologies activity, advancing the operation of the DEISA HPC infrastructure to a turnkey solution for the future European HPC ecosystem by improving the operational model and integrating new sites.

Technologies Services cover monitoring of technologies, identifying and selecting technologies of relevance for the project, evaluating technologies for pre-production deployment, planning and designing specific sub-infrastructures to upgrade existing services or deliver new ones based on approved technologies.

Applications Services cover the areas of applications enabling, extreme computing projects, environment and user related application support, and benchmarking. Applications enabling focuses on enhancing scientific applications from the DEISA Extreme Computing Initiative (DEISA 2010c), Virtual Communities and EU projects. Environment and user related application support addresses the maintenance and improvement of the DEISA application environment and interfaces, and DEISA-wide user support in the applications area. Benchmarking refers to the provision and maintenance of a European Benchmark Suite for supercomputers.

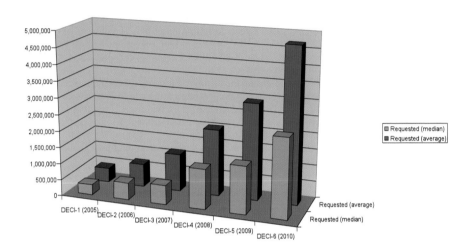

Figure 1.   Average and median amount requested during the lifetime of DECI.

4.   DEISA Extreme Computing Initiative and Astrosciences

DECI, the DEISA Extreme Computing Initiative (DEISA 2010c) was launched in May 2005 by the DEISA Consortium, as a way to enhance its impact on science and technology. The main purpose of this initiative is to enable a number of "grand challenge" applications in all areas of science and technology. These leading, ground-breaking applications must deal with complex, demanding and innovative simulations that would not be possible without the DEISA infrastructure, and which benefit from the exceptional resources provided by the Consortium. The DEISA applications are expected to have requirements that cannot be fulfilled by the national services alone.

In DEISA2, the activities oriented towards single projects have been qualitatively extended towards persistent support of Virtual Science Communities. This service extension benefits from and build on the experiences of the DEISA scientific Joint Research Activities where selected computing needs of various scientific communities and a pilot industry partner were addressed. Examples of structured science communities with which close relationships have been established are the European fusion community, the European climate community, and the science community within the Virtual Physiological Human project. DEISA2 provides a computational platform for them, offering integration via distributed services and web applications, as well as managing data repositories.

DECI enables European computational scientists to obtain access to the most powerful national computing resources in Europe regardless of their country of origin or work, and to enhance DEISA's impact on European science and technology at the highest level. Through an annual call, a number of capability computing projects are selected by peer-review on the basis of innovation and scientific excellence. The consortium has designed and deployed, and operates a complex, distributed, heterogeneous supercomputing environment with an aggregate peak performance in excess of two PetaFLOPS.

Successful projects are given access to the exceptional resources in the DEISA infrastructure (on an HPC architecture selected for its suitability to run the project's codes efficiently) and are offered applications support to enable them to use it productively. The number of proposals received has grown, from 51 at DECI-1 (in 2005) to 122 at DECI-6 (in 2010), with particularly rapid growth over the past three years. The number of CPU cycles requested per project has grown steadily since DECI's inception.

Figure 1 shows how both the median and the average amount of CPU requested by projects has grown year on year. The growing divergence between the average and the median CPU requested reflects the increasing number of larger, collaborative projects applying to DEISA for computational and applications enabling resources. The following four case studies taken from DECI astrosciences projects presented in Alessandrini et al. (2008) and DEISA (2010a) illustrate the great benefit of DEISA for the astrosciences research community.

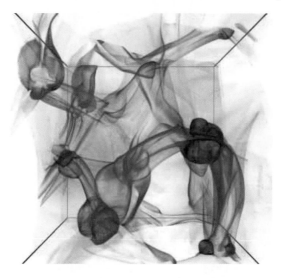

Figure 2.   Magnitude of the vorticity in an AMR simulation of forced supersonic turbulence. The sheet-like structures indicate the formation of strong shock fronts while turbulence is still developing.

4.1.   Modeling of Turbulent Astrophysical Flows

Turbulence in engineering applications and atmospheric sciences has frequently been modeled by large eddy simulations (LES). In LES, the dynamics of turbulent eddies is computed on large scales, while a subgrid scale model approximates the influence of smaller eddies. However, in astrophysics, phenomena such as supersonic turbulence in star-forming gas clouds challenge the LES approach. The self-similarity hypothesis employed in LES fails to be applicable over a wide range of disparate scales.

The adaptive mesh refinement (AMR) method proposes to insert computational grids of higher resolution into flow regions where turbulent structures such as eddies or shock fronts are forming. A major problem is to find the criteria for the generation of refined grids based on various fluid dynamic processes. The computational resources granted by the DEISA to perform highly resolved AMR simulations of supersonic turbulence were used. Depending on the size of the computational grid, 16 to 126 CPUs of the SGI Altix supercomputer at SARA, The Netherlands, were required for each simulation.

The basic idea in Schmidt (2008) was to trigger the refinement by monitoring flow properties such as the vorticity (the rotation of the velocity field) and the rate of gas compression (due to shocks or gravity). This is illustrated by the simulation of a three-dimensional visualization of the isosurfaces of the vorticity. The refined grids are mostly generated in the vicinity of sheet-like structures arising from shocks. The tube-like structures in Figure 2 indicate the centers of eddies.

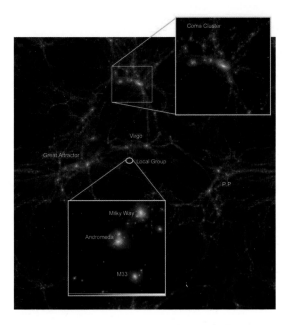

Figure 3.   Dark matter density distribution of the Local Universe simulation. The boxsize is 160 Mpc. The circle shows the position of the Local Group. A blown up panel shows the structure of the simulated Local Group. Resolution of this small region of 2 Mpc radius is equivalent to 4096^3 (70 billion) particles in the whole box, which translates into a dynamical mass range of more than 10^6.

4.2. Simulating the Local Universe

The SIMU-LU project (Yepes, G. et al. 2008) in 2008 has carried out the most accurate representation of the formation of the Local Universe performed to date, starting from cosmological initial conditions compatible with the most recent astronomical observations. To this end, it has been running a set of high resolution N-body cosmological simulations from initial conditions which include observational constraints on the mass distribution and velocity fields derived from nearby galaxies and clusters around us. They were able to run these simulations with up to 1 billion dark matter particles in different computational volumes.

By zooming in to the region where we are supposed to live, the Local Group, the project partners have been able to simulate the formation and evolution of a system pretty similar to our own galaxy and its closest neighbor: the Andromeda Galaxy. The outcome of these numerical experiments will provide a deep insight into the dynamics of our local environment and will constitute the starting point for more realistic simulations, in which ordinary matter (i.e. gas and stars) will be included together with the more exotic, yet dominant, components of the universe: Dark Matter and Dark Energy.

The purpose of project was to perform very large cosmological simulations of 1 billion particles in a computational cubic volumes ranging from 64 to 160 Mpc (i.e. 200 to ~ 500 million light years) on a side, with results demonstrated in Figure 3. The simulations need to resolve objects that are expected to be formed if dark matter is in the form of cold, weakly interacting massive particles (WIMPS) with differences in mass over many orders of magnitude: the largest superclusters of galaxies have masses of up to 10^{16} solar masses or more, while the tiniest dwarf galaxies orbiting around the normal ones have masses less than 10^{10} solar mass.

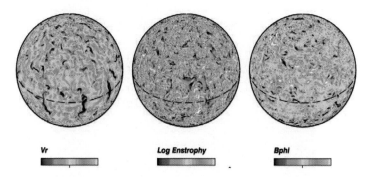

Figure 4.   Snapshot of radial velocity, log of enstrophy and toroidal magnetic field in the bulk of the highly turbulent convection zone. Highly intermittent convection and magnetic fields are observed in this low Pm simulation of the solar convective envelope. We note the high degree of vorticity present in the down flows. ©Allan Sacha Brun, CEA, France.

4.3.   Turbulent, Active, and Rotating Stars

The STARS project (Sacha Brun 2008) aims at modelling the complex, time dependent, and nonlinear dynamics present in the Sun and stars, and in particular, understanding stellar magnetic activity. Depending on the spectral type of the star considered, such activity can be cyclic, irregular, or for stars with stellar mass greater than 2 solar masses, without any activity or simply possessing a modulated signal.

The mechanism thought to be at the origin of the magnetism seen in solar type stars or in low mass stars is likely to be linked to the dynamo action in the upper convective layers of such stars. The simultaneous existence of convective turbulent motions, and of the rotation and its associated differential rotation and shear layers in stars, favors the emergence of a small and/or large scale magnetic field through induction. For more massive stars, possessing a convective core, understanding the interaction between the dynamo generated magnetic field and the probable fossil magnetic field of their radiative envelope constitute a major challenge in stellar fluid dynamics. To study in great details the interaction between convection, rotation, and magnetic field in stars was the main scientific goal of this project.

The researchers have computed several models in order to reach the dynamo threshold while keeping a solar-like differential rotation profile. The first model had a magnetic Reynolds number of around 300 (resolution Nr=256, Ntheta=512 and Nphi=1024), the threshold determined by previous studies. This model did not succeed and the seed magnetic finally decayed away. They then progressively increased the level of turbulence while keeping Pm=0.8, to reach about Rm=400 (and a resolution of 256 × 784 × 1568) to get a successful dynamo. This seems to indicate an increase of about 30 per cent of the dynamo threshold with respect to the high Pm cases, and confirms that turbulence is actually making it harder to get a successful dynamo rather than easier.

To illustrate the richness of the simulations, Figure 4 shows a snapshot of the convective radial velocity, of the log10 of the enstrophy (square of the vorticity) and the associated dynamo toroidal magnetic field in the bulk of the modeled convection zone. It can be seen how turbulent the convective patterns are and how small scale and intricate the magnetic field can be. This indicates that the magnetic fields generated by dynamo processes in stellar plasma are likely to be disorganized, and that in order to get a large-scale organized field (mostly toroidal in nature), one needs to include an omega effect not only in the convection zone but also in a sheared stable layer like the tachocline.

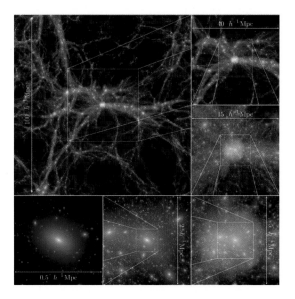

Figure 5.   The cosmic web of dark matter. A zoom of the large-scale structure of the cosmos, from the scale of superclusters to that of galaxies.

4.4.   Simulating the Formation of the Milky Way

The AQUILA project (MacDonald 2010) has sought to confront the grand challenge of simulating the formation of a realistic galaxy. It brought together over 30 researchers from all over the world under the auspices of the Virgo Consortium for cosmological simulations. Thanks to recent advances in cosmology, which have revolutionized our view of the cosmos and opened up some of the most interesting questions in modern science, we now know that only some 4%

of it is made up of ordinary matter; 20% is "dark matter", and the remainder is formed by "dark energy". Galaxies like our own Milky Way are thought to have arisen from quantum fluctuations generated immediately after the Big Bang. These notions, together with the view that our universe has a flat geometry, form the core of the current paradigm for the formation of galaxies, known as the cold dark matter (CDM) model. This model provides a hypothesis for the statistical properties of dark matter structures at early epochs when the universe was almost uniform.

As the Virgo Consortium is an international collaboration, it was natural that they turned to the international infrastructure DEISA in order to carry out their project. The resources made available by DEISA were crucial, not only because they represented substantial amounts of computing power, but they also facilitated the transfer of data amongst the participating Virgo nodes. Over 1.2 million DEISA core hours were used at the RZG supercomputing center in Garching, Germany, to carry out two series of simulations including the largest simulations of galaxy formation to date, see Figure 5.

Acknowledgments.   This work is supported by the European DEISA project, funded through European Commission contracts RI-508830, RI-031513 and RI-222919. The author wants to thank the whole DEISA Team for their continuous support and contributions. I also want to thank Ian Evans from the Smithsonian Astrophysical Observatory for his valuable support with publishing this paper.

References

Alessandrini, V., Lederer, H., Streit, S., & Värttö, G. J. (eds.) 2008, DEISA: Advancing Science in Europe
DEISA (ed.) 2010a, DEISA Digest: Extreme Computing in Europe. URL http://www.deisa.eu/news_press/media
DEISA 2010b, Distributed European Infrastructure for Supercomputing Applications, project website. URL http://www.deisa.eu
— 2010c, e-Science in a Collaborative, Secure, Interoperable and User-Friendly Environment DEISA Extreme Computing Initiative, DECI website. URL http://www.deisa.eu/deci
ESFRI 2006, European Strategy Forum on Research Infrastructures, ESFRI website. URL http://cordis.europa.eu/esfri
Gentzsch, W., Girou, D., Kennedy, A., Lederer, H., Reetz, J., Reidel, M., Schott, A., Vanni, A., Vasquez, J., & Wolfrat, J. 2011, J. Grid Comp., in press
Gentzsch, W., Kennedy, A., Lederer, H., Pringle, G., Reetz, J., Riedel, M., Schuller, B., Streit, A., & Wolfrat, J. 2010, in Proceedings of the e-Challenges Conference e-2010
Lederer, H. 2008, J. Phys. Conf. Ser., 125, 011003
MacDonald, E. 2010, in DEISA Digest: Extreme Computing in Europe, edited by The DEISA Consortium. URL http://www.deisa.eu/news_press/DEISA_DIGEST2010.pdf
Sacha Brun, A. 2008, in DEISA: Advancing Science in Europe, edited by V. Alessandrini, H. Lederer, S. Streit, & G. J. Värttö. URL http://www.deisa.eu/news_press/Media/DEISA-AdvancingScienceInEurope.pdf
Schmidt, W. 2008, in DEISA: Advancing Science in Europe, edited by V. Alessandrini, H. Lederer, S. Streit, & G. J. Värttö. URL http://www.deisa.eu/news_press/Media/DEISA-AdvancingScienceInEurope.pdf
Yepes, G., Gottlöber, S., Klypin, A., & Hoffman, Y. 2008, in DEISA: Advancing Science in Europe, edited by V. Alessandrini, H. Lederer, S. Streit, & G. J. Värttö. URL http://www.deisa.eu/news_press/Media/DEISA-AdvancingScienceInEurope.pdf

Astronomical Data Analysis Software and Systems XX
ASP Conference Series, Vol. 442
Ian N. Evans, Alberto Accomazzi, Douglas J. Mink, and Arnold H. Rots, eds.
© *2011 Astronomical Society of the Pacific*

The JSA and the Grid: How "Infinite" Computing Power Enables a New Archive Model for PI-led Observatories

Frossie Economou,[1] Tim Jenness,[1] Brad Cavanagh,[1] Sharon Goliath,[2] Russell Redman,[2] Dustin Jenkins,[2] David Berry,[1] Antonio Chrysostomou,[1] Brian Chapel,[2] Pat Dowler,[2] Séverin Gaudet,[2] John Ouellette,[2] and David Schade[2]

[1]*Joint Astronomy Center, 660 N. A'ohōkū Place, Hilo, HI, 96720, USA*

[2]*Canadian Astronomy Data Center, Herzberg Institute of Astrophysics, National Research Council of Canada, 5071 West Saanich Road, Victoria, BC V9E 2E7, Canada*

Abstract. The JCMT Science archive has adopted an agile "reduce-often release-always" model that continuously processes data and exposes it to the user. This paper describes the reasons for, and advantages of, this model.

1. The JCMT Science Archive

The James Clerk Maxwell Telescope (JCMT) is the world's largest single-dish sub-millimeter facility. Its instrumentation includes a heterodyne back end (the ACSIS auto-correlation spectrometer; Buckle et al. 2009) and a continuum back end (the sub-mm imager SCUBA-2; Holland et al. 2006). As a result of its heterogeneous instrumentation, its diverse scientific program (ranging from small PI projects to large legacy surveys), its flexible operating mode (Economou et al. 2002; Adamson et al. 2004) and the complications of sub-mm data reduction (Jenness et al. 2011), the JCMT data presents a challenging candidate for science archiving and VO publication.

The JCMT Science Archive (JSA; Gaudet et al. 2008) is a collaboration between the JCMT and the Canadian Astronomy Data Center (CADC). The JCMT's interest lies in increasing productivity by providing its community with science-ready data products, and enhancing the value of the JCMT Legacy Survey Program (Economou et al. 2008). The CADC's interest lies in the challenges of dealing with a dataset that is far removed from the more common 2-D optical/IR datasets found in astronomical archives, the infrastructure required to produce automatically generated science products for data of this complexity, as well as the challenges of its integration into the Virtual Observatory.

At the time of writing, the JCMT Science Archive is now entering its second year of deployment, offering automatically generated multi-night science products from ACSIS and SCUBA-2 data. This puts us in a position to draw some lessons that may be of value to future projects of similar nature.

2. The JSA as an Agile "Reduce-and-Release" Archive Model

Traditionally, astronomical archives fall into one of two categories: raw data archives where the end user is expected to process the data themselves after retrieval (perhaps with provided software), or science archives where a heavily curated science product is made available in relatively infrequent official releases.

In contrast, at the JSA data is processed every night using JCMT's ORAC-DR-based data reduction pipeline (Cavanagh et al. 2008; Jenness et al. 2008) on the CADC's computing hybrid.

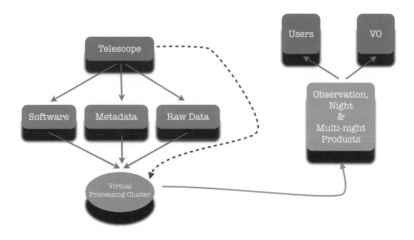

Figure 1.   Conceptual diagram of the JSA. The telescope makes raw data, meta-data and archive available to the data center. It then triggers processing on the data center's processing nodes. Products are immediately exposed.

The pipeline produces science-ready calibrated products for each observation and for all the observations on that given night for a given field (Fig. 1). Additionally, in frequent intervals, multi-night (whole project) fields are also generated. The products are immediately exposed to the proprietary user, the public user and/or the VO, as appropriate.

There are numerous advantages to such a model. The user has immediate access to products that allow them to assess their observing strategy while their program is still underway. Because the user does not have to do a lot of reduction before trying to establish whether their scientific goals are being met, they engage more rapidly with their data. Because the data is constantly being re-reduced, improvements in the reduction software immediately propagate (instead of waiting for fixed releases and hope that the users will upgrade their installations and reprocess their data).

There is a potential disadvantage, in that while data is immediately exposed, by definition any errors are also immediately exposed. However it appears that in this model (which we call "agile" in analogy to the Agile software development philosophy; Larman & Basili 2003), the users are willing to take speed and convenience over guaranteed accuracy. Moreover, a facility like JCMT that lacks the resources to publish a highly curated data product gets real value from the ability to harness its extensive community of users in reviewing their products.

A serendipitous benefit of this model has been the positive effect of the agile archive model to observatory operations. Normally the kinds of data rates that we are now seeing ($\sim$ 100 TB/year) would make the wholesale processing of data for internal reasons quite prohibitive. Thanks to the considerable resources of the CADC computing infrastructure as well as the ease of interaction with it, the observatory can benefit from reprocessing years of data in a matter of days. This can aid observatory functions such as monitoring long term calibration and instrument stability, as well as providing a comprehensive testing platform for the data reduction pipelines.

Of course, all this pre-supposes that the observatory has a fully data-driven pipeline that has access to all the metadata that is required to automatically process the data. In our experience, this requires not only sophisticated data processing, but the kind of observation management system that tracks important metadata from observation preparation, through data acquisition to observatory management.

3.　Planned and Possible Future Improvements

There are a number of planned improvements for the JSA, such as the publication of 4-D products (RA, Dec, frequency, polarisation), the VO publication of "clump" (extended asymmetrical sources) catalogs (e.g. Berry et al. 2007), and the migration to the new CANFAR platform (Gaudet et al. 2011).

More ambitiously, our experience with the system so far deployed is that there is significant potential in further developing the traditional model of science archives. To date, data flow from the archive to the user has been mono-directional: the user downloads the data. We anticiapte making this bi-directional by providing an API for the upload and ingestion of user-generated products. These will not be recomputed along with the standard pipeline products, but will be stable copies suitable for reference in publications. By re-using our existing ingestion process, we will provide a mechanism for astronomers to publish their data in parallel with their papers analyzing the data

However the ultimate extension of this process would be for the archive to not only facilitate archive-to-user data flow, but also user-to-user data flow, in order for the archive to become not just a data repository, but a platform for scientific collaboration.

4.　The Telescope-Data Center Partnership

The JSA was a project of modest financial resources that was only made possible by the extensive prior software infrastructures at JCMT and CADC. Its successful deployment highlights the potential for partnerships between established observatories and advanced data centers. The observatory can bring domain knowledge and the data processing software that encodes it, while the data center can shield the working observatory from the complexities of large scale infrastructures and VO publication.

While one has to remain vigilant about the problems that can beset such collaborations in some circumstances, such as a lack of focus and communication amongst the distributed teams (which happily was not the case here), the outcome can be a significant increase in the effective scientific productivity of the observatory, and an enrichment of the VO by the incorporation of rich and diverse data sets.

References

Adamson, A. J., Tilanus, R. P., Buckle, J., Davis, G. R., Economou, F., Jenness, T., & Delorey, K. 2004, in Optimizing Scientific Return for Astronomy through Information Technologies, edited by P. J. Quinn, & A. Bridger (Bellingham, WA: SPIE), vol. 5493 of Proc. SPIE, 24

Berry, D. S., Reinhold, K., Jenness, T., & Economou, F. 2007, in Astronomical Data Analysis Software and Systems XVI, edited by R. A. Shaw, F. Hill, & D. J. Bell (San Francisco, CA: ASP), vol. 376 of ASP Conf. Ser., 425

Buckle, J. V., et al. 2009, MNRAS, 399, 1026

Cavanagh, B., Jenness, T., Economou, F., & Currie, M. J. 2008, Astron. Nach., 329, 295

Economou, F., Jenness, T., Chrysostomou, A., Cavanagh, B., Redman, R., & Berry, D. S. 2008, in Astronomical Data Analysis Software and Systems XVII, edited by R. W. Argyle, P. S. Bunclark, & J. R. Lewis (San Francisco, CA: ASP), vol. 394 of ASP Conf. Ser., 450

Economou, F., Jenness, T., Tilanus, R. P. J., Hirst, P., Adamson, A. J., Rippa, M., Delorey, K. K., & Isaak, K. G. 2002, in Astronomical Data Analysis Software and Systems XI, edited by D. A. Bohlender, D. Durand, & T. H. Handley (San Francisco, CA: ASP), vol. 281 of ASP Conf. Ser., 488

Gaudet, S., Dowler, P., Goliath, S., & Redman, R. 2008, in Astronomical Data Analysis Software and Systems XVII, edited by R. W. Argyle, P. S. Bunclark, & J. R. Lewis (San Francisco, CA: ASP), vol. 394 of ASP Conf. Ser., 135

Gaudet, S., et al. 2011, in Astronomical Data Analysis Software and Systems XX, edited by I. N. Evans, A. Accomazzi, D. J. Mink, & A. H. Rots (San Francisco, CA: ASP), vol. 442 of ASP Conf. Ser., 61

Holland, W., et al. 2006, in Millimeter and Submillimeter Detectors and Instrumentation for Astronomy III, edited by J. Zmuidzinas, W. S. Holland, S. Withington, & W. D. Duncan (Bellingham, WA: SPIE), vol. 6275 of Proc. SPIE, 62751E

Jenness, T., Berry, D., Chapin, E., Economou, F., Gibb, A., & Scott, D. 2011, in Astronomical Data Analysis Software and Systems XX, edited by I. N. Evans, A. Accomazzi, D. J. Mink, & A. H. Rots (San Francisco, CA: ASP), vol. 442 of ASP Conf. Ser., 281

Jenness, T., Cavanagh, B., Economou, F., & Berry, D. S. 2008, in Astronomical Data Analysis Software and Systems XVII, edited by R. W. Argyle, P. S. Bunclark, & J. R. Lewis (San Francisco, CA: ASP), vol. 394 of ASP Conf. Ser., 565

Larman, C., & Basili, V. R. 2003, Computer, 36, 47

Astronomical Data Analysis Software and Systems XX
ASP Conference Series, Vol. 442
Ian N. Evans, Alberto Accomazzi, Douglas J. Mink, and Arnold H. Rots, eds.
©*2011 Astronomical Society of the Pacific*

Distributed GPU Volume Rendering of ASKAP Spectral Data Cubes

A. H. Hassan,* C. J. Fluke, and D. G. Barnes

Center for Astrophysics and Supercomputing,
Swinburne University of Technology,
PO Box 218, Hawthorn, Australia, 3122.

*ahassan@swin.edu.au

Abstract.   The Australian SKA Pathfinder (ASKAP) will be producing 2.2 terabyte HI spectral-line cubes for each 8 hours of observation by 2013. Global views of spectral data cubes are vital for the detection of instrumentation errors, the identification of data artifacts and noise characteristics, and the discovery of strange phenomena, unexpected relations, or unknown patterns. We have previously presented the first framework that can render ASKAP-sized cubes at interactive frame rates. The framework provides the user with a real-time interactive volume rendering by combining shared and distributed memory architectures, distributed CPUs and graphics processing units (GPUs), using the ray-casting algorithm. In this paper we present two main extensions of this framework which are: using a multi-panel display system to provide a high resolution rendering output, and the ability to integrate automated data analysis tools into the visualization output and to interact with its output in place.

1. Introduction

Upcoming radio observation facilities such as the Australian Square Kilometer Array Pathfinder (ASKAP), the MeerKAT Karoo Array Telescope, the Low Frequency Array (LOFAR), and ultimately the Square Kilometer Array (SKA) will pose a significant challenge for current astronomical data analysis and visualization tools. The expected data product sizes (e.g. 2.2 TB per 8 hours of observation for the proposed WALLABY all-sky HI survey[1]) are orders of magnitude larger than astronomers, and existing astronomy software, are accustomed to dealing with.

In Hassan et al. (2011), we presented a distributed GPU framework to interactively volume render larger-than-memory astronomical data cubes. Throughout this work, we demonstrated how volume rendering offers an alternative to standard 2D visualization techniques and provided a way to overcome the technological barrier caused by the computational requirement of volume rendering for large data cubes. The presented framework utilizes a heterogeneous CPU and GPU hardware infrastructure, combining shared- and distributed-memory architectures, to yield a scalable volume rendering solution, capable of volume rendering image cubes larger than a single machine memory limit, in real-time and at interactive frame rates. The usage of GPUs as the main processing backbone for the system, and the remote visualization architecture adopted in our design for this framework enables further enhancement for the astronomer's visualization experience.

In this paper we present an extension to this framework to provide: better visualization output by integrating external quantitative information with the volume rendering output (see section 2), and high resolution output via multi-panel displays (see section 3).

---

[1]See http://www.atnf.csiro.au/research/WALLABY for details.

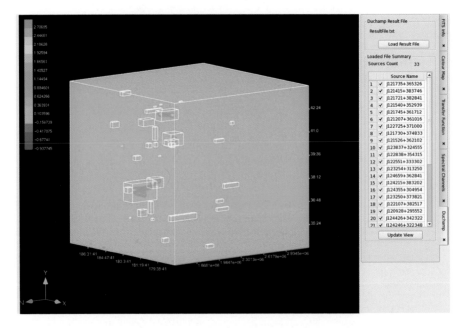

Figure 1.   Illustration of the process of integerating Duchamp catalog with the volume rendering output. The data cube used is a neutral hydrogen (21cm) observation of part of the Ursa Major galaxy cluster, made with the Lovell radio-telescope at Jodrell Bank, Manchester. Data courtesy Virginia Kilborn (Swinburne).

2.   Integrating Duchamp Output

While the main processing burden is moved to the back-end cluster of GPUs, the rendering client is free to do some relatively smaller processing to enhance the interactivity and the scientific outcomes of the visualization output. The usage of remote visualization enables the viewer application to integrate different graphical primitives with the volume rendering output without affecting the rendering processes [see Figure 2 in Hassan et al. (2011) for details of the framework's main components]. We already utilized this ability to present simple primitives such as the color map, and orientation arrows in the original implementation [see Figure 1, 5, and 6 in Hassan et al. (2011)]. We have extended our work to enable the overlay of a source finder output catalog on the volume rendering output. We use the Duchamp source finder (Whiting 2008), the selected source finder for the proposed ASKAP survey pipeline, to demonstrate this integration.

First, the user loads a Duchamp output file, an ASCII file that contains a set of detected astronomical sources represented by their bounding cubes. The viewer application parses the file and extracts the sources' bounding cubes, which are superimposed as wireframe boxes on the volume rendering output. The user interaction with the displayed volume rendering output is automatically applied to the displayed source cubes. The user can select one of the source cubes to display further available information about that source. See Figure 1 for an illustration of this process. The user can also show or hide a group of the displayed sources to enable better understanding of the output. We anticipate other applications of this extension such as comparing multiple source finder outputs (by assigning different colors for the source cubes), and overlaying radio/optical catalogs.

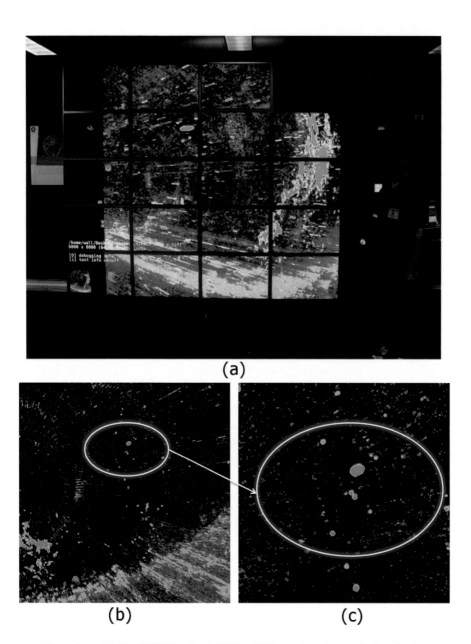

Figure 2.   (a) The HIPASS cube in 8000 × 8000 pixel resolution displayed using
the CSIRO OptIPortal ATNF, Marsfield (9600 × 5400 pixel). Credit: C. J. Fluke.
(b) A portion of the HIPASS volume rendering with 3000 × 3000 output resolution.
(c) The same portion of the HIPASS cube with 8000 × 8000 otuput resolution.

3. OptIPortal Integeration

Another benefit of using GPUs is the ability to render high resolution output in a relatively short time. As a further extension of the current viewer implementation, we present support for high resolution tiled display systems (such as the OptIPortal display[2]). We used the OptIPortal display facility operated by the Australian Commonwealth Scientific and Research Organization (CSIRO) at Marsfield (Sydney, Australia) to demonstrate this integeration (see Figure 2-a). The current CSIRO OptIPortal consists of 5×5 high definition screens (1920×1080 pixels) with an overall output size of 9600×5400 pixels. The viewer uses the back-end GPU cluster to render a static image of the current data file with the same orientation information but with a different output size (defined by the user). The rendering output is stored in TIFF format, which is converted to the pyramidal tiled TIFF format.[3]

Figures 2-b and 2-c illustrate the importance of high resolution output with the HIPASS southern sky cube[4] as an example. With the expected increase in the output pixel resolution within the different main ASKAP HI surveys (e.g. WALLABY with its expected output cube size of $6144 \times 6144 \times 16384$), it will be useful to utilize such high resolution output facilities to provide the astronomers with a better level of detail for both qualitative and quantitative data visualization.

On the other hand, such facilities may be useful to support collaborative visualization if a suitable data interaction mechanism is provided. We think the current framework can support an interactive OptIPortal display of the volume rendering output, but with a lower frame rate.

4. Conclusion

We demonstrate the ability to extend our GPU volume rendering framework to offer better visualization outcomes. Two main features were presented: integrating source finder output, and utilizing multi-panel displays. Both of these features demonstrate two main strong points in the framework design, namely the usage of GPU cluster as the main processing back-bone, and the separation between the rendering and the result display. We anticipate the ability to further enhance this for integrating more quantitative visualization tools and a better utilization of the GPU processing power.

Acknowledgments.   We thank Russell Jurek (ATNF — CSIRO), and Virginia Kilborn (Swinburne) for providing sample data cubes.

References

Hassan, A. H., Fluke, C. J., & Barnes, D. G. 2011, New Astron., 16, 100
Whiting, M. T. 2008, Galaxies in the Local Volume, 343

---

[2]http://research.ict.csiro.au/research/labs/information-engineering/ie-lab-projects/optiportal

[3]See http://iipimage.sourceforge.net/documentation/images/

[4]The southern sky cube was generated by Russell Jurek (ATNF) from 387 individual cubes. Its dimensions are $1721 \times 1721 \times 1025$ with a data size of 12 GB.

Astronomical Data Analysis Software and Systems XX
ASP Conference Series, Vol. 442
Ian N. Evans, Alberto Accomazzi, Douglas J. Mink, and Arnold H. Rots, eds.
©2011 Astronomical Society of the Pacific

The Marriage of Mario (NHPPS) and Luigi (OGCE)

Francisco Valdes[1] and Suresh Marru[2]

[1]*NOAO Science Data Management, P.O. Box 26732, Tucson, AZ 85732*
[2]*Pervasive Technology Institue, Indiana University, Bloomington, IN 47408*

Abstract. Pipeline systems, like the plumbing characters of video game fame, have their strengths and weaknesses. This is generally due to the different niches they occupy in the landscape of pipeline systems. People frequently discuss the potential interchange of application elements (the modules and algorithms) between pipeline systems. However, pipeline systems themselves are applications in their own right. This paper describes an interesting architectural marriage where one pipeline system orchestrates another to take advantage of the strengths and niches of both. This marriage is the result of a (proposed) project between NOAO and Indiana University to provide a Pipeline, Portal, and Archive (PPA) system for the WIYN One-Degree Imager (ODI) which is briefly introduced here.

1. Introduction

The WIYN Observatory is building a forefront mosaic camera with a one-degree field of view, hence the name One-Degree Imager (ODI), and a pixel scale of 0.1 arcseconds for its 3.5-meter telescope (Jacoby et al. 2002). This gigapixel camera means observers will obtain large volumes of data which most will be unable to store, calibrate, and analyze with their own resources. Therefore, Indiana University (IU) and NOAO have partnered to propose a Pipeline, Portal, and Archive (PPA) system to support this instrument. In this contribution, we describe the plans and synergies within this partnership for developing the standard ODI calibration pipeline.

The pipeline framework, arrived at through design and prototype studies, is a marriage of two pipeline/workflow systems. The NOAO High Performance Pipeline System (NHPPS) is good at orchestrating host commands into multi-process workflows that effectively make use of a dedicated cluster of multi-core machines. The Open Gateway Computing Environment (OGCE) (Pierce et al. 2009) Workflow Suite is designed to wrap command line-driven science applications and make them into robust, network-accessible services. The resulting services are orchestrated as workflows moving data and executed as jobs on computational resources ranging from local workstations to high performance computational grids and clouds including the TeraGrid (Catlett 2005). For the ODI project we will wrap NHPPS pipeline applications with OGCE for orchestration into distributed Teragrid workflows.

Before going into more detail on this marriage, we introduce the IU/NOAO partnership. The collaborating groups are the Pervasive Technology Institute (PTI) at IU and the Science Data Management (SDM) group at NOAO. PTI is dedicated to the development and delivery of innovative information technology while SDM is a traditional astronomical data management group. Difference in expertise between these groups is one of the characteristics of this partnership. SDM brings experience with pipeline processing of astronomical image data similar to ODI and managing a principle investigator and community archive. PTI brings experience with the IT problems of data transport, storage, and computations on a large scale as well as multidisciplinary science gateways. The blending of expertise is needed to address the astronomical needs of the ODI users in working with the volume of ODI data which is larger than that from other current NOAO instruments.

2.   NOAO High Performance Pipeline System (NHPPS)

The NOAO High Performance Pipeline System (NHPPS), internally code named "Mario" at NOAO, has been described in previous ADASS proceedings (see Scott et al. 2007) and in a more detailed paper (Valdes et al. 2006). Briefly, NHPPS is an event-driven executor of host commands (aka modules), where any number of these may be executed at the same time. The commands are submitted to the host operating system which is where the processing scheduling and management takes place. Because many processes are running in an asynchronous, time-shared manner, the compute resources, such as multiple cores, can be efficiently utilized by the operating system even when the processes are not threaded or MPI applications.

NOAO uses this framework to build pipeline applications structured as a number of pipeline services. A pipeline service is a sequence of (possibly parallel) steps running on one type of data object and on a single node to perform a piece of the overall application. These pipeline services interact with each other in a workflow. The workflow can be executed on a single node or distributed across peer nodes in a dedicated cluster. In NOAO applications the individual steps in the pipeline services consist primarily of IRAF-shell scripts calling standard IRAF data processing tools. One of the main advantages of NHPPS in these applications is the ability to run many of these IRAF modules in parallel to fully utilize compute resources.

In a high performance application the pipeline services are deployed on a set of compute nodes. The workflow is structured into a hierarchy such that higher level pipeline services operate on larger data objects and call multiple instances of lower level services to process smaller data objects in parallel. The calling pipeline service waits for the results of the lower level pipelines. This type of workflow execution is typically called hierarchal orchestration and the workflow structure generally follows the map/reduce paradigms.

To understand the marriage with OGCE, a key point is that the NHPPS framework basically consists of a server, called the Node Manager, which runs on each compute node and responds to events by executing modules on particular dataset objects. The efficiency occurs because it manages many processes at the same time so that the operating system can keep the CPU cores busy.

When NHPPS is configured as a distributed pipeline system, the Node Managers communicate with each other and manage multiple instances of the same pipeline or different pipelines on different nodes depending on the configuration design. The need to communicate means that the compute nodes be part of the cluster providing peer-to-peer connections.

3.   Open Gateway Computing Environment (OGCE)

The Open Gateway Computing Environment (OGCE), dubbed "Luigi" in this tongue-in-cheek marriage analogy with "Mario", provides science gateway software. It includes several middleware components that can be used by themselves or integrated to provide more comprehensive solutions. The toolkit wraps command line scientific applications into web services, deploys and executes them on the TeraGrid or other Grid and Cloud Infrastructures, and orchestrates them as a workflow applications. There are a variety of components and aspects of OGCE with details provided in Pierce et al. (2009).

A key point is that the Grid infrastructure, while providing good features of uniform access to heterogeneous resources, add some overhead to perform meta scheduling and unified security. The overhead is ignorable for long running applications but significant enough for very small running applications. OGCE as a general purpose toolkit abstracts the utilization of multiple cores on a grid node and hands it over to the application for efficient threading or other parallelization techniques. So primary it launches a single command line application on a grid node or nodes. The command may be multi-threaded, an MPI application, a simple shell script, etc. The utilization of the node resources, such as multiple cores, resides in the application. In the cases like astronomical data processing the application can be a pipeline itself. When a pipeline is wrapped as a web service, the OGCE toolkit orchestrates the pipeline applications onto grid nodes and leaves the details of internal node orchestration to the pipeline itself.

4. The Marriage

The fundamental technical question for the NOAO/IU partners with regards to the Tier 1 pipeline was how NOAO developed science pipeline software could be integrated (married) with the IU grid workflow system. Another goal, which we only mention here, is to enable observer workflows, called Tiers 2/3, using the standard data products, pieces of the standard pipeline and other tools such as IRAF.

Primary considerations were to utilize the NOAO data reduction and pipeline expertise and to capitalize on the large body of already developed software. This means use of the IRAF toolkit and NOAO pipeline applications. This leads to using IRAF-shell modules; with reuse of modules from similar NOAO pipeline applications and addition of new modules as appropriate. It also means using the same high performance map/reduce workflow design.

A secondary consideration was to have the potential to run the pipeline application on both the TeraGrid and on a dedicated cluster of machines. This provides a development framework, a fallback in case of grid limitations, and increased flexibility for deployment. This again leads to using the NOAO pipeline applications model of NHPPS pipeline services.

The first step was to develop an ODI data flow design (Valdes 2009) which identifies the map/reduce strategies and estimates for the number of modules ($\sim$ 150) and pipeline services ($\sim$ 20) required. Then the first integration concept was to use the individual modules in an OGCE workflow. After studying this option there were a number of problems. These include the large number of modules, the heterogeneity of function and parameters, and the inefficiency in deploying many light-weight modules as grid jobs.

Then came the "epiphany" that NHPPS pipeline services (the collection of modules that address a particular map or reduce operation) would satisfy all of the considerations. The number of services is relatively small, the services have consistent input and output parameters, are not too-fine grained, efficiently utilize node resources, and are easily orchestrated both by the dedicated cluster configuration of NHPPS and the TeraGrid framework of OGCE.

The main challenge was that a pipeline service uses NHPPS as the execution framework that glues the modules together efficiently. Normally the NHPPS framework is a daemon on the nodes where the pipeline services are deployed. However, it was realized that with some minor changes a captive NHPPS framework can be started for one or more pipeline services within a single host application. This host application is then what is wrapped as an OGCE service.

So here is the core of this contribution: The marriage of NHPPS with OGCE consists of wrapping NHPPS pipeline services as single host applications. Because these pipeline services have a standard structure a single basic wrapper is all that is needed. This wrapper accepts input parameters and produces output parameters that OGCE builds services around. The wrapper also takes care of starting the single node NHPPS framework, supplying the triggers that start the pipeline, wait for the pipeline service to complete, and shutdown the framework.

For OGCE, one web service wrapper is created for each of the pipeline services. These web services are then orchestrated by OGCE as a Grid workflow. OGCE provides the workflow monitoring and collecting and managing of the pipeline input and output. In addition it handles the typical submission and security aspects of a Grid application.

There was one other detail to resolve. An NHPPS pipeline service implements the hierarchal orchestration and map/reduce structure by calling other pipelines and waiting for results. The OGCE workflows must be described by a directed acyclic graph (DAG) which means there is no "return" to a service. The solution was to automatically split pipeline services at their call (map) and return (reduce) steps and promote this to the OGCE workflow framework as a list of call requests and the the collection of map results.

An organizational benefit of this design is that the interface, and resulting division of work, between NOAO and IU is simple and clear. NOAO can concentrate on the NHPPS pipelines knowing the wrapper strategy is simple and clear and IU can concentrate on the map/reduce workflow aspects with little need to know about the modules and algorithms.

References

Catlett, C. 2005, in Cluster Computing and the Grid, 2002. 2nd IEEE/ACM International Symposium on (IEEE), 8

Jacoby, G. H., Tonry, J. L., Burke, B. E., Claver, C. F., Starr, B. M., Saha, A., Luppino, G. A., & Harmer, C. F. W. 2002, in Survey and Other Telescope Technologies and Discoveries, edited by J. A. Tyson, & S. Wolff (Bellingham, WA: SPIE), vol. 4836 of Proc. SPIE, 217

Pierce, M., et al. 2009, Open Grid Computing Environments (TeraGrid 2009)

Scott, D., Pierfederici, F., Swaters, R. A., Thomas, B., & Valdes, F. G. 2007, in Astronomical Data Analysis Software and Systems XVI, edited by R. A. Shaw, F. Hill, & D. J. Bell (San Francisco, CA: ASP), vol. 376 of ASP Conf. Ser., 265

Valdes, F. 2009, ODI Pipeline Data Flow Design, SDM Pipeline Document PL013, NOAO/SDM. URL http://chive.tuc.noao.edu/noaodpp/Pipeline/PL013.pdf

Valdes, F., Cline, T., Pierfederici, F., Thomas, B., Miller, M., & Swaters, R. 2006, The NOAO High-Performance Pipeline System, SDM Pipeline Document PL001, NOAO/SDM. URL http://chive.tuc.noao.edu/noaodpp/Pipeline/PL001.pdf

Astronomical Data Analysis Software and Systems XX
ASP Conference Series, Vol. 442
Ian N. Evans, Alberto Accomazzi, Douglas J. Mink, and Arnold H. Rots, eds.
©*2011 Astronomical Society of the Pacific*

Synergy Between Archives, VO, and the Grid at ESAC

Christophe Arviset,[1] Ruben Alvarez,[1] Carlos Gabriel,[1] Pedro Osuna,[1] and Stephan Ott[2]

[1]*ESA-ESAC, PoBox 78, 28691 Villanueva de la Canada, Madrid, Spain*

[2]*ESA-ESTEC, Keplerlaan 1, NL-2201 AZ Noordwijk, The Netherlands*

Abstract. Over the years, in support to the Science Operations Centers at ESAC, we have set up two Grid infrastructures. These have been built: 1) to facilitate daily research for scientists at ESAC, 2) to provide high computing capabilities for project data processing pipelines (e.g., Herschel), 3) to support science operations activities (e.g., calibration monitoring). Furthermore, closer collaboration between the science archives, the Virtual Observatory (VO) and data processing activities has led to an other Grid use case: the Remote Interface to XMM-Newton SAS Analysis (RISA). This web service-based system allows users to launch SAS tasks transparently to the GRID, save results on http-based storage and visualize them through VO tools. This paper presents real and operational use cases of Grid usages in these contexts

1. Grid at ESAC

ESA's European Space Astronomy Center (ESAC), near Madrid, is the default center for ESA astronomy and planetary missions' science operations and their respective scientific archives (e.g. XMM-Newton, ISO, Integral, Herschel, Planck, Mars Express, Venus Express, Rosetta, Huygens, Soho). As such, ESAC can be considered as a Data Center for European space science missions. Building Grid infrastructure extends this concept for ESAC to be as well a Data Processing Center, ensuring a natural link between science archives, the Grid, and the VO.

The first objective for the Grid at ESAC is to provide a shared and powerful Grid computing environment for projects enabling them to carry more efficiently projects main tasks such as standard data processing (from raw data to science data), on-the-fly reprocessing facilities from the science archives and instruments calibration monitoring and trend analysis. An other objective was for the ESAC scientists to use the Grid for their daily science tasks, providing them access to more processing capacities than just their standard desktop. To complete this goal, access to the Grid must be easy, ensuring user's home directories and standard scientific software is available on the Grid as well as providing user support in porting user's software to the Grid. These objectives could have been met by just building a cluster, but we had in mind to be able to cooperate with other institutions having Grid infrastructure, hence our decision from the beginning to use Grid middleware. This has enabled us to participate to the EGEE (Enabling Grid for E-scienceE), and later EGI (European Grid Infrastructure) framework, although we must admit that we did not manage to foster collaboration as much as we had initially hoped.

To fulfill these objectives and due to the ESA network and firewall constraints, we had to build two Grid infrastructures. One, ESACGRID is located on the external network (DMZ), close to the science archives, part of EGEE/EGI (hence based on the gLite middleware), and can be used for external collaboration. It consists of 56 blades with more than 400 cores and around 1.4TB of RAM. The second, SCIGRID is located on the Intranet, on the same network as the ESAC users' workstations and disk storage. Users can transparently launch their tasks to SCIGRID or even run an interactive session on one of the SCIGRID node through a simple

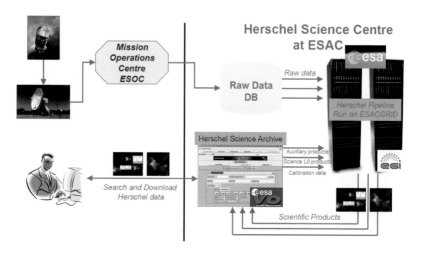

Figure 1.   Herschel Data Processing workflow.

command ("gridlogin"). It is based on the Globus middleware and consists of 35 blades with more than 250 cores and around 450GB of RAM.

2.   Grid Operational and Science Use Cases

2.1.   Herschel Data Processing

The Herschel Science Center at ESAC is responsible (amongst other tasks) for the pipeline data processing to convert the raw data into science data to be ingested into the Herschel Science Archive. Data processing can be run in several modes: 1) systematic daily processing when receiving spacecraft telemetry, 2) on-the-fly reprocessing from archive requests and 3) bulk reprocessing of all raw data at regular intervals, with improved versions of the pipeline software. Each observation can be processed independantly on individual Grid worker nodes, hence the usage of the Grid offers a good framework to run many observation processings in parallel, with regular interactions from the Grid and the Herschel Science Archive (see Fig. 1). Grid processing queues enable to assign different priorities to various type of processings (systematic, on-the-fly, bulk reprocessing), as well as assigning specific worker nodes for observations requiring higher RAM to be processed. The Grid is also used to perform daily test runs of automatic regression testing on different release tracks.

2.2.   XMM-Newton Remote Interface to Science Analysis (RISA)

The XMM-Newton RISA provides remote analysis of XMM-Newton data while making use of GRID computing technology. This framework (see Fig. 2) provides a more user friendly access to XMM-Newton Science Analysis Software, without the need to download and install SAS on one's local desktop, and by offering predefined data analysis workflows. It gives free access to large ESAC Grid hardware resources. Accessing and processing of the data is done at ESAC, where the data resides (in XMM-Newton Science Archive), avoiding long transfer of data. XMM-RISA is making use of various Virtual Observatory (VO) standards and tools, bringing closer together the best of all worlds. This has been a very good example of real Grid usage with collaboration with the University of Cantabria, Spain. Another potential benefit of

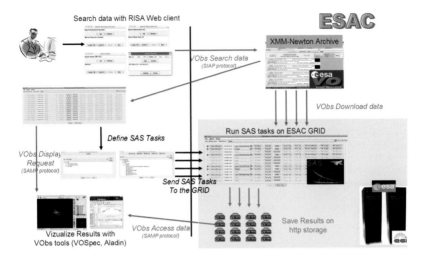

Figure 2.   XMM-Newton Remote Interface for Science Analysis.

this framework is the reduction of XMM-Newton SAS software maintenance effort, as fewer operating systems would need to be supported, once the system becomes fully operational.

2.3. XMM-Newton Calibration Analysis

XMM-Newton calibration scientists typically study the variation of some instrument calibration parameters through the mission lifetime. An example is the monitoring since XMM-Newton launch (1999) of the EPIC-pn X-Ray imager Long-Term CTI (Charge Transfer Inefficiency). That usually requires the analysis of about 900 calibration observations (corresponding to 7.2 Ms or 83 days of observing time). Through several iterations, various parameters are derived and tweaked against a calibration model of source line energies. Processing of one dataset requires around 3 hours, hence for 900 datasets would require almost 4 months. Usage of Grid with its full capacity (160 processors) allows to perform the same tasks within less than 1 day! The Grid hence offers the possibility to repeat similar runs very easily and exploring calibrations trends for many other parameters.

2.4. Science Case Measuring Transverse Motion of 730000 Stars

Scientists at ESAC are also using the Grid to do their daily science. A very good example is the processing of 520Gb of wide field images obtained from various space and ground based telescopes (Subaru, HST, CFHT, Isaac Newton Telescope) over a 12 years period to calculate the transverse motion of around 730.000 stars. The usage of the SCIGRID was critical, as it requires vast amount of storage, fast multi-threaded computers to extract the source photometry and astrometry, fast multi-threaded computers, and vast amounts of RAM to cross-match the multi-epoch catalogs and derive the kinematics of each star. Moreover these jobs had to be run on a parallel environment and SCIGRID proved to be the proper infrastructure. The result is a vector point diagram of the motion of all 730000 stars located in the Pleiades cluster, showing that the Field and background objects are distributed randomly around (0, 0) mas/yr, while the Pleiades members are all co-moving and form the locus near (15, −35). This allows unambiguous identification of several thousand members–when less than 1000 were known to date– down to the planetary mass regime. The scientific outcome will be extremely valuable and rich, from refining the mass function of the cluster, identifying planetary mass objects, and detailed studies of the internal dynamics. The same technique could be re-used for other regions of the sky, the limit being the number of Grid worker nodes as this was the highest usage of SCIGRID to date.

3.   The Cloud and the Grid

The Cloud has recently become the new "revolution" in the IT world. Naturally, building on an existing Virtual Infrastructure, we are investigating the Cloud concepts. Keeping in mind that we want to continue using our operational Grid infrastructure, we are looking at synergy between both computer infrastructures. Building an internal Cloud at ESAC will make easier and more flexible the redeployment of existing hardware resources for Grid and non Grid usage. An idea could be to deploy on-demand worker nodes on the ESAC internal cloud when more Grid resources are required. In parallel, we're exploring the usage of a public Cloud where we could deploy full applications (like for Gaia) or again deploying on-demand extra Grid worker nodes through a VPN.

4.   Conclusion

Two Grid infrastructures have been built at ESAC, serving complementary purposes: support to science mission operational tasks (such as data processing, calibration monitoring) and offering high computing capabilities for science use. Combining science archives, Grid and VO technology, this computing framework makes ESAC a unique European Data Center and Data Processing Center. The Grid for operational use still has a bright future, although it's obvious that close synergy with the emerging Cloud technology is to be investigated.

Acknowledgments.   The authors want to thank ESAC Computer and GRID Support Group, the ESAC Science Archives and VO Team, the XMM-Newton SAS team, the Herschel Data Processing Team, M. Smith for the XMM-Newton calibration work and H. Bouy and B. Merin for their science work using the Grid.

Astronomical Data Analysis Software and Systems XX
ASP Conference Series, Vol. 442
Ian N. Evans, Alberto Accomazzi, Douglas J. Mink, and Arnold H. Rots, eds.
© 2011 Astronomical Society of the Pacific

A Distributed Datacube Analysis Service for Radio Telescopes

Venkat Mahadevan and Erik Rosolowsky

The University of British Columbia, Okanagan Campus,
3333 University Way, Kelowna, BC Canada V1V 1V7

Abstract. Current- and next-generation radio telescopes are poised to produce data at an unprecedented rate. We are developing the cyberinfrastructure to enable distributed processing and storage of FITS data cubes from these telescopes. In this contribution, we will present the data storage and network infrastructure that enables efficient searching, extraction and transfer of FITS datacubes. The infrastructure combines the iRODS distributed data management with a custom spatially-enabled PostgreSQL database. The data management system ingests FITS cubes, automatically populating the metadata database using FITS header data. Queries to the metadata service return matching records using VOTable format. The iRODS system allows for a distributed network of fileservers to store large data sets redundantly with a minimum of upkeep. Transfers between iRODS data sites use parallel I/O streams for maximum speed. Files are staged to the optimal host for download by an end user. The service can automatically extract subregions of individual or adjacent cubes registered to user-defined astrometric grids using the Montage package. The data system can query multiple surveys and return spatially registered data cubes to the user. Future development will allow the data system to utilize distributed processing environment to analyze datasets, returning only the calculation results to the end user. This cyberinfrastructure project combines many existing, open-source packages into a single deployment of a data system. The codebase can also function on two-dimensional images. The project is funded by CANARIE under the Network-Enabled Platforms 2 program.

1. Introduction

The goal of the CyberSKA project is to develop the cyberinfrastructure for the Square Kilometer Array (SKA). Within the context of this overall project, we are currently developing the cyberinfrastructure to enable distributed storage and processing of data in the form of FITS data cubes. Our primary motivations are to provide transparent access to cloud computing resources and to provide users with access to their data via a web based science portal so they can manage and analyze large data sets. To this end, a web portal has been established at www.cyberska.org that allows users to access various web-based data analysis applications and data sets. The web-based service presented as part of this work is one of several available on the portal.

The remainder of this paper will focus on the system architecture we have developed for enabling the ingestion, querying, storage, and transfer of FITS data cubes across a data grid powered by the integrated Rule-Oriented Data-management System or iRODS for short (Data Intensive Cyber Environments 2010). Further, the web-based service which can be used to query for and perform various processing operations on FITS files is presented. The paper concludes with a discussion of possible future developments and enhancements.

2. System Architecture

The service is designed around a three-layer architecture comprised of a Data Layer, a Metadata Layer, and a Web Services Layer. The end user will interact with the system via the Web

Services Layer, either using a custom web-based application as shown in this work or via other clients that will communicate with the Web Service Layer using a RESTful API (Fielding 2000).

2.1. Data Layer

At this level, FITS files are stored in the data grid based on iRODS. iRODS is a "hands off" distributed data system i.e., a data grid management system that supports data replication and cross-site backups, abstraction of the data storage location from the end user, and high speed data transfers using multiple TCP/IP streams. For our initial test environment, two iRODS servers were configured: one at the University of British Columbia, Okanagan Campus and the other at the University of Calgary, our partner institution. Eventually there will be multiple sites each housing various collections of data.

2.2. Metadata Layer

This layer consists of a PostgreSQL database with spatial capabilities based on the pgSphere extension (pgSphere Dev Team 2010). FITS header data is ingested into the metadatabase, which has a schema based on the IVOA Resource Metadata recommendation (Hanisch et al. 2007). The schema has been modified to directly support FITS files as a resource type. Figure 1 shows the data ingestion process into the Metadata and Data Layers.

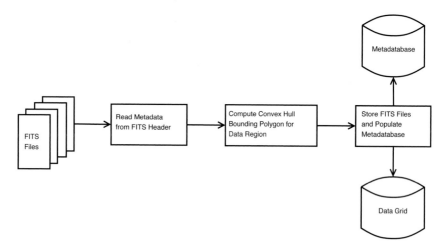

Figure 1.   Ingesting FITS files into the Metadata and Data Layers.

Three types of metadata are stored for querying purposes, namely: spatial metadata (the bounding polygon of the spatial region encompassing each FITS file), spectral metadata (frequencies, bandwidths, stokes codes), and temporal metadata (observation dates). Queries for FITS files can be done via any logical combination of spatial, spectral, and temporal queries. A GiST (Generalized Search Tree) index can be created on spatial metadata in pgSphere to speed up operations such as contains and overlaps when there are large numbers of rows in the database table containing spatial information.

FITS files in the metadatabase are organized into *resources*, where each resource is comprised of a collection of FITS files with common attributes (for example, all the files from a particular survey). FITS files in a resource share the same *capabilities*, i.e., the set of processing analysis options which can be applied to the individual files. Two-dimensional images have capabilities like subregion extraction, mosaicking, and remapping using the Montage (Laity et al. 2005) and WCS Tools (Mink 2006) packages. Three-dimensional images add frame extraction to the aforementioned capabilities.

2.3. Web Services Layer

End users interact with the service via a RESTful API. Custom web-based applications such as the one presented here can utilize this API to query the metadatabase for FITS files matching various parameters as described earlier. The query and return of files from the data grid is a two-stage process. The initial query checks the metadatabase for the names of FITS files matching the given query parameters and returns all matching files in a list, which is in a custom XML format. The initial result from an API query returns lists of matching files organized into resources. This XML list can be parsed for direct download of all matching files or the list can be based to a web-based application for workflow processing.

The web-based application shown in figure 2 uses the notion of workflows for data processing. Users can build workflows by selecting the FITS files they want to process, dragging and dropping processing options (such as subregion extraction, mosaicking, spatial reprojection, pixel scale, etc.), and staging the output to a directory that will be web accessible once processing is completed. The file list from step one mentioned above is used to determine the list of files to process and other operations that happen in sequence, such as spatial registration/alignment. Leveraging the iRODS distributed data storage facilities in the background, data is shipped to where the processing occurs automatically. Once processing is completed, results are returned to the user in VOTable format (Ochsenbein & Williams 2009) with FITS serialization of data.

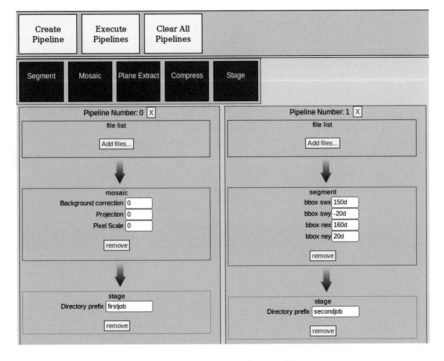

Figure 2.   Web-based Data Processing Workflow Builder.

3.   Conclusion and Future Developments

This work explored the core foundations of an infrastructure for analyzing FITS datacubes which combined a distributed data grid, a custom spatially-enabled PostgreSQL database, and a web services API. In the near term, the infrastructure will be expanded to cater for distributed processing in conjunction with the existing distributed data storage for analyzing large data sets. The workflow builder will be expanded to include operations such as: image convolution, object identification, image statistics, Fourier transforms/spectral analysis, and basic pixel array manipulation. We also hope to build an API for user contributed data processing modules.

Acknowledgments.   The authors would like to thank our project collaborators at the University of Calgary and the developers of the various open-source software packages utilized by our system. This research made use of Montage, funded by the National Aeronautics and Space Administration's Earth Science Technology Office, Computation Technologies Project, under Cooperative Agreement Number NCC5-626 between NASA and the California Institute of Technology. Montage is maintained by the NASA/IPAC Infrared Science Archive. This research has been made possible by funding from CANARIE under the Network-Enabled Platforms 2 program.

References

Data Intensive Cyber Environments 2010, IRODS:Data Grids, Digital Libraries, Persistent Archives, and Real-time Data Systems. http://www.irods.org

Fielding, R. T. 2000, Ph.D. thesis, University of California, Irvine

Hanisch, R., et al. 2007, Resource Metadata for the Virtual Observatory Version 1.12. URL http://www.ivoa.net/Documents/latest/RM.html

Laity, A. C., Anagnostou, N., Berriman, G. B., Good, J. C., Jacob, J. C., Katz, D. S., & Prince, T. 2005, in Astronomical Data Analysis Software and Systems XIV, edited by P. Shopbell, M. Britton, & R. Ebert (San Francisco, CA: ASP), vol. 347 of ASP Conf. Ser., 34

Mink, D. 2006, in Astronomical Data Analysis Software and Systems XV, edited by C. Gabriel, C. Arviset, D. Ponz, & S. Enrique (San Francisco, CA: ASP), vol. 351 of ASP Conf. Ser., 204

Ochsenbein, F., & Williams, R. 2009, VOTable Format Definition, Version 1.2. http://www.ivoa.net/Documents/VOTable/

pgSphere Dev Team 2010, pgSphere. http://pgfoundry.org/projects/pgsphere/

Part VI

Large Observatory Challenges

Astronomical Data Analysis Software and Systems XX
ASP Conference Series, Vol. 442
Ian N. Evans, Alberto Accomazzi, Douglas J. Mink, and Arnold H. Rots, eds.
© *2011 Astronomical Society of the Pacific*

First Year In-Flight and Early Science with the Herschel Space Observatory

P. García-Lario

Herschel Science Center, ESAC/ESA, Madrid, Spain

Abstract. Herschel, an ESA space observatory equipped with science instruments provided by European-led Principal Investigator consortia with important participation from NASA, was launched on 14 May 2009. With its 3.5m diameter primary mirror, Herschel is the largest telescope ever launched into space. Herschel carries three science instruments whose focal plane units are cryogenically cooled inside a superfluid helium cryostat. The PACS and SPIRE instruments provide broadband imaging photometry in six bands centered on 75, 100, 160, 250, 350, and 500 μm and imaging spectroscopy over the range 55–672 μm. The HIFI instrument provides very high-resolution heterodyne spectroscopy over the ranges 157–212 and 240–625 μm. The prime science objectives of Herschel are intimately connected to the physics of, and processes in, the interstellar medium (ISM) in the widest sense. Near and far in both space and time, they stretch from solar system objects and the relics of the formation of the sun and our solar system, through star formation in the ISM and the feedback material returned by evolved stars to the ISM, to the star formation history of the universe, galaxy evolution, and cosmology. The very first observational results from Herschel already show that it will have strong impact on research in all of these fields, as exemplified by the few observational results presented here, These are just the tip of the iceberg of what is yet to come in the remaining 2 years of operations.

1. Introduction

ESA's Herschel space observatory (Pilbratt et al. 2010) was successfully launched, together with the Planck satellite, on board an Ariane 5 ECA launcher from Kourou, the European's Spaceport in French Guiana, on 14 May 2009. Equipped with a passively cooled 3.5m primary mirror, Herschel is the largest telescope ever launched into space, orbiting around the second Lagrangian point of the Sun-Earth system (L2) at a mean distance of 1.5 million km in the anti-sun direction.

With its ability to observe across the far infrared and sub-millimeter wavelengths (55–672 μm), Herschel is bridging the gap between earlier infrared space missions and ground-based facilities observing in the sub-millimeter range. Designed to observe the 'cool and distant universe', Herschel's primary science objectives are to:

- Study the formation of galaxies in the early Universe and their subsequent evolution;
- Investigate the creation of stars and their interaction with the interstellar medium;
- Observe the chemical composition of the atmospheres and surfaces of comets, planets and satellites in our solar system; and
- Examine the molecular chemistry of the universe.

For this, Herschel is equipped with three main instruments: HIFI, PACS and SPIRE, housed in a superfluid helium cryostat. The PACS and SPIRE instruments provide broadband imaging photometry in six bands centered at 75, 100, 160, 250, 350, and 500 μm and imaging spectroscopy over the range 55–672 μm, while HIFI provides very high-resolution heterodyne spectroscopy over the ranges 157–212 and 240–625 μm.

Herschel is operated as an observatory facility designed to provide a minimum of 3 years of routine science operations, with an estimated total mission lifetime of 3.5 years. As an observatory, it is available to the worldwide scientific community, with roughly two thirds of the observing time considered 'open time', allocated through standard competitive calls for observing proposals.

2.   Early Mission Phases

Exactly one month after the launch, Herschel's crycover was successfully opened. The first light images of an astronomical source (M51) were obtained with the PACS instrument, which demonstrated perfect optical performance, fully according to specifications.

Building on the experience from this PACS 'sneak preview', and making use of time initially allocated to 'thermal stabilisation' of the whole spacecraft, all three instruments performed their initial test observations. These included SPIRE images of nearby galaxies, HIFI spectroscopy of a star forming region and PACS imaging spectroscopy of a planetary nebula. These very first attempts provided spectacular data, and were followed by a two-month 'commissioning phase', mainly used for instrument and spacecraft functional tests. During three additional months of 'performance verification', the different planned instrument observing modes were optimised according to the results obtained in-orbit.

As a consequence of the experience gained during this three-month period, various observing modes were declared 'ready to use' in a gradual fashion. This led to the so-called 'science demonstration phase', in which small snippets of the various observing programs were executed for verification that the users were getting what they wanted. If not, they had to update and optimize their observing programs.

The first 'science demonstration phase' observations were delivered to users on 28 September 2009. Since then all remaining observing modes have been released, adapted to in-flight circumstances. In this process, some observing modes were discarded, while others had to be fully revamped.

This sliding transition took longer than initially expected, as HIFI became unavailable on 2 August 2009, affected by an anomaly, and was only recovered for science in April 2010. Despite this adversity,we can say that PACS and SPIRE have routinely been operated since mid-December 2009 and HIFI since mid-April 2010. At the present moment, the three instruments are being operated at full speed.

3.   Early Science Results

In the first year of routine science phase operations, Herschel has started to do the science that it will continue doing for as long as the observatory will work. From the very beginning, Herschel has demonstrated that its scientific results will have an enormous impact on essentially all fields of astronomy. From solar system studies to cosmology, from the analysis of star formation, included the origin of our own sun, to the models which describe the formation of the first galaxies in the very early days of our universe. This will include the physical processes that take place in the interstellar medium and the feedback material that is returned by evolved stars, all of which makes the universe look the way it looks now.

3.1.   Solar System Studies

One of the most interesting results obtained in the early days of Herschel is addressed in Figure 1, where we can see the detection with SPIRE of the dwarf planet Makemake. With a diameter of arounc 1500 km, it is the third largest dwarf planet known. With a surface temperature of only 30 K it is one of the coldest objects in the solar system, and thus very hard to detect. By taking images 44 hours apart and subtracting the 'before' and 'after' image, the background sky is removed. What is left are positive and negative images of Makemake, showing a much fainter emission at sub-millimeter wavelength than model predictions (9.5 ± 3.1 mJy at 250 μm). This suggests that the object is much more complex than expected, that Makemake is more reflective (i.e., has a higher albedo) than we thought, implying that its size might not be as large as has been derived from optical data (Lim et al. 2010).

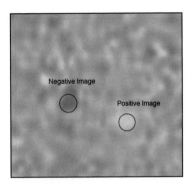

Figure 1.   SPIRE differenced image of the dwarf planet Makemake (Key Program: TNOs are cool!; P.I.: Thomas Müller).

Another interesting result obtained with Herschel in this field was the detection of very high levels of stratospheric carbon monoxide (CO) in the atmosphere of Neptune (Key Program: "Water and Related Chemistry in the Solar System"; P.I. Paul Hartogh). The only explanation for this result seems to be that a cometary impact may have happened about two centuries ago (Lellouch et al. 2010).

3.2. Protoplanetary Disks

Herschel can also see the late stages of the formation of planetary systems like our own solar system. Several Key Programs are searching for extended emission around nearby stars which may be indicative of the presence of protoplanetary disks. Kuiper belt-like structures, made up of icy objects ranging in size from micron-sized grains to comets many kilometers in diameter, have been detected by Herschel (see Fig. 2; Matthews et al. 2010)

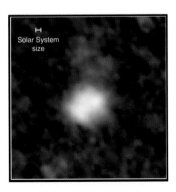

Figure 2.   The star Eta Corvi as seen by Herschel. An outer ring of icy, comet-like bodies is seen, much like the Kuiper belt in our solar system. This star in particular is peculiar as the system seems to contain a second warmer, dusty belt. The size of our solar system is shown for comparison (DEBRIS; P.I.: Brenda Matthews).

3.3.   New Views of the Galactic Plane

Hi-GAL, the Herschel infrared GALactic Plane Survey, is providing spectacular images of the Milky Way, demonstrating the large area mapping capabilities of Herschel. The survey covers a region of 120 degrees around the Galactic center, and it is intended that this coverage will be expanded in future open time calls to a similar area in the anti-center region. In Figure 3 we show one of the first tiles corresponding to this survey (Molinari et al. 2010).

Figure 3.   PACS/SPIRE three-color composite image (blue=70 μm, green=160 μm, red= 350μm) which unveils our own Milky Way galaxy around $l = 59°$ as one giant nursery where generations of new young stars are continuously born (Hi-GAL: the Herschel infrared GALactic Plane Survey; P.I.: Sergio Molinari).

This image is taken in the constellation of Vulpecula and shows the entire assembly line of newborn stars. The diffuse glow reveals the widespread cold reservoir of raw material which our Galaxy has in stock for the production of new stars. Large-scale turbulenc, possibly due to giant colliding Galactic flows, causes this material to condense into the web of filaments that we see throughout the image. These filaments act as "incubators" where the material becomes colder and denser. Eventually gravitational forces will take over and fragment these filaments into chains of stellar embryos that can finally collapse to form infant stars.

3.4.   Star Formation

In the area of star formation Herschel has also contributed with the systematic mapping of some nearby star forming regions in the Gould Belt (André et al. 2010) and giant molecular clouds (Motte et al. 2010). These observations are providing us with a new view of the physical processes which lead to the formation of new stars.

The most extraordinary feature observed in all Herschel maps is the ubiquitous pattern of filaments in the ISM structure. The compact sources detected at 250 μm are preferentially distributed along these filaments (see Figure 4). The high degree of association between bright filaments and dense pre-stellar cores suggests a column density threshold for the appearance of these cores, and a formation scenario that starts with the condensation of diffuse clouds into long filaments. As the column density increases, the filaments become gravitationally unstable and fragment into condensations which will become the embryos of future stars. In less dense regions, at higher Galactic latitudes, prestellar cores are located in clumps and are not gravitationally bound, so they will likely not be able to produce new stars.

Figure 4.   SPIRE and PACS images have been combined into a single composite (blue=70μm, green=160μm, red=combined SPIRE emission from all three SPIRE bands at 250/350/500 μm). The composite image makes it easy to locate the star-forming filaments that would be very difficult to isolate from a map made at a single far-infrared or submillimeter wavelength. The image contains an incredible network of filamentary structures with surprising features indicative of a chain of near-simultaneous star-formation events.

Other observations taken with Herschel in the vicinities of some well-known OB associations show clear indications of triggered massive star formation. The observations obtained e.g., around the Rosette molecular complex (Motte et al. 2010) or RCW 120 (Zavagno et al. 2010), are good examples of this (see Figure 5). With Herschel we can easily identify a new population of highly embedded stars around ionized regions. These are formed in the gas and dust cocoons but are completely invisible at optical wavelengths.

3.5.   Feedback Material from Evolved Stars

Among Herschel's results concerning evolved stars, the discovery of more than 60 spectral lines from warm water vapor in the circumstellar environment around the aging carbon star IRC +10216 (Decin et al. 2010) is certainly one of the highlights. These results from PACS and SPIRE spectroscopy seem to indicate that the clumpy structure of these circumstellar shells allows UV photons from the ISM to penetrate deeply enough to trigger a set of reactions leading to the production of water.

The circumstellar shells of evolved stars are enormous molecular laboratories which allow very complex molecular chemistry. Hundreds of lines are observed in some of these sources. Some of them are complex organic molecules, considered to be the building blocks of life. In Figure 6 we show, as an example, the rich molecular spectrum from PACS of the circumstellar shell around the red supergiant star VY CMa (Royer et al. 2010). This oxygen-rich star is in an extreme evolutionary state and could explode as supernova at any time. The SPIRE spectrum (not shown) is dominated by prominent features coming from carbon monoxide (CO) and water (H_2O).

3.6.   The Molecular ISM

Herschel's heterodyne instrument, HIFI, is ideally suited to study the physical and dynamical processes that take place in the interstellar medium, with unprecedented spectral resolution.

Figure 5.   PACS/SPIRE three-color composite image (blue=70 μm, green=160 μm, red=250 μm) of the Rosette molecular complex, a good example of triggered massive star formation (HOBYS: the Herschel imaging survey of OB Young Stellar objects; P.I.: Frédérique Motte).

Some regions have been observed in detail with HIFI by different projects. In Orion, Herschel has gotten the most complete spectrum of molecular gas at high spectral resolution ever (Bergin et al. 2010), with more than 100,000 lines in one single spectrum scan (see Figure 7). Among the organic molecules identified in this spectrum are water, carbon monoxide, formaldehyde, methanol, dimethyl ether, hydrogen cyanide, sulphur oxide, sulphur dioxide and their isotope analogues. It is expected that new organic molecules will also be identified.

This spectrum is just the first glimpse at the spectral richness of the sources that Herschel will observe. It harbors the promise of a deep understanding of the chemistry of the interstellar space once this and other complete spectral surveys are available.

3.7. Extragalactic Astronomy

Herschel PACS and SPIRE integral field spectroscopic data is providing the most detailed images ever obtained of dusty nearby galaxies in the infrared and submilimeter range, as well as a new view of the more distant galaxies in the Universe.

Near our Milky Way, Herschel/PACS has been able to provide for the first time spatially resolved spectroscopic images of the ISM in the nearest starburst galaxy M82. The line ratio [O III]/[C II], a diagnostic of ionized gas vs. neutral gas drops rapidly going outwards from the galaxy center along the disk. In contrast, this ratio does not drop so significantly when going outward in the super-wind direction. On the other hand, the SPIRE spectrum of M82 shows strong emission lines from CO over the whole spectral range. This can be used with emission lines from atomic carbon and ionized nitrogen to constrain the fundamental properties of the gas (Panuzzo et al. 2010).

With the help of Herschel we can now study the process of star formation and nuclear activity in infrared bright galaxies at practically all redshifts. For each redshift range we can observe a larger number of galaxies, going fainter than any other previous studies made with other infrared telescopes such as Spitzer. The results confirm that star formation was several times more active and efficient in the early Universe, compared to current rates.

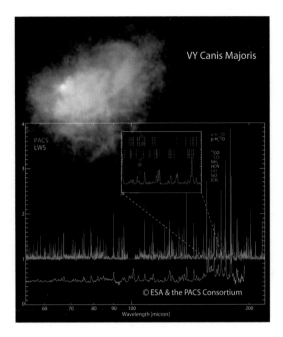

Figure 6.   PACS spectrum of the red supergiant VY CMa between 57 and 210 μm. For comparison, depicted in gray, and offset by −0.5, is the observation from the ISO Long Wavelength Spectrometer (LWS). The inset shows a zoom into the 156 to 172 μm, containing 44 different identified molecular lines (MESS: Mass-loss of Evolved StarS; P.I. Martin Groenewegen).

Some Herschel Key Programs are obtaining deep images of cosmological fields such as GOODS North and South, the Hubble Deep Field, and COSMOS (see Figures 8 and 9). The results indicate that with Herschel we can now resolve more than half of the Cosmic Infrared Background (CIB) into individual galaxies. The CIB is a relic, isotropic emission distinct from the Cosmic Microwave Background (CMB) associated with the formation of galaxies. Predicted in the mid-1960s, it was first detected 30 years later with the Cosmic Background Explorer (COBE). The CIB is particularly hard to probe, as all galaxies at all redshifts contribute to it. Hence it has no characteristic signature. Since the CIB peaks around 100–200 microns, the majority of the galaxies contributing to this emission remained unidentified in the pre-Herschel era.

The images obtained on these cosmological fields are providing us with a much clearer idea of how star formation has progressed throughout the history of the Universe. Studying galaxies at this early stage of the Universe will allow astronomers to test their models of star and galaxy formation.

3.8.   Much More is Yet to Come

A lot of people all around the world have been working over the last 20 years to make this mission possible. Thanks to all of them, Herschel is now a fully working observatory with its three science instruments working at full speed, providing plenty of new food for thought every day. Not only are the observatory and the instruments working very well, but it is already clear that in this unexplored region of the spectrum the Universe is even more interesting that we ever thought. Some of the initial findings have been put together in two special issues of Astronomy & Astrophysics (vols. 518 and 521). They contain in total more than 200 refereed

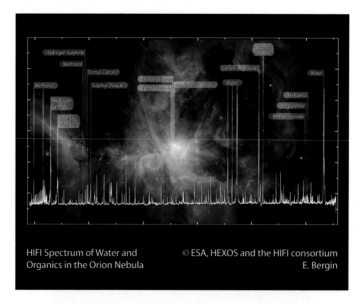

Figure 7.   HIFI spectrum of the Orion Nebula, superimposed on a Spitzer image of Orion, showing the spectrum richness of this source (HEXOS: Herschel/HIFI Observations of Extraordinary Sources; P.I.: Ted Bergin).

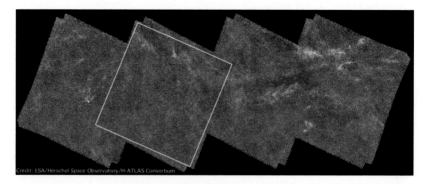

Figure 8.   Herschel-ATLAS survey (P.I. Steven Eales) composite color images in bands 250, 350, and 500 μm of a field covering roughly 4×4 degrees on the sky (more than 60 times larger than the full moon). This took 16 hours to observe and contains more than 6,000 galaxies in the inner 14 deg^2, which represents only 2.5% of the total survey area.

papers describing just the first 3 months of scientific operations with Herschel. This is certainly only the beginning of what Herschel will be able to provide us in the coming two years. Herschel will undoubtfully expand our knowledge on the formation and evolution of stars and galaxies significantly, showing us our Universe as we have never seen it before.

Acknowledgments.   Herschel is an ESA space observatory with science instruments provided by European-led Principal Investigator consortia and with important participation from NASA.

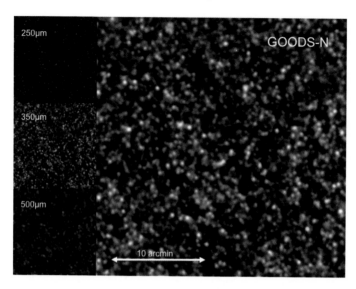

Figure 9.   SPIRE image of GOODS-North, an area of sky devoid of foreground objects, such as stars within our Galaxy, or any other nearby galaxies. It is a little larger than the area of the full moon as observed from Earth. The image is made from the three SPIRE bands, with red, green and blue corresponding to 500 μm, 350 μm and 250 μm, respectively. The image took just 14 hours of observations. Every fuzzy blob in this image is a very distant galaxy, seen as they were 3 to 10 thousand million years ago when star formation was very widely spread throughout the Universe (HERMES survey; P.I.: Seb Oliver).

References

André, P., et al. 2010, A&A, 518, L102
Bergin, E. A., et al. 2010, A&A, 521, L20
Decin, L., et al. 2010, Nat, 467, 64
Lellouch, E., et al. 2010, A&A, 518, L152
Lim, T. L., et al. 2010, A&A, 518, L148
Matthews, B. C., et al. 2010, A&A, 518, L135
Molinari, S., et al. 2010, A&A, 518, L100
Motte, F., et al. 2010, A&A, 518, L77
Panuzzo, P., et al. 2010, A&A, 518, L37
Pilbratt, G. L., et al. 2010, A&A, 518, L1
Royer, P., et al. 2010, A&A, 518, L145
Zavagno, A., et al. 2010, A&A, 518, L81

Astronomical Data Analysis Software and Systems XX
ASP Conference Series, Vol. 442
Ian N. Evans, Alberto Accomazzi, Douglas J. Mink, and Arnold H. Rots, eds.
© 2011 Astronomical Society of the Pacific

What do Astronomers Really Want?

Richard Mushotzky

Astronomy Department, University of Maryland

Abstract. After 10 years of XMM and Chandra as well as many other astronomical space missions (Swift, Spitzer, Suzaku, Hubble, etc.), it is clear that the present system of data analysis is in need of review. I will discuss some of the advantages and disadvantages of several of the present high energy data systems and suggest a path to the future. Many observers now routinely utilize many different data sets to understand the physics of the objects under study, but the data comes in many forms with different analysis systems, different data archive systems, and different calibration architectures, each of which has its own pluses and minuses. While it is far too late to have a uniform system for all NASA and ESA space missions, it is still possible to insure that the data in the archives can be analyzed by one software package that has a uniform look and feel, yet is still capable of handling the different types of data and the properties of the instruments. I will discuss some of the properties of such a system.

1. Introduction

The fundamental purpose of astronomical software is to *allow* science to be done, either now or in the future. In this regard the astronomical community has expressed many desires for its software, many of which are incompatible with each other. Thus we have the question "what does the community **really** want?"

This is directly connected to why we are astronomers. We primarily choose a job in astronomy because we really want to do it — it is a career/profession, not a job. In this regard we would like our "work" to be as straightforward as possible. It's fair to say that we get really annoyed by having to do unnecessary work. Of course, the key word is "unnecessary".

There are many things in our profession that appear to be necessary, and there is a wide range of opinions about what is unnecessary. In this regard a little personal story is enlightening. When I was a graduate student my first thesis advisor asked me to scan the spectra that she had taken on film. The standard approach was to stare into a microscope and align a vertical marker on the apparent features and then hit a button which punched a card with the mechanical value of where the "hair" was. Then, one went back and forth four times to average out measurement error. I thought this was a bit unnecessary since a newfangled digitizer was available and all one had to do was digitize the spectrum and write a little computer program to analyze it. This horrified my advisor who told me bluntly not to do all this "unnecessary" work. Clearly opinions have changed. I would argue that we are now in the middle of a change in concept of what software is and what it should do.

Despite not being a software developer or manager, I think I am qualified to comment on this. I have been on five NASA/JAXA/ESA science working groups (ASCA, Chandra (IDS), XMM (Mission scientist), Suzaku and Astro-H) and have been deeply involved in several others (HEAO-1, Einstein, Swift, BBXRT etc). I have been the US Project Scientist for XMM in charge of the US guest observer support facility (GOF; > 120 GOs/yr for 10 years). I also do science — I have been a guest observer on over 14 astronomical facilities (from IUE to Hubble) and have been author or co-author of more than 340 refereed papers with > 22,000 citations. I

235

have supervised more than 12 PhD students and 15 post-docs. **I know software issues when I see them.**

To repeat, the purpose of software is to allow us, the users, to do our jobs, which is to discover new and interesting things about the universe. Thus, software is necessary (but not sufficient) to accomplish this goal. However, the nature of the software that is required is changing. Astronomy is now 'panchromatic' — we use data across the entire electromagnetic spectrum and thus must analyze and interpret a very broad range of types of data. However, most software is still chromatic — caring about where and how the data are created. Of course, this is important: without detailed knowledge of how the instruments perform and the conditions under which the data were taken, the data are not very useful. However, this should not drive the final formats nor the means by which the data are accessed, archived, or, in a perfect world, analyzed. We care about the objects and their properties, not the details of the data. Thus, what we need is "objects" oriented software, not, as at present, data oriented software.

Of course astronomical software is only a tiny part of the software world. What, if anything, is special about astronomy (astronomical software)? I believe it is breadth — to be an excellent astrophysicist it is useful to know a little about a lot and our software framework has to recognize that. We are specialists trying to understand complex objects and thus have to be generalists as well. This is similar to physics 75 years ago, when someone like Fermi understood electronics, quantum mechanics, as well as how to get funding.

So who is the astronomical community that is telling us what they want? The vast majority of the astronomical community are users of data and software and **not** generators (see Figure 1). Most are in universities/colleges and not only are not software experts, they do not have the support structure to call on software experts. At the present time **most** are not involved in large surveys/data sets. We work on Objects, in multi-wavelength studies.

We have a major frustration — there are roughly 11 different wavelength regimes but more than 30 different data and analysis systems. A short summary of the range of possible data sets indicates the complexity: Long wavelength radio (LOFAR), classical radio (VLA), Millimeter (ALMA, Planck), far IR (Herschel), Mid-IR (Spitzer), Optical-IR (Keck, Hubble, JWST), UV (Hubble, FUSE, EUVE), X-rays (Chandra, XMM), Hard X-rays (Swift, Integral), Gamma-Rays (Fermi), TeV gamma-rays (Veritas, HESS). Each band has a wide variety of data and its own special issues with respect to spectra, timing, images, surveys, observing strategies, nature of the data (photon counting, Fourier transforms, etc.).

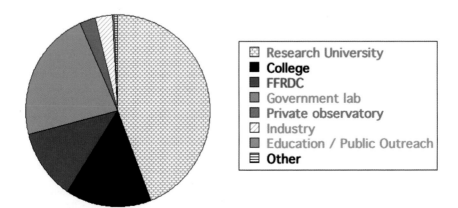

Figure 1.   Distribution of astronomers by employment.

2. Sources of Data

There are three main sources of data in the U.S. I will only briefly mention the situation in Europe, Australia and Japan due to lack of space, a vast increase in complexity and my relative ignorance of the situation. The situation is even more complex because of the availability of data from one continent to researchers everywhere. I apologize to my international friends.

1) NASA missions. These range from WMAP to Fermi, covering a factor of 10^{14} in frequency and a wide variety of instrumentation. Access is fully open via proposals and peer review. All of the data becomes public (eventually) and there have been attempts to have robust data analysis systems, user help, data access and open archives with frequent updates and new versions of both the data (as processing improves), and software. However, there has been no attempt at uniformity in either data, archives or analysis tools. Thus while the data are easy to access, one must master very different analysis systems and data archives and there are no "non-mission specific" general tools to analyze or access the data. A similar situation occurs with ESA missions and data systems.

2) NSF facilities. These fall into 3 big categories: optical/near-IR, radio and Millimeter and some smaller bits such as TeV gamma-rays. Access is usually fully open via proposals and peer review. There are a wide range of policies: some data go public (VLA, Gemini) and some do not. The archives are very variable in quality and content. Often there is instrument specific software available — frequently with cookbooks. There are general user analysis tools (AIPS, IRAF) which are fairly uniform for a given type of data and data reduction pipelines frequently exist. I believe that the situation with the VLT is much better with respect to archives and data systems, but I do not know the situation with other sources of data such as the EVN or HESS.

3) Private Observatories (e.g., Keck, Magellan, etc.). These tend to be focused on optical telescopes. Many do not have archives and for those that do, most are not open to the public. As in the case of NSF archives, software is different at each site and for each instrument and there is a wide range in policies. Some of them do not have pipelines or much user support. Access is usually restricted to members of the consortium but for some there are complex rules that allow more open access. This is like the US health care system. Caveat emptor.

Archival data analysis is rapidly increasing in importance. There has been a vast increase in archival research — based mostly on space-based data. However, due to the relatively poorer condition of ground archives in the US, this is not yet a major activity, but it is coming soon. However, to combine Chandra/HST/Spitzer data you need to become "expert enough" in three data systems which are mutually incompatible; a similar situation exists for Herschel and XMM in Europe.

3. What is "Unnecessary" Work?

1) The data are hard to find — perhaps the VO will fix this! They are in a vast variety of non-compatible formats and need to be accessed through a large number of non-compatible tools. Then comes the problem of "Once you get it what do you do?" The analysis tools differ in look, feel, and procedure, but, in general, do the same things. In astronomy if you want to turn the same type of screw you may need six types of screwdrivers! Thus the user must learn, to a fairly high degree of competence many mutually incompatible systems. Many of these are rather "arcane" with poor interfaces and difficult to use systems.

How did this happen? Primarily because of the lack of "forced" collaborations: each team did what they perceived to be best for their data set (e.g., XMM compared to Chandra) and many decided to use a data analysis system of their own creation which was optimal for their data set. There was no reason to be uniform. That is, there was no benefit to the program/project to produce data formats/archives/software compatible with what came before or new more uniform standards. I fully sympathize with the projects, which are often under great financial and time pressures, but it does not help the general observer. We need some sort of feedback into this system.

Is there a solution? It is not really possible to "force" compatibility except at the agency level (e.g., NASA/ESA), but then one has to include NSF, private observatories and the whole European/International ground based community), and it seems as if neither NSF nor private observatories will ever agree. Can "we" (whoever that is) re-do all the formats ex post facto? It is lots of work and probably expensive. Who will pay? What is the gain? Because archival data is vastly increasing in importance this issue is becoming more urgent.

2) Software un-organization — there are no software "indexes." The user does not know what the software is capable of and how to do what they want to do. The user wants to do X — but the documentation says that program P does this that and the other, and one has to go through all the programs to find the one that does the right thing.

Frequently doing X requires chaining together multiple programs in *various* locations. There are two frequent solutions to this problem: *threads* — but when the number of threads gets large, this does not work well unless the threads are also organized (e.g., spectral extraction threads, image analysis threads); and *cookbooks* — a good solution but deadly for the "attention deficit" generation, which expects simple directions such as: "to make a loaf of bread, buy frozen bread and then microwave it" rather than "buy flour, milk," etc.

3) One needs different commands to do the exact same things inside a given system and between system and system. This is just madness! Instead, there should be a uniform set of commands to do common tasks (e.g., plot, extract spectrum, make a light curve). Again, one does not need new software, just an overlay program that converts from one simple system into the complex set of available software. While this is far from trivial (there is often not a one to one translation) it is much easier than requiring all the systems to use the same tools.

4) Frequently the software does not 'know' the relevant data processing parameters even though they are all in the FITS header for common tasks. For example, in IRAF and Suzaku software one has to modify FITS headers by hand to perform standard functions! FITS headers should be in English(!), not in code. The software should "know" how to obtain the relevant parameters (they are all in the FITS header after all) for common tasks.

4. Goal

The primary goal is to be able to do multi-wavelength research as easily as possible (consistent with accuracy) **now**. With new systems (ALMA, JWST, NuStar, etc) coming on line, we need to do something **ASAP**. Unless something is done **now** it will be very difficult to retro-fit. The future is **now**.

In order to achieve this goal one needs to make the input into the data and the analysis packages as simple and straightforward as possible e.g., take the Macintosh vs. the IBM JCL model. In this conception there is no reason for the average user to ever see the complexity that underlies the data and the software. However, astronomers are not the average person, we are smarter and more knowledgeable (at least I hope), so there has to be a path to "seeing" the underlying complexity as well (as opposed to blindly following the Mac model).

The system should allow the user to get something useful quickly and easily on the first try. The top level web page must be obvious, clear and not require the manual — (RTFM is not the goal). The system must assume the user is naive and knows just a little (but to avoid tedium have an expert web page also available at the very top level). Also if there is a functional pipeline it must produce useful results. In many cases a huge effort has gone into the pipeline but because the results are not trustworthy the user ignores the outputs.

Let's consider 3 models for system design.

1) An old fashioned radio: It did a few things very well. It was simple, worked well, was reliable, cheap, and had a very long life. It was easy to repair, open access, one could build it oneself. However, it had limited capability and was hard to modify or update.

2) The space shuttle model: Very powerful, very capable, very expensive, hard to repair and maintain, very complex, and hard to modify. This is similar to many present data analysis systems.

3) The iPad model: Simple, works well, very capable, reliable, moderate cost, life span unknown, (the working hypothesis is to replace not repair), very easy to modify, but controlled access (hard to get at the innards).

The problem is to find the right combination of these eigenvectors — e.g., moderate to low cost, but powerful, easy to maintain, having a long life, and being open access. I think the best option is pairing the iPad model with a bit of the old fashioned radio thrown in.

5. Present Examples of "Bad" Practice

I am most familiar with the software in the high energy community and thus will use it as an example. This is not to say that high energy software is any better or worse than that found in the rest of our field. However, in general it is produced by teams funded by the space agencies (NASA/ESA/JAXA) and thus tends to have a more secure and larger funding base than some other fields.

In X-ray astronomy we have at present three "full" data analysis systems (each of which runs on many platforms) CIAO for Chandra, SAS for XMM, FTOOLS for Suzaku, RXTE, and Swift. They all use FITS (sort of) as the primary means of data storage and transfer.

Each of these systems does very similar things: 1) clean data, calculate background; 2) make images in sky coordinates with a precise aspect correction; 3) extract spectra and light curves in standard formats; and 4) provide files for spectral, temporal and spatial analysis. In addition, the Chandra project has also developed its own analysis tools (e.g., Sherpa) and the Chandra PI teams have developed tools that others can use that are not part of any other system (ISIS, ACIS Extract, etc.). Finally each mission (Swift, Suzaku, etc.) also has its own data archive, organized in its own way. Each of these systems is well documented and robust. However, each has its own interface, set of commands, help system, and style. Knowledge of one does not help very much with learning the others. Some of these systems can also analyze the data from other missions (e.g., FTOOLS can analyze XMM data). Given the way that our proposal system works, it is not at all clear that if one obtains Chandra data one year, one will obtain more data in following years; one is just as likely to have XMM or Suzaku or Swift data. Thus by the time you learn CIAO well you will have to attempt to learn SAS!

Thus, the problem is an your average X-ray astronomer will need (attempt) to use data from all these missions but does not have the time to learn them all well nor the memory to remember what one learned well three years ago. One can assume that an X-ray astronomer (assuming that this is not a archaic term in this era of multi-wavelength science) will be more expert in at least one of these systems then the average astronomer. So, what is a poor radio astronomer to do (other than find a X-ray astronomer as a collaborator)?

This raises several sets of "Big Issues:" Which one of these systems should an individual astronomer learn and use? Which one(s) of these systems should the agencies continue to support in the long term? The archives will be useful for many years to come and without a supportable data system may not be accessible (it took a major effort on the part of the HEASARC to "restore" older data sets). In addition, which ones (if any) of these systems should future missions (e.g., NuStar, Astro-H, GEMS) choose and why? Unfortunately there is a well known sociological problem: when one has too many choices, one becomes demotivated from making any choice at all.

6. Are Things Getting Better? — Herschel

To quote from Ott (2010), "Developing the Herschel Data Processing system is a major project, with over 200 contributors and currently 60 full-time equivalents working on calibration, coding, documentation, quality control, testing and tutoring. Development started in 2002 for a 2009 launch."

It appears from the documentation to be well designed: it has both a script driven, command line based environment (suited to developers and experts needs) and a GUI based end user-oriented environment. This interface is data-centric rather than language-centric, providing astronomers who are not experienced in Java a state of the art interface. The same framework can be used to download, reprocess, analyze and compare data from all three instruments on Herschel simultaneously. But despite these great concepts and years of work, the main analysis tool, HIPE 2.0 (Science Routine Phase version), was updated three times this year and updates are anticipated every three months. In addition, use is not simple: the "How To" Guide is 208 pages long with lots of screen shots and instrument specific documentation ~ 100 pages/instrument. As with almost all other systems there is no mention of other wavelengths. As of today, reading the first page of the PACS quick look document you find ... *"It will take a while to get used to HIPE and to reducing PACS data, so allow yourself a lot of patience"* — not exactly what a user wants to read. I think that we will have to wait and see if Herschel has produced a better system. Certainly it is not the system we need for multi-wavelength research.

7.   The Need for a Re-focus of Software Efforts has been Recognized

The Virtual Astronomical Observatory science council has said, in its report this spring (VAO-Science Council 2010), "We strongly endorse and support the refocusing of VAO activities on ease of use and science productivity. This focus should include paying immediate attention to user-interface design, tool and data discoverability, interoperability among tools, and visualization strategies".

And they also said: "A Change of Emphasis from Infrastructure to Science Tools is needed: The NVO has made admirable progress in 'laying the pipes' for convenient user access to complex distributed astronomical databases. However, early use of the system for top quality science has been weak, pointing to a lack of value-added intermediate software products that assist the user in extracting knowledge and understanding from the extracted data".

In other words the software should become "Objects oriented:" We have the road, but still need the cars to be designed and built and the drivers trained.

Continuing the quotes from the VO advisory group: "Existing user interfaces tend to be specific to a particular database (e.g., SDSS), putting the burden of learning the intricacies of all these interfaces on the user who wants to engage in multi-wavelength research. Moreover, the casual or beginner user should not be required to be expert in SQL (or any other arcane systems (RM added)). We urge the VAO to provide a user-friendly search interface that could cross-match several VO-published databases."

Unfortunately this statement is true of the majority of our fields software! This is probably due to the old dictum, "The best thing about programmers is they like to program. The worst thing about programmers is they like to program." As stated in one of the other ADASS presentations, "It turns out that all good examples: were developed either by professional astronomers with very strong IT/CS background or by IT/CS professionals working closely with astronomers for years and understanding astronomy. One cannot simply hire an industrial software engineer to develop astronomical software and/or an archive and/or a database" (Chilingarian & Zolotukhin 2011). My personal experience strongly supports these findings. Without direct hands-on interaction between astronomers and software professionals one ends up with junk; without software professionals, astronomer-written code is slow, hard to maintain and frequently buggy.

There are a huge number of available packages: At the NRAO web sites over 170 software packages and systems were linked to, from ASSIST to XASSIST! — and this is surely an incomplete list. Most of them do more or less the same things. I am far from the first person to comment on this (see Weiner et al. 2009).

Much of the requirements of the various astronomical software systems are *the same* — e.g., spectral extraction, source detection, time series, background analysis. But the input data are various — frequently each instrument has its own special requirements and sometimes the desired outputs are very different (but the number of categories is rather small).

I propose a simple process: 1) Take what we have now and find simple, cheap, fast ways to make it better — KISS. Fancy new systems may not be necessary. What is required are simple, robust, easy to understand functional systems. While revolutionary changes will certainly come, they take time, effort, and money and will only occur if people see the short and long term benefit. Perhaps the astronomical community should not lead in new software development but utilize the vast output from the commercial world — take what is best for us and leave the rest.

2) Test the web site with post-docs and grad students — not experts! If an expert is needed to access the system at the first try, it is a failure, unless you do not want the general astronomer to be able to use it.

3) Have overlay software that *converts* the *unusual* format data into a standard one *on the fly* (based on user choice). This requires *real units* or conversion factors (e.g., cts/pixel/ readout, photons/cm^2/sec or ergs/m^2/Å) and must have correct and up to date comprehensible documentation.

We must remember that better software can be equivalent to bigger telescopes and new high tech instruments. One gets more out of the data and more of the data can be useful. I believe that if software were easier to use and more robust we would get a lot more science out of our present instruments. A good analogy is the thousands of features in your video recorder that you never use (or even know about) while you often tried and failed to figure out how to use the functions you really needed. If proper attention had been paid, the effort to develop the unused features would have gone into making the frequently used ones easier and more robust.

Finally, having to deal with complex data does necessarily mean being stuck with a complex user interface (e.g., Google).

Acknowledgments. The author wishes to thank the organizers of the meeting for allowing me to present this opinionated presentation and to Peter Teuben for making it possible for this presentation to be printable.

References

Chilingarian, I., & Zolotukhin, I. 2011, in Astronomical Data Analysis Software and Systems XX, edited by I. N. Evans, A. Accomazzi, D. J. Mink, & A. H. Rots (San Francisco, CA: ASP), vol. 442 of ASP Conf. Ser., 471

Ott, S. 2010, in Astronomical Data Analysis Software and Systems XIX, edited by Y. Mizumoto, K.-I. Morita, & M. Ohishi (San Francisco, CA: ASP), vol. 434 of ASP Conf. Ser., 139

VAO-Science Council 2010, Recommendations of the VAO-Science Council. URL http://www.aui.edu/vao.php?f=vao_files/public/science_council/ Report%20from%20the%201st%20meeting%2026-27%20March%202010%20and% 20Directors%20response.pdf

Weiner, B., et al. 2009, in astro2010: The Astronomy and Astrophysics Decadal Survey, ArXiv Astrophysics e-prints. 0903.3971

Astronomical Data Analysis Software and Systems XX
ASP Conference Series, Vol. 442
Ian N. Evans, Alberto Accomazzi, Douglas J. Mink, and Arnold H. Rots, eds.
©*2011 Astronomical Society of the Pacific*

Data Challenges for the Gaia Science Alerts System

Ross Burgon,[1] Lukasz Wyrzykowski,[2] and Simon Hodgkin[2]

[1]*Planetary and Space Sciences Research Institute, The Open University, Walton Hall, Milton Keynes, MK7 6AA, United Kingdom*

[2]*Institute of Astronomy, University of Cambridge, Madingley Road, Cambridge, CB3 0HA, United Kingdom*

Abstract.   Gaia is a European Space Agency (ESA) cornerstone mission due to launch late 2012. Its mission is to precisely survey over one billion sources to create an accurate three-dimensional map of the sky. The Gaia Science Alerts (GSA) System, based in the Institute of Astronomy (IoA) at Cambridge University in the UK, aims to use the daily data stream from Gaia to look for and report on transient events both from within and beyond our galaxy. The data stream will be processed in near real-time in order to provide rapid alerts to facilitate ground-based follow-up. This paper provides an overview of the Gaia Science Alerts System and highlights the data processing and storage challenges from data ingestion and event-detection to event classification and the eventual publication mechanism.

1.   Introduction

ESA's Gaia mission aims to create a 3-dimensional map of the sky to a parallax accuracy $< 7\,\mu$as, a photometric accuracy $< 10\,$mmag and a radial velocity accuracy of $< 1\,$km s^{-1} (Perryman et al. 2001). In order to achieve this Gaia will utilize the largest CCD focal plane ever to be flown in space (Gare et al. 2010). Over 100 individual CCDs will make up the astrometric, photometric and radial velocity instruments on the spacecraft resulting in over a billion pixels that, when read in time-delay integration (TDI) mode will create a data rate of 7 Mb per second.

Gaia will perform its observations from the L2 Lagrange point of the Sun-Earth system. It will operate for 5 years, spinning constantly at 60 arcseconds/sec. The two astrometric fields of view will scan across all objects located along a great circle perpendicular to the spin axis. Both fields of view will be registered on one focal plane, but (almost) the same part of the sky will transit the second FOV 106.5 minutes after the first. The spin axis precesses slowly on the sky resulting in multiple observations of the whole sky over the lifetime of the spacecraft. For a spin rate of 60 arcseconds/sec and a solar aspect angle of 45 degrees, the precessional period will equal 63 days.

The data rate transmitted to the ground stations will be approximately 50 GB per day. Pipeline processing for this data will be managed by the Gaia Data Processing and Analysis Consortium (DPAC), a European-wide organization tasked with creating the final Gaia catalog by 2021.

2.   Gaia Science Alerts

The Gaia Science Alerts (GSA) System is based at the Institute of Astronomy (IoA) at the University of Cambridge in the UK. The aims of the system are to:

1. Detect unexpected and rapid changes in the flux, spectrum or the position or appearance of new objects,

2. Release alerts to the community and trigger ground-based follow-up,

3. Monitor a 'watch list' of known potentially eruptive sources.

The GSA system will achieve this by running in near real time, using the photometric, spectroscopic and astrometric data available from the Gaia data stream and cross-matching sources against existing information. As Gaia is primarily an astrometry mission the design presents many obstacles to performing transient detection and classification. The GSA system will therefore be performed on a best effort basis and as such has only limited manpower and hardware resources.

2.1. Triggers, Contaminants and Cadence

Objects of interest to the GSA system include supernovae, micro-lensing events, GRB afterglows, M-dwarf flares, R CrB-type stars, classical and recurrent novae and unknown transients. However there will be triggers, including variable stars, AGNs and asteroids that will be viewed as contaminants as other units within DPAC are tasked with processing these objects (O'Mullane et al. 2007). Figure 1 illustrates some of the types of objects of interest defined in the magnitude change / transient duration parameter-space accessible by Gaia. The large range in transient duration detection is partly due to the design of the focal plane and partly due to the Gaia scanning law (Jordan 2008).

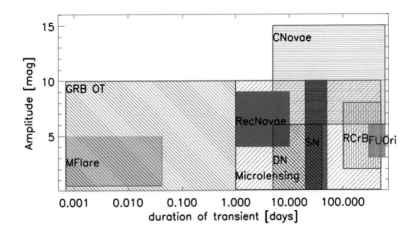

Figure 1.   Illustration of the objects that reside within the magnitude change / transient duration parameter space potentially accessible by Gaia

2.2. Science Alerts Pipeline

The processing pipeline (AlertPipe) is illustrated in Figure 2. New data will arrive at the pipeline having already been preprocessed at the Science Operations Center in Madrid with an expected mean rate of 13.5 million observations per day. The data may only be a few hours old at this point but could be up to 36 hrs old. It is then fed into the pipeline where it is ingested and processed.

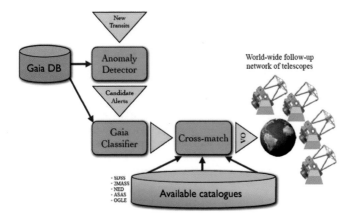

Figure 2.   Illustration of the Gaia Science Alerts processing pipeline.

2.2.1.   Data Storage

The end of mission data volume to be assessible by AlertPipe could be up to 100 TB so it is essential that the storage model is robust and efficient enough to operate within a real-time processing environment. The current implementation for the data storage involves multiple PSQL databases. The entire sky is split into HealPix(els) (Górski et al. 2005) and a separate database is used for all sources within each HealPix. This sub-division of the data storage model is to control the size of each the database (which will be subjected to a large number of I/O operations) and should aid multi-threading within AlertPipe (though only a single-threaded implementation is used at present).

2.2.2.   Anomaly Detector

The detection will be two-fold: (i) known sources exhibiting anomalous behavior and (ii) the appearance of new sources. Both types of alert detection will rely solely on the fluxes observed. Anomaly detection with an 'old source' will rely on the history of that source, ideally incorporating both Gaia observations collected so far and historic ground-based data. New sources are defined as not existing in the Gaia database despite previous observation. Naturally, triggers on new sources will only be valid after enough data is gathered by the mission and the whole sky observed at least once or twice.

2.2.3.   Object Classification and Cross-matching

Classification of objects that are flagged by the anomaly detector will be achieved in three ways: (i) light curve classification, (ii) spectral classification and (iii) position cross-matching with existing catalog sources.

- Classification on the morphology of the light curve will be performed using historical Gaia observations of the source and the new observations that tripped the anomaly detector. The current implementation for this algorithm is based on the Bayesian Classifier with Gaussian Mixture technique Debosscher et al. (2007).

- Spectral classification will be based on the spectrum at the epoch the anomaly detector was tripped. The current implementation for this is through a Self Organized Map (SOM).

- Cross match classification involves data-mining astronomical catalogs around the observed position of the source and identifying any links between the Gaia observation and catalog observations. Catalog access is currently through 'http get' requests however it is likely that local copies of each catalog will be maintained at the IoA to decrease network latency issues.

2.2.4. Alert Publication

The alerts will be published in the VOEvent format which offers a standardized but flexible way to convey alert information that is both human and machine-readable. Alert dissemination will be based on three mechanisms. The primary mechanism will be through a subscription e-mail service offered by the IoA which simply e-mails the VOEvent to subscribers. Alert information will also be available via the Gaia Science Alerts website, that in addition to the VOEvent, will also offer ancillary information and images/charts for each alert. Finally, it is intended to stream VOEvent to a VOEvent repository where it can be used for data-mining through services like SkyAlert.org.

3. Further Information

Interested readers are directed to the wiki[1] for further information and contact details.

References

Debosscher, L. M., Sarro, L. M., Aerts, C., Cuypers, J., Vandenbussche, B., & Solano, E. 2007, A&A, 475, 1159
Gare, P., Sarri, G., & Schmidt, R. 2010, ESA Bull., 137, 51
Górski, K. M., Hivon, E., Banday, A. J., Wandelt, B. D., Hansen, F. K., Reinecke, M., & Bartelmann, M. 2005, ApJ, 622, 759
Jordan, S. 2008, Astron. Nach., 329, 875
O'Mullane, W., et al. 2007, in Astronomical Data Analysis Software and Systems XVI, edited by R. A. Shaw, F. Hill, & D. J. Bell (San Francisco, CA: ASP), vol. 376 of ASP Conf. Ser., 99
Perryman, M. A. C., et al. 2001, A&A, 369, 339

---

[1]http://www.ast.cam.ac.uk/research/gsawg/

Astronomical Data Analysis Software and Systems XX
ASP Conference Series, Vol. 442
Ian N. Evans, Alberto Accomazzi, Douglas J. Mink, and Arnold H. Rots, eds.
©*2011 Astronomical Society of the Pacific*

Goodbye to WIMPs: A Scalable Interface for ALMA Operations

Joseph Schwarz,[1] Emmanuel Pietriga,[2] Marcus Schilling,[1] and Preben Grosbol[1]

[1]*European Southern Observatory (ESO), 85748 Garching, Germany*

[2]*Institut national de recherche en informatique et en automatique (INRIA), Paris, France*

Abstract. The operators of the ALMA Observatory will monitor and control more than 50 mm/submm radio antennas and their associated instrumentation from an operations site that is separated from this hardware by 35–50 km. Software that enables them to identify trouble spots and react to failures quickly in this environment will be critical to the safe and efficient functioning of the observatory. Early commissioning of ALMA uses a operator interface implemented with a standard window, icon, menu, pointing device (WIMP) toolkit. Early experience indicates that this paradigm will not scale well as the number of antennas approaches its full complement. Operators lose time as they manipulate overlapping or tabbed windows to drill-down to detailed diagnostic data, losing a feeling for "where they are" in the process. The WIMP model reaches its limits when there is so much information to present to users that they cannot focus on details while maintaining a view from above. To simplify the operators' tasks and let them concentrate on the real issues at hand rather than continually re-organizing their use of screen space, we are replacing the existing top-level interface with a multi-scale interface that takes advantage of semantic zooming, dynamic network visualization and other advanced filtering, navigation and visualization features. Following the first of several planned participatory design workshops, we have developed prototypes to show how users' needs can be met with the kinds of navigation that become possible when the restrictions of the WIMP model are lifted. Cycles of design and implementation coupled with active user feedback will characterize this project up through deployment.

1. Introduction/The Current Operator Interface

The ALMA Operator Monitoring and Control (OMC) application provides, as its name indicates, an interface for operators and astronomers on duty (AoDs) to monitor and control the ALMA observatory and all its subsystems and devices. As the 50-plus antennas, the correlator and nearly all supporting hardware, at 5000 m altitude, are ≥ 35 km distant from the Operations Support Facility (OSF) at 2900 m, the OMC plays a critical role in how well the operations staff can react to abnormal situations, and their ability to maintain the observing efficiency of ALMA.

Typical monitoring scenarios include viewing the status of several antennas at once, or using anomalous baseline data to identify a single malfunctioning antenna. In the OMC version currently used for commissioning and science verification, this requires the user to manage several overlapping windows via many mouse clicks to arrive at a screen that resembles that shown in Figure 1. In this case, the user must rely on text fields to know which antenna or device is being examined. There are no visual clues to help the user know where an antenna is located, much less assistance in changing focus from one antenna to another, which requires uncovering one tabbed window and suppressing another. In general, managing screen space and windows distracts the user from the task at hand, *e.g.*, system verification or troubleshooting. As ALMA grows towards its full complement of > 50 antennas, this type of bookkeeping chore will present an ever-larger problem.

248   *Schwarz et al.*

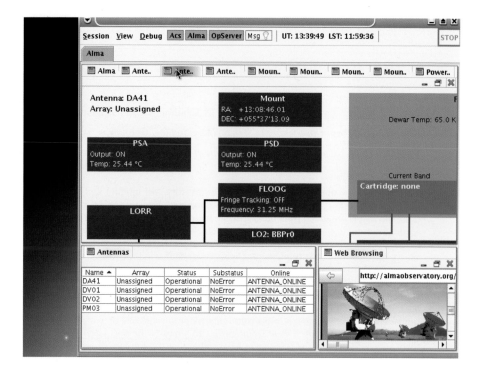

Figure 1.   A typical view of the current OMC; the user has brought up status information for two antennas, one of whose devices is displayed, with all other details hidden behind the several tabs shown.

2.   Towards a Scalable Interface

Well-designed graphical user interfaces are more than just pretty pictures or crutches for beginners; a difficult-to-use UI will result in more errors and longer reaction times, even for experts. In general, an effective interface should enable the user to a) create and maintain a mental model, and therefore retain context (*i.e.*, "Where am I?" and "What route did I take to get here?"); b) leverage gut-level cognitive abilities such as spatial memory and spatial orientation, *e.g.*, by the judicious use of pictures to enhance or replace text; c) minimize *extraneous* (*e.g.*, window management) in favor of *germane* cognitive load.

We have started a project to take account of these criteria in the OMC, beginning with a two-day workshop involving users, Java Swing developers and HCI experts, in order to identify the areas of the OMC that need to be reworked to help operators and astronomers deal with obvious scalability and efficiency issues. Several days of prototyping followed, and the results were shown to users in a follow-on workshop to solicit feedback and plan future development.

Users at the OSF will not normally see the hardware they command and monitor. During the year, antennas will be moved between array configurations in which antennas are as little as a few hundred meters or as much as 15 km apart. Without visual support, operators will find it hard to locate a particular antenna. As the full ALMA array may also be subdivided into smaller arrays, it is important to be able to concentrate on the antennas in this "sub-array," or determine whether some anomaly concerns a group of antennas that are near to each other. Our prototype solution is a pseudo-geographical map and a chessboard to let users select antennas either by hardware ID or by location in the array (see Figure 2). Selecting an array from the

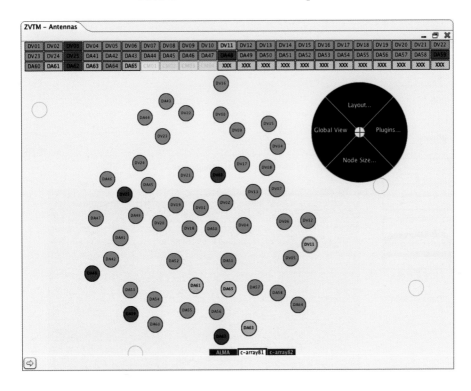

Figure 2. Geographical antenna layout. The chessboard at the top of the window shows antennas ordered by their IDs, while the antenna positions are mapped to either a true or a distorted scale that preserves antenna order but, *e.g.*, reduces the separation of distant antennas so as to display all antennas within limited screen space. The example pie menu can bring up any of the antenna's device monitoring panels.

mini-chessboard at the bottom of the figure highlights the antennas in this array and grays out the others. At the user's option, all views can be synchronized, so that clicking on one antenna will update all relevant panels to show the status of *that* antenna. A semantic zoom facility lets the user drill down into an antenna's details without losing all connection to the antenna's location.

While presenting the status of $n_{ant} \geq 50$ antennas at once is a challenge, it is dwarfed by the need to monitor the $n_{ant}(n_{ant} - 1)/2 \geq 1475$ baselines that characterize the array. Figure 3 demonstrates the approach adopted in our prototype. An $n_{ant} \times n_{ant}$ adjacency matrix lets us identify cells (i, j) and (j, i) with the baseline joining antennas i and j. Using colors such as red and black, the matrix can highlight baselines that require attention. The user can then semantically zoom in to the cell to see whatever details are appropriate. A single malfunctioning antenna, which affects $n_{ant} - 1$ baselines, will show up as two intersecting red or black lines in each half of the matrix. User reaction to a demonstration of a modified OMC incorporating these prototyped features has been overwhelmingly positive. The next phase of the project will deliver a working version for testing at the OSF.

Acknowledgments. We thank Romain Primet of INRIA and Pierre-Henri Cubaud of CNAM for their collaboration on this project.

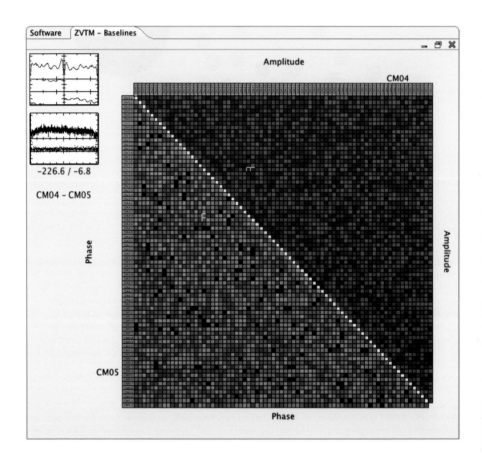

Figure 3.   Adjacency matrix displaying baselines (phase status in lower left, amplitude status in upper right). When the user's mouse is over a cell of the matrix, the graphs in the upper left corner can show amplitude and phase of the corresponding baseline.

Astronomical Data Analysis Software and Systems XX
ASP Conference Series, Vol. 442
Ian N. Evans, Alberto Accomazzi, Douglas J. Mink, and Arnold H. Rots, eds.
©2011 Astronomical Society of the Pacific

The Cherenkov Telescope Array: The Project and the Challenges

Daniel Ponz,[1] José Luis Contreras,[2] Aitor Ibarra,[1] Ignacio de la Calle,[1] and
Irene Puerto-Giménez[3]

[1]*INSA, ESA-ESAC, E-28691 Villanueva de la Cañada, Madrid, Spain*

[2]*Dpto. Física Atómica, Universidad Complutense, E-28040 Madrid, Spain*

[3]*Instituto de Astrofísica de Canarias, E-38200 La Laguna, Tenerife, Spain*

Abstract.   The Cherenkov Telescope Array (CTA) is a new observatory for γ-ray
astronomy at very high energies — between tens of GeV to hundreds of TeV, with
unprecedented sensitivity, energy coverage and angular resolution. This contribution
describes the current status of the project and the challenges involved in this international facility, including the scientific aspects of this new window to the Universe and
technical issues associated to the design of the observatory.

1.   Introduction

In the last decades we have seen major developments in γ-ray astronomy, both from ground-
based and space observatories. Space instruments are now able to observe at energies up to some
tens of GeV, where they become statistics-limited due to their small collecting areas (Pinkau
2009). In the very high energy range, above 100 GeV, Imaging Atmospheric Cherenkov Telescopes (IACTs) are highly successful for the detection of γ-rays from ground, since the first
measurements of TeV emissions from the Crab nebula at the Whipple observatory (Weekes
et al. 1989).

IACTs observe Extensive Air Showers (EAS); γ-ray photons striking the atmosphere will
produce a cascade of relativistic electrons, positrons and secondary γ-rays via pair production and bremsstrahlung. Among those, charged relativistic particles generate a wide pool of
Cherenkov light that reaches the ground in a short pulse of a few nanoseconds. To detect these
faint Cherenkov bursts IACTs have a large tessellated primary mirror focussing the light into
a photomultiplier camera with fast electronics. Stereoscopic methods, using several telescopes
that observe the same shower on coincidence, allow the determination of the γ-ray direction and
improve the sensitivity and resolution of the observations. Major observatories based on IACT
currently in operation are: MAGIC, HESS, VERITAS and Cangaroo (Völk & Bernlöhr 2009).

Astronomy at this very high energy γ-rays range probes the non-thermal Universe, where
mechanisms other than thermal emission by hot bodies are needed to concentrate very large
amounts of energy into a single quantum of radiation. The current catalog includes 110 sources,
encompassing 40 extragalactic sources, mainly AGNs of different types, and 70 galactic objects:
Supernovae Remnants, Pulsar Wind Nebulae, binary systems and unidentified sources in the
Galactic plane without counterpart at other wavelengths (Weekes 2008).

2.   The Project

The CTA observatory project is conceived as an open facility to serve several science communities including high energy astrophysics, cosmology and fundamental physics. Due to its unique
features, CTA has been ranked as one of the top priorities in major roadmaps that highlight

the trends in future science infrastructures (ASPERA,[1] ASTRONET,[2] ESFRI,[3] US National Academy Decadal Survey[4]). Although born in Europe, the project is now fully international with the participation of more than 700 scientists from over 20 countries worldwide. Structured as an international consortium, CTA has completed the initial design study (The CTA Consortium 2010) and started, in October 2010, the preparatory phase which includes prototyping of key systems.

In order to provide full sky coverage, the CTA observatory consists of two large arrays, in the Southern and Northern hemisphere respectively. The Southern site, devoted to observe the central part of the Galactic plane and the dominant part of the Galactic sources, will be designed to cover the full spectral range, while the Northern site, specialized in extragalactic astronomy, will not cover the highest energy range.

Due to the wide energy range covered by CTA, a unique telescope design would not be efficient. To optimize the observatory three energy ranges have been identified and specific telescope designs are being developed for each energy range to improve the sensitivity:

Low-energy range: Below primary energies of 100 GeV the detection of the faint Cherenkov light becomes the limiting factor. In the current design, a few very large, closely packed, telescopes of about 20 to 30 m dish diameter are foreseen. Based on the experience with MAGIC the telescopes will have a parabolic primary mirror (Acciari et al. 2009).

Core energy range: The technology to observe primary γ-rays from 100 GeV to 10 TeV is well understood, based on HESS and VERITAS experience. The appropriate solution seems to be a grid of telescopes in the 10 to 15 m class with an spacing between in the 100 m range. This part of the array will be the workhorse of the observatory. A Davies-Cotton reflector (Davies & Cotton 2007) is foreseen; although this design is not isochronous, it provides very good off-axis properties and a wide field of view.

High-energy range: Above primary energies of 10 TeV the key limitation is the number of detected γ-ray showers, therefore, the array will have to cover a large area. As the Cherenkov light yield is large at high energies, an option is to implement a large number of small size telescopes covering an area of about 10 km^2. Telescopes based on Davies-Cotton or Schwarzschild-Couder (Vassiliev et al. 2007) designs are being considered. This last design, with a secondary mirror, presents two major advantages: it is isochronous and has a small focal field, suitable to incorporate small cameras.

The CTA observatory is defined as an open facility promoting collaboration and data sharing. Scientists will issue observing proposals that will go through a selection process evaluated by an external committee. Selected proposals will be scheduled for observation depending on priorities and on observing conditions within a flexible schedule. After execution of the observations the data will be reduced and the results will be delivered to the scientist and also stored in the project archive. Finally, after a guarantee period, data will be available to the science community via standard Virtual Observatory facilities.

It is foreseen that the complete cycle will be implemented in three main centers:

1) **Science Operations Center** responsible for the scientific outcome. It receives the proposals, performs the science planning and evaluation.

---

[1] http://www.aspera-eu.org/

[2] http://www.astronet-eu.org/

[3] http://ec.europa.eu/research/infrastructures

[4] http://sites.nationalacademies.org/bpa/BPA_049810

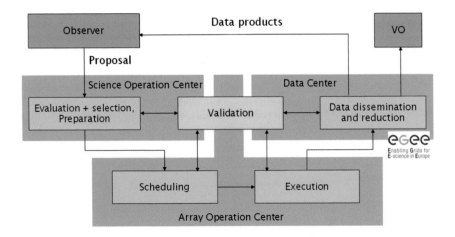

Figure 1.   Data flow diagram of the CTA observatory.

2) **Array Operations Center** responsible for the execution of the observing plan, performs the scheduling and execution of the observations. The actual control system will be based on the ALMA architecture.

3) **Science Data Center** performs the data reduction procedures and distribute the results. Processing pipelines implemented in grid architectures, data distribution and archive based on Virtual Observatory technology.

The selection of the site is a critical task. The requirements aim at sites with very good sky quality at latitudes around 30° North and South, respectively, and an altitude between 1500 and 4000 m in a flat area of about 10 km^2 for the Southern array and about 1 km^2 for the Northern site. In addition, the site must have adequate infrastructure available in terms of access roads, power grid, high speed internet connectivity and safety conditions. A critical aspect of the project is to minimize the environmental impact of the telescope arrays.

The following sites are currently being considered for the Southern array: (1) Khomas Highlands in Namibia (23° S, 1800 m) close to HESS observatory, (2) North of La Silla in Chile (29° S, 2400 m) and (3) in Argentina El Leoncito in Argentina (32° S, 2600 m), or Puna Highlands (30° S, 3700 m).

Pre-selected sites for the Northen array are (1) Canary islands observatories in Spain (26° N, 2400 m) and (2) San Pedro Mártir, Baja California in Mexico (31° N, 2800 m).

3.   Goals and Challenges

CTA will offer a new way to explore the Universe at very high energies with many important scientific goals. These include astrophysics questions as the understanding of the origin of cosmic rays and their impact on the constituents of the Universe or the study of particle accelerators such as pulsars, supernova remnants and γ-ray binaries. In addition, we expect that CTA will substantially contribute to the understanding of the ultimate nature of matter and to the search for new physics beyond the standard model, including dark matter through its possible annihilation signatures.

At the technical level the project builds on the experience of current IACT observatories, therefore, no major risks are expected. However, to fulfill the scientific objectives, the project aims at unprecedented capabilities in performance with an increase of a factor of 10 in sensitivity compared to current observatories and an angular resolution in the arc-min range over the wide energy coverage from some tens of GeV to beyond 100 TeV. The foreseen performance requires an optimal design of the telescopes and a major effort is required concerning the industrial production of the large telescope arrays. Fast electronics and photon detectors with improved quantum efficiency are being investigated to fulfill the required sensitivity.

As a conclusion, CTA will advance the state of the art in astronomy at the highest energies. For the first time it will bring together all the groups working in Europe in the field, integrating as well partners from USA, Japan and other countries to form a worldwide collaboration aimed at building the first open Cherenkov observatory.

Acknowledgments. This contribution has been done on behalf of the CTA project consortium. The support of the Spanish MICINN is gratefully acknowledged.

References

Acciari, V. A., et al. 2009, ApJ, 703, 169
Davies, J. M., & Cotton, E. S. 2007, Sol. Ener. Sci. Eng, 1, 16
Pinkau, K. 2009, Exp. Astron., 25, 157
The CTA Consortium 2010, ArXiv e-prints. 1008.3703
Vassiliev, V., Fegan, S., & Brousseau, P. 2007, Astroparticle Phys., 28, 10
Völk, H. J., & Bernlöhr, K. 2009, Exp. Astron., 25, 173
Weekes, T. C. 2008, in High Energy Gamma-Ray Astronomy, edited by F. A. Aharonian, W. Hofmann, & F. Rieger, vol. 1085 of AIP Conf. Ser., 3
Weekes, T. C., et al. 1989, ApJ, 342, 379

Part VII

Pipelines (Ground-Based Instruments)

Astronomical Data Analysis Software and Systems XX
ASP Conference Series, Vol. 442
Ian N. Evans, Alberto Accomazzi, Douglas J. Mink, and Arnold H. Rots, eds.
© *2011 Astronomical Society of the Pacific*

Sky Subtraction for the MUSE Data Reduction Pipeline

Ole Streicher,[1] Peter M. Weilbacher,[1] Roland Bacon,[2] and Aurélien Jarno[2]

[1]*Astrophysikalisches Institut Potsdam, An der Sternwarte 16, D-14482 Potsdam, Germany*

[2]*Centre de recherche astrophysique de Lyon, Observatoire de Lyon, 9 av Charles André, 69561 Saint Genis Laval Cedex, France*

Abstract.   The Multi Unit Spectroscopic Explorer (MUSE) is a second-generation integral-field spectrograph currently in development for the Very Large Telescope (VLT), consisting of 24 Integral Field Units (IFU). This paper presents the methodology for sky subtraction to be used in the MUSE data reduction pipeline. The method is based on the parameterized simulation of the night sky emission processes and the instrument response. In most cases, this method makes a separate sky exposition obsolete. The approach is described and illustrated here using simulated data.

1. Introduction

One of the key science drivers of MUSE is the observation of deep fields similar to the Hubble Deep Fields. Such deep fields contain very faint emission-line galaxies that will not be detected in a single exposure, but only when a large series is summed up. For such observations the accuracy of sky subtraction is essential and classical methods will not work. This was the motivation to developed an advanced sky subtraction method. In the past, several advanced sky subtraction methods were developed (Content 1996; Kelson 2003; Davies 2007).

In 3D spectroscopy, we can take advantage of the huge number of spectra recorded simultaneously to greatly improve the accuracy of the sky spectrum as well as the instrumental properties.

The MUSE instrument consists of 24 individual IFUs, spanning a total field of view of $1' \times 1'$ (wide field mode) or $7\rlap{.}''5 \times 7\rlap{.}''5$ (narrow field mode) in the visible and near infrared wavelength range (4650–9300 Å). Each IFU has an image slicer that divides the light into 48 slices to be fed into the spectrograph, resulting in a total number of 1152 spectral slices. Each slice has a width of about 78 usable pixels. The final data cube has a size of about 300×300 spatial bins (spaxels) and 4000 spectral bins for one exposure (Bacon et al. 2010).

The optical properties of the instruments are simulated by a numerical model including the whole acquisition chain, including atmosphere, optical light path, and electronic effects (Jarno et al. 2010). This simulation is used to generate input images for the sky subtraction testing.

2. Procedure

The sky reduction will be part of the standard MUSE data reduction pipeline (Weilbacher et al. 2009). The procedure consists of several steps:

1. Selection of sky spaxels,

2. Determination of the emission line fluxes,

3. Calculation of the continuous residual spectrum,

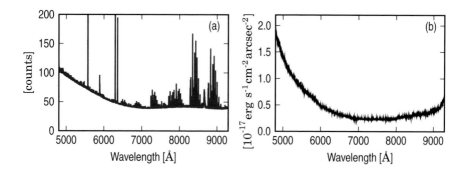

Figure 1.   (a) Sample spectrum of selected sky spaxels (simulated data). (b) Residual spectrum of the sky after the removal of the fitted emission lines. This spectrum is used as continuous background emission.

4. Determination of the individual instrumental properties,

5. Subtraction of the constructed sky spectrum.

The sky spaxel selection is done by choosing the spaxels with the lowest intensity summed over the whole spectrum (usually 5–10 % of all spaxels). All selected spaxels are combined into one spectrum with a fixed binning of $\Delta\lambda = 2.5$ Å. Figure 1a shows this sky spectrum for simulated data.

For the determination of the emission line fluxes, the known atmospheric emission lines (van der Loo & Groenenboom 2007; Osterbrock et al. 1996) are grouped by the originating molecules (OH, [O I], Na I, O_2) and the upper transitional level into twelve groups. Within each group, the emission flux ratio is fixed. The line fluxes for each group are then taken into a fit routine, minimizing the error of the differences between neighboring bins:

$$s^2(p) = \sum_\lambda \left(\frac{\Delta I(\lambda, p)}{\Delta\lambda} - \frac{\Delta I_0(\lambda)}{\Delta\lambda} \right)^2 .$$

Here, $\Delta I(\lambda, p)$ is the computed intensity difference between two neighboring bins (with fit parameters p), $\Delta I_0\lambda$ the measured intensity difference, $\Delta\lambda$ the bin width, and $s(p)$ the error. The intensity difference between neighboring bins is used to minimize the influence of the continuous background to the fit.

If the spaxels used for the emission line flux determinations do not contain object contributions, the residuum of the sky spectrum after subtraction of the fitted sky lines consists only of the continuous night sky background as shown in Figure 1b. However, if the original image does not contain sky regions, the background from a designated sky exposure may be used instead since the continuous background varies only slowly in time. Due to fast variations of the emission line spectrum, the line fluxes should always be estimated directly in the object exposure. Since most emission line groups contain many lines that are fitted simultaneously, their flux estimation works well also within a continuum object spectrum.

3. Instrument Model

Since in MUSE the spectra are undersampled, a very accurate modelling of the instrument parameters and their variations across the instrument (24 IFUs, 48 slices per IFU) is essential to avoid artifacts in the final spectrum. Therefore, for each slice, an individual fit of the instrument

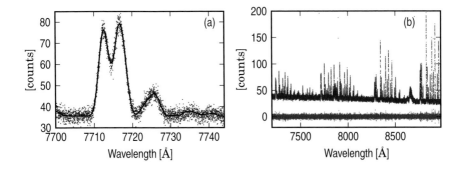

Figure 2.   (a) Modelling the LSF in the sky subtraction. The black line shows the LSF, the blue dots are the original data of one slice. (b) Residuals after the sky subtraction for one slice. The blue dots show the original simulated data from the Instrument Numerical Model, the green dots the residual flux after subtraction.

line spread function (LSF) is done. The LSF shape is parameterized as a damped Gauss-Hermite function (Zhao & Prada 1996):

$$LS F(x) = e^{-x^2/2w^2} + e^{-x^2/w^2} \cdot \sum_{i=3}^{6} k_i H_i \left(\frac{x}{w} \right),$$

with the coefficients k_i of the Hermitean polynomials $H_i(x)$ and the width w as slice and wavelength dependent fit parameters ($x = \lambda - \lambda_0$; λ_0 is the wavelength of the emission line). This LSF is then convolved analytically with rectangular functions representing the slit and the binning. To suppress the influence of the signal, the fit is done using the intensity differences between each data point and the data point with a wavelength difference closest to $\Delta\lambda = 2.5$ Å as in the determination of the original spectrum in step 2. The result of the fit compared with the original data is shown in Figure 2a.

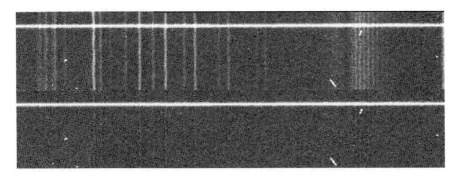

Figure 3.   Spectra of one slice with one continuum object spectrum before (top) and after (bottom) sky subtraction in the wavelength range between 8300 and 8750 Angstrom. This range contains the OH (6–2) band as well as the O_2 (0–1) band as main sky line contributions.

Finally, the slice-dependent sky emission line fit and the continuous sky background are used to reconstruct the sky spectrum and subtract it from the original data. The continuous sky background is here applied without the LSF since its variations over the wavelength are small compared to the LSF width.

4. Results and Discussion

The method is tested on simulated data from the Instrument Numerical Model (INM). The INM incorporates a simulation of the continuous and line sky emission as well as a detailed optical simulation of the complete light path in the instrument. The remaining residuals are shown in Figure 2b. Figure 3 shows the spectra of all spaxels of one slice before and after the sky subtraction.

The model used describes the simulated data well. The resulting spectrum has almost no remaining flux rising above the statistical limit from the night sky emission. However, the method depends strongly on the correct description of the detector parameters (LSF, throughput), so further tests with real exposures have to be done.

Acknowledgments.   We acknowledge support by the German Verbundforschung through the MUSE/D3Dnet project (grant 05A08BA1).

References

Bacon, R., et al. 2010, in Ground-based and Airborne Instrumentation for Astronomy III, edited by I. S. McLean, S. K. Ramsay, & H. Takami (Bellingham, WA: SPIE), vol. 7735 of Proc. SPIE, 773508
Content, R. 1996, ApJ, 464, 412
Davies, R. I. 2007, MNRAS, 375, 1099
Jarno, A., Bacon, R., Ferruit, P., Pécontal-Rousset, A., Pandey-Pommier, M., Streicher, O., & Weilbacher, P. 2010, in Modeling, Systems Engineering, and Project Management for Astronomy IV, edited by G. Z. Angeli, & P. Dierickx (Bellingham, WA: SPIE), vol. 7738 of Proc. SPIE, 77380A
Kelson, D. D. 2003, PASP, 115, 688
Osterbrock, D. E., Fulbright, J. P., Martel, A. R., Keane, M. J., Trager, S. C., & Basri, G. 1996, PASP, 108, 277
van der Loo, M. P. J., & Groenenboom, G. C. 2007, J. Comp. Phys., 126, 4314
Weilbacher, P. M., Gerssen, J., Roth, M. M., Böhm, P., & Pécontal-Rousset, A. 2009, in Astronomical Data Analysis Software and Systems XVIII, edited by D. A. Bohlender, D. Durand, & P. Dowler (San Francisco, CA: ASP), vol. 411 of ASP Conf. Ser., 159
Zhao, H., & Prada, F. 1996, MNRAS, 282, 1223

Astronomical Data Analysis Software and Systems XX
ASP Conference Series, Vol. 442
Ian N. Evans, Alberto Accomazzi, Douglas J. Mink, and Arnold H. Rots, eds.
© *2011 Astronomical Society of the Pacific*

Reflex: Scientific Workflows for the ESO Pipelines

Pascal Ballester,[1] Daniel Bramich,[1] Vincenzo Forchi,[1] Wolfram Freudling,[1] Cesar Enrique Garcia-Dabó,[2] Maurice Klein Gebbinck,[1] Andrea Modigliani,[1] and Martino Romaniello[1]

[1]*European Southern Observatory — Karl-Schwarzschild-Str. 2,*
D-85748 Garching bei München, Germany

[2]*FRACTAL SLNE, Tulipán 2, portal 13, 1. A E-28231 Las Rozas de Madrid, Spain*

Abstract.   The recently released Reflex scientific workflow environment supports the interactive execution of ESO VLT data reduction pipelines. Reflex is based upon the Kepler workflow engine, and provides components for organizing the data, executing pipeline recipes based on the ESO Common Pipeline Library, invoking Python scripts, and constructing interaction loops. Reflex will greatly enhance the quick validation and reduction of the scientific data. In this paper we summarize the main features of Reflex, and demonstrate as an example its application to the reduction of echelle UVES data.

1.   Reflex: a Data Processing Workflow Environment Based on Kepler

The ESO Recipe Flexible Execution Workbench (Reflex) is an environment which allows an easy and flexible way to execute VLT pipelines. It is built using the Kepler workflow engine (https://kepler-project.org). Reflex allows the users to process their scientific data in the following steps:

- Associate scientific files with required calibrations,

- Choose datasets to be processed,

- Execute several pipeline recipes in a logical sequence,

- Inspect intermediate products, and

- Interactively change recipe parameters.

A workflow accepts science and calibration data, as delivered to Principal Investigators (PIs) in the form of PI-Packs or as downloaded from the archive, and organizes them into groups of files called LoSOs (List of Science Observations), where each LoSO contains one science object observation and all associated raw calibrations required for a successful data reduction. The data organization process is fully automatic, which is a major time-saving feature provided by the software. The LoSOs selected by the user for reduction are fed through the workflow which executes the relevant pipeline recipes (or stages) in the correct order, providing optional user interactivity at key data reduction points with the aim of enabling the iteration of certain recipes in order to obtain better results. Full control of the various recipe parameters is available within the workflow, and the workflow deals automatically with optional recipe inputs via built-in conditional branches. Additionally, the workflow stores the reduced final data products in a logically organized directory structure and employing user-configurable file names.

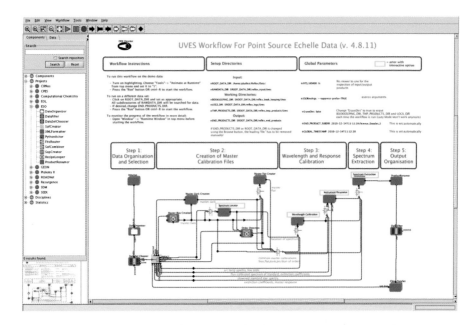

Figure 1.   The main workflow window for the UVES pipeline. Annotations and top-level parameters of the workflow are shown to the user, while the logic pertaining to individual reduction steps is hidden in composite subworkflows.

2.   Data Organization, Selection, and Routing

The processing components in Kepler workflows are called actors. The Reflex package currently includes twelve actors for the development of data reduction workflows. Moreover, all other existing Kepler actors can be used. Four Reflex actors are usually involved before the actual execution of data reduction recipes:

The **DataOrganizer** organizes and classifies raw and reduced data: it takes as inputs all the files contained in a given directory, classifies them and organizes them into Lists of Science Observations. Each list contains science frames, all the associated calibrations needed to reduce them, and custom parameters for the reduction recipes. The user can check and modify the selection.

The **DataSetChooser** allows the user to view and select the groups of files created by the DataOrganizer: it displays a list of LoSOs and provides the possibility of selecting, deselecting and analysing them. It produces one token on the output port per selected LoSO.

The **FitsRouter** sorts files based on their category and routes them to the corresponding output port. The **SofCombiner** groups related set of files into a common channel. For example, the averaged master bias together with the master flat, the order location table, the arc lamp exposure and the arc line catalog are routed to the wavelength calibration recipe.

3.   CPL Recipe Executer

The RecipeExecuter executes a CPL recipe from the VLT pipelines (Fig. 2). In the workflow, each parameter of the recipe is initially set to the default value. The user can change any of the parameters, or provide calculated parameter values to a special input port. The RecipeExecuter supports a variety of execution modes, facilitating the processing of large data sets.

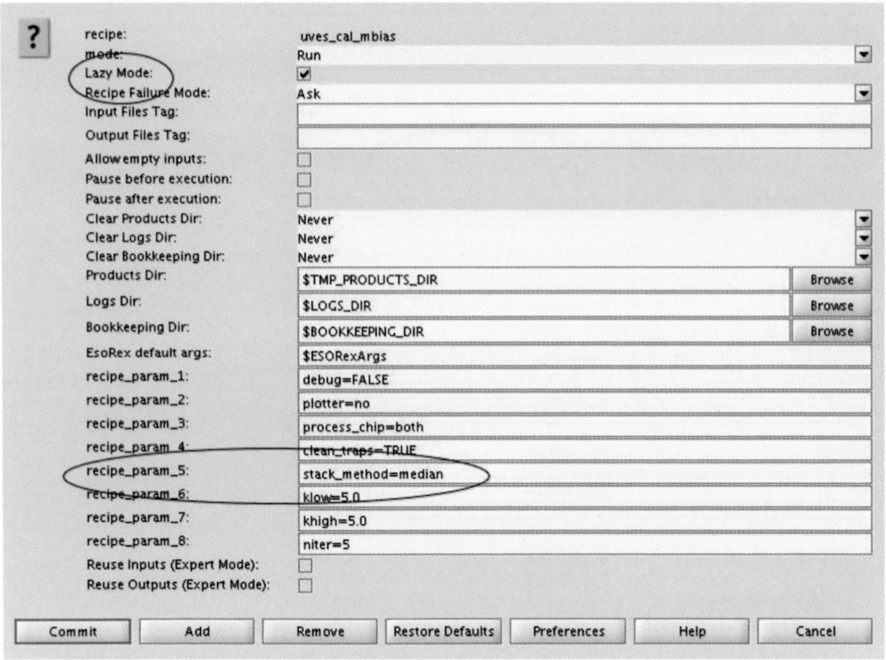

Figure 2.   The configuration panel for the RecipeExecuter actor. The list of parameters is created for a given pipeline recipe at the time the actor is instantiated in the workflow. All parameters of a recipe can be modified at this level. In addition the RecipeExecuter supports several execution modes, giving the user control over the input, output, and conditions of execution of the recipe.

The Lazy mode is particularly useful: if true the RecipeExecuter will check whether the pipeline recipe has already been executed with the same input files and with the same recipe parameters. If this is the case then the recipe will not be executed, and instead the previously generated products (which are the same as those that would have been generated if the recipe were executed, except for a timestamp) will be broadcast to the output port.

Kepler supports subworkflows, which make it possible to display the data processing tasks at different levels of detail. Examples of subworkflows are the iterative subworkflows: it is sometimes useful to be able to visualise the products of a recipe execution, tweak some recipe parameter and execute the recipe again, until the products are as expected. This can be easily achieved in Reflex by means of the RecipeLooper and the PythonActor, embedded in a Kepler subworkflow.

4.   Python Actor

The PythonActor executes custom Python scripts. A module included in the Reflex release provides the functions necessary to insert scripts into the workflows. Upon selecting a script, input and output ports are automatically created, together with a parameter for each output port. Python actors are used to create plotting windows, graphical user interface, or to invoke

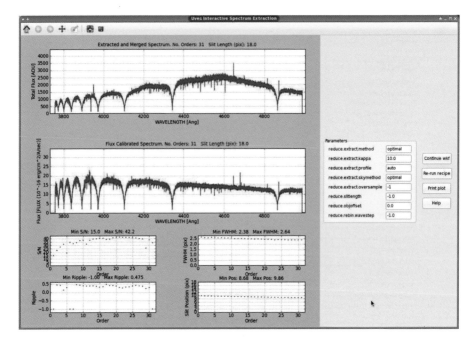

Figure 3.   The Python actor can be used to develop interactive windows that display intermediate results, and let the user fine-tune a set of recipe parameters.

tasks from external data reduction systems. The released UVES workflow comes with several examples of interactive Python actors (Fig. 3).

5.   Where Can I Find Reflex?

Reflex is distributed on the ESO Web site http://www.eso.org/reflex. The pipeline specific workflows, like the UVES workflow shown in this article are available from the ESO pipelines Web site: http://www.eso.org/pipelines, together with documentation, installation procedures, and test data.

Astronomical Data Analysis Software and Systems XX
ASP Conference Series, Vol. 442
Ian N. Evans, Alberto Accomazzi, Douglas J. Mink, and Arnold H. Rots, eds.
© *2011 Astronomical Society of the Pacific*

Closing the Observing Loop Across Continents: Data Transfer Between Chile and Europe

R. Hanuschik,[1] S. Zampieri,[1] M. Romaniello,[1] C. Cerón,[2] A. Wright,[2] C. Ledoux,[2] and F. Comerón[1]

[1]*European Southern Observatory, Karl-Schwarzschild-Strasse 2, D-85748 Garching bei München, Germany*

[2]*European Southern Observatory, Alonso de Córdova 3107, Vitacura, Santiago, Chile*

Abstract. The different tasks supporting the operations of ESO telescopes in Chile are carried out at widely separated locations, with some taking place at the observatory site and others in Europe at ESO's headquarters. A fundamental requirement to make such an operations scheme viable is a stable way of transferring the large amounts of data generated by the telescopes to Europe on the shortest possible timescale. We review technical progress that has allowed ESO in the last years to move from the transfer of data on physical media, with delays of up to two weeks, to the current transfer of most of the data stream through the internet within minutes. We also describe the possibilities that will be open in the near future with EVALSO, a European Union co-funded project to provide full-fiber connectivity from the ESO observatory site on Paranal, and the nearby Cerro Armazones hosting the telescopes of the University of Bochum, all the way to Europe. Given the recent choice of Cerro Armazones as the future location of the European Extremely Large Telescope, upgrades in the communications infrastructure of Paranal and Cerro Armazones are very relevant to the ability to operate the current and future facilities there.

1. Introduction

The operation of world-class ground-based astronomical facilities is nowadays a global enterprise, in which operations processes can take place at locations thousands of kilometers away from the telescope. Underpinning the capability of managing operations in such geographically distributed manner is the ability to transfer quickly the vast amounts of data generated by current astronomical instrumentation across the operations sites, so that the various processes composing the operations scheme of the facility can feed back on each other in a timely manner.

The ESO observatory of Cerro Paranal (Chile), hosting the Very Large Telescope (VLT), the VLT Interferometer (VLTI), the Visible and Infrared Survey Telescope for Astronomy (VISTA) and, in the near future, the VLT Survey Telescope (VST) is no exception to these demands. The operations facilities on Paranal are complemented by the ESO Headquarters in Garching (Germany), where user support, data processing and permanent storage takes place. During the initial 10 years of operations, limitations of the available bandwidth coupled with the growing overall data rates coming from the increasing number of telescopes and instruments on Paranal forced ESO to transfer the data from Chile to Europe using physical media, with the subsequent time lag of up to two weeks between the generation of the data and their detailed quality analysis. Recently it has become possible to use the Internet through the microwave link connecting the observatory to the outside world to transfer this data stream within hours or less of it being produced. However, qualitative leaps lie ahead with the arrival to Paranal in the period 2011–2014 of the VLT/VLTI 2nd generation instruments and of the VST. Furthermore, the communications infrastructure on Paranal will be used to transfer the data from the

265

future European Extremely Large Telescope (E-ELT), to start operations on the nearby Cerro Armazones around 2020.

2.   VLT End-to-End Operations and its Communication Needs

Since its initial design in the mid 1990s (Quinn et al. 1998), VLT operations were conceived as an integrated system (Péron 2008) strongly relying on flexible scheduling. Underlying this end-to-end model is a system with tools to manage the flow of data and information. A central element in this scheme is the ESO archive, located at the ESO Headquarters in Garching, where all observations collected at the observatory are stored and made available to operations groups in charge of quality control, instrument trending and health check, and data package preparation and distribution (Eglitis & Suchar 2010), as well as to the users. The essential role played in operations by the archive thus makes the operations workflow critically dependent on the speed with which the data obtained at the observatory can be ingested into it. The end-to-end model has been validated by its successful implementation for over 10 years, and is now applied with minor variations to the operation of the La Silla facilities and to the other facilities on Cerro Paranal. Many elements of this model have been adopted by the planning of operations of the ALMA millimeter and submillimeter array (Andreani & Zwaan 2008). Conveniently evolved in the coming years, it will also form the basis of the operation of the E-ELT (Spyromilio et al. 2008).

3.   Current Demands

The transfer of data through the Internet (Zampieri et al. 2009) from Paranal to Garching became operational in late 2009. At present, the transfer of data from Paranal (and also from the ESO La Silla observatory near La Serena, Chile) to the ESO Archive in Garching takes place online. Data are also copied at the observatories onto USB disks, which are shipped to Garching on a weekly basis. The double channel is designed to ensure the prompt online transfer of high priority data without saturating the available bandwidth, in which case data can be made to overflow to the disk channel.

3.1.   Why Fast Data Transfer?

Having the data available in the ESO archive in Garching within a short time from acquisition has positively impacted both the user experience and the observatory operations. Investigators can access their proprietary data in almost real time to speed up their scientific exploitation. More than 100 datasets are downloaded per month on the average. Data handling operations are simplified on both sides of the Atlantic by eliminating most of the logistics associated with the transfer and handling of physical media. Finally, the Data Processing and Quality Control group in Garching can perform in-depth quality control (Hanuschik 2007) in quasi real time using the most recent files and feed the results back to Paranal typically within one hour. This removes most of the needs for similar activities on the mountain.

3.2.   Current Data Volumes and Data Transfer Capabilities

The current average production rate of VLT-VLTI instruments is 20–25 GB of data (all data volumes given in this paper are after compression) per 24-hour period. The bandwidth currently reserved for data transfer from Paranal is currently 9.12 Mbit/s, out of a total available bandwidth of 11.2 Mbit/s, corresponding to about 800 Gb over 24 hours. The total bandwidth from Santiago de Chile to Garching is 50 Mbits/s. The indication from the first year of operations of the Scientific Data Transfer is that the nominal available bandwidth should be over-dimensioned by at least a factor of two with respect to the average data rate to be transferred.

4. VLT/VLTI Second Generation, Survey Telescopes, and Beyond

4.1. VLT/VLTI Second Generation Instruments and Survey Telescopes

With the new telescopes and instruments scheduled to start operating at Paranal in the next few years Moorwood (2009), the amount of data that will be produced, and that will have to be transported to Garching, will significantly increase. The most significant contributor is VISTA, which has already started operations in 2010 and produces on average 75 GB/night of compressed data. Others will be VST (2011, 35 GB/night compressed), SPHERE (2011, 40 GB/night) and MUSE (2012, 30 GB/night). The expected data volumes of VLTI 2nd generation instruments are smaller, adding another 20 GB/night to the data volume. The total average data production of Paranal in 2014+ will thus be about 225 GB/night, requiring a sustained bandwidth of about 21 Mbits/s for their transfer to Garching within 24 hours. The current bandwidth from Paranal to Garching will thus fall short by a factor 2 from that needed.

4.2. The E-ELT

The nightly data volume to be produced by E-ELT instrumentation is very difficult to predict one decade in advance, but the instrument concepts that have been developed have provided already a crude estimate of the average nightly data production to be set at the level of 1–2 TB (uncompressed) per night, with large variations depending on the instruments and modes actually used on a given night. A rough estimate of the data transfer requirements of the Paranal observatory, including the E-ELT on Armazones, is thus 0.5–1 TB/night (compressed). The quantity may increase further through the upgrade of planned Paranal instruments with larger format detectors.

5. Dealing with Future Demands: EVALSO

The daily transfer of the Paranal and Armazones data production over the internet will require a bandwidth increase of roughly one order of magnitude over the next decade with respect to the capabilities of the current setup supported by the existing microwave link. Furthermore, new applications such as the implementation of a remote interaction mode involving the transfer of data just acquired from the observatory to a user located on another continent, would require Gbit/s-level capabilities. The EVALSO project (for "Enabling Virtual Acces to Latin South American Observatories", Filippi et al. 2010), funded by the European Union under its Framework Program 7, is expected to provide such capabilities for Paranal and Armazones already in the very near future.

The main goal of EVALSO is to create the missing parts of the physical infrastructure to connect the Paranal and Cerro Armazones observatories to Europe with a high capacity link. To this end, a consortium was formed in 2007 by seven European institutions (the GARR consortium, the University of Trieste, and the Astronomical Observatory of Trieste in Italy, Queen Mary University of London, NOVA in the Netherlands, the Astronomical Observatory of the Ruhr University of Bochum, and ESO, plus the REUNA and RedCLARA networks in Chile). The project will use the ALICE/ALICE2 research network infrastructure within South America and transatlantic connection to European National Research Networks (NREN) via GEANT2. ESO has procured the infrastructure needed to connect Paranal to the existing networks linking Santiago with Europe, and in particular a 75 km-long fiber link between Paranal and the access point to the Chilean backbone. The EVALSO infrastructure is now completed and undergoing commissioning. In the EVALSO implementation currently being put in place, the capacity of the path between Paranal and Armazones is limited by that of the transatlantic link, which with the planned ALICE2 upgrade is expected to exceed 1 Gbit/s. Tests on this segment using the current ALICE infrastructure have already achieved a sustained transfer rate exceeding 100 Mbit/s between Santiago and Garching using UDP-based file transfer tools. Considering that all existing links have higher nominal capacity and that the planned upgrade within ALICE and the trans-Atlantic link will even increase such limits, even faster transfer rates are foreseen in the near future.

The new infrastructure should thus provide sufficient data transfer capacity for at least the next decade on Paranal and Armazones, enabling data files produced by the scientific instruments to be stored in the Garching archive within seconds from being produced and making data processing to start in a practically instantaneous manner. The high capacity of already existing NREN would make it possible the extension of fast data transfer to many other locations in Europe, which may be used in the future to enable remote interaction with the facility as described above (Comerón et al. 2008).

References

Andreani, P., & Zwaan, M. 2008, in Observatory Operations: Strategies, Processes, and Systems II, edited by R. J. Brissenden, & D. R. Silva (Bellingham, WA: SPIE), vol. 7016 of Proc. SPIE, 701611

Comerón, F., Filippi, G., & Emerson, J. 2008, in Observatory Operations: Strategies, Processes, and Systems II, edited by R. J. Brissenden, & D. R. Silva (Bellingham, WA: SPIE), vol. 7016 of Proc. SPIE, 70161U

Eglitis, P., & Suchar, D. 2010, in Ensuring long-term preservation and adding value to scientific and technical data, ESA, in press

Filippi, G., Jaque, S., Liello, F., Chini, R., Utreras, F., Wright, A., Lemke, R., & Heissenhuber, F. 2010, in Software and Cyberinfrastructure for Astronomy, edited by N. M. Radziwill, & A. Bridger (Bellingham, WA: SPIE), vol. 7740 of Proc. SPIE, 77401G

Hanuschik, R. 2007, in Astronomical Data Analysis Software and Systems XVI, edited by R. A. Shaw, F. Hill, & D. J. Bell (San Francisco, CA: ASP), vol. 376 of ASP Conf. Ser., 373

Moorwood, A. 2009, in Science with the VLT in the ELT era, Springer-Verlag

Péron, M. 2008, in The 2007 ESO Instrument Calibration Workshop, Springer-Verlag

Quinn, P. J., Albrecht, M. A., Ballester, P., Banse, K., Chavan, A. M., Grosbol, P., Péron, M., & Silva, D. R. 1998, in Observatory Operations to Optimize Scientific Return, edited by P. J. Quinn (Bellingham, WA: SPIE), vol. 3349 of Proc. SPIE, 2

Spyromilio, J., Comerón, F., D'Odorico, S., Kissler-Patig, M., & Gilmozzi, R. 2008, The Messenger, 133, 2

Zampieri, S., Forchi, V., Gebbinck, M. K., Moins, C., & Padovan, M. 2009, in Astronomical Data Analysis Software and Systems XVIII, edited by D. A. Bohlender, D. Durand, & P. Dowler (San Francisco, CA: ASP), vol. 411 of ASP Conf. Ser., 540

Astronomical Data Analysis Software and Systems XX
ASP Conference Series, Vol. 442
Ian N. Evans, Alberto Accomazzi, Douglas J. Mink, and Arnold H. Rots, eds.
©*2011 Astronomical Society of the Pacific*

The DASCH Data Processing Pipeline and Multiple Exposure Plate Processing

Edward Los,[1] Jonathan Grindlay,[1] Sumin Tang,[1] Mathieu Servillat,[1] and Silas Laycock[2]

[1]*Harvard College Observatory*

[2]*Department of Physics, University of Massachusetts at Lowell*

Abstract.   Digital Access to a Sky Century @ Harvard (DASCH) is a project to digitize the collection of approximately 525,000 astronomical plates held at the Harvard College Observatory. This paper presents an overview of the DASCH data processing pipeline, with special emphasis on the processing of multiple-exposure plates. Such plates extended the dynamic range of photograph emulsions and improved photometric accuracy by minimizing variations in plate development procedures. Two approaches are explored in this paper: The repetitive use of `astrometry.net` and local correlation searches. Both procedures have yielded additional quality control checks useful to the pipeline.

1. Introduction

The Harvard College Observatory plate collection consists of approximately 525,000 photographs produced by over 80 telescopes spanning over 100 years from about 1885 to 1992. The goal of the Digital Access to a Sky Century @ Harvard[1] (DASCH) project (Grindlay et al. 2009) is to digitize this entire collection and provide photometry measurements for all objects. To date, we have digitized over 10,000 plates and extracted an average of 80,000 objects per plate. The analysis of this digitized data presents a number of challenges which are no longer encountered with modern CCD photographic techniques. One of these challenges is the presence of many plates which have multiple instances of the same objects.[2]

1.1. Types of Multiple Exposure Plates

Plates with multiple exposures were produced to extend the limited dynamic range of photographic emulsions, to account for variations in development procedures, and to include known standards such as the Harvard North Polar Sequence. Figure 1 shows examples of successive exposures of the same field and of ghost objects generated by a Pickering or Racine Wedge (Leavitt 1917). A third technique for producing ghost objects made use of coarse gratings (King 1931).

1.2. Number of Multiple Exposure Plates

Now that 217,000 of the 525,000 logbook entries have been transcribed and entered into a MySQL database, a simple query shows that 5,842 or 2.6% of the transcribed plates have multi-

---

[1]See http://hea-www.harvard.edu/DASCH/.

[2]See http://hea-www.harvard.edu/DASCH/papers/P033f.pdf for a full-length version of this paper.

Figure 1.   Left: Plate mc05077 showing multiple exposures of the M44 field. Nine exposures were taken with exposure times decreasing by 50% for each successive exposure. Right: Plate i31090 showing examples of Pickering Wedge objects.

ple exposures. Unfortunately, no mention of the use of Pickering Wedge plates has been found in the logbooks. Logbook transcriptions suggest that less than 100 plates used coarse gratings, although the total may be as high as 8,000 because of issues with the logbook plate classification system.

2.   DASCH Pipeline

An overview of the DASCH digitizer and pipeline appears below. The two key steps involved in the processing of multiple exposure plates are the "Pickering Wedge Filter" and the "Multiple Exposure Loop".

2.1.   Plate Preparation

Before any scanning can occur, the relevant entries in the logbook must be transcribed and entered into the MySQL scanner database. Both plate jackets and plates with ink annotations are photographed with a Nikon D200 camera. All ink annotations on the reverse side of the plate from the emulsion must then be cleaned to avoid confusion with astronomical objects.

2.2.   Mosaic Generation, WCS Fitting, and Source Extraction

The digitizer (Simcoe et al. 2006) generates 60 tiles to cover a typical 20×25 cm plate with 10×6 images using half-steps in width to assure two exposures for every plate object. The mosaicing process registers and combines these tiles with flatfield tiles to produce a single mosaic of approximately 780 megapixels.

The SExtractor program (Bertin & Arnouts 1996) next generates object lists. The WCS fitting procedure first begins with `astrometry.net` (Lang et al. 2010) and moves to successively accurate fits using WCSTools (Mink 1999) and then a polynomial fit. A companion paper (Servillat 2011) describes this procedure in greater detail.

2.3.   Pickering Wedge Filter

The Pickering Wedge Filter is used to flag ghost objects by performing a spatial correlation within a limited region around bright stars. The procedural steps are as follows. (1) Select the 300 brightest objects on the plate. (2) For each of these stars, superimpose all of the objects at fixed distances from the primary objects. If wedge objects are present, there will be a peak in the secondary object distribution. Two peaks indicate a grating plate. (3) If a Pickering Wedge plate identification has been made, a plot of the difference of instrumental magnitudes of both the wedge and primary objects against the instrumental magnitude of the primary object will show two distributions: normal stars and the wedge objects. (4) Finally, go through the entire SExtractor population and flag as Pickering Wedge objects all objects which meet the above position and magnitude criteria.

2.4. Star Matching, Defect Filter, Photometry Calibration, and Magnitude Calculation

These steps are described in more detail in Laycock et al. (2010), Tang et al. (2010), S. Tang, in preparation, and on the DASCH website.[3] After wedge filtering, objects are matched either to the GSC 2.3.2 catalog (Lasker et al. 2009) or the Kepler Input Catalog (for calibration of the Kepler satellite field). A defect filter removes emulsion defects, dust, and development defects by comparing PSF characteristics of matched objects with unmatched objects. Next, the plate is divided into nine annular bins and a colorterm algorithm estimates the spectral response of the plate. For each annular bin, a LOWESS curve fitting algorithm is used to generate a calibration between the SExtractor instrumental magnitudes and the blue catalog magnitudes. Finally, a local correction algorithm is used on a 50×50 grid to account for variations in emulsion and/or sky conditions.

2.5. Multiple Exposure Loop

The multiple exposure loop removes the objects successfully matched to the calibration catalog from the SExtractor source list and submits the remainder source list to `astrometry.net`. At the completion of the WCS fitting procedures, the new WCS parameters are applied to all objects in the original SExtractor dataset and the pipeline proceeds normally from Pickering Wedge filtering to completion.

Two ambiguities arise: (1) If a catalog object from one solution coincides with a different catalog object from another solution, then the result is considered a "Multiple Exposure Blend". (2) If an object is not matched to any catalog object, these objects receive a special flag because it is impossible to assign the object to a particular exposure.

To prevent infinite loops, the algorithm is terminated if a new solution is close to or overlaps a previous solution.

2.6. Photometry Database

Magnitude measurements are stored in a set of binary files optimized for performance. Supporting star-specific and plate-specific data appear in MySQL tables. A detailed database design document will appear in a future publication.

3. Results and Discussion

Results from a sample of 87 multiple exposure plates with full logbook transcriptions show successful detection of 31% of the double exposures and 18% of the triple exposures. There were 481 non-Pickering-Wedge plates which showed more solutions than described in the transcribed logbook entries.

The multiple exposure loop provided additional quality control checks by detecting at least 292 plates that can not provide good photometry because they are grating plates or plates with bad astrometric solutions.

The Pickering Wedge filter flagged 857 of 11461 scanned plates. Use of this filter has been extended to detect a possible optics misalignment in the "dsy" telescope series. These results suggest that a general near-neighborhood search algorithm would be useful for better detection of grating images and double-peaked star PSF's.

---

[3]See `http://hea-www.harvard.edu/DASCH/photometry.php`.

Acknowledgments. This work is funded by National Science Foundation grants AST-0407380 and AST-0909073. Additional contributers to the DASCH project are the Harvard College Plate Stacks Curator, Alison Doane; hardware engineer, Robert Simcoe; and our transcribers, plate cleaners, and scanners.

References

Bertin, E., & Arnouts, S. 1996, A&AS, 117, 393

Grindlay, J., Tang, S., Simcoe, R., Laycock, S., Los, E., Mink, D., Doane, A., & Champine, G. 2009, in Preserving Astronomy's Photographic Legacy: Current State and the Future of North American Astronomical Plates, edited by W. Osborn & L. Robbins (San Francisco, CA: ASP), vol. 410 of ASP Conf. Ser., 101

King, E. S. 1931, A Manual of Celestial Photography (Boston, MA: Eastern Science Supply Co.), 1st ed.

Lang, D., Hogg, D. W., Mierle, K., Blanton, M., & Roweis, S. 2010, AJ, 139, 1782

Lasker, B. M., et al. 2009, AJ, 136, 735

Laycock, S., Tang, S., Grindlay, J., Los, E., Simcoe, R., & Mink, D. 2010, AJ, 140

Leavitt, H. S. 1917, Ann. Harvard College Obs., 71, 106

Mink, D. 1999, in Astronomical Data Analysis Software and Systems VIII, edited by D. Mehringer, R. Plante, & D. Roberts (San Francisco, CA: ASP), vol. 8 of ASP Conf. Ser., 498

Servillat, M. 2011, in Astronomical Data Analysis Software and Systems XX, edited by I. N. Evans, A. Accomazzi, D. J. Mink, & A. H. Rots (San Francisco, CA: ASP), vol. 442 of ASP Conf. Ser., 273

Simcoe, R. J., Grindlay, J., Los, E. J., Doane, A., Laycock, S. G., Mink, D. J., Champine, G., & A., S. 2006, in Applications of Digital Image Processing XXIX, edited by A. G. Tescher (Bellingham, WA: SPIE), vol. 6312 of Proc. SPIE, 631217

Tang, S., Grindlay, J., Los, E., & Laycock, S. 2010, ApJ, 710, L77

Astronomical Data Analysis Software and Systems XX
ASP Conference Series, Vol. 442
Ian N. Evans, Alberto Accomazzi, Douglas J. Mink, and Arnold H. Rots, eds.
© 2011 Astronomical Society of the Pacific

Correcting the Astrometry of DASCH Scanned Plates

M. Servillat,[1] E. J. Los,[1] J. E. Grindlay,[1] S. Tang,[1] and S. Laycock[2]

[1]*Harvard College Observatory, 60 Garden St., Cambridge, MA 02138, USA*

[2]*Department of Physics, University of Massachusetts, Lowell, MA, USA*

Abstract.   We describe the process implemented in the DASCH pipeline which applies a reliable astrometric correction to each scanned plate. Our first blind astrometric fit resolves the pointing, scale, and orientation of the plate in the sky using `astrometry.net` code. Then we iteratively improve this solution with WCSTools *imwcs*. Finally, we apply a 6th order polynomial fit with SCAMP to correct the image for distortions. During a test on 140 plates, this process has allowed us to successfully correct 79% of the plates. With further refinements of the process we now reach a 95% success rate after reprocessing all our scanned plates ($\sim 11,000$ in Nov. 2010). We could extract a lightcurve for 2.85 times more objects than with the previous pipeline, down to magnitude 17. The resulting median RMS error is $0\rlap{.}''13$ for objects with mag. 8 to 17.

1. Introduction

The DASCH project (Digital Access to a Sky Century @ Harvard) is a project that aims to digitize the $\sim 525,000$ photographic plates stored at the Harvard College Observatory that were exposed on various telescopes from 1885 to 1992 (Grindlay et al. 2009). The plates cover the whole sky and provide typically 500 to 1000 images of any object brighter than the detection limit, of typically 14 to 17 B magnitude.

We developed a specific pipeline (Grindlay et al. 2009; Laycock et al. 2010) to process the plates and store those measurements in a database (see Los et al. 2011, for the description of the pipeline). In order to extract the long-term lightcurves (over 100 years!) of an object without confusing it with a neighboring object, we need to obtain a good astrometric solution for each plate.

An accuracy of 1 arc second ($''$) or lower generally allow us to associate an object with its entry in the GSC 2.3.2 catalog uniquely, or to classify it securely as a new transient event. In practice, the scale of the plates varies from sub-arc-second to about $6''$ per pixel depending on the plate series and we expect to obtain positional accuracy lower than a 3 pixels limit (radius used for cross-correlations). Distortions from the original telescope optics can have dramatic effects, with offsets of up to few arc minutes on the edges.

2. Implementation

We implemented a 3 step procedure that allows us to blindly find the position of the plate center and ultimately correct the distortions on the plates.

First guess.   It is first necessary to find the position of the plate in the sky with the right scale and orientation. The transcriptions of the hand-written logbooks could not reliably give this information. We thus need to find a solution blindly, which is possible by using pattern

273

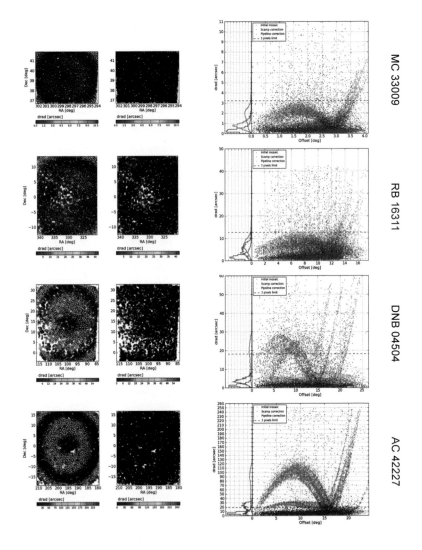

Figure 1.   Distortions observed for 4 plates. *Left*: Distortion maps showing each reference stars in RA/Dec with a color corresponding to the position error (drad: radial distance from reference to detection). Maps before (left) and after (right) SCAMP correction. *Right*: Position error as a function of the offset from the plate center for each reference star. Red: before correction. Green: previous pipeline. Blue: Pipeline with SCAMP. Some scattered points with high drad correspond to mismatches.

matching techniques. `astrometry.net` procedures (Lang et al. 2010) are optimized for this, and are integrated since June 2008 in our pipeline. In order to speed up the process, the catalog of detections is generated with SExtractor (Bertin & Arnouts 1996) and a high threshold on a reduced image with a 16 pixels binning. This procedure leads to a 99.75% success rate.

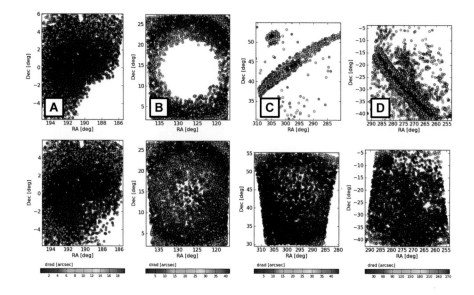

Figure 2. Erratic distortion maps before SCAMP correction. Top: results from the test on 140 plates. Bottom: latest results after refinement of the process. A: B 71955, B: BM 00389, C: RH 06526, D: AM 00538.

Refining the solution. The first guess is too crude to match the detected objects efficiently. We thus use WCSTools (Mink 2002) *imwcs* to iteratively reach a more precise solution. The Tycho 2 catalog is used with coordinates corrected for proper motions, and generally provides a positioning of the plate with a precision of 10–20″.

Fitting the distortions. For each plate, a low threshold detection of objects is performed with SExtractor. In order to correct the distortion, we select the 10 000 brightest sources and a reference catalog of 10 000 is extracted from UCAC3 (Zacharias et al. 2010). This reference catalog has indeed the best astrometric references to date (0.″015 to 0.″100) as well as the best proper motions estimates, which are important on 100 year timescales. UCAC3 is filtered to remove its known biases: we keep objects with a 2MASS counterpart, and keep only reference stars with a M magnitude in the range 8–16. With those inputs, we run SCAMP (Bertin 2006) to obtain a 6th order polynomial fit stored in the header of the plate image file for use in the following pipeline steps. This fit was previously performed with IRAF *ccmap* and the GSC catalog.

3. Test and Results

We performed a test on 140 plates chosen randomly from different plate series. We took care to respect the proportion of already scanned plates for each series so the results could be extrapolated to the larger number of plates now scanned.

This resulted in 79% of the plates correctly processed, with a mean error well below the 3 pixels limit and close or lower than 1″ (depending on the series). We note that 44% have even better accuracy than with the previous version of the pipeline. Distortion maps and plots for plates with different scales are shown in Figure 1.

The other 21% of the plates showed various problems. The problems discovered during this test do not seem to be linked to SCAMP, and they generally could be solved by refining the parameters. We present examples of erratic maps in Figure 2. For map A, a cloud in the sky was reported in the logbook, seen as a hole in the map. Plates with spotting or uneven development will produce similar holes. For map B, the plate is possibly over-exposed and the background emission is particularly high in the center of the plate due to vignetting. By changing the detection parameters, we could recover more matches with reference stars and in both cases improve the astrometry.

For 15% of the plates, the SCAMP correction step could not even be applied due to an insufficient number of matches with reference stars that are only locally distributed (see Figure 2, C and D). This showed us that the initial astrometry is too uncertain for SCAMP to work, with scaling issues in RA and Dec. We reviewed the code used for the second step to provide a better estimate of the astrometry before using SCAMP.

We now reach a success rate of ~ 95% after reprocessing all the scanned plates to date (more than 11 000). We extracted photometry for a subset of ~ 3300 plates covering one field and obtained 2.85 times more objects with a lightcurve, down to magnitude 17 (i.e., 2 magnitudes fainter than with the previous version of the pipeline). The median RMS error on the position is about $0\rlap{.}''13$ for objects with magnitude 8 to 17.

Acknowledgments.   DASCH is supported by NSF grants AST-0407380 and AST-0909073. MS gratefully thanks A. Doane for her comments and for sharing her knowledge on the plate collection.

Visit the DASCH website at `http://hea-www.harvard.edu/DASCH`

References

Bertin, E. 2006, in Astronomical Data Analysis Software and Systems XV, edited by C. Gabriel, C. Arviset, D. Ponz, & S. Enrique (San Francisco, CA: ASP), vol. 351 of ASP Conf. Ser., 112
Bertin, E., & Arnouts, S. 1996, A&AS, 117, 393
Grindlay, J., Tang, S., Simcoe, R., Laycock, S., Los, E., Mink, D., Doane, A., & Champine, G. 2009, in Preserving Astronomy's Photographic Legacy: Current State and the Future of North American Astronomical Plates, edited by W. Osborn & L. Robbins (San Francisco, CA: ASP), vol. 410 of ASP Conf. Ser., 101
Lang, D., Hogg, D. W., Mierle, K., Blanton, M., & Roweis, S. 2010, AJ, 139, 1782
Laycock, S., Tang, S., Grindlay, J., Los, E., Simcoe, R., & Mink, D. 2010, AJ, 140, 1062
Los, E., Grindlay, J., Tang, S., Servillat, M., & Laycock, S. 2011, in Astronomical Data Analysis Software and Systems XX, edited by I. N. Evans, A. Accomazzi, D. J. Mink, & A. H. Rots (San Francisco, CA: ASP), vol. 442 of ASP Conf. Ser., 269
Mink, D. J. 2002, in Astronomical Data Analysis Software and Systems XI, edited by D. A. Bohlender, D. Durand, & T. H. Handley (San Francisco, CA: ASP), vol. 281 of ASP Conf. Ser., 169
Zacharias, N., et al. 2010, AJ, 139, 2184

Astronomical Data Analysis Software and Systems XX
ASP Conference Series, Vol. 442
Ian N. Evans, Alberto Accomazzi, Douglas J. Mink, and Arnold H. Rots, eds.
©2011 *Astronomical Society of the Pacific*

TELCAL: The On-line Calibration Software for ALMA

Dominique Broguière,[1] Robert Lucas,[2] Juan Pardo,[3] and Jean-Christophe Roche[1]

[1]*Institut de Radioastromie Millimetrique (IRAM), Grenoble — France*

[2]*ESO ALMA, Santiago — Chile*

[3]*Laboratorio de Astrofisica Molecular, Madrid — Spain*

Abstract. The ALMA on-line calibration regroups all the operations needed to maintain the ALMA interferometer optimally tuned to successfully execute the planned observations. The results of the calibrations are used in quasi-real time by the ALMA Control System. Since the first ALMA antennas were put into operation in 2009, TELCAL has been used for all the basic calibration operations and is still being improved following the project advancement. We describe here the calibrations done by TELCAL, its relationships with the other ALMA software subsystems and, briefly, the architecture of the software based on CORBA.

1. Introduction

The Atacama Large Millimeter/submillimeter Array (ALMA) is a major new facility for world astronomy. The ALMA array will on completion total 66 12 m antennas with baselines up to 16 km and state-of-the-art receivers that cover all the atmospheric windows from 30 GHz to 1 THz. The ALMA project is an international collaboration between Europe, East Asia and North America in cooperation with the Republic of Chile. ALMA is still under construction in northern Chile's Atacama desert, at an altitude of 5000 meters. As of today, 8 antennas have been put into operation and are intensively used for the commissioning of the instrument. On-line calibration, performed by the TELCAL software, groups all the operations needed to maintain the ALMA interferometer optimally tuned to successfully execute the planned observations, in particular:

1. Measurement of the atmospheric absorption and of the phase radiometric correction,
2. Pointing and focus offsets computation,
3. Delay measurements,
4. Solution of antenna position,
5. Monitoring of phase and amplitude on an astronomical calibrator source.

We describe here the calibrations done by TELCAL, its relationships with the other ALMA software subsystems and, briefly, the architecture of the software based on CORBA.

2. Architecture

TELCAL (see Fig. 1) is a component of the ALMA instrument operation. It receives data from several instruments (Correlators, Total Power detectors, Water Vapor radiometers), performs the required calibrations and delivers the results to the control system and to the archive. The ALMA Common software (ACS) is based on object oriented CORBA middleware. TELCAL uses the ACS infrastructure and deploys its code into CORBA components which process in

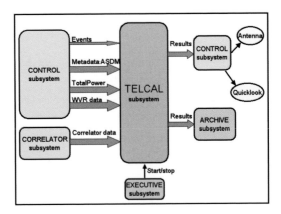

Figure 1.   Sketch of the ALMA data flow showing the relationship of TELCAL
with respect to the other subsystems.

parallel the data from several arrays and exchange messages with other CORBA components.
Metadata which describe the observing process, and results, are stored in XML format. The
binary data issued from the correlator are sent to TELCAL via an audio-video stream, because
the high data rates (up to 60MBytes/s) are not supported by CORBA.

3.   Calibration Procedures

3.1.   Pointing and Focus Measurements

Pointing sessions are required every $\sim$ 20 minutes to obtain and keep the required pointing
accuracy (0″.6). Several methods are available (e.g., five-points, cross scanning). Similar measurements are done to determine focus offsets in the X, Y or Z directions.

3.2.   Atmospheric Calibration

At specific intervals (1–10 min), we need to compute atmospheric-corrected system temperatures, which are used later by the quick-look and science pipelines to scale raw data onto the
antenna temperature scale. At the same time, receiver temperatures, optical depths and other
atmospheric parameters are also computed. The Water Vapor Radiometer data are used to compute the quantity of water in the atmosphere and the path length correction coefficients. These
coefficients are applied to the raw visibilities in real time by the Correlator subsystem, at 1.0s
intervals, prior to integration. These calculations use the atmospheric library (ATM) developed
at IEM (Madrid). Figure 2 show an example of the gain induced by this real-time correction.

3.3.   Delay Measurements and Antenna Position Determinations

The delays must be measured each time an antenna is moved or when something is changed in
the signal path. The delay measurement can be done by processing the spectral data of a strong
continuum source (phase) in interferometric mode. The antenna position (Fig. 3) determination
requires the observation of several sources across the sky.

3.4.   Phase and Amplitude Calibration

At regular intervals, phase or amplitude calibrators are observed, in order to monitor the overall
system: during a phase calibration, the rms phase fluctuations are measured as a function of

ASDM:uid___A002_X9f64b_X1

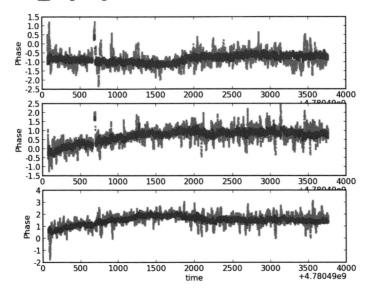

Figure 2. Phase versus time on 3 baselines. The red curve corresponds to the uncorrected data and the green curve to the corrected data, using the path length correction coefficients.

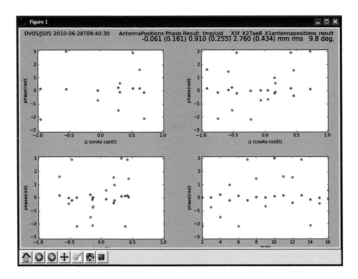

Figure 3. Antenna position using phases. Red: raw data (phase difference between two sources observed in sequence); Green: residuals of fit. Top left: plotted against a change of $X = \sin(Az)\cos(El)$; top right: against a change of $Y = \cos(Az)\cos(El)$; bottom left: against a change of $Z = \sin(El)$.

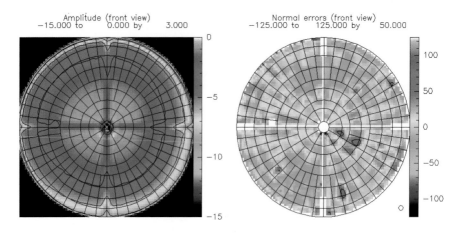

Figure 4.   Antenna Holography.

baseline length to qualify the atmosphere and determine a seeing parameter. For an amplitude calibration, the antenna efficiencies are measured. An efficiency drop can be an indication of bad pointing or focus, or another problem.

3.5.   Sideband Ratio Measurement

The sideband ratio is measured with a strong continuum source and is used in the atmospheric calibration, to compute the system temperature.

3.6.   Holography

Figure 4 shows the result of an holography done on an ALMA antenna. The goal of the holography is to measure the quality of the surface of the dish and to identify the panels which need to be adjusted. In this example, the surface rms is about 17μm.

4.   Conclusion

The TELCAL software has been mainly developed at IRAM, using the institute's large expertise in millimeter interferometry. Today, TELCAL is used every day for the commissioning of the ALMA project. It is still evolving, following the experience that we acquire on the site, especially in the domain of atmospheric calibration, which is a critical point at millimeter and sub-millimeter frequencies.

Astronomical Data Analysis Software and Systems XX
ASP Conference Series, Vol. 442
Ian N. Evans, Alberto Accomazzi, Douglas J. Mink, and Arnold H. Rots, eds.
© *2011 Astronomical Society of the Pacific*

SCUBA-2 Data Processing

Tim Jenness,[1] David Berry,[1] Ed Chapin,[2] Frossie Economou,[1] Andy Gibb,[2] and Douglas Scott,[2]

[1]*Joint Astronomy Center, 660 N. A'ohōkū Place, HI, 96720, USA*

[2]*Department of Physics & Astronomy, University of British Columbia, 6224 Agricultural Road, Vancouver, BC V6T 1Z1, Canada*

Abstract. SCUBA-2 is the largest submillimeter array camera in the world and was commissioned on the James Clerk Maxwell Telescope (JCMT) with two arrays towards the end of 2009. A period of shared-risks observing was then completed and the full planned complement of 8 arrays, 4 at 850 μm and 4 at 450 μm, are now installed and ready to be commissioned. SCUBA-2 has 10,240 bolometers, corresponding to a data rate of 8 MB/s when sampled at the nominal rate of 200 Hz. The pipeline produces useful maps in near real time at the telescope and often publication quality maps in the JCMT Science Archive (JSA) hosted at the Canadian Astronomy Data Center (CADC).

1. SMURF Iterative Map-Maker

The Sub-Millimeter Common-User Bolometer Array 2 (SCUBA-2; Craig et al. 2010; Holland et al. 2006) is a direct-detection bolometer array and so measures the temperature variation of the sky as well as the astronomical signal. Data are taken by scanning the telescope over the source using a pattern designed such that the time taken to return to the same place on the sky is not a fixed interval. The telescope software has been modified to use a number of different patterns called "pong", "lissajous" and "daisy" that have this property (Kackley et al. 2010). This allows the map-maker to separate time-varying signals from those that are fixed in a particular location on the sky. We have developed the SMURF software package (Sub-Millimeter User Reduction Facility, Chapin et al. 2010) to process SCUBA-2 data. The SMURF map-maker works by iteratively fitting a collection of models to the time stream in turn, and subtracting them, leaving the astronomical signal. All model components are refined at each iteration, resulting in improved astronomical image estimates. However, before the iterations may begin, we must repair several problems with the bolometer time series. For example, the SQUID readout electronics can introduce steps which need to be fixed. We have developed a robust algorithm for correcting these problems and an example is shown in Fig. 1.

1.1. Data Models

After glitch repair, the time series data in Fig. 1 still have a periodic structure that is dominated by a 25 second oscillation in the fridge along with low-frequency variations due to changes in the sky power and instrument drifts. These signals are common to all the bolometers and so can be removed, albeit with a loss of sensitivity to signals larger than the array footprint. We reject any bolometers that do not exhibit this strong common-mode signal. In addition, the relative amplitudes of the common-mode in different bolometers may be used to refine the flatfield.

Other models include a Fourier filter to remove low and high frequencies from the data based on the scan speed of the telescope and the wavelength, correction for atmospheric extinction, and an alternative to the Fourier filter that removes a median value from a rolling box.

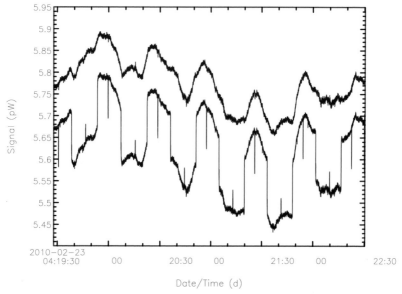

Figure 1.   The lower curve shows some flatfielded data containing many steps along with some spikes. The upper curve is the same data with the steps and spikes removed (including an offset of 0.15 pW to make this easier to see).

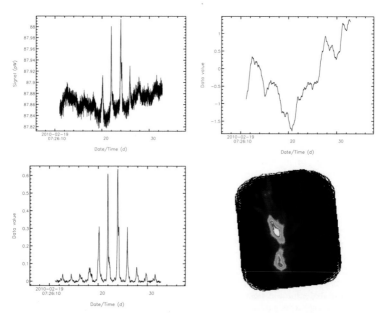

Figure 2.   Top left shows some raw flatfielded time-series data from a single bolometer for a 20 second observation of a source in Orion. Top right is the common-mode signal and the bottom left is the astronomical signal determined from the time series after removing all the models. The final map is shown bottom right and covers an area of about 4 arcmin × 4 arcmin.

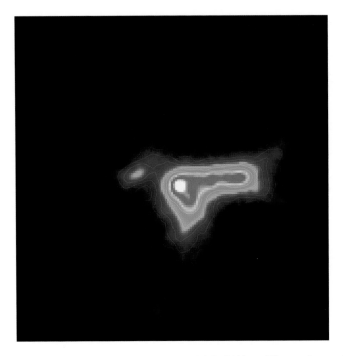

Figure 3.   Commissioning image of IRAS 18592+0108 at $450\,\mu$m taken on 2009 October 25th with contours from an $850\,\mu$m SCUBA scan map commissioning observation from 1997 August 19th (see also Gear et al. 1996).

At the end of each iteration, once the low-frequency noise and astronomical signal components have been estimated and removed, the residual signal is considerably flatter making it easy to identify smaller spikes. We have also implemented a map-based despiker that identifies outliers in the data that land in each map pixel. This procedure is repeated until the RMS of the residuals do not change appreciably. Fig. 2 shows an example of some models for a source in the Orion Molecular Cloud and Fig. 3 shows a commissioning observation compared with some commissioning data from the first SCUBA taken in 1997.

2.   Pipeline and the JCMT Science Archive

At the telescope and at the JSA (hosted at CADC) we use the ORAC-DR data reduction pipeline (Cavanagh et al. 2008) to run the SMURF map-maker and to perform mosaicking, pointing corrections and data analysis. PiCARD (Jenness et al. 2008) is used for off-line data analysis.

At the summit there are two pipelines (one for each wavelength) for Quick Look, and two pipelines for science processing, all running on dedicated machines. The Quick Look pipeline (Gibb et al. 2005) processes single observations as quickly as possible to provide instant feedback to the observer and also to reduce pointing and focus observations. The science pipelines have a little more time to process the data and can demonstrate observation progress by waiting for more data to accumulate and mosaicking multiple observations.

In the JSA time is not an issue, so the pipeline can run on the CADC grid processing cloud (Economou et al. 2011) using more complex models and many more iterations. It is also possible to run the pipeline in the new CANFAR cloud computing infrastructure (Gaudet et al. 2011).

References

Cavanagh, B., Jenness, T., Economou, F., & Currie, M. J. 2008, Astron. Nach., 329, 295

Chapin, E., Gibb, A. G., Jenness, T., Berry, D. S., & Scott, D. 2010, Starlink User Note 258, Joint Astronomy Centre

Craig, S. C., et al. 2010, in Millimeter, Submillimeter, and Far-Infrared Detectors and Instrumentation for Astronomy V, edited by W. S. Holland, & J. Zmuidzinas (Bellingham, WA: SPIE), vol. 7741 of Proc. SPIE, 77411K

Economou, F., et al. 2011, in Astronomical Data Analysis Software and Systems XX, edited by I. N. Evans, A. Accomazzi, D. J. Mink, & A. H. Rots (San Francisco, CA: ASP), vol. 442 of ASP Conf. Ser., 203

Gaudet, S., et al. 2011, in Astronomical Data Analysis Software and Systems XX, edited by I. N. Evans, A. Accomazzi, D. J. Mink, & A. H. Rots (San Francisco, CA: ASP), vol. 442 of ASP Conf. Ser., 61

Gear, W. K., Holland, W. S., Cunningham, C. R., & Lightfoot, J. F. 1996, in Submillimetre and Far-Infrared Space Instrumentation, edited by E. J. Rolfe & G. Pilbratt, vol. 388 of ESA Special Publication, 135

Gibb, A. G., Scott, D., Jenness, T., Economou, F., Kelly, B. D., & Holland, W. S. 2005, in Astronomical Data Analysis Software and Systems XIV, edited by P. Shopbell, M. Britton, & R. Ebert (San Francisco, CA: ASP), vol. 347 of ASP Conf. Ser., 585

Holland, W., et al. 2006, in Millimeter and Submillimeter Detectors and Instrumentation for Astronomy III, edited by J. Zmuidzinas, W. S. Holland, S. Withington, & W. D. Duncan (Bellingham, WA: SPIE), vol. 6275 of Proc. SPIE, 62751E

Jenness, T., Cavanagh, B., Economou, F., & Berry, D. S. 2008, in Astronomical Data Analysis Software and Systems XVII, edited by R. W. Argyle, P. S. Bunclark, & J. R. Lewis (San Francisco, CA: ASP), vol. 394 of ASP Conf. Ser., 565

Kackley, R., Scott, D., Chapin, E., & Friberg, P. 2010, in Software and Cyberinfrastructure for Astronomy, edited by N. M. Radziwill, & A. Bridger (Bellingham, WA: SPIE), vol. 7740 of Proc. SPIE, 77401Z

Astronomical Data Analysis Software and Systems XX
ASP Conference Series, Vol. 442
Ian N. Evans, Alberto Accomazzi, Douglas J. Mink, and Arnold H. Rots, eds.
©2011 Astronomical Society of the Pacific

New Control System Software for the Hobby-Eberly Telescope

Tom Rafferty,* Mark E. Cornell, Charles Taylor III, and Walter Moreira

McDonald Observatory, University of Texas at Austin, 1 University Station C1402, Austin, TX, USA 78712-0259

*rafferty@astro.as.utexas.edu, thomasrafferty@gmail.com

Abstract. The Hobby-Eberly Telescope at the McDonald Observatory is undergoing a major upgrade to support the Hobby-Eberly Telescope Dark Energy Experiment (HETDEX) and to facilitate large field systematic emission-line surveys of the universe. An integral part of this upgrade will be the development of a new software control system. Designed using modern object oriented programming techniques and tools, the new software system uses a component architecture that closely models the telescope hardware and instruments, and provides a high degree of configuration, automation and scalability. Here we cover the overall architecture of the new system, plus details some of the key design patterns and technologies used. This includes the utilization of an embedded Python scripting engine, the use of the factory method pattern and interfacing for easy run-time configuration, a flexible communication scheme, the design and use of a centralized logging system, and the distributed GUI architecture.

1. Overview

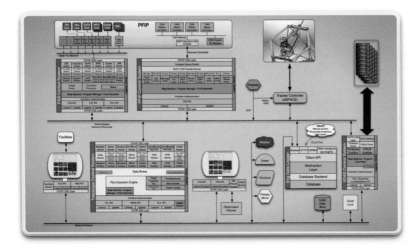

The new control system architecture consists of a closely coupled group of distributed systems. Each system is responsible for specific functions based on type or proximity to hardware, and is designed to be run autonomously. A simple but flexible messaging scheme allows communications between the systems.

2. Control System Architecture

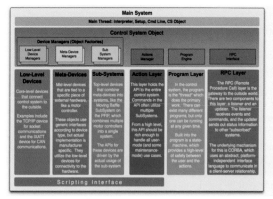

Phased Startup:

1. Main thread starts the server, reads command-line args, creates the context object and interpreter
2. Control System object creation, managers are created and setup with device via script
3. Pre-wiring: All managers instantiate their devices and device objects get pre-wiring setup
4. Wiring: Managers "wire" all their devices
   - All low-level devices establish communications with input/output hardware
   - Meta-devices are configured to talk to their respective low-level devices
   - Sub-Systems are configured with Meta-devices
   - Post-wiring configuration of devices
5. Actions layer enabled
6. CORBA/RPC layer enabled
7. Default program is loaded and starts to run

All major control systems use the same basic architecture, highlighted above. Each control system has a collection of managers, or object factories, whose job is to create and destroy all device objects in the system. They are configured at run-time by the embedded scripting engine. The API to the control system consists of a set of actions, which are controlled by a high-level, state-machine based program called actions processing. This allows asynchronous control of the underlying hardware.

3. Design Principles

Flexibility, along with reliability, were two of the primary drivers of the design. We are using modern, yet proven, programming techniques, utilizing a common tool chain and widely available libraries.

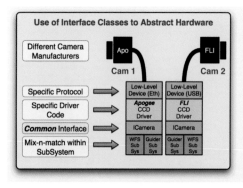

- Component oriented design
- Object oriented programming techniques
- Physical world is composed of components and modules that are connected using defined interfaces
- Common code throughout
  - Provides base classes for all objects in system
  - Similar hardware drivers implement the same generic interface class
  - Shared algorithms, routines, utilities, etc
- Scriptable interface built in at device level
- Each individual control system:
  - Controls low-level device drivers to provide I/O to the hardware
  - Provides abstraction interface to the devices so changing hardware manufacturer is simple
  - Defines it's API using Actions with a marshaled interface to the action processing system
  - Run programs that use actions to manipulate the device sub-systems
  - State machine driven

4. Scripting

At the heart of the control system lies an embedded Python scripting interface. All devices, sub-systems, and actions are available to scripting engine through the "magic" of SWIG wrappers. The control system is configured entirely from within scripting layer which executes shortly after startup. This allows the capability to completely change the configuration of your system simply by using a different or modified configuration script. Testing/debugging is simplified because you can easily isolate any driver or change its properties. Program code can start out as a prototyping script to allow agile development without changing any of the core driver code or algorithms. The use of Python as interpreter brings in a very rich and well tested set of modules for extending the base functionality of the control system, with little or no modification to the base code.

5. Communications and Messaging

In order to keep the communications and messaging between components and major systems as flexible as possible, we use JSON documents. JSON (JavaScript Object Notation) is a lightweight, data-interchange format that is easy for humans to read and write. By using JSON as the payload in our communications, we can easily add or modify commands without having to re-factor the interface.

6. GUI

The design of the GUI system allows multiple individual GUI screens to exist at any given time. Each can be configured differently based on the role of the user. The data model for the GUI resides in the control systems; therefore there is no disconnect between different GUIs that are running simultaneously.

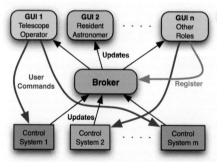

Data Flow

- ◆ Data model in control systems, not GUI.
- ◆ Each control system sees only one GUI (a broker), simplifying their operation.
- ◆ Broker distributes data to multiple GUIs as it is updated (no use for redundant model).
- ◆ GUIs send commands to control systems; control systems respond to broker; and broker distributes back to client GUIs.
- ◆ Each GUI has a common dashboard at top, but the remaining components can be customized according to the role of the user.

User Interface

- ◆ Qt/PyQt with custom widget library.
- ◆ Mayavi using OpenGL to render images and 3D visualizations in real time.
- ◆ Image visualizations allow standard set of operations in a DS9-like fashion. User can send images to DS9 for further processing, if necessary.
- ◆ User configurable set of modules (tabs) and visual elements at a given time or for a given job role.
- ◆ Fully interactive stripcharts to display data. Smaller versions (sparklines) displayed side by side with important values to quickly show trends in a glance.

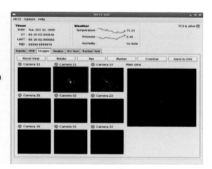

7. Logger

The logging system is intended to be an obervatory-wide system available 24/7 to a wide variety of data-producing systems using a variety or protocols. Command-line as well as web-based clients can be used to extract data from the system.

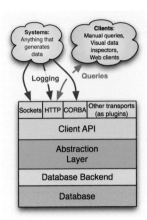

Network Server:

♦ 24/7 availability allows high number of concurrent connections.
♦ All major system connect to it via sockets, http, etc.
♦ Uniform API for all the protocols.
♦ Separation of network server and database engine processes through a fast internal synchronized queue.
♦ Uses JSON format for logging and querying, and for returning set of results.
♦ Supports raw unformatted strings for clients with limited computing power.

Database Backend:

1. Uses schema-less design (NOSQL).
2. Database engine can be changed by writing a thin layer to adapt to a defined interface.
3. Currently uses MongoDB allowing high insertion speed and easy distributed replication.
4. Separation of logging and querying OS processes to minimize interference in case of overloading. Each of them can run on separate machines if necessary.

Abstraction Layer:

1. Handles the logic between network and dB backend.
2. Does stemming analysis on text to allow fuzzy queries (plurals, conjugated verbs, etc.).
3. Adds data about the client and timestamps if necessary.

8. Development Environment

The development team uses the usual list of open source tools and libraries as described in the following table.

Primary Platform	RHEL 64-bit
Languages Used	C/C++, Python, tcl/tk
Toolchain	gcc, gnu make, bash, lapack, libtool, automake
Technologies	SWIG, omniORB, omniORBpy, SLALIB, pySlalib, CFITSIO, pyFits, libedit, sqlite, cJSON, log4cplus, Qt4/pyQt, Mayavi/OpenGL, Chaco, NumPy, SciPy, VTK, SetupDocs, Pyrex, Sphynx, Greenlet, GSL, MongoDB, Stemming
Continuous Integration	Nightly builds using Hudson
Version control	git with a central 'master' repository
Other	Automatic building of auxiliary libs; use of bugzilla for bug tracking; wiki used to capture specs, brainstorm amongst developers, document processes

9. Further Information

To obtain the original ADASS XX (2010) poster in pdf form, please use the following URL: http://www.philomather.com/ADASS_XX_P060.pdf. The author can be contacted via email at rafferty@astro.as.utexas.edu or thomasrafferty@gmail.com.

Astronomical Data Analysis Software and Systems XX
ASP Conference Series, Vol. 442
Ian N. Evans, Alberto Accomazzi, Douglas J. Mink, and Arnold H. Rots, eds.
© *2011 Astronomical Society of the Pacific*

A New Approach to Presenting Ancillary Data at the Hobby-Eberly Telescope

Sergey Rostopchin and Matthew Shetrone

Mcdonald Observatory, University of Texas, Mcdonald Observatory, TX, USA

Abstract. A new approach of managing and presenting ancillary data is developed at the Hobby-Eberly Telescope (HET). We have switched from using an ASCII form of saving all our auxiliary observation data, such as weather information, seeing data and flux from guiders software, to RDBMS. This information will be available to our users as an addition to their science data.

1. Introduction

The HET (Ramsey et al. 1998) is a 9.2 meter telescope located at McDonald Observatory at West Texas. It is an Arecibo-type optical telescope and is operated by McDonald Observatory on behalf of the University of Texas at Austin, the Pennsylvania State University, the Ludwig-Maxmilians Universität München, the Georg-August-Universität Göttingen, and Stanford University.

In our previous paper (Rostopchin et al. 2009) we described our efforts to create or improve software tools for the planning and scheduling of observations at the HET. These software tools are combined into the Night Operations Software System (NOSS). Moreover, we are upgrading all our existing programs in order to have a uniform programming language, and switching from keeping data in tab-delimited ASCII flat files to RDBMS.

Traditionally, all ancillary and engineering data at HET were kept in plain ASCII files. We are now using an RDBMS approach not only for science observation related information such as proposals and object information, but also for ancillary and some of engineering data.

We are going to make all relevant ancillary data available to our users as a supplement to their science data.

2. Ancillary Data

Ancillary data which can be useful to astronomers includes all available weather information, seeing from our differential image motion monitor and guider data.

2.1. Image Quality

The intrinsic HET site seeing is measuring using a Differential Image Motion Monitor (DIMM, Barker et al. 2003). It was somewhat a challenge to pull data from the DIMM in real time because the computer controlling it is antiquated and was intended to run stand-alone.

2.2. Weather Information

A central station collects our basic weather information. We have developed scripts which pull data every few minutes into our database. These includes: temperature, humidity, dew points and wind speed and direction. In addition, we have a cloud sensor whose information we parse into the database.

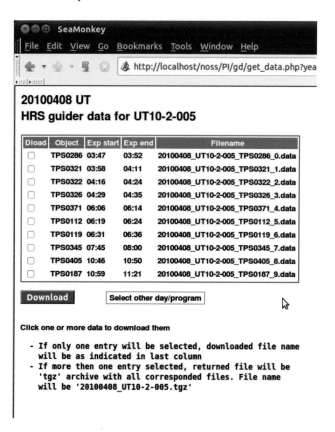

Figure 1.   This figure shows an example of the user interface to HRS guider data. With this interface it is possible to get guider data for a specific date, program and science exposures.

2.3. Guide Star Data

Guide star flux behavior during an exposure can give valuable information. For example, for planet search programs it can help to determine the flux weighted center of the observation and hence more precise radial velocities determination. The data we are storing in our database are FWHM, raw and Gauss estimated fluxes. Scripts pull the data every minute. Fig. 1 shows an example of retrieving data from the HRS guider.

At the present time only the HRS guider data are available. Future additions will make data from the other guide cameras available.

References

Barker, E. S., Adams, M. T., Deglman, F., Riley, V., George, T., Booth, J. A., Rest, A., & Robinson, E. L. 2003, in Large Ground-based Telescopes, edited by J. M. Oschmann & L. M. Stepp (Bellingham, WA: SPIE), vol. 4837 of Proc. SPIE, 225
Ramsey, L. W., et al. 1998, in Advanced Technology Optical/IR Telescopes VI, edited by L. M. Stepp (Bellingham, WA: SPIE), vol. 3352 of Proc. SPIE, 34

Rostopchin, S., Shetrone, M. C., M. E., Umbarger, J., & Mason, C. 2009, in Astronomical Data Analysis Software and Systems XVIII, edited by D. A. Bohlender, D. Durand, & P. Dowler (San Francisco, CA: ASP), vol. 411 of ASP Conf. Ser., 474

Astronomical Data Analysis Software and Systems XX
ASP Conference Series, Vol. 442
Ian N. Evans, Alberto Accomazzi, Douglas J. Mink, and Arnold H. Rots, eds.
© 2011 Astronomical Society of the Pacific

The GBT Dynamic Scheduling System: An Update

M. H. Clark,[1] D. S. Balser,[2] J. Braatz,[2] J. Condon,[2] R. E. Creager,[1] M. T. McCarty,[1] R. J. Maddalena,[1] P. Marganian,[1] K. O'Neil,[1] E. Sessoms,[3] and A. L. Shelton[1]

[1]*National Radio Astronomy Observatory, Green Bank, WV 24944-0002, USA*

[2]*National Radio Astronomy Observatory, Charlottesville, VA 22903-2475, USA*

[3]*Nub Games Inc., Carrboro, NC 27510, USA*

Abstract. The Robert C. Byrd Green Bank Telescope's (GBT) Dynamic Scheduling System (DSS), in production use since September 2009, was designed to maximize observing efficiency while maintaining the GBT's flexibility, improving data quality, and minimizing any undue adversity for the observers. Using observing criteria, observer availability and qualifications, three-dimensional weather forecasts, and telescope state, the DSS software is capable of optimally scheduling observers 24 to 48 hours in advance on a telescope having a wide-range of capabilities in a geographical location with variable weather patterns. Recent improvements for the GBT include an expanded frequency coverage (0.390–90 GHz), proper treatment of fully sampled array receivers, increasingly diverse observing criteria, the ability to account for atmospheric instability from clouds, and new tools for scheduling staff to control and interact with generated schedules and the underlying database.

1. Introduction

For the Fall 2009 observing trimester, we placed into production a Dynamic Scheduling System (DSS) for the Robert C. Byrd Green Bank Telescope (GBT). The goal was to maximize observing efficiency — especially for high frequencies — while not sacrificing the GBT's flexibility, data quality, or generate undue adversity for the observers. An overview of the system is covered in O'Neil et al. (2009) and Marganian et al. (2010); further details may be obtained from Sessoms et al. (2009), McCarty et al. (2009), Marganian et al. (2009), Clark et al. (2009), Braatz et al. (2009), Balser et al. (2009), and at our website `http://science.nrao.edu/gbt/scheduling/dynamic.shtml`.

In short, observing projects are divided into sessions, each of which is described by a set of parameters such as observing frequency, sky coordinates, and time allocated. Using a formula based on telescope efficiency, scheduling pressures, weather conditions, and administrative needs (Balser et al. 2009; Maddalena 2010a), it is possible to generate for every session a numeric score for every 15-minute time period available for scheduling. A schedule is then generated with a packing algorithm that maximizes the score for the 48-hour scheduling period. Before being finalized, the schedule is inspected and possibly modified by a human scheduler by trading off overall telescope performance to achieve specific project goals.

Over the last year the experience we have gained using the system has generated new requirements leading to enhancements which are described here.

2. Expanded Frequency Coverage

New instrumentation, new receivers, and improved antenna pointing in 2010 have made it practical to observe with the GBT at higher frequencies than before. The first release of the DSS in 2009 provided accurate scheduling only for frequencies between 2 and 50 GHz. With the 2010 release, the DSS can handle frequencies up to 120 GHz and provides better scheduling below 2 GHz.

3. Fully-Sampled Array Receivers

The advent of filled-array receivers that fully sample the sky in one pointing makes the GBT less vulnerable to the effects of wind. Modifications to the ranking algorithm account for use of this class of receiver.

4. New Atmospheric Stability Limits

Scheduling of continuum observing (as opposed to spectral line) requires defining and predicting the atmospheric stability.

Hydrosols are water droplets in clouds. They are not well mixed in the atmosphere, so they cause short-term fluctuations in the total power detected by the antenna that are proportional to frequency squared and can degrade single-dish continuum observations at frequencies higher than about 2 GHz. These relatively slow, broadband fluctuations have little effect on pulsar and spectral-line observations. Continuum observations at frequencies above about 2 GHz should normally be scheduled only when there are no hydrosols; that is, when the sky is clear. We can use the forecast downward atmospheric irradiance at long wavelengths (4.5–40 microns) to estimate cloud cover and hence atmospheric stability.

We do not consider scheduling continuum observations above 2 GHz when the downward atmospheric irradiance exceeds $330\,\text{W}/\text{m}^2$, in accordance with the recommendation of Balser (2010). These limits are also now reflected in the weather information available through the GBT page Maddalena (2010b).

5. New Scheduling Tools

A rigorous application of observing efficiencies and the knapsack algorithm are used daily to generate the optimal schedule for the following two days. However, there are times when the human scheduler needs to sacrifice observing efficiency to meet other needs. The tools available in the DSS scheduler's interface provide the information and means to make these tradeoffs. In addition, tools are provided to aid the scheduler in updating observer qualifications, session observing criteria, observing window specifications, receiver schedule, and time accounting.

Monitoring projects that require regularly spaced observations, such as pulsar timing experiments, can be a challenge for scheduling and require special consideration within the DSS. These projects must be scheduled in such a manner that conflicts are avoided. Currently 40% of GBT time is consumed by windowed observations.

The approach for scheduling windowed observations is conceptually simple. All monitoring windows are selected a priori along with an associated default telescope (or observing) period near the end of the window. The default period is when the observing will take place regardless of weather forecasts, unless a period with a better ranking score wins a time earlier in the window. This process guarantees observations within the window, while providing the scheduling algorithms an opportunity to provide appropriate weather. Movement of the telescope period to an earlier slot occurs about half the time.

6. Extended Weather History

Solitary weather forecasts are insufficient to generate realistic scheduling measures. Any metric must be weighed against the best possible weather and the likelihood of a specific weather measurement occurring. This information is drawn from historical weather. Previously, we extracted these measures from 2004–2007 forecasts. We now have added weather forecasts since 2007 providing a more reliable baseline.

7. Unexpected Benefits

The use of dynamic scheduling has provided additional benefits not originally targeted.

7.1. Observer's Ability to Disable Sessions

The observer has the option of disabling any subset of her sessions, i.e., removing them as candidates for scheduling. It was understood such a feature would be needed to table sessions until scripts or other prepatory needs were completed. What was not foreseen was the need to iteratively enable and disable sessions, e.g., for disabling of pulsar sessions to allow adjustment to unexpected timing parameters.

7.2. Adjustment of the Schedule Because of Unexpected Events

Because the schedule is not fixed weeks in advance, unexpected events, such as equipment failures, become just another factor in scheduling preventing a domino effect which results in many projects being affected.

7.3. Minimum Downtime for Major Repair/Maintenance Activities

Before dynamic scheduling, allotted time for non-observing activities had to encompass worst-case scenarios to ensure sufficient time, but by generating a schedule via a two-day moving window, we can track progress on major activities and schedule time accordingly.

7.4. Seamless Handling of Target-of-Opportunity Projects

Because of the domino effect of inserting new projects into a predetermined schedule, the impact of a target-of-opportunity period goes beyond the sessions being replaced since they in turn must be inserted into the schedule somewhere. A target-of-opportunity project becomes simply a very high-priority project.

8. Things to Come

- Correctly handle the scheduling of non-guaranteed monitoring sessions.

- Provide increased options for elective/pre-scheduled sessions, e.g., VLB and other observing requiring coordination in scheduling with other telescopes.

- Continue to improve the selection and analysis of weather information.

- Implement a sensitivity calculator to aid in observation planning.

Acknowledgments. The National Radio Astronomy Observatory is a facility of the National Science Foundation operated under cooperative agreement by Associated Universities, Inc.

References

Balser, D. S. 2010, Forecasting Clouds in Green Bank, Tech. Rep. 13, NRAO, https://safe.nrao.edu/wiki/pub/GB/Dynamic/DynamicProjectNotes/dspn13.0.pdf

Balser, D. S., et al. 2009, in Astronomical Data Analysis Software and Systems XVIII, edited by D. A. Bohlender, D. Durand, & P. Dowler (San Francisco, CA: ASP), vol. 411 of ASP Conf. Ser., 330

Braatz, J., et al. 2009, in Astronomical Data Analysis Software and Systems XVIII, edited by D. A. Bohlender, D. Durand, & P. Dowler (San Francisco, CA: ASP), vol. 411 of ASP Conf. Ser., 334

Clark, M., et al. 2009, in Astronomical Data Analysis Software and Systems XVIII, edited by D. A. Bohlender, D. Durand, & P. Dowler (San Francisco, CA: ASP), vol. 411 of ASP Conf. Ser., 338

Maddalena, R. J. 2010a, in BAAS, vol. 42 of BAAS, 406

— 2010b, High Frequency Weather Forecasts. URL `http://www.gb.nrao.edu/~rmaddale/Weather/`

Marganian, P., Clark, M., McCarty, M., Sessoms, E., & Shelton, A. 2009, in Astronomical Data Analysis Software and Systems XVIII, edited by D. A. Bohlender, D. Durand, & P. Dowler (San Francisco, CA: ASP), vol. 411 of ASP Conf. Ser., 342

Marganian, P., Clark, M., Shelton, A., McCarty, M., & Sessoms, E. 2010, in Astronomical Data Analysis Software and Systems XIX, edited by Y. Mizumoto, K. I. Morita, & M. Ohishi (San Francisco, CA: ASP), vol. 434 of ASP Conf. Ser., 47

McCarty, M., Clark, M., Marganian, P., O'Neil, K., Shelton, A., & Sessoms, E. 2009, in Astronomical Data Analysis Software and Systems XVIII, edited by D. A. Bohlender, D. Durand, & P. Dowler (San Francisco, CA: ASP), vol. 411 of ASP Conf. Ser., 346

O'Neil, K., et al. 2009, in Astronomical Data Analysis Software and Systems XVIII, edited by D. A. Bohlender, D. Durand, & P. Dowler (San Francisco, CA: ASP), vol. 411 of ASP Conf. Ser., 147

Sessoms, E., Clark, M., Marganian, P., McCarty, M., & Shelton, A. 2009, in Astronomical Data Analysis Software and Systems XVIII, edited by D. A. Bohlender, D. Durand, & P. Dowler (San Francisco, CA: ASP), vol. 411 of ASP Conf. Ser., 351

Astronomical Data Analysis Software and Systems XX
ASP Conference Series, Vol. 442
Ian N. Evans, Alberto Accomazzi, Douglas J. Mink, and Arnold H. Rots, eds.
©2011 *Astronomical Society of the Pacific*

Technologies for High Speed Data Handling in the ATST

Bruce Cowan and Steve Wampler

National Solar Observatory, Tucson, AZ, USA

Abstract. The Advanced Technology Solar Telescope (ATST) has some very demanding data handling requirements. Sixteen high speed camera lines are available, each capable of transmitting data at up to 1 gigabyte per second. Camera lines must be fully isolated from each other in order to cope with the high data volume, yet be defined dynamically on a per-experiment basis, providing the ability to route data through various real-time data processing/reduction and quality assurance nodes as needed. This paper discusses the investigation and technologies chosen to deliver on those requirements.

1. Introduction

The ATST[1] is a world-class, 4-meter optical telescope being built atop the Haleakala volcano on the Hawaiian island of Maui. It is designed to serve the solar physics community to the middle of the 21st century. It hosts an array of instruments—many using technologies still in the laboratory.

The ATST Data Handling System (DHS) facilitates the delivery of data from the observatory instruments to the observer. The DHS provides high-speed bulk data transport along with intermediate storage space. Each instrument camera is attached to an associated *camera line* (Wampler & Goodrich 2009) that manages the data flow for the duration of the observation. The camera line provides any necessary storage, data processing, or interface to external systems. Some instruments have their data routed through processing nodes for real time data reduction. For data quality assurance, the DHS provides both a quick-look display for raw data and a detailed display that routes the data through a processing node prior to its display. The DHS is scalable architecture, so as new instruments are added, the DHS adds more camera lines to accommodate them.

The primary requirements of the system focus on two main features:

- **Bandwidth:** Initially targeting 4K × 4K cameras running at 30 fps, plus other instrument data required for processing nodes within the pipeline, each camera line must be capable of transmitting data between nodes at up to 1 gigabyte per second. Future instruments may run at even higher data volumes, so an upgrade path beyond this must also be considered.

- **Flexibility:** In order to cope with the high data volume, camera lines must be fully isolated from each other, yet be defined dynamically on a per-experiment basis. This provides the the ability to route data through various real-time processing, data reduction and quality assurance nodes as needed. This level of flexibility also facilitates the easy addition of future camera lines, with minimal impact to operations

---

[1]http://atst.nso.edu/

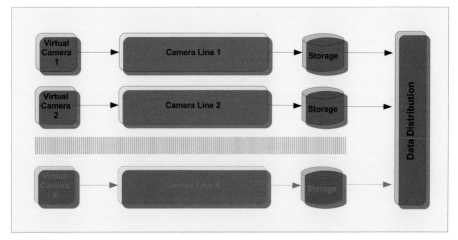

Figure 1.   Camera lines are isolated from each other, and new ones can be added as required.

2.   Technology Choices

Hardware technology selection for projects with long lead times come with a higher level of risk. By the time the ATST is completed, it is reasonable to expect that today's cutting edge offerings may be a generation or two old, and some may become completely obsolete. To minimize the risk of investing time building on a limited technology, associated industry trends need to be considered to get an understanding of what future technologies are likely to become reality, and to identify which current technologies are likely to offer the lowest impact upgrade potential. In these areas, 10 Gigabit Ethernet (10GbE) offered the most promise. Its theoretical maximum bandwidth seemed to solve the immediate 1 gigabyte per second target, and the IEEE was a few years into working on 40GbE and 100GbE standards (since ratified in June 2010) to define the next generation. Widespread 10GbE adoption has been relatively slow, but the recent availability of a wide range of reasonably priced hardware has had a big effect. It is now becoming entrenched in the corporate data center.

Hardware that is physically capable of the bandwidth target must be driven by a software architecture that uses that capability efficiently. An issue was presented by a situation shown in the camera line use cases: multiple data paths within the line. If a camera line is running at near maximum bandwidth, standard unicast transmission would make it unable to send the data from a source node to more than one recipient node, because it would have to send the same data multiple times. Going to a multicast implementation solves this. A multicast message is sent once and received by multiple recipients, so the camera line can run at full speed with multiple paths. RTI's DDS[2] implementation allows multicast transmission, and also brings another feature that fulfills the dynamic definition requirement. Its publish/subscribe model allows the processing nodes to "discover" each other using multicast without having a data path predefined for each sender. A sender simply declares it will "publish" data of a certain topic, and all receivers "subscribing" to that topic automatically receive any messages that are published.

---

[2]http://www.rti.com/products/dds/index.html

3. Hardware Notes

Our initial test platform consists of two rack-mounted computers, each running 64-bit CentOS LINUX. Both machines have a single Intel Xeon processor. One has an X5650 (6 cores), and the other a X3450 (4 cores). The machines are equipped with a PCI Express V2.0 bus. Connectivity is via Intel X520-DA2[3] 10GbE cards, attached with an SFP+ direct attach cable. The MTU for the Ethernet ports are configured for a size of 16,110 bytes, which is the highest jumbo frame size supported by the X520-DA2 cards. The most current driver for the Intel cards was installed, and configured to enable the card's special I/OAT (I/O Acceleration Technology) high-performance feature.

We are in the process of expanding this to a third machine, identical to the existing X3450 box. They will be connected through a Fujitsu XG2600[4] 10GbE switch. For the DHS, the feature set is an important consideration in switch selection. Efficient routing of multicast packets requires the switch to perform *IGMP Snooping* (Internet Group Management Protocol[5]) and also to act as an *IGMP querier*. This allows the switch to build a map of which clients have specifically requested certain types of multicast traffic, and route the packets only to them. Without this feature, a switch's default behavior is to flood multicast traffic to all machines on the VLAN, whether they contain an interested listener or not. This would be catastrophic for a high-bandwidth application like the DHS.

Other important switch features are its ability to process "jumbo frames" and the sizes of jumbo frames it supports. Typically, Ethernet frames have a payload of 1,500 bytes, which means that a 32 MB message within the DHS gets sliced up and transmitted from one node to another in over 22,000 pieces, each causing additional processing overhead. Jumbo frames are technically anything over the 1,500 bytes, but almost all switches that support this feature limit the size to 9,000 bytes. The XG2600 supports jumbo frame sizes up to 16K, which will allow us to continue to run the X520-DA2 cards at their maximum frame size, and transmit that 32 MB DHS data message in just over 2,000 pieces.

4. Tweaks

When we first tested DDS using Java, the initial experiments were quite disappointing. Data transmissions were plagued with lost packets, dropping the effective throughput of 32 MB data messages to zero, unless the sending logic throttled the rate back to a fraction of the line speed. Investigation revealed that the default LINUX kernel network settings are only optimized for 100 Mb connections. Once these were identified and adjusted, data transmission became reliable, but still fell far short of our bandwidth requirement. We were now bottle-necked at the CPU, which was running at 100%, but only utilizing a single core.

A closer examination of marshaling code generated by an RTI utility showed it to contain inefficiencies that required only minor tweaks to greatly reduce CPU usage. Data arrays were copied element-by-element instead of using much faster methods that copied the entire array all at once. Also, examination of the generated code revealed that some data types are handled more efficiently than others. Using a list instead of an array might be more convenient, but it came at the price of extra handling overhead. Using the patched marshaling code and a slightly adjusted adjusted data structure, we were running reliably at 725 MB per second, but still maxing out a single CPU core.

The final step was to adjust how our code used RTI DDS. The RTI documentation said that each publisher, or subscriber, would run on a separate thread. To take advantage of this, we

---

[3]http://www.intel.com/Assets/PDF/prodbrief/322217.pdf

[4]http://www.fujitsu.com/us/services/computing/peripherals/ethernet/xg2600-switch.html

[5]http://www.networksorcery.com/enp/protocol/igmp.htm

added the concept of multiple "channels" to our publisher and subscriber classes, where each channel equates to a separate RTI publisher and subscriber. This allowed high-bandwidth topics to specify more than one channel, thereby using more than a single CPU core. Configuring our test application to use two channels instead of one, resulted not only in sustained and reliable data transmission at our target rate of 1 GB per second, but also a comfortable margin above it.

5. Summary

Although some instruments transmit much smaller data sizes, and some of those at much higher frame rates, we chose to simulate a 4K × 4K camera running at 30 fps for our technology evaluation because it represented the maximum bandwidth load in our requirements. Once it was established that DDS could handle that, additional experiments were conducted to explore the performance of smaller sizes, and it continued to meet expectations.

DDS and multicast has proven to be a viable vehicle for high-performance distribution of large data sets in the ATST DHS. The DHS API available to the other systems within the ATST is based on the concepts of a publish/subscribe model, but doesn't expose any of RTI's DDS implementation lying below. This not only greatly simplifies usage of the DHS, it also insulates the other systems from change should it ever become necessary to move DHS to different middleware.

References

Wampler, S., & Goodrich, B. 2009, in Astronomical Data Analysis Software and Systems XVIII, edited by D. A. Bohlender, D. Durand, & P. Dowler (San Francisco, CA: ASP), vol. 411 of ASP Conf. Ser., 527

Astronomical Data Analysis Software and Systems XX
ASP Conference Series, Vol. 442
Ian N. Evans, Alberto Accomazzi, Douglas J. Mink, and Arnold H. Rots, eds.
© *2011 Astronomical Society of the Pacific*

The Archive and Digitizer Facility at the ROB

Jean-Pierre De Cuyper,[1] Georges de Decker,[1] Lars Winter,[2] and Norbert Zacharias[3]

[1] *Royal Observatory of Belgium, Ringlaan 3, B-1180 Ukkel, Belgium*

[2] *Hamburg, Germany*

[3] *US Naval Observatory, 3450 Massachusetts Ave, NW, Washington, DC 20392-5420, USA*

Abstract. The building of a 2D-digitizer facility of high geometric and radiometric resolution and precision, started under the D4A pilot-project (2002–2005), has been continued during the DI/07 production-project financed by the Belgian Science Policy for the digitization of the astro photographic archives of the Royal Observatory of Belgium (ROB), the aerial photographic archives of the Belgian National Geographic Institute (NGI) and of the Belgian Royal Museum of Central Africa (RMCA) and the photographic archive of the Belgian Royal Institute for Cultural Heritage (RICH). An international collaboration has been set up between the US Naval Observatory (USNO) in Washington DC, the IMCCE of Paris Observatory (OBSPM) and the Royal Observatory of Belgium (ROB) to make a new astrometric reduction of archival photographic plates of planetary satellites.

1. The ROB Digitizer

In autumn 2007 the ROB digitizer has been installed and housed in a temperature and humidity stabilised clean room with adjacent archive room. The ROB digitizer can digitize photographic images up to 342mm (14inch) wide on glass plates, film sheets and film rolls. A cast aluminum counter pressure plate and plate trays with a central opening corresponding to the actual image size are used for the automatic loading into focus (de Cuyper et al. 2004; de Cuyper & Winter 2005, 2006; de Cuyper et al. 2009). The positioning repeatability of the XY table is better than $0.01\mu m$ and the fitting of stellar images is stable to better than $0.07\mu m$ on the plate. This corresponds to the proposed submicron accuracy (Zacharias et al. 2004) for digitizing astronomical photographic plates. The XY-table is an adapted Aerotech ABL3600 air bearing system. The mechanical subsystem, designed and build to our specifications by Aerotech, Pittsburgh (Schmidt et al. 2010), includes: an automatic plate holder assembly and a plate change robot with plate tray magazine and turntable for the photographic glass plates and film sheets; and a full automatic film roll transport system. These custom made devices are necessary for a rapid change of the photographs to be digitized, without manual intervention.

The heart of the optical system is a BCi4, 12bit CMOS camera from C-Cam Vector International, mounted on a Schneider Xenoplan telecentric 1:1 objective. The back light illumination system consists of very bright LED's that are powered by a precision power supply.

Software was developed at the ROB for the automation of the step-and-stare digitization process, (i.e., autofocus, auto exposure bracketing for adjusting the exposure of each individual sub-image of the aerial photographic images, high dynamic range enhancement of the camera resolution is used for digitizing the art photographic images), the determination and correction of the optical distortion caused by the object/detector unit of the digitizer, the post-treatment of the raw images (i.e., flat and dark correction, and for aerial images also dodging and vignetting correction) and the creation of digital images in TIFF and FITS format.

301

2. Astro Plates

An international collaboration has been set up between the US Naval Observatory (USNO), the IMCCE of Paris Observatory and the Royal Observatory of Belgium (ROB) to make a new astrometric reduction of the USNO archival photographic plates collection of the planetary satellites in order to provide a better knowledge of their orbital motions. Hereto the photographic plates, taken with the 26-inch Yale telescope at the Union Observatory in Johannesburg, South Africa from 1926 till 1947 and with the 26-inch telescope at the US Naval Observatory in Washington DC, USA by Dan Pascu between 1967 till 1998 (Pascu 1977, 1979, 1994) will be digitized with the new generation digitizer at the ROB, providing accurate fitted positions of better than 0.1μm on the plates. Specific procedures and algorithms are used to obtain accurate sky position of each moon using the Tycho2, UCAC2 (20–30 mas) and later the UCAC4 (15 mas) catalogs.

The ESO Schmidt plates archive will also be stored at the ROB. At first the ROB and ESO plates will be digitized on specific demand. The aim is to also make available tables containing the extracted and distortion corrected object positions in mm on the glass plates as well as calibrated astrometric positions on the sky and photometric data together with object identifications. This extracted objects information will be stored in a database.

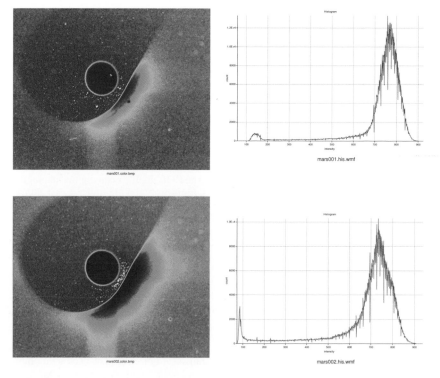

Figure 1.   On the left two false color images obtained from digitizing two USNO plates of Mars and his moons taken with different integration times. On the right the corresponding transmission histograms.

Acknowledgments.   This digitization project is financed by the Belgian Federal Science Policy Office (Project DI/07).

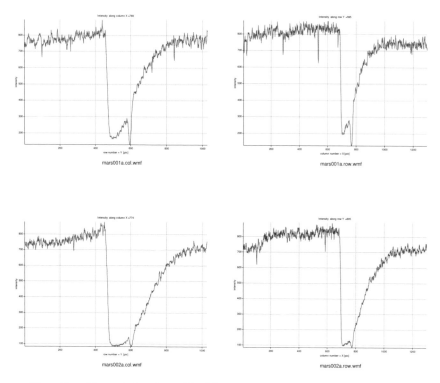

Figure 2.   Transmission plots of the Mars images passing through the inner moon. Notice the non-linear sky background in the vicinity of Mars.

References

de Cuyper, J., & Winter, L. 2005, in Astronomical Data Analysis Software and Systems XIV, edited by P. Shopbell, M. Britton, & R. Ebert (San Francisco, CA: ASP), vol. 347 of ASP Conf. Ser., 651

— 2006, in Astronomical Data Analysis Software and Systems XV, edited by C. Gabriel, C. Arviset, D. Ponz, & S. Enrique (San Francisco, CA: ASP), vol. 351 of ASP Conf. Ser., 587

de Cuyper, J., Winter, L., de Decker, G., Zacharias, N., Pascu, D., Arlot, J., Robert, V., & Lainey, V. 2009, in Astronomical Data Analysis Software and Systems XVIII, edited by D. A. Bohlender, D. Durand, & P. Dowler (San Francisco, CA: ASP), vol. 411 of ASP Conf. Ser., 275

de Cuyper, J., Winter, L., & Vanommeslaeghe, J. 2004, in Astronomical Data Analysis Software and Systems XIII, edited by F. Ochsenbein, M. G. Allen, & D. Egret (San Francisco, CA: ASP), vol. 314 of ASP Conf. Ser., 77

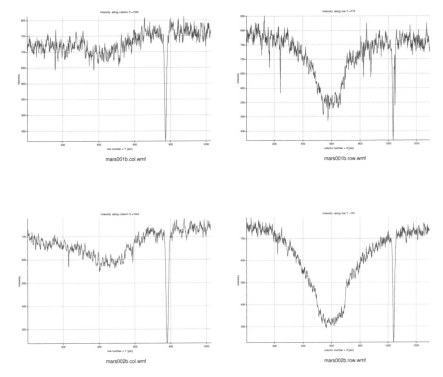

Figure 3.   Transmission plots of the Mars images passing through the outer moon.

Pascu, D. 1977, in IAU Colloq. 28: Planetary Satellites, edited by J. A. Burns, 63
— 1979, in Natural and Artificial Satellite Motion, edited by P. E. Nacozy & S. Ferraz-Mello, 17
— 1994, in Galactic and Solar System Optical Astrometry, edited by L. V. Morrison & G. F. Gilmore, 304
Schmidt, E., De Cuyper, J.-P., Winter, L., Ciez, A., & Ludwick, S. 2010, in Proceedings of the 10th international conference of the european society for precision engineering and nanotechnology, edited by H. Spaan, P. Shore, H. Van Brussel, & T. Burke (Bedford, UK: euspen), 372
Zacharias, N., Urban, S. E., Zacharias, M. I., Wycoff, G. L., Hall, D. M., Monet, D. G., & Rafferty, T. J. 2004, AJ, 127, 3043

Astronomical Data Analysis Software and Systems XX
ASP Conference Series, Vol. 442
Ian N. Evans, Alberto Accomazzi, Douglas J. Mink, and Arnold H. Rots, eds.
© *2011 Astronomical Society of the Pacific*

Data Pipelines for the TRES Echelle Spectrograph

Douglas J. Mink

Smithsonian Astrophysical Observatory, 60 Garden St., Cambridge, MA 02138

Abstract.   An IRAF-based processing pipeline has been written to reduce spectra from the Smithsonian Astrophysical Observatory's TRES (Tillinghast Reflector Echelle Spectrograph) two-fiber echelle spectrograph on Mt. Hopkins in Arizona. A modular system was written in IRAF so that the same software could be used at the telescope for quick-look processing and later with more accuracy for scientific processing. IRAF tasks developed for the SAO FAST long-slit and Hectospec multi-fiber spectrographs, as well as NOAO-developed echelle tasks, were adapted for this instrument.

The TRES Spectrograph

The Tillinghast Reflector Echelle Spectrograph (TRES, Szentgyorgyi & Furész 2007) is a two-fiber (object and sky spectra) echelle spectrograph on the 1.5-meter Tillinghast telescope at the Smithsonian Astrophysical Observatory's Fred L. Whipple Observatory on Mt. Hopkins in Arizona. It covers optical and near-IR wavelengths from 3850 to 9100 Ångstroms at a resolution which varies from 0.03 to 0.07 Ångstroms/pixel over the spectrograph's 51 orders.

Design Strategy

Our two design goals were to keep the maximum wavelength precision and to use the same software for both quick-look processing at the telescope and more accurate processing for scientific analysis.

To maintain accuracy, we fit a two-dimensional (order and pixel) dispersion function to each spectrum and maintain that function in the header through further processing using formats developed in IRAF for dealing with multi-order echelle data. Using IRAF made it possible to adapt and re-use tasks written to process data from other SAO spectrographs such as FAST (Tokarz & Roll 1997), Hectospec (Mink et al. 2005, 2007), and Hectochelle, as well as IRAF tasks written at NOAO (Valdes 1992) which process images and extract multiple-order Echelle spectra. For portability, all code is written in IRAF CL or SPP.

Data Structure

The process needed to be easy enough for scientists to use and even harder for them to misuse. This was accomplished by keeping all processed data in a different directory than raw data and enforcing that separation in code.

Each night's raw data resides in a directory designated by *[rawdir]/yyyy/yyyy.mmdd*, with filenames beginning with a three-digit sequence number, followed by an object name or calibration type (BIAS, DARK, FLAT, or COMP). These filenames are kept through all processing.

Corresponding processed data is in a directory *[procdir]/yyyy/yyyy.mmdd*. When composite files, such as sums and medians, are created, they are automatically given sequence numbers beyond those used for observed data.

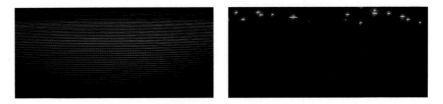

Figure 1.   Processed flat field and flat field images.

Data Processing

One master IRAF CL script, **tproc**, runs the entire data processing pipeline. In the past, while processing data from other instruments, it has been difficult to work with IRAF scripts which called one master program with different arguments to do different processing steps. When tasked with writing a multiple purpose pipeline, however, it became apparent that having one main program made it easier to process all of the different data types identically. It works on a single spectrum or a group of spectra of the same object with the same fiber configuration and exposure time.

First, the standard IRAF **ccdproc** task is used to remove the bias levels from the overscan region of the chip and trim that region from the image file. Then the two amplifiers are corrected to the same gain and merged into a single image. Bad pixels are fixed in that merged and trimmed image because they can only be discovered in images combined to remove the many bright spots caused in each image by radiation from the anti-reflection coating on the dewar window. Each image is then corrected by the aperture flattening image to minimize fringing and pixel to pixel variation. If a median image is requested, it is made next, before particle hits and cosmic rays are removed from a group of similar images by a locally-written task which intercompares them. At this point a sum of all of the similar images may be made and added to the processing list along with a requested median file. Then a model of the scattered light is fit to the background of each image, with the spectrum or spectra masked out, and subtracted.

For ThAr wavelength calibration and flat field lamps, one spectrum is then extracted from each image. For object plus sky images, two spectra are extracted. Thorium Argon (ThAr) wavelength calibration spectra are compared to a reference spectrum, and a wavelength to pixel mapping is fit for each one. For each object or sky spectrum, wavelength solutions are found by combining the closest or most recent ThAr solutions. If the simultaneous sky spectrum is going to be removed from the object spectrum, it is first rebinned to the same wavelength solution as the object spectrum, and then corrected for fiber throughput before being subtracted.

Processing Data

Initalization

To make the separation of raw and processed data easy, the **trsdate** yyyy.mmdd command sets the processing and raw data directories based on the date. After this, all of the tasks can find the raw data and put the processed data only in the processing directory. Tasks that write data always check to make sure that they are writing to the processing directory.

Pipeline Processing

Quick Look

qtres reduces one raw spectrum image to one to three multi-order echelle spectra using **tproc**. Default flat fields and ThAr spectra can be used because night-to-night instrumental shifts are less than a pixel.

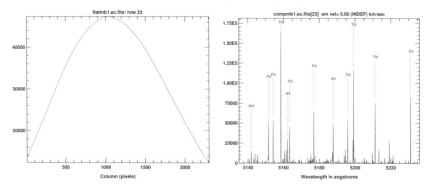

Figure 2.   Single order of extracted flat field and ThAr spectra.

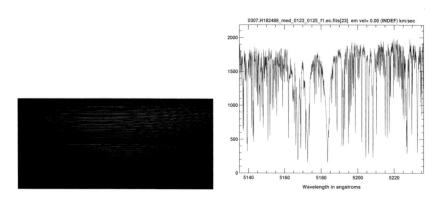

Figure 3.   Raw stellar spectrum and one order of extracted spectrum.

Quick Look with Cosmic Ray Removal

ctres is a quick-look task for multiple images with the same configuration reducing a list of raw spectrum images to dispersion- corrected multi-order echelle spectra with cosmic rays removed by tcosmic.

Setting Up and Processing a Full Night's Data

trsgroup makes lists all of the TRES files to be processed, grouped by object name, fiber configuration and exposure time *[obj][s m l][b if binned][fiber(s)]x[exposure in seconds].list*. Multiple groupings of exposures of the same object and configuration are split into separate lists. A master list of lists is made for object data. Processing then proceeds with optional interaction.

   btres processes and takes the median of all bias images (those with BIAS as the object name), plotting a histogram of the combined image as an instrument check.

   If dark field images are taken, **dtres** makes a median of all of them and plots a histogram of the combined image as an instrument check.

   Images of flat field spectra (with FLAT as the object name) are used to create extraction functions, aperture flattening masks, scattered light masks, and throughput ratios. **ftres** is run on lists of short-exposure images from each fiber for each observation configuration, creating the necessary masks and functions.

Thorium Argon spectra (with COMP as the object name) are processed using **ttres**, which reads a list of raw ThAr spectrum images taken with the same fiber size and binning and extracts them to dispersion-corrected multi-order echelle spectra. Each extracted spectrum (see Figure 1) is cross-correlated to a reference spectrum in pixel space using **rvsao.xcsao** (Kurtz & Mink 1998). The resulting pixel shift is added to the line positions in the reference spectrum database file which is then refit to the lines identified in the ThAr spectrum.

Object spectra are then processed by the **trsproc** task which runs **otres** on each list of similar object observations read from the list file created by **trsgroup**.

Acknowledgments. I would like to thank Gabor Furész for designing the instrument and suggesting processing possibilities, Lars Buchave and Dave Latham for suggesting additional processing steps, Nathalie Martimbeau and Perry Berkind for using the pipeline and suggesting improvements, and Susan Tokarz for helping me clarify this description of how the pipeline works.

References

Kurtz, M. J., & Mink, D. J. 1998, PASP, 110, 934
Mink, D. J., Wyatt, W. F., Caldwell, N., Conroy, M. A., Furesz, G., & Tokarz, S. P. 2007, in Astronomical Data Analysis Software and Systems XVI, edited by R. A. Shaw, F. Hill, & D. J. Bell (San Francisco, CA: ASP), vol. 376 of ASP Conf. Ser., 249
Mink, D. J., et al. 2005, in Astronomical Data Analysis Software and Systems XIV, edited by P. Shopbell, M. Britton, & R. Ebert (San Francisco, CA: ASP), vol. 347 of ASP Conf. Ser., 228
Szentgyorgyi, A. H., & Furész, G. 2007, in The 3rd Mexico-Korea Conference on Astrophysics: Telescopes of the Future and San Pedro Mártir, edited by S. Kurtz, vol. 28 of Rev. Mexicana Astron. Astrofis. Conf. Ser., 129
Tokarz, S. P., & Roll, J. 1997, in Astronomical Data Analysis Software and Systems VI, edited by G. Hunt & H. Payne (San Francisco, CA: ASP), vol. 125 of ASP Conf. Ser., 140
Valdes, F. 1992, in Astronomical Data Analysis Software and Systems I, edited by D. M. Worrall, C. Biemesderfer, & J. Barnes (San Francisco, CA: ASP), vol. 25 of ASP Conf. Ser., 417

Astronomical Data Analysis Software and Systems XX
ASP Conference Series, Vol. 442
Ian N. Evans, Alberto Accomazzi, Douglas J. Mink, and Arnold H. Rots, eds.
© 2011 Astronomical Society of the Pacific

The Challenge of Data Reduction for Multiple Instruments on the Stratospheric Observatory for Infrared Astronomy (SOFIA)

M. V. Charcos-Llorens,[1] R. Krzaczek,[2] R. Y. Shuping,[3] and L. Lin[1]

[1]*Universities Space Research Association, NASA Ames Research Center, Moffett Field, CA 94035, USA*

[2]*Chester F. Carlson Center for Imaging Science, RIT, 54 Lomb Memorial Drive, Rochester NY 14623, USA*

[3]*Space Science Institute, 4750 Walnut Street, Boulder, Colorado 80301, USA*

Abstract. SOFIA, the Stratospheric Observatory For Infrared Astronomy, presents a number of interesting challenges for the development of a data reduction environment which, at its initial phase, will have to incorporate pipelines from seven different instruments developed by organizations around the world. Therefore, the SOFIA data reduction software must run code which has been developed in a variety of dissimilar environments, e.g., IDL, Python, Java, C++. Moreover, we anticipate this diversity will only increase in future generations of instrumentation. We investigated three distinctly different situations for performing pipelined data reduction in SOFIA: (1) automated data reduction after data archival at the end of a mission, (2) re-pipelining of science data with updated calibrations or optimum parameters, and (3) the interactive user-driven local execution and analysis of data reduction by an investigator. These different modes would traditionally result in very different software implementations of algorithms used by each instrument team, in effect tripling the amount of data reduction software that would need to be maintained by SOFIA. We present here a unique approach for enfolding all the instrument-specific data reduction software in the observatory framework and verifies the needs for all three reduction scenarios as well as the standard visualization tools. The SOFIA data reduction structure would host the different algorithms and techniques that the instrument teams develop in their own programming language and operating system. Ideally, duplication of software is minimized across the system because instrument teams can draw on software solutions and techniques previously delivered to SOFIA by other instruments. With this approach, we minimize the effort for analyzing and developing new software reduction pipelines for future generation instruments. We also explore the potential benefits of this approach in the portability of the software to an ever-broadening science audience, as well as its ability to ease the use of distributed processing for data reduction pipelines.

1. Introduction

SOFIA is an airborne observatory designed primarily to carry out observations at infrared and sub-millimeter wavelengths that cannot be carried out from ground-based facilities. SOFIA will host a variety of instruments observing in wavelength ranges from 0.3 to 600 microns which will be upgraded over time. This will produce a large diversity of data types which will likely increase as new generations of instruments are operated.

The SOFIA Data Cycle System (DCS)[1] is a collection of tools and services that support both the General Investigator (GI) and the Science and Mission Operations staff from observation and mission planning, through observation execution on-board the aircraft, to data archiving and processing post-flight and distribution to the GI and the scientific community. The DCS will provide a uniform, extensible and supportable framework for all aspects of this data cycle.

The DCS will support data processing for both facility and Principal Investigator-class instruments, including archiving and pipelining of raw (Level 1), processed (Level 2), flux calibrated (Level 3), and higher level data products (e.g., mosaics and source catalogs). Data processing includes all steps required to obtain good quality flux calibrated data for spectroscopy, imaging, fast-acquisition, polarimetry, etc. Processing each data type requires a sequence of unique or common algorithms with specific parameters to be tuned. The DCS will incorporate, improve and maintain these algorithms which are provided by the instrument teams and developed in a variety of environments. In addition, these algorithms may require user-interaction or fine tuning of input parameters in order to return good quality data

2. Concepts and Associations

The DCS uses an Astronomical Observation Request (AOR) concept to collect up all needed information required to carry out an observation. AORs are produced by the GI and SMO staff during the observation planning stage and then passed to the SI during flight for execution. In addition, the AOR is the link between science and calibration data of the same observation type and defines the parameters necessary for post-processing. Therefore, it will identify the reduction pipeline and its parameters. For each Level 2 product, the Pipeline Pedigree (PP) records the pipeline generating the data, the parameters, the processing date and the data involved in the process. AOR and PP concepts has been implemented and are operative in DCS. A similar concept will be necessary to track calibration activities. DCS will include a Flux Calibration Parameter (FCP) which will support the calibration process in order to document and reproduce the same results as needed. AOR, PP and FCP are characterized by unique key numbers that identify them as well as information about the data involved in the process.

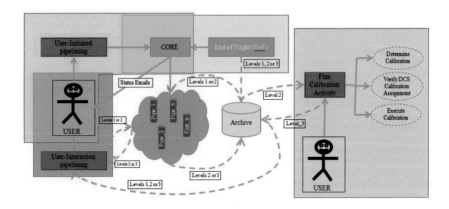

Figure 1.   Data processing scenarios.

---

[1]See http://dcs.sofia.usra.edu.

3. Architecture

The DCS will provide the framework for both automatic pipelining and human-in-the-loop processing. The DCS will host automatic pipelining at End-of-Flight (EoF), user-initiated pipelining, and user-interactive processing and analysis. Figure 1 illustrates how these scenarios relate to each other. The CORE is in charge of data processing within DCS (green actors). User can perform data processing outside DCS (red actors) and use DCS tools to extract and archive data. We show data fluxes as dashed arrows and process requests as plain arrows. We explain below the four main scenarios defining the data processing scenarios illustrated in Fig. 1:

- **EoF automatic pipelining producing immediate Level 2 products** [green area]
  Flight data, as for example raw observations or flight-processed products, are ingested at EoF. DCS calls pipelines automatically after ingestion of data observed during flight operation. Products from data reduction, typically Level 2 data, are automatically archived as the data are processed, making them quickly available for scientific analysis.

- **Flux calibration producing Level 3 products** [purple area]
  Outstanding scientific results can be obtained only with flux calibrated data. Flux calibration is a complicated processes that is difficult to automate — especially for an airborne observatory. The difficulty of defining a metric of the data quality makes necessary intervention of experienced scientists. Final Level 3 products can be archived in SOFIA database as well as their associated FCP.

- **User initiated processing and inspection (all levels)** [gray area]
  Pipelining can also be manually initiated by SMO scientists. For example, they may re-pipeline the data with modified parameters which would improve the quality of the final results, or when a new version of a pipeline becomes available.

- **User interactive processing (all levels)** [orange area]
  Likely, human intervention is often needed to verify results at any of the data levels. DCS will provide an interaction interface to extract data from the archive and run locally the same algorithms used during automatic pipelining. This allows the user to analyze the data at any step of the process, eliminate undesirable data and fine-tune parameters of the reduction. This step will result on the data validation or the appropriate parameters required to re-pipeline data in order to improve the quality of the final product.

4. Pipelines Approaches

A pipeline is a collection of algorithms which are run in a particular order. The DCS will host pipelines coded in IDL, Python and other languages which are delivered by the instrument teams with a description specified as XML. With the appropriate pipeline specification, DCS can currently run pipelines in any language with no modification of the code as soon as the pipeline is delivered as an executable, likely the same that runs in SMO machines outside DCS. Because DCS does not have a knowledge of the details of the pipeline execution after it is called, we name this pipeline "blackbox". Level 2 blackboxes are applied based on the specifications of the AOR — which is detailed before the flight as part of the observation planning process — and the details of the process are recorded on the PP — which is created after pipelining (§ 2). This approach is currently implemented in the DCS and embraces both automatic and user-initiated pipelining within the same framework. This answers the need for re-pipelining with the goal of improving the quality of the Level 2 data by fine tuning pipeline parameters after manual inspection or applying an improved version of the pipeline. Although, this approach represents an enormous cost saving on the implementation and maintenance of the pipelines it lacks the advanced functionalities that the DCS could offer including parallel execution of processes of a single pipeline, status report, and intermediate user intervention.

We plan to complement the current functionality with another approach allowing human interaction. User interaction is required for step-by-step data processing and intermediate data analysis. These will be performed using a DCS graphical interface tool which runs user-interaction data process and analysis tools locally (outside DCS) after downloading updated algorithms from DCS. As a long term goal, DCS will integrate user-interaction pipelining within the same framework as automatic pipelining. For that purpose, pipelines will be delivered as a collection of functions (modules) performing a portion of the pipeline and XML files describing them. The pipeline recipe (another XML file) will describe how modules are executed, the order of execution and how data is transfered between modules. Technically, the pipeline manager objects (pipe_man) are in charge of executing specific modules (module->process method) or the whole pipeline (pipe_man->run method). This new approach fits in the actual black box structure by calling run method as the pipeline executable. When implemented within DCS, pipe_man will be able to process modules in parallel, control their execution, and allow user data analysis. In addition, pipe_man will manage modules in different computer languages for the same pipeline thus reducing the number of algorithms in the system. Instrument teams will be encouraged to use existing algorithms when developing their pipelines, resulting in a common library of algorithms which will decrease the efforts of the instrument teams for developing pipelines and of the DCS team for maintaining and upgrading them.

5. Conclusion

Combining automatic pipelining and user interaction of processing algorithms which are developed in various languages presents an important challenge to the SOFIA DCS — especially when trying to minimize efforts required for long-term maintenance and upgrade of the code. We divide the problem in four distinct cases of interaction with the data. These scenarios can be developed independently but are based on a common architecture. The case of automatic pipelining, either at EoF or user-initiated, is already implemented and has been demonstrated with FLITECAM data. User-interactive pipelining is in its design phase but we have shown its feasibility using a prototype implemented in IDL. Flux calibration is not included in current DCS development plans due to resource/schedule constraints, but we provide the required tools for the user to ingest human validated data.

Acknowledgments.   RYS is suported by USRA Contract to the Space Science Institute. For more information about SOFIA visit http://www.sofia.usra.edu.

Astronomical Data Analysis Software and Systems XX
ASP Conference Series, Vol. 442
Ian N. Evans, Alberto Accomazzi, Douglas J. Mink, and Arnold H. Rots, eds.
© *2011 Astronomical Society of the Pacific*

The LOFAR Transients Pipeline

John Swinbank

Astronomical Institute "Anton Pannekoek", University of Amsterdam,
Postbus 94249, 1090 GE Amsterdam, The Netherlands

Abstract.   LOFAR is a new radio telescope being commissioned in the Netherlands and across Europe. It provides a uniquely wide-field, high-sensitivity view of the low frequency radio sky. This capability will be exploited to carry out a large-scale transients monitoring program. This has necessitated the development and application of a number of new techniques and technologies. Here, some of the approaches which have been adopted are highlighted, and their applications within the LOFAR project are discussed.

1.   Introduction

LOFAR, the LOw Frequency ARray, is a next-generation 'software' radio telescope: mechanical dishes and analogue correlators have been replaced by dipole antennae, high-bandwidth digital data transport, and powerful computing facilities. This new system combines an unprecedented wide-field, high-sensitivity view of the low frequency (30–240 MHz) radio sky with a remarkably flexible data-processing system. Centered on a core in the Netherlands, LOFAR is currently undergoing construction and commissioning across Europe, with stations in the UK, France, Germany and Sweden.

The LOFAR Transients Key Science Project (TKP, Fender et al. 2006) will exploit LOFAR's capabilities to explore the low-frequency transient radio sky. As illustrated in Figure 1, LOFAR will use a 'multi-beaming' capability to tile out a large area of observation on the sky (varying from 65.8 square degrees per beam at 30 MHz to 4.0 square degrees per beam at 240 MHz), while simultaneously using individual beams to monitor noteworthy objects. Images will be made on a logarithmic range of integration times, with transients and variable sources

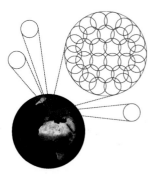

Figure 1.   The Radio Sky Monitor concept: multiple beams from the LOFAR core (at left) tile out a large area on the sky, while other beams can simultaneously target specific sources.

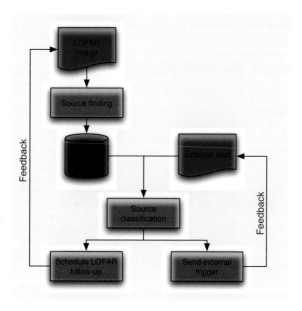

Figure 2.   A simplified outline of data flow through the TKP pipeline.

being identified by a combination of image differencing and statistical analysis of lightcurves. This mode of observation is referred to as the "Radio Sky Monitor", or RSM, and will be capable of surveying the majority of the visible sky to a depth of tens of mJy (depending on frequency) in a 24 hour period.

The TKP has been developing a sophisticated pipeline system to process and respond to the large amounts of data which will be generated in this mode. A schematic overview is shown in Figure 2. In brief, the Transients Pipeline (TP) is tightly coupled with an optimised version of LOFAR's Standard Imaging Pipeline (SIP). This provides data flagging, compression, calibration and imaging, eventually delivering an "image cube" (a group of simultaneous images of the same area of sky at different frequencies) to the TP. The images are then searched for sources, and the results fed into a database which will automatically associate them with known objects and generate lightcurves. Interesting lightcurves are extracted from the database and fed to a source classification and response system, which can then arrange for appropriate follow-up actions to be taken.

2.   Pipeline Framework

The various tools which are used as part of the TP do not all present a uniform interface. However, they must all interface with each other, and with the overall LOFAR control system. A framework has therefore been developed to provide a uniform interface to all the various components. Each tool is wrapped in standard way, exporting its functionality to other pipeline components through a consistent interface. The framework also provides a number of useful services which may be exploited by the pipeline, such as distribution of jobs across the LOFAR processing cluster.

The pipeline developer can connect the tasks via a straightforward Python script, making use of arbitrary logic (it is not adequate to simply chain them sequentially, as pipelines can be responsible for decision making, looping, and so on). Pipelines can then be automatically started when data is available, and will ensure that results and processing logs are fed back in a consistent way to the scientist.

Although designed initially to meet the requirements of the TP, the framework is now being used for several LOFAR science pipelines, including the aforementioned SIP.

3. Source Finding

Fast and accurate source identification and measurement is critical to the TP. This is used both for the direct identification of new transient sources (by searching difference images) and for the addition of new measurements to the lightcurve database. After evaluating available packages, the TKP has implemented a new source finding system as a Python module (Spreeuw 2010). This combines a false detection rate algorithm with a rigorous treatment of the correlated noise properties of radio maps, deblending using a multi-thresholding technique and fitting with elliptical Gaussians.

In the longer term, the TKP plans to make all its source finding routines available to the community as a standalone package. It is likely that future development will supplement the Python-based system with a faster implementation of the same algorithms in C++.

4. Database

The LOFAR transients database is a repository of lightcurve information: it stores (and make available for data-mining) information on all the sources observed by the RSM over its lifetime. While the RSM is running, up to 10 MB/s of data is being being added to the archive (the growth rate per year obviously depending on how much observing time is allocated for RSM observations). To store this data volume, the TKP has been collaborating with the Centrum Wiskunde & Informatica (CWI) in Amsterdam, to deploy the high-performance MonetDB database.[1] MonetDB innovates at all levels of the database stack, from a column-oriented data storage model to a vectorized query execution system.

This foundation makes it possible to extend pipeline processing into the database. When a new source measurement is inserted into the database, an automatic routine will determine what other detections are of the same object, and combine them all to build a lightcurve. The database keeps track of all lightcurves, monitoring them for variability, and providing notification of scientifically interesting events.

5. Classification and Response

Classification of discoveries is essential not only for data-mining of the lightcurve archive, but also to make it possible to respond to ongoing events in real-time. Responses could include re-running the pipeline with a different configuration, scheduling a follow-up observation, or broadcasting a notification of the event to the community at large.

As lightcurves are stored and updated in the database, they are automatically classified. Classification is based on a list of simple parameters which can be derived from the lightcurve: flux, variability, dispersion measure, spectral index, and so on. Based on these parameters, a number of classification techniques are applied. For maximum performance and generality, a machine learning approach is being investigated. However, there will always be a requirement for identifying events of particular interest to specific astronomical cases. Therefore, astronomer-defined classification steps may also be added to the pipeline.

---

[1] http://monetdb.cwi.nl/

Classifications and other calculated parameters are stored in the database, and can thus be referred to in future pipeline runs. As more data becomes available, the classification will become increasingly refined.

6. Notifications and VOEvent

Since it will explore a new parameter space, it is hard to predict the rate of transient discovery by the RSM. Estimates in the range of tens to hundreds per day are likely conservative. Manual followup of all detections will quickly become impractical. Further, given the software-driven nature of LOFAR it is possible to respond to events in near-real-time, potentially catching the most scientifically valuable results. In order to best take advantage of this, however, it is obviously necessary to remove the requirement for human intervention.

An automatic system for both generating and receiving alerts of transient events as quickly as possible is being developed. This is based on the IVOA VOEvent standard which provides a structured way of representing information about transient events. We anticipate private communication channels being developed with partner facilities as well as broadcasting LOFAR-derived events to as wide a community as possible.

7. Conclusion

The LOFAR RSM will regularly monitor the sky for transients. In order to make this possible, a number of new technologies have been developed which may be relevant to other projects. In particular, the pipeline framework has already been adopted by other LOFAR science pipelines, the source finding code will be publicly available, and the database systems are potentially useful for many astronomical catalogs.

References

Fender, R. P., et al. 2006, in Proceedings of the VI Microquasar Workshop: Microquasars and Beyond, edited by T. Belloni (PoS). PoS(MQW6)104

Spreeuw, H. 2010, Ph.D. thesis, University of Amsterdam

Astronomical Data Analysis Software and Systems XX
ASP Conference Series, Vol. 442
Ian N. Evans, Alberto Accomazzi, Douglas J. Mink, and Arnold H. Rots, eds.
© *2011 Astronomical Society of the Pacific*

GALFACTS Data Processing Pipeline

S. S. Guram, M. Andrecut, S. J. George, and A. R. Taylor

Institute for Space Imaging Science,
University of Calgary, Alberta, T2N 1N4, Canada

Abstract. The Galactic ALFA Continuum Survey (GALFACTS) is a large-area spectro-polarimetric survey being carried out with the Arecibo Radio Telescope in Puerto Rico. It uses the seven-beam focal plane feed array receiver system (ALFA) recently installed at Arecibo to carry out an imaging survey project of the 12,700 square degrees of sky visible from Arecibo at 1.4 GHz. The raw data produced by the spectrometer creates fifty-six digital data streams (seven beams, four polarization states and two frequency bands) each with 4096 spectral channels sampled at 1 millisecond. The aggregate data rate is 875 MB/s. The data processing pipeline for the GALFACTS observations consists of two parts: (i) the derivation of parameters necessary for processing of the GALFACTS main run data using the calibration observations, and (ii) processing of the main run data to create calibrated spectral image cubes converted to Stokes parameters (I, Q, U, V). Thus, the imaging process consist of mapping the time-frequency observations to sky coordinate-frequency data cubes. This process is computationally expensive, and involves many calibration and transformation steps, which will be discussed here. Being the first such survey ever attempted, the data processing offers significant challenges right from dealing with a very large data-set to the fact that new techniques need to be developed to reduce such a data-set. The multi-beam nature of observations requires significant modifications of the traditional data reduction techniques for single feed, single dish data.

1. Introduction

The observations for Galactic Arecibo L-band Feed Array Continuum Transit Survey (GALFACTS) are currently underway. It is using the Arecibo Radio Telescope in Puerto Rico to produce a full Stokes spectro-polarimetric cube of all the sky visible from the Arecibo Observatory (about 12,700 square degrees). A description of the technical specifications and science goals of the project can be found elsewhere (Guram & Taylor 2009; Taylor & Salter 2011). Being the first such wide bandwidth spectro-polarimetric survey, new techniques and algorithms need to be developed for processing the GALFACTS data. Currently, work is being carried out on the development of data processing pipeline for GALFACTS at the Institute of Space Imaging Science, University of Calgary.

2. The Data Processing Pipeline

Firgure 1 shows schematic of the GALFACTS Data Processing Pipeline. The data for each set of GALFACTS observations amounts to about 5 TB (about 12% of the whole survey). This data is first run through a quick-look program. This program serves as a quick heuristic to check if the observations went alright. The program produces plots of the band averaged and time averaged data. If there is a problem with the data we can submit a makeup observation request before the LST range for the observations is unavailable for this year. After this, the data is converted to the format used in the main processing pipeline. The main reason for this has been the commissioning of a new spectrometer at Arecibo, which outputs data in a different

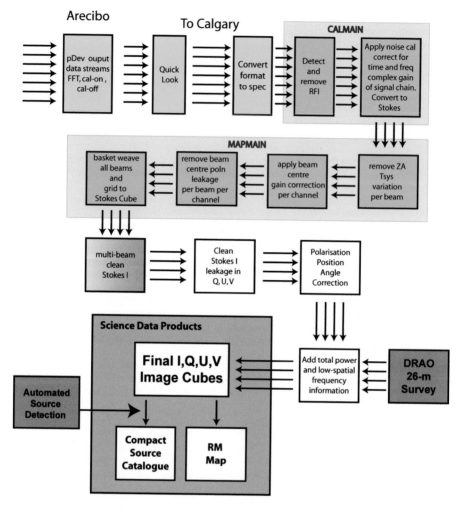

Figure 1.   The Block Diagram for the GALFACTS data processing pipeline. Implementation exists for the blocks in yellow. The development work is under progress for the white blocks.

format. Once the data is converted the main processing programs can be used. The bulk of the data processing is handled by CALMAIN and MAPMAIN. We will now discuss the details of each.

2.1. CALMAIN

The CALMAIN program performs radio frequency interference (RFI) detection and removal, as well as correcting for the complex gain of the electronic signal chain using the pulsed noise diode signal. This signal is injected into the receiver chain during the observations. The RFI detection algorithm is described in more detail elsewhere in these proceedings (Andrecut et al. 2011). The final step performed by this program is to transform the X and Y auto and cross-

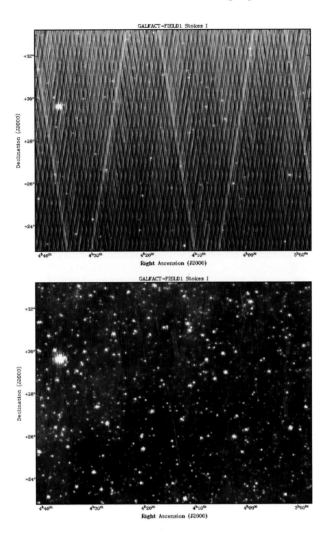

Figure 2. Results of MAPMAIN. The increase in the background system temperature at higher declination can be seen in the top image. The striping pattern in image on the top is due to the day to day variations in the system temperature. These have been removed in the image on the bottom.

correlation values into the four Stokes Parameters I, Q, U, and V. All of these corrections can be applied independently for each day and each of the seven beam observations. This allows us to parallelize the processing of the data by dedicating each core on the machine to one day and one beam of observations. The only limitation being the fact that this process needs large amounts of RAM to load all the data into its memory.

2.2. MAPMAIN

The output of CALMAIN produces "raw" Stokes values which can in principal be used to generate a FITS cube of the dataset. MAPMAIN performs a number of corrections on the "raw" Stokes image to remove artifacts in the images. The first step is to subtract a zenith angle (ZA) dependent polynomial representing the background from the data. This has to be done because of the peculiarity of the Arecibo Telescope Structure which leads to a rise in the background system temperature with increasing declination. This is caused by the spilling of the beam over the side of the dish, which picks up ground radiation. This correction has to be applied separately for each beam since the spillover is dependent on the location of the beam in the feed horn array. After this each beam is normalized for its peak gain. This is done because the off-axis beams have a lower peak antenna gain due to the presence of side-lobes. This correction is essential to put all the data on a consistent normalized scale before imaging. The next step is called "basket-weaving" (Haslam et al. 1970). This process helps achieve internally consistent zero levels for all the data and gets rid of the scanning artifacts from the image. Figure 2 shows the effect of corrections applied by MAPMAIN on the image. Most of the artifacts in the data have been removed.

Current work on the data processing pipeline is focusing on Multi-beam CLEAN (Guram 2007) and polarization leakage correction. Working prototypes exist for these but the algorithms are computationally very intensive. Significant work needs to be done to achieve a parallelized and fast implementation. Following this shall be the implementation of polarization position angle correction and using data from the Dominion Radio Astrophysical Observatory GMIMS Survey (Wolleben et al. 2009) to add the low spatial frequency information. This is required since the ZA dependent background removal takes out some of this information from the data.

References

Andrecut, M., S., G. S., George, S. J., & Taylor, A. R. 2011, in Astronomical Data Analysis Software and Systems XX, edited by I. N. Evans, A. Accomazzi, D. J. Mink, & A. H. Rots (San Francisco, CA: ASP), vol. 442 of ASP Conf. Ser., 115

Guram, S. S. 2007, Master's thesis, University of Calgary (Canada)

Guram, S. S., & Taylor, A. R. 2009, in The Low-Frequency Radio Universe, edited by D. J. Saikia, D. A. Green, Y. Gupta, & T. Venturi (San Francisco, CA: ASP), vol. 407 of ASP Conf. Ser., 282

Haslam, C. G. T., Quigley, M. J. S., & Salter, C. J. 1970, MNRAS, 147, 405

Taylor, A. R., & Salter, C. J. 2011, in The Dynamical ISM: A celebration of the Canadian Galactic Plane Survey, edited by R. Kothes, T. L. Landecker, & A. G. Willis (San Francisco, CA: ASP), ASP Conf. Ser., in press

Wolleben, M., et al. 2009, in Cosmic Magnetic Fields: from Planets, to Stars ancd Galaxies, edited by K. G. Strassmeier, A. G. Kosovichev, & J. Beckmann, vol. 259 of IAU Symp., 89

Astronomical Data Analysis Software and Systems XX
ASP Conference Series, Vol. 442
Ian N. Evans, Alberto Accomazzi, Douglas J. Mink, and Arnold H. Rots, eds.
© *2011 Astronomical Society of the Pacific*

The MMT-POL Instrument Control System

C. Warner,[1] C. Packham,[1] T. J. Jones,[2] F. Varosi,[1] S. S. Eikenberry,[1] K. Dewahl,[2] and M. Krejny[2]

[1] *University of Florida, Gainesville, FL*

[2] *University of Minnesota, Minneapolis, MN*

Abstract. Instrument control system (ICS) suites are a continually evolving class of software packages that are highly dependent upon the design choices and application programming interfaces (Apis) of the observatory control system (OCS), as well as the hardware choices for motors and electronics. We present the ICS for MMT-POL, a 1–5 μm polarimeter for the MMT telescope, in the context of being a transitional step between the software packages developed for facility class instruments at the University of Florida (UF), such as Flamingos-II and CanariCam, and in preparation for 30 m-class instruments. Our goals for improving ICS suites are to make them (a) portable (compile once, run anywhere), (b) highly modular and extensible (through the re-use of common libraries), (c) multi-threaded (to allow multiple tasks to be performed in parallel), (d) smart, and (e) easy to use and maintain. An ICS should also be well-defined and use mature languages (we choose Java and Python) and common standards (such as XML and the FITS file format). We also note that as hardware moves away from serial communications to ethernet, the use of TCP sockets makes communication faster and easier. Below, we present our design choices for the MMT-POL ICS and discuss our reasons for these choices and potential issues that must be addressed for future ICS suites ready for thirty meter class instruments.

1. Introduction

MMT-POL (Packham et al. 2010) is an adaptive optics optimized imaging polarimeter for use at the 6.5 m MMT (Fig. 1). By taking full advantage of the adaptive optics secondary mirror of the MMT, this polarimeter will offer diffraction-limited polarimetry with very low instrumental polarization. This instrument will permit observations as diverse as protoplanetary discs, comets, red giant winds, galaxies and AGN.

The MMT-POL ICS (Fig. 2) is a transitional step in UF instrumentation software. Previous software packages developed for facility class instruments had many inherent shortcomings that must be addressed in preparation for instrumentation in the 30 m class Giant Segmented Mirror Telescope (GSMT) era. Because

Figure 1.   MMT Observatory in Arizona.

MMT-POL is a PI-class instrument, we are able to explore new ideas without the restrictions placed on facility-class instruments with the goal of building new common libraries that can be incorporated into future proposals for 30 m class instruments.

Below, we discuss some of the shortcomings of previous generation ICS suites and our goals for the MMT-POL ICS and common libraries for future instruments.

2.   Goals of the MMT-POL ICS

2.1.   Portable

Previous software packages for facility-class instruments were developed mostly in C/C++ and required specific hardware architectures such as old Sun SPARC systems and VMS VAX machines. Furthermore, these older packages commonly used non-standard data formats and protocols and slow serial communications. Because hardware quickly becomes outdated or obsolete, it is important for ICS software to be portable and hardware independent. The use of Java and Python allow for code to be compiled once and run on any architecture. The use of common standards such as Ethernet TCP sockets allow for fast, future-proof communication and allow the software agents and GUIs to be run from anywhere and connect to the servers and electronics running in the instrument rackmount.

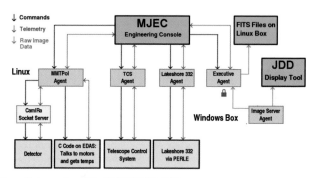

Figure 2.   Schematic of the MMT-POL ICS.

2.2.   Modular and Extensible

Older ICS suites often became bloated and cluttered, and solutions to common tasks were often reinvented over and over. An important goal of the MMT-POL ICS is to organize common tasks and data structures into libraries that can be used by both software agents and GUIs. These libraries should be extensible and reusable for future instruments, with well-documented APIs. This allows for the rapid development of new ICS suites with minimal effort.

Figure 3.   Thread diagram for Executive Agent.

2.3. Multi-Threaded

All agents and GUIs in the MMT-POL ICS are multi-threaded so that multiple tasks can be performed in parallel. This results in a much higher efficiency than older, single-threaded code. However, synchronization between threads must be carefully implemented to avoid collisions or deadlock.

2.4. Smart

Agents and GUIs should work together to automate complex tasks, such as observation sequences. The executive agent, for instance, has 14 active threads running during an observation sequence to coordinate the various agents and GUIs (Fig. 3). An observation sequence entails dithering the telescope to X positions (via the TCS agent), cycling through four half-wave plate positions Y times at each (via the MMT-POL agent), commanding the detector to take an image Z times at each position (also via the MMT-POL agent), and collecting telemetry data from all of the agents to be added to the image header of each exposure. This is all performed with one simple click of the **Observe** button after setting a few parameters.

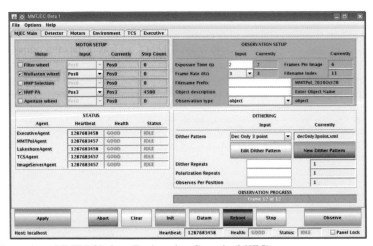

Figure 4.   MMT-POL Java Engineering Console (MJEC)

2.5. Easy to Use and Maintain

Transparency and simplicity are very important for any modern software package. Astronomers observing on an instrument cannot be troubled to spend hours or days learning complex tasks and commands. Older software suites sometimes required issuing commands from a terminal and configuration changes sometimes required recompiling of code and/or rebooting VMS VAX machines. Newer ICS suites must be intuitive to operate and reconfigure. Everything in the MMT-POL ICS is run through the MMT-POL Java Engineering Console (MJEC) GUI. Its main tab (Fig. 4) displays all commonly changed parameters along with important health, status, and telemetry data. The use of XML for all configuration files allows configuration changes, such as creating or modifying dither patterns, to be performed live from MJEC without the need to recompile or restart any software. The astronomer must also be able to examine their data in real-time. This is accomplished through a quick look display tool (Java Data Display, JDD), which can connect to the image server agent from anywhere over a TCP socket and provde a real-time view of the data. It can store multiple buffers and offers tools such as line cuts, statistics, arithmetic between buffers, zooming, and various scaling algorithms and color maps.

3. Conclusions and Issues for Future ICS Suites

The MMT-POL ICS is a transitional step between software packages developed for previous facility-class instruments and future instrumentation in the GSMT era. It provides common, extensible libraries with well-documented APIs that can serve as the backbone for a future GSMT class instrument. Agents and GUIs could be built upon these existing libraries combined with observatory-defined APIs to communicate with the Observatory Control System (OCS). Fig. 5 shows a potential design for a GSMT ICS based on the MMT-POL ICS. The two largest remaining issues for GSMT instrumentation software are (1) implementing observatory-defined APIs to communicate with the OCS through a middleware layer (sofware communications backbone)

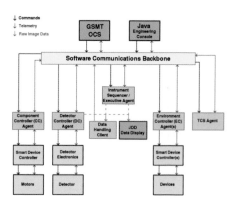

Figure 5.   Potential design for a GSMT ICS.

and (2) the Data Handling Client must be able to process vast amounts of data quickly (10 Gbit/s ethernet would be optimal) and use relational database management systems (RDBMS) to mine huge databases of meta-data.

References

Packham, C., Jones, T. J., Krejny, M., Dewahl, K., Warner, C., & Lopez Rodriguez, E. 2010, in Ground-based and Airborne Instrumentation for Astronomy III, edited by I. S. McLean, S. K. Ramsay, & H. Takami (Bellingham, WA: SPIE), vol. 7735 of Proc. SPIE, 77356J

Astronomical Data Analysis Software and Systems XX
ASP Conference Series, Vol. 442
Ian N. Evans, Alberto Accomazzi, Douglas J. Mink, and Arnold H. Rots, eds.
© *2011 Astronomical Society of the Pacific*

CARMA Correlator Graphical Setup

Dalton Wu, Ben Shaya, and Marc W. Pound

Astronomy Department, University of Maryland, College Park, MD 20742

Abstract. CARMA Correlator Graphical Setup (CGS) is a Java tool to help users of the Combined Array for Research in Millimeter-wave Astronomy (CARMA) plan observations. It allows users to visualize the correlator bands overlaid on frequency space and view spectral lines within each band. Bands can be click-dragged to anywhere in frequency and can have their properties (e.g., bandwidth, quantization level, rest frequency) changed interactively. Spectral lines can be filtered from the view by expected line strength to reduce visual clutter. Once the user is happy with the setup, a button click generates the Python commands needed to configure the correlator within the observing script. CGS can also read Python configurations from an observing script and reproduce the correlator setup that was used. Because the correlator hardware description is defined in an XML file, the tool can be rapidly reconfigured for changing hardware. This has been quite useful as CARMA has recently commissioned a new correlator. The tool was written in Java by high school summer interns working in UMD's Laboratory for Millimeter Astronomy and has become an essential planning tool for CARMA PIs.

1. Why

The Correlator Graphical Setup (CGS) was created to enhance the scientific utility of CARMA by providing astronomers with a tool to investigate and understand the capabilities of the CARMA correlator. CGS was important to create because a new correlator was coming online in 2009. Since the new correlator was going to be much more flexible and contain many more bands than the previous correlator, we wanted users to be able to successfully take advantage of its capabilities. One way to do this is to allow astronomers to visualize the spectral lines and spectral windows in frequency space, and to modify the properties of the spectral windows in a natural way. With CGS, users can easily transfer the visual representation of their desired correlator set up into the actual Python commands that configure the correlator for observations.

2. How

The capabilities of the correlator are specified in XML. These include the receiver IF ranges, the number of bands available, the bandwidth modes, the number of channels per mode, polarization mode, and variations of some these parameters as a function of quantization bit level. Specifying the capabilities in XML allows easy modification of the CGS tool as the telescope hardware evolves. For instance, not all of the new bands came online simultaneously and so we were able to have the CGS tool accurately track the current state of the commissioned bands by simple edits. The spectral line characteristics are read in from a tabular version of the Lovas 3 mm and 1 mm catalogs. Using such a table also makes it easy to add spectral lines in the future, which is particularly important for planning high red-shift observations.

3. Who

CGS was developed entirely by high school students working in UMD's Laboratory for Millimeter Astronomy. They wrote all of the software in Java with the Netbeans IDE under the guidance of Marc Pound. Ben Shaya started CGS during the summer following his 10th grade year at Montgomery Blair High School. He created a functional tool that contained all of the necessary initial features. Later on, fellow student Dalton Wu made additional modifications to CGS during the summer following his 12th grade year at Montgomery Blair High. He fixed many bugs that were reported by users, and added additional features to meet the needs of current users. For example, many astronomers use Macbooks, and the older versions of CGS had a screen layout that did not fit on Macbooks! Dalton implemented this change as well as many others including: flexible bit configuration options that could be set for each individual band (thereby changing the bands properties), spectral line frequencies that adjust for the Doppler sources, an additional panel to display band velocity properties, and support for new CARMA-23, dual-polarization, and full-Stokes modes.

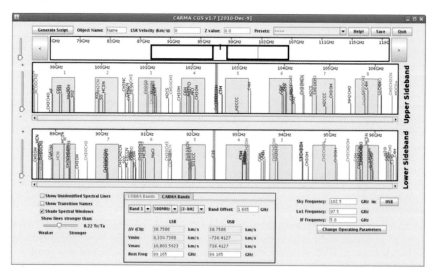

Figure 1.   This is the default view of CGS. The vertical lines represent spectral lines of known (blue) and unknown (red) intensity. Each of the pink rectangles represents one of the two spectral windows of each band. The properties of each band can be modified with the pull-down boxes inside the lower center box or with right-mouse clicks. Bands can be moved in frequency space with a mouse drag or by the Band Offset text box. Band edge locations in LSR velocity space are updated in real-time as a band is moved.

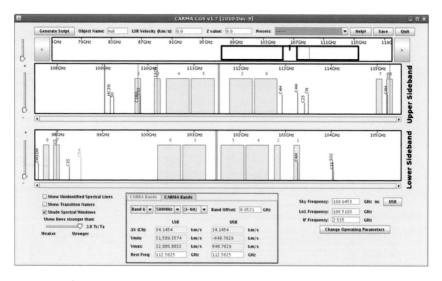

Figure 2. This is an example of what CGS looks like after it has been configured by the user. This set-up covers 4 spectral lines: 12CO(1–0), 13CO(1–0), C18O(1–0), CS(2–1). The user has tagged each of these lines with its rest frequency shown by the vertical green lines, so that each band has its own velocity range. Note the feature to show only the strongest spectral lines (slider in the lower left), which reduces visual clutter.

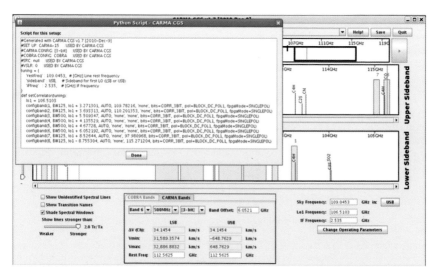

Figure 3. After the user clicks "Generate Script", CGS creates this popup window containing the Python commands for this set up. The user can now copy and paste the commands directly into an observing script to configure the correlator for the observations. A user may also save the script to a file which can be loaded back into CGS later via a file read or by pasting into an input text box. Commonly used set ups are available under the Presets menu.

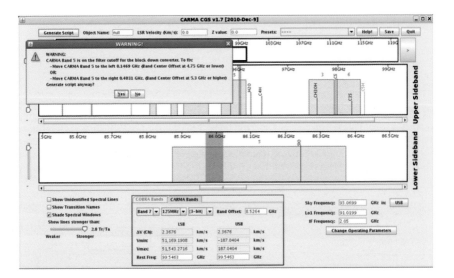

Figure 4.   This is an example of CGS firing a warning to indicate that the desired
set up has problems. The problem with this specific set up is that one of the bands is
on the edge of the block down converter filter, which is marked by the light gray zone.
This would cause some degradation of data from that band, so CGS warns the user.
This figure also shows the zooming feature of CGS: the "Lower Sideband" panel
is zoomed in more than the "Upper Sideband" panel. This can help in visualizing
narrower bands or spectral lines which are closely spaced in frequency.

Astronomical Data Analysis Software and Systems XX
ASP Conference Series, Vol. 442
Ian N. Evans, Alberto Accomazzi, Douglas J. Mink, and Arnold H. Rots, eds.
© *2011 Astronomical Society of the Pacific*

A Framework for End to End Simulations of the Large Synoptic Survey Telescope

Robert R. Gibson,[1] Zarah Ahmad,[2] Justin Bankert,[2] Deborah Bard,[3,4]
Andrew J. Connolly,[1] Chihway Chang,[5,4] Kirk Gilmore,[5,4] Emily Grace,[2]
Mark Hannel,[2] J. Garrett Jernigan,[6] Lynne Jones,[1] Steven M. Kahn,[5,4]
K. Simon Krughoff,[1] Suzanne Lorenz,[2] Stuart Marshall,[5,4] Satya Nagarajan,[2]
John R. Peterson,[2] James Pizagno,[1] Andrew P. Rasmussen,[5,4] Marina Shmakova,[3,4]
Nicole Silvestri,[1] Nathan Todd,[2] and Mallory Young[2]

[1] *University of Washington*

[2] *Purdue*

[3] *Kavli Institute for Particle Astrophysics and Cosmology*

[4] *SLAC*

[5] *Stanford*

[6] *SSL/UC Berkeley*

Abstract.   As observatories get bigger and more complicated to operate, risk mitigation techniques become increasingly important. Additionally, the size and complexity of data coming from the next generation of surveys will present enormous challenges in how we process, store, and analyze these data. End-to-end simulations of telescopes with the scope of LSST are essential to correct problems and verify science capabilities as early as possible. A simulator can also determine how defects and trade-offs in individual subsystems impact the overall design requirements. Here, we present the architecture, implementation, and results of the source simulation framework for the Large Synoptic Survey Telescope (LSST). The framework creates time-based realizations of astronomical objects and formats the output for use in many different survey contexts (i.e., image simulation, reference catalogs, calibration catalogs, and simulated science outputs). The simulations include Milky Way, cosmological, and solar system models as well as transient and variable objects. All model objects can be sampled with the LSST cadence from any operations simulator run. The result is a representative, full-sky simulation of LSST data that can be used to determine telescope performance, the feasibility of science goals, and strategies for processing LSST-scale data volumes.

1. A "Cosmological Database"

We have assembled a database of sky sources that we may query to generate catalogs. The database allows us to simulate multiple overlapping observations of the LSST sky with realistic source positions, colors, morphologies, proper motion, and variability.

There are 8 billion stars down to $r = 28$ mag in our simulated southern sky. The Milky Way simulation (Jurić et al. 2008) currently includes FGKMLT, RGB, BHB, and RR Lyrae stars and is flux-limited using a 3D Milky Way dust model. Figure 1, *Left*, shows a stellar number density map for a region of sky that includes the dusty Galactic plane.

Galaxies in a 4.5×4.5 deg^2 patch are generated from the semi-analytic model of De Lucia et al. (2006). This patch is tiled across the full sky.

Figure 1, *Right*, shows the distribution of asteroids in the LSST simulations as a function of eccentricity (or inclination) and semi-major axis. We store 11 million asteroids in the database at a resolution of one position per night.

Several different types of variability are currently simulated. AGN activity is assigned to appropriate host galaxies and given "Damped Random Walk" light curves (e.g., Kelly et al. 2009). Supernova are assigned to galaxies and given supernova template light curves. RR Lyrae stars are generated in the star simulations and given RR Lyrae template light curves.

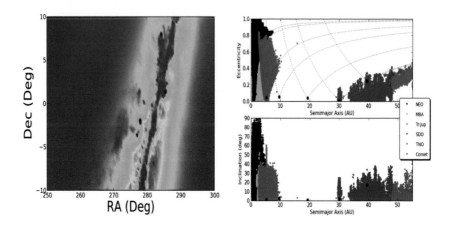

Figure 1.   (Left panel:) Number density map of stars in an equatorial strip that crosses the Galactic plane; the structure represents simulated thin/thick disk and halo stellar components extincted by a Galactic dust model. Red pixels have up to ten million stars per square degree. (Right panel:) Distribution and types of simulated asteroids. Black dots indicate Solar System planets, for reference.

2.   LSST Cadence

The LSST Operations Cadence Simulator (OpSim) simulates an entire 10-year survey to meet LSST science requirements with realistic observing constraints. The LSST image and catalog simulations are time-aware: for each telescope exposure, we query the OpSim to determine time, filter, and observing conditions that are used to affect moving source positions, variability, and image quality.

3.   Simulation Flow

The simulator generates a variety of different catalog types from the Cosmological Database. These catalog types include: 1) Image catalogs that are used to generated raytraced images using an LSST telescope model; 2) Truth catalogs that describe the sources in a raytraced image, for testing the results of image-processing pipelines; 3) Science catalogs that are used for direct scientific investigations (e.g., source-classification algorithms) with no images involved.

The catalogs themselves are very large (currently about 500 MB compressed per focal-plane exposure), and require significant data/processing resources to generate, store, and maintain. The processing flow for user input, catalog generation, and image generation is shown in Figure 2.

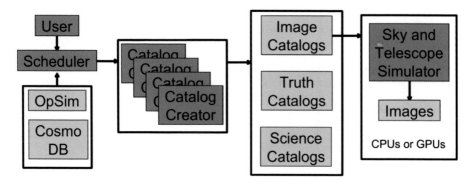

Figure 2.   Flow diagram for LSST image simulations showing the processing structure used to generate catalogs, a variety of catalogs that may be produced, and image generation using an LSST telescope model.

4. Simulated Images

Each image catalog is ray-traced through a model of the atmosphere and LSST telescope (Connolly et al. 2010). The processing, data storage, and IO requirements are even more severe at this stage; besides the 500 MB (compressed) catalog of sources, each instance of the ray-tracer requires a large library of template spectra, currently 25 GB. Each photon is drawn from the reddened spectrum of a source, then raytraced from the sky to the detector and simulated readout.

There are 189 imaging chips in the LSST focal plane. Figure 3, *Left*, shows a simulated focal plane; at this resolution only the amp structure (16 per chip) is noticeable. Figure 3, *Right*, shows a simulated false-color image of a *single* chip with no simulated background. Fainter sources in this image are "washed out" when the background is applied, but provide important structure to the background for testing image-processing pipelines.

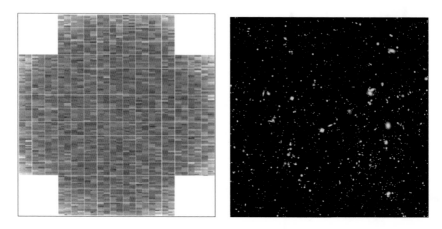

Figure 3.   (Left panel:) Simulated LSST focal plane showing vignetting and amp structure. (Right panel:) 3-color (*gri*) simulated image for one chip with no simulated background.

5. Conclusion

Simulated LSST images are currently being used in large numbers to test LSST software development, and will evolve to add new features as required for testing. The simulations make use of incredibly detailed databases of source properties and telescope characteristics. This level of detail permits us to test real-world scenarios at image depths, size scales, and time scales that are not available in existing survey catalogs. Of course, we can also mock up contrived scenarios to test specific parts of the image-processing pipelines, or large samples of "outlier cases" that occur infrequently but must be handled correctly during data collection. The simulation framework is also designed to generate large amounts of image data to test LSSTs ability to process data in real time and generate alerts for interesting sources.

Acknowledgments.   Support for this work was provided by the LSST Corporation and NSF grant IIS-0844580.

References

Connolly, A. J., et al. 2010, in Modeling, Systems Engineering, and Project Management for Astronomy IV, edited by G. Z. Angeli, & P. Dierickx (Bellingham, WA: SPIE), vol. 7738 of Proc. SPIE, 77381O
De Lucia, G., Springel, V., White, S. D. M., Croton, D., & Kauffmann, G. 2006, MNRAS, 366, 499
Jurić, M., et al. 2008, ApJ, 673, 864
Kelly, B. C., Bechtold, J., & Siemiginowska, A. 2009, ApJ, 698, 895

Part VIII

Pipelines (Space-Based Instruments)

Astronomical Data Analysis Software and Systems XX
ASP Conference Series, Vol. 442
Ian N. Evans, Alberto Accomazzi, Douglas J. Mink, and Arnold H. Rots, eds.
© *2011 Astronomical Society of the Pacific*

Processing Chains for Characterization and Data Analysis of TES X-ray Detectors

M.T. Ceballos,[1] R. Fraga,[1] B. Cobo,[1] J. van der Kuur,[2] J. Schuurmans,[2] I. González,[1] and F. Carrera[1]

[1]*Instituto de Física de Cantabria (CSIC-UC), Avda Los Castros s/n, 39005 Santander, Spain*

[2]*Netherlands Institute for Space Research, Sorbonnelaan 2,3584 CA Utrecht, NL*

Abstract.   The EURECA (EUropean-japanese Calorimeter Array) project[1] was aimed at developing a prototype for an X-ray Imaging detector based on TES (Transition Edge Sensor) technology. The software development group in this project is responsible for the elaboration of a full set of **processing chains** to characterize the instrument and to process the real data it can collect (SRON and IFCA [CSIC-UC] institutes) as well as to create the graphical tools to display the results (ISDC institute). We present here a description of the processing chains (purpose, component tasks, output information) as well as the Test Harness and the additional pipeline created to fully process the input data according to their type and some fixed user's specifications in an automatic, un-managed way. They are being developed by the IFCA-SRON collaboration.

1. The Processing Chains

The processing chains are a set of interactive tasks written in C++[2] that make use of the ISDC Data Access Layer Library (Jennings et al. 1998). The development of these tasks is version controlled through a Subversion,[3] to which the developers can access via a Web interface (Web-SVN[4]).

1.1. Chains Structure

We have developed four different processing chains to account for TES characterization and data analysis:

- XRAYCHAIN: It gets energy resolution of the instrument through pulse analysis

    - Input: Current (I) versus Time (t)

    - Output: energy resolution of TES

---

[1]See de Korte et al. (2008).

[2]GCC, the GNU compiler collection (http://gcc.gnu.org).

[3]See http://subversion.apache.org.

[4]See http://websvn.tigris.org.

- Tasks: `trigger` (pulse finding) + `pulseshape` (pulse quality) + `filter` (pulse template creation) + `energyresol` (convolution with template) + `holzgauss` (energy resolution)

- IVCHAIN: this chain gets the IV (current-voltage) curve characteristics analyzing the behavior of the current while the voltage is changing in ascending and descending ramps.

  - Input: input voltage V(t) and measured current I(t)
  - Output: critical current, dissipated power, TES resistance in normal and superconducting states
  - Tasks: `ivrepr` (IV definition) + `ivproc` (IV curve fitting)

- Z(POST)CHAIN: the chain ZCHAIN calculates TES complex impedance (CI) and then ZPOSTCHAIN gets TES properties from this CI (bias power, heat capacity, thermal constant, etc.).

  - Input: input noise voltage and measured current I(t)
  - Output: CI and then TES properties
  - Tasks: `zcmplx` (calculates CI) + `polezero` (find Gain and Poles and Zeros) + (POST)`zpars` (gets physical parameters)

- TES(POST)CHAIN: TESCHAIN finds instrument noise characteristics for a given bias voltage and the TESPOSTCHAIN fits this noise components to obtain relevant physical parameters.

  - Input: input current I(t)
  - Output: fits to noise components
  - Tasks: `tesps` (calculates current noise spectral density) + `polezero` (finds gain, poles and zeros) + (POST)`calcnoise` (calculates Johnson and phonon noise)

1.2. Running the Tasks

The individual tasks that are part of a processing chain can be (independently) run through the command line or using the ISDC Guest User Interface (GUI). The command line for each task is a list of `parameter=value` combination for each of the parameters specified in the parameter files. An example of the command line for the task `trigger` is as follows:

```
> trigger inFile=in.fits outFile=out.fits nbins=1000 n=1.E-5
tauFall=3.E-5 ntaus=20 numBitsQuality=16 writePulse=y ql=n
nameLog=trigger.log verbosity=3
```

In addition, all the tasks that make up a given chain can be run sequentially with a single command, as in this example for the IV chain:

```
#!/bin/sh
IVchain chain.groupFile="groupIV.fits" chain.logfile="chainIV.log"
ivrepr.inFile="iv.fits" ivrepr.ivrFile="a_ivr.fits"
ivrepr.nbins=1000 ivrepr.tauFall=3.E-5 ivrepr.ql=no
ivrepr.nameLog="ivp_error.log" ivrepr.verbosity=1 ivrepr.ntaus=20
ivrepr.n=1E-5 ivproc.ivrFile="a_ivr.fits" ivproc.chisqrLimit=1.5
ivproc.ivpFile="a_ivp.fits" ivproc.step=10 ivproc.ql=no
ivproc.nameLog="ivp_error.log" ivproc.verbosity=1
```

Both the individual tasks and the complete chains can also be run using the ISDC provided GUI. See Fig. 1 for an example of a task that is run through the user interface.

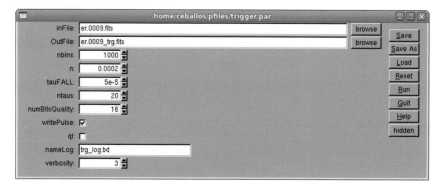

Figure 1.   Running trigger task using the ISDC GUI.

2. Test Harness

The software package also includes a suite of test perl scripts (based on the Perl Test::Harness module[5]) designed for each task in every chain, aimed at checking the output results against input (known) simulated values (TES simulator `eur_simulate` developed at SRON). These tests run the task installed and, in addition to screen output, generate an output summary file (log file) with the results given by the tasks, the comparable simulated values and the results of the tests.

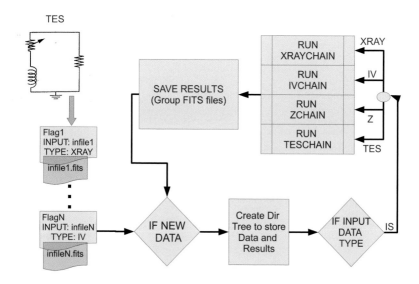

Figure 2.   Automatic pipeline flowchart

---

[5]See http://search.cpan.org.

3. Automatic Pipeline

The processing chains can be run manually during the development process and for testing purposes. However, once the software package is mature enough it should be able to run continuously, in an un-managed way over a large set of input files. This is the purpose of the Automatic Pipeline, a set of perl scripts that automatically process a continuous flow of input data (in FITS format) according to their data type. The flowchart of this pipeline can be seen in figure Fig. 2. The pipeline is continuously looking for new data to be produced. Once a new data file and its partner "flag" file (which describes the type of processing the input file requires) are produced, the pipeline creates the directory tree to store the calculations and runs the appropriate processing chain. Once the file is created, the FITS index files are updated with these new data and the pipeline returns to its initial stage.

Acknowledgments. Funding provided by the Ministry of Science and Innovation (MICINN) under project ESP2006-13608-C02-01 and AYA2009-0859 as an in-kind contribution to the EURECA project.

References

de Korte, P. A. J., et al. 2008, in Space Telescopes and Instrumentation 2008: Ultraviolet to Gamma Ray, edited by M. J. L. Turner, & K. A. Flanagan (Bellingham, WA: SPIE), vol. 7011 of Proc. SPIE, 701122

Jennings, D., Borkowski, J., Contessi, T., Lock, T., Rohlfs, R., & Walter, R. 1998, in Astronomical Data Analysis Software and Systems VII, edited by R. Albrecht, R. N. Hook, & H. A. Bushouse (San Francisco, CA: ASP), vol. 145 of ASP Conf. Ser., 220

Astronomical Data Analysis Software and Systems XX
ASP Conference Series, Vol. 442
Ian N. Evans, Alberto Accomazzi, Douglas J. Mink, and Arnold H. Rots, eds.
©*2011 Astronomical Society of the Pacific*

Overview of CeSAM Tools and Services for the CoRoT Mission

P.-Y. Chabaud, F. Agneray, J.-C. Meunier, P. Guterman, R. Cautain, C. Surace, and M. Deleuil

Laboratoire d'Astrophysique de Marseille, Université de Provence, CNRS, France

Abstract. CoRoT (Convection Rotation et Transits planétaires) is a space mission of astronomy led by French space agency (CNES) in association with French laboratories (CNRS). One of CoRoT goal is to detect exoplanets by transit method. Since the beginning of the CoRoT project, the Centre de donnéeS Astrophysique de Marseille (CeSAM) is involved in many fields of activities: database, information system, software development and scientific analysis. We present some CeSAM implications.

1. CeSAM

The Centre de donnéeS Astrophysiques de Marseille (CeSAM) at the Laboratoire Astrophysique de Marseille (LAM) has been set up to provide access to quality controlled data via web based applications, tools, pipelines developments and VO compliant applications to the astrophysical community (see Fenouillet et al. 2011).

2. Mission Definition

CoRoT is a 27-cm diameter telescope and a 4-CCD wide field camera operating in the visible loaded on a PROTEUS platform. Launched on the 27th of December 2006, the telescope is expected to survey and explain the inner structure of the stars, as well as to detect many extrasolar planets, using the transit method brightness. With its high photometric performance and its observing runs, one of the goals is to discover telluric extra-solar planets and rocky planets.

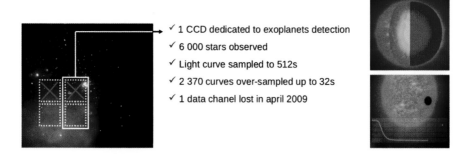

✓ 1 CCD dedicated to exoplanets detection
✓ 6 000 stars observed
✓ Light curve sampled to 512s
✓ 2 370 curves over-sampled up to 32s
✓ 1 data chanel lost in april 2009

Figure 1.   Mission definition.

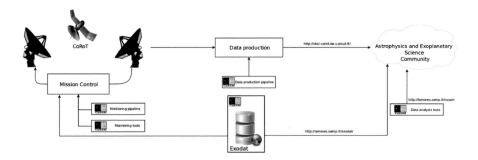

Figure 2.   CoRoT project.

3.   Projects

3.1.   Exodat

The Exodat project aims at providing the information needed for the selection of the best targets and statistical analysis.

The Exodat archive consists of a database (80 GB) under PostgreSQL for storage and two data access methods, web application and web services. These data are supplied to the entire scientific community and for the operation of the satellite.

The web application was built from the SiTools framework. Users can find information in astronomical catalogs of data through a set of forms, e.g., search by star ID, by spectral type.

Recently, the Exodat team has developed a set of RESTful web services to query the database via URIs. Exodat users can access information through the HTTP protocol, using their web browser or an API in any programming language.

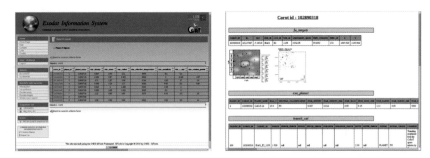

Figure 3.   Exodat page on `http://lamwws.oamp.fr/exodat`.

3.2.   Monitoring Pipeline

The alarm pipeline was designed to detect planetary transit signals. It works every week on Day+10 CoRoT data.

Alarms are checked by a scientist team which can raise the on-board sampling rate of validated targets up to 32s. At the same time, complementary observations are performed at ground observatories.

The alarm pipeline is composed of several filter stages: cosmic detection, low frequencies correction, etc, and one transit detection stage using the BLS algorithm.

The computation process is managed by the SCPC system (Shared Cores Process Coordinator) which organizes and executes processing on available stations all over the laboratory.

3.3. Monitoring Tools

The CRT application is a step by step detection pipeline which enables visualisation and access at each filter stage of the alarm pipeline through a GUI. This application is used to investigate process mistakes which can occur on specific target data. It is also used to design new filter stages.

The PAD application is a visualisation tool created in order to search quickly through a huge amount of pipeline results. Detection transit and target parameters are displayed. Star physical parameters are provided by Exodat web services.

SAMP compliance enables inter-operability between monitoring tools.

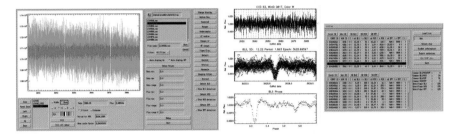

Figure 4.   CoRoT tools.

3.4. Data Production Pipeline

The CoRoT Exo-planets channel provides high precision photometric data. The on-board software computes photometry before sending it to the CoRoT Mission Center in Toulouse.

The N1 → N2 pipeline, developed at LAM, provides usable data without instrumental effects and cleared from environmental effects. Data are free from SAA influence, satellite motion effects (jitter) and proton impacts. The pipeline produces a light-curve for every target in FITS format data. Star physical parameters are retrieved from ExoDat and included in header files.

LAM is also involved in data efficiency improvements such as systematic noise attenuation and photometry computation on saturated stars.

References

Fenouillet, T., et al. 2011, in Astronomical Data Analysis Software and Systems XX, edited by I. N. Evans, A. Accomazzi, D. J. Mink, & A. H. Rots (San Francisco, CA: ASP), vol. 442 of ASP Conf. Ser., 17

Astronomical Data Analysis Software and Systems XX
ASP Conference Series, Vol. 442
Ian N. Evans, Alberto Accomazzi, Douglas J. Mink, and Arnold H. Rots, eds.
© 2011 Astronomical Society of the Pacific

First Simulation and Data Reduction of a JWST/NIRSpec Observation

Bernhard Dorner,[1] Pierre Ferruit,[2] Laure Piquéras,[1] Emeline Legros,[1] Arlette Pécontal,[1] Aurélien Jarno,[1] Aurélien Pons,[1] Xavier Gnata,[3] and Camilla Pacifici[4]

[1]*Université de Lyon, Lyon, F-69003, France;*
Université Lyon 1, Observatoire de Lyon,
9 Avenue Charles André, Saint-Genis Laval, F-69230;
CNRS, UMR 5574, Centre de Recherche Astrophysique de Lyon;
Ecole Normale Supérieure de Lyon, Lyon, F-69007, France

[2]*European Space Agency, Dep. RSSD,*
Keplerlaan 1, 2200 AG Noordwijk, Netherlands

[3]*EADS Astrium GmbH, P.O., 81663 Munich, Germany*

[4]*Institut d'Astrophysique de Paris, 98bis Boulevard Arago, 75014 Paris, France*

Abstract. The James Webb Space Telescope (JWST), a joint project by NASA, ESA, and CSA, is the successor mission to the Hubble Space Telescope. One of the four science instruments of the observatory is the multi-object spectrograph NIRSpec. It will be able to measure the spectra of more than 100 objects simultaneously and will cover the near infrared wavelength range from 0.6 to $5.0\,\mu$m at various spectral resolutions. Due to the instrument complexity, it was seen as necessary to create an instrument simulator for studies of the instrument performance, optical and geometrical effects, as well as the creation of realistic calibration and science exposures to develop and test data analysis tools. The Centre de Recherche Astrophysique de Lyon (CRAL), as subcontractor to EADS Astrium GmbH, is developing this Instrument Performance Simulator (IPS) software for NIRSpec. One of the key objectives of the IPS is to generate realistic simulated JWST/NIRSpec exposures of astrophysical sources, providing a check of NIRSpec in-orbit performance and inputs for the definition of the best observation strategies. We briefly summarize how the different input data is used for the instrument model, and present a first spectral extraction pipeline tailored to the IPS. Following this, we show the simulated exposure of an observation of a typical NIRSpec target, a modeled redshifted galaxy, and compare the finally extracted spectrum with the input.

1. Introduction

The James Webb Space Telescope (JWST, Gardner et al. 2006), scheduled for launch in 2014, will be the successor observatory to the Hubble Space Telescope. One of the science instruments, the Near-Infrared Spectrograph NIRSpec (Bagnasco et al. 2007), is designed for low- and mid-resolution spectroscopy in the near-infrared from 0.6 to $5\,\mu$m. A team at the Centre de Recherche Astrophysique de Lyon (CRAL) is developing the NIRSpec Instrument Performance Simulator (IPS) software (Piquéras et al. 2010), which provides an end-to-end optical and performance simulation of the telescope and the instrument, including the detector readout. The software package is written in C/C++ and is close to delivery to EADS Astrium.

343

We defined a simple file interface to feed science target data into the IPS, and simulated a typical science case, a sample galaxy at redshift $z = 2$. In addition we created a first data processing pipeline for IPS output. The extracted spectrum of the exposure is compared with the input galaxy spectrum and is found to match very well.

2.   IPS Sky Simulations

2.1.   IPS Data Flow

The instrument model in the IPS is based on data from different subsystems, as the optics, microshutters and slits, dispersers, and detectors. The available light sources comprise ground and on-board calibration lamps and sources on the sky. Combining the instrument data with one of the source modules, an "IPS model" is created which describes the selected instrument mode. From the source description, a source file is computed and used in combination with the IPS model to calculate a noiseless electron count rate map, a first main simulation output. The tool to create the NIRSpec MULTIACCUM data cube with the sequential non-destructive readouts is a separate module. It uses the detector description and the selected readout parameters to generate a realistic raw data file from the electron rate maps.

2.2.   Sky Scene Interface

The native IPS source file format is very general to accommodate various source types, but not well suited to exchange data with external users. Besides, the source placement for observations is only possible on the sky. In order to test the instrument performance in certain scenarios, it is more convenient to place the sources directly in shutters or slits. Therefore, we established a set of input FITS files that contain common data types: single spectra, an intensity image and a spectrum, and a data cube. Furthermore, we defined a scene description that allows users to place the sources relative to single shutters, slits, slices, or on the sky. A set of IDL scripts and Python classes were prepared, that also give examples how to construct exposure scenes from external data.

3.   Data Processing and Results

3.1.   Extraction Pipeline Workflow

The final IPS output is a FITS datacube with raw data. To ease the comparison with the input, we created a first processing pipeline for NIRSpec. It follows the classical approach of long-slit spectroscopy, adjusted to the multi-object capabilities of the instrument. The pipeline inputs are the fitted slopes from the readout cube, providing gain and linearity corrected electron rate, variance, and quality flags for each pixel. They can be generated from the readout simulation with external pre-processing software, or by adding analytic noise and pixel crosstalk to the electron rate maps.

Each target exposure is divided by a simulated flatfield taken with the same spectrograph configuration and the internal calibration lamp. The pipeline then uses the instrument model data from the IPS to predict the spectrum location and extract the traces for all the targets in the observation. It rectifies the spectra to a uniform spectral and spatial sampling, creates the target spectra, and subtracts the sky background.

For the absolute radiometric calibration, there is a set of simulated reference star exposures, probing the field of view and different positions inside shutters. These exposures are processed the same way as science targets and constitute the photometric reference spectrum set. For each observed target, the local calibration spectrum is interpolated from these reference spectra according to the position on sky and inside the slit. Applied to the background subtracted target spectrum, the final calibrated spectrum is obtained. Depending on the extraction parameters, this is a one- or two-dimensional spectrum with uniform wavelength sampling, containing the spectral photon rate, variance, and quality flags.

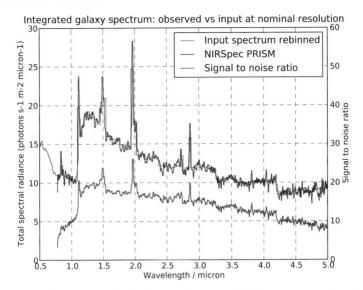

Figure 1.   Comparison of a simulated and extracted NIRSpec spectrum (blue) and input galaxy spectrum (red) for the prism mode, including the derived signal to noise ratio (green).

3.2.   Extraction Pipeline Framework

All the spectrum extraction has been coded in Python in a very modular approach. The main class encompasses the top-level methods of the pipeline, as determination of the exposure parameters, data extraction and processing, data file I/O, and logging. These methods make use of classes for single pipeline models, which contain characterization data of the instrument modules read from the IPS database. Those classes again use low-level classes from the framework to handle the IPS input data. They offer methods to read IPS data files and for basic calculations directly related to the instrument module. In addition, there is a set of data file classes for the different data products (count rate maps and spectra), providing file I/O, data arithmetic, and graphical output. All the pipeline classes use a dedicated calculation library that provides generic mathematical functions, partly extended to handle pixel quality flags. Scripting the pipeline is therefore possible on a high level using the main class' methods only.

3.3.   Example Simulation

One of the major science cases of NIRSpec is the multi-object observation of high-redshift galaxies. We used a simulated galaxy spectrum at redshift $z = 2$ (Pacifici et al., in prep.), and put it as a point source at the center of a minislit (1×3 shutters). In addition, we set an average Zodiacal spectrum as a background source covering the complete field of view. The observation was simulated with the prism at a spectral resolution of $R \approx 100$. We added synthetic noise to the electron rates corresponding to a standard exposure duration of 902 s. With the same spectrograph configuration we also simulated a flatfield exposure with the internal calibration light source.

   The spectrum was processed as described in subsection 3.1 and finally collapsed in the spatial dimension. We rebinned the input galaxy spectrum to the nominal spectral resolution of the instrument to compare it with the simulated data. Figure 1 shows the plot of the extracted spectrum (blue) and the rebinned input spectrum (red). Both curves match very well. The obtained result is slightly better resolved, as the spectral resolution element of a point source is smaller than the nominal 2.2 detector pixels. The signal to noise ratio was determined from the propagated variances in each pixel and is plotted in the green curve, confirming a good spectrum quality beyond wavelengths of $1.1\,\mu$m.

4. Conclusion and Outlook

The IPS provides a realistic simulation of NIRSpec of in-orbit observations. We designed a simple data interface for astronomical users and started the first on-sky simulations with as-built instrument models. In addition, we created a spectrum extraction pipeline for IPS output. So far, point sources have been successfully extracted and the pipeline is now extended to other sources and instrument modes. The software will also be used during the upcoming instrument cryo tests and can serve as a testbed for different extraction techniques.

Acknowledgments. BD and CP are funded by the European Community's Seventh Framework Program (FP7/2007-2013) under grant agreement PITN-GA-2008-214227 - ELIXIR, and thank Stephane Charlot for the network lead and all the encouragement. All authors would like to thank the engineers at EADS Astrium and the ESA JWST science team members for the great cooperation and inspiring discussions.

References

Bagnasco, G., et al. 2007, in Cryogenic Optical Systems and Instruments XII, edited by J. B. Heaney, & L. G. Burriesci (Bellingham, WA: SPIE), vol. 6692 of Proc. SPIE, 66920M
Gardner, J. P., et al. 2006, Space Sci. Rev., 123, 485
Piquéras, L., et al. 2010, in Modeling, Systems Engineering, and Project Management for Astronomy IV, edited by G. Z. Angeli, & P. Dierickx (Bellingham, WA: SPIE), vol. 7738 of Proc. SPIE, 773812

Astronomical Data Analysis Software and Systems XX
ASP Conference Series, Vol. 442
Ian N. Evans, Alberto Accomazzi, Douglas J. Mink, and Arnold H. Rots, eds.
© *2011 Astronomical Society of the Pacific*

HIPE, HIPE, Hooray!

S. Ott, on behalf of all contributors[1] to the Herschel mission

Herschel Science Center, European Space Agency

[1]`http://herschel.esac.esa.int/HerschelPeople.shtml`

Abstract. The Herschel Space Observatory, the fourth cornerstone mission in the ESA science program, was launched 14th of May 2009. With a 3.5 m telescope, it is the largest space telescope ever launched. Herschel's three instruments (HIFI, PACS, and SPIRE) perform photometry and spectroscopy in the 55–671 micron range and will deliver exciting science for the astronomical community during at least three years of routine observations. Starting October 2009 Herschel has been performing and processing observations in routine science mode. The development of the Herschel Data Processing System (HIPE) started nine years ago to support the data analysis for Instrument Level Tests.[1] To fulfil the expectations of the astronomical community, additional resources were made available to implement a freely distributable Data Processing System capable of interactively and automatically reducing Herschel data at different processing levels. The system combines data retrieval, pipeline execution, data quality checking and scientific analysis in one single environment. HIPE is the user-friendly face of Herschel interactive Data Processing. The software is coded in Java and Jython to be platform independent and to avoid the need for commercial licenses. It is distributed under the GNU Lesser General Public License (LGPL), permitting everyone to access and to re-use its code. We will summarise the current capabilities of the Herschel Data Processing system, highlight how the Herschel Data Processing system supported the Herschel observatory to meet the challenges of this large project, give an overview about future development milestones and plans, and how the astronomical community can contribute to HIPE.

1. Introduction

The Herschel Space Observatory,[2] the fourth cornerstone mission in the ESA science program, was successfully launched 14th of May 2009. With a 3.5 m Cassegrain telescope it is the largest space telescope ever launched (Pilbratt et al. 2010). Herschel's three instruments (HIFI, PACS and SPIRE) perform photometry and spectroscopy in the 55–671 micron range and will deliver exciting science for the astronomical community during at least three years of routine observations (de Graauw et al. 2010; Poglitsch et al. 2010; Griffin et al. 2010). One month after launch, halfway through its commissioning phase and thermalisation period and while enroute to its operational orbit around L2, the Lagrange point located 1.5 million km away from the Earth, the cryostat lid was opened, and the first observational tests were conducted. All of Herschel's performance verification and science demonstration phase activities have been completed. While PACS and SPIRE have been performing routine science observing since late

---

[1]See Ott (2010).

[2]Herschel is an ESA space observatory with science instruments provided by European-led Principal Investigator consortia and with important participation from NASA.

2009, HIFI was initially delayed due to an anomaly which has since been resolved, and as of spring 2010 Herschel has been performing routine science operations phase with its full payload complement.

2. Current Status of HIPE

The system combines data retrieval, pipeline execution and scientific analysis in one single environment. All tools for data reduction and analysis, and also the expert applications for "Instrument Calibration", "Trend Analysis" and "Quality Control" systems are part of the Herschel Data Processing System. Therefore the community has access to the same system as the instrument experts. Also the Standard Product Generation (SPG) software which automatically generate data products are a subset of the Herschel Data Processing System.

Since the last ADASS XIX meeting four more major HIPE versions were made available to the astronomical community. HIPE 4.1 was used to bulk-reprocess all observations taken by Herschel up to now. Currently HIPE 5.1 is used for standard product generation. Generated products are ingested in the Herschel Science Archive (Osuna et al. 2010). The same version is offered for interactive data reduction. HIPE 5.1 can be downloaded, as all official HIPE releases, via http://herschel.esac.esa.int/HIPE_download.shtml. HIPE 6.0 is currently undergoing its validation process, with an expected release date of March 2011.

In addition the latest HIPE developer releases are available to the astronomical community via http://herschel.esac.esa.int/CIB_disclaimer.html.

Formal support is provided for Windows XP, Vista and Windows 7, Linux, Mac OS X 10.5 ("Leopard") and Mac OS X 10.6 ("Snow Leopard").

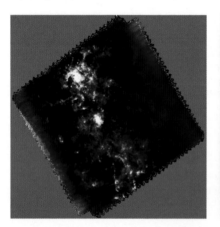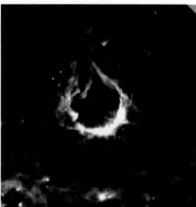

Figure 1.   Results of Herschel data reduced with HIPE.
Left: SPIRE Parallel Mode Observations of the Milky Way at 30° galactic latitude. Public data taken from Hi-GAL proposal (PI: S. Molinari). Credits D. Coia and L. Conversi
Right: PACS Photometer Scan Mapping Observation of RCW 120. Credits F. Motte, A. Zavagno, B. Gonzalez and the HOBYS Consortium.

3. Future Development Milestones

It is foreseen that major HIPE versions will be released regularly; during the coming year that will be around each three months. Twice a year a bulk reprocessing exercise will take place to repopulate the Herschel Science Archive. For those future HIPE versions we will concentrate on the following improvements:

- Refinement of data reduction algorithms, instrument calibration and product quality assessment criteria, specially for PACS and SPIRE spectroscopy,

- Improvement of the user friendliness of HIPE, e.g., by adding more example scripts and simplifying the data input/output,

- Enhancement of performance and stability by identifying memory leaks and the reported root causes of HIPE crashes and hangs,

- Upgrade from Jython 2.1 to Jython 2.5, the latest Jython release,

- Harmonisation of the workflow between Herschel products/HIPE and the Virtual Observatory,

- Combination of data from the different Herschel instruments,

- Production of publication quality plots,

- Facilitation of contributions via a plug-in mechanism,

- Improvement of the code quality so that the system will be maintainable with the reduced manpower one can expect during Herschel's post-operational phase.

4. Contributions to and Re-Use of HIPE

We are welcoming all contributions from the astronomical community to HIPE. From `http://herschel.esac.esa.int/Data_Processing.shtml` you can subscribe to Herschel Data Processing Interest Lists. The following mailing lists have been created to allow user-to-user sharing of information about specific Herschel data reduction needs and problems:

- PACS point source photometry,
- SPIRE point source and small maps photometry,
- Large maps and point source extraction for PACS and SPIRE,
- PACS Spectroscopy,
- SPIRE Spectroscopy,
- Spectral maps for PACS, SPIRE and HIFI,
- HIFI Point sources and spectral scan,
- HIPE General,
- HIPE Contribution.

The Herschel Data Processing system is coded in Java/Jython to be license free and portable for different operating systems. Its source code is freely available under the GNU lesser general public license for modifications, and reuse.

Acknowledgments. The Herschel Science Center (ESA), the Instrument Control Centers (HIFI, PACS and SPIRE) and the NASA Herschel Science Center jointly manage and contribute to the Herschel Data Processing System.

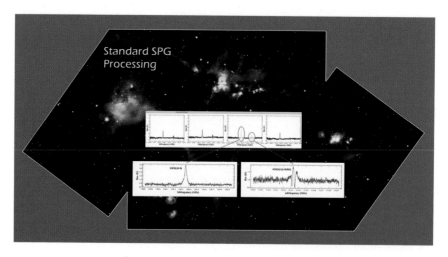

Figure 2.   HIFI Mapping Dual Beam Switch Raster data reduced by the standard processing pipeline: Fast chopping observation of DR21. Credits F. Helmich, F. van der Tak, M. Marseille, A. Abreu and the HIFI Consortium.

References

de Graauw, T., et al. 2010, A&A, 518, L6
Griffin, M. J., et al. 2010, A&A, 518, L3
Osuna, P., et al. 2010, in Astronomical Data Analysis Software and Systems XIX, edited by Y. Mizumoto, K.-I. Morita, & M. Ohishi (San Francisco, CA: ASP), vol. 434 of ASP Conf. Ser., 3
Ott, S. 2010, in Astronomical Data Analysis Software and Systems XIX, edited by Y. Mizumoto, K.-I. Morita, & M. Ohishi (San Francisco, CA: ASP), vol. 434 of ASP Conf. Ser., 139
Pilbratt, G. L., et al. 2010, A&A, 518, L1
Poglitsch, A., et al. 2010, A&A, 518, L2

Astronomical Data Analysis Software and Systems XX
ASP Conference Series, Vol. 442
Ian N. Evans, Alberto Accomazzi, Douglas J. Mink, and Arnold H. Rots, eds.
© 2011 Astronomical Society of the Pacific

Gaia: Processing to Archive

William O'Mullane, Uwe Lammers, and Jose Hernandez

European Space Astronomy Center of ESA, Madrid, Spain

Abstract. Gaia is ESA's ambitious space astrometry mission with a foreseen launch date in late 2012. Its main objective is to perform a stellar census of the 10^9 brightest objects in our galaxy (completeness to $V = 20$ mag) from which an astrometric catalog of μas level accuracy will be constructed. We update the viewer briefly on the status of the Astrometric Global Iterative Solution for Gaia. The results of AGIS feed in to the Main Database (MDB) which will be described here also. All results from Gaia processing in fact are in the MDB which is governed by a strict Interface Control Document (ICD). We describe the Distributed Data Model tool developed for Gaia, the Data Dictionary. Finally we mention public access to Gaia data in the archive. We present current plans and thinking on the archive from the ESA/DPAC perspective.

1. The Astrometric Global Iterative Solution (AGIS)

AGIS is the mathematical and numerical scheme selected by Gaia's Data Processing and Analysis Consortium (DPAC) for constructing the astrometric catalog from all available observation data gathered during Gaia's 5 year operational lifetime. It has been presented in a couple of previous ADASS conferences now, (e.g. Lammers et al. 2009; O'Mullane et al. 2007, and references therein); only a short status update is given here.

AGIS approximates the astrometric solution through a step-wise adjustment process in which the optimal match, in a global, least-square sense, between all observation data and an observational model expressed in terms of the desired unknowns is sought. These unknowns are the 5×10^9 astrometric **S**tellar parameters, a few Million **A**ttitude (satellite orientation), some 10,000 **C**alibration, and a handful of **G**lobal (General Relativity) parameters. The least-square problem is cast into an algebraic form with Normal equations and solved block-iteratively with one of two available solution methods: Simple Iterations (SI) or Conjugate Gradients (CG; Bombrun et al. 2010).

CG is superior to SI in a number of ways, notably the rate at which the unknowns are converging. In the current test scenarios with selected (very conservative) initial noise levels and outlier-free simulated input data we see that the difference is about a factor of 2. In about 20 iterations CG converges to a solution approximation that with SI is not found in less than 40 iterations. Recently, all four algorithmic blocks, S, A, C, G, have been brought into the CG framework and are working reliably, including the Calibration, which is formulated in a flexible, very generic way (see Joliet & Lammers 2010). This is regarded as a major milestone in the AGIS development.

Fig. 1 shows two results from a recent full 4-block (S-A-C-G) AGIS run: The spatial distribution of parallax error (left) shows an expected structure and remaining error level. The right curve shows the evolution of parallax adjustments and mean errors as a function of iteration number.

A last central remaining problem in AGIS is a needed detection and elimination of outlier observations. A robust algorithm has recently been formulated and is being implemented now with first results expected over the next couple of months. Outlier rejection is a surprisingly complex process and tackled in AGIS via downweighting with the downweighting factor being determined as part of the iterative loop. While this is no problem in SI, changing weights cause

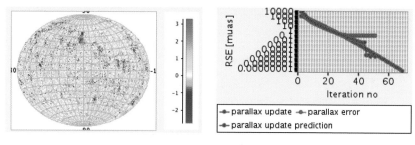

Figure 1.   Nearly-converged spatial parallax error map (left in μas) and mean parallax error versus iteration number from a full 4-block AGIS CG run.

CG to diverge as the least-square problem becomes ill-defined. This means, that it will not be possible to employ CG from the beginning but a "hybrid" solution scheme with alternating phases of SI and CG iterations will be needed: AGIS starts with SI iterations up to a point where the weights have stabilized to a given degree, then activate CG, followed by perhaps another SI phase to refine the weights further, then again CG, etc.

The AGIS-NanoJASMINE collaboration is progressing nicely and we are looking forward to seeing results from this Japanese Gaia-precursor mission processed with AGIS. See (Yamada et al. 2011) in this volume for a detailed status update.

2.   The Main Database (MDB)

The processing of Gaia's data will be distributed across six European data processing centers, each one of them taking care of different tasks which are interdependent on each other. The Gaia DPAC consortium in charge of developing the data processing system involves more than 300 individuals geographically distributed in 20 countries.

Each data processing center will receive and send data to ESAC where we will host a central database where data will be collected together and integrated.

One of the issues to be tackled is the definition and agreement of the data model of the central database. We have undertaken a novel approach with the development of an on-line Dictionary Tool (Fig. 2) which allows the contributors to define the form and structure of their data. From a single database we can then generate the data model entity classes, database schemas, ICD and documentation, data model releases, etc. The Data Processing Centers also use the Dictionary Tool to indicate what pieces of data from the main database they need to receive in order to perform their part of the data processing.

Furthermore all DPAC software is being developed in Java, the Dictionary Tool fully supports the Object Oriented approach so we can extend and contain data model entities and exploit the commonalities existing in our subsystems and develop common libraries when relevant.

The Dictionary Tool and the data model are reaching a mature level now with more than 10 data model releases, 100 contributors, the main database data model is made up of over 1000 data entities containing something like 8000 parameters.

Since 2009, changes to the data model have been reviewed and controlled by a CCB involving one member from each DPAC coordination unit. So far we've only needed a couple of physical meetings and regular telecoms to manage the data model evolution.

3.   The Gaia Archive

Finally after much processing data will be published to the Gaia archive from the Main Database. This year we start the Gaia Archive Preparation Working Group to gather further requirements for the archive and to pool ideas on the implementation. More explicitly GAP should:

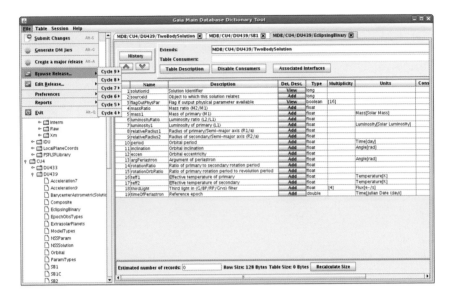

Figure 2.   Main GUI of the Gaia Main Database Dictionary Tool.

- Coordinate a response to the ESA AO for the Gaia Archive,
- Define an Archive plan for activities in each Coordination Unit (CU),
- Solicit interest and intent from parties currently external to DPAC,
- Define a mode of work between existing CUs and possible new partners,
- Coordinating any externally funded activities/investigations considered as potentially useful to the Gaia Archive.

There were many interesting talks in ADASS XX on relevant topics. Clearly the Virtual Observatory is coming as seen in presentations from CADC and CyberSka, developments of interest for Gaia. On the front end we also saw ASCOT which is developing the needed more configurable user experience. These are only seeds today; they need to all come together to make a truly great archive.

References

Bombrun, A., Lindegren, L., Holl, B., Hobbs, D., Lammers, U., & Bastian, U. 2010, A&A, in prep.
Joliet, E., & Lammers, U. 2010, in Astronomical Data Analysis Software and Systems XIX, edited by Y. Mizumoto, K.-I. Morita, & M. Ohishi (San Francisco, CA: ASP), vol. 434 of ASP Conf. Ser., 301
Lammers, U., Lindegren, L., O'Mullane, W., & Hobbs, D. 2009, in Astronomical Data Analysis Software and Systems XVIII, edited by D. A. Bohlender, D. Durand, & P. Dowler (San Francisco, CA: ASP), vol. 411 of ASP Conf. Ser., 55
O'Mullane, W., et al. 2007, in Astronomical Data Analysis Software and Systems XVI, edited by R. A. Shaw, F. Hill, & D. J. Bell (San Francisco, CA: ASP), vol. 376 of ASP Conf. Ser., 99
Yamada, Y., Lammers, U., & Gouda, N. 2011, in Astronomical Data Analysis Software and Systems XX, edited by I. N. Evans, A. Accomazzi, D. J. Mink, & A. H. Rots (San Francisco, CA: ASP), vol. 442 of ASP Conf. Ser., 367

Astronomical Data Analysis Software and Systems XX
ASP Conference Series, Vol. 442
Ian N. Evans, Alberto Accomazzi, Douglas J. Mink, and Arnold H. Rots, eds.
©2011 Astronomical Society of the Pacific

Machine Learning: Quality Control of HST Grism Spectra

Felix Stoehr, Jeremy Walsh, Harald Kuntschner, Piero Rosati, Robert Fosbury,
Martin Kümmel, Jonas Haase, Richard Hook, Marco Lombardi, Kim Nilsson, and
Michael Rosa

ST-ECF/ESO, Karl-Schwarzschild-Str. 2, 85748 Garching, Germany

Abstract.   The Pipeline for Hubble Legacy Archive Grism data (PHLAG) had
been used to extract more than 70000 wavelength and flux calibrated 1D spectra. They
were obtained from 153 fields observed in G800L grism spectroscopy mode with the
Advanced Camera for Surveys on the Hubble Space Telescope. This number of spectra
is far too large to allow detailed visual inspection for quality control on reasonable
time-scales. As a solution, we use machine learning techniques to classify spectra into
"good" and "bad" based on a careful visual inspection of only about 3% of the full
sample. A final visual skim through the set of "good" spectra was made to remove
catastrophic failures. The remaining 47919 spectra form the largest set of slitless high-
level spectroscopic data products publicly released to date.

1. Introduction

The Hubble Legacy Archive (HLA) is a collaboration between the Space Telescope Science
Institute (STScI), the Canadian Astronomy Data Center (CADC), and the Space Telescope -
European Coordinating Facility (ST-ECF). As part of the HLA project, the ST-ECF aims to
provide calibrated spectra for objects observed with HST slitless modes (left-hand side of Fig.
1).

At the ST-ECF the Pipeline for Hubble Legacy Archive Grism data (PHLAG) had been
developed (see Freudling et al. 2008, and references therein) to allow for automatic extraction of
slitless spectra. This pipeline was first applied to observations with the grism G141 of the Near
Infrared Camera and Multi-Object Spectrometer (NICMOS) camera onboard Hubble (Freudling
et al. 2008) and is now used to extract spectra from observations with the grism G800L of
the Advanced Camera for Surveys (ACS). In both cases, the dispersing elements are the most
popular ones for the respective instruments.

Details of the extraction process and a description of the data can be found in M. Kümmel
et al. (submitted).

2. Quality Control

In slitless spectroscopy a number of effects can lead to flawed spectral measurements. The
contaminating light trace from one source overlaying the dispersed light of the source to be
extracted is by far the most common corrupting effect (right-hand side of Fig. 1). Less common
effects are pixel-saturation at the centers of the sources, spectra from the borders of the chips or
spectra which contain pixels with with insufficient cosmic ray detection.

Whereas quality control of the 4825 NICMOS spectra was tedious but doable, and the
flawed ("bad") spectra could be identified by visual inspection, such a scenario is impossible to
do on a reasonable time-scale for the 73581 spectra produced by the PHLAG pipeline for ACS.

However, the pipeline computes for each spectrum a large number of quality parameters.
These include

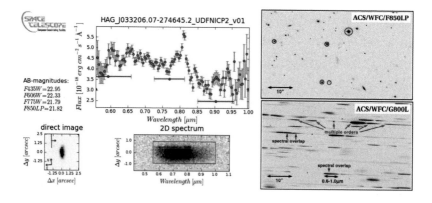

Figure 1.   Left-hand side: Preview image of a extracted spectrum taken in slitless spectroscopy mode with the ACS camera on the HST. On top the spectrum with error estimates and tree photometric measurements from the direct (undispersed) image are shown. That direct image is shown in the lower left corner and the corresponding 2D spectrum in the lower right corner. The blue box shows the region that was used to extract the spectrum. Right-hand side: The basic dataset of one slitless spectroscopic image (bottom) and its corresponding direct image (top).

- Two estimations of the signal-to-noise ratio,

- The magnitude of the source in different bands,

- The total exposure time,

- Two estimates of the the contamination,

- The position of the maximum of the light with respect to the expected position,

- The length of the spectrum in pixels,

- The source type (point or extended source),

- The difference of the magnitudes computed from the spectrum and from the separate image(s) taken without dispersive element.

Correlations between these parameters and visual classification of a few spectra are weak and thus automatic classification using simple cuts in the parameter space are inadequate. Therefore a more powerful classification methodology needs to be employed. We use the open-source machine-learning package Weka (Hall et al. 2009) to perform the classification.

3.   Methodology

In a first step, a relatively small subsample is inspected visually. 2020 spectra are classified independently by three of us into "good" and "bad". All spectra that are not classified in the same way in that process (about 12% of the spectra in the sample) are looked-at again by the team and consolidated classifications are produced. This process results in a very clean and homogeneously classified training sample, key to high automatic classification success rates.

Although all spectra are classified, it turns out that the number of spectra that are border-line cases between "good" and "bad" is surprisingly large, i.e. of the order of perhaps 15%. Many spectra which have some kind of problem still contain a lot of scientific value which

training_sample_001

Select the **bad** spectra from the group that has been classified **good** by clicking on the preview image. Once the bottom is reached, click on the link to submit the classification.
Hovering over the image number shows the corresponding preview in full size.

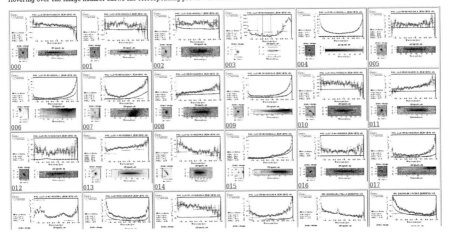

Figure 2.   Webpage used to remove flawed spectra that were not picked up by the automatic classification.

makes the decision difficult. We opt for a scheme that when in doubt the spectra were included into the "good" sample.

The visually classified set is randomized and split into 2/3 and 1/3 of the spectra for training and testing, respectively. All available classification algorithms of Weka are trained with the larger sample and the classification accuracy is measured with the smaller one. Several Weka algorithms provide classification success rates of 89% or better with ClassificationViaRegression using weka.classifiers.trees.M5P being the best performing one for the test set at hand. It yielded 90.5% of correct classifications and a particularly low rate of false negatives (i.e. "good" spectra classified "bad") of only 2.7%. This classification quality is comparable to that a single astronomer can achieve.

The ClassificationViaRegression algorithm trained with the training set is then used to classify the remaining 97% of the spectra. The 21416 spectra falling in the "bad" category are discarded. We use a semi-supervised classification scheme and passed the 50145 spectra from the "good" category on to post-classification analysis.

4.   Post-Classification Analysis

While analyzing the training sample, it turned out that there are two classes of spectra that are clearly flawed but are classified by the algorithm as "good". One such class of spectra is composed of spectra that have their centers saturated (67 spectra). A second class of spectra has the curve of the spectrum raise at two ends of the spectral range (1628 spectra). The reason for this rise is that the size of the object could not be estimated well enough within the pipeline. In order to remove them from the "good" sample we developed detection algorithms which are simple given that these classes are easily identifiable. Flagged spectra are then inspected visually again, a quick task given the small number.

As these data are part of the Hubble Archive Legacy Project, which aims at providing fully calibrated science-ready data, at all of the spectra marked "good" are given a quick look.

This is executed in a mode where 100 previews are concatenated into a poster-like web interface for very rapid inspection (Fig. 2). This procedure takes roughly as much time as the careful classification of the 2020 spectra used for training although the sample is about 24 times larger.

1951 spectra (4% of the "good" sample where the saturated-centers and rising-ends had been removed) were discarded through this procedure resulting in a total 47919 published spectra.

5. Conclusions

We use machine learning techniques in Weka for quality classification of slitless spectra. We find that this classification can be done very satisfactorily. Key ingredients are a large set of relevant metadata parameters describing each dataset as well as a statistically relevant sample of carefully visually classified spectra. Two classes of flawed spectra were removed with dedicated detection algorithms. We choose to skim visually through the set of remaining spectra to make sure that no third class of flawed spectra slips through and to remove a small number of failed spectra.

The remaining sample of 47919 "good" spectra can be accessed through web-interfaces at

archive.eso.org/hst/science
cadcwww.dao.nrc.ca/hst
hla.stsci.edu

References

Freudling, W., et al. 2008, A&A, 490, 1165
Hall, M., Frank, E., Holmes, G., Pfahringer, B., & Reutemann, P. 2009, SIGKDD Explorat., 11

Astronomical Data Analysis Software and Systems XX
ASP Conference Series, Vol. 442
Ian N. Evans, Alberto Accomazzi, Douglas J. Mink, and Arnold H. Rots, eds.
©2011 *Astronomical Society of the Pacific*

HILTS: The Herschel Inspector and Long-Term Scheduler

Pedro Gómez-Alvarez,[1,2] Jon Brumfitt,[1] Rosario Lorente,[1] and Pedro García-Lario[1]

[1]*Herschel Science Center, European Space Astronomy Center (ESAC), European Space Agency (ESA), P.O. Box 78, Villanueva de la Caada, 28691, Madrid, Spain*

[2]*Ingeniería y Servicios Aeroespaciales, S.A. (INSA), Paseo Pintor Rosales 34, 28008, Madrid, Spain*

Abstract. Visualization, querying, statistical analysis and mid-long-term scheduling are common concerns for any observatory. HILTS is a Java tool developed for the Herschel project to address all these issues in a unified way.

1. Introduction

Herschel (Pilbratt et al. 2010) is an ESA cornerstone observatory mission launched on 14 May 2009. Herschel covers the range from 55 to 672 microns with three instruments: HIFI, PACS and SPIRE, see García-Lario, P. G. (2011) in these proceedings. If planning, visualization and inspection capabilities are important in any observatory, cryogenic space observatories such as Herschel, with a estimated lifetime of 3.5–4 years, call for additional efforts to maximize the observatory scientific return.

HILTS was initially conceived to assist Herschel medium and long-term planning. The tool is also useful to assess the mission's past, present and future status. Short-term mission planning for a given Operational Day (OD) is executed using the Herschel Scientific Mission Planning System (SMPS; Brumfitt 2005), which generates satellite telecommands that are up-linked to Herschel on a daily basis. HILTS has been developed sharing the common object-oriented framework.

2. The Mission

The Herschel operational database is populated by more than 50,000 observation requests pertaining to 1,300 proposals, from around 500 astronomers (see Figure 3 for their geographic distribution). There are several factors impacting Herschel scheduling: helium optimization, slews minimization, proposal completion, scientific grades, Targets of Opportunity (ToOs), and operational issues. Herschel also has thermal and communication constraints. The observatory attitude is constrained by the (anti-) Sun, Earth, Moon and some planets. The observatory also needs to communicate with the ground station every 24 hours, during the so-called Daily Tele-Communication Period (DTCP).

3. Tool Description

HILTS is a Java tool, whose main screen is divided into a set of panels (see Figure 1):

- *Time panel:* It is composed of a set of horizontal sub-panels: a simple Gregorian calendar; a time selector, where the current time is selected; operational days, where the OD

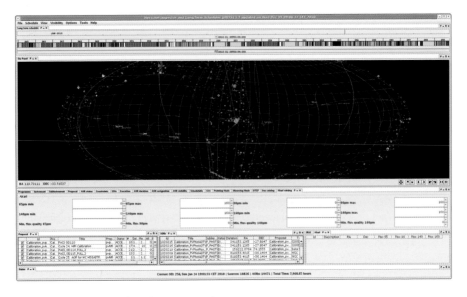

Figure 1.   HILTS main screen.

divisions are represented; scheduled observations, where the scheduled and already executed observations are represented; current observing block restrictions (groups of ODs
preallocated to a given instrument mode) and the available scheduling interval.

- *Sky panel:* Visible observations and current constraints are represented in this panel. The
satellite pointing history for the current OD can also be plotted.

- *Query panel:* Composed of multiple tabs, arbitrary complex selections are allowed using these available criteria: observation programs, instruments and instrument modes,
request status, Solar System Objects (SSOs), duration, etc.

- *Proposal panel:* Current selected proposals are listed and can also be (de)selected.

- *Requests panel:* Current selected observations are listed.

- *Catalogs panel:* By default, IRAS and AKARI catalogs are available. User catalogs can
also be loaded.

- *Status panel:* General status information is displayed.

All panels are interconnected with each other. For instance, when a new time is selected in the
time panel, constraints and visible observations are updated simultaneously in the sky panel,
while visible proposals and catalog objects are also updated in their respective panels. The
default visibility selection is objects visible at a given time. Other available alternatives are
(in)visibility during a time interval, during the DTCP, always visible, etc.

3.1. Scheduling

HILTS supports both manual and automatic scheduling. The former, by simple drag-and-drop
from the observation panel to the scheduled observations sub-panel. The tool automatically
places the observation at the earliest time within the dropped OD, taking into account observing
blocks, observations duration, configuration and slews, amongst other factors. The latter (see
Figure 2) is attained by first selecting a suitable interval of typically several months and one of
the set of pluggable strategies. For instance, if the "remaining visibility" strategy is selected,
the tool will assign each of the visible observations by order of remaining visibility. HILTS
scheduling is typically an iterative process: starting with one of the available "filler" strategies

Figure 2.   Before and after an automatic scheduling run.

and finishing with an optimization phase (a simulated annealing optimization is being developed). Once a satisfactory schedule is obtained, it can be exported to XML and loaded into the SMPS.

3.2.   Statistics

HILTS can generate detailed statistics (see Figure 3) focused on several mission aspects, which help evaluating current mission status and thus retrofit observation strategies. Among the reports HILTS is capable to generate are: Execution reports where the completion of each program, proposal and instrument mode is displayed; duplication studies where possible collisions between proposals are identified helping to avoid redundant science; scheduling reports, where several figures of merit and reports are generated as a result of a long-term schedule generation or an already executed period. The AJAX Google presentation API[1] has been extensively used to implement this functionality.

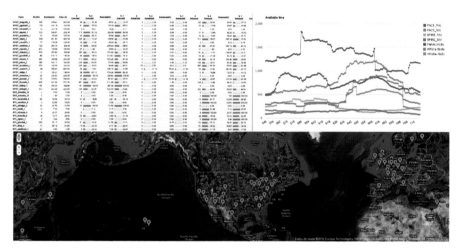

Figure 3.   Some examples of HILTS statistical capabilities: from upper left and clockwise, proposal completion report, available time per instrument mode during a long-term schedule and Herschel PIs geographical distribution.

3.3.   Catalogs and Virtual Observatory

HILTS is also able to interact with on-line catalogs from Vizier (Genova et al. 2006). Specially relevant catalogs such as IRAS and AKARI can be also filtered by flux in the query panel.

---

[1]http://code.google.com/apis/visualization/documentation/gallery.html

A synthetic catalog of IR sources for the selection of candidate "filler observations" is also included. HILTS can also inter-operate with VO tools such as Aladin (Boch et al. 2011) using the SAMP protocol (see Figure 4).

Figure 4.   Some screen-shots illustrating HILTS catalogs and VO interoperability: From left to right, IRAS sources available during DTCP, Vizier information of a given source and a joint HILTS-Aladin session centered at M42.

References

Boch, T., Oberto, A., Fernique, P., & Bonnarel, F. 2011, in Astronomical Data Analysis Software and Systems XX, edited by I. N. Evans, A. Accomazzi, D. J. Mink, & A. H. Rots (San Francisco, CA: ASP), vol. 442 of ASP Conf. Ser., 683
Brumfitt, P. J. 2005, in 5th NSSA Australian Space Science Conference (Melbourne: RMIT University), 358
García-Lario, P. G. 2011, in Astronomical Data Analysis Software and Systems XX, edited by I. N. Evans, A. Accomazzi, D. J. Mink, & A. H. Rots (San Francisco, CA: ASP), vol. 442 of ASP Conf. Ser., 225
Genova, F., et al. 2006, in BAAS, vol. 38 of BAAS, 1003
Pilbratt, G. L., et al. 2010, A&A, 518, L1

Astronomical Data Analysis Software and Systems XX
ASP Conference Series, Vol. 442
Ian N. Evans, Alberto Accomazzi, Douglas J. Mink, and Arnold H. Rots, eds.
© *2011 Astronomical Society of the Pacific*

Science Validation of the Spitzer Source List

H. I. Teplitz, P. Capak, T. Brooke, D. W. Hoard, D. Hanish, V. Desai, I. Khan, and R. Laher

Spitzer Science Center, MS 100-22, Caltech, Pasadena, CA 91125, USA

Abstract. The Spitzer Science Center will produce a source list (SL) of photometry for a large subset of imaging data in the Spitzer Heritage Archive (SHA). The list will enable a large range of science projects. The primary requirement on the SL is very high reliability — with areal coverage, completeness and limiting depth being secondary considerations. The SHA at the NASA Infrared Science Archive (IRSA) will serve the SL as an enhanced data product. The SL will include data from the four channels of IRAC (3–8 microns) and the 24 micron channel of MIPS. The Source List will include image products (mosaics) and photometric data for Spitzer observations of about 1500 square degrees and include around 30 million sources. We describe ongoing science validation of the Spitzer Source List, and discuss the range of use cases which will be supported

The Spitzer Science Center (SSC) will produce a source list (SL) of photometry for a large subset of imaging data obtained by the IRAC and MIPS instruments onboard the *Spitzer Space Telescope* (Werner et al. 2004) during its recently-completed cryogenic mission. The SL will enable a large range of science projects, but is not intended to meet the standards of a mission wide catalog (e.g. 2MASS). The primary requirement on the SL is very high reliability, even at the cost of completeness. The Spitzer Heritage Archive (SHA) at the NASA/IPAC Infrared Science Archive (IRSA) will serve the SL as an enhanced data product and will ensure that appropriate caveats and warnings are prominently placed. The SL products are planned for public release in late 2011. Details of the SL requirements and construction were given in the proceedings of the 2009 ADASS.

This SL will ensure high reliability by including only a subset of Spitzer data that is well behaved and can be processed and verified autonomously. This means observing programs must meet a minimum set of requirements for processing by the SL pipeline. These minimum requirements include data obtained with IRAC channels 1–4 (3–8 microns) in high-dynamic range or mapping mode and MIPS channel 1 (24 microns) in scan or photometry mode. In these proceedings, we present possible science use cases that may be used for SL validation.

1. Galactic Science I. Dust Extinction in Dark Clouds from UKIDSS and Spitzer

Knowledge of the wavelength dependence of interstellar dust grain extinction gives a constraint on the dust size and composition, and improves estimates of the true colors of objects behind dust. Studies of dust extinction in galactic molecular clouds typically show changes in the wavelength dependence of the extinction compared to dust in the diffuse interstellar medium, usually ascribed to grain growth (e.g. Flaherty et al. 2007; Chapman et al. 2009).

The UKIDSS (Lawrence et al. 2007) includes a Galactic Plane Survey about 3 mag deeper than 2MASS at K (2.2 μm). By combining UKIDSS data with Spitzer data one can get the dust extinction law deep into dark clouds through the IRAC 8 μm band (to MIPS 24 μm in some cases).

As an example of the kind of science one might do with the Spitzer source list, we chose a patchy dark cloud, L673, seen against a dense stellar background (galactic coordinates $l = 46°$,

$b = -1°$). Parts of the cloud were observed with Spitzer by the "c2d" Legacy survey (Evans et al. 2007). For this example, we used the objects identified in their catalog as likely to be photospheres and matched these to UKIDSS sources.

The extinction at K in mag, A(K), for each object was estimated from the UKIDSS data by assuming a power-law for A(λ) between J (1.25 μm), H (1.63 μm), and K (2.2 μm), and assuming a mean intrinsic color for the background stars of J − H = 0.7 and H − K = 0.3 from off-cloud regions, in a similar manner as Lombardi & Alves (2001). The relative extinctions in IRAC bands were then calcuated assuming photospheres. Values in separate bins in A(K) were fit with a robust linear fit and mean fit values in low and high extinction bins are plotted in Figure 1.

While strong conclusions are not possible from just this small region, the trend appears consistent with what's been seen in other dense clouds: flatter extinction in IRAC bands relative to K when looking towards the high extinction areas, compared to lower extinction regions, which are more like the diffuse ISM. Though ice bands contribute to this effect, grain growth is probably dominant, according to McClure (2009).

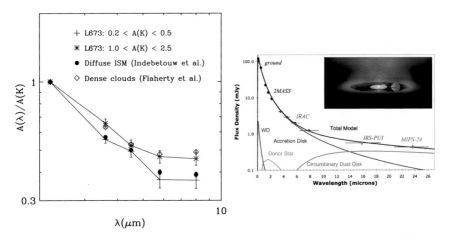

Figure 1.   **Left:** Dust extinction law in low and high extinction regions of L673, compared to other lines of sight. The Indebetouw et al. (2005) values are those adjusted by Flaherty et al. (2007) and do not include the uncertainty in A(H)/A(K). **Right:** Spectral energy distribution of the bright, nearly face-on cataclysmic variable V592 Cas, along with a multi-component model. Infrared data from pointed observations with the Spitzer Space Telescope were crucial to discovering the presence of a circumbinary dust disk that surrounds this close binary star (see artistic depiction in inset panel). For details, see Hoard et al. (2009).

2.   Galactic Science II. Cataclysmic Variables in the Infrared

Recent infrared observations, particularly from the Spitzer Space Telescope, of white dwarfs, cataclysmic variables (CVs), and other interacting compact binaries, have revealed the presence of dust in many systems (for example, see Figure 1). Cataclysmic variables, in particular, are important astrophysical laboratories for the study of mass transfer and accretion processes. These play key roles in a wide range of astrophysical scenarios, from the formation of stars and planets to the central engines of quasars and AGN. The recent, and surprising, discovery of dust in CVs has also highlighted the importance of these systems for expanding the understanding of post-main sequence stellar evolution, as well as the eventual fate of planetary systems.

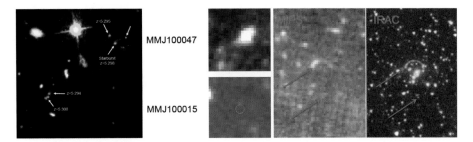

Figure 2.   **Left:** The core of a proto-cluster at z=5.3 with spectroscopically confirmed sources marked. The source list photometry was confirmed to be accurate enough to measure masses for these objects. **Center:** IRAC Ch1 cutouts around optically faint sub-mm bright galaxies are shown (Aravena et al. 2010) with the mm positions marked by a 1″ radius green circle. Note the lack of a clear detection for either source. **Right:** Source list images for an extended gravitational lens and counter image (indicated with red arrows) reported by Wuyts et al. (2010). Note the 24μm emission is resolved in this $z = 1.7$ lensed galaxy.

Up to now, the Spitzer observations of CVs have targeted only about two dozen systems (out of 2500+ known CVs) in pointed programs. We will use the Spitzer Source List, with a recently updated, comprehensive target list for all known cataclysmic variables (CVs), to complete the census of mid-infrared photometry of CVs.

Prior to our observational confirmation of their existence, circumbinary dust disks had been proposed as an additional angular momentum loss mechanism to help rectify theories and observational constraints for the secular evolution of CVs. We have already shown that the total masses of dust appear to be many orders of magnitude too small to agree with the proposed evolutionary models. However, many important questions remain; for example, what is the origin and formation mechanism for the dust? The Spitzer observations to date have given clues to the answer for this question, such as the general lack of a 10-micron emission feature (indicating that the dust grains are likely relatively large), but a complete answer awaits a full exploration of parameter space for the dust in CVs. Even such a basic question as the true ubiquity of dust in CVs remains unanswered. The data to support such an undertaking will be found in the Spitzer Source List. In addition, when combined with upcoming results from the all sky infrared survey by WISE, we will be able to examine the infrared spectral energy distributions of CVs for variability on time scales of several years. This will allow us to address questions relating to the ongoing formation and longevity of dust in these systems.

3.   Extragalactic Science

Statistical surveys of the sky at all wavelengths have become the backbone of modern extragalactic astronomy and are set to become even more essential in the next decade as new survey facilities come online. Novel re-analysis of this survey data often results in valuable and surprising findings, especially if the data are cross correlated with existing data at other wavelengths. Here we give three examples of such science that made use of the SL pipeline. In all cases, a more careful analysis by the science team confirmed the SL results.

A massive proto-cluster was recently confirmed around a sub-mm system at z=5.3 (Capak et al. 2011, Figure 2 Left). The mass of the galaxies in the system was determined using IRAC 3.6 & 4.5μm data obtained in both the Cryogenic and Warm missions. The IRAC images and photometry for this cluster were analyzed using the source list pipeline. The results were checked with a manual re-reduction of the mosaics and Point Spread Function (PSF) fitting of the IRAC source photometry. The images were essentially identical and the photometry was consistent for isolated sources.

In a recent paper (Aravena et al. 2010) the Sub-Millimeter Array (SMA) was used to localize two potentially high-redshift sub-mm systems found with the IRAM 30m telescope. Neither of these sources have significant optical or near-Infrared counterparts. Even in very deep IRAC observations reduced with the SL pipeline no clear counterpart is visible (Figure 2 Center).

Finally, the Red Sequence Cluster Search (RCS) recently reported the discovery of a very extended strongly lensed galaxy (Wuyts et al. 2010, Figure 2 Right). Analysis of the lensing arcs shows the 24μm emission is isolated to one side of the extended arc, and is significantly weaker in a similarly magnified, but un-extended counter-arcs. This indicates the 24μm emission is coming from a very compact region within the galaxy that is more apparent when the emission is resolved by the lensing.

References

Aravena, M., Younger, J. D., Fazio, G. G., Gurwell, M., Espada, D., Bertoldi, F., Capak, P., & Wilner, D. 2010, ApJ, 719, L15
Capak, P. L., et al. 2011, Nat, accepted
Chapman, N. L., Mundy, L. G., Lai, S., & Evans, N. J. 2009, ApJ, 690, 496
Evans, N. J., et al. 2007, Final delivery of data from the c2d legacy project: Irac and mips. URL http://ssc.spitzer.caltech.edu
Flaherty, K. M., Pipher, J. L., Megeath, S. T., Winston, E. M., Gutermuth, R. A., Muzerolle, J., Allen, L. E., & Fazio, G. G. 2007, ApJ, 663, 1069
Hoard, D. W., et al. 2009, ApJ, 693, 236
Indebetouw, R., et al. 2005, ApJ, 619, 931
Lawrence, A., et al. 2007, MNRAS, 379, 1599
Lombardi, M., & Alves, J. 2001, A&A, 377, 1023
McClure, M. 2009, ApJ, 693, L81
Werner, M. W., et al. 2004, ApJS, 154, 1
Wuyts, E., et al. 2010, ApJ, 724, 1182

Astronomical Data Analysis Software and Systems XX
ASP Conference Series, Vol. 442
Ian N. Evans, Alberto Accomazzi, Douglas J. Mink, and Arnold H. Rots, eds.
© 2011 Astronomical Society of the Pacific

Application of Gaia Analysis Software AGIS to Nano-JASMINE

Yoshiyuki Yamada,[1] Uwe Lammers,[2] and Naoteru Gouda[3]

[1]*Department of Physics, Kyoto University,*
Oiwake-cho Kita-Shirakawa Kyoto, 606-8502, Japan

[2]*European Space Astronomy Center, P.O. Box 78,*
28691 Villanueva de la Canada, Madrid Spain

[3]*JASMINE project office, National Astronomical Observatory Japan,*
Osawa Mitaka Tokyo 181-8588 Japan

Abstract. The core data reduction for the Nano-JASMINE mission is planned to be done with Gaia's Astrometric Global Iterative Solution (AGIS). Nano-JASMINE is an ultra small (35 kg) satellite for astrometry observations in Japan and Gaia is ESA's large (over 1000 kg) next-generation astrometry mission. The accuracy of Nano-JASMINE is about 3 mas, comparable to the Hipparcos mission, Gaia's predecessor some 20 years ago. It is challenging that such a small satellite can perform real scientific observations. The collaboration for sharing software started in 2007. In addition to similar design and operating principles of the two missions, this is possible thanks to the encapsulation of all Gaia-specific aspects of AGIS in a Parameter Database. Nano-JASMINE will be the test bench for the Gaia AGIS software. We present this idea in detail and the necessary practical steps to make AGIS work with Nano-JASMINE data. We also show the key mission parameters, goals, and status of the data reduction for the Nano-JASMINE.

1. Nano-JASMINE and JASMINE Series

We have three astrometric missions in Japan as shown in Table 1. Nano-JASMINE is a rotating satellite with two beams and is similar to the large European astrometry satellite, Gaia. Small JASMINE and JASMINE are pointing satellites with one beam and have different observational strategies from Nano-JASMINE.

Nano-JASMINE is the first Japanese astrometric satellite (Yamada 2008; Kobayashi et al. 2010) and the second one in the world. The launch contract is completed: Nano-JASMINE will be launched at around Aug 2011 from Alcantara Space Port at Brazil by a Cyclone 4 Ukrainian launch vehicle. In the summer of 2010, the flight model shown in Figure 1 was completed; it is now being tested.

2. What Is AGIS

Astrometric Global Iterative Solution (AGIS, O'Mullane et al. 2011) is the name of the software developed by Gaia team. Astrometric data reduction is essentially a large least squares problem.

$$\begin{pmatrix} S & U^T & V^T \\ U & A & W^T \\ V & W & C \end{pmatrix} \begin{pmatrix} \Delta\vec{s} \\ \Delta\vec{a} \\ \Delta\vec{c} \end{pmatrix} = \begin{pmatrix} \vec{b}_s \\ \vec{b}_a \\ \vec{b}_c \end{pmatrix}$$

The vector $\vec{s}$ is stellar parameters, $\vec{a}$ is satellite attitude parameters, and $\vec{c}$ is the calibration parameters. The number of components of $\vec{s}$ is 10^7 because each star has 5 astrometric

	Nano-JASMINE	Small JASMINE	JASMINE
D	5cm	30cm	1m class
Size/weight	$(50cm)^3$, 35kg	400kg	1500kg
accuracy	3mas at z < 7.5	10μas at H_W < 11	10μas at K_W < 11
survey	whole sky	several sqr. deg.	200 sqr deg.
launch	Aug 2011, Alcantara Cyclone Space, cyclone 4	Will submit small science satellite program in ISAS/JAXA	TBD
operation	2011~2013	2016	
		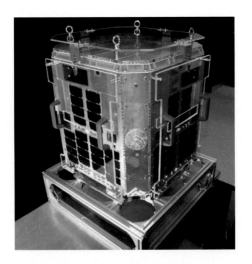	

Table 1.   Space astrometry program in Japan.

Figure 1.   Nano-JASMINE flight model.

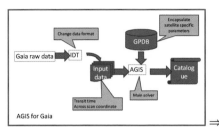 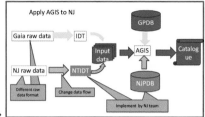

Figure 2. Application of AGIS to Nano-JASMINE. By replacing PDB and IDT for Gaia with those for Nano-JASMINE, we can apply AGIS to Nano-JASMINE.

parameters, and the number of stars observed by Nano-JASMINE is about 2×10^6. For satellite attitude, we apply a statistical model. According to the Gaia attitude model, we also apply the cubic spline of each quaternion component as the first step. AGIS can solve the best fit parameters from the observed data iteratively.

In AGIS, the input data is only the transit time and cross-scan coordinate in the observed stellar position. The part of calculating the two input parameters from the observational data (IDT) is separated from the main solver / iterator. The satellite-specific parameters are encapsulated within a database(PDB). We can apply AGIS to Nano-JASMINE by replacing IDT and PDB from the Gaia-specific version to the Nano-JASMINE-specific version as shown in Figure 2.

3. Implementation Status

For IDT, we convert downlinked data of Nano-JASMINE which is 9×5 pixel stellar images (shown in Figure 3) to the transit time and cross-scan coordinate. The photo center is not the real center of the stellar image. The principal components as shown in Figure 3 have the information of the image center. The first component denotes the shape of stellar images. The second and third ones are the shift of the image center in each direction. By sampling, the coefficients of these two components, the normalization which is the PSF dependent values, will be estimated. Cross-scan coordinates are written in the unit "pixel". For the coordinate along the scan, we have GPS time of each TDI line in on-board data and can convert to baricentric time.

For PDB, we already picked up the parameters which are used in AGIS. The Gaia team has almost completed modification of parameter handling Java classes in AGIS.

Due to effects of the radiation, the PSF changes, and the centroid will be shifted. The NJ team will measure the change of CTI, and within two years operation, CTI will be about 10^{-4} when the mission ends. Gaia will be affected in a similar way. The radiation exposure test was performed using the proton beam at the Tandem Accerelator in the Department of Physics of Kyoto University. From the analysis of the Gaia team, the shift will be written as the third polynomial of the magnitude. In that case, the AGIS "calibration model" can be applied. The coefficients of the polynomial will be estimated simultaneously as other parameters.

In Nano-JASMINE and also in Gaia, proper motion accuracy is improved by combining the Hipparcos data and using the long time baseline (more than 20 years with 1 mas accuracy). One of our team members in ESA has modified the AGIS to do this.

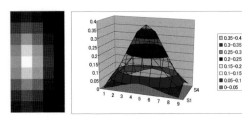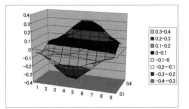

Figure 3.   Centroiding will be performed by Principal value analysis. The left figure is the sample stellar image. The central and right ones are the first two principal components.

4.   Future Plan

We are planning small-JASMINE after the Nano-JASMINE. The observational strategy is different, so we cannot apply AGIS to small-JASMINE easily. We are starting to implement a more general least squares solver.

Each new implementation supports flexibility for replacing models. For example in Nano-JASMINE, we can replace the attitude model from a statistical (cubin spline) model to a physical model. This will be helpful for the design of the next small scientific satellite.

By replacing the attitude model and observation model, we will able to apply the same software to small-JASMINE data.

Acknowledgments.   This work is supported by Coordination Funds for Promoting Space Utilization, Ministry of Education, Culture, Sports, Science and Technology, Japan.

References

Kobayashi, Y., Yano, T., Gouda, N., Niwa, Y., Murooka, J., Yamada, Y., Sako, N., & Nakasuka, S. 2010, in Space Telescopes and Instrumentation 2010: Optical, Infrared, and Millimeter Wave, edited by J. M. Oschmann Jr., M. C. Clampin, & H. A. MacEwen (Bellingham, WA: SPIE), vol. 7731 of Proc. SPIE, 77313Z

O'Mullane, W., Lammers, U., & Hernandez, J. 2011, in Astronomical Data Analysis Software and Systems XX, edited by I. N. Evans, A. Accomazzi, D. J. Mink, & A. H. Rots (San Francisco, CA: ASP), vol. 442 of ASP Conf. Ser., 351

Yamada, Y. 2008, in Astronomical Data Analysis Software and Systems XVIII, edited by D. Bohlender, D. Durand, & P. Dowler (San Francisco, CA: ASP), vol. 411 of ASP Conf. Ser., 39

Astronomical Data Analysis Software and Systems XX
ASP Conference Series, Vol. 442
Ian N. Evans, Alberto Accomazzi, Douglas J. Mink, and Arnold H. Rots, eds.
© *2011 Astronomical Society of the Pacific*

Recording the History of Herschel[1] Data Processing in the Data Products

J. A. de Jong,[1] R. Huygen,[2] J. Bakker,[3] E. Wieprecht,[1] R. Vavrek,[3] M. Wetzstein,[1] J. Schreiber,[4] E. Sturm,[1] and S. Ott[3]

[1]*Max Planck Institute for Extra-Terrestial Physics,
Garching bei München, Germany*

[2]*Institute of Astronomy, KU Leuven, Belgium*

[3]*ESA Herschel Science Center, Madrid, Spain*

[4]*Max Planck Institute for Astronomy, Heidelberg, Germany*

Abstract.   We present how the history is recorded during data processing with the Herschel Common Science System (HCSS), and how users can inspect this in the Herschel Interactive Processing Environment (HIPE). The Herschel DP software records after the execution of each (pipeline) task all information needed to redo the task. This includes the name of the task, the used software version, input parameter names and their values. In case an input parameter is a product, the complete history of that product and if available a human readable identifier (such as a calibration filename) is recorded. Also, the final product contains the complete chain of tasks needed to reproduce that product from the raw data. The history information is stored in two binary tables when the product is saved as a FITS file. Users can inspect this dataset with a dedicated viewer in HIPE, or get it as a Python script to redo the reduction with optional changes to the parameters.

1. Introduction

For the scientific assessment of any products generated by a Data Processing (DP) system it is important to know how the product has been generated. Therefore, such a DP system should record the history of tasks/recipes used to generate the product. Many DP systems store processing keywords in the FITS headers (like the ESO data flow, see e.g., Ballester et al. 2001), but that is not enough to completely reproduce a data reduction.

Therefore, we decided for the Herschel (Pilbratt et al. 2010) Data Processing System (Ott 2010) to record the processing history into a data format which can hold a complete tree of reduction steps from several raw products to a final product. We basically store this tree into two binary tables which are linked to each other by an ID, like in a relational database (see § 3). This tree can be converted to a Python script with which users can redo the reduction. However, this full recording only works when all pipeline tasks accept only products (with a history) or simple parameters (see § 4) and return a product.

---

[1]Herschel is an ESA space observatory with science instruments provided by European-led Principal Investigator consortia and with important participation from NASA.

We describe following sections how users can inspect and use this data in HIPE (Ott 2011), what the data format looks like, and how the software has been designed. The code is available as open source in the HIPE distribution which can be downloaded from the Herschel website.[1]

2. User Interfaces

Users can inspect the history in HIPE in several ways:

- On the command line, type `print product.history` to show all of the tasks, the software version used, and the special products, such as calibration files, used.
- Get a complete script to redo the reduction by means of `product.history.saveScript("script.py")`.
- Inspect the history in a dedicated viewer in HIPE. This viewer consists of two tables: The first table shows all the tasks, the used software versions and special products. The second table shows the details of a task after selecting it in the first table. In case of one of the parameters is a product then one can click on a button in the second table to jump to the task which generated the product. Fig. 2 shows this viewer with the history of a PACS spectrometer pipeline (Schreiber et al. 2009) reduction.

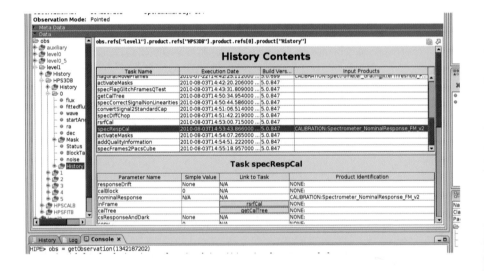

Figure 1.   History viewer in HIPE with history of PACS spectrometer pipeline.

3. Data Format

The history is stored in three tables, which are stored as binary tables in FITS files:
HistoryTasks: (see Table 1) Contains the task ID, name, execution date and used software version (build).
HistoryParameters: (see Table 2) Contains all the information about the parameters used in

---

[1]`http://herschel.esac.esa.int/HIPE_download.shtml`

the task. The rows in this table are linked to a task by the task ID.
HistoryScript: Contains just the generated python script for reference. Provided for users who quickly want to check the history outside of HIPE.

Table 1.   HistoryTasks table

Column	Type	Description
ID	Long	Unique ID which links the tasks and parameters
Name	String	Name of the task
ExecDate	Long	Timestamp of execution
BuildVersion	String	The software version which was used (e.g., 5.0.1576)

Table 2.   HistoryParameters table

Column	Type	Description
TaskID	Long	ID which links the parameter to its task
Name	String	Name of the parameter
Type	String	Type of parameter (simple STRING, INTEGER, etc., PRODUCT, OBJECT (generic))
Value	String	The simple value of the parameter (when available)
IsDefault	Boolean	Was the default parameter value used?
IncTaskId	Long	If applicable: The ID of the task which generated this product. Used to merge the history of products.
UserInput	Boolean	Needs user input? (neither a product nor a simple value)
Class	String	The Java class of the value object
ProductType	String	Type of Product (e.g., CALIBRATION)
ProductId	String	A human readable identification (when available)

4.   Software Design

The code is written in Java, like most of HCSS. The most important class is the **HistoryDataset**. This class contains the complete history of the product and the code to read/write this in the tables described in § 3. The history is represented in the API by an array of **TaskHistory** objects, which in turn each contain an array of **ParameterHistory** objects. These arrays contain all the tasks in sequence of execution with their parameter descriptions. The parameter API provides also two interfaces to define simple parameters and special products: The **SimpleParameter-Handler** defines how to represent an arbitrary object by a string and how to convert this string back to the object. This is used to represent short arrays of numbers by a string. These handlers are registered in the **SimpleParameterRegistry** with the classes to which they can be applied. The **HistoryIndentifiable** interface can be implemented in a product to provide a human readable indentifier for special products (like calibration products). Users will see these identifiers in the overview table of the history. The history module is called after successfully executing a task in the HCSS task framework. This framework invokes the addHistoryLine() method of **HistoryManager** with the Task object. This method then reads the task parameters, if needed merges the history of any products in the parameters and adds the lines for the task and its parameters to the tables. Merging a history is simply done by appending the history tables of the input product to that of the output product before adding the new task and parameter lines. All the classes and their dependencies are shown in Fig. 2. They are located in the packages herschel.ia.dataset.history and herschel.ia.task.history. Apart from the classes discussed here there is also a HIPE history viewer (see § 2) located in herschel.ia.toolbox.history.

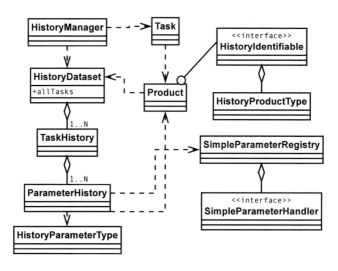

Figure 2.   Class diagram of history package.

Acknowledgments.   The Herschel Science Center (ESA), the Instrument Control Centers (HIFI, PACS and SPIRE) and the NASA Herschel Science Center jointly manage and contribute to the Herschel Data Processing System

References

Ballester, P., et al. 2001, in Astronomical Data Analysis, edited by J.-L. Starck & F. D. Murtagh (Bellingham, WA: SPIE), vol. 4477 of Proc. SPIE, 225
Ott, S. 2010, in Astronomical Data Analysis Software and Systems XIX, edited by Y. Mizumoto, K.-I. Morita, & M. Ohishi (San Francisco, CA: ASP), vol. 434 of ASP Conf. Ser., 139
— 2011, in Astronomical Data Analysis Software and Systems XX, edited by I. N. Evans, A. Accomazzi, D. J. Mink, & A. H. Rots (San Francisco, CA: ASP), vol. 442 of ASP Conf. Ser., 347
Pilbratt, G. L., et al. 2010, A&A, 518, L1
Schreiber, J., et al. 2009, in Astronomical Data Analysis Software and Systems XVIII, edited by D. A. Bohlender, D. Durand, & P. Dowler (San Francisco, CA: ASP), vol. 411 of ASP Conf. Ser., 478

Astronomical Data Analysis Software and Systems XX
ASP Conference Series, Vol. 442
Ian N. Evans, Alberto Accomazzi, Douglas J. Mink, and Arnold H. Rots, eds.
© 2011 Astronomical Society of the Pacific

An Automated Release Manager for the Fermi Large Area Telescope Software Systems

Thomas E. Stephens and Navid Golpayegani
for the Fermi LAT Collaboration

Wyle IS/NASA Goddard Spaceflight Center

Abstract. The Fermi Gamma-ray Space Telescope (Fermi) Large Area Telescope (LAT) collaboration maintains a large software system that covers all aspects of the instrument operation from simulations of the instrument response to event reconstruction and data analysis. Much of this software is supported and developed across a variety of operating systems and platforms (Windows, Linux and Mac OS X, both 32 and 64 bit). In order to ensure that the software works across the full range of supported systems, the LAT collaboration has developed an automated Release Manager system to checkout, compile and test any new code across all these systems regardless of which system it was developed on. This poster describes the newest version of this Release Manager system developed in conjunction with the move by the collaboration to the use of Scons as our build tool of choice. Built upon the Qt framework, the Release Manager leverages the batch submission system at the SLAC National Accelerator Laboratory (SLAC) to build and test any new code changes on all relevant platforms. Here we describe the design of the system as well as issues encountered in its implementation.

1. Software Systems

The LAT collaboration software is contained in two major and two minor software systems or release packages:

- **Science Tools** — The Science Tools contain all the collaboration software related to the scientific analysis of the Fermi LAT data. This is the package used by the individual scientists to do data analysis and by tools in various automated pipelines in the collaboration (flaring source detection, catalog generation, etc.)

- **GlastRelease** — Named before the mission name was changed to Fermi, the GlastRelease package contains all of the simulation and data reconstruction software for the mission. It contains a high fidelity spacecraft model and physics simulations used to study the instrument response. It contains all the software and algorithms used to reconstruct the data received from the spacecraft (or simulation) into useful scientific data. This package is used by the data reconstruction pipeline to process the data as it arrives from the spacecraft and prepare it for the data archive and use by the Science Tools.

- **Command, Health and Safety (CHS)** — This package contains the software responsible for generating commands sent to the instrument to control operation as well as receiving and analyzing telemetry data downloaded from the instrument during operation.

- **TMineRelease** — This package contains a classification tree data mining package that is used as part of the data analysis and reconstruction done by the software in the GlastRelease package. It was split out into its own package to facilitate rapid development without encumbering the much larger GlastRelease package.

375

2. Supported Operating Systems

The LAT collaboration software is supported across a variety of operating systems and environments including systems used by developers, end users within the LAT collaboration, and the various environments the production software needs for simulation and data processing.

Currently we support development and/or operation of the various software systems on the following operating systems: Redhat Enterprise Linux (RHEL) 4 (32 & 64 bit), RHEL 5 (32 & 64 bit), Windows (VS 2003 and VS 2008) and Mac OS X versions 10.4 (Tiger) and 10.6 (Snow Leopard).

3. Build Types

The Release manager supports three basic build types: Integration, Release Candidate, and Release.

- **Integration Builds** — This build type is automatically triggered when a software component receives a new tag in the CVS repository. The Release Manager Daemon regularly checks the repository looking for new tags on sub-packages that make up each release package. When one or more new tags are discovered, a new Integration build is triggered. Debug versions of the packages are built automatically but optimized versions can be triggered manually if desired. The Integration builds are primarily designed to provide rapid feedback to developers on changes made to the code and to verify that changes made work on all supported operating systems.

- **Release Candidate Builds** — These builds are triggered by a specific tag that is manually applied by the release package owner. They are triggered in preparation for a release build to verify that the selected tagged versions of the sub-packages build and work together properly. They contain the appropriate tags for the release in a combination that may or may not have existed in the Integration builds. Once the tag is applied to the appropriate sub-packages, debug versions of this build are created for each supported OS.

- **Release Builds** — These are the builds intended for distribution to the collaboration and for use in the automated systems run by the LAT team (data processing, catalog analysis, etc.). These builds are triggered by the existence of the appropriate release tag in CVS which is applied manually by the package owner. The existence of the appropriate tag causes the Release Manager to build debug and optimized versions of the software for each supported operating system.

4. Release Manager Components

The Release Manager consists of three main components: the batch submission system, the workflow system, and the release manager proper. Each of these three components are supported by a series of database tables to drive the processes and store state and metadata about the pending, running and completed builds.

- **Batch Submission System** — The Release Manager utilizes the LSF batch submission system at the SLAC National Accelerator Laboratory. Through this system we have access to hardware (either purchased by the Fermi mission or as a shared resource) running all of our supported operating systems that we can use for build and testing of the various software packages.

  This is the lowest level component of the system and manages the individual processes of the builds and provides the status information to the other portions of the system.

- **Workflow System** — The Workflow system is a rule based script execution system. Each script or program is considered a stage in the workflow. The workflow moves from one stage to the next by evaluating rules set forth for each stage. The rules are stored in the database as a series of conditions and steps to execute if the conditions are met. Each stage of the workflow consists of a script or program that is passed to the Batch Submission System for execution.

- **Release Manager System** — The Release Manager System consists to two main parts. At the very top is the Release Manager Daemon that runs and monitors the CVS repository for new tags and triggers the appropriate workflows to actually execute the builds.

At the lower level, this system is composed of the actual programs executed by the Workflow system to check out code, build the software, test the softwarei, and package it for distribution and download. It also includes command line tools for deleting and triggering builds.

5. Building a Release

The process of building, testing, and packaging a release consists of several steps that are managed by the Workflow System and executed by software that is part of the Release Manager System. The basic steps are:

- **Checkout** — Selects all the sub-packages that are to be part of the build via the appropriate tag in the CVS repository and places them in a central build location

- **Compile** — SCons is invoked to build the package. Output is recorded and stored in the database.

- **Test** — Nearly all of the sub-packages have unit and validation tests that are run to verify that all the code is working properly. Each test is run independently and the output is stored in the database.

- **Package** — Running in parallel with the testing, each build has a source, user and developer distribution package created to allow users and developers to download and work with the specific version of the code built and tested.

- **Cleanup** — Once all other processing is done, a script is run to clean up the build process and store any final metadata in the database.

6. Using Qt

Qt is used as a framework to build the various programs that comprise the build system in order to provide parallel and asynchronous execution of the various parts of the system. The main highlights are discussed here.

The Release Manager Daemon leverages the QTimer class to set up an asynchronous polling system to check each of the 12 possible builds (4 Packages and 3 build types) on a configurable polling interval to look for new builds that need to be started.

Once the need for a new build has been determined, the software leverages the QProcess class to launch each of the up to 16 build variations (8 OSes each with a Debug and/or Optimized build) in its own thread for processing.

In addition to allowing the entire system to be multi-threaded, the QProcess class allows the system to be robust against hung processes and other unexpected failures as we utilize the ability to limit the time the process remains active. Processes and build stages that exceed the configured (generous) time limit are cancelled and errors are reported.

Finally, all of the classes and tools associated with the build system make heavy use of the QSqlQuery class to provide easy access to the MySQL databases that hold all of the configuration parameters, build status information and logging.

7. Issues and Lessons Learned

No large system is constructed and works without issues. Here we highlight some of the hurdles we had to overcome and lessons learned along the way.

- **LSF on multiple operating systems** — Since the batch queuing system was at the center of the build system, understanding its operation and peculiarities was essential to correct operation. There were subtle differences between the way the LSF system worked with the underlying operating system on the target machines, especially between the Unix-like OSes and Windows.

- **General Windows Support** — Supporting Windows has been both a boon and a thorn in our side. On the one hand, the Windows tools and compilers are high quality and having to support builds with both Visual Studio and GCC compilers has resulted in a strong, robust code base. On the other hand, the Windows environment is very different from the Unix-like OSes and special care has to be taken in many of the configuration issues to account for the differences.

- **Windows Network Storage** — In addition, the use of AFS network mounted storage on Windows resulted in very slow performance of the entire system. Since the builds are run through the batch system and different portions may run on different machines, all the code, libraries, etc. need to be on shared disks. The performance on Windows is so bad (8–12 hours instead of ~1) that we are moving all of our Windows builds to a single multi-core machine with a large local disk.

Astronomical Data Analysis Software and Systems XX
ASP Conference Series, Vol. 442
Ian N. Evans, Alberto Accomazzi, Douglas J. Mink, and Arnold H. Rots, eds.
© 2011 Astronomical Society of the Pacific

Maintaining Software for Active Missions: A Case Study of Chandra's Instrumentation Over Time

D. G. Gibbs II,[1] J. C. Chen,[1] K. J. Glotfelty,[1] G. E. Allen,[2] D. P. Huenemoerder,[2] and F. A. Primini[1]

[1] *Smithsonian Astrophysical Observatory,*
60 Garden Street, Cambridge, MA 02138, USA

[2] *MIT Kavli Institute for Astrophysics and Space Research,*
77 Massachusetts Avenue, 37-287, Cambridge, MA 02139, USA

Abstract. During eleven plus years of operation, ample knowledge has been gained regarding the Chandra X-ray Observatory's scientific instruments and how they are performing over time. In this paper we will summarize the significant software changes related to the performance and knowledge gained about the observatory's instrumentation specifically targeting the last five years of the mission. With this knowledge, numerous upgrades to the Chandra processing software have taken place to correct issues that have developed on orbit (ACIS CTI effects, LETG/ACIS rotation), to correct for hardware issues (HRC timing), and for our better understanding of the evolution of the instrumentation (temporal gain shifts in both instruments). In addition, we will discuss the challenges in maintaining software when the calibrations for different operating modes are incrementally made available. This includes challenges in maintaining an archive with a mixture of data products with different calibrations applied. This paper roughly covers the era of the 3rd reprocessing of the Chandra archive (Repro-3).

1. Introduction

Chandra has enabled many cosmological discoveries (Weisskopf et al. 2002). In addition to scientific discoveries, ample knowledge has been gained regarding the instruments themselves and how they are performing over time. With this knowledge, numerous upgrades to the Chandra processing software have taken place to correct issues that have developed on orbit, to correct for hardware issues by adapting the software to meet the changes, and for our better understanding of the evolution of the instrumentation. What follows are some examples of software updates made in response to our evolving understanding of the instruments.

2. ACIS

2.1. CTI Correction

The charge transfer inefficiency (CTI) is a known issue that has been addressed before (Masters et al. 2004). However, as the knowledge of the CTI has evolved, methods to incorporate graded mode and time-dependent CTI corrections into the software processing have occurred (release December 2010).

2.2. Destreak

There is a flaw in the serial read-out of the ACIS chips, causing a significant amount of charge to be randomly deposited along pixel rows as they are read out.[1] ACIS-S4 (ccd_id=8) is significantly affected by this problem. The destreak CIAO tool detects coincidence of events in adjacent pixels along a row (i.e., the serial read), flags probable streak events, and (optionally) removes them.

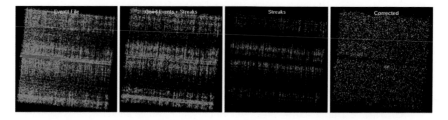

Figure 1.   Example of ACIS destreak algorithm. Starting at the left, the panels show all Level 1 events, the good events plus the streaks, the streaks, and the last panel shows the destreak correction.

2.3. Afterglow Detection

A cosmic-ray "afterglow" is produced when a large amount of charge is deposited on a CCD by a cosmic ray. Most of the charge is clocked off of the CCD in a single frame. However, a small amount can be captured in charge traps, which release the charge relatively slowly. As a result, a sequence of events can appear in a single detector pixel over a few frames as the trapped charge is released.

To date, two algorithms have been used by the CXC to identify cosmic-ray afterglows. The first algorithm was implemented in the CIAO tool acis_detect_afterglow. In an attempt to minimize the loss of source events, another algorithm was developed and implemented in the CIAO tool acis_run_hotpix.

A third afterglow-detection algorithm is currently being implemented. The principal change being the third algorithm searches for afterglows using the events in a short, sliding time window instead of using the events from the entire duration of an observation (i.e., the algorithm searches in three dimensions instead of two). Like the second algorithm, it is designed to avoid discarding events associated with real astrophysical sources. It is also designed to enhance the detection efficiency for afterglows that have as few as four events (Allen 2010).

3. HRC

3.1. HRC Time Varying Gain

The new time varying gain correction algorithms for HRC-I and HRC-S have been implemented in the CIAO tool hrc_process_events to support the new calibration files. The new gain files for both instruments are based on the sum of amplitudes (SAMP) metric. Unlike the previous pulse height amplitude based gain, the SAMP based gain is spatially invariant. For the HRC-S instrument, the primary purpose of the new gain map application is to reduce background. The figure below has two plots for HRC-S arlac in the observation ID 11930. The plots were created from processing the events with the old gain file (solid) and with the new time gain map (dash). The background count rates are dropped as desired after applying the new gain file.

---

[1]http://cxc.harvard.edu/ciao/why/destreak.html

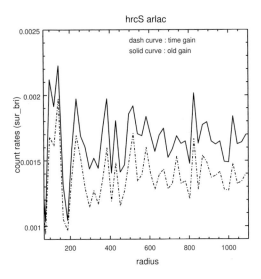

Figure 2.   Two plots from processing the events with the old gain file (solid) and with the new time gain map (dash).

3.2. HRC-S Timing

The CIAO tool hrc_calc_dead_time is used to compute the dead time factor (DTF) for a record as a function of the Secondary Science (SS) counts (total and valid), the Primary Science (PS) Counts, and the telemetry saturation case (in which there are many more SS counts than Primary Science).

For HRC-S, when the event amplitude is sufficiently large, the SS valid event rate, as reported by the HRC hardware scalars, is double counted and that affected the DTF result. To correct this timing problem for HRC-S, we have implemented a new algorithm in hrc_calc_dead_time. Based on the thresholds set from the calibration, we computed how many of the SS valid events were double counted (DD) and divided it with PS counts (FF) then correct the DTF value by the factor of $1/(1 + FF)$.

4.   Transmission Gratings

Pixlib, the Chandra coordinate library, has been updated to support a new format geometry file which allows independent grating angles for each grating arm as seen by each instrument (ACIS, HRC-I, and HRC-S). Calibration data are different for ACIS LETG compared to HRC LETG. Figure 3 below shows the before and after of the ACIS LETG for observation ID 4148. Both images are in the diffraction coordinates tg_r and tg_d with the old and the new geometry calibration files. The spectrum on the left becomes horizontal after applying the new geometry file, as depicted on the right.

5.   Conclusion

Numerous upgrades to the Chandra processing software have taken place to correct issues that have developed on orbit (ACIS CTI effects), to correct for hardware issues (HRC timing), and

Figure 3.   Before and after of the ACIS LETG with new geometry calibration files.

for our better understanding of the evolution of the instrumentation (temporal gain shifts in both instruments). Priority for upgrades is given to the most popular instrumental configurations. In addition to learning about the evolution of the instrumentation, we have learned from our own software and algorithms. As an example the algorithm to detect afterglows is currently in its third iteration.

Some of the keys to success have been the implementation of flexible, multi-dimensional software, a consistent style of programming (to coordinate multiple programmers/developers), robust testing for incremental changes, a single code base for operations and user analysis tools (CIAO) and CIAO being based on a mission independence design. With each step in our understanding we learn better and more efficient ways to enhance and upgrade the software for one of the most complex scientific instruments ever to fly.

Acknowledgments.   Support of the development and maintenance of Instrument tools is provided by National Aeronautics and Space Administration through the Chandra X-ray Center, which is operated by the Smithsonian Astrophysical Observatory for and on behalf of the National Aeronautics and Space Administration contract NAS8-03060.

References

Allen, G. E. 2010, Afterglow spec, Revision 1.9, Tech. rep., MIT Kavli Institute for Astrophysics and Space Science. URL `http://space.mit.edu/ASC/docs/afterglow_spec_1.9.pdf`

Masters, J. S., He, H., McLaughlin, W., Glotfelty, K., & Allen, G. 2004, in Astronomical Data Analysis Software and Systems XIII, edited by F. Ochsenbein, M. G. Allen, & D. Egret (San Francisco, CA: ASP), vol. 314 of ASP Conf. Ser., 800

Weisskopf, M. C., Brinkman, B., Canizares, C., Garmire, G., Murray, S., & Van Speybroeck, L. P. 2002, PASP, 114, 1

Astronomical Data Analysis Software and Systems XX
ASP Conference Series, Vol. 442
Ian N. Evans, Alberto Accomazzi, Douglas J. Mink, and Arnold H. Rots, eds.
© *2011 Astronomical Society of the Pacific*

Changing Horses in Midstream: Fermi LAT Computing and SCons

J. R. Bogart[1] and Navid Golpayegani[2]
for the *Fermi* LAT Collaboration

[1]*SLAC National Accelerator Laboratory*
[2]*Wyle IS/NASA Goddard Space Flight Center*

Abstract. Several years into GLAST (now Fermi) offline software development it became evident we would need a replacement for our original build system, the Configuration Management Tool[1] (CMT) developed at CERN, in order to support Mac users and to keep pace with newer compilers and operating system versions on our traditional platforms, Linux and Windows. The open source product SCons[2] emerged as the only viable alternative and development began in earnest several months before Fermi's successful launch in June of 2008. Over two years later the conversion is nearing completion. This paper describes the conversion to and our use of SCons, concentrating on the resulting environment for users and developers and how it was achieved. Topics discussed include SCons and its interaction with Fermi code, GoGui, a cross-platform gui for Fermi developers, and issues specific to Windows developer support.

1. Background

The LAT (Atwood et al. 2009) is essentially a small orbiting HEP detector; the raw data consist of events triggered by interactions with the detector's active elements. The ones of interest after filtering typically correspond to a single photon or cosmic ray particle. Considerable analysis (commonly known as *reconstruction*) is required to identify the best candidate particle and infer trajectory energy, etc.

1.1. Offline Software

The bulk of offline software is organized into two large, partially-overlapping collections of *packages* (see 1.1.3): **GlastRelease** and **ScienceTools**. Both are used in automated simulations at SLAC and Lyon CC-IN2P3 and in automated flight data processing at SLAC, as well as by individual developers and end-users, often running on personal laptops and desktops at remote locations.

1.1.1. GlastRelease. Event reconstruction is handled by Gleam, the primary build product of GlastRelease. Gleam can also simulate events from astronomical source through digitization in the detector; that is, to the input to reconstruction

Gleam simulations were used years before launch as an aid in determining design parameters for the instrument and developing reconstruction algorithms. Gleam continues to evolve as lessons learned from real data are fed back into simulation and reconstruction. Because of its compute-intensive nature and dependence on externals libraries for key functions, such as

---

[1]http://www.cmtsite.org/

[2]http://www.scons.org/

Geant for physics simulation. GlastRelease is written almost entirely in C++. The group of active developers is small but crucial to the continued success of the mission.

1.1.2.   ScienceTools.   Members of the Collaboration use ScienceTools for science analysis of real and simulated data to produce analysis products, chief among them exposure maps and light curves. ScienceTools code is written in a mixture of C++ and python.

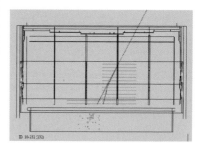

Gleam simulated photon event

1.1.3.   Packages.   Both GlastRelease and ScienceTools are organized into packages. Build products for a typical package include a library (static or shared), one or more test programs or other applications linked against the library, and program inputs: data files, and configuration files of various kinds. ScienceTools packages also often provide python modules and applications. The package organization facilitates concurrent development by different developers but does not always map well to the dependency hierarchy since an application program in package P may depend on components from more packages than does the library of package P.

1.2.   Supported Platforms

Much of the initial LAT offline code development was done on Windows. There is still a substantial and critical contribution from Windows developers, particularly for GlastRelease, but the SLAC batch farm machines used for automated processing run Linux. Hence there has always been a requirement to support both operating systems. Our old build system, CMT, would not build our software correctly — or at all — on 64-bit Linux systems, on newer versions of the kernel (anything beyond Redhat Enterprise 4), nor on versions of Visual Studio beyond VS 2003. This has caused consternation among users and developers working on remote systems for some time. More recently, the standard SLAC batch machines were upgraded to Redhat 5, so we can no longer use them to build our software with CMT.

As ScienceTools began to mature sufficiently to be of interest to end-users, it became clear that many preferred to do their analysis on Macs, another platform not supported by the old build system.

2.   SCons

SCons is an open-source build tool which supports all platforms of interest to LAT Offline. It is written in python as are all local user customization, configuration, etc. This imposes welcome regularity in syntax and behavior and undoubtedly contributes to its impressive extensibility.

2.0.1.   Builders and other functions.   Rules for building a target from its sources are specified in Builder methods. SCons comes with several Builders (e.g. for compilation, making libraries, making Java archives, making tar archives, installing files, etc.). SCons also provides several other functions, including AddOption (add command-line options) and Alias (define new targets in terms of existing ones, similar to phony targets for Make). It is straightforward to extend SCons with custom builders and functions.

2.0.2.   Construction environments.   SCons facilitates fine-grained control of dependencies and other aspects of the build process through construction environments. One can, for example, independently manipulate compile or link options for each construction environment.

3. Fermi Use and Customization

Standard compiler options and other generic settings are added to an initial environment, baseEnv. Most packages clone baseEnv twice (once for an environment to build its library, again for building applications) and customize the clones as appropriate.

Our local extensions include additional command-line options (e.g. to specify non-default compiler), added or modified Builders (for Doxygen, dynamic ROOT libraries, VS project files etc.) and added Tools to encapsulate common operations.

3.0.3. Environment Set-up. SCons, particularly when combined with our decision to install all files needed at run-time in centralized locations, allowed us to streamline the distribution for remote users (easy to exclude intermediate build products and, for end-user distribution, package source not needed at run time) and simplify set-up, especially for end-users. End-users of an SCons build can establish a process with environment suitable for running all applications by running a single set-up file; they do not need a local installation of SCons.

3.0.4. Windows Project and Solution Files. The largest stumbling block in adapting SCons to make Windows builds has been the creation of adequate project and solution files. SCons native builders produce files which cannot be usefully modified in the Visual Studio environment. Generated files include an encoded version of dependencies and procedures, then invoke SCons to interpret it. The (substantial) advantage of this approach is its robustness: the build proceeds identically whether invoked from Visual Studio or directly from SCons. The disadvantage is that it is impossible to change anything about how the build proceeds from Visual Studio.

The fundamental differences between Windows and Linux make it impossible for a build system to do "the same thing" with both. In our use of CMT, the first step was to generate project and solution files, then use them for the build. Hence automated builds made by the Release Manager and builds made interactively by developers were created the same way, but there was a continual maintenance problem in keeping Windows builds similar to Linux builds of the same code base.

SCons does a better job of automating similarity of behavior on Windows and Linux by invoking compilers, linker, etc. directly on both, but at the cost of producing unidiomatic project and solution files. Our approach has been for the Release Manager to produce Windows builds in the standard SCons manner (sans Visual Studio) and to use additional build targets to generate custom VS project files for our developers. However, capturing all the information of SCons builds and translating, bit by bit, for VS consumption is labor-intensive, unlikely to ever be perfect, and difficult to maintain.

3.1. Accompanying Applications

Each of the tools in the large suite developed in support of CMT had to be adapted or entirely rewritten for SCons, including the Release Manager, Installers to install RM builds remotely, and a new developer gui, GoGui.

The gui, written from scratch in C++ and based on the Qt library, runs on all supported platforms. Functions available from GoGui include (CVS) repository checkout, update, commit and tagging; concurrent package-centric and full-hierarchy views; building targets; and running and debugging programs.

4. Conclusions

Conversion of the build system for a mature project was, as we expected, a considerable amount of work. In spite of the numerous shortcomings of our old system, we would not have seriously contemplated switching (nor should we have) if not forced by the platform support issue. Particularly troublesome were

- **Windows developer support** We knew it would be difficult, but nevertheless seriously underestimated the time involved. We're still not quite done with GlastRelease support on Windows. Overall, Windows issues have probably added a year to the project. A substantial fraction of that is for developer support.

- **Change to installation strategy** Installing all files needed at run- or build-time has significant advantages, however it was not required as part of the conversion, it added to the work, and it takes some getting used to for developers.

- **Tendency for other upgrades to get dragged in** New external library versions, new CVS tagging convention and install policy above are examples.

- **Cost of supporting parallel systems** Most developers and end-users are loathe to switch before they have to. Some minimal communication between systems is necessary but the implementation is inevitably imperfect: since the dual-system configuration is temporary there is little incentive to do a better job.

Acknowledgments. The *Fermi* LAT Collaboration acknowledges support from a number of agencies and institutes for both development and the operation of the LAT as well as scientific data analysis. These include NASA and DOE in the United States, CEA/Irfu and IN2P3/CNRS in France, ASI and INFN in Italy, MEXT, KEK, and JAXA in Japan, and the K. A. Wallenberg Foundation, the Swedish Research Council and the National Space Board in Sweden. Additional support from INAF in Italy and CNES in France for science analysis during the operations phase is also gratefully acknowledged.

References

Atwood, W. B., et al. 2009, ApJ, 697, 1071

Astronomical Data Analysis Software and Systems XX
ASP Conference Series, Vol. 442
Ian N. Evans, Alberto Accomazzi, Douglas J. Mink, and Arnold H. Rots, eds.
© *2011 Astronomical Society of the Pacific*

Two Years Before the Mast: Fermi LAT Computing Two Years After Launch — Eight to Go!

R. Dubois

For the Fermi LAT Collaboration
SLAC National Accelerator Laboratory, Stanford, CA, USA

Abstract.   The Fermi Observatory was launched on June 11, 2008 and the Large Area Telescope (LAT)[1] was activated on June 25. Some 13 GB of data is downlinked daily, transformed into 750 GB in the event reconstruction process, spread out over approximately 8 contacts per day. Each data run is farmed out to several hundred computing cores and results merged back together in our processing pipeline. The pipeline is designed to execute complex processing trees defined in xml and to handle multiple tasks simultaneously, including prompt data processing, simulations and data reprocessings. Our system has a pair of Oracle servers at its core to maintain all the state and dataset bookkeeping. Batch processing is centrally dispatched to the SLAC LSF and Lyon (France) BQS batch farms with more than 5000 shared cores. The xrootd cluster filesystem is used for high throughput and management of large disk pools. Nagios and Ganglia are used for problem alerts and tracking resource usage. The HEP-like instrument event reconstruction lives in a Root world, while high level science is done in FITS. LAT Collaboration users have access to the data via web query engines that slice and dice the data to their needs, also executing the queries in the processing pipeline. The data is now public, so we have the issues of new development vs stability for an outside user base. Two years later, we are dealing with the issues of long term support — how to keep a complex operation alive and vital for 10 years, and how to deal with dependencies on external packages whose support is out of our control.

1. Principal Computing Resources

Our processing system supports generic pipelines (Flath et al. 2009) of complex graphs, running in parallel. We apply this to prompt processing and monitoring; reprocessing and simulations.

We downlink about 15 GB/day, delivered on average every 3 hours. With an event reconstruction rate of about 4 Hz and and a downlink rate of about 500 Hz, we need to apply 125 computing cores to keep up. Input raw data is broken up into many pieces and reconstructed in parallel. In practice, with all the other activities going on (monitoring, file merging etc) we routinely use about 300 cores on average.

The system needs to scale well, since we are running thousands of jobs per day, with hundreds of thousands of files. Ten years of operations is envisaged. The overall system must handle all tasks in parallel with about 45,000 jobs per day being our peak usage to date.

The data arrives from the Mission Operations Center (MOC, at Goddard Space Flight Center, GSFC) via the FastCopy commercial application, which is the standard file transfer tool between all the Fermi ground elements. There is an automated response to the delivery which decodes, repackages and archives the incoming data, then triggers the science ("Level 1", L1) processing task.

---

[1]See Atwood et al. (2009).

L1 does full event reconstruction, with about ×50 expansion of the data size yielding some 750 GB per day of reconstruction output files. These highly detailed data are used for instrument monitoring. The high level science summary files, occupying some 200 MB/day, are sent on to the Fermi Science Support Center (FSSC, at GSFC). Finally L1 triggers the high level science application, ASP or Automated Science Processing.

ASP's job is to refine the parameters of any onboard triggered GRBs, perform blind searches for untriggered GRBs, and to search the sky for flaring sources. The Data Release Plan stipulates that the ASP light curves for 23 sources are released to the public as well as those flaring sources above a flux of 10^{-6} ph/cm^2/s.

Asynchronously to these activities, the system also supports running pipelines for the Science working groups, data reprocessings and sky and instrument simulations.

Primarily the collaboration makes use of the Linux batch farm at SLAC, in which usage is determined by per-project allocations against the total number of available cores controlled by the LSF batch system. As of this writing, Fermi's allocation is 1600 cores, with dedicated disk space of about 1.1 PB. We were loathe to dedicate a cluster of 300 cores to L1, since many cpu cycles would fall on the floor. This was solved elegantly in LSF by configuring a much larger number of cores (800) with a special queue wherein if all job slots were taken, then non-L1 jobs (from any source on the SLAC farm) running on those would be suspended and the L1 job started immediately. Since the L1 jobs were typically about one elapsed hour long, this did not cause tremendous disruption to other users and gave us an on-demand cluster with no lost cycles.

The bookkeeping heart of the system is a pair of "Niagara" Oracle servers, which provide 64 execution threads per server. At present the second server is operated as a hot spare.

In addition, the processing system has been extended to the Lyon Computing Center in France. The system was designed with a base class interface to generic batch systems. We have implemented two: LSF[2] at SLAC and bqs at Lyon. All files are transferred back to SLAC from Lyon; we have an allocation of about 600 cores and primarily use them for simulations.

Data processing cpu usage has been steady at 10 CPU-yrs and 25 TB per month. We have some 600,000 data files and counting with about 72 million log files! We have consumed 600 CPU-yrs of processing (data+simulations) since 2008.

2. System Monitoring, Tuning and Reliability

We use the open source products Nagios[3] and Ganglia[4] for much of our system monitoring. Nagios provides alerts for server aliveness and disk space, while Ganglia monitors cpu, disk and network loads vs time. In addition, we maintain a trending system (dynamic web plots of database quantities) for specialised information from the pipeline.

We run a disk-resident operation, and find NFS (at least as configured at SLAC) to be insufficiently reliable when hundreds of clients are hammering a given server. We have placed most of our files into the xrootd[5] cluster file system: it is more reliable, has much higher throughput and naturally spreads files across many servers. It also relieves us of the need to manage space on individual servers. We have used AFS as a repository for source code and scripts. We recently converted our dedicated AFS L1 I/O buffers (1.2 TB total) to xrootd: we no longer see the servers getting overloaded.

We have configured our processing system to allow automatic reruns of failed jobs. We still find that most failures are transient glitches which run fine the next time. With this "auto-rollback" feature, we find the failure rate in L1 to be about 0.03%. Finally, we noticed that sick

---

[2]http://www.platform.com/Products/platform-lsf

[3]http://www.nagios.org/

[4]http://ganglia.info/

[5]http://xrootd.slac.stanford.edu/

batch machines are magnets for failing jobs, since they fail quickly and hence are always available. We now monitor the machines for excessive resource usage and reboot them automatically when they become sick.

3. Issues and Lessons Learned

No large system is constructed and works without issues. Here we spotlight some of the hurdles we had to overcome and lessons learned along the way.

- Millions of log files overwhelming nfs file servers: the only solution so far is to keep a pool of servers and switch when they fill up.

- Keep finding new bottlenecks in processing: we have not been able to anticipate them all and have to address some on the fly (with delays until fixed). We still need more throttles in the pipeline system.

- Slow web pages: some of the pipeline task monitoring pages have become very slow with the millions of jobs run. We have no fixes yet, but we will probably have to scale back on what is quickly seen. The slow page loading times are getting painful.

- Monitoring the kitchen sink: 120k science quantities are being tracked (some different representations of the same detector quantities). These are in canned histograms and (eg) 90k numbers stored in Oracle every 5 mins for dynamic trending. This is much larger than we planned on and required us to significantly increase the Oracle server disk space.

- Ever expanding definition of "usable photon": as our understanding of LAT data deepens, more of it becomes useful. We are now storing 10× as many useable photons as planned. This has made resource planning tricky, but expansion trays for the Oracle servers have done the trick for now.

- PB of disk files: our storage model has been to keep latest the latest versions of files all on disk. xrootd provides a clustered file system with connection to hierarchical storage (tape). File load balancing and transparent hierarchical storage management (HSM) has required some effort on the client side.

- "Corporate Memory" and maintaining experts' interest: A thorny issue is how to keep a complex system humming in a stable experiment. We are leveraging several of the LAT tools for use by other experiments at SLAC, which should keep them under active development.

Acknowledgments. The Fermi LAT Collaboration acknowledges generous ongoing support from a number of agencies and institutes that have supported both the development and the operation of the LAT as well as scientific data analysis. These include the National Aeronautics and Space Administration and the Department of Energy in the United States, the Commissariat à l'Energie Atomique and the Center National de la Recherche Scientifique / Institut National de Physique Nucléaire et de Physique des Particules in France, the Agenzia Spaziale Italiana and the Istituto Nazionale di Fisica Nucleare in Italy, the Ministry of Education, Culture, Sports, Science and Technology (MEXT), High Energy Accelerator Research Organization (KEK) and Japan Aerospace Exploration Agency (JAXA) in Japan, and the K. A. Wallenberg Foundation, the Swedish Research Council and the Swedish National Space Board in Sweden.

Additional support for science analysis during the operations phase is gratefully acknowledged from the Istituto Nazionale di Astrofisica in Italy and the Center National d'Etudes Spatiales in France.

References

Atwood, W. B., et al. 2009, ApJ, 697, 1071

Flath, D. L., Johnson, T. S., Turri, M., & Heidenreich, K. A. 2009, in Astronomical Data Analysis Software and Systems XVIII, edited by D. A. Bohlender, D. Durand, & P. Dowler (San Francisco, CA: ASP), vol. 411 of ASP Conf. Ser., 193

Astronomical Data Analysis Software and Systems XX
ASP Conference Series, Vol. 442
Ian N. Evans, Alberto Accomazzi, Douglas J. Mink, and Arnold H. Rots, eds.
© *2011 Astronomical Society of the Pacific*

Recent Improvements to COS Calibration Pipeline

Philip E. Hodge

Space Telescope Science Institute, 3700 San Martin Dr, Baltimore, MD 21218

Abstract. The pipeline calibration program (calcos) for the Cosmic Origins Spectrograph (COS) was originally written based on requirements established in 2002. While the basic outline remains the same, much has been learned about the instrument since then, and extensive testing of calcos has helped reveal shortcomings (aka bugs). This paper describes some of the more significant changes, such as for wavelength calibration, keywords for the world coordinate system, and improving the data quality flags.

1. Introduction

COS is an ultraviolet spectrograph designed for very high sensitivity. Both the far ultraviolet (FUV) and near ultraviolet (NUV) channels use microchannel plate detectors, but with very different designs. The FUV channel has two 16k x 1k detectors ("segments") mounted end to end, and the spectrum runs along both. The NUV channel has a 1k × 1k MAMA detector, and three camera mirrors focus separate portions of the spectrum onto nearby regions of the detector. The preferred operating mode is TIME-TAG, i.e. the raw data file is a table of the arrival times and positions of detected photons. The output products include a corrected TIME-TAG events list (the corrtag file), detector images created by binning the corrtag table over pixels (the flt and counts files), and 1-D extracted spectra (the x1d file).

2. Wavelength Calibration

The offset of a spectrum from the nominal location is determined by taking a wavelength calibration spectrum (an exposure of a Pt-Ne lamp) and comparing it with a lamp template spectrum. The offset in each axis is then subtracted from the pixel coordinate of each event to move the image on the detector to its nominal location.

χ^2 is used to compare the wavelength calibration and lamp template spectra, but first the data must be scaled because the exposure times and background levels of the wavelength calibration spectrum and template spectrum may differ. These data are emission-line spectra, so there are sometimes many local minima of χ^2, and the lowest minimum is not always the correct one. The following algorithm is currently used: (1) find the 10 deepest local minima in the RMS difference between the wavelength calibration spectrum and the (scaled) template; (2) find and reject thos local minima which are just noise by fitting a quadratic to five points around each minimum and rejecting those that don't have positive curvature in the fit, (3) use χ^2 to select the best of the remaining minima.

For those cases where the algorithm doesn't give the correct shift, e.g. if the wavelength calibration spectrum is too faint or the background too high, the user can specify the shift via an input text file, and calcos will use those values. In order to do this, it must be possible for the user to independently determine the shift, e.g. from the science data.

3. WCS Keywords for Spectroscopic Data

Wavelengths for the 1-D extracted spectra are computed from polynomial expressions developed by the COS instrument team. These are written to the WAVELENGTH column in the x1d file and are the wavelengths that are intended to be used when one is working with the extracted spectra. The values in the WAVELENGTH column in the corrtag file are computed from the same polynomial expressions. Computation of these wavelengths is independent of the world coordinate system (WCS) header keywords described in this section. The WCS keywords are written to images (flt and counts files) and to corrected TIME-TAG tables (corrtag files) to make these files more useful to observers.

The COS aperture is 2.5 arcseconds in diameter, which is large compared with a stellar PSF. If the target has resolved spatial structure, each point in the spectrum will have three relevant coordinates: wavelength, right ascension, and declination. Since the detector image is only 2-D, there is an inherent ambiguity in a mapping from pixel coordinates to these world coordinates. In some cases, however, the mapping can be meaningful, for example if the target is extended and has narrow, separated emission lines, as in a planetary nebula. The spectrum would show a separate image of the nebula for each emission line. Within one such image (i.e. one wavelength), it should be possible to convert between pixel coordinates and celestial coordinates. Or if one were interested in the spectrum at a given point within the astronomical source, the celestial coordinates at that point would be specified, and then the wavelength could be computed at each pixel that the spectrum passes through. There would, however, be overlapping spectra from other parts of the source.

To support conversions between pixels and world coordinates, COS images (flt and counts files) now include the keywords for a 3-D world coordinate system (WCSAXES = 3). The corrtag files have only a 2-D WCS because that is what is supported for the FITS pixel list format; in this case the coordinates are wavelength and angle perpendicular to the dispersion. The FUV channel uses Rowland gratings, so the dispersion should be linear, so keyword CTYPE1 is 'WAVE'. This works well for the medium dispersion gratings, but the low resolution grating actually needs a quadratic term which is currently not included. The NUV channel uses flat gratings, so keyword CTYPE1 is set to 'WAVE-GRI' to indicate that the wavelengths should be computed from the grating equation. Other keywords give the reference pixel number, wavelength at the reference pixel, dispersion, groove density, spectral order, and the incident angle onto the grating. Since the NUV channel has three science spectra ("stripes") on the image, the FITS option of an alternate axis description is used to give a different coordinate mapping for each of the three spectra. The spectral stripes are called NUVA, NUVB, and NUVC, so the WCS keywords for those stripes end in the letters "A", "B", and "C" respectively. The primary WCS is for stripe NUVB. The FUV channel has just one science spectrum on the detector, so only the primary WCS is used.

4. DQ Array

There is a data quality (DQ) image extension in the flt and counts FITS files. This image is primarily used to flag regions within which the data should not be used, although some flags are just for information to the user. Figure 1 shows corresponding portions of the science array (left half of the figure) and data quality array (right half) of an NUV image. The left ends of the three science spectra (stripes NUVA, NUVB, NUVC) are shown, and one emission line from the wavelength calibration spectrum for NUVA is visible near the upper left corner. (The large, green, nearly rectangular area toward the left side of the DQ array flags a vignetted region. Since the vignetting is corrected via the flat field, data within this region are taken as valid and are flagged just for information.) Accurate alignment of the DQ array with the science data is essential when combining files taken at different offsets, so that regions flagged as bad in one file can be filled in with data from another. Two- or three-pixel offsets in the horizontal (dispersion) direction can be seen near the left edge of the DQ image. These correspond to the different offsets of the three science spectra from their positions in the raw data; the spectra

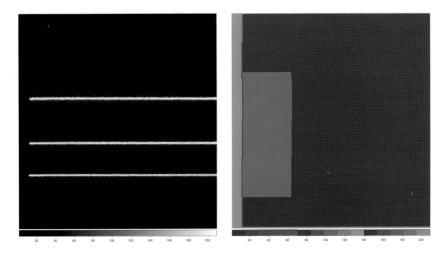

Figure 1.   Lower left portions of SCI and DQ extensions.

have been shifted by amounts determined during wavelength calibration. A recently fixed bug involved an error in the handling of these offsets at the edge of the detector, which resulted in artifacts that looked like narrow absorption lines in the combined spectrum. This observation (la9g01luq) of the white dwarf G-191B2B is used by permission of David J. Sahnow, Johns Hopkins University.

5.   hstcos Package in PyRAF

The hstcos package in PyRAF contains three tasks: calcos, splittag, and x1dcorr. calcos is an interface to the Python module calcos.py, which performs the COS pipeline calibration. The input to calcos can be either a raw file or a corrtag file.

The splittag task splits a corrtag file into multiple files, depending on the times of events in the input file. This can be used for a target that varied during the exposure, or to exclude a portion of the exposure, e.g. during orbital day because of airglow lines. The times may be uniformly spaced, specified by start time, increment, and end time; or an explicit list of arbitrary times may be given. The input to splittag should be a corrtag file rather than raw so that wavelength correction will already have been done. The raw data table will by default include both science spectrum and a wavelength calibration spectrum ("tagflash" mode), the latter with the lamp on for only one or a few short periods of time, so splitting the raw data would likely result in some output files without lamp flashes. The wavelength calibration will be more accurate if done using all the lamp flashes in the original data, and the corrtag file will have been so corrected.

The x1dcorr task extracts a 1-D spectrum, starting with a corrtag file as input. One could use this task on the output of splittag, for instance, or the data quality flags in a corrtag file might have been modified, perhaps to change the pulse height filtering. When x1dcorr is run, it will recreate the flt and counts files and extract the 1-D spectrum (x1d file). There is an option to specify the extraction height, but at the time of writing this is not fully supported because it will affect the flux calibration. The location of the target in the direction perpendicular to the dispersion may be taken from a reference file (this is the default), or the user may explicitly specify the location, or the task can find the location by collapsing the data along the dispersion direction and fitting a quadratic to the brightest pixels.

A task to combine spectra (x1d files) should be added to the hstcos package in the near future. Calcos does combine spectra for data in an association, but there are limitations (e.g. same grating and central wavelength) on what can be included in an association. Also, the user may have run splittag or modified the corrtag file and run x1dcorr; it would therefore not be possible to generate the same x1d file or files directly by running calcos. Having a separate task which will call the same function used by calcos will add useful flexibility.

Astronomical Data Analysis Software and Systems XX
ASP Conference Series, Vol. 442
Ian N. Evans, Alberto Accomazzi, Douglas J. Mink, and Arnold H. Rots, eds.
© *2011 Astronomical Society of the Pacific*

WFC3RED: A HST Wide Field Camera 3 Image Processing Pipeline

Daniel K. Magee,[1] Rychard J. Bouwens,[2] and Garth D. Illingworth[1]

[1]*University of California Observatories / Lick Observatory*

[2]*Leiden University*

Abstract. WFC3RED is a pipeline for automatically processing imaging data taken with the Wide Field Camera 3 instrument on the Hubble Space Telescope (HST). The pipeline currently supports processing of imaging data from both the IR and UVIS channels and is written in Python and C. The automated processing steps include cosmic-ray removal (UVIS), super-sky subtraction, user defined artifact masking, robust alignment and registration for large mosaics, weight map generation, and drizzling onto a final image mosaic. WFC3RED can combined data across different HST observations, visits and proposals without the need for any pre-defined associations. WFC3RED can create imaging products with a signal-to-noise ratio that matches the most careful step-by-step manual WFC3 reductions.

1. Introduction

The Wide Field Camera 3 (WFC3) was installed during HST Servicing Mission 4 (SM4) by the Space Shuttle astronauts in May 2009. After several months of on-orbit verification and other preparatory work, WFC3 began science operations late in the summer of 2009. Some of the first observations to be taken were for the Early Release Science (ERS) and the Hubble Ultra Deep Field 2009 (HUDF09) programs. The HUDF09 project was awarded 192 orbits of HST WFC3/IR observations over three fields in the GOODS South field with deep optical HST ACS observations. More than half of the WFC3/IR orbits were acquired over the Hubble Ultra Deep Field ACS pointing (Figure 1).

In order to quickly process the large amounts of imaging data provided by the HUDF09 and ERS programs we developed an automatic image reduction pipeline required for processing all HST WFC3 observations over these four areas of the sky. The HUDF09 project pipeline-processed data will made publicly available through the Multi-Mission Archive at STScI (MAST) as high level science products.

2. Pipeline Modules

The WFC3RED pipeline includes eight processing steps from basic calibration to generating final co-added registered mosaics. See Table 1 for a short description of each module.

3. Running WFC3RED

As input, the WFC3RED pipeline requires the full set of calibrated data products and best reference files for each observation in the input image set. These files can be readily obtained through the MAST HST archive. For most image sets, WFC3RED can be run in the default

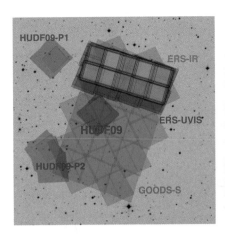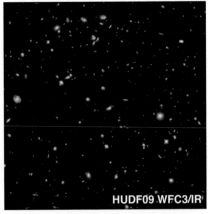

Figure 1.   *Left*: The three HUDF09 WFC3/IR fields in red, the ERS WFC3/IR and UVIS in orange and purple and the GOODS South field in green. *Right*: A color composite image of the HUDF09 WFC3/IR data. The HUDF09 WFC3/IR image is the deepest image of the universe ever taken in the near infrared.

Table 1.   A listing of WFC3RED modules

WFC3RED Modules	
setup	Ingest raw data and builds a SQLite database containing fits header data
medsub	Removes residual instrument signatures by subtracting a median stacked *super-sky* image
flatten	Uses bicubic spline to flatten the background
definemask	Apply user-defined masks (optional)
align	Uses object matching algorithms to improve image alignment and registration
weightmap	Creates accurate rms maps for use with MultiDrizzle
mdrizzle	Creates final CR-cleaned, distortion-free drizzled image mosaics using MultiDrizzle
refineshift	Refines the final output image WCS (if needed)

configuration with no additional input from the user. However, if the default settings are insufficient for processing a particular data set, there are a number of run-time options which can be applied to help improve the reductions:

- Modules can be run step-by-step (i.e., manually) allowing for the inspection of the output at each step,

- Modules can be skipped,

- A pipeline run can be stopped, restarted or rerun at any stage of the reduction after the initial setup,

- External reference images can be used to improve the internal alignment of the reduced WFC3 frames,

- Single image CR cleaning data before alignment (UVIS only),
- Group and combine observations by visit for CR cleaning then align (UVIS only).

```
Usage: wfc3red.py run_name flt_in_dir

Options:
  --version             show program's version number and exit
  -h, --help            show this help message and exit
  --setup               run setup only
  --definemask          interupt to run definemask
  --flatten             run flatten only
  --medsub              run medsub only
  --align               run align only
  --weightmap           run weightmap only
  --mdrizzle            run mdrizzle only
  --refineshift         run refineshift only
  -a ALIGNREF, --alignref=ALIGNREF
                        external reference image used for alignment (absolute
                        path)
  -n MODULE, --rerun=MODULE
                        rerun wfc3red from MODULE on
  -r, --restart         restart wfc3red
  -s SKIP, --skip=SKIP  Skip running module/s
  -t TARGNAME, --targname=TARGNAME
                        run wfc3red only on images with name TARGNAME
  -f FILTER, --filter=FILTER
                        run wfc3red only on images with filter name FILTER
  --pixel_scale=PIXEL_SCALE
                        pixel scale (in arcsec/pixel) for final drizzled
                        images (default=0.06 for IR 0.0396 for UVIS)
  --pixel_frac=PIXEL_FRAC
                        pixel fraction for final drizzled images (default=0.7)
  --drz_kernel=KERNEL   drizzel kernel for final drizzled images
                        (default=square)
  --nomasking           turn off masking
  --noskysub            turn off multidrizzle sky subtraction
  --nodrzcr             turn off multidrizzle cosmic ray rejection
  --brightobj           take several actions to better cope with bright
                        objects
  -d DRZREF, --drzref=DRZREF
                        external reference image used for final drizzle
                        (absolute path)
  --tshift=TSHIFT       external reference shift file used for final drizzle
                        (absolute path)
  -c, --crclean         cosmic ray clean UVIS images before running align
                        (helpful for UV data)
  -e, --sparse          skip align and cosmic ray clean UVIS images by visit
                        (helpful for UV data)
```

3.1. Masking Artifacts

User defined mask can be generated with the help of SAOImage DS9:

- Images are displayed in DS9,

- The user marks artifacts with DS9 polygon region tool,

- A script is run that saves the DS9 region files for each image that has a marked artifact,

- A second script is run that applies the masks region in each region file to the associated images data quality.

The definemask module is useful for masking artifacts such as satellite trails and unwanted features due to persistence afterglow in pixels that have been saturated in earlier exposures.

Acknowledgments.   We would like to thank NASA and the Space Telescope Science Institute for their support with this project.

Astronomical Data Analysis Software and Systems XX
ASP Conference Series, Vol. 442
Ian N. Evans, Alberto Accomazzi, Douglas J. Mink, and Arnold H. Rots, eds.
© *2011 Astronomical Society of the Pacific*

The Data Handling and Analysis Software for the Mid-Infrared Instrument (MIRI) on JWST

Jane Morrison

Steward Observatory, University of Arizona, Tucson, Arizona

Abstract. The Mid-Infrared Instrument on the James Webb Space Telescope (JWST) provides imaging, coronography, and integral field spectroscopy over the 5–28 μm wavelength range. The camera module provides wide-field broadband imagery. The Medium Resolution Spectrograph (MRS) obtains simultaneous spectral and spatial data on a relatively compact region of the sky using four integral field units. MIRI is jointly developed by the US and a nationally funded consortium of European institutes, working with ESA and NASA. The MIRI Data Handling and Analysis Software (DHAS) provides a software package to support the MIRI Focal Plane Electronics testing at Jet Propulsions Lab and to support the MIRI assembly, integration, and verification at Rutherford Appleton Lab in the UK. It will also be used to support the the JWST Telescope integration and verification at Goddard Space Flight Center. In addition, it is being used to test various processing algorithms and will be delivered to Space Telescope Science Institute with the recommended procedure for processing MIRI data based on pre-flight data.

1. MIRI Overview

MIRI is one of four instruments being built for JWST (Wright et al. 2008) The MIRI imager, coronograph, and low-resolution (R ~ 100) grism spectrometer are combined onto one focal plane. The imaging module uses a 1024×1024 pixel arsenic-doped silicon sensor chip assembly (SCA). The MRS splits the 5 to 28 μm wavelength range into four channels. An Integral Field Unit (IFU) then slices the rectangular field of view into several slices. Each channel has its own IFU. The image slices are then dispersed and imaged onto one of two MRS focal planes. Since each channel uses a set of three grating, the entire wavelength range can be split in 12-sub channels. Each of the MRS focal planes, like the imager, uses a 1024×1024 pixel As:Si SCA. On each SCA there are four additional reference pixels at the beginning of each row and four at the end of each row that are unconnected "pixels" with no light sensitivity. All these pixels (including the reference pixels) are read out through four data outputs. There is an additional "reference output" that is essentially a clocked DC voltage source. The reference output is read out with a 5th output line. The read out of the reference output line will occur simultaneously with the four readout lines (Ressler et al. 2008). This reference output and additional reference pixels on each rows may be used when processing the data to reduced the noise.

The basic readout mode for the MIRI detectors is a simple sample-up-the-ramp pattern (see Figure 1). The pixels are continuously addressed at a time interval of 10 micro-seconds. For MIRI a "frame" is defined as a single clocking scan through the

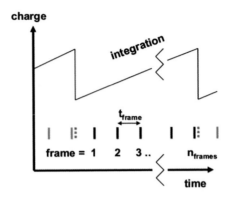

Figure 1.   Detector readout pattern for MIRI: sample-up-the-ramp.

array. While an "integration" is the number of non-destructive readout frames where photons are are allowed to integrate (i.e. there are no resets). The integration length is an integer number of frames times the time required to read a frame.

2.   MIRI DHAS Software

The MIRI DHAS software is comprised of two-components: a C++ processing pipeline and an IDL viewing and analysis tool set. The DHAS operates in non-real time on FITS-formatted science data captured earlier over one or more exposures. An overview of the MIRI DHAS software in given by the flowchart in Figure 2.

The DHAS C++ Pipeline consists of three programs. The first routine, miri_sloper, converts the measured charge sample-up-the-ramps into slope values for each pixel. This program can apply a bad pixel mask to the data by assigning a NaN to the resulting slope. Initial or final frames and saturated data can be rejected from the slope fit. In addition using an algorithm based on the difference between a pixel's adjacent frames, cosmic-rays and noise spikes can also be rejected. There a various schemes to use the reference pixels and reference output to assist in reducing noise. The program also applies a non-linearity correction.

The second routine in the pipeline suite, miri_caler, takes the output from miri_-sloper and applies a dark or flat calibration image to the data. If the data is from the MRS a pixel fringe flat can be applied to reduce/eliminate the fringing. The third routine, miri_cube, is limited to MRS data and takes MRS data that has been run through miri_sloper and/or miri_caler and creates a spectral cube. The optical distortion in the MRS image slicing and dispersive optics leads to distortions in the spatial and spectral information on the detector plane (Glauser et al. 2010). A set of transformation maps which corrects for the distortions are used in the cube building software to map each detector pixel onto a pseudo sky cube (see Figure 3 A.) The overlap area of this mapped pixel onto the various cube pixels is determined using the Sutherland-Hodgman polygon clipping algorithm (Sutherland & Hodgman 1974, Figure 3 B). This overlap area is used as a weight in determining flux contained in sky cube pixel.

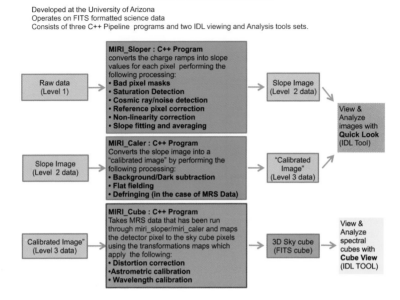

Figure 2.   Flow Chart of MIRI DHAS Software tool.

The transformation maps are determined using the best currently available astrometric and wavelength calibrations. The strategy for obtaining these transformation maps depends on the phase of the project. During the Verification Model testing the ray-tracing software ZEMAX was used to produce the transformation maps. During Flight Model Testing (occurring in the Spring of 2011) we are planning an end-to-end measurement of the instrumental distortions.

The Viewing and Analysis tool set is a group of IDL software routines for displaying and analyzing the MIRI science frames, reduced images and spectral cubes. This tool set is comprised of two packages: the Quick-Look (ql) tool to display and analyze the science frames and reduced data and the Cube-View (cv) tool for displaying and analyzing spectral cubes.

3.   Conclusion

With the MIRI DHAS software we have been able to analyze image quality, characterize the PSF, test various methods to process the data, characterize the detectors, and test the astrometric calibration of the spectrograph. The combination of the miri_sloper and the QL have allowed us to test different processing algorithms to find the optimal one.

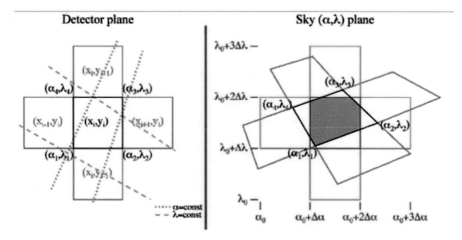

Figure 3.   A (left side): A diagram showing the mapping of detector pixels (x,y) to the sky cube coordinates (α, β) using the transformation maps. B (right side): The shaded area is the overlap between the detector pixel and the central cube pixel.

Acknowledgments.   The author thanks Adrian Glauser for his discussions with designing and testing the cube building software, as well as, creating the detector to cube transformation maps. The author also thanks the Mike Ressler, Alistair Glasse, Tim Grundy, Scott Friedman and Karl Gordon who gave valuable suggestions on improving th DHAS software.

References

Glauser, A. M., Glasse, A., Morrison, J. E., Kelly, B. D., Wells, M., Lahuis, F., & Wright, G. S. 2010, in Space Telescopes and Instrumentation 2010: Optical, Infrared, and Millimeter Wave, edited by J. M. Oschmann Jr., M. C. Clampin, & H. A. MacEwen (Bellingham, WA: SPIE), vol. 7731 of Proc. SPIE, 77313K

Ressler, M. E., et al. 2008, in High Energy, Optical, and Infrared Detectors for Astronomy III, edited by D. A. Dorn, & A. D. Holland (Bellingham, WA: SPIE), vol. 7021 of Proc. SPIE, 70210O

Sutherland, I. E., & Hodgman, G. W. 1974, Commun. ACM, 32

Wright, G. S., et al. 2008, in Space Telescopes and Instrumentation 2008: Optical, Infrared, and Millimeter, edited by J. M. Oschmann Jr., M. W. M. de Graauw, & H. A. MacEwen (Bellingham, WA: SPIE), vol. 7010 of Proc. SPIE, 70100T

Part IX

Scientific Computing

Astronomical Data Analysis Software and Systems XX
ASP Conference Series, Vol. 442
Ian N. Evans, Alberto Accomazzi, Douglas J. Mink, and Arnold H. Rots, eds.
© *2011 Astronomical Society of the Pacific*

Amdahl's Laws and Extreme Data-Intensive Scientific Computing

Alexander S. Szalay

Dept. of Physics and Astronomy, The Johns Hopkins University, Baltimore, MD 21218

Abstract.   Scientific computing is increasingly revolving around massive amounts of data. In astronomy, observations and numerical simulations are on the verge of generating many Petabytes of data. This new, data-centric computing requires a new look at computing architectures and strategies. The talk will revisit Amdahl's Law establishing the relation between CPU and I/O in a balanced computer system, and use this to analyze current computing architectures and workloads. We will discuss how existing hardware can be used to build systems that are much closer to an ideal Amdahl machine. Scaling existing architectures to the yearly doubling of data will soon require excessive amounts of electrical power. We have deployed various scientific test cases, mostly drawn from astronomy, over different architectures and compare performance and scaling laws. We discuss an inexpensive, yet high performance multi-petabyte system currently under construction at JHU.

1. Introduction

Modern science is approaching the point where novel computational algorithms and tools, combined with computational thinking, will become as indispensable as mathematics. The fact that scientific data sets across a wide range of fields are doubling in size every year and thus create complex challenges for traditional analysis techniques is an important driver of this trend. Analyses of the information contained within these data sets have already led to revolutionary breakthroughs, in fields ranging from Genomics to High Energy Physics, encompassing every scale of the physical world. Much more remains as-yet-undiscovered.

The availability of large experimental datasets coupled with the potential to analyze them computationally is changing the way we do science (Szalay & Gray 2001). In many cases however, our ability to acquire experimental data outpaces our ability to process them leading to the so-called data deluge (Bell et al. 2009). This data deluge is the outcome of three converging trends: the recent availability of high throughput instruments (e.g., telescopes, high-energy particle accelerators, gene sequencing machines), increasingly larger disks to store the measurements, and ever faster CPUs to process them.

Not only experimental data are growing at a rapid pace; the volume of data produced by computer simulations, used in virtually all scientific disciplines today, is increasing at an even faster rate. The reason is that intermediate simulation steps must also be preserved for future reuse as they represent substantial computational investments. The sheer volume of these datasets is only one of the challenges that scientists must confront. Data analyses in other disciplines (e.g., environmental sciences) must span thousands of distinct datasets with incompatible formats and inconsistent meta-data. Overall, dataset sizes follow a power law distribution and challenges abound at both extremes of this distribution.

Astronomy provides a particularly good example of this trend. Large, uniform sky surveys, like the Sloan Digital Sky Survey (SDSS) have completely transformed the way we think about astronomy. More and more exciting and novel science can be undertaken by "observing" the Universe through a new kind of instrumental facility, namely the publicly available large database which contains tens of terabytes of well-calibrated and well-documented data. Astronomers can now go from the formulation of an idea to its detailed analysis in a matter of days, compared to the standard year-long (or more) cycle of submission of proposal for telescope time, acceptance (not always, due to high over-subscription), observation (good weather permitting), analysis, etc. The idea of the Virtual Observatory is compelling: we should be able to integrate all the astronomy data in the world into a single, federated system.[1,2]

Adapting to this new data-intensive paradigm is far from trivial. As data sets become larger, many aspects of their analysis become increasingly harder. Even previously straightforward tasks, like extracting the data from databases and downloading them to individual researchers can be challenging, since there is not enough bandwidth to a typical university desktop to move tens to hundreds of terabytes at will. Cutting-edge science requires full, unlimited access to these refined, high-resolution data sets. Even simulations, used to compare the experimental results to first-principle, theory-driven models, suffer from our inability to easily move the data — the largest simulations are typically analyzed "on-the-fly", by the scientists creating them, with only a small fraction of the output retained for later analyses. As simulations become larger, it becomes impractical to immediately rerun them, should the limited stored snapshots of the simulation hint at new, unexpected results.

Large experimental or observational data sets often reside in external databases, far from the computers where the simulations are performed, thus further complicating a joint analysis. At the same time, revolutionary science increasingly requires combined, complex comparisons. We are at a point where the whole science community understands the magnitude of the problem but lacks a clear-cut, solution template. We have yet to come up with a new, innovative approach to handle such challenges.

2. Scientific Data Analysis Today

While improvements in computer hardware have enabled this data explosion, the performance of different architecture components increases at different rates. CPU performance has been doubling every 18 months, following Moore's Law. The capacity of disk drives is doubling at a similar rate, somewhat slower that the original Kryder's Law prediction, driven by higher density platters. On the other hand, the disks' rotational speed has changed little over the last ten years. The result of this divergence is that while sequential IO speeds increase with density, random IO speeds have changed only moderately. Due to the increasing difference between the sequential and random IO speeds of our disks, only sequential disk access is possible — if a 100TB compu-

---

[1]Virtual Astronomical Observatory, `http://usvao.org/`

[2]International Virtual Observatory Alliance, `http://ivoa.org/`

tational problem requires mostly random access patterns, it cannot be done. Finally, network speeds, even in the data center, are unable to keep up with the doubling of the data sizes. Said differently, with petabytes of data we cannot move the data where the computing is — instead we must bring the computing to the data.

The typical analysis pipeline of a data-intensive scientific problem starts with a low level data access pattern during which outliers are filtered out, aggregates are collected, or a subset of the data is selected based on custom criteria. The more CPU-intensive parts of the analysis happen during subsequent passes. Such analyses are currently implemented in academic Beowulf clusters that combine compute-intensive but storage-poor servers with network attached storage. These clusters, often placed to rooms resembling "broom closets", can handle problems of a few tens of terabytes, but they do not scale above hundred terabytes, constrained by the very high costs of PB-scale enterprise storage systems. Furthermore, as we grow these traditional systems to meet our data needs, we are hitting a "power wall" (Szalay et al. 2010), where the power and space requirements for these systems exceed what is available to individual PIs and small research groups.

Existing supercomputers are not well suited for data intensive computations either; they maximize CPU cycles, but lack IO bandwidth to the mass storage layer. Moreover, most supercomputers lack disk space adequate to store PB-size datasets over multi-month periods. Finally, commercial cloud computing platforms are not the answer, at least today. The data movement and access fees are excessive compared to purchasing physical disks, the IO performance they offer is substantially lower (20MBps), and the amount of provided disk space is woefully inadequate (e.g. 10GB per Azure instance).

Based on these observations, we posit that there is a vacuum today in data-intensive scientific computations, similar to the one that lead to the development of the BeoWulf cluster: an inexpensive yet efficient template for data intensive computing in academic environments based on commodity components. This situation is not scalable, and not maintainable in the long run.

As data sets are growing at or faster than Moore's Law, outpacing the speedup of our computers, it becomes increasingly harder to tackle computationally challenging data analyses. Imagine that we have an algorithm whose execution time is quadratic in the number of data points. As our data doubles by next year, we need to perform four times as much CPU time on the analysis, while our computers are only a factor of two faster. Soon, we will only be able to run analyses which scale linearly with the number of data points (N), or at worst scale as $N \log N$. We can make up for the $\log N$ factor through parallelism.

2.1. How Long Does the Data Growth Continue?

Once we build the world's largest detector, and turn it on, it will generate data at a constant rate, with the aggregate data volume growing at a linear rate. This is happening in High Energy Physics, with the Large Hadron Collider, where even the detectors are enormous and cost hundreds of millions of dollars. So, the experimental data volume in HEP is not going to double every year.

In astronomy we see old telescopes getting new instruments. The Dark Energy Survey (DES) is building a huge mosaic to be placed on an older telescope in Chile. In genomics, the generated data volume is growing whenever a new, higher resolution CCD camera is placed in the high throughput sequencers. Our satellites have ever higher resolution, and more pixels. Inexpensive digital cameras have new generations coming out every six months. In this world, successive generations of exponentially more capable sensors at the same cost are the reason for the data explosion. All of this can be traced back to semiconductor technology and ultimately, to Moore's Law.

However, not every domain of science has such growth areas. One could argue that optical astronomy will soon reach the point when increasing CCD mosaic sizes will become impractical, and the atmospheric resolution will constrain the reasonable pixel size, causing a slow-down our data collection. But time domain astronomy is emerging, and by taking images every 15 seconds, even a single telescope (LSST) can easily generate data that can reach 100 petabytes in a decade.

New instruments and new communities will also emerge. Radio astronomy, with focal plane arrays on the horizon, is likely to undergo a paradigm shift in data collection, resembling the time when CCDs replaced photographic plates. Amateur astronomers already have quite large, cooled CCD cameras. When a community of 100,000 people will start collecting high resolution, sensitive images, the aggregate data may easily outgrow the professional astronomy community.

Human behavior on the Internet always obeyed the "long tail", originally noted by Pareto, as the 80-20 rule. On the Internet today, 90% of the people only look at 10% of the pages. A similar rule applies to scientific data: 90% of scientists only look at approximately 10% of the data. Hierarchical data replication of very large data sets, following the preferences of the 90 percent of the user community is a very reasonable way to optimize how we spend our scarce resources. This principle is already adopted by the HEP community, who organize their data into a multi-tiered system. CERN contains the Tier0 data, the Tier1 centers replicate data by experiment, the Tier2 and Tier3 centers by analysis project. This also provides a healthy impedance match between the many petabytes of raw data and the thousands of users.

2.2. Numerical Simulations

Numerical simulations are becoming another new way of generating enormous amounts of data. This has not always been the case. Traditionally, gravitational N-body simulations have been analyzed while the simulation was running, since checkpointing and saving the snapshots was overly expensive. This fact has substantially limited the widespread use of simulations — an astronomer had to have access to a high end parallel computer to use these data sets.

Even when a few snapshots have been saved and made public, downloading large files over slow network connections made the analysis highly impractical, once the simulations reached the terabyte range. The Millennium simulation (Springel et al. 2005) has changed all that by creating a remotely accessible database with a collaborative environment (Lemson & the Virgo Consortium 2006), following the example of the SDSS SkyServer. The Millennium database drew hundreds, if not thousands of astronomers into analyzing simulations as easily as if it were public observational data.

The emerging challenge in this area is scalability. The Millennium has 10 billion particles. The raw data is about 30 terabytes, but the database does not contain the individual dark matter particles, only the halos, subhalos and the derived galaxies. Newer simulations are soon going to have a trillion particles, where every snapshot is tens of terabytes, so the data problem becomes much worse. At the same time, there is an increasing demand by the public to get access to these best and largest simulations. It is inevitable that the Millennium database model is going to proliferate. We need a Virtual

Observatory of the Virtual Universe, that can provide adequate access and the ability to do analysis, visualization and computations of these large simulations remotely.

As data become increasingly unmovable, the only way to analyze them is "in place". We need new mechanisms to interact with these large simulations — we cannot simply download the raw files. For interactive visualizations, it will be easier to send a high-definition 3D video stream to every astronomer in the world than moving even a single snapshot of a trillion particle simulation from one place to another.

In a project related to isotropic turbulence (Li et al. 2008) we have taken a 1024^3 simulation, with a Reynolds number of 470. The simulation output is over a regular grid where every point contains the three components of the fluid velocity and the pressure. The data is partitioned along a space filling curve (z-index) into cubes of 64^3, as a 6MB Binary Large Object (BLOB), stored in a separate row of the database.

The data are accessed via a web service where users can submit a set of about 10,000 particle positions and times and then can retrieve the interpolated values of the velocity field at those positions. This can be considered as the equivalent of placing small sensors into the simulation instead of downloading all the data or significant subsets of it. This service is public and is typically delivering about 108 particles per day world-wide.[3] Several papers appearing in the top journals (Phys. Rev. Letters, etc.) have used this facility. Currently we are adding a 70TB simulation of a magneto-hydrodynamic system and next year we will add a 50TB simulation of a channel flow. We are currently experimenting with different blob sizes, overlap regions and partitioning schemes across servers. Visualization services are around the corner and we are also considering enabling users to easily grab a sub-domain of the data.

How can this "immersive" paradigm be used in astronomy? One of the most interesting discoveries over the last decade was the "tidal streams" in the Milky Way Galaxy, found in the SDSS data (Belokurov et al. 2006). Would it not be nice, if we had a dynamic simulation of the Milky Way, stored at a high temporal resolution in a database, and we could interactively "shoot" dwarf galaxies on different trajectories at it, and watch them as they are disrupted. Or, what if we could plug in our detailed particle physics prescription for the energy dependence of the dark matter annihilation cross section and see the emissivity map for a Milky Way-like galaxy.

The "Silver River" simulation, currently running at the Oak Ridge Jaguar system, involves 50B dark matter particles in a 5 Mpc/h box containing a Milky Way-like galaxy. By storing more than 800 snapshots, the resulting uncompressed raw data will exceed 2.6PB. Using today's technologies, it is not impossible to turn this into an interactive public experiment. In collaboration with the Silver River team we have proposed such an experiment, the Milky Way Laboratory, where users would be able to launch test particles into the precomputed simulation. We will see more of such numerical experiments, creating new kinds of immersive analysis templates to deal with extreme amounts of data.

2.3. Typical Data-Intensive Scientific Workloads

Over the last few years we have implemented several eScience applications, in experimental data-intensive physical sciences applications such as astronomy, oceanography and water resources. We have been monitoring the usage and the typical workloads

---

[3]Turbulence simulation services website, `http://turbulence.idies.jhu.edu/`

corresponding to different types of users. When analyzing the workload on the publicly available multi-terabyte Sloan Digital Survey SkyServer database (Gray et al. 2002), it was found that most user metrics have a 1/f power law distribution (Singh et al. 2007).

Of the several hundred million data accesses most queries were very simple, single row lookups in the data set, which heavily used indices such as on position over the celestial sphere (nearest object queries). These made up the high frequency, low volume part of the power law distribution. On the other end there were analyses which did not map very well on any of the precomputed indices, thus the system had to perform a sequential scan, often combined with a merge join. These often took over an hour to scan through the multi-terabyte database. In order to submit a long query, users had to register with an email address, while the short accesses were anonymous.

We have noticed a pattern in-between these two types of accesses. Long, sequential accesses to the data were broken up into small, templated queries, typically implemented by a simple client-side Python script, submitted once in every 10 seconds. These "crawlers" had the advantage of returning data quickly, and in small buckets. If the inspection of the first few buckets hinted at an incorrect request (in the science sense), the users could terminate the queries without having to wait too long. The "power users" have adopted a different pattern. Their analyses involved complex, multi-step workflows, where the end result was approached in a multi-step, hit-and-miss fashion. Once the workflow was finalized, they executed it over the whole data set, by submitting a large job into a batch queue.

Most scientific analyses are done in a exploratory fashion, where "everything goes", and few predefined patterns apply. Users typically want to experiment, try innovative things that often do not fit preconceived notions, and would like to get rapid feedback on the momentary approach.

2.4.   Computations Inside the Database

Many of the typical data access patterns in science require a first, rapid pass through the data, with relatively few CPU cycles carried out on each byte. These involve filtering by a simple search pattern, or computing a statistical aggregate. Such operations are quite naturally performed within a relational database, and expressed in SQL. So a traditional relational database fits these patterns extremely well.

The picture gets more complicated when one needs to run more complex algorithms on the data, not necessarily easily expressed in a declarative language. Examples of such applications can include complex geospatial queries, processing time series data, or running the BLAST algorithm for gene sequence matching.

The traditional approach of bringing the data to where there is an analysis facility is inherently not scalable, once the data sizes exceed a terabyte, due to network bandwidth, latency, and cost. It has been suggested (Szalay & Gray 2006), that the best approach is to bring the analysis to the data. If the data are stored in a relational database, nothing is closer to the data than the CPU of the database server. It is quite easy today with most relational database systems to import procedural (or object oriented) code and expose their methods as user defined functions within the query.

This approach has proved to be very successful in many of our reference applications, and while writing class libraries linked against SQL was not always the easiest coding paradigm, its excellent performance made the coding effort worthwhile.

3. Data-Intensive Computing

3.1. Amdahl's Laws

Amdahl has established several laws for building a balanced computer system (Amdahl 2007). These were reviewed recently (Bell et al. 2006) in the context of the explosion of data. The paper pointed out that contemporary computer systems IO subsystems are lagging CPU cycles. In the discussion below we will be concerned with two of Amdahl's Laws:

- A balanced system needs one bit of IO for each CPU cycle,
- A balanced system has 1 byte of memory for each CPU cycle.

These laws enumerate a rather obvious statement — in order to perform continued generic computations, we need to be able to deliver data to the CPU, through the memory. Amdahl observed that these ratios need to be close to unity and this need has stayed relatively constant.

Table 1.   Amdahl number of different platforms

System	CPU [count]	GIPS [GHz]	diskIO [MBps]	Amdahl IO
BeoWulf	100	300	3,000	0.08
Desktop	2	6	150	0.20
Cloud VM	1	3	30	0.08
SC1	213,000	150,000	16,900	0.001
SC2	2090	5000	4,700	0.008
GrayWulf	416	1107	70,000	0.506

One might ask, why would such a number, based on computer performance metrics originating in the 60's be relevant today? In the eighties and nineties, as the cache hierarchies have increased system performance enormously, people argued against the relevance of Amdahl's Laws. However, once we have to scan through petabytes of data, orders of magnitude beyond the size of the main memory in the system, in order to look at the data, we have to move the data from disk to CPU.

Thus, for large data sets the only way we can even hope to accomplish the analysis if we follow a maximally sequential read pattern. The sequential IO rate has grown somewhat faster as the density of the disks has increased by the square root of disk capacity. For commodity SATA drives the sequential IO is typically 60MBps, compared with 20MBps 10 years ago. Nevertheless, compared to the increase of the data volumes and the CPU speedups, this increase is not fast enough to conduct business as usual. Just loading a terabyte at this rate takes 4.5 hours. Given this sequential bottleneck, the only way to increase the disk throughput of the system is to add more and more disk drives and to eliminate obvious bottlenecks in the rest of the system. Excessive CPU capacity, beyond what is needed to have a quick look at the data will be wasted, and just burns excessive power. Thus Amdahl's Laws are reemerging today as an extremely relevant metric for efficient data intensive computing. In the next paragraphs we look at several architectures for scientific High Performance Computing and calculate their Amdahl number for comparison. The Amdahl IO number is computed by dividing the aggregate sequential IO speed of the system in Gbits/sec by the GIPS (Giga Instructions Per Second) value. A ratio close to 1 indicates a balanced system in the Amdahl sense.

We consider first a typical BeoWulf cluster, with 50 3GHz dual-core nodes, each with one SATA disk with 60MBps, and compare it to a typical desktop used by the average scientist. Today such a machine has 2 CPUs, and an aggregate sequential IO of 150MB/sec. A virtual machine in a commercial cloud would have a single CPU, but a lower IO speed of about 30MB/sec per instance.

Let us consider two hypothetical supercomputers. An approximate configuration "SC1" for a typical BlueGene-like machine was obtained from the web. The sequential IO performance of an IO-optimized BlueGene/L configuration with 256 IO nodes has been measured to reach 2.6 GB/sec peak. A simple scaling to the 1664 IO nodes in the selected system gives us the nominal 16.9 GB/sec used in the table. The other hypothetical supercomputer, "SC2," has been modeled on a Cray XT3-like system. We have also attempted to get accurate numbers from several of the large cloud computing companies — our efforts have not been successful, unfortunately. The Graywulf IO numbers have been measured as part of our entry for the SC-08 Data Challenge, based upon sequential IO performance during typical DB workloads.

The Amdahl number can also be defined for data sets. for a given simulation one can take the ratio of final data size divided by the compute cycles needed to create the simulation. For database queries, one can estimate the size of the data that had to be queried to deliver the result, and the CPU time for producing the result set. When we do this analysis, we find an interesting trend. The largest simulations, like Aquarius and Via Lactea-2 have Amdahl numbers between 10^{-5} and 10^{-4}, several hundred times lower than even the otherwise relatively slow IO of the supercomputers allow. BeoWulf-class computations, like typical astronomical image processing applications are a factor of 50–100 lower in their Amdahl number than the hardware would allow. Database queries at the same time seem to use as much IO as the hardware allows.

What is the reason behind this trend? Once we plot the total data set sizes involved, a very obvious fact emerges: all the data sets involved are in the range of 20–100TB. In hindsight, if the large simulations used a 100 times more IO, their data sets would have been in the 2PB range. It is the lack of disk space that primarily limits the full utilization of these large simulations. The magic 100TB seems to be the practical limit today for these data-intensive scientific computations.

4.   Extreme Data-Intensive Computing

4.1.   DISC Needs Today

Reflecting on the conclusion from the previous section, the primary need of Data-Intensive Scientific Computing (DISC) is disk space, everything else is secondary. While we use supercomputers costing several hundred million dollars, it is interesting to consider that our computations are often limited by the lack of an other 100TB of disks, which can be purchased for less than $10,000. Extreme computing is about tradeoffs. When we build systems with performance characteristics that focus on certain parameters, we do extreme computing. It is also clear that for HPC there are different criteria than for commercial cloud computing. Below we attempt to define a ranked list of priorities for data-intensive scientific computing:

- Total amount of storage,
- Total cost,
- Sequential IO performance,

- Stream processing capability,
- Low power.

The total amount of storage is critical. Also, if we want some level of redundancy, data must be stored more than once. At the petabyte scales, traditional RAID systems do not seem to be performing very well, and the large commercial cloud-storage companies are experimenting with redundant BLOB stores. The biggest factor in the total cost is the ratio of the system price over the cost of the raw disks. Most commercial storage vendors sell enterprise class drives with their systems, often at five times higher cost than a commodity drive. Yet, there is little evidence that these drives have a much longer lifetime or much lower failure rate. Furthermore, these days any computer system has a practical 5 year lifecycle. If a commodity drive comes with a 5 year warranty, there is little sense to try to extend its life beyond that.

There is an interesting discussion on the Internet about the cost of a petabyte.[4] In 2009, when the website was built, buying a PB of commodity drives cost $81,000. Putting these into home-brewed servers resulted in a total cost of $117,000. A PB set of servers from different commercial vendors had prices between $826K–$1.7M. Putting this much data in the cloud is $2.8M. Buying a high-end, redundant, hierarchical storage system was also in the range of $2.8M. This shows why assembling custom systems is still a viable alternative today. Disk prices have fallen since; at the time of this paper, the street price of 1PB of disks is $40K. High end storage systems, if anything, have become even more expensive in the mean time.

We have already discussed the importance of sequential IO. The implication for system design is that the best and cheapest way to get good sequential IO speeds is through locally attached disks instead of network drives. Once we can read the data fast, do the first pass and create a statistical sample, we usually have to do a much heavier computation on the resulting data stream. For CPU heavy applications scientists are increasingly turning to GPUs, often enabling two orders of magnitude speedups in applications that can be run in parallel. However, the bottleneck of data-intensive computations on GPUs lies in the difficulty of getting the data into the GPUs. A GPU cluster connected through a slow network fabric is not going to solve the problem — we need the GPUs also local to the data servers for high enough throughput.

Amdahl's Laws tell us that in order to control the basic low level data storage servers, it is enough to use much fewer CPU cycles than those of a typical BeoWulf, maybe even by an order of magnitude. Indeed, our experiments have shown that low-power motherboards, based on the Intel Atom processor and containing GPUs for HDTV rendering, can provide a very attractive solution. Our small cluster of 36 Atom boards has reached a sequential read performance of 18 GBps, at a power consumption of 1kW, with 70TB of disks.

We see an increasing diversification in high end scientific computing today — one shoe does not fit all! This diversity is emerging naturally, as scientists are designing and tuning systems to get optimal performance for their shoestring budgets and huge computational problems. We see large floating point calculations moving to GPUs, large data moving into high-density storage servers with locally attached disks, and calculations with high random IO needs are moved to solid state disks (SSDs).

---

[4]"How to Build a Cheap Petabyte?"   `http://blog.backblaze.com/2009/09/01/`
`petabytes-on-a-budget-how-to-build-cheap-cloud-storage/`

Fast online stream processing is emerging, since soon we will not be able to store the incoming data stream (e.g. SKA), rather it has to be processed on the fly. Even in software we see a huge diversity emerging. Whereas a few years ago we had only three large database vendors, today there is a diversity of many NOSQL solutions, like MapReduce, Hadoop, LINQ, Dryad, column stores (Vertica), and open source databases aimed at the large scientific data sets of the future are emerging (SciDB).

Responding to this diversification, at JHU we are building the Data-Scope, a system with over 5PB of storage, with an extreme IO performance of 450GBps, faster than any supercomputer in the world today. Each of the 100 server nodes will have one or more GPU cards, and each server can pump 4.5GB of data per second into the GPUs, almost at the limit of the PCI bus. The projected cost of the system will be slightly over $1M, making it quite cost effective. One of the most exciting part of the Data-Scope story is that as we started to write the proposal, 20 groups of scientists across all disciplines at JHU came forward with large-data-related problems, each data set in excess of 100TB. They did not realize before this, that these challenges can be realistically addressed today. The Data-Scope should be thought of as a new kind of instrument, a microscope and a telescope for data. Large data sets of several hundred TBs can stay on the system for many months, while they are analyzed in detail, something difficult to do anywhere in the world today.

5. Summary

We live at a time when an interesting transition is happening in science, similar to the BeoWulf revolution of the 90's. Computation is increasingly centered around data, which in turn is becoming increasingly un-movable. This new world of data-centric computing requires us to rethink from the ground up how we build our computers, where we do our computations, how we do our statistics, and ultimately, how we do our science.

Acknowledgments.   The author would like to acknowledge support for the Gordon and Betty Moore Foundation, and NSF grants ITR-AST-0428325, OCI-104114 and OCI-106256.

References

Amdahl, G. M. 2007, IEEE Solid State Circuits Society News, 12, 4
Bell, G., Gray, J., & Szalay, A. S. 2006, IEEE Computer, 39, 110
Bell, G., Hey, T., & Szalay, A. S. 2009, Science, 323, 1297
Belokurov, V., et al. 2006, ApJ, 642, L137
Gray, J., Szalay, A. S., Thakar, A. R., Kunszt, P. Z., Stoughton, C., Slutz, D., & vandenBerg, J. 2002, ArXiv Computer Science e-prints. arXiv:cs/0202014
Lemson, G., & the Virgo Consortium 2006, ArXiv Astrophysics e-prints. arXiv:astro-ph/0608019
Li, Y., et al. 2008, J. Turbulence, 9, 31
Singh, V., Gray, J., Thakar, A., Szalay, A. S., Raddick, J., Boroski, B., Lebedeva, S., & Yanny, B. 2007, ArXiv Computer Science e-prints. arXiv:cs/0701173
Springel, V., et al. 2005, Nat, 435, 629
Szalay, A., & Gray, J. 2001, Science, 293, 2037
— 2006, Nat, 440, 413
Szalay, A. S., Bell, G. C., Huang, H. H., Terzis, A., & White, A. 2010, SIGOPS Oper. Syst. Rev., 44, 71

Astronomical Data Analysis Software and Systems XX
ASP Conference Series, Vol. 442
Ian N. Evans, Alberto Accomazzi, Douglas J. Mink, and Arnold H. Rots, eds.
© *2011 Astronomical Society of the Pacific*

Semantic Interlinking of Resources in the Virtual Observatory Era

Alberto Accomazzi and Rahul Dave

Harvard-Smithsonian Center for Astrophysics, 60 Garden Street, Cambridge, MA 02138, USA

Abstract. In the coming era of data-intensive science, it will be increasingly important to be able to seamlessly move between scientific results, the data analyzed in them, and the processes used to produce them. As observations, derived data products, publications, and object metadata are curated by different projects and archived in different locations, establishing the proper linkages between these resources and describing their relationships becomes an essential activity in their curation and preservation. In this paper we describe initial efforts to create a semantic knowledge base allowing easier integration and linking of the body of heterogeneous astronomical resources which we call the Virtual Observatory (VO). The ultimate goal of this effort is the creation of a semantic layer over existing resources, allowing applications to cross boundaries between archives. The proposed approach follows the current best practices in Semantic Computing and the architecture of the web, allowing the use of off-the-shelf technologies and providing a path for VO resources to become part of the global web of linked data.

1. Introduction

The explosion of content on the web has been partially tamed by the availability of services that aim to organize and link resources in ways that allow end-users to locate, filter, and rank the available resources. The enormous success of Google and its page-rank algorithm is mainly due to its capability of using the architecture of the web to organize this content, thus demonstrating that successful web-based information systems need not only take into account the content of the resources it knows about, but also the kinds of connections between them.

In the commercial world, there are a number of popular websites that provide extremely useful services based on organizing and presenting information in novel ways which enhance the discovery process. Some of the enabling techniques used by such sites are auto-suggest services, display of "facets" to allow narrowing or broadening of search results, ranking by different criteria, personalization and recommendations. When locating information on the web through one of these services, the current user expectation is that it not only be available through an intuitive interface, but also that it be organized in an efficient way, and that relevant content be only one click away.

These expectations are understandably also present when a scientist uses web-based services to access resources and data for research activities. With the proliferation of scientific digital data becoming available from different web-based science archives, it is essential for information providers to think of their content and services as being part of a network of interconnected science products. As such, their effective discovery and re-use will be enhanced by portals and search engines that index and expose the context and properties of these products through the appropriate interfaces.

Any system supporting resource discovery in astronomy will need to be built upon our community's distributed environment. Publications, now completely in digital format, are published worldwide, but their metadata are collected and indexed in one single database, the ADS. Similarly, metadata characterizing Astronomical Objects is collected by three projects, SIMBAD, Vizier and NED. While these projects provide centralized, well-curated access to their comprehensive databases of literature and object metadata, the same is not true for observational data repositories. Observational data and their basic metadata are stored in a number of archives and are usually partitioned based on their observational wavelength or the observatory which was used to collect them. Given the fact that these data are stored in heterogeneous archives and are accessible through interfaces which are very much tied to the underlying data model, no effective discovery mechanism exists today for this body of data. While services have been built implementing federated positional searches over the contents of data archives, the challenge of providing a single search paradigm over such an heterogeneous set of data products has proven difficult to solve in a satisfactory way.

In addition to the problem of ubiquitous discovery and access to datasets, data preservation principles require that we capture, curate, and connect all of the activities and digital data products which are part of the typical research workflows in astronomy. In order to support the principle of repeatability of the scientific process, it is critical that all artifacts created during a scientist's research activity be properly preserved and described (Pepe et al. 2010). In addition, provenance of data used, both between publications and data, and also between high-level data products and raw datasets is critical to the reproduction of scientific results by others. Documenting provenance of evidence and conclusions has been done sporadically and in ad-hoc ways at best, but the coming flood of multi-terabyte per night datasets require that we adopt best practices and frameworks that help us do this efficiently and automatically.

This paper presents work currently being carried out within the US Virtual Astronomical Observatory (VAO) Data Curation and Preservation efforts to create an infrastructure supporting curation, discovery and access to VAO resources. The two main objectives of the project are to capture and describe the linkages between data and publications and to capture and describe as much as possible the lifecycle of the research process, thus enabling us to track the provenance of both data and publication assets produced by researchers. Both of these goals contribute to achieving our end goal: creating services enabling discovery of Virtual Observatory (VO) resources via a seamless search over bibliographic and observational metadata.

2. Semantics

In order to provide the proper infrastructure for our project, we rely on the current best practices and technologies used in semantic computing (Heflin et al. 1999). These provide us with formal models to uniquely name resources, concepts, and their relationships; frameworks to represent and store them in databases; and standard languages to query and infer over this knowledge base. In this section we describe how our project takes advantage of these well-established techniques to achieve its goals: first we identify the resources in our research lifecycle, then we model their relationships, and finally we describe them in a formal way.

2.1. Resources

The linkage between astronomical data and publications is complex. Data may be used to reach conclusions, and this process is published in papers. But data are also measured in order to identify and characterize the celestial objects which generated the observed signal. These Astronomical Objects are then studied by other papers, and additional data are taken to reach further conclusions about their nature. Thus, there is a triangle of concepts to consider: Publications, Data, and Objects (see Figure 1). Given any instance from one of these concepts, one would like to be able to describe (and later discover) all the possible linkages to the other two, across all known datasets, papers, and astronomical objects.

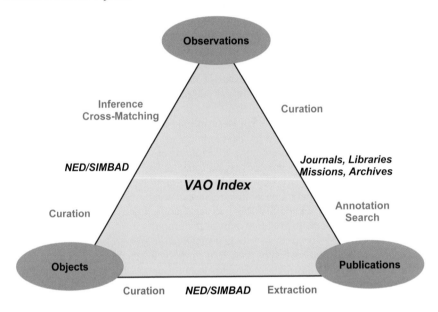

Figure 1. Relationships between Publications, Objects, Observations and the corresponding major actors in the curating process and their activities (in red).

For example, assume we want to know all papers written about a particular galaxy, say M31, and all datasets known about it. Or, given the Chandra COUP dataset, we want to know all known astronomical objects in the footprint of the dataset, as well as all papers written using COUP. There are further products of these linkages: all datasets sharing overlapping footprints, and all papers written about objects in these footprints.

As mentioned earlier, the linkages between publications and Astronomical Objects are well curated. The curation for the linkages between data and objects, and between data and publications currently relies on the heroic efforts of individual librarians and archivists working at a number of different institutes. Our efforts will leverage on their work to provide a centralized repository of these links *across* multiple missions and archives. At the same time, we intend to make their work easier by creating an infrastructure to simplify the curation process. Eventually we hope to leverage on other VO efforts and encourage direct participation from researchers in identifying linkages between their publications and datasets described therein.

2.2.   Ontologies

In order to capture the research lifecycle of astronomical research projects, we make use of formal tools to model the activities and artifacts involved in this process. These include: writing a proposal applying to a grant, securing funding, making observations, analyzing datasets, creating high-level data products, finding and characterizing objects, and writing papers. We do so in layers, at each level creating one or more **Ontologies** to describe the concepts and activities within the layer. We first start with the fundamentals of the Scientific process, creating an ontology called **VAOBase**. We build on that an observational ontology, **VAOObsv**, which describes observations and their associated datasets. We also build an ontology for publications, **VAOBib**, which relates to the other two ontologies.

An ontology is a formal representation of the concepts within a domain of knowledge (Heflin et al. 1999), and of the relationships between these concepts. For example, an *Observation* is a subtype of a *ScienceProcess* **Class**, and it results in a *DataProduct* **Class**. We represent this linkage as a property named *hasDataProduct*. We then say that an instance of the *Observation* class *hasDataProduct* an instance of the *DataProduct* class. We define ontologies in a formal language known as OWL (Ontology Web Language, McGuinness & van Harmelen (2004)), which is itself defined in a simpler formal language called RDF (Resource Description Framework.[1]) RDF is widely used on the web, and its use has led to the development of a parallel web of resources that can be linked to each other, and whose descriptions are machine readable, called the Semantic Web.[2] Since RDF provides typed links between resources, every site that publishes RDF contributes to a large, world-wide graph over which computations can be performed. Such computations include relational database-like queries on the graph using an analog of SQL called SPARQL, as well as the inferring of relationships between resources from existing relationships in the graph.

We have chosen to use industry standard RDF and OWL technologies since these are widely deployed, and have very good tool support. Furthermore, we can make use of a number of existing excellent ontologies to build upon. These include the Provenance, Authoring, and Versioning ontology from the SWAN Project[3] which provides a basis for all provenance related activity in our ontologies. We also use the FABio and CiTO ontologies from the Semantic Publishing and Referencing Ontologies[4] which provide a way for typing the different kinds of publications and citations respectively. For Astronomy semantics, we utilize the IVOA SKOS vocabularies for astronomical keywords,[5] as well as the CDS vocabulary for Astronomical Objects and their variability types (Derriere et al. 2007).

Our model of Observations and Data Products follows that of the Common Archive Observation Model (CAOM, Dowler et al. 2008). Wherever possible, we have chosen to track existing, deployed standards. Datums, datasets, and their associated observations are described by metadata properties such as position, URI, flux data, band,

---

[1] http://www.w3.org/RDF/

[2] http://en.wikipedia.org/wiki/Semantic_Web

[3] http://swan.mindinformatics.org/ontology.html

[4] http://opencitations.wordpress.com/

[5] http://www.ivoa.net/Documents/latest/Vocabularies.html

Instruments used, etc. We have chosen the metadata properties we wish to model and their names following the ObsCore specification from the ObsTAP project.[6] ObsCore is rapidly gaining steam amongst archives as a minimal, simple standard to provide ADQL[7] compatible querying of data product metadata, and we intend to ride on its coattails.

2.3. Representing the Research Lifecycle in Astronomy

In this section we illustrate, by way of example, how the formal tools described above can be used to represent scientific assets, their relationships, and research activities performed on them. A schematic representation of the main concepts and relationships can be found in Figure 2, and a narrative of some of these activities is given below.

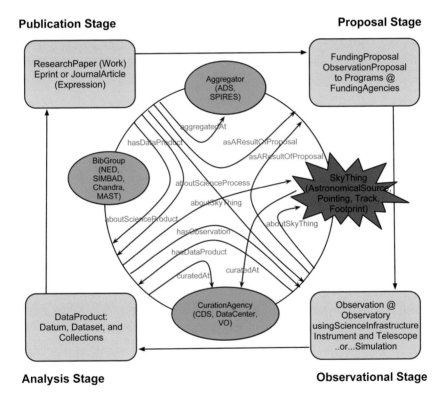

Figure 2.   A model of the Research Lifecycle in Astronomy, showing some of the classes in our three ontologies, and some of the links between instances of these classes (created as ObjectProperties in OWL). For example, an instance of an Observation may (or may not) have the property *asAResultOfProposal* whose range is an instance of the class *ObservationProposal*.

---

[6]http://www.ivoa.net/cgi-bin/twiki/bin/view/IVOA/ObsDMCoreComponents

[7]http://www.ivoa.net/Documents/latest/ADQL.htm

We submit *Proposal*s for funding and *ObservationProposal*s for time to *Program*s and *ObservationProgram*s at *FundingAgency*s and *ScienceInfrastructureAgency*s respectively. Upon the granting of observing time to the proposals we carry out a type of *ScienceProcess* called *Observation* at *ScienceInfrastructure* such as *Observatory*s using *Instrument*s and *Telescope*s. We then carry out *Analysis* of the observations leading to the production of *DataProduct*s. Further analysis and possibly *Simulation*s, also examples of science processes, are carried out leading to the production of *WrittenProduct*s such as reports or papers.

Observations taken on the sky may be known *AstronomicalSource*s at known *Position*s, as identified by one or more *CurationAgency*s such as the CDS, or of a random *Pointing, Track,* or *FootPrint* on the sky. Observations may be *SimpleObservation*s, which correspond to photons collected in one time interval or *ComplexObservation*s such as multi-point skews, grid observations, etc. A piece of data from a simple observation is called a *Datum*, e.g., the FITS file corresponding to a single exposure. Multiple simple observations (more precisely their data) may be combined into a *Dataset*, such as a mosaic, or light curve. *ComplexObservation*s too are represented by datasets. Both datum and dataset are types of *SingularDataset*s, which might be combined together to create *CompositeDataset*s such as cartouches of all files associated with a given astronomical source.

Publications are described in our ontologies using FABIO's support for FRBR (Functional Requirements for Bibliographic Records.[8]) FRBR advocates tracking *Work* through its various *Expression*s, and the *Manifestation*s of these expressions. For example, the work in case may be a *ResearchPaper* on spiral galaxies. This paper is expressed as a *JournalArticle*, and before this article is ever published, as an *Eprint* on the arXiv site.[9] Manifestations of this paper are the various formats in which the article is available, at various online *Aggregator*s, or in printed form. Such a research paper represents a *WrittenProduct* about the data products, observations and analysis.

The links from Publications to Data, and from Publications and Data to Proposals are maintained by *BibGroup*s at various institutions. These links are captured in our ontologies by properties such as *aboutScienceProcess, aboutScienceProduct, underProgram, asAResultOfProposal* and *hasDataProduct* whose domain is the Work or Expression, or even the *Observation* or *AstronomicalSource* at hand. These linkages constitute the key part of our project.

It is probably obvious by now than any such database of such resources and linkages is incomplete. Here the usage of semantic technology shines when compared to relational technology: we only need to assert the properties we know about. Nevertheless, our framework has been designed so that all the crucial entities and their properties can be captured according to the model at any point in time. As an example, we intend to use text mining techniques at a later date to search the full-text literature for grant numbers, program names, and organizations. There are many other terms defined in our ontologies and the ontologies that they depend upon. These can be examined in more detail in our code repository.[10]

---

[8] http://archive.ifla.org/VII/s13/frbr/frbr_current_toc.htm

[9] http://arxiv.org

[10] https://github.com/rahuldave/ontoads

3. Infrastructure and Applications

In the previous section we discussed the concepts that our ontologies capture, and the languages we represent these concepts in, RDF and OWL. The reason for using these languages is the vast infrastructure available as open source software for the semantic web. The purpose of our server and database infrastructure is to: provide a linked data endpoint to various astronomical resources and the relationships between them; enable the querying and inferencing on this graph of resources and relationships; index certain key resources and relationships in order to provide a fast query interface over selected properties of publications, datasets, and astronomical objects; enable applications such as search and discovery engines, and faceted browsers of astronomical resources to be built, so as to deliver services to end users; enable applications to be built which will help future identification of data-publication linkages, and provide these services to bibliographic groups at different astronomy institutions, as well as directly to astronomers.

We intend to create an indexed database of publications, datasets, and their relationships to provide an effective infrastructure for resource discovery, leveraging on ADS's expertise in metadata and full-text indexing. Bibliographic metadata will be incorporated into the knowledge base from the ADS database. Integration of object metadata and linkages will be accomplished utilizing the astronomical object databases maintained by NED and SIMBAD. Observational metadata will be incorporated from a number of collaborators at the CDS, Chandra, NED and MAST who maintain curated connections between datasets and publications.

3.1. Server Infrastructure

To store RDF statements, we use a database system called a triplestore, and have selected the open source Sesame[11] as the DBMS. The triplestore stores statements, creates indices on some subjects, objects, and predicates, and provides SPARQL and RESTian[12] interfaces to resources and simple queries. Additionally, Sesame stores triples with a context, which may then be used to track transactional additions and removals from the database.

However, because a triplestore has no knowledge of the structure of relationships in the data, it provides slow performance in the common search cases, such as finding the datasets associated with a publication, for example. To provide fast results which can then be faceted, we use SOLR[13] as an indexing server in front of the triplestore. This allows us to have a two-tier system, where complex SPARQL or subject/object/predicate queries are handed over to Sesame, while SOLR serves the more common search cases with real fast indices. Furthermore, since SOLR provides faceting out of the box, we can write user interfaces for our application, once we index the properties we wish to filter upon.

Finally, a web service written in Python is used to make choices as to which server to query and proxy, manage authentication, run federated searches to SIMBAD and NED, and handle any additional features that a user-facing application requires. The

---

[11]http://www.openrdf.org/index.jsp

[12]http://en.wikipedia.org/wiki/Representational_State_Transfer

[13]http://lucene.apache.org/solr/

triplestore is currently accessible via the SESAME API and SPARQL query language, using our Python library code. We use it in our data pipeline to populate the SOLR server, and to inferentially add data into it. We are planning to use this infrastructure in the core pipeline for ingesting publications from the ADS to normalize author and organization names, to keep track of the linkages between papers and proposals, and to track the provenance of publications.

The triplestore has been populated with a select subset of bibliographic data from ADS and makes use of an object cache automatically populated as SIMBAD and NED are queried. For observational data, our strategy is to ingest metadata from larger archives (starting from Chandra and MAST), and make our way to the smaller ones. Chandra data are complex and we have collaborated with the Chandra Archive team to convert their metadata into RDF. Because our observation model is based on ObsCore and CAOM, we will be able to ingest data from any mission which publishes metadata in ObsCore compatible tables. This is how we will be tackling most of the data from MAST.

3.2. Applications

A first prototype user interface is being developed in javascript with jquery, AJAXSolr and our own custom code which talks to the backend SOLR indexing server, Sesame triplestore, and the python web service. This user interface makes use of AJAX to pull metadata from the server in the background while the interface is being manipulated.

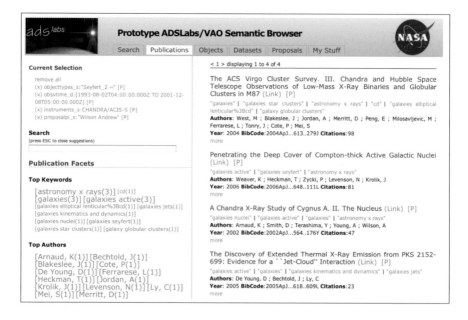

Figure 3.   A prototype of a faceted search on publications, with filtering via observational and object metadata.

In the screenshot depicted in Figure 3, publications are being faceted by various metadata belonging to the datasets used in them, the objects described within, and the proposals used to fund the research and make observations. Clicking on any facet link will filter the publication set by that facet in addition to the facets already chosen; clicking a P (or pivot) link will change to a view in which the publications are filtered by that facet only. In the figure, we have faceted by Seyfert Objects, observation time, the CHANDRA ACIS-S instrument, and selected a particular proposal PI (Andrew Wilson). These simple filtering activities lead us to find papers associated with Seyfert research proposed by Andrew Wilson in a particular timeframe and with a particular instrument. Interestingly, only one of the papers that result from this selection is co-authored by him, indicating that these observations have had impact beyond the original intent of the proposal, a result that would have been difficult to conclude without the support of this knowledge base.

This interface is being extended to facet datasets, objects, and proposals in order to provide a generic search and bookmarking capability over all these resources. It will be made available as part of the VAO toolset, the ADS "Labs" experimental search interface, and possibly integrated in the upcoming VAO portal.

4. Conclusions and Future work

Our backend server infrastructure and javascript prototype experiments accomplish a first goal: exposing the linkages between objects, datasets, and publications in a natural way, thus making it easier for astronomers to explore the space of astronomical concepts and phenomena using an iterative process through an interface which exposes key relationships among them. The knowledge base and infrastructure we are building is meant to provide support for a variety of applications, some of which we will develop ourselves, with others being contributed by collaborators. A list of potentially useful applications that we have envisioned include:

- **The APOD Browser**: A 3-pane search and exploration browser which will allow users to simultaneously browse Astronomical Publications, Objects, and Datasets (APOD). The contents of each pane view will change depending on selections in the other panes. It will also be possible to pivot on any asset in any pane and see what resources are available for the other two. Any search will be bookmark-able and will act as a live search, so that additions to our and other mission and archival databases will be immediately reflected in the search through a process of notification. Thus APOD will serve as a research portfolio tool for graduate students and seasoned astronomers alike. By linking APOD into the VAO portal, we will be able to provide one-stop service to users of the VAO.

- **Annotation Server**: The working of our tool depends largely on the mostly unsung efforts of bibliographic groups maintained by multiple archives such as Chandra, MAST, ESO, NED, CDS, and ADS. By combining our triple store with semantic annotation technology and the ADS literature full text search, we are in a position to provide infrastructure to the curators who maintain bibliographic information, allowing them to carry out their annotation of literature-data and object-literature connections in a more efficient manner, simplifying their curation efforts.

- **Metrics Tool**: By leveraging the efforts of bibliographic groups across multiple missions, and by full-text mining of publications, we are also capable of providing a queryable infrastructure that links publications to proposals and observations. This allows the computation of metrics on the efficacy of observing and funding programs, as well as the output of researchers. This is invaluable information for both funding agencies and mission directorates. Thus user interfaces can be developed which make such metric extraction as easy as the faceted browsing of astronomical concepts.

- **Paper of the Future**: Leveraging on the database of the connections from any given publication to the objects studied therein, the datasets used, and the proposals that went into the production of the paper, we will be able to provide a more wholistic view of the paper, with direct linking to (and in some case inline depiction of) datasets, catalogs, objects, SEDs, etc. In conjunction with full text searching, the extraction of table, figure, and equation assets from the paper, and added encouragement to users to provide enhanced publication-data linking themselves, we will be able to provide a very rich view of the paper itself. In addition, we will be able to provide to the users links to relevant resources and recommendations based on a variety of criteria, such as citations, usage of data products, objects studied, etc.

We have emphasized earlier the dawn of a new age of data-intensive astronomy, which will require a paradigm shift in the way research is conducted in our discipline. The work we have presented in this paper is part of the effort to automate and make easier the characterization and indexing of scientific resources and their relationships. Additionally, by capturing and formally describing the linkages from published research to data used, we will make progress towards the creation of a digital environment enabling the repeatability of the scientific process.

Acknowledgments. We are grateful to a number of individuals and groups who have provided us with the metadata currently indexed in our knowledge base, in particular Sherry Winkelman (Chandra), Karen Levay (MAST), and the SIMBAD, NED and ADS teams. Sherry and members of the VAO collaboration, in particular Doug Burke, Matthew Graham and Brian Thomas offered suggestions on a number of topics related to the development of our Ontologies and technical infrastructure. We thank Alyssa Goodman and Michael Kurtz for inspiring us to pursue this effort. This work was supported by the Astrophysics Data System project which is funded by NASA grant NNX09AB39G, Microsoft Research WorldWideTelescope, and the Virtual Astronomical Observatory, funded under NSF and NASA grants.

References

Derriere, S., Richard, A., & Preite-Martinez, A. 2007, Highlights of Astronomy, 14, 603
Dowler, P., Gaudet, S., Durand, D., Redman, R., Hill, N., & Goliath, S. 2008, in Astronomical Data Analysis Software and Systems XVII, edited by R. W. Argyle, P. S. Bunclark, & J. R. Lewis, vol. 394 of ASP Conf. Ser., 426
Heflin, J., Hendler, J., & Luke, S. 1999, SHOE: A Knowledge Representation Language for Internet Applications, Tech. rep., University of Maryland
McGuinness, D. L., & van Harmelen, F. 2004, OWL Web Ontology Language Overview, W3C recommendation, W3C
Pepe, A., Mayernik, M., Borgman, C. L., & Van de Sompel, H. 2010, Journal of the American Society for Information Science and Technology, 61, 567

Astronomical Data Analysis Software and Systems XX
ASP Conference Series, Vol. 442
Ian N. Evans, Alberto Accomazzi, Douglas J. Mink, and Arnold H. Rots, eds.
© *2011 Astronomical Society of the Pacific*

What Python Can Do for Astronomy

Perry Greenfield

*Space Telescope Science Institute, 3700 San Martin Dr., Baltimore,
MD 21218, USA*

Abstract. Python continues to see increased use in astronomy, both by institutions developing software for new instruments, telescopes and missions, and by astronomers for use with analyzing their data. I will describe the advances in Python-based tools for astronomy in the past few years and the current effort to bring those tools together into a common repository. What are the most serious challenges facing the use of Python in astronomy and how should these be addressed? Finally, I'll discuss the non-technical contributions that Python can bring to astronomy and illustrate with an example and suggestions.

1. Suitability of Python for Astronomy

The positive aspects of Python as a language or its associated libraries are generally well known and will not be detailed at length here. It is a very accessible language for casual programmers, particularly astronomers, while also being a very powerful language. This enables many of the same tools to be shared by both researchers and for production quality programs. Python has clearly become the scripting language of choice for nearly all new large astronomy projects, and is gaining wider use among astronomers as well. While many consider it to be only a scripting language, there is also increasing usage of Python as an applications language.

The major downside of Python is its speed. Primitive operations are many tens of times slower than corresponding operations in low-level compiled languages such as C, C++, or Fortran. But this is not nearly as serious an obstacle as it may appear at first glance. In practical applications, relatively small fractions of the code need to be efficient. And where it needs to be efficient, there are often high level operations available that run efficiently in Python because they perform a lot of work underneath in C. For example, array operations in Python are efficiently carried out by the numpy package so long as operations are performed on large parts of the array with numpy primitives. And when no language libraries or constructs are available, it is not difficult to call code written in C (or other languages) to perform operations that must be done efficiently at a low level.

Another issue viewed a liability is Python's Global Interpreter Lock. This prevents multiple Python threads from running simultaneously. Thus, multi-threaded applications do not exploit the now common multi-core processors. The Python community generally frowns on the use of threads for such performance gains (for more reasons than performance) and generally suggests the use of multiple processes for parallel processing. In any event, the efficient use multiple processors with Python is a topic that has many in the community investigating the best ways to do so.

In our view, the ease of installation is probably our greatest concern. Since the tools needed to process and visualize data in Python have many third-party dependencies, each with its own peculiar means of installation, installations can consist of several independent steps, each having possible snags leading to more difficulties than most users wish to deal with. This is an area that we are investing significant effort that will be described briefly later.

The suitability of Python as an applications language has been enabled by great progress in the past decade in the improvement of the tools that permit it to be used effectively to manipulate astronomical data. Both numpy and matplotlib now generally surpass the capabilities that are present in IDL for array manipulation and 2-d plotting and in some areas far surpass the equivalent capabilities. Likewise the same is true for PyFITS. No other language has the equivalent functionality to PyRAF. These combined with such tools as ipython, mayavi, and scipy give Python a wealth of basic tools to analyze and visualize data. STScI is pleased to have played an important part for many of these.

Nevertheless, there is no question that Python does not yet have the equivalent functionality of IRAF or IDL in the astronomical applications area, or in specialized astronomical libraries. This is clearly an area for improvement. It is worth noting that the past two years have seen a blossoming of activity in contributed astronomical packages such as:

APLy Astronomical plotting utilities. `http://aplpy.sourceforge.net/`

Asciitable Read and write ascii tables.
`http://cxc.harvard.edu/contrib/asciitable/`

astlib Astronomy utilities. `http://astlib.sourceforge.net/`

astronomical utilities `http://www.astro.washington.edu/users/rowen/`

astropysics Astrophysics utilities. `http://packages.python.org/Astropysics/`

ATpy generic Table module. `http://atpy.sourceforge.net/`

cosmology calculator `http://www.astro.ucla.edu/~wright/CC.python`

cosmolopy Cosmology tools. `http://roban.github.com/CosmoloPy/`

IDLSave Module to read IDL save files. `http://idlsave.sourceforge.net`

Kapteyn package Plotting, coordinate, tables utilities.
`http://www.astro.rug.nl/software/kapteyn/`

Pandora `http://cosmos.iasf-milano.inaf.it/pandora/software.html`

PyEphem Ephemeris tools. `http://rhodesmill.org/pyephem/`

pygalkin Tools for spectral cubes. `http://code.google.com/p/pygalkin/`

Python-Montage Python interface to Montage.
`http://astrofrog.github.com/python-montage`

Python scripts for astronomy
http://www.atnf.csiro.au/people/Enno.Middelberg/python/python
.html

pytpm Python interface to the TPM library. http://phn.github.com/pytpm

pywcsgrid2 Astronomical image grid overlay and labeling tools.
http://leejjoon.github.com/pywcsgrid2

sampy Python SAMP implementation.
http://cosmos.iasf-milano.inaf.it/pandora/sampy.thml

Most of these have been contributed by individuals, not institutions. In addition to these we have significant Python contributions from larger institutional efforts that have been around awhile (e.g., STScI, NRAO, Chandra/SAO, ESO/ST-ECF). There is good reason to believe we will see much improvement in this area over the next few years.

2. Future STScI Community-Oriented Efforts for Python

We have three main areas of effort planned for the immediate future.

2.1. Addressing Installation Ease

The most important is developing an easy-to-install release of AURA software. We have been working with Gemini (and particularly with James Turner) over the past year to package all AURA distributed analysis and reduction software into a "one-step" install. Such a "unified release" will contain Python, IRAF/STSDAS/TABLES, stsci_python, Gemini packages, and all needed dependencies, as well as other third party libraries and tools that may be needed by common add ons. Our goal is that this should support all popular platforms, e.g., Mac OS X and Linux (but given the multiplicity of Linux variants, universal support is unlikely). It should be understood that there is no universal, easy-to-use, foolproof solution to the installation issues of software distribution, nor is one expected in the near future.

The distribution we are planning will allow users to update various software components, most easily with end-level libraries and applications (updating core items such as Python or numpy may require updating the many tools depending on them and therefore may be impractical for most users). Making such a distribution will allow us to start using the many tools available in packages such as scipy or mayavi that we previously have avoided because of installation issues. Users will also be able to install additional packages on top of this distribution (Python packages most easily). Nevertheless, the more the core distribution is updated or modified, the greater likelihood problems may be encountered. We are hoping to balance flexibility with ease of installation and provide users with the choices between the two. We expect to make an early version of this "unified release" available in early 2011.

Part of this distribution system is a nightly testing framework that ensures that new versions of all software components continue to work properly together to ensure its consistency. Such nightly testing highlights potentially disruptive changes in packages from all contributing institutions. This feedback will encourage greater interdependencies between the institutions than permitted by a traditional release cycle where fixing such changes becomes much more difficult given the frequency of such releases.

2.2. Replacing IRAF Functionality

In order to support the JWST mission and reduce dependencies on IRAF we are considering an effort to begin replacing important core IRAF functionalities. This is likely to focus first on spectral data calibration and processing capabilities, then on still useful STSDAS packages such as isophote and restore, and finally on generic image processing tools.

2.3. Fostering Community Python Contributions

We have set up a code repository at `http://astrolib.org`. The repository is intended to help merge the various community contributions into a more integrated and coherent whole. By putting different packages in the same repository it should help drive such packages to a more consistent documentation system and style (based on sphinx, the new standard system of the Python community), and more consistent interfaces to make their interoperability easier. Finally, it should make regular regression testing easier since there will be an existing framework running on many platforms for contributors to take advantage of.

3.   What Else Python Can Contribute

Python has more to offer than just a language or libraries. The first Python conference I attended was about 12 years ago. It was small as such conferences go, with only about 90 attendees. Yet it made a great impression on me because I saw a significantly different culture at work than I had seen in the astronomical software community.

3.1.   The Python Culture and Mindset

There are aspects of the Python developer culture that are fairly common in many other open source projects, and aspects that aren't. The characterizations that I'll describe are of course generalizations, and one must understand that these are all a matter of degree rather than absolutes. The culture I'll attempt to describe is primarily that of those that contribute to the language itself, its standard libraries, and to a somewhat lesser extent, the heavily used libraries in the community that aren't part of the standard library, and not of the wider community that uses Python.

I'll first note the aspects that are more peculiar to the Python community. These include:

- A steady temperament that is resistant to fads.

- A judicious restraint to adding new features to the language.

- But at the same time, recognizing that evolution is necessary.

- While backward compatibility is important, it isn't an absolute requirement.

- A good balance between practicality and purity.

- Great helpfulness to newcomers in the community.

There isn't much doubt that these reflect the characteristics of the language's originator, Guido van Rossum. The real success of Python goes well beyond the technical merits of the language itself. It is much more due to the community that has grown up to support the language. Without that community the language would not have been as successful as it has been. Guido van Rossum's great contribution has been in creating a large and productive community that leverages the advantages of the language he created.

The Python developer community is large and quite active. It has attracted many very talented and enthusiastic developers. The community is loosely and primarily self-organized. There is very little top-down control or coordination. Guido van Rossum still holds the final say on many decisions, where he chooses between various alternatives, and vetoes or approves various proposals (Guido has been awarded the title "Benevolent Dictator For Life" by the community). But he hasn't done much development himself on the core system over the past several years. This isn't to say that the development process is always orderly. It isn't. Often there are false starts, or strong disagreements on the best approach, leading to various competing alternatives that are tried before there is a clear outcome. Nevertheless, the community is quite cooperative and generally makes great progress. The process exhibits a lot of detailed technical communication. The software produced is usually of very high quality. Most of it is well out in the open, warts and all.

There are other interesting aspects of the developer community. Much of the philosophy is reflected in the "Zen of Python" (which can be displayed by typing "import this" at the python interpreter prompt). Their rapid engagement with different approaches appears to lead to quick and good evaluations of new technologies. Technologies that are viewed as having problems or which are overhyped are fairly quickly discarded or de-emphasized. Their weight in decision making is heavily weighted towards those that actually contribute effort (either in development or use of the new tools). They have little patience for those who want to tell others how to do things without contributing work themselves.

3.2. Contrasted with the Astronomical Software Development Community

On the other hand, much of the software development in the astronomical software community is fragmented, despite sharing many of the same needs. The organizations and individuals in that community usually are protective of their code and slow to share. Cooperative efforts, when they do occur, are committee- and consensus-driven, often lead to very feature-rich goals, and often fail to achieve them, taking a long time in the process. These cooperative efforts are, ironically given the nature of the science, more oriented towards talk and analysis than experimentation. Very often, such efforts are top-down and centrally planned. The community is slow to discard obsolete or failed technologies. In comparison to the Python developer community, the astronomical community appears quite slow moving and not nearly as productive. Is there anything that the astronomical software community can learn from the Python developer community?

3.3. Reasons for Difference

Before trying to answer that question it is important to try to understand the possible reasons for the differences between the two communities. I'll go through various possibilities starting with those I consider least important.

Difference in nature of work or funding
I do not consider this a serious reason. Many in the Python community have similar issues and yet are able to work around them.

Lack of Guidos
I suspect the lack of leaders like Guido is due more to our culture rather than our culture being due to the lack of leaders like Guido.

Lack of theme comedy troupe/enthusiasm
It does appear to me that the astronomical community takes itself more seriously than the Python community, and that there is more enthusiasm in the Python community. This may be due in part to software being secondary to astronomy, and thus getting less respect and correspondingly less enthusiasm (e.g., it isn't necessarily good to be seen as enjoying the software aspect too much as that may not be beneficial to one's career). Likewise, the seriousness may arise out the academic culture taking itself more seriously.

Academic culture
There are possibly a number of aspects to this that are significant. A couple were already mentioned for the previous item. Other elements of the academic culture that may be relevant are:

- Committee/consensus orientation to decision making.
- Territorial views towards software as being part of a competitive research or organizational advantage.
- A view of software issues as having final, long-term (and thus static) solutions much like physics or math problems.

Lack of external competition
This is perhaps the most important reason. The Python community is well aware that it is in competition with other languages. If they wish to see their language thrive, they know they must make good progress in comparison to its competitors. This is a strong motivating force for cooperation. If they don't they may see the language they favor fail to attract new users and eventually die, and with it, all the tools they have invested in it. The astronomical software community has no corresponding motive. If astronomical software is less efficient or more fragmented, astronomy will not cease, and correspondingly, the need for astronomical software will not end.

Lack of a Guido
The Python community has an advantage that when occasions arise in which there is no clear technical winner when multiple alternatives are presented, someone must make a decision. They have a clear leader who commands much respect and who can make that decision.

3.4.   Different for Good Reasons

There are clear reasons why the astronomical software community has not developed a culture similar to that of the Python community. But does that mean the astronomical software community cannot learn from the beneficial aspects of the Python culture? I do not believe that it is clear that it cannot. There does not appear to have been any attempt yet to try. Perhaps the forces that prevented it from forming naturally would prevent its adoption. But we don't really know that.

4. Case Study

To this point the comparisons made have been very high level and general. It is more interesting (and risky) to illustrate with a specific example. There are many examples that could be given, but I'll use one that I think is most important.

4.1. Backward Compatibility in the Python Community

While the Python developer community views backward compatibility as important, it will break it when the benefits are seen to outweigh the costs. Some good examples of this are:

1. Python 3 seriously breaks backward compatibility. Yet the changes were made regardless since these were viewed as necessary in fixing some of Python's intrinsic problems or drawbacks.

2. The Python documentation system underwent major changes when a newer and more maintainable technology (sphinx) became available. It meant redoing all the standard documentation formatting, yet that was accomplished fairly quickly despite the work required. Most of the Python world has fallen in line with the new standard quickly, as well.

3. Transition from Numeric to numarray to numpy made various backward incompatible changes to the array manipulation tools (twice!). Despite the pain, what resulted is far superior to the original package.

Breaking backward compatibility obviously causes problems. On the other hand, it does allow the community to evolve their software to keep up with the rest of the software world. There are costs in not evolving that must be considered along with the costs of revising existing software to deal with incompatible changes.

4.2. Backward Compatibility in the Astronomical Software Community

The astronomical software community generally appears much more resistant to change, and much more unwilling to change standards even when there are great benefits to doing so.

One such example is with the FITS standard. The FITS standard has been a great success. But arguably too great. What is the basis for this shocking claim? The standard originated in the very early 1980s when the software (and hardware) environment was quite different. And even though extensions have been made to it, much of it still reflects very outdated limitations.

I argue that these limitations are making our software harder to develop, understand, and maintain. Furthermore, the data files being generated are less consistent than they could be due to the fragmentation of various approaches used to work around the limitations of the format.

A good example of how FITS limitations have impeded progress is the World Co-ordinate System standards for FITS. It took about 10 years for the basics aspects to become part of the standard once the effort started. There are aspects still in limbo 16 years after this effort started with no end in sight (i.e., FITS "Paper IV" that deals with distortion representations sees no prospects of near-term acceptance). The existing standard is complex yet inflexible and has arbitrary restrictions imposed by the keyword size restriction (e.g., how many polynomial terms are permitted). Neverthe-less, even with the Paper IV additions, it is inadequate for HST data, where we must resort to non-standard representations to fully describe the WCS. The existing WCS ap-proach illustrates the contortions that one must go through to adapt coordinate system transformations to the FITS standard.

There are much better ways to deal with such issues these days, yet the issue of updating or replacing the FITS standard is not really entertained in the community. Adopting or changing standards is so painful that it is a topic widely avoided. While the principle that "once FITS, always FITS" is a sensible requirement, many in the FITS community take that further in requiring any future FITS standard be accessible at some level (e.g., being able to interpret headers) by older FITS readers. Any such restrictions either make it nearly impossible to make significant improvements to the FITS capabilities, or make such improvements convoluted in order to fit the current header restrictions. There is no question that the FITS format is "Turing complete" in a data sense. It does not mean that it is sensible to layer enhancements on top of it.

VOTable was a potential chance to come up with a new standard, yet it was fa-tally flawed in its inability to deal with binary data efficiently. So now we have two inadequate data formats to support now. In some respects, we are even worse off for it.

Perhaps HDF5 is a suitable replacement. It is an existing standard, and has many features, but its drawback is that it is complex, making software re-implementations of supporting libraries expensive.

There are many possible alternatives that would be clear improvements. It isn't so much the issue of picking the perfect optimum, but rather settling on a solution that is a significant improvement for most cases. There are times to replace standards when they have outlived their usefulness (and many examples exist where standards have changed in the commercial world despite the high cost of switching standards). We are well overdue for such an improvement.

The cost of switching standards is likely overrated. Given the increasing growth of data, the amount of data in older formats will quickly become a minor component of the data available, and correspondingly, a small amount that needs conversion. Secondarily, all of the existing FITS data is on media that is comparatively non-permanent, due to the physical nature of the physical media itself (e.g., magnetic tapes) or the fact that the media may not have devices to read it indefinitely. If such data must be copied anyway to preserve it, inserting a conversion step in the copying process is likely an easy thing.

I don't believe the Python community would have found itself in a similar situa-tion. They are much more aware of the changing nature of the software landscape and much more willing to deal with such issues more quickly. The astronomy community appears to prefer isolating itself more, resulting in a more static environment for them-selves, but leading to greater discontinuities in the future with the external software world.

4.3. Conclusions

For those who use Python, look beyond the language and libraries and see what can be learned from the Python community itself. Even non-users of Python may find it beneficial to look at how its developer community works (or one of any number of other successful open source communities) for ideas of how to better collaborate with other projects. Finally, if you produce Python libraries that have potentially greater applicability within astronomy, consider hosting them on `http://astrolib.org`.

Astronomical Data Analysis Software and Systems XX
ASP Conference Series, Vol. 442
Ian N. Evans, Alberto Accomazzi, Douglas J. Mink, and Arnold H. Rots, eds.
© *2011 Astronomical Society of the Pacific*

Automated Morphometry with SExtractor and PSFEx

E. Bertin

Institut d'Astrophysique de Paris, UMR 7095 CNRS,
*Université Pierre et Marie Curie, 98*bis *Boulevard Arago,*
F-75014 Paris, France

Abstract.   Variable blurring of astronomical images by seeing and the instrumental Point Spread Function (PSF) makes it challenging to obtain accurate photometric and morphometric measurements under changing observing conditions. I show how the new PSFEx PSF modeling software and the latest version of the SEXTRACTOR source extraction tool can be combined to perform fully automated, PSF-corrected source measurements. An implementation of these techniques in the Dark Energy Survey Data Management (DESDM) pipeline is presented.

1.   Introduction

The purpose of the SEXTRACTOR software package (Bertin & Arnouts 1996) is to create lists of sources in astronomical images in an efficient and fully automated way. Until recently, measurements of features detected by SEXTRACTOR has been limited to rather basic quantities, partly because of processing time constraints. Recent increases in computer performance now allow for more sophisticated measurements to be carried out in a reasonable amount of time,

One of these features is two-dimensional model-fitting. The fitting of two-dimensional models of galaxies convolved with a model of the instrumental Point Spread Function (PSF) has been proposed as an effective way of measuring shape parameters of faint galaxy images (see, e.g. Peng et al. 2002, and references therein). Several efforts have recently been made to automate the process on the scale of complete images (Barden et al. 2009; Vikram et al. 2010). These tools have the disadvantage of relying on a heterogeneous collection of codes, which impacts their efficiency. In the following, I introduce the PSFEx companion software and describe an optimized, fully automated, two-dimensional model-fitting code implemented in C directly within SEXTRACTOR. I also describe the extra step taken in the Dark Energy Survey Data Management (DESDM) pipeline, to homogenize PSFs prior to image stacking and model-fitting.

2.   Modeling the PSF with PSFEx

The modeling of the PSF itself is performed by the newly released software package PSFEx.[1]   Briefly, PSFEx starts by identifying detections that are likely to be point-

---

[1]Available at http//astromatic.net.

sources using an empirical recipe which includes finding the position of the stellar locus in a magnitude vs half-light-radius diagram (Kaiser et al. 1995). PSFEx models the PSF as a linear combination of basis vectors. The basis vectors are rendered as small images at a resolution chosen to minimize aliasing, which makes it possible to recover the PSF even with severely undersampled data. The vector basis may be the pixel basis, the Gauss-Laguerre basis (Massey & Refregier 2005), the Karhunen-Loève basis derived from a set of actual point-source images, or any user-provided basis. PSFEx fits the image of every point-source $\vec{p}_s$ with a projection on the local pixel grid of the linear combination of basis vectors $\vec{\psi}_b$ by minimizing the χ^2 function of the coefficient vector $\vec{c}$:

$$\chi^2(\vec{c}) = \sum_s \left[\vec{p}_s - f_s \mathbf{R}(\vec{x}_s) \sum_b c_b \vec{\psi}_b \right]^T \mathbf{W}_s \left[\vec{p}_s - f_s \mathbf{R}(\vec{x}_s) \sum_b c_b \vec{\psi}_b \right], \tag{1}$$

where f_s is the flux within some reference aperture, and $\mathbf{W}_s$ the inverse of the pixel noise covariance matrix for point-source s. $\mathbf{R}(\vec{x}_s)$ is a resampling operator that depends on the image grid coordinates $\vec{x}_s$ of the point-source centroid:

$$\mathbf{R}_{ij}(\vec{x}_s) = h\left(\vec{x}_j - \eta.(\vec{x}_i - \vec{x}_s) \right), \tag{2}$$

where h is a 2-dimensional interpolating function, $\vec{x}_i$ is the coordinate vector of image pixel i, $\vec{x}_j$ the coordinate vector of model sample j, and η is the image-to-model sampling step ratio (oversampling factor). PSFEx is able to model smooth PSF variations by making the c_b coefficients (equation 1) themselves a linear combination of polynomial functions of the source position within the image (Fig. 1).

χ^2 minimization is fast, but restricts the current modeling process to images with noise in the Gaussian regime. The point-source selection and modeling process is iterated several times to minimize contamination of the sample by image artifacts, multiple stars and compact galaxies. More details about the working of PSFEx can be found in Bertin et al. (in preparation).

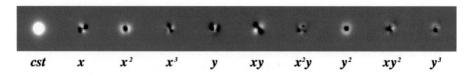

| cst | x | x^2 | x^3 | y | xy | x^2y | y^2 | xy^2 | y^3 |

Figure 1.   Examples of PSF image components recovered using PSFEx on a CCD image. PSF variations are modeled as a 3rd degree polynomial in image coordinates. The final PSF at a given position is the sum of all image components, each of which is weighted by the associated polynomial term.

3. Morphometry Measurements with SEXTRACTOR

Like GALFIT, SEXTRACTOR's two-dimensional model-fitting procedure relies on the Levenberg-Marquardt minimisation algorithm. It uses a modified version of the LEVMAR library (Lourakis 2004). Minimization is carried out on a modified χ^2 of the residuals:

$$\chi_g^2(\vec{q}) = \sum_i g^2 \left(\frac{p_i - m_i(\vec{q})}{\sigma_i} \right), \tag{3}$$

where $\vec{q}$ is the vector of parameters to fit, p_i the background-subtracted value of galaxy image pixel i, $m_i(\vec{q})$'s the associated model sample (2-D galaxy model convolved with the local PSF model and resampled to image resolution), σ_i the uncertainty, and $g(u)$ a derivable function that reduces the influence of large deviations:

$$g(u) = \begin{cases} u_0 \log\left(1 + \dfrac{u}{u_0}\right) & \text{if } u \geq 0, \\ -u_0 \log\left(1 - \dfrac{u}{u_0}\right) & \text{otherwise.} \end{cases} \tag{4}$$

Setting $u_0 \approx 10$ makes the solution more immune against occasionally large non-Gaussian deviations such as contamination by neighbor sources or artifacts, while essentially preserving the convergence properties for regular objects.

Galaxy models tested so far include linear combinations of concentric Sérsic (1963), exponential and delta functions. Best fitting parameters, as well as estimates of uncertainties derived from the approximate Hessian matrix, are directly available as standard SExtractor measurements, both in pixel or world coordinates. Figure 2 shows examples of galaxy models fitted on deep imaging data.

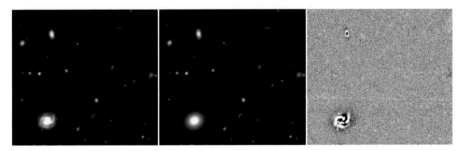

Figure 2.   Examples of bulge+disk galaxy models fitted to deep imaging data. *Left*: $1' \times 1'$ fragment of an image (CFHTLS D1-deep field, i-band). *Middle*: best-fitting galaxy models, convolved with the local PSF model. *Right*: residuals of the fit; residual features in late-type galaxies are dominated by spiral arms and star formation regions.

4. PSF Homogenization

One of the challenges faced by the Dark Energy Survey data management (Mohr et al. 2008) is to provide homogeneous photometry and morphometry for 300 million galaxies over 5000 square degrees of imaging data. Observations are ground-based and use a large, regular tiling pattern. Image stacks involve several tiles and therefore images with different shifts and different PSFs. A PSF homogenization procedure has been set up to avoid "jumps" in the PSF from tile to tile (Darnell et al. 2009). The procedure generates variable noise correlations at the scale of about 1 arcsec, but these are much easier to track than composite PSF variations. Briefly, PSFEx computes a set of convolution kernels $\vec{\kappa}_l$ which, when applied to the variable model PSF $\vec{\phi}(\vec{x})$, minimizes (in the χ^2 sense) the difference with a constant target PSF. The resulting PSF $\vec{\phi}^{(H)}$ is:

$$\vec{\phi}^{(H)} = \sum_l X_l(\vec{x})\vec{\kappa}_l * \vec{\phi}(\vec{x}), \tag{5}$$

where the $X_l(\vec{x})$'s are terms of a polynomial in pixel coordinates $\vec{x}$. The kernels $\vec{\kappa}_l$ are computed as combinations of the first ~ 60 vectors of the Gauss-Laguerre basis (e.g. Massey & Refregier 2005). All overlapping images, each convolved with its own combination of kernels, now deliver the same well-defined PSF everywhere and can be stacked before applying model-fitting.

4.1. Current Developments

I am currently concentrating my efforts on adding new model ingredients in SExtractor and allowing multiple sources to be fitted simultaneously in order to improve the photometric and morphometric accuracy in dense environments such as galaxy clusters.

Acknowledgments. Part of this work has been supported by grant 04-5500 ("ACI masse de données") from the French Ministry of Research. The author acknowledges the DESDM team for continuous and fruitful interactions, as well as support from the Dark Energy Survey consortium and hospitality of the University of Illinois department of Astronomy, where some of these developments were completed.

References

Barden, M., Haüßler, B., Peng, C. Y., McIntosh, D. H., & Guo, Y. 2009, A&A, submitted
Bertin, E., & Arnouts, S. 1996, A&AS, 117, 393
Darnell, T., et al. 2009, in Astronomical Data Analysis Software and Systems XVIII, edited by D. A. Bohlender, D. Durand, & P. Dowler (San Francisco, CA: ASP), vol. 411 of ASP Conf. Ser., 18
Kaiser, N., Squires, G., & Broadhurst, T. 1995, ApJ, 449, 460
Lourakis, M. I. A. 2004, LevMar: Levenberg-marquardt nonlinear least squares algorithms in C/C++. URL http://www.ics.forth.gr/~lourakis/levmar/
Massey, R., & Refregier, A. 2005, MNRAS, 363, 197
Mohr, J. J., et al. 2008, in Observatory Operations: Strategies, Processes, and Systems II, edited by R. J. Brissenden, & D. R. Silva (Bellingham, WA: SPIE), vol. 7016 of Proc. SPIE, 70160L
Peng, C. Y., Ho, L. C., Impey, C. D., & Rix, H. 2002, AJ, 124, 266
Sérsic, J. L. 1963, Boletin de la Asociacion Argentina de Astronomia La Plata Argentina, 6, 41
Vikram, V., Wadadekar, Y., Kembhavi, A. K., & Vijayagovindan, G. V. 2010, MNRAS, 409, 1379

Astronomical Data Analysis Software and Systems XX
ASP Conference Series, Vol. 442
Ian N. Evans, Alberto Accomazzi, Douglas J. Mink, and Arnold H. Rots, eds.
©*2011 Astronomical Society of the Pacific*

pyblocxs: Bayesian Low-Counts X-ray Spectral Analysis in Sherpa

Aneta Siemiginowska,[1] Vinay Kashyap,[1] Brian Refsdal,[1] David van Dyk,[2] Alanna Connors,[3] and Taeyoung Park[4]

[1] *Smithsonian Astrophysical Observatory, Cambridge, MA 02138*

[2] *University of California, Irvine, CA 92697*

[3] *Eureka Scientific, Oakland, CA 94602*

[4] *Yonsei University, Seoul, South Korea*

Abstract.   Typical X-ray spectra have low counts and should be modeled using the Poisson distribution. However, χ^2 statistic is often applied as an alternative and the data are assumed to follow the Gaussian distribution. A variety of weights to the statistic or a binning of the data is performed to overcome the low counts issues. However, such modifications introduce biases or/and a loss of information. Standard modeling packages such as XSPEC and *Sherpa* provide the Poisson likelihood and allow computation of rudimentary MCMC chains, but so far do not allow for setting a full Bayesian model. We have implemented a sophisticated Bayesian MCMC-based algorithm to carry out spectral fitting of low counts sources in the *Sherpa* environment. The code is a Python extension to *Sherpa* and allows to fit a predefined *Sherpa* model to high-energy X-ray spectral data and other generic data. We present the algorithm and discuss several issues related to the implementation, including flexible definition of priors and allowing for variations in the calibration information.

1. Introduction

Standard spectral modeling packages provide a library of physical models, sets of statistics and optimization methods to fit spectral data (e.g. *Sherpa*, XSPEC, or ISIS). Two classes of statistics are available: (1) Many flavors of χ^2 statistics with different weights to allow for fitting low counts X-ray spectra. However, even these statistics can lead to biased results when applied to the non-Gaussian X-ray data (see Arnaud et al. 2011); (2) Poisson based likelihood statistics provide unbiased results, e.g. *cash* (derived by Cash 1979) or *C*, a slightly modified form of *cash*. When using this approach, the background and source data have to be modeled simultaneously (which is not trivial) and there is no simple goodness-of-fit test, so often the various modifications of χ^2 have been used.

Poisson likelihood methods appropriate for low counts data require techniques for checking model selections and assessing "goodness-of-fit" that involve sampling from the posterior probability distribution. Available software packages contain the Poisson likelihood and standard optimization methods. However, there is no generally available software to probe the posterior probability and check the applied models using the Bayesian methods which include prior. Markov Chain Monte Carlo (MCMC) methods

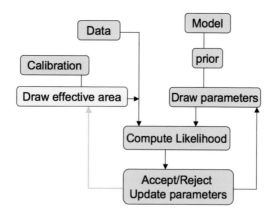

Figure 1.   Data flow diagram for the PyBLoCXS algorithm: Draw parameters from a "proposal distribution," calculate likelihood and posterior probability of the "proposed" parameter value given the observed data, use a Metropolis-Hastings criterion to accept or reject the "proposed" values. The step "draw effective area" to account for calibration uncertainties in the simulations is marked in yellow.

explore the posterior probability in Bayesian analysis. They are more reliable than the standard downhill optimization algorithms which can get stuck in local minima and are highly sensitive to stopping rules, especially for complex likelihood surfaces. The MCMC provides the full view of the posterior, and gives a direct way to calculate parameter uncertainties and p-values (and ppp-values).

We have developed a Bayesian model for exploring the posterior probability (van Dyk et al. 2001). The method has been implemented in the Python based package pyblocxs which can be used in *Sherpa* modeling and fitting application.

2. PyBLoCXS

PyBLoCXS is a sophisticated MCMC based algorithm designed to carry out Bayesian Low-Count X-ray Spectral (BLoCXS) analysis in the *Sherpa* environment. The code is a Python extension to *Sherpa* that explores parameter space at a suspected minimum using a predefined *Sherpa* model. It includes a flexible definition of priors and allows for variations in the calibration information. It can be used to compute posterior predictive p-values for the likelihood ratio test (see Protassov et al. 2002).

pyblocxs is based on the methods described in van Dyk et al. (2001) but employs a different MCMC sampler than the one presented in that article. In particular, pyblocxs has two sampling algorithms. The first uses a Metropolis-Hastings jumping rule that is a multivariate t-distribution with user specified degrees of freedom centered on the best spectral fit and with multivariate scale determined by the *Sherpa* function, covar(), applied to the best fit. The second algorithm mixes this Metropolis-Hastings jumping rule with a Metropolis jumping rule centered at the current draw, also sampling according to a t-distribution with user specified degrees of freedom and a multivariate scale determined by a specified scalar multiple of covar() applied to the best fit.

A general description of the MCMC techniques we employ along with their convergence diagnostics can be found in Appendices A.2 – A 4 of van Dyk et al. (2001) and in more detail in Chapter 11 of Gelman et al. (2004)

3. Applications

`pyblocxs` is open source code. It can be used to perform several important statistical tasks such as:

- Exploring parameter space and summarizing the full posterior or profile posterior distributions,
- Computing parameter uncertainties that can include calibration errors,
- Simulating data from the posterior predictive distributions,
- Testing for added spectral components by computing the Likelihood Ratio Statistic on replicate data and the ppp-value (posterior-predictive-p-values).

3.1. Calibration Uncertainties

Instrument calibration measurements such as an effective area of a telescope have known uncertainties. These uncertainties are often non-linear and cannot simply be added to the statistical uncertainties. A standard approach is to just ignore these uncertainties, mainly because there have been no methods to account for them in the analysis software. However, these uncertainties are important as they limit the final parameter constraints given by the observations. Also, their impact is more significant in the high signal to noise spectra (see Drake et al. 2006; Kashyap et al. 2008; Lee et al. 2011).

PyBLoCXS MCMC methods can take into account calibration uncertainty, by including an additional "update calibration" step in the MCMC loop (e.g. 'draw effective area' step in Fig. 1). The new calibration data (e.g.effective area) is drawn just before each computation of the likelihood and is used in the model evaluation and the final acceptance of the parameters in the loop. Lee et al. (2011) discuss the model that includes the calibration uncertainties and which was applied to Chandra spectra. They also compare several methods to account for these uncertainties and discuss some implications on the overall data analysis. Figure 2 shows an impact of the calibration errors on the uncertainties. The departure between the statistical and total errors is larger for the data with the highest signal to noise indicating the limit in the constraints that can be put on model parameters, implying that we cannot improve our knowledge about these sources with a larger number of counts.

4. Summary

`pyblocxs` is used to analyze astronomical counts data. It provides the MCMC simulations to explore parameter space of models applied to Poisson data. It requires *Sherpa* and was only tested on applications to simple one component models, while a parameter space can be complex for composite models. It is available as a *Sherpa* Python extension at `http://hea-www.harvard.edu/AstroStat/pyBLoCXS/index.html`.

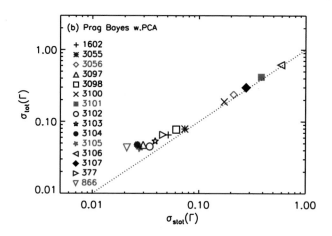

Figure 2.   Comparison of the statistical and total error which accounts for calibration uncertainties.   Points are results of a `pyblocxs` fit to the data including the calibration step in Fig.1.   The dotted line represents equality between the statistical(x-axis) and total (y-axis) errors (Lee et al. 2011). Note that for high counts sources where the effect of calibration uncertainty is most prominent, the overall error reaches a minimum even as the statistical error continues to decrease with increasing data quality.

Acknowledgments.   This project was supported by CXC NASA contract NAG8-39073 and NASA AISRP grant NNG06GF17G.

References

Arnaud, K., Smith, R., & Siemiginowska, A. 2011, Handbook of X-ray Astronomy (Cambridge: Cambridge University Press), 1st ed.

Cash, W. 1979, ApJ, 228, 939

Drake, J. J., Ratzlaff, P., Kashyap, V., Edgar, R., Izem, R., Jerius, D., Siemiginowska, A., & Vikhlinin, A. 2006, in Observatory Operations: Strategies, Processes, and System, edited by D. R. Silva, & R. E. Doxsey (Bellingham, WA: SPIE), vol. 6270 of Proc. SPIE, 62701I

Gelman, A., Carlin, J. B., Stern, H. S., & Rubin, D. B. 2004, Bayesian Data Analysis (London: Chapman & Hall), 2nd ed.

Kashyap, V. L., et al. 2008, in Observatory Operations: Strategies, Processes, and Systems II, edited by R. J. Brissenden, & D. R. Silva (Bellingham, WA: SPIE), vol. 7016 of Proc. SPIE, 70160P

Lee, H., et al. 2011, ApJ, submitted

Protassov, R., van Dyk, D. A., Connors, A., Kashyap, V. L., & Siemiginowska, A. 2002, ApJ, 571, 545

van Dyk, D. A., Connors, A., Kashyap, V. L., & Siemiginowska, A. 2001, ApJ, 548, 224

Astronomical Data Analysis Software and Systems XX
ASP Conference Series, Vol. 442
Ian N. Evans, Alberto Accomazzi, Douglas J. Mink, and Arnold H. Rots, eds.
© 2011 Astronomical Society of the Pacific

Using the Browser for Science: A Collaborative Toolkit for Astronomy

A. J. Connolly, I. Smith, K. S. Krughoff, and R. Gibson

Department of Astronomy, University of Washington, WA 98160

Abstract. Astronomical surveys have yielded hundreds of terabytes of catalogs and images that span many decades of the electromagnetic spectrum. Even when observatories provide user-friendly web interfaces, exploring these data resources remains a complex and daunting task. In contrast, gadgets and widgets have become popular in social networking (e.g. iGoogle, Facebook). They provide a simple way to make complex data easily accessible that can be customized based on the interest of the user. With ASCOT (an AStronomical COllaborative Toolkit) we expand on these concepts to provide a customizable and extensible gadget framework for use in science. Unlike iGoogle, where all of the gadgets are independent, the gadgets we develop communicate and share information, enabling users to visualize and interact with data through multiple, simultaneous views. With this approach, web-based applications for accessing and visualizing data can be generated easily and, by linking these tools together, integrated and powerful data analysis and discovery tools can be constructed.

1. Introduction

The last decade has witnessed a change in the way astronomers work with data. Two forces have driven this: multispectral surveys have been undertaken that cover many decades of the electromagnetic spectrum and initiatives are underway to develop a Virtual Observatory (VO) that will provide seamless access to this panchromatic view of the night sky. Federation of these massive data streams through the VO will provide an opportunity for astronomers to study the origin, structure and evolution of the universe with unprecedented accuracy. It does, however, come with many associated challenges: how do we provide access to data in a way that will enhance scientific returns, how do we link different functionality to provide an easy and intuitive interface to data and services, how do we account for the fact that not all astronomers wish to interact with data in the same way and how do we enable scientists to share their ideas and results with each other. With a 1000-fold increase in the data rate expected from the next generation of surveys, resulting in Petabyte databases containing 10^{10} sources each with hundreds of measured attributes, the need for a new way to explore and interact with these multiple data streams is becoming paramount.

There are a number of different ways to take applications and services, each aimed at different audiences, and make them accessible and useful to a broad community. One way is to create portals that handcraft a specific set of services designed to satisfy the needs of a typical user (e.g. querying catalog databases, displaying footprints of different survey geometries, extracting postage stamp images). This has the advantage of being able to address the needs of the average user by providing a simple interface

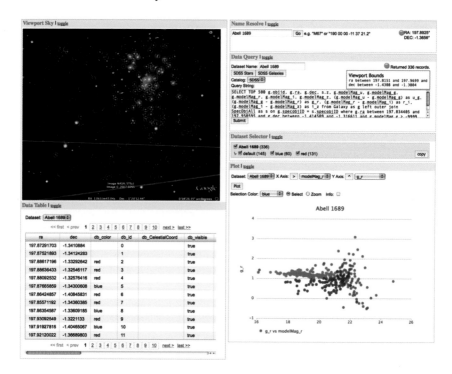

Figure 1.   A screenshot from the ASCOT framework showing a series of astronomy gadgets embedded within a webpage. A data gadget handles the communication
between these gadgets through the use of events registered to global variables. In this
way, when the name resolver returns the position of a source on the sky the viewport
is triggered to pan and zoom to that place. Sources returned by a database query are
displayed in a scatter plot and superimposed on the sky viewer. Interaction in one
gadget (e.g. coloring of the points) is mirrored in the others.

that can hide complex interactions between different data sets. It does, however, come at
a substantial cost. Different users have different priorities and interests. Questions can
use the same data resources (one using catalogs and the other images) but the interfaces
to accomplish these tasks are often substantially different. A single static portal cannot
easily capture the diversity of users requests. It cannot evolve with the user as his or
her experience and expertise grows and it does not allow users to bring in new services
if their scientific questions change with time (i.e. it limits our access to those questions
that have been decided ahead of time).

A second approach is to utilize emerging technologies for providing customizable
and personalized views of services and data. For example, in social networking sites
(e.g. Facebook), there has been a fundamental change in the way information is presented through the inclusion of atomic services that can be tailored to the needs of a
user. These services go by many names: widgets, gadgets, and portlets to name a few.
For consistency, we will refer to them as gadgets throughout. From news pages (e.g.
http://www.bbc.co.uk) to web pages (http://www.google.com/ig) to phones

(http://www.apple.com) it is now recognized that we need to create interfaces that users can change and adapt over time, rather than imposing a fixed view. Users need to have access to a common and extensive set of functionality but this must include the ability to customize and personalize the appearance of their interfaces and the ability to share these interfaces with others in their community.

2. ASCOT: A Framework for Gadget-Based Astronomy

ASCOT (an AStronomical COllaborative Toolkit) is a browser-based framework that enables the creation of gadgets designed for astronomical applications. Providing a browser-based interface ensures that applications are cross-platform and easily accessible. Ease of access also results in a low cost of buy-in for users (browser-based client style applications typically get $\sim 10\times$ the usage of clients that require a download). AS-COT is written in Javascript and is built upon the open-source, OpenSocial container Shindig. As shown in Figure 1, the initial implementation of ASCOT includes: visualization (image and catalog), data access, and data selection with all gadgets capable of inter-gadget interaction. Communication between gadgets is provided through the use of triggers or events associated with shared variables.

2.1. A Common Communication Layer

While gadget frameworks are fairly prevalent on the internet (from iGoogle to the BBC homepage) these gadgets are typically standalone applications. There is no communication between gadgets; you cannot link a weather and map gadget so that the weather forecast changes as you browse from city to city. For science applications, and in particular data discovery applications, interactivity is essential. We require a communication layer that is fast and has low latency. To accomplish this the inter-gadget communication model for ASCOT is resident within the browser and is based on triggered events. When a new gadget is loaded into the framework it registers itself with a central "data gadget." This data gadget contains a set of global variables to which functions can be assigned (or registered). In response to a change in a variable (for example a user requests a new position on the sky) any gadget that has registered a function with that variable will have its function triggered and will respond appropriately. The API for this data gadget is simple, requiring functions to add and remove variables and to register and unregister a function to a variable. In-browser communication removes the need for client-server polling which can introduce substantial latency in the response times. Figure 1 shows an example of the use of the ASCOT framework.[1] In this example there are five separate gadgets: an image viewer, a name resolver, a SQL database query gadget, a table viewer, and a scatter plot gadget. Each of these gadgets can be embedded within a webpage and work as an independent entity or loaded within the ASCOT framework and act as part of a coherent analysis tool. In the example given here, the name resolver has used the SIMBAD service to return the position of the galaxy cluster Abell 1689. The change in the variables associated with the sky position triggers an action on the image viewport that pans and zoom to this position. The data query gadget then uses a SQL query to search the SDSS archive for galaxies within the field-of-view of the image viewport and these sources are returned to a table viewer (the

---

[1]see http://ssg.astro.washington.edu/research.shtml

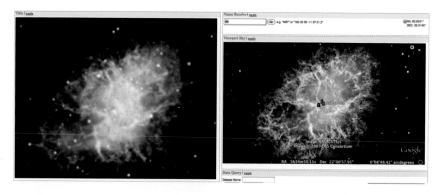

Figure 2.   In this example, different viewports are interlinked with the Google Sky gadget showing an image of the Crab Nebula taken by the Hubble Space Telescope and the World Wide Telescope gadget displaying the same region of the sky but with data from the Palomar Optical Sky Survey.

data table gadget), overlaid on the viewport, and plotted within the plot gadget. Plotting the color-magnitude diagram for Abell 1689, data sources within the scatter plot can be selected or "brushed" and the points in the viewport and table gadgets are highlighted in response.

Selection of which gadgets are available and how they should interact is specified within the webpage and, with gadget servers, can be defined directly by the user (see Figure 2.1 where Google Sky and World Wide Telescope gadgets show two separate data sets for the same region of the sky).

3. Conclusions

We demonstrate with the initial release of ASCOT how we can build simple, browser-based interactions to astronomical data sets. With the development of ASCOT we will generate granular tools which interact, can be moved, deleted, minimized, and added (customization). This will enable individual users to tailor their interface to particular tasks of interest to them (personalization). Once created, an interface with constituent granular tools will be savable and viewable by others (collaboration). These three aspects are at the core of our attempt to offer an alternative to traditional methods for data discovery in astronomy.

Acknowledgments.   We acknowledge support from NASA AISRP program NNX-09AK48G and from the University of Washington Return Royalty Fund.

Astronomical Data Analysis Software and Systems XX
ASP Conference Series, Vol. 442
Ian N. Evans, Alberto Accomazzi, Douglas J. Mink, and Arnold H. Rots, eds.
© *2011 Astronomical Society of the Pacific*

Automatic QSO Selection Using Machine Learning: Application on Massive Astronomical Database

Dae-Won Kim,[1,2] Pavlos Protopapas,[1] Charles Alcock,[1] Yong-Ik Byun,[2] and Roni Khardon[3]

[1]*Harvard-Smithsonian Center for Astrophysics, Cambridge, MA, USA*

[2]*Department of Astronomy, Yonsei University, Seoul, South Korea*

[3]*Department of Computer Science, Tufts University, Medford, MA, USA*

Abstract. We present a new QSO (Quasi-Stellar Object) selection algorithm using Support Vector Machine (SVM), a supervised classification method, on a set of multiple extracted times series features such as period, amplitude, color, and autocorrelation value. We train a model that separates QSOs from variable stars, non-variable stars and microlensing events using the richest possible training set consisting of all known types of variables including QSOs from the MAssive Compact Halo Object (MACHO) database. We applied the trained model on the MACHO Large Magellanic Cloud (LMC) dataset, which consists of 40 million lightcurves, and found 1,620 QSO candidates. During the selection none of the 33,242 known MACHO variables were misclassified as QSO candidates. In order to estimate the true false positive rate, we crossmatched the candidates with astronomical catalogs including the Spitzer Surveying the Agents of a Galaxy's Evolution (SAGE) LMC catalog. The results further suggest that the majority of the candidates, more than 70%, are QSOs.

1. QSO Selections

In order to separate QSOs from non-variable stars and variable stars, we quantify the variability characteristics of lightcurves using 11 time series features. These 11 features were independently proposed to quantify certain types of variability features such as amplitudes, periods, colors and distribution of data points. They can complement each other because they pick out different variability features. Thus, by using these multiple features, we can identify various types of variability characteristics (e.g. non-varying sources, periodic variables and non-periodic variables). Note that we selected these time series features not only for characterizing QSO time series but also for characterizing other types of variable sources or non-variable sources because we want to identify QSOs while excluding the other types of sources at the same time. We will give details about the 11 time series features in Kim et al. (2010). In brief, we use seven time series features that have been previously used for astronomical variability searches ($\sigma/\bar{m}$; color; period, Lomb 1976; period S/Ns, Scargle 1982; Stetson L, Stetson 1996; η, von Neumann 1941; and Con, Woźniak 2000) and four other time series features we developed for this work. The four other features are based on the autocorrelation function and cumulative sum (Ellaway 1978) of individual light curve.

We then used SVM, a supervised classification algorithm, to train a classification model for QSO selection. In general, SVM defines a linear hyperplane that separates a training set of two classes. To select a unique hyperplane among nearly infinite numbers of possible hyperplanes, SVM chooses the hyperplane which maximizes the margin between the training set of the two classes and the hyperplane, which is called *maximum margin* SVM. However, in many cases, it is not possible to find the optimal hyperplane that can perfectly separate two classes with the maximum margin. In other words, a training set of the two classes cannot be separated without errors. In order to solve this problem, a *soft-margin* SVM which allows errors in a training set (i.e. mislabeled samples) was proposed (Cortes & Vapnik 1995). The soft-margin SVM uses a constant, $C > 0$, to control trade-offs between maximizing the margin and minimizing the errors of a classification model. The parameter C should be manually determined to balance the margin with the errors. Small C allows a large margin between two classes and thus tends to ignore mislabeled samples. On the other hand, large C allows a small margin and tries to separate even mislabeled samples. Thus models with large C can over-fit the training data. For details about the soft-margin SVM, see Cortes & Vapnik (1995).[1]

Using the 11 times series features and SVM, we trained a classification model for QSO selection. We selected the richest possible training set consisting of all known variable types such as RR Lyrae, Cepheids, eclipsing binaries, long-period variables, microlensing events, Be stars and QSOs from the MACHO database (Alcock et al. 1996). We also included about five thousand non-variable stars in the training set. Finally, we applied the trained model on the whole MACHO LMC database and selected 1,620 QSO candidates.

2. Crossmatching Results

In order to estimate the true false positive rate without spectroscopic confirmation, we crossmatched our 1,620 QSO candidates with other astronomical catalogs. In the following subsections, we present the crossmatching results and the false positive rate estimated based on the crossmatched counterparts.

2.1. Spitzer SAGE Counterparts

It is known that mid-IR color selection is efficient at separating AGNs from other galaxies or stars because the spectral energy distribution of these types are substantially different from each other (Laurent et al. 2000; Lacy et al. 2004; Trichas et al. 2010; Kalfountzou et al. 2010). Based on the characteristics, Lacy et al. (2004) and Stern et al. (2005) introduced a mid-IR color cut to separate AGNs using the the Spitzer SAGE (Surveying the Agents of a Galaxy's Evolution; Meixner et al. 2006) catalog. Kozłowski & Kochanek (2009) (hereinafter KK09) employed the mid-IR color cut and selected about 5,000 AGN candidates from the Spitzer SAGE catalog.

To check whether our candidates are inside the mid-IR selection cut that KK09 used, we crossmatched them with the Spitzer SAGE LMC catalog containing 6 million mid-IR objects and found 1,239 counterparts. Among the crossmatched counterparts,

---

[1] We used radial basis kernel function to deal with non-linearly separable classes and cross-validation test to find the best C and γ. However, due to the page limit we are not able to provide the details in this paper. For more information, please see Kim et al. (2010).

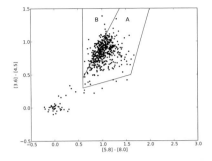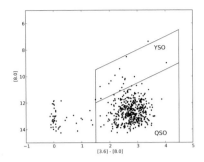

Figure 1.   Mid-IR color-color and color-magnitude diagrams of the Spitzer SAGE counterparts crossmatched with the QSO candidates. Each axis of the figure is either Spitzer magnitude or color. All sources inside the region A, B, QSO and YSO are potential QSOs (Kozłowski & Kochanek 2009). The majority of the candidates are inside the region A and QSO, which are thought to be the most promising QSO regions.

about 500 counterparts had been observed with at least three Spitzer IRAC (InfraRed Array Camera) bands. Note that we need a minimum of three Spitzer IRAC magnitudes to apply the mid-IR color cut. Figure 1 shows the color-color and color-magnitude diagrams of these counterparts. The solid line in the figure shows the mid-IR color selection cut. KK09 suggested that the sources inside region B could either be AGNs or black bodies such as stars, while the sources inside region A are likely AGNs (left panel). In the color-magnitude diagram (right panel), there are two regions as well. The region labeled as YSO is thought to be dominated by *young stellar objects* (YSO) while the region QSO is thought to be dominated by QSOs. Nevertheless, all the sources inside these four regions (AGN region) are potential QSOs. Among them, the sources inside the QSO and A regions are the most promising QSO candidates. As the figure clearly shows, most of the crossmatched QSO candidates are inside the QSO (88.2%; 480 out of 544) and A regions (76.9%; 407 out of 529), which implies that most of the candidates are likely true QSOs. The number of QSO candidates that are in both the QSO and A regions are 391 out of 529 (73.9%). Under the assumption that all the 391 candidates are QSOs, the false positive rate is 26.1%, which is the upper bound of false positive rate. There are only about 9% of the candidates outside the AGN region (9.3% outside A and B region, 9.0% outside YSO and QSO region), which is the lower bound of false positive rate.

3.   Summary and Future Work

We presented a new QSO selection algorithm based on the 11 time series features and a supervised classification. We first introduced 11 time series features to quantify a variety of variability characteristics of lightcurves. We then used Support Vector Machine (SVM) to train a classification model which separates QSOs from other types of variable stars and non-variable stars. Using the training set of the MACHO variables, 4,288 non-variables and the 58 known MACHO QSOs, we trained the classification model.

We applied the model on the whole MACHO LMC database consisting of 40 million lightcurves (i.e. 20 million from each MACHO band) to select QSO candidates. As a result, we found 1,620 candidates from the MACHO LMC database. To estimate the true false positive rate of the QSO candidates, we crossmatched the candidates with astronomical catalogs including the Spitzer SAGE LMC catalog. The crossmatching results confirmed that most of our candidates are promising QSO candidates. For instance, the majority of candidates with Spitzer counterparts are inside the AGN region that is defined by a mid-IR color cut and is known to be effective in confirming QSO candidates.

We will observe the QSO candidates with spectroscopic instruments to check if they are QSOs or not. Based on the projection of the models and the crossmatching results, we expect at least several hundred candidates would turn out to be QSOs. Using the confirmed QSOs and the false positives, we will improve our model. The current model is constructed based on the relatively small number of known QSOs (58 known MACHO QSOs) which may be too few to represent the true variability characteristics of all QSOs in the MACHO database. Thus using a large number of QSOs (i.e. more than a few hundred) would help improve the models.

In addition, our model is effective selecting not only QSOs but also other types of variable sources. Preliminary tests showed that selection efficiencies for periodic variables such as RR Lyrae, Cepheids and eclipsing binaries, were almost 100%. Even for long-period variables, microlensing events and Be stars which generally show non-periodic and irregular variability, selection efficiencies were about 80%.

References

Alcock, C., et al. 1996, ApJ, 461, 84
Cortes, C., & Vapnik, V. 1995, Machine Learning, 20, 273
Ellaway, P. H. 1978, Electroencephalography and Clinical Neurophysiology, 45, 302
Kalfountzou, E., Trichas, M., Rowan-Robinson, M., & et al. 2010, ArXiv:1005.4353
Kim, D.-W., Protopapas, P., Alcock, C., Byun, Y.-I., & Khardon, R. 2010, in Preparation
Kozłowski, S., & Kochanek, C. S. 2009, ApJ, 701, 508
Lacy, M., Storrie-Lombardi, L. J., Sajina, A., & et al. 2004, ApJS, 154, 166
Laurent, O., Mirabel, I. F., Charmandaris, V., Gallais, P., Madden, S. C., Sauvage, M., Vigroux, L., & Cesarsky, C. 2000, A&A, 359, 887
Lomb, N. R. 1976, Ap&SS, 39, 447
Meixner, M., Gordon, K. D., Indebetouw, R., & et al. 2006, AJ, 132, 2268
Scargle, J. D. 1982, ApJ, 263, 835
Stern, D., Eisenhardt, P., Gorjian, V., & et al. 2005, ApJ, 631, 163
Stetson, P. B. 1996, PASP, 108, 851
Trichas, M., Rowan-Robinson, M., Georgakakis, A., & et al. 2010, MNRAS, 405, 2243
von Neumann, J. 1941, Ann. Math. Statist., 12, 367
Woźniak, P. R. 2000, Acta Astron., 50, 421

Astronomical Data Analysis Software and Systems XX
ASP Conference Series, Vol. 442
Ian N. Evans, Alberto Accomazzi, Douglas J. Mink, and Arnold H. Rots, eds.
© *2011 Astronomical Society of the Pacific*

Fitting Galaxies on GPUs

Benjamin R. Barsdell, David G. Barnes, and Christopher J. Fluke

Swinburne University of Technology,
PO Box 218, Hawthorn VIC 3122 (Mail H39), Australia

Abstract. Structural parameters are normally extracted from observed galaxies by fitting analytic light profiles to the observations. Obtaining accurate fits to high-resolution images is a computationally expensive task, requiring many model evaluations and convolutions with the imaging point spread function. While these algorithms contain high degrees of parallelism, current implementations do not exploit this property. With ever-growing volumes of observational data, an inability to make use of advances in computing power can act as a constraint on scientific outcomes. This is the motivation behind our work, which aims to implement the model-fitting procedure on a graphics processing unit (GPU). We begin by analysing the algorithms involved in model evaluation with respect to their suitability for modern many-core computing architectures like GPUs, finding them to be well-placed to take advantage of the high memory bandwidth offered by this hardware. Following our analysis, we briefly describe a preliminary implementation of the model fitting procedure using freely-available GPU libraries. Early results suggest a speed-up of around 10× over a CPU implementation. We discuss the opportunities such a speed-up could provide, including the ability to use more computationally expensive but better-performing fitting routines to increase the quality and robustness of fits.

1. Introduction

Recent trends in commodity computing hardware have seen a dramatic shift first from single-core processors to multi-core and then to accelerated platforms like graphics processing units (GPUs). GPUs were originally designed to speed up 3D graphics calculations for video games, but their immense memory bandwidth and arithmetic capabilities have seen them re-purposed for the needs of scientific computing. While unquestionably powerful, their radically different, massively-parallel architectures have shaken up the software community. Astronomy is one of many fields trying to adapt to these changes in computing hardware.

While the area is still in its infancy, GPUs have already been shown to provide significant speed-ups across a range of astronomy problems. These include direct N-body simulation (e.g., Hamada et al. 2009), adaptive mesh refinement hydrodynamics (e.g., Schive et al. 2010), galaxy spectral energy density calculations (Jonsson & Primack 2010), gravitational microlensing (Bate et al. 2010), correlation for radio telescopes (e.g., Wayth et al. 2009) and coherent pulsar dedispersion (van Straten & Bailes 2010). The approach taken in each of these cases has, however, been *ad hoc* in nature — the transition to the GPU has been guided largely by hardware-specific documentation, code samples and simple trial and error. While such an approach has proven very successful for these early adopters, it is not clear that it will remain effective when it comes to more complex algorithms. Furthermore, in some cases the cost of re-implementing a code may be too large to gamble on a return (i.e., a speed-up) of unknown magnitude.

In this paper we discuss the potential for accelerating the process of galaxy fitting using GPUs. Rather than tackling the challenge blind, we instead use a generalised method based on algorithm analysis as outlined in Barsdell et al. (2010). The galaxy fitting process is described in Section 2, which is followed by a full analysis of the problem in Section 3. A preliminary implementation and results are described in Section 4 before our summary discussion in Section 5.

2. Galaxy Fitting

A common problem in astronomy is to fit analytic surface brightness profiles to observations of globular clusters or galaxies in order to extract structural parameters such as the effective radius, ellipticity or integrated flux magnitude. The fitting procedure is typically non-linear and of high dimensionality, demanding the use of powerful optimisation routines. Many codes have been developed to perform this task, including, e.g., ISHAPE (Larsen 1999), GALFIT (Peng et al. 2002) and GALPHAT (Yoon et al. 2010), which use the downhill simplex, Levenberg Marquardt and Markov-Chain Monte-Carlo methods respectively. While there is a variety of optimisation techniques in common use, most follow a similar pattern:

1. Evaluate a model on a grid using the current set of parameters (guessed initially).

2. Convolve the model with the point spread function (PSF) of the observation.

3. Compare the model and observation.

4. Adjust the model parameters.

5. Check finishing criteria and return optimised parameters if complete, otherwise repeat from Step 1.

Step 4 is where the specifics of a particular fitting routine come into play, while steps 1–3 generally remain unchanged between methods. Given the pixel-counts of modern astronomical observations, and the fact that most fitting routines require a very large number of iterations, the optimisation process can be highly computationally intensive. The quantity and quality of science results are thus tied to the available computing power and a code's ability to take advantage of it.

Computationally-limited problems like galaxy fitting are ideal candidates for acceleration. The fact that steps 1–3 are common to a large number of fitting routines allows us to study the problem with significant generality. Additionally, the image-based nature of the operations immediately suggests suitability for GPUs.

3. Algorithm Analysis

In order to determine whether galaxy fitting is a suitable application for GPU acceleration, we use an approach based around algorithm analysis as described in Barsdell et al. (2010). We begin by identifying known algorithms within the steps in the problem outline presented in Section 2:

1. Evaluation of a model on a grid is an example of a **transform** algorithm.

2. Convolution with the PSF is best done in Fourier space, requiring the **fast Fourier transform (FFT)** and regular **transform** algorithms.

3. Comparison of a model with an observation typically involves computation of the "sum of squared differences," which involves the **transform** and **reduce** algorithms.

Note that the algorithms required during optimisation of the model parameters in step 4 will depend on the chosen fitting routine.

It is thus seen that the fitting procedure makes use of only the transform, FFT and reduce algorithms, all of which are known to be very efficient on many-core architectures like GPUs (Barsdell et al. 2010).

The next step in the analysis is to look at the global characteristics of the computation. Both the transform and reduce algorithms have a work complexity of $O(N)$, indicating that a constant number of operations is performed for every image pixel. The FFT algorithm has a work complexity of $O(N \log N)$, indicating that $O(\log N)$ operations are performed for each of the N image pixels. The convolution step is thus expected to consume the majority of the processing time for large images.

The fitting procedure for an image of N pixels therefore requires reading $O(N)$ values, repeatedly performing $O(N \log N)$ arithmetic operations $O(N_{iter})$ times, and writing out $O(1)$ optimised parameter values (where N_{iter} is the number of iterations required to obtain a good fit). In the best case scenario then, the problem has a ratio of memory to compute operations, or *arithmetic intensity*, of $O(N_{iter} \log N)$.

The ability to achieve this arithmetic intensity depends on the memory access patterns of the component algorithms. Because the FFT algorithm requires an all-to-all communication pattern (i.e., each output value depends on every input value), it is necessary to have the entire image globally accessible during each iteration of the fitting routine. This rules out storing the image data in very small caches (e.g., the *shared memory* on NVIDIA GPUs) between iterations. However, there is generally more than enough main memory on a GPU to hold an entire image. This means that the data may be left on the device for all N_{iter} iterations, with no need to go back to the host's memory or disk until the final results have been obtained. The limiting factor will instead be the internal memory bandwidth of the device. This is a good result, as current GPUs have significantly more memory bandwidth than CPUs, and one can expect a corresponding speed-up.

4. Implementation Results

Given the positive results of the algorithm analysis in Section 3, a prototype implementation of the galaxy fitting problem was deemed worthwhile. NVIDIA's CUDA[1] platform was used to interface to the GPU. FFTs were performed using the CUFFT

---

[1]http://www.nvidia.com/object/cuda_home_new.html

library[2], and the Thrust[3] C++ library was used for its efficient implementations of the transform and reduce algorithms.

Given the subtleties of mature codes like GALFIT (Peng et al. 2002), performing an accurate comparison with our prototype GPU code is not yet possible. Preliminary results, however, suggest a speed-up in the main computations of around 10× when using a single NVIDIA Tesla C1060 GPU versus an Intel Nehalem CPU. Profiling results also indicate that the GPU hardware is being used efficiently by all of the algorithms in the code. These results support the conclusions of our analysis in the previous section.

5. Discussion

Many-core architectures like GPUs are now an important part of the computing landscape. While many software challenges remain, a generalised approach to analysing astronomy problems has proven very useful in tackling new GPU codes.

Galaxy fitting looks to be a promising application of GPU technology. Significant speed-ups present the opportunity to perform faster fits, which may be crucial for the next generation of galaxy surveys. Alternatively, the additional processing speed could be fed back into the fitting routine to provide fits of much better quality in the same length of time, helping to overcome common problems such as local minima and unphysical results.

While useful as a profiling tool, our prototype GPU code requires significant further development before it can be considered a viable alternative to other galaxy fitting codes in use by the astronomy community. Future work will address such development.

Given the generality of our analysis, it is likely that other fitting problems in astronomy would also benefit from GPU acceleration. If one allows flexibility in the dimensionality of the problem, procedures such as spectral line or cube fitting become possible. Such problems will also be the subject of future work.

References

Barsdell, B. R., Barnes, D. G., & Fluke, C. J. 2010, MNRAS, 408, 1936
Bate, N. F., Fluke, C. J., Barsdell, B. R., Garsden, H., & Lewis, G. F. 2010, New Astron. Accepted for publication June 2010, 1005.5198
Hamada, T., et al. 2009, Comput. Sci. Res. Devel., 24, 21. URL http://www.springerlink.com/content/j2881042547v4403
Jonsson, P., & Primack, J. R. 2010, New Astron., 15, 509
Larsen, S. S. 1999, A&AS, 139, 393
Peng, C. Y., Ho, L. C., Impey, C. D., & Rix, H. 2002, AJ, 124, 266
Schive, H., Tsai, Y., & Chiueh, T. 2010, ApJS, 186, 457
van Straten, W., & Bailes, M. 2010, ArXiv e-prints. 1008.3973
Wayth, R. B., Greenhill, L. J., & Briggs, F. H. 2009, PASP, 121, 857
Yoon, I., Weinberg, M., & Katz, N. 2010, ArXiv e-prints. 1010.1266

---

[2]http://developer.nvidia.com/object/cuda_archive.html

[3]http://code.google.com/p/thrust

Astronomical Data Analysis Software and Systems XX
ASP Conference Series, Vol. 442
Ian N. Evans, Alberto Accomazzi, Douglas J. Mink, and Arnold H. Rots, eds.
© *2011 Astronomical Society of the Pacific*

The Gemini Recipe System: A Dynamic Workflow for Automated Data Reduction

Kathleen Labrie, Paul Hirst, and Craig Allen

Gemini Observatory, 670 N. A'ohoku Pl, Hilo, HI 96720, USA

Abstract.   Gemini's next generation data reduction software suite aims to offer greater automation of the data reduction process without compromising the flexibility required by science programs using advanced or unusual observing strategies. The Recipe System is central to our new data reduction software. Developed in Python, it facilitates near-real time processing for data quality assessment, and both on- and off-line science quality processing. The Recipe System can be run as a standalone application or as the data processing core of an automatic pipeline. Building on concepts that originated in ORAC-DR, a data reduction process is defined in a Recipe written in a science (as opposed to computer) oriented language, and consists of a sequence of data reduction steps called Primitives. The Primitives are written in Python and can be launched from the PyRAF user interface by users wishing for more hands-on optimization of the data reduction process. The fact that the same processing Primitives can be run within both the pipeline context and interactively in a PyRAF session is an important strength of the Recipe System. The Recipe System offers dynamic flow control allowing for decisions regarding processing and calibration to be made automatically, based on the pixel and the metadata properties of the dataset at the stage in processing where the decision is being made, and the context in which the processing is being carried out. Processing history and provenance recording are provided by the AstroData middleware, which also offers header abstraction and data type recognition to facilitate the development of instrument-agnostic processing routines. All observatory or instrument specific definitions are isolated from the core of the AstroData system and distributed in external configuration packages that define a lexicon including classifications, uniform metadata elements, and transformations.

1. Dynamic Workflow

The Gemini facility instruments are highly diverse in wavelength regimes and modes of observation. The instrument suite spans the optical, near-infrared, and mid-infrared wavelength regimes, each requiring significantly different data processing procedures and tools.

Within each wavelength regime, both imaging and spectroscopy are offered, including long-slit, multi-object and integral-field spectroscopy. Additionally, in the optical regime, all three spectroscopy modes can be used with an electronic "nod-and-shuffle" technique, and in the mid-infrared polarimetry is offered.

The extreme diversity of the data and scientific objectives of the programs makes significant demands on the design and implementation of a data processing software suite, especially on the automation and pipeline features. It is clear that a static automated data reduction system, a standard linear pipeline, would not be sufficient to obtain reliable data quality metrics and eventually science quality products. A multi-purpose, smart system is required.

The system must be able to make decisions on the fly based on the headers, the pixels statistics, and on the availability, or non-availability, of the calibration data. Those decisions must be made at the stage of processing where they are needed. Nothing is pre-scripted; the sequence of events is dynamically adapted during processing. This is drastically different from most existing pipelines which usually process the same type of data over and over again, the same exact way every time.

The Gemini Recipe System builds heavily on the concepts and experiences gained in building and using previous generations of astronomy data reduction software. It combines the flexibility of interactive data reduction as provided by PyRAF (Greenfield & White 2000) and the Gemini IRAF tasks, with the on-the-fly automatic flow control and generic data reduction module approach of the ORAC-DR pipeline (Cavanagh et al. 2003).

2. Instrument-Agnostic Programming

At the core of the Recipe System is AstroData, a Python meta-class that serves as an active abstraction for a dataset. As designed, AstroData provides the abstraction necessary for instrument-agnostic programming through header abstraction, data type recognition, validation, metadata propagation, history and provenance. All observatory or instrument-specific definitions are isolated from the core of the AstroData system and distributed in external configuration packages.

The configuration packages define a lexicon that includes: the AstroDataType libraries for the classification, the Descriptors libraries for the uniform metadata elements (e.g. header keyword mapping), and the Primitives libraries for the transformations.

3. A Scientific Language

Recipes are reduction sequences meant to function as a science level interface to the reduction system. They consist of lists of sequential instructions. These instructions are in general Primitives, though in actual fact these steps can be other Recipes as well. Recipes do not contain computer language artifacts such as conditionals or variables. They are not written in Python, but in their own very simple syntax. The flexibility of a Recipe is not in any explicit conditional behavior within it, but rather in its implicit conditional behavior based on the dataset's AstroDataType.

Our guidelines call for Recipes and Primitives to bear names that are meaningful within astronomical data reduction semantics. Given the sequential nature of the Recipe, upon viewing a Recipe it should be clear to a typical science user what steps it will run on the data and the intent of those steps. For example, the "flatfieldCorrect" Primitive instruction in a Recipe means, "do the appropriate type of flat field correction for the input data." If that can be done in a generic manner, a common Primitive can be used for all types of data. However, if some type of dataset requires a special version of the code to do this semantically identical step, then that specialized version is automatically loaded and used based on the type of dataset at that point in the processing.

Primitives are associated together in sets defined by the dataset type on which they run. These sets are collected as Python classes and since Python supports both hierarchical and multiple inheritance, there is ample flexibility to share Primitives through object oriented inheritance arrangements.

4. The Recipe System Infrastructure

The Recipe System is launched via a controlling application. In pipeline mode this application is named "reduce". A great strength of the system is that the controlling application can also be a PyRAF task, transforming a pipeline system into an interactive system that allows users to experiment and optimize their data reduction procedure. Figure 1 illustrates broadly how the various components described in this section relate to one another.

When the application is launched, the RecipeManager takes charge. Its task is to build a ReductionObject instance that will encapsulate everything about the reduction. The controlling application, such as "reduce," gives the RecipeManager either a dataset from which data type can be inferred, or a specific AstroDataType for which a ReductionObject will be built.

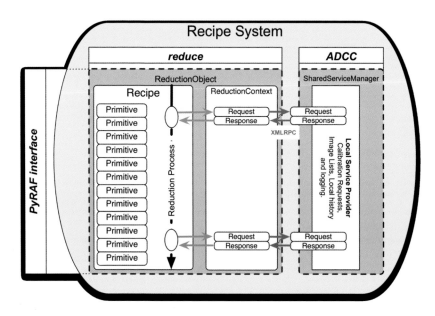

Figure 1.   The "reduce" application or a PyRAF task initiates the Recipe System. The reduction occurs within the ReductionObject that contains the Recipe, the Primitives and the ReductionContext. Resources are accessed by the AstroData Control Center and communicated through an interface provided by the ReductionContext.

The ReductionObject is composed piece by piece by the RecipeManager via the RecipeLibrary which serves as the interface to the files containing the Recipes, Primitives, and default input parameters to the Primitives. The Recipes, Primitives and parameters attached to the ReductionObject are used to create a ReductionContext instance that will handle communications with the control loop. The Recipe is translated to Python code, compiled into a function object and dynamically attached to the ReductionObject. Once the ReductionObject is fully composed, the Recipe is executed by the control loop.

The ReductionContext object exists primarily to store rather than act on information. Though it has an interface to make requests for calibrations and report statistics, it does not contain the code to make the request itself, but merely stores the information so that it is available to the control system that will handle the request appropriately, depending on the context. The ReductionContext stores the following information: input and output data URIs, input parameters, recipe status and commands, calibration interfaces, reduction history, and data quality metrics.

The Primitives share a particular uniform interface, accepting only the 'self' argument for the instance and a ReductionContext instance that contains all the required input information. The end result is that use of the Reduction Context as a 'single argument' is really a variable argument list passing mechanism.

The Astrodata Control Center (ADCC) is a shared-service provider for Reduction Objects. It is multi-threaded and multi-process safe. It communicates with the ReductionContext with XMLRPC. The controlling application registers and unregisters with the ADCC. The ADCC is responsible for access to local and external service providers.

5.   Closing Remarks

The Recipe System is Gemini's new data processing software platform for end-users reducing data on their desktop. It is also designed to form the heart of an automated data reduction pipeline that will provide the night observer with near-real time data and site quality assessment measurements from the delivered data and allow for automated instrument performance monitoring.

Acknowledgments.   The Gemini Observatory is operated by the Association of Universities for Research in Astronomy, Inc., under a cooperative agreement with the NSF on behalf of the Gemini partnership: the National Science Foundation (United States), the Science and Technology Facilities Council (United Kingdom), the National Research Council (Canada), CONICYT (Chile), the Australian Research Council (Australia), Ministério da Ciência y Tecnología (Brazil), and Ministerio de Ciencia, Tecnología e Innovación Productiva (Argentina)

References

Cavanagh, B., Hirst, P., Jenness, T., Economou, F., Currie, M. J., Todd, S., & Ryder, S. D. 2003, in Astronomical Data Analysis Software and Systems XII, edited by H. E. Payne, R. I. Jedrzejewski, & R. N. Hook (San Francisco, CA: ASP), vol. 295 of ASP Conf. Ser., 237

Greenfield, P., & White, R. L. 2000, in Astronomical Data Analysis Software and System IX, edited by N. Manset, C. Veillet, & D. Crabtree (San Francisco, CA: ASP), vol. 216 of ASP Conf. Ser., 59

Astronomical Data Analysis Software and Systems XX
ASP Conference Series, Vol. 442
Ian N. Evans, Alberto Accomazzi, Douglas J. Mink, and Arnold H. Rots, eds.
© *2011 Astronomical Society of the Pacific*

A Pattern Recognition System for the Automated Tracking and Classification of Meteors Using Digital Image Data

Kenny McGarvey,[1] Ricardo Vilalta,[1] Marilia Samara,[2] and Robert Michell[2]

[1]*Department of Computer Science, University of Houston, Houston Texas 77204-3010, USA*

[2]*Department of Space Science, Space Science and Engineering Division, Southwest Research Institute, San Antonio Texas 78228-0510, USA*

Abstract.   In this project, we aim to develop an efficient and reliable method to search through vast amounts of digital image data of the sky (at video rates of 30 fps) to identify meteors. We describe a pattern recognition system that accomplishes this goal in two steps. The first step keeps track of luminous objects that follow a straight path through a set of frames; multiple features become then available for each moving luminous object. During the second step, the system automatically labels the moving object as either a meteor or not (e.g., or satellite). The benefit of such a pattern recognition tool is twofold: 1) it obviates the process of searching through image data by eye, which is infeasible for producing a database with reliable statistics; 2) derived information can provide a better understanding of the mass distribution of meteors, which is crucial for determining the total mass flux incident into the upper atmosphere.

1. Introduction

There is broad and far reaching interest in studying meteor trail physics. For example, understanding the properties of the incoming meteoroid provides information about the interplanetary dust environment around Earth and how it has evolved during the evolution of the solar system. There are currently two main methods for this type of study: radars and imagers. The use of radars is more mature, but the techniques for deriving mass are not without limitations and assumptions about the radar scattering cross section of the plasma trail. The use of television imaging for meteor detection is relatively young, and has the main advantage of incurring fewer errors during mass calculations because of simpler and fewer assumptions than radar techniques. The main disadvantage is the tedious process of sorting through the television data; this has been a major drawback to setting up consistent, long term observations for detecting small meteors. Additionally, the large amount of data generated from an imager running at 30 or 60 frames per second (fps) can be around 15 GB per hour.

The goal of this study is to develop an algorithm to search through large volumes of television imaging data in order to identify and quantify meteors in an accurate and consistent manner. We describe a pattern recognition system for the automatic detection of meteors using image processing and classification. Our system will be tailored to the identification of meteors by exploiting the precise signature of these objects along several frames (e.g., trajectories are linear, few frames to cross the sky, trajectory displays a rectangular shape, etc.). Such singular behavior leaves patterns that can be quickly identified. The general mechanism for our proposed software tool can be divided into two main tasks: 1) removal of background and noise, and 2) meteor detection. We describe each of these tasks next.

2. Background and Noise Removal

Our current system receives as input a video capturing part of the night sky; videos are usually 8 minutes long and contain around 16,000 frames. Each frame is a pixel matrix of size 256×256. A pixel at coordinate (x, y) contains the value of gray intensity at that point in the sky.

For a new frame F_n, a first step is to remove some background noise by making all pixels binary-valued as follows. Let $V_{x,y} = I(x, y)$ be the value of gray intensity at entry (x, y) in the matrix. We assign a value of minimum intensity (black) if $V_{x,y} < \theta$, and a value of maximum intensity (white) if $V_{x,y} \geq \theta$ (we currently set θ empirically). This enables us to work with binary-valued pixels only, getting rid of faint sources, and highlighting sources bright enough for analysis. Figure 1 illustrates this operation.

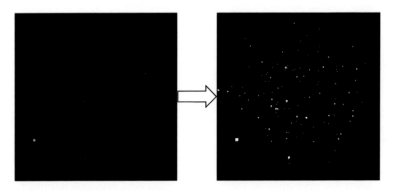

Figure 1.   Our first step is to simplify the image to enable us to work with two values only, getting rid of tenuous sources, and highlighting bright sources.

The next step is to build an average background that we will use to subtract from our current frame F_n to eliminate all luminous objects that are static on the sky during a short time interval. To proceed, we create a queue Q of 50 frames that immediately precede current frame F_n. When we move to the next frame, the current frame will be added to the queue, while the last frame will be removed. In short, we keep a window of size 50 containing those frames that precede current frame F_n. Pixels in all frames in Q are averaged to produce a single average frame F_Q. This creates a background with no noise or fast moving objects. Figure 2 illustrates the process of averaging over several frames.

To remove noise corresponding to flickering stars, the average background image is dilated by one pixel (i.e., every white pixel is increased in size by making white its nearest neighbors). This increases the radius of all luminous objects by one pixel. The averaged background image F_Q is then subtracted from the current frame F_n. This has

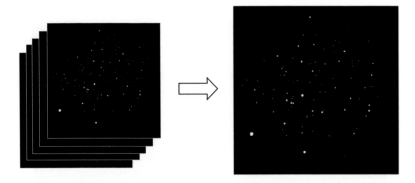

Figure 2.　A queue of 50 frames previous to current frame F_n is used to produce a new image by averaging over all pixel values.

the effect of removing objects that appear in both F_n and F_Q (e.g., stars). Figure 3 shows an example of the effect of background removal. The last frame shows a few luminous objects corresponding to those that have just appeared in the sky.

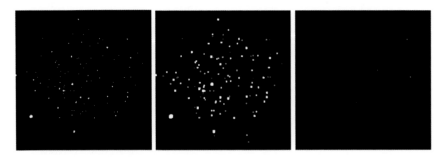

Figure 3.　Left: Current frame F_n. Middle: Averaged background F_Q image after dilation. Right: The effect of subtracting F_Q from F_n.

3.　Meteor Detection

We now explain how to identify meteors on an image where background has been already removed. We identify luminous structures using a technique known as blob detection (Treiber 2010). An 8-neighborhood algorithm is applied to each white pixel enlarging the size of the blob as long as white pixels remain connected. To eliminate faint sources, any blob with a size less than a minimum threshold ϑ is ignored (in our experiments we set $\vartheta = 3$). Our blob detection mechanism helps us extract several properties of the object under analysis, such as size, centroid, and shape. Due to the exposure of each frame, a particular signature of all meteors is that they will appear elongated; we thus eliminate all non-elongated blobs from the image using object properties. We infer that any object not eliminated by the previous step corresponds to a meteor. Figure 4 shows an example of the blob detection mechanism during meteor identification.

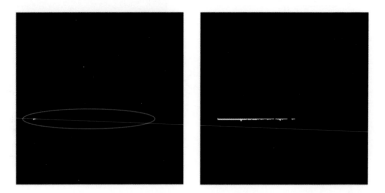

Figure 4.   After applying the blob detection mechanism, it is possible to identify meteors by searching for long elongated bright signals.

4.   Conclusions and Future Work

Our pattern recognition system provides a preliminary approach to meteor identification by isolating and extracting bright linear trajectories from video images. Many components of the system are amenable to refinement. For example, during the removal of background and noise, it is unclear what the correct value of θ is to transform the initial frame into a binary-valued frame. There is a possibility of eliminating faint sources corresponding to our target class (meteors). We plan to learn this value using machine learning techniques (Duda et al. 2001; Hastie et al. 2009). Our goal is to approximate the probability distribution of meteors and other luminous objects according to their brightness, which would enable us to predict the posterior probability of the class (meteor or other source) conditioned on brightness.

In addition, the background removal operation can be improved in several ways. The window size that is used to average over previous frames can be selected dynamically to increase computational efficiency (i.e., to decrease CPU time). The process of dilating luminous objects to avoid the effect of variability of the light source can also be adjusted to spread over a larger region. Such adjustments depend on the characteristics of the background during video recording, and can be optimized using a history of previous recordings.

Acknowledgments.   This work was supported by National Science Foundation under grants IIS-0812372 and IIS-0448542.

References

Duda, R., Hart, P., & Stork, D. 2001, Pattern Classification (New York, NY: Wiley-Interscience), 2nd ed.
Hastie, T., Tibshirani, R., & Friedman, J. 2009, The Elements of Statistical Learning: Data Mining, Inference and Prediction (Dordrecht: Springer), 2nd ed.
Treiber, M. 2010, An Introduction to Object Recognition (Dordrecht: Springer), 1st ed.

Astronomical Data Analysis Software and Systems XX
ASP Conference Series, Vol. 442
Ian N. Evans, Alberto Accomazzi, Douglas J. Mink, and Arnold H. Rots, eds.
©2011 *Astronomical Society of the Pacific*

ℒIRA — The Low-Counts Image Restoration and Analysis Package: A Teaching Version via R

A. Connors,[1] Nathan M. Stein,[2] David van Dyk,[3] Vinay Kashyap,[4] and Aneta Siemiginowska[4]

[1] *Eureka Scientific, 2452 Delmer St. Suite 100, Oakland, CA 94602*

[2] *Harvard Statistics, 1 Oxford St., Cambridge, MA 02138*

[3] *UCI Statistics, Bren Hall 2019, Irvine, CA 92697*

[4] *Harvard-Smithsonian Center for Astrophysics, 60 Garden St., Cambridge, MA 02138*

Abstract. In low-count discrete photon imaging systems, such as in high energy astrophysics, the spatial distribution of a very few (or no!) photons per pixel can indeed carry important information about the shape of interesting emission. Our Low-counts Image Restoration and Analysis package, ℒIRA, was designed to: 'deconvolve' any unknown sky components; give a fully Poisson 'goodness-of-fit' for any best-fit model; and quantify uncertainties on the existence and shape of unknown sky components. ℒIRA does this without resorting to χ^2 or rebinning, which can lose high-resolution information. However, running it thoughtfully requires understanding of several key areas, since it combines a Poisson-specific multi-scale model for the sky with a full instrument response, within a (Bayesian) probablility framework, sampled via MCMC. To this end, we have created and are releasing a 'teaching' version of ℒIRA. It is implemented in R. The accompanying tutorial and R-scripts step through all the basic analysis steps, from simple multi-scale representation and deconvolution; to model-testing; setting quantitative limits; and even simple ways of incorporating uncertainties in the instrument response.

1. Intro: Wonder, Glee, Skepticism, and ℒIRA

As one confronts beautiful, beautifully processed, astronomical images — such as many in these proceedings — who does not feel the pull of wonder? As well, when one recognizes that a newly visible feature appears to match one's theory, isn't there a sharp pull of glee? Yet, in this paper, we advocate doubt: "where are the error bars?"

ℒIRA was developed precisely to quantify this doubt for low-count Poisson data. To do this, ℒIRA brings together several different kinds of machinery, from Multi-scale (MS) models to Markov chain Monte Carlo (McMC) in a Bayesian framework (Esch et al. 2004; Connors & van Dyk 2007; van Dyk et al. 2006). Although made for Poisson counts, our schema that consists of a flexible or non- or semi-parametric model, including a background or Null model, and operating within a full likelihood framework, can serve as a model for more general data. The combination can at first feel non-intuitive for even seasoned researchers. Hence, we have created a 'teaching' version, with many

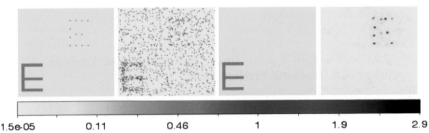

1.5e-05   0.11   0.46   1   1.9   2.9

Figure 1.   SkyTruth(unconvolved); Simulated Data; Null Model(unconvolved); *LIRA* result.

examples, within the easy-to-use public statistical package 'R'.[1] *LIRA* is available from: nathanmstein at gmail.com or aconnors at eurekabayes.com.

Here, we briefly exhibit parts of one of the teaching examples. It is based on a hypothetical 'skytruth' of a diffuse component (a broad letter E) and a cluster of point sources (also forming a letter E) on a flat background, in 128×128 bins, as shown in the first figure. The instrument smearing, or point-spread function (PSF), is assumed to be a circular Gauss-Normal distribution with $\sigma = 1.5$ bins. Simulated Poisson data D based on these is shown in the 2nd panel. We display a 'Null Model' of the diffuse emission based on hypothetical measurements and theory: a broad 'E' — in the 3rd panel. The simulated data, PSF, and Null model, are inputs to *LIRA*; one of the outputs is the mean 'mismatch' between the data and theory, shown in the last panel of the 1st figure.

2. *LIRA* Mechanics

LIRA can be termed a 'forward-fitting' likelihood-based method, built under a 'Bayesian umbrella'. That is, we use a Bayesian framework to successively add 'spokes' to the total likelihood: Poisson likelihood of the data D (red); given a Null Model with parameters θ, designated by $M(\theta)$ (blue); and the Instrument Response by IR (brown). Then, using Bayes' theorem, the posterior probability can be written as in the first panel of the second figure, where @ designates a convolution.

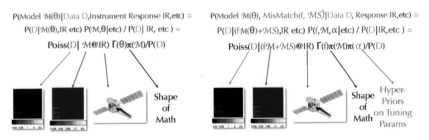

Figure 2.   Bayes Umbrellas: Adding a spoke (right panel; green).

In this form, it is easy for us to add a *model/data mis-match* 'spoke' (green) to our Bayesian umbrella. In our low-count regime, we formulate the *mode/data mis-match*

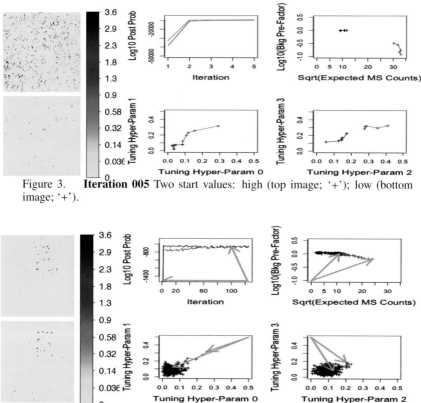

Figure 3.   **Iteration 005** Two start values: high (top image; '+'); low (bottom image; '+').

Figure 4.   **Iteration 125.** Two start values: high (top image, '+'); low (bottom image, '+'). Orange arrows roughly indicate burn-in range for high starting values.

term to be a *prefactor* times the null model, plus a Poisson-tailored *multi-scale model* (Esch et al. 2004; Connors & van Dyk 2007; Kolaczyk & Nowak 2005) that will handle both fine details and broad features. But now there are a great many parameters: rates at each successively finer multi-scale level, given the previous level; tuning (or smoothing or regularization) hyper-parameters, for each level; the Null Model prefactor. Hence rather than e.g., a Powell or Levenberg-Marquardt method for finding a mode, we use Markov chain Monte Carlo to map out the full probability space. This allows us to get both a 'best fit', and a way to express uncertainties on any feature from the data/model mis-match.

3.   Running $\mathcal{L}IRA$

In the next several figures, we illustrate McMC in action, mapping out the shape of our posterior likelihood (or Bayesian Umbrella from the second figure). It shows both a 'burn-in' phase and a converged phase. Finally we illustrate that, in order to get full quantitative limits, we must perform the same $\mathcal{L}IRA$ analysis on a handful of simulated data sets based on the Null Model (convolved with the instrument response). We then use a small subset of the parameters — in this case, the total counts inferred to be in

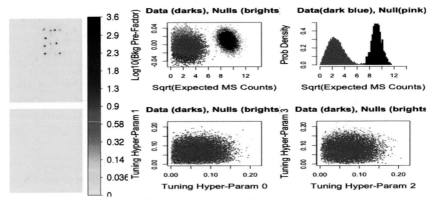

Figure 5.   *£IRA* **Results, after burn-in.** Left: Mean Images from Data (top) vs. Simulated Nulls (bottom). Right: Distributions of Data (dark colors) vs. Simulated Nulls (bright colors).

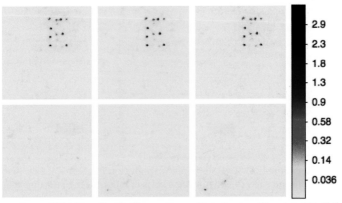

Figure 6.   *£IRA* **Results, limits on shape.** Data (top) vs Simulated Null (bottom): Left: lower 5% limit; Middle: Mean; Right: upper 95% limits.

the multi-scale (*MS*) component — as a *summary statistic* of the 'distance' between the data and the null of the summary statistic give the upper and lower bounds on the *shape* of the Data/Null-Model mis-match.

Acknowledgments.   This work was supported by NSF Awards DMS 04-06085, DMS-09-07522, and DMS-09-07185. A. C. and N. M. S. were supported by NASA AISRP grant NNG06GF17G. *£IRA* is based on EMC2, and is an out-growth of CHASC, the California Harvard Astrostatistics Collaboration.

References

Connors, A., & van Dyk, D. A. 2007, in Statistical Challenges in Modern Astronomy IV, edited by G. J. Babu & E. D. Feigelson (San Francisco, CA: ASP), vol. 371 of ASP Conf. Ser., 101
Esch, D. N., Connors, A., Karovska, M., & van Dyk, D. A. 2004, ApJ, 610, 1213
Kolaczyk, E. D., & Nowak, R. D. 2005, Biometrika, 92, 119
van Dyk, D. A., et al. 2006, Bayesian Analysis, 1, 189

Astronomical Data Analysis Software and Systems XX
ASP Conference Series, Vol. 442
Ian N. Evans, Alberto Accomazzi, Douglas J. Mink, and Arnold H. Rots, eds.
©2011 Astronomical Society of the Pacific

On-Line Access and Visualization of Multi-Dimensional FITS Data

Pavol Federl, Arne Grimstrup, Cameron Kiddle, and A. R. Taylor

University of Calgary,
2500 University Drive NW, Calgary, AB T2N 1N4, Canada

Abstract. As astronomical data sets continue to grow in size, the challenges involved in data access and visualization increase in complexity. Obtaining entire data sets is time consuming, awkward and in many cases even impractical. Furthermore, visualization of data is traditionally accomplished off-line, requiring specialized software that is often operating system or library dependent and cannot be easily installed everywhere. In this paper we describe two interactive web-based tools we implemented for accessing and visualizing multidimensional FITS data. The intent in offering on-line tools is to provide astronomers with easy access to data from any computer with a modern web browser. The described tools were developed by the CyberSKA project and are accessible on its web portal.

1. Introduction

The goal of the CyberSKA project is to develop a scalable and distributed cyberinfrastructure platform to meet evolving science needs of the Square Kilometer Array (SKA). We plan to deploy the CyberSKA platform as a distributed system, initially consisting of sites from several North American universities. Each participating site will host a variety of data management, processing, visualization and other services. Users will be provided with transparent access to these distributed services via a collaborative web portal enhanced with social networking features (Kiddle et al. 2011).

As a starting point we are creating a cyberinfrastructure to support the current large-scale astrophysical data needs generated by GALFACTS, PALFA and other high data volume SKA Pathfinder projects. The current web portal for the CyberSKA project[1] is implemented on top of the Elgg[2] open source social networking platform. We have developed two web-based tools for accessing and visualizing multidimensional FITS data and integrated them into this portal. These tools and some of the most relevant implementation details are described below.

2. FITS Viewer Tool

One of Elgg's important features is the ability to share files of any type. Since users often wish to visualize FITS files uploaded by other users, we have implemented an interactive visualization module that can display FITS files directly inside a browser.

---

[1]http://www.cyberska.org

[2]http://www.elgg.org

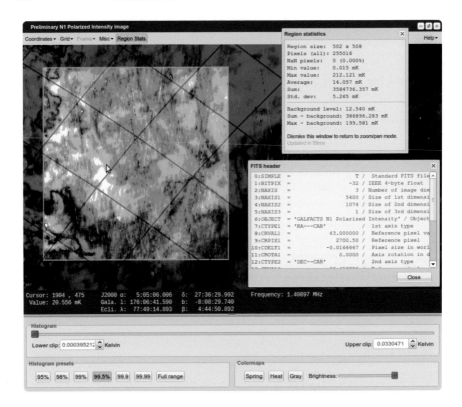

Figure 1.   Screenshot of the FITS Viewer Tool.

Without this FITS Viewer Tool, users intending to display a FITS file would first have to download the data and then invoke some local application. Our tool affords the same functionality without the need to download any files or install additional software. The only requirement is a modern web browser.[3]

The FITS viewer, implemented as an Elgg plugin on the CyberSKA web portal, allows users to interactively visualize any 2+ dimensional FITS file uploaded to the site, provided the user was granted read access to the file. A sample screenshot of the tool is given in Figure 1. When the user opens the file using the viewer tool, a graphical representation of the data is shown in the browser. The user can then interactively zoom and pan around the image using the mouse, in a manner similar to the Google Maps interface. The tool displays the FITS data value at the cursor location, as well as the actual cursor position in pixels and in up to 4 different coordinate systems. Additionally, a grid can be rendered over the image in any of the supported coordinate systems. The grid dynamically adjusts based on the current zoom and pan. The user can also interactively select and display simple statistics about a rectangular region, such as min/max values, average and standard deviation. If the displayed FITS file has more

---

[3]Browser has to support HTML5 Canvas element, e.g. Google Chrome, Mozilla Firefox, Apple Safari and Opera. Microsoft Internet Explorer 8 does not currently support HTML5.

Figure 2.   Screenshot of the Data Selector Tool.

than two dimensions, a different frame (cross section) of the cube can be selected for inspection. The histogram of the rendered image can be adjusted either by using a dual thumb slider, or by using buttons with some predefined values, such as 95% or 99%. Another noteworthy functionality is the ability to interactively adjust the colormap of the displayed data. Currently we support three predefined colormaps: heat, spring and grey.

3.   Data Selector Tool

One of the user groups of the CyberSKA portal is the GALFACTS consortium project,[4] which regularly produces data sets they wish to make available to their community for download. Each data set is normally composed of five 3D FITS cubes, corresponding to stokes I, Q, U and V, as well as a weight cube used to generate the 4 stokes cubes. Data size depends on the particular survey run, but in general ranges between 15 to 75 gigabytes per cube. Files of this size are not well suited for download over the Internet at today's download speeds. Further, users are often interested in only a small portion of the data set. For this purpose we have implemented a data access tool that allows users to interactively select the desired subset of data to be extracted for download. A sample screenshot of the Data Selector Tool is displayed in Figure 2.

The user interface is similar to the FITS Viewer Tool described in the previous section, however this tool is streamlined for selecting a subset of multiple FITS cubes

---

[4]GALFACTS uses the Arecibo L-band Feed Array to carry out a spectro-polarimetric survey of the sky.

for download. The user can interactively select a rectangular region in the first two dimensions by drawing a rectangular region using the mouse, while the range in the 3rd dimension is selected by a dual thumb slider. The user can also select from which of the 5 data cubes to extract the data for download. Once all selections are chosen, the download is submitted for processing using the submit button at the bottom of the GUI. The user is then presented with a status of all submitted download requests. When the data extraction is completed, the result can be downloaded.

4. High-Level Implementation Details

The implementation of the tools described above can be divided into two parts: client side code and server side code. On the client side, the tools utilize JavaScript and HTML5 canvas element[5] to render the FITS data, to display the GUI and to handle user interaction. When the tools are launched, they request one frame (2D cross-section) of the raw FITS data from the server through an AJAX call. Once received, the data is transformed into a format suitable for display by the canvas element. We chose the canvas element for its ability to efficiently draw arbitrary graphics on the screen through a low level JavaScript API, allowing us to implement features like interactive zooming, panning, grid drawing, histogram and colormap operations.

On the server side we use PHP to handle all incoming AJAX requests from the tools, such as submitting a new download request, or selecting a different frame for visualization. These requests are then handled by spawning appropriate jobs on our high-performance cluster, where we use Condor[6] for job scheduling. The data extraction from the actual FITS files is performed by hand-optimized C++ code.

5. Limitations and Future Work

The most significant limitation of the current approach to the FITS Viewer Tool is that it does not scale to very large images. At resolutions higher than about 4000×4000 pixels the JavaScript performance is not sufficient to maintain interactive user experience. It also takes longer to launch as it needs to download more information.

We anticipate the frame sizes of the FITS files produced by the SKA telescope to be orders of magnitude larger than the current limit of our FITS viewer. To address this we are working on a new approach, where all computation and rendering will be performed on the server side, and only the necessary information will be transferred to the browser for visualization.

References

Kiddle, C., et al. 2011, in Astronomical Data Analysis Software and Systems XX, edited by I. N. Evans, A. Accomazzi, D. J. Mink, & A. H. Rots (San Francisco, CA: ASP), vol. 442 of ASP Conf. Ser., 669

---

[5]http://www.whatwg.org/specs/web-apps/current-work

[6]http://www.cs.wisc.edu/condor

Astronomical Data Analysis Software and Systems XX
ASP Conference Series, Vol. 442
Ian N. Evans, Alberto Accomazzi, Douglas J. Mink, and Arnold H. Rots, eds.
© *2011 Astronomical Society of the Pacific*

The True Bottleneck of Modern Scientific Computing in Astronomy

Igor Chilingarian[1,2] and Ivan Zolotukhin[3,2]

[1]*CDS, Observatoire Astronomique de Strasbourg, Université de Strasbourg, CNRS UMR 7550, 11 rue de l'Université, 67000 Strasbourg, France*

[2]*Sternberg Astronomical Institute, Moscow State University, 13 Universitetsky prospect, Moscow, 119992, Russia*

[3]*Observatoire de Paris, LERMA, UMR 8112, 61 Av. de l'Observatoire, 75014 Paris, France*

Abstract. We discuss what hampers the rate of scientific progress in our exponentially growing world. The rapid increase in technologies leaves the growth of research result metrics far behind. The reason for this lies in the education of astronomers lacking basic computer science aspects crucially important in the data intensive science era.

1. Motivation

Present-day astronomical instruments and large surveys produce data streams increasing exponentially in time. The CPU power required to analyze these data is also growing at the same pace following Moore's law; the same applies to data storage per price unit. However, in astronomy we do not see a corresponding avalanche of scientific results produced with this computational power. This suggests the presence of *a bottleneck* somewhere in the loop: *if we consider the system containing three modules "A", "B", and "C" so that "A" is connected to "C" via "B," then optimizing features in module "A" or "C" will not produce a change in the performance of the system until the performance problems in module "B" are addressed.*

Where is the true bottleneck of the scientific computing? Just like many other scientists, astronomers prefer to develop their computational codes and software systems (including database solutions) themselves, often having no coding skills, insufficient background in algorithms and computational science.

2. Code Writing: Astronomers vs. Software Engineers

2.1. Scientific Software by Scientists

Most computer programs developed by astronomers without computer science background, regardless of their purposes (numerical modeling or simulations, data reduction or visualization, etc.), often have some specific common features.

(a) They are usually written in *Fortran-95, -90, -77* (or even prehistoric *Fortran-4* and *-66*). Sometimes high-level languages (e.g., *IDL, MATLAB*) are used. Primitive building scripts are used instead of *Makefiles* or more advanced building solutions (e.g., *ant*, or *maven* for *Java*). Code is non-portable.

471

(b) They often contain the GOTO statement every 10–20 lines; names of variables do not follow any conventions, i.e. *a1, a2, aa1*; the code is unreadable: poor or no indentation, very long function bodies and/or source files. There is a lot of hard-coding of file and device names, file system paths.

(c) They are undocumented and full of "intuitive" algorithmic solutions, such as "re-invented" sorting and search algorithms, which sometimes end up are quite far from the standard solutions that computer science students learn at school.

(d) The "multi-layered" code structure is another typical feature. When the author is returning to the same program after several months or years, he/she often finds that the existing procedure/function calls do not satisfy his/her needs, however he/she is not willing to modify them to keep backward compatibility. Then, a wrapper routine is created which calls some underlying procedures/functions in a slightly different way. As a result, after several such periods of development, one can find multiple (undocumented) interfaces to the same functional blocks.

(e) However, in the end the program does what it is supposed to, because the author knows exactly what it should do, even though it may sometimes crash during run-time or have very poor performance.

2.2. Scientific Software by IT Engineers

The software developed by IT engineers in research is notably different. Here the quality of the final product strongly depends on the job of a project manager.

(a) Usually it is done using a "real" programming language: *C/C++/Java*, primarily because it is virtually impossible to find an IT professional developing in *Fortran*.

(b) All necessary solutions for computational algorithms are conventional because the developer has at least heard about "The Art of Computer Programming" (Knuth 1978).

(c) The code is usually well organized and structured; correct indentations and variable naming conventions are used; sometimes the author follow one of the coding styles (e.g., GNU). Therefore, the code becomes readable and comprehensible.

(d) The quality and completeness of the documentation strongly depends on the project manager's competence. It can be from *none* to nearly perfect.

(e) However, the author often does not understand the physical principles behind the algorithm or particular features of the instrumentation that generated the data, therefore some bad surprises are possible. For example, arithmetic bugs leading to results which are wrong by many orders of magnitude cannot be spotted by a software engineer because for him/her these are "just numbers." This may dramatically slow down development.

2.3. Databases by Scientists

The worst class of software solutions is probably DBs developed by researchers.

(a) Often they contain custom implementation in *Fortran* or *IDL* of re-invented indexing solutions and primitive requests to the data. Indices and data tables are stored in a proprietary undocumented binary format.

(b) If an existing database management system (DBMS) is used, then the DB usually contains one or several flat tables without mutual links, i.e. no data model.

(c) DB constraints are not used for consistency checks. In some rare cases they are implemented externally in a DB management interface (often written in *Fortran*).

(d) User interfaces, both application programming interface (API) and web front-ends, are undocumented, have very low usability and terrible design.

3. Bad and Good Examples

For obvious reasons, we will name directly the projects in the list of bad examples. The list of good examples is neither exhaustive nor complete.

3.1. Bad Example #1: an Unnamed Galaxy Catalog

The project is very interesting scientifically and recognized in the community. But,
(a) There is no access interface on the web.
(b) The data are distributed as a set of dozens of FITS tables with a total volume >10Gb and *IDL* access routines to perform queries on these tables. One has to download nearly everything in order to study just a handful of objects.
(c) Therefore, huge memory requirements if one uses the whole catalog at once.
(d) Therefore, very slow and inefficient data access and selection.

3.2. Bad Example #2: an Unnamed Database Using *PostgreSQL*

(a) DB administration and ingestion interface (implemented in *Fortran*) has a function with over 250 arguments
(b) Inside the DB restore script, to delete a record from a table, instead of
  DELETE FROM table1 WHERE field1=value1
the authors do:
  pg_dump -t table1 mydb | grep -v value1 | pg_restore -c mydb
(c) One of the stored procedures which is triggered on *INSERT* connects externally to the same DB and makes some selections. Obviously this new connection cannot see the changes introduced before the trigger is fired because the transaction has not been committed.

3.3. Good Examples #1: Technologically Advanced Projects

1. HLA - the Hubble Legacy Archive (`http://hla.stsci.edu/`). Innovative solutions implemented inside HLA include: (a) Virtual Observatory standard interfaces (Simple Image Access Protocol) as a hidden middleware; (b) XSLT transformation of *VOTables* into *AJAX*-enabled HTML pages; (c) advanced visualization tools.
2. SDSS CasJobs (`http://cas.sdss.org/CasJobs`, Szalay et al. 2002). Efficient and easy-to-use access to a large DB featuring user management, user table upload, I/O of tabular data in different formats, comprehensive SQL query builder.
3. GalexView (`http://galex.stsci.edu/GalexView/`) — a Flash-based interactive web-access to the GALEX satellite images.
4. Millennium Simulation (`http://www.mpa-garching.mpg.de/millennium/` by G. Lemson) — access to the DB containing the results of large cosmological simulations with a comprehensive data model and full SQL access.
5. GalMer (`http://galmer.obspm.fr/`, Chilingarian et al. 2010) — a DB to access numerical simulations of merging and interacting galaxies. The project implements a set of Virtual Observatory (VO) standards, features efficient interactive preview visualization of the datasets on the server side and complex on-the-fly data analysis algorithms. The *JavaScript*-powered web-interface working in most modern browsers is integrated with VO tools in order to visualize complex datasets (Chilingarian & Zolotukhin 2008; Zolotukhin & Chilingarian 2008).

3.4. Good Examples #2: Computations, Data Analysis and Visualization

1. GADGET-2 by V. Springel (2005), a cosmological simulation code that is well documented and easily extensible: there are numerous third-party add-ons implementing different physical phenomena, e.g., radiative transfer, metallicity evolution in galaxies.
2. SExtractor (Bertin & Arnouts 1996): a software to perform object extraction and photometry from CCD images has very intuitive configuration, although outdated documentation. It is relatively clearly coded.
3. TOPCAT/STILTS (Taylor 2005, 2006) — the best available platform-independent table manipulation software integrated with VO services and resources.
4. CDS Aladin (Bonnarel et al. 2000) — a VO data browser for images and catalogs.
5. SAOImage DS9 (Joye & Mandel 2003) — probably the most frequently used desktop FITS visualization software in astronomy implementing some VO access methods.

4. The Main Message and a Possible Solution

It turns out that all "good examples" were developed either by professional astronomers with very strong IT/CS background or by IT/CS professionals working closely with astronomers for years and understanding astronomy. One cannot simply hire an industrial software engineer to develop astronomical software and/or an archive and/or a database.

A possible solution is *to change the teaching paradigm for students in astronomy*. Basic courses in algorithms, programming, software development and maintenance have to be made mandatory in the education of modern astronomers and physicists; advanced courses should be recommended to some of them. The *Fortran* language is now obsolete and we have to accept this. Instead of teaching research students *Fortran* programming, one should teach how to interface legacy *Fortran* code in *C/C++*.

As soon as this *bottleneck* is resolved, the avalanche of discoveries will loom.

References

Bertin, E., & Arnouts, S. 1996, A&AS, 117, 393
Bonnarel, F., et al. 2000, A&AS, 143, 33
Chilingarian, I., & Zolotukhin, I. 2008, in Astronomical Data Analysis Software and Systems XVII, edited by R. W. Argyle, P. S. Bunclark, & J. R. Lewis (San Francisco, CA: ASP), vol. 394 of ASP Conf. Ser., 351
Chilingarian, I. V., Di Matteo, P., Combes, F., Melchior, A., & Semelin, B. 2010, A&A, 518, A61
Joye, W. A., & Mandel, E. 2003, in Astronomical Data Analysis Software and Systems XII, edited by H. E. Payne, R. I. Jedrzejewski, & R. N. Hook (San Francisco, CA: ASP), vol. 295 of ASP Conf. Ser., 489
Knuth, D. E. 1978 (Reading, Massachusetts: Addison-Wesley)
Springel, V. 2005, MNRAS, 364, 1105
Szalay, A. S., Gray, J., Thakar, A. R., Kunszt, P. Z., Malik, T., Raddick, J., Stoughton, C., & vandenBerg, J. 2002, ArXiv Computer Science e-prints. arXiv:cs/0202013
Taylor, M. B. 2005, in Astronomical Data Analysis Software and Systems XIV, edited by P. Shopbell, M. Britton, & R. Ebert (San Francisco, CA: ASP), vol. 347 of ASP Conf. Ser., 29
— 2006, in Astronomical Data Analysis Software and Systems XV, edited by C. Gabriel, C. Arviset, D. Ponz, & S. Enrique (San Francisco, CA: ASP), vol. 351 of ASP Conf. Ser., 666
Zolotukhin, I., & Chilingarian, I. 2008, in Astronomical Data Analysis Software and Systems XVII, edited by R. W. Argyle, P. S. Bunclark, & J. R. Lewis (San Francisco, CA: ASP), vol. 394 of ASP Conf. Ser., 393

Astronomical Data Analysis Software and Systems XX
ASP Conference Series, Vol. 442
Ian N. Evans, Alberto Accomazzi, Douglas J. Mink, and Arnold H. Rots, eds.
© *2011 Astronomical Society of the Pacific*

Bayesian Inference of Stellar Parameters and Interstellar Extinction with Heterogeneous Data

C. A. L. Bailer-Jones

Max Planck Institute for Astronomy, Königstuhl 17, 69117 Heidelberg, Germany. Email: calj@mpia.de

Abstract.   I outline a probabilistic method for estimating stellar parameters. It uses not only the spectral energy distribution but also the apparent magnitude, parallax (if available) and the strong prior information provided by the Hertzsprung-Russell Diagram. This (a) improves the accuracy and precision over the use of just the spectrum, and (b) ensures that the inferred parameters (e.g. effective temperature, interstellar extinction and absolute magnitude) are both physically realistic and are consistent with the distance and apparent magnitude. The method — which gives covariate probability distributions over the parameters — is applied to some 85 000 Hipparcos-2MASS stars.

1. The Problem

Traditionally we estimate stellar astrophysical parameters (APs) — such as T_{eff}, $\log g$ and [Fe/H] — using only spectroscopy. If we are lucky enough to have high resolution spectra, then we can often achieve good performance. But increasingly we are interested in massive scale (10^9 objects or more) parametrization using low resolution spectroscopy or multiband photometry. This is the case with surveys such as SDSS, Pan-STARRS and LSST (five band photometry), and Gaia (very low resolution spectrophotometry).[1] The AP accuracy we can achieve with such data alone is limited.

What other information is available to help improve performance? The Hertzsprung-Russell Diagram (HRD) describes the location of stars in the (M_V, T) (absolute magnitude, effective temperature) plane, and for virtually any stellar population it is very sparsely and non-uniformly populated (see Fig. 1). A priori we can place plausible constraints on the relative probability of the stellar parameters.

Stellar parametrization in large, deep surveys faces another problem, namely interstellar extinction (A_V). In principle this can also be estimated from the photometry, but it is frequently highly degenerate with T_{eff}. This problem is sometimes ignored in survey projects by assuming that the stars have negligible extinction (e.g. at high Galactic latitudes), or by using an extinction map. The first solution is inadmissible for surveys near the Galactic plane (or near molecular clouds), and extinction maps often have very low spatial resolution or are not three-dimensional (i.e. they only give the integrated extinction to the edge of the modelled Galaxy).

---

[1]Gaia will measure positions, parallaxes and proper motions with an accuracy of up to 10 microarcseconds for almost all 10^9 stars in our Galaxy brighter than G=20. It will also obtain low resolution optical spectrophotometry, as well as radial velocities.

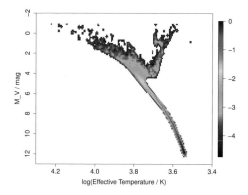

Figure 1.   HRD prior. The color scale shows $\log P(M_V, T)$ normalized to have zero at its maximum. Unoccupied areas are shown in white.

Extinction is a major issue for the all-sky Gaia survey. Yet herein also lies an opportunity. Gaia measures parallaxes (ϖ) and apparent magnitudes (V).[2] *If* we knew A_V then we could estimate the absolute stellar magnitude (M_V), a fundamental stellar property, via the relation

$$V + 5 \log \varpi = M_V + A_V - 5.\tag{1}$$

However, we also have to estimate A_V from the data. How can we do this?

2. The Solution

The solution is to approach the problem probabilistically. Where we have noise we have uncertainties; these are best represented by probability density functions (PDFs). The Bayesian approach allows one to include all available information as PDFs in a self-consistent manner, and to propagate these PDFs through the calculation to provide not only parameter estimates but confidence intervals on these estimates.

Let us consider the problem of estimating just the two APs $\phi = (A_V, T)$. We have three pieces of information:

1. The spectrum (p), which constrains T and A_V;
2. The quantity $q = V + 5 \log \varpi$, which constrains $M_V + A_V$ from equation 1;
3. The HRD, which constrains M_V and T (Fig. 1).

The goal is to determine $P(\phi|p, q)$. We can predict the spectrum given ϕ using a *forward model*, which is the result of a fit to a set of labelled data ("training" in machine learning speak; see Bailer-Jones 2010b). Combined with a suitable photometric noise model, this gives $P(p|\phi)$. Adopting a noise model for the apparent magnitude and parallax measurement allows us to write item (2) as $P(q|\phi, M_V)$. Then, applying Bayes's theorem we achieve an expression for $P(\phi|p, q)$ in terms of these quantities. It involves marginalizing over the unknown M_V to give a (non-parametric) two-dimensional PDF over ϕ for given measurements p and q.

---

[2]Gaia actually measures in a broad G band rather than the V band, but this is a detail here.

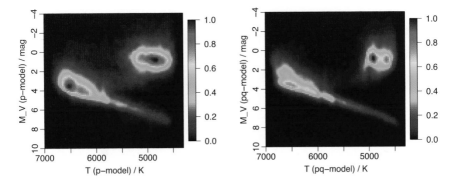

Figure 2.   HRD for the Hipparcos-2MASS stars derived from the p-model (left) and pq-model (right) shown as a density plot (achieved via smoothing with a Gaussian kernel). The number of stars per unit area is normalized to a value of 1.0 at the maximum density (separate normalization in each plot).

3.   Application to 85 000 Hipparcos-2MASS Stars

This method has been tested by using it to estimate A_V and T for 5280 FGK stars with known "true" parameters based on their $BVJHK$ photometry and Hipparcos parallaxes. These data are derived from a set of 880 stars with T estimated from high resolution spectroscopy by Valenti & Fischer (2005), to which I have applied artificial redenning to provide variance in A_V. The parameter accuracy (mean of absolute residuals) using just the four colors — at the mean of $P(p|\phi)$ — is 5.5% in T and 0.3 dex in A_V (the *p-model*). When introducing the parallax, apparent magnitude and an HRD prior (to give the *pq-model*), these errors are reduced to 3.5% and 0.2 dex respectively, an increase in accuracy of around 40%. (We can also apply the method using just the colors and the HRD prior but no measurement of q. Even this improves accuracy by 13% over the p-model.) We can also determine 90% confidence intervals on the parameters (or estimate the corresponding Gaussian 1σ uncertainties), and these are also reduced by introducing the new information.

I then applied the method to a set of 85 000 Hipparcos stars for which I obtained a reliable astrometric cross match with 2MASS (to give $BVJHK$ photometry). Many of these stars (42%, it turns out) are not FGK stars, so their APs cannot be estimated reliably by this method. (I identify these stars from their inferred PDF peaking at or very close to the edge of the parameter space.) Once we have estimated A_V and T we can estimate M_V (or rather a PDF over it) from equation 1 and so plot the stars in an HRD: see Fig. 2. These results are discussed in more detail in Bailer-Jones (2010a). As the Hipparcos sample covers the whole sky, we can also combine the individual extinction measurements to produce an extinction map; a 2D map is shown in Fig. 3. The median distance to these stars is 170 pc (90% have distances between 40 and 730 pc).

As an additional test, I identified 137 stars in my sample in the list of 218 Hipparcos Hyades members from Perryman et al. (1998). The HRD diagram (pq-model) for these objects is plotted in Fig. 4. As expected, the majority of these have very low extinctions, yet a significant number of the cooler stars have relatively large extinctions. Further investigation of this is beyond the space available in this paper.

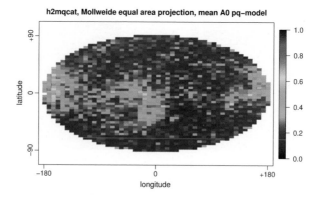

Figure 3.   The mean extinction (A_V) from the pq-model along the line of sight to the Hipparcos stars, plotted in Galactic coordinates.

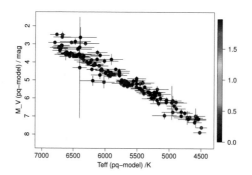

Figure 4.   HRD for 137 Hyades stars derived using the pq-model. Individual stars are colored according to their estimated extinction, A_V.

A catalog of parameter estimates (plus uncertainties) from both the p-model and pq-model for 46 900 stars is available from the author's website[3] and from the CDS in Strasbourg.[4] More results and discussion can be found in Bailer-Jones (2010a).

References

Bailer-Jones, C. A. L. 2010a, MNRAS, in press (arXiv:1009.2766)
— 2010b, MNRAS, 403, 96
Perryman, M. A. C., et al. 1998, A&A, 331, 81
Valenti, J. A., & Fischer, D. A. 2005, ApJS, 159, 141

---

[3]http://tinyurl.com/qmethod

[4]http://vizier.u-strasbg.fr/viz-bin/VizieR?-source=J/MNRAS/411/435

Astronomical Data Analysis Software and Systems XX
ASP Conference Series, Vol. 442
Ian N. Evans, Alberto Accomazzi, Douglas J. Mink, and Arnold H. Rots, eds.
© *2011 Astronomical Society of the Pacific*

Measuring the Physical Properties of Galaxy Components Using Modern Surveys

Steven P. Bamford,[1] Boris Häußler,[1] Alex Rojas,[2] and Andrea Borch[2]

[1]*School of Physics and Astronomy, University of Nottingham, University Park, Nottingham, NG7 2RD, UK*

[2]*Carnegie Mellon University in Qatar, PO Box 24866, Doha, Qatar*

Abstract. Most galaxies are multicomponent stellar systems, comprising a spheroid and disk with largely independent origins. Separating the properties of these components can put powerful constraints on models of galaxy formation. However, current decomposition methods are not sufficiently developed for routine use. We identify areas where substantial improvements may be made, including fully utilizing multiband imaging, and incorporating non-parametric components. These issues will be addressed by the MegaMorph project, which aims to produce a tool for robustly and accurately measuring galaxy components in large surveys.

1. Galaxy Internal Components

The primary division in the galaxy population is commonly considered to be that between spiral and elliptical galaxies. However, there are good reasons for considering the disk and spheroid stellar components of galaxies as the more fundamental dichotomy (e.g., Allen et al. 2006; Benson & Devereux 2010). These components sometimes occur separately, in pure-disk and elliptical galaxies, but more often appear together within the same galaxy. In our simplest models, galaxies grow through two principal mechanisms: the gradual accretion of gas, which cools and settles into a thin disk in which stars form; and the merging of existing stellar systems, which result in a spheroid. The two components thus form through disparate mechanisms, and represent disjoint periods in a galaxy's history (e.g., Cole et al. 2000; Cook et al. 2009; Benson 2010).

Despite this contrast, large surveys usually consider only integrated properties of galaxies or simplified indicators of structure. Summing the components in this way discards much of their physical information. One would prefer to observationally separate the components and measure their individual properties; enabling a deeper understanding of the galaxy population and more effective comparison with theoretical models.

2. Today's Solutions

A number of galaxy decomposition solutions exist (e.g., GALFIT, Peng et al. 2002; GIM2D, Simard et al. 2002, etc.), all of which fit smooth, parametric models to a single image of each galaxy. Applying these methods to relatively large samples of galaxies has met with some success (e.g., Allen et al. 2006), and has confirmed the value of the approach. However, significant problems remain regarding the robustness and physical meaning of the fitted model parameters. The fit may fail to reproduce the light profile of the galaxy under consideration, but when the issues of masking, deblending and sky estimation are properly handled these 'catastrophic' failures are rare. Of more concern are cases where the best-fitting model well represents the light profile, but does not correspond to a realistic combination of physical components. Allen et al. find that ~ 15% of their attempted bulge+disk fits result in unphysical parameters. These difficulties arise for two principal reasons: (1) strong degeneracies between the parameters of the two galaxy components in a single wavelength-band model, and (2) the presence of features in the data (e.g., star-formation regions, spiral arms, bars, etc.) which are not included in the model.

3. Improving the Reliability of Galaxy Decompositions

3.1. Efficient Use of Multiband Data

Galaxy surveys now produce imaging with comparable quality across many photometric bands. For example, the SDSS (Abazajian et al. 2009) and UKIDSS LAS (Lawrence et al. 2007) surveys together provide imaging spanning nine optical and near-infrared bands. Currently available bulge-disk decomposition solutions may be applied to these data in a limited manner. One can treat each band independently, but the resulting set of parameters are of little scientific use. The recovered components vary in size and shape as a function of wavelength, because of the varying features and noise properties. They, therefore, do not correspond to the same population of stars at each wavelength, and thus the inferred component colors are largely meaningless. More commonly one performs a full fit in a single band, and then fits the resulting model to each additional band, allowing only the flux of each component to vary. This results in more meaningful representations of the true colors of the physically distinct components. However, their quality depends upon the reliability of the initial single-band fit. As we have explained above, a significant fraction of these fits will be physically unrealistic.

We propose that one can greatly increase the reliability of the decomposition process, while ensuring physically meaningful component colors, by using all of the available multiband imaging simultaneously to constrain a single wavelength-dependent galaxy model. Firstly, this increases the signal-to-noise of the data used to constrain the fit, for a comparatively small increase in the number of fitted parameters. Secondly, and more importantly, multi-wavelength imaging provides information which is not available to techniques which operate on only a single band at a time: the wavelength dependence of the luminosity of each component. This is critically important to overcoming the degeneracies inherent in multicomponent fitting, particularly in the usual case of an exponential disk and Sérsic bulge with free Sérsic index. Finally, this approach enables the inclusion of spectral energy distribution constraints for each component, thereby further reducing the effective number of free parameters.

3.2. Non-Parametric Components

For most galaxies, the majority of their stellar mass may be identified with a small number of relatively smooth, approximately azimuthally-symmetric components. However, the observed surface brightness distribution, particularly at blue wavelengths, is strongly affected by both star formation and dust, both of which have complex spatial distributions. Spiral arms, bars and rings provide additional complicating features.

It is rarely wise to fit data with a model which is unable to adequately represent the behavior of that data and attempt to interpret the resulting parameters in a meaningful way (e.g., see Hogg et al. 2010). Unfortunately, this is the case when fitting optical images of real, messy, galaxies with an idealized, smooth model. Instead, we propose to add non-parametric components to the galaxy model to account for features that are not represented by the usual parametric components. Clearly, a penalty must be imposed to ensure that any non-parametric components contribute the minimum flux necessary to ensure a good fit, such that the parametric aspects of the model accurately reflect their respective galaxy components. We expect that this approach will substantially increase the reliability of the recovered parameters (and their uncertainties) for the spheroid and disk components.

Galaxy bars often contain significant stellar mass and are fairly well described by a parametric model. Therefore, it is usually preferable for bars to be included in the galaxy model as specific, parametrized components. However, there may be times when it would be more convenient to account for them via a non-parametric component. Note that the latest version of GALFIT supports non-axis-symmetric components (Peng et al. 2010). These may also be useful in ensuring robust and accurate fits, but fulfill a different purpose to non-parametric components.

4. Preparation

To develop the improvements described above we have embarked on a project named MegaMorph. Our first priority has been to select and construct a baseline system, to use as the basis for our development and the benchmark against which our improvements will be evaluated, via application to a number of test datasets.

4.1. Baseline System

We choose to modify an existing, tried-and-tested system which will take care of the numerous details that are essential for reliable galaxy fitting, but which do not require significant alteration for our purposes. We have selected GALAPAGOS (Häussler et al. 2007), which takes care of detection, deblending and preliminary object measurements using SExtractor (Bertin & Arnouts 1996); as well as image extraction, masking, fitting each galaxy using GALFIT (Peng et al. 2002), and collating the results.

In addition to the changes regarding multiband data and non-parametric components, we are also making several modifications to GALAPAGOS to enable its use on large ground-based surveys. These include a position dependent point spread function, efficiency tweaks, and adaptations to a supercomputer environment.

4.2. Test Datasets

In order to assess the performance of our modifications to GALAPAGOS and GALFIT we have prepared three test datasets:

Nearby galaxy sample — We have collected a sample of nearby galaxies with measured structural parameters, often by multiple independent studies, and SDSS imaging. Using the FERENGI software tool (Barden et al. 2008), we have simulated how these galaxies would appear at a range of redshifts. By fitting these images with our baseline and development versions of GALFIT, we can compare our structural measurements with those of independent studies, and examine the reliability of the method for galaxies viewed at different distances.

GAMA real sample — The Galaxy And Mass Assembly survey (GAMA, Driver et al. 2009, 2010) brings together a spectroscopic survey with many multi-wavelength datasets to create a powerful resource for studies of galaxy evolution and cosmology. Amongst many other things, GAMA provides homogenized imaging over nine optical to near-infrared bands, which is ideal for both testing our modified code on real data and subsequently for scientific studies. GAMA also includes useful visual morphological data from Galaxy Zoo (Lintott et al. 2010).

GAMA simulation sample — We are creating simulations of the GAMA imaging following the method of Häussler et al. (2007). These will be used to quantify the performance of the full structural fitting code with all the problems of real data, but without the complication of whether or not the model allows a good fit to individual objects. Comparing with the results obtained for real data will help us understand the observational limitations of the decomposition technique.

5. Next Steps

Our priorities for the next year are to implement and quantify the advantages of both multiband fitting and non-parametric components. We will then focus on the issue of model selection: deciding which combination of components is supported by the data for each galaxy, and finally tune the software for application to the full GAMA survey. After producing some initial science results to demonstrate the value of our improvements, we plan to release the MegaMorph software as a public tool.

Acknowledgments.   This work was made possible by a NPRP grant from the Qatar National Research Fund (a member of the Qatar Foundation). The statements made herein are solely the responsibility of the authors. SPB is supported by an STFC Advanced Fellowship.

References

Abazajian, K. N., et al. 2009, ApJS, 182, 543
Allen, P. D., Driver, S. P., Graham, A. W., Cameron, E., Liske, J., & de Propris, R. 2006, MNRAS, 371, 2
Barden, M., Jahnke, K., & Häußler, B. 2008, ApJS, 175, 105
Benson, A. J. 2010, Phys. Rep., 495, 33
Benson, A. J., & Devereux, N. 2010, MNRAS, 402, 2321
Bertin, E., & Arnouts, S. 1996, A&AS, 117, 393
Cole, S., Lacey, C. G., Baugh, C. M., & Frenk, C. S. 2000, MNRAS, 319, 168
Cook, M., Lapi, A., & Granato, G. L. 2009, MNRAS, 397, 534
Driver, S. P., et al. 2009, Astron. Geophys., 50, 5.12
— 2010, MNRAS, in press. arXiv:1009.0614
Häussler, B., et al. 2007, ApJS, 172, 615
Hogg, D. W., Bovy, J., & Lang, D. 2010, ArXiv e-prints. arXiv:1008.4686
Lawrence, A., et al. 2007, MNRAS, 379, 1599
Lintott, C., et al. 2010, MNRAS, in press. arXiv:1007.3265
Peng, C. Y., Ho, L. C., Impey, C. D., & Rix, H. 2002, AJ, 124, 266
Peng, C. Y., et al. 2010, AJ, 139, 2097
Simard, L., et al. 2002, ApJS, 142, 1

Part X

Software Tools

Astronomical Data Analysis Software and Systems XX
ASP Conference Series, Vol. 442
Ian N. Evans, Alberto Accomazzi, Douglas J. Mink, and Arnold H. Rots, eds.
© 2011 Astronomical Society of the Pacific

FASE — Future Astronomical Software Environment: How to Include Tools and Systems into the FASE Environment

Y. Granet,[1] L. Paioro,[2] C. Surace,[1] B. Garilli,[2] P. Grosbøl,[3] D. Tody,[4] and the
FASE Consortium

[1]*Laboratoire d'Astrophysique de Marseille, OAMP, CNRS,
Université de Provence*

[2]*INAF — IASF Milano*

[3]*European Southern Observatory*

[4] *National Radio Astronomy Observatory / VAO*

Abstract. The new OPTICON Network 9.2 (FP7) is working on a concrete prototype of shared astronomical software environment for scalable and desktop systems (FASE). A prototype of packaging system has been defined that allows to include new tools and programs as well as major legacy systems (e.g. AIPS, CASA, IRAF/PyRAF, Starlink and ESO Common Pipeline Library) within FASE environment.

1. Future Astronomical Software Environment

1.1. FASE

For several years the astronomical community has used different data reduction and analysis tools. The Future Astronomical Software Environment (FASE) project (Grosbol 2010) aims to create a new astronomical software framework based on Virtual Observatory standards. The main concept targets are:

- To allow the reuse of the most important legacy software (e.g., IRAF, MIDAS, AIPS, GIPSY) within a modern framework;

- To make easier the support and development of new interoperable and distributed applications or simple computational tasks;

- To increase software sharing and astronomical software development collaboration;

- To define stable, controlled and open software interfaces.

To achieve these goals, we built up a first prototype (which aims to become a first basic reference implementation) with the purpose of demonstrating the concepts and fesability. In order to facilitate the integration of astronomical software within the FASE framework, we provide also a packaging system. The core implementation is defined in Python and ANSI-C. The packaging system has been provided with a Wizard coded in Java to be used whatever the Operating System is.

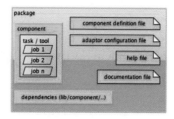

Figure 1.   Package logical organization.

1.2. The FP7 Network

The high level requirements and the main architectural design have already been defined within the OPTICON Network 3.6 in collaboration with NRAO/USVAO. OPTICON Network 9.2 (FP7) has been set up to make available to the community a FASE prototype leading to an eventual reference implementation of the basic core system, as well as a packaging system. Several papers have been published during these years providing information on evolution and activities of this framework (Grosbol et al. 2008)

2. Packaging System

The FASE packaging is defined as the way to package an existing or a new component in order to be distributed and easily integrated in the FASE environment. Its purpose is to ensure an easy implementation of any kind of software, with the minimization of the efforts of component developers as a guideline. Another guideline is to avoid any mixing between the astrophysical developments (Business part) and any specific framework (e.g., SAMP, ICE), to be able to switch to any other bus system, if necessary. And finally, the goal is to provide a developer friendly framework (keep it simple for high level applications). Based on the Package Manager and component-container interfaces document (Grosbol et al. 2010; Tody 2010), this packaging includes the component itself, the container system, a file containing dependencies information and documentation. A component can be a task or a tool (tasks are stateless while tools maintain their state). A task (resp. tool) can have several functions. These functions are defined as "jobs" in the packaging system. Each job could be launched by a specific direct command.

Each package is a compressed tar file including a component definition file, an adaptor configuration file, a directory named as the component, the packaged component inside this directory, and internal dependencies.

2.1. Standards

In order to make the packaging easier, the packaging system has been split in two different steps leading to two different configuration files: the "component definition file" and the "adaptor configuration file". Indeed, while the component definition file needs a good knowledge of the component itself to fill in the necessary input information, the adaptor configuration file needs a good knowledge of the bus protocols, for example SAMP (Taylor et al. 2010). These files can be created by two different people, at dif-

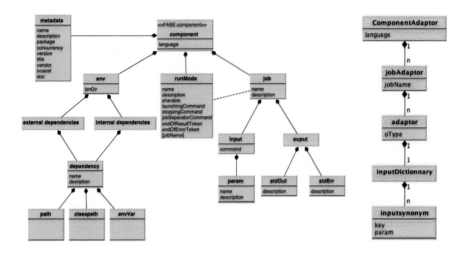

Figure 2.   Architecture of the component definition file and adaptor configuration file.

ferent times. Moreover, changing the bus access protocols will lead to a modification of only one file.

- The component definition file is filled in by the component developer and contains component metadata. The purpose of the definition file is to describe the functionalities of the component, including its calling sequence with arguments and parameters and the result sequence of such component. It describes also the component itself, its characteristics, its name, its version, etc. It lists, as well, the running modes, the parameters it can accept, the input output streaming, and the dependencies, both internal (within the packages) and external (that have to be present on the host machine). This file format is standardized and is framework independent. The component definition file is built as described in Figure 2. Serialization is performed under XML format.

- The adaptor configuration file is filled in by either the component or FASE developer. It describes the way the component should respond to the bus infrastructure that is specific to the communication protocol.

The packaging system is based on these two files that describe the component to include and the relationship with the FASE framework. These files are XML files that can be edited manually, but to ensure a faster and easier integration of tools, we developed a FASE Packaging Wizard.

2.2. FASE Packaging Wizard

A packaging wizard has been built in order to facilitate the creation of packages. The packaging wizard can be downloaded at: `http://lamwws.oamp.fr/fase/`. The wizard guides the programmer through the creation of a new package for a component and helps fill out the necessary fields for both description files. The packaging system wizard is able to list and load the package into the FASE environment.

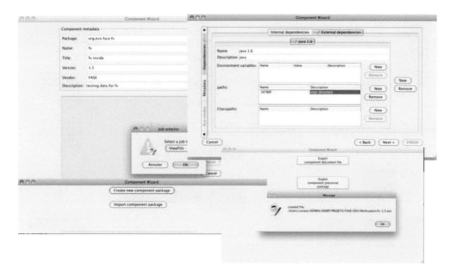

Figure 3.   Packaging Wizard for FASE Environment.

Documentation on how to use the FASE Packaging wizard can be found at: `http://lamwws.oamp.fr/fase/Site/FASE-Packaging.html`.

3. Conclusion

Over the past few years, the FASE project has defined the concepts of an architecture for a future shared astronomical software environment, taking into account the previous and still widely used reduction and analysis legacy software. A packaging system prototype has been developed in order to demonstrate the concepts defined in the FASE architecture. Such a packaging system prototype allows legacy programs to easily be plugged into the FASE framework. Many tests using commonly used legacy software and developing new software using the framework facilities are underway. This software is freely available on request from the authors.

Acknowledgments.   The OPTICON activities are funded by the European Commission. The Network 9.2 on Future software environments for processing and analysis of astronomical data was started in spring 2009. This work has been started during OPTICON Network FP6 WP-3.6 funded by the European Commission (contract: RII3-Ct-2004-001566) and FP5 (grant: 226604).

References

Grosbol, P. 2010, `https://www.eso.org/wiki/bin/view/Opticon/WebHome`
Grosbol, P., et al. 2008, `https://www.eso.org/wiki/bin/view/Opticon/GeneralDocuments`
— 2010, White paper on `http://archive.eso.org/opticon/twiki/pub/Main/WebHome/WPArch.pdf`
Taylor, M., et al. 2010. URL `http://www.ivoa.net/Documents/latest/SAMP.html`
Tody, D. 2010, AppFramework `http://trac.us-vo.org/project/nvo/wiki/VOAppFramework`

Astronomical Data Analysis Software and Systems XX
ASP Conference Series, Vol. 442
Ian N. Evans, Alberto Accomazzi, Douglas J. Mink, and Arnold H. Rots, eds.
©2011 Astronomical Society of the Pacific

ASPRO2: A Modern Tool to Prepare Optical Interferometry Observations

G. Duvert,[1,2] L. Bourgès,[1] G. Mella,[1,2] and S. Lafrasse[1,2]

[1]*Center Jean-Marie Mariotti, Observatoire de Grenoble, BP 53, F–38041 Grenoble, France*

[2]*Laboratoire d'Astrophysique, Observatoire de Grenoble, BP 53, F–38041 Grenoble, France*

Abstract.   We present ASPRO2, a Java observation preparation tool developed and maintained by the Jean-Marie Mariotti Center for Expertise in Interferometry (JMMC). ASPRO2 allows to prepare observations for optical interferometers, in particular for the VLTI and CHARA arrays. It automates the writing of VLTI and CHARA Observing Blocks. ASPRO2 persistent metacode can be used to exchange observation lists between observers or even instruments. Relying on Virtual Observatory techniques, AS-PRO2 is interoperable with other applications, in particular those developed at JMMC for optical interferometry.

1. Introduction

ASPRO2 is the second version of the Astronomical Software to PRepare Observations created by the JMMC.[1] It is quickly replacing its predecessor, ASPRO (Duvert et al. 2002; Mella & Duvert 2004). ASPRO was also a complete observation preparation tool for preparing interferometric observations with the ESO/VLTI. It was based on a client-server model, with a light Java display interface on the client side and a complex server side, relying on a special "network-aware" version of the GILDAS[2] software suite, a series of FORTRAN and C programs and SIC scripts.

Initially intended as a demonstrator only, ASPRO had a long and useful life (10 years), but is now difficult to maintain and improve due to its dependency to obsolete components. Based on the numerous positive returns and evolution requests from the community, the JMMC Scientific Council started the ASPRO2 project in September 2009.

ASPRO2 is a Java standalone program improving on all the functionality of ASPRO and adding a dynamic graphical interface, the abilities to be used off line, to load and save observation settings, to generate Observing Blocks, and much more. ASPRO2 is developed in close relationship with a panel of users, in an AGILE-like development environment.

---

[1]http://www.jmmc.fr

[2]http://iram.fr/IRAMFR/GILDAS

2. Functional Description

ASPRO2 is designed to prepare observations with a variety of long baseline optical interferometers. Although some prior knowledge of interferometry is preferable when dealing with such a specialized tool, we have tried to design an intuitive user interface (GUI). We also insist on keeping the GUI as simple as possible and dealing only with the parameters really needed to prepare an observation, using interoperability with other VO tools to fulfill additional user needs.

Preparing an observation in our case is very much alike simulating the whole observational process, since in the absence of a directly interpretable result (such as an image would be, for example), the astronomer needs to use data interpretation tools such as model-fitting or image-reconstruction programs to estimate the feasibility of the observations, something which is not readily conveyed by a simple "exposure time calculator". Thus, ASPRO2 is internally designed as simulator and produces simulated observables in the data format, data format!OI-FITS OI-FITS (Pauls et al. 2005), used by the optical interferometry community.

2.1. Data Model

ASPRO2 is based on a Data Model to easily maintain and update the configuration for interferometers, instruments and observations:

- The interferometer itself, with its constituents and relevant parameters: telescope(s) size(s), stations positions, optical path lengths, delay lines throw, atmospheric conditions, etc. The interferometer "section" of the data model is named from the actual interferometer (e.g., CHARA, VLTI) and roughly corresponds to the OI_ARRAY table of the OI-FITS format.

- The instrument used, with its related properties: transmissivity, bandwidth, number of spectral channels and resolution, detector properties, etc. The supported instruments at the time of writing are AMBER, MIDI, VEGA, CLIMB, CLASSIC, MIRC.

- The instrument noise figures. At this time we have an all-purpose "generic" noise model valid for any fibered recombiner (monoaxial/multiaxial).

- The science target itself, with several basic properties (RA, DEC, magnitude, proper motion, etc., usually retrieved on-line from the CDS database), and a structural model based on a collection of simple parametric models, such as point source, elliptical disk, Gaussian, etc. In the future, we will provide support for user-defined models (a functionality already present in ASPRO).

Adding new interferometers or instruments is just a matter of editing our simple XML configuration file.

2.2. ASPRO2 as a Workflow

Due to its "simulator" properties, ASPRO2 has a workflow-like structure. It structures and sequences the different steps needed to simulate the observation and produce the interferometric observables:

- Retrieve the object or object list basic parameters.

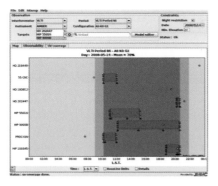 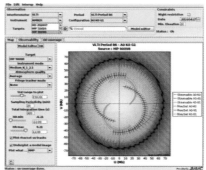

Figure 1.   Some aspects of the ASPRO2 GUI. Left: the GUI with the "Observability" tab selected. It summarizes the observability of all the objects included in the observation preparation, and serves as a chart for planning a night of observation. Right: with the "UV Coverage" tab selected. It permits to assign a model to an object, find the best telescope configuration for observation of this object etc... The image in the background is the Fourier transform of the object's model (in this case a giant star with a faint companion).

- Select the interferometer and instrument.

- Compute an **Observability Chart** based on the interferometer configuration, the night restriction for the observation date chosen, the minimum elevation, the delay line compensation for the selected base lines, telescope shadowing (for VLTI), zenithal constraints, etc.

- Select model for each object. ASPRO2 shares its model editor with our model-fitting program LITpro (Tallon-Bosc et al. 2008).

- Compute for each object the **UV Coverage** from the object's observability, the object's model, the instrumental configuration, the geometrical delays and constraints.

2.3.   The Graphical User Interface

We have tried to keep the GUI of ASPRO2 simple. Figure 1 shows two examples of the GUI panel. The upper part of the GUI groups the information or entry points common to the whole observational project. Below, a tabbed panel groups different kind of "views" needed by the preparation process. At the moment only three tabs are present, the Map of the Interferometer, the Observability and the UV Coverage. This will be completed in a future release by another tab with the OI-FITS visualization panel, that we are building as an independent, interoperable tool.

2.4.   Interoperability

ASPRO2 uses SAMP to interact with other VO-compliant, "SAMPified" applications, in particular two JMMC tools:

- SearchCal (Bonneau et al. 2006), the JMMC tool to find calibrators for the planned observations (and get back the star list in ASPRO2).

- `LITpro` (Tallon-Bosc et al. 2008), the JMMC tool to fit models in interferometric observables, applied here to the OI-FITS simulated by `ASPRO2`.

Besides, `ASPRO2` is also able to pass VOTables to, e.g. `Aladin`.

Acknowledgments. We wish to acknowledge the use in `ASPRO2` of the following components:

- The `JMCS`[3] Shared library providing GUI and common features.

- `JskyCalc`[4] which is to our knowledge the only Java library available dealing with ephemeris, astronomical coordinates conversions, etc.

- `nom.tam.fits`[5] which provides FITS file handling. We added the support of single and double-precision Complex values and handling of the COMMENT and UNIT keywords

- `JFreeChart`[6] handles all vector plots and provides science-grade exports in SVG or PDF (meaning that they can be used directly as illustrations in proposals or publications, an important feature frequently absent from JAVA-based graphical tools).

- `JSAMP`[7] for the SAMP VO query protocol.

- And SIMBAD[8] to retrieve many of the needed object's information (position, magnitude, proper motions, etc.) with a simple name query.

References

Bonneau, D., et al. 2006, A&A, 456, 789

Duvert, G., Bério, P., & Malbet, F. 2002, in Observatory Operations to Optimize Scientific Return III, edited by P. J. Quinn (Bellingham, WA: SPIE), vol. 4844 of Proc. SPIE, 295

Mella, G., & Duvert, G. 2004, in Advanced Software, Control, and Communication Systems for Astronomy, edited by H. Lewis, & G. Raffi (Bellingham, WA: SPIE), vol. 5496 of Proc. SPIE, 582

Pauls, T. A., Young, J. S., Cotton, W. D., & Monnier, J. D. 2005, PASP, 117, 1255

Tallon-Bosc, I., et al. 2008, in Optical and Infrared Interferometry, edited by M. Schöller, W. C. Danchi, & F. Delplancke (Bellingham, WA: SPIE), vol. 7013 of Proc. SPIE, 70131J

---

[3]by S. Lafrasse and G. Mella, JMMC

[4]by J. R. Thorstensen, Dartmouth College

[5]by Dr. Thomas A. McGlynn, HEASARC

[6]see http://www.jfree.org/jfreechart

[7]by Mark Taylor, AstroGrid

[8]Centre de Données Astronomiques de Strasbourg, http://simbad.u-strasbg.fr/simbad

Astronomical Data Analysis Software and Systems XX
ASP Conference Series, Vol. 442
Ian N. Evans, Alberto Accomazzi, Douglas J. Mink, and Arnold H. Rots, eds.
© *2011 Astronomical Society of the Pacific*

A New Compression Method for FITS Tables

William Pence,[1] Rob Seaman,[2] and Richard L. White[3]

[1] *NASA Goddard Space Flight Center, Greenbelt, MD 20771*

[2] *National Optical Astronomy Observatories, Tucson, AZ 85719*

[3] *Space Telescope Science Institute, Baltimore, MD 21218*

Abstract.   As the size and number of FITS binary tables generated by astronomical observatories increases, so does the need for a more efficient compression method to reduce the amount disk space and network bandwidth required to archive and download the data tables. We have developed a new compression method for FITS binary tables that is modeled after the FITS tiled-image compression convention that has been in use for the past decade. Tests of this new method on a sample of FITS binary tables from a variety of current missions show that on average this new compression technique saves about 50% more disk space than when simply compressing the whole FITS file with gzip. Other advantages of this method are (1) the compressed FITS table is itself a valid FITS table, (2) the FITS headers remain uncompressed, thus allowing rapid read and write access to the keyword values, and (3) in the common case where the FITS file contains multiple tables, each table is compressed separately and may be accessed without having to uncompress the whole file.

1.   Overview

Modern astronomical observatories continue to produce greater volumes of data in FITS binary table format, for example in the form of photon event lists and in mega- and giga-sized object catalogs. Currently, the only widely available method for reducing the storage size of these data tables is to compress the entire FITS file using an external file compression utility such as gzip or bzip2. We have developed a new compression method specifically designed for FITS binary tables that generally produces significantly higher compression. In addition, this new method has the advantage that the FITS headers remain uncompressed, which allows rapid read and write access to the keywords. Also, in cases where the FITS file contains multiple binary tables, each table is compressed separately and may be directly accessed without having to first uncompress the entire FITS file.

This new binary table compression method is modeled after the FITS tiled-image compression convention (see the FITS Support Office site at `http://fits.gsfc.nasa.gov/registry/tilecompression.html`) that has been in use for about a decade. This present paper provides a high-level description of this compression method and some preliminary performance results on a sample of FITS binary tables. Further technical details about this proposed FITS table compression method can be found in a draft specification document also available on the FITS Support Office web site.

2. Description of the Compression Method

The algorithm for compressing a FITS binary table is described in the following sub-sections. Only the tabular data values are compressed, and the header keywords remain uncompressed for rapid access. Almost all the keywords in the uncompressed table are copied verbatim into the header of the compressed table with only a few exceptions (such as the NAXISn and TFORMn keywords which necessarily define the structure of the compressed table itself).

2.1. Divide the Table into Tiles (Optional)

In order to limit the amount of data that must be manipulated at any one time, large FITS tables may be optionally divided into blocks, or tiles, where each tile contains an equal number of rows, except for the last tile which may contain fewer rows. Each tile is compressed in sequence, and is stored in a row in the output compressed table which is itself a valid FITS binary table. There is no fixed upper limit on the tile size, but it is recommended that FITS tables larger than about 10 MB in size be divided into multiple tiles so as to not impose too large of a memory resource burden on software that must uncompress the table.

2.2. Transpose the Rows and Columns

The data cells in a FITS binary table are stored in row by row order, so that the values in the first row of every column are given in order, followed by the values for the second row, and so on. It can be difficult to efficiently compress this native stream of data values, however, because binary tables can contain very heterogeneous types of data in different columns. For this reason, almost all FITS tables can be better compressed by first internally transposing the rows and columns in the table so that all the values for the first column occur first, followed by all the values for the second column, and so on. Then the data values within each column can be compressed separately.

2.3. Compress Each Column

After transposing the rows and columns in the table, each column of data is compressed with a suitable compression algorithm. Our prototype table compression utility program uses the gzip compression algorithm by default for all columns, but in principle each column could be compressed using a different algorithm that is optimized for that particular type of data (e.g., using the Rice algorithm for integer columns).

Our prototype compression code also preprocesses numeric column values (integer and floating-point) using a "byte shuffling" technique which rearranges the bytes so that the most significant byte of every column element appears first, followed by the next most significant byte of every element, and so on. Since the most significant byte(s) of each column value often contains nearly the same bit pattern, these bytes tend to compress very well when reordered in this way. Then when the table is uncompressed, the inverse shuffling is applied to restore the original byte order of all the column values.

In most cases a lossless compression algorithm will be required in order to exactly preserve the values in each column. In specialized cases, however, a lossy compression algorithm could be used to achieve higher compression if the values in a particular column only need to be approximately represented.

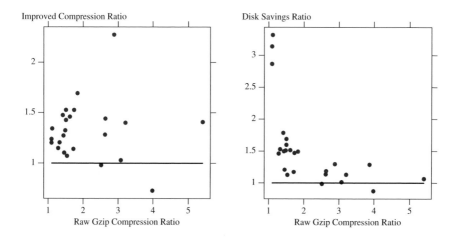

Figure 1. (Left): The points show the factor by which the compression ratio is improved when using the new compression method (y-axis), as compared to simply compressing the whole file with gzip (x-axis) for the 24 tables in our sample. (Right): The ratio of how much more disk space is saved when using the new compression method, as compared with just using gzip.

2.4. Store the Compressed Bytes

The compressed stream of bytes for each column is stored in the corresponding column in the output table, which is itself a valid FITS binary table with variable-length array columns. Each row in the output compressed table corresponds to one tile of rows in the uncompressed table. If the input table is compressed as a single tile, then the output table will only contain one row.

3. Example Results

We have implemented a prototype version of this compression method and applied it to a sample of 24 different types of FITS binary tables produced by 9 different missions (Chandra, Dark Energy Camera, Fermi, Herschel, PanSTARRS, ROSAT, SDSS, Swift, and XMM). For these tests we used the gzip algorithm to compress every column, after applying the previously described byte shuffling algorithm to the numeric column values. We compared the compression ratio of the tables when using this new method to that of the standard technique of simply compressing the whole FITS table with the gzip utility program. All the sample FITS tables are very large, so the fact the FITS headers remain uncompressed using our new method has no significant effect on the overall table compression ratio.

Figure 1a shows that in 22 out of the 24 cases, the technique of transposing the rows and columns in the table and shuffling the bytes in the numeric columns before compressing the column with gzip, produces better compression than simply using gzip to compress the raw FITS binary table. On average, our new method produced 1.3 times more compression then just using gzip alone. In only 1 case did the table compress significantly less well after shuffling the bytes in the numeric values. A more sophisticated approach would be to only apply the byte shuffling technique in cases where it actually improves the subsequent compression ratio. The CPU times needed to compress or uncompress these tables using the new method is not much greater (less than a factor of 2) than when simply using gzip, because the additional time needed to transpose the table and shuffle the bytes is relatively small compared to the time needed to actually compress the bytes with the gzip algorithm.

Of more direct practical interest, Figure 1b shows the ratio of how much more disk space is conserved by using our new method compared to just gzipping the whole file. On average, the new method saves about 50% more disk space than when just using gzip. The biggest disk space gains are for the files that compress by less than a factor of 2.

4. Summary

These tests demonstrate that this new compression method can significantly reduce the network bandwidth and disk space needed to transfer and archive the large volumes of FITS binary tables produced by current and future astronomical missions. In addition, the fact that the header keywords remain uncompressed, and that individual tables within a multi-table FITS file can be directly accessed without uncompressing the whole file should make this an attractive on-line data analysis format. We plan to conduct additional tests on a more comprehensive set of astronomical FITS tables to further optimize the compression strategy for different types of tables.

Astronomical Data Analysis Software and Systems XX
ASP Conference Series, Vol. 442
Ian N. Evans, Alberto Accomazzi, Douglas J. Mink, and Arnold H. Rots, eds.
©2011 Astronomical Society of the Pacific

NOAO E2E Integrated Data Cache Initiative Using iRODS

Irene Barg, Derec Scott, and Erik Timmermann

National Optical Astronomy Observatory, Science Data Management, Tucson, AZ 85719

Abstract. The NOAO Mass Storage System (MSS) holds astronomical data collected from about two dozen different scientific instruments at eleven telescopes on three mountain tops in two different countries both north and south of the Equator. Data are transferred via the net, from each mountain to Data Centers in La Serena, Chile and Tucson, Arizona. Then replicated across both hemispheres. A third copy is saved on tape at NCSA. This system is collectively called the End-to-End system (E2E). The data flow and file repository management is accomplished using a collection of custom code built on top of the Storage Resource Broker (SRB) developed by Data Intensive Cyber Environments (DICE) research group at the University of North Carolina at Chapel Hill, and the Institute for Neural Computation at the University of California, San Diego. iRODS, the Integrated Rule-Oriented Data System, is a data grid software system developed by DICE and collaborators and the successor to SRB. Both SRB and iRODS provide the ability to manage large amounts of data which can be distributed across data centers. This paper will describe why NOAO Science Data Management (SDM) chose iRODS as the next generation file repository management system for the E2E data management system.

1. Introduction

SRB was nearing its end of life status when the first release of iRODS appeared in 2008. At the same time, the Science Data Management Operations (SDM) group had planned to move away from the Apple XSAN architecture for our mass storage system (MSS). The two NOAO data centers, Tucson, Arizona and La Serena, Chile, provide the frontend to the MSS for all services that required access to the Archive. These frontend systems were Apple XServers running XSan client software. It was getting increasingly difficult to maintain both Mac OS X upgrades and the SRB due to compatibility issues. It was time to not only retire SRB but to migrate our MSS frontend systems to Linux.

2. Why iRODS?

The SRB had been a key component of the DCI since it's inception (August 2004). Since iRODS was the successor to SRB, we decided to take a look at iRODS. After downloading and installing iRODS in our test environment, it felt familiar, and the functionality we had grown to rely on were still there.

2.1. Ease of Migration

Before migrating to iRODS the following requirements needed to be met:

Table 1.   iRODS Migration Requirements

Migration requirement	iRODS	Comment
Retain existing physical and logical resources	yes	
Retain collection hierarchy	yes	
Retain existing mdasCollectionHome	yes	
Retain existing groups (noao, smarts, etc)	no	re-implement manually
Retain access control (ACL)	no	re-implement manually
Retain user metadata (md5sum)	no	not necessary, iRODS checksum is md5

The three requirements not met had acceptable workaround. Discussions with two members of the iRODS team suggested that the best migration path would be as follows:

1. Recreate the physical and logical resources in iRODS;

2. Bulk register the existing holdings;

3. Recreate your institutional groups + users;

4. Recreate the access control.

There were pros/cons to this approach.

- The pros were:

    - Clean installation;

    - Allowed us to make changes;

    - Could be implemented in parallel to live system.

- The cons were:

    - Time comsuming.

It took one week to migrate over 2 million records. The 'time consuming' disadvantage was mitigated by the ability to run an iRODS system in parallel to the SRB in order to implement a phased migration.

2.2. Phased Migration from SRB to iRODS

The migration to iRODS was implemented in phases. There were three drivers to a phased approach:

- We had to keep the current SRB-MCAT system running to support the production Archive.

- We needed all pre-existing MCAT metadata available at the time we switched to iRODS.

- We needed to time the transition to full iRODS with an Archive release that supported iRODS. The NOAO Archive uses Jargon, the Java client API to interface with iRODS.

A brief description of this hybrid system:

1. Two new iRODS scripts were written to take advantage of bundled transfers.

2. A modified version of the DCI client was written to register files in both SRB and iRODS.

3. A process to migrate MCAT-to-ICAT was implemented.

Phasing the migration this way allowed us to make the switch with the least amount of downtime to the production system as possible.

2.3. Reduction in Code

The original DCI code used the Perl API for SRB. This code needed to be re-written to utilize the iRODS implementation. However, this was a good thing, not bad. At the iRODS core is a Rule Engine that interprets the Rules to decide how the system is to respond to various requests. Micro services are small, well-defined procedures/functions that perform a certain task. These micro-services and the rule engine interact to provide you with control over what happens when one performs a macro-level functionality. Functionality that was originally implemented in the API, are now implemented using a set of rules. For example we can set the access control (ACL) for each file using a set of pre-defined rules.

2.4. Improved Performance

Improved performance was achieved in many ways:

1. Simpler code — moving repetitive tasks at the rule and micro-service level that get compiled into the iRODS sever code results in faster run-time application. We replaced Perl API with more efficient *icommands*. With finer control being done at the rule level, most of the functionality of the previous DCI is now implemented with 2-3 simple *icommands*.

2. Bundle file operations — bundling many small files into a large 350MB tar file then using iRODS *iget* to upload and transfer the tar bundle improved file transfer efficiency overall by 30 percent.

3. Improved file transfer performance — was achieved by using multi-threaded, concurrent, parallel TCP/IP connections.

The current iDCI data flow is illustrated in Figure 1.

3. Summary

iRODS is easy to install and configure. iRODS rule base engine and use of micro-services gives the end user greater flexibility and control over your file repository system. We have just started learning the capabilities of iRODS. We created an auto-retire

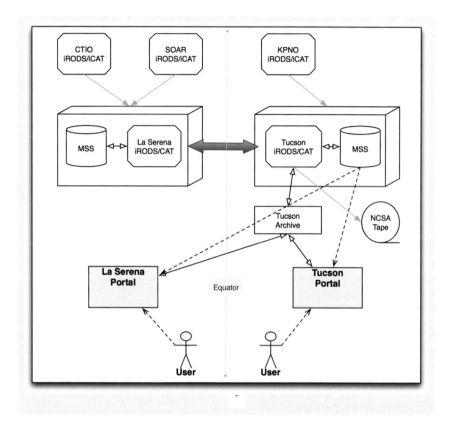

Figure 1.   The current iDCI data flow.

program that uses a *md5checker* to verify a night of data and if successful, adds a comment to the iRODS collection for the corresponding night. iRODS collections which have the verified comment and are greater than 9 weeks old are purged from the mountain caches automatically. This task was previously done manually.

Acknowledgments.   We wish to acknowledge the following:

- iRODS^{TM}, the Integrated Rule-Oriented Data System, is a data grid software system developed by the Data Intensive Cyber Environments (DICE) research group (developers of the SRB, the Storage Resource Broker), and collaborators. We wish to thank the DICE team for their support and feedback and for providing such a fine product as iRODS.

- Jargon, the iRODS Java^{TM} client is distributed by the University of California under the BSD License.

Astronomical Data Analysis Software and Systems XX
ASP Conference Series, Vol. 442
Ian N. Evans, Alberto Accomazzi, Douglas J. Mink, and Arnold H. Rots, eds.
©2011 Astronomical Society of the Pacific

Tile-Compressed FITS Kernel for IRAF

Rob Seaman

National Optical Astronomy Observatory

Abstract. The Flexible Image Transport System (FITS) is a ubiquitously supported standard of the astronomical community. Similarly, the Image Reduction and Analysis Facility (IRAF), developed by the National Optical Astronomy Observatory, is a widely used astronomical data reduction package. IRAF supplies compatibility with FITS format data through numerous tools and interfaces. The most integrated of these is IRAF's FITS image kernel that provides access to FITS from any IRAF task that uses the basic IMIO interface. The original FITS kernel is a complex interface of purpose-built procedures that presents growing maintenance issues and lacks recent FITS innovations. A new FITS kernel is being developed at NOAO that is layered on the CFITSIO library from the NASA Goddard Space Flight Center. The simplified interface will minimize maintenance headaches as well as add important new features such as support for the FITS tile-compressed (*fpack*) format.

1. Tile-Compressed FITS as a Runtime Image Format

Data compression has been a topic in the astronomical community for many years. In recent years, the increasing deployment of rapid readout large format cameras, automated pipeline software, and federated science archives have led to the anticipation of a looming "tsunami" of data (Seaman et al. 2007).

The FITS tile-compression convention (White et al. 2009) has emerged as the standard compressed data format for several "big data" astronomical archives and surveys. Project requirements may lead to either lossless (Pence et al. 2009) or lossy (Pence et al. 2010b) solutions. Tile-compressed FITS supports both options. As with FITS itself, tile-compressed FITS images can be accessed on the command line via standalone compression and decompression tools like *fpack* and *funpack* (http://heasarc.gsfc. nasa.gov/fitsio/fpack).

Ideally, support for FITS format data whether tile-compressed or not would be built directly into astronomical image processing software and systems. Rather than viewing data compression as a separate step to be applied to more fundamental data formats, it is more accurate to see it as an extension of the entire notion of the efficient representation of data. This is precisely what is done with non-astronomical formats such as JPEG, GIF, and PNG which have compression codecs built directly into these formats. Compression is a question of optimizing workflow throughput, not just minimizing storage footprint. FITS tile-compression also provides this option of run-time access, for instance through support built into the CFITSIO library from NASA Goddard Space Flight Center (http://heasarc.gsfc.nasa.gov/fitsio).

We discuss a project to add such runtime FITS tile-compression support to a major astronomical software package. In the future a similar sequence of projects, from library support to command line tools and then integrated runtime access are likely to build on top of the foundation of the proposed tiled-table compression convention (Pence et al. 2010a) for FITS binary tables applied to purposes such as astronomical catalog data.

2. Discussion of Options

The National Optical Astronomy Observatory (NOAO) was instrumental in the creation of both the Image Reduction and Analysis Facility (IRAF) and the Flexible Image Transport System. Both have also benefited from significant community development and oversight since their beginnings. While IRAF (http://iraf.net or http://iraf.noao.edu) has supported FITS from the very start, direct access to FITS files as a runtime format was only added after the fact through collaboration with the Space Telescope Science Institute (STScI) in the original FITS Image Kernel project.

The IRAF FITS kernel (Zarate & Greenfield 1996) has been a very successful addition to IRAF on the one hand and in providing community access to the burgeoning trade in archival FITS files on the other. It is unsurprising that it is showing its age, however, and when the idea emerged of adding the ability to access tile-compressed FITS files to IRAF, NOAO considered the best strategy for accomplishing this.

Two possibilities presented themselves. First, to modify the current FITS kernel to add the features required to read and write tile-compressed FITS. Second, rather to develop another FITS kernel layered on the CFITSIO library that already understands tile-compressed FITS. For the motivating reasons and features listed below we chose the second route.

2.1. Motivation

- The original FITS kernel was a complex interface of purpose-built procedures,

- that presents growing maintenance issues, and

- lacks recent FITS innovations.

2.2. Features

- A new FITS kernel is being developed at NOAO,

- layered on CFITSIO library from NASA/GSFC,

- that minimizes maintenance headaches, and

- supports the FITS tile-compression (*fpack*) format.

Pending the availability of the new tile-compression aware FITS kernel, the IRAF FITSUTIL package has been updated (Seaman 2010) with the ability to access tile-compressed FITS. This was accomplished by layering the IRAF package on the CFITSIO *fpack* and *funpack* tools, thus providing useful experience for the larger project. The intent is to first deploy the new tile-compression aware FZF image kernel in parallel with the original IRAF FXF FITS image kernel.

3. IRAF Image Kernel Interface

The IRAF Image I/O library (IMIO) is layered on the Image Kernel Interface (IKI), described by Tody (1986), that provides compatibility with several disk image formats. For each of the eight IKI procedures below, a matching format-specific procedure translates the disk format. This interface also permits adapting to idiosyncrasies of the different formats such as the ubiquitous Multi-Extension FITS (MEF) format. The logistics of accommodating the different formats may generate internal complexity in the individual interfaces.

3.1. Self-Explanatory Procedures

* iki_access

* iki_copy

* iki_delete

* iki_rename

3.2. Complexity Under the Hood

* iki_open

* iki_updhdr — update the image header

* iki_opix — open or create the pixel file

* iki_close

References

Pence, W. D., Seaman, R., & White, R. L. 2009, PASP, 121, 414
— 2010a, A tiled-table convention for compressing fits binary tables. URL http://fits.gsfc.nasa.gov/tiletable.pdf
Pence, W. D., White, R. L., & Seaman, R. 2010b, PASP, 122, 1065
Seaman, R. 2010, Fits tile compression using fpack. URL http://archive.noao.edu/doc/SDM_fpack_usernotes.html
Seaman, R., Pence, W., White, R., Dickinson, M., Valdes, F., & Zárate, N. 2007, in Astronomical Data Analysis Software and Systems XVI, edited by R. A. Shaw, F. Hill, & D. J. Bell (San Francisco, CA: ASP), vol. 376 of ASP Conf. Ser., 483
Tody, D. 1986, Image kernel interface (iki), source code, & documents such as: iraf$sys/imio/iki/readme
White, R. L., Greenfield, P., Pence, W., Tody, D., & Seaman, R. 2009, Tiled image convention for storing compressed images in fits binary tables. URL http://fits.gsfc.nasa.gov/registry/tilecompression/tilecompression2.1.pdf
Zarate, N., & Greenfield, P. 1996, in Astronomical Data Analysis Software and Systems V, edited by G. H. Jacoby & J. Barnes (San Francisco, CA: ASP), vol. 101 of ASP Conf. Ser., 331

Astronomical Data Analysis Software and Systems XX
ASP Conference Series, Vol. 442
Ian N. Evans, Alberto Accomazzi, Douglas J. Mink, and Arnold H. Rots, eds.
© *2011 Astronomical Society of the Pacific*

MAGIX: A Generic Tool for Fitting Models to Astrophysical Data

I. Bernst,[1] P. Schilke,[1] T. Moeller,[1] D. Panoglou,[1] V. Ossenkopf,[1] M. Roellig,[1] J. Stutzki,[1] and D. Muders[2]

[1]*Physical Institute, University of Cologne, Zuelpicher Str. 77, 50937 Koeln, Germany*

[2]*Max Planck Institute for Radio Astronomy, Auf dem Huegel 69, 53121 Bonn, Germany*

Abstract. MAGIX (Modelling and Analysis Generic Interface for eXternal numerical codes) is a model optimizer developed under the framework of the CATS (Coherent set of Astrophysical Tools for Spectroscopy) project, funded by the European ASTRONET. The goal is to use existing astrophysical models, and optimize their parameters to fit a given astrophysical data set. A number of algorithms can be used to explore the parameter space, to find the best-fitting values of the parameters to be optimized, and the corresponding confidence intervals. A GUI based frontend is being developed to make the registration of new models and creation/editing of new instances of initial conditions user-friendly. MAGIX complies with the data structures and reduction tools of ALMA (Atacama Large Millimeter Array), and is developed to be a tool accompanying observations assembled with the ALMA interferometer, but can be used on any, even non-astronomical, data sets.

1. About MAGIX

MAGIX[1] is a model optimizer developed under the framework of the CATS project. It is an ASTRONET[2] funded German-French-Swedish Project that will provide common tools and databases for astrophysical applications. (M)any theoretical models can plug into it, and its goal is to provide the best-fit parameters, within the framework of the model, to a particular data set, including confidence intervals for the parameter values. It consists of a GUI based frontend to register new models (registration), to create model instances, i.e. set up a model with initial conditions (instantiation), the fitting engine, and an output module.

MAGIX is able to read a variety of model data formats, including FITS, and has a number of algorithms available for finding the best fit: Levenberg-Marquardt, Simulated Annealing, Particle Swarm Optimization, Bees algorithm, Genetic algorithm, Nested Sampling.

It is basically written in Python, whereas various algorithm packages are written in Fortran. MAGIX requires the following packages: python 2.6 (or later), numpy 1.3 (or later), pyfits 1.0 (or later), gfortran 4.3 (or later), matplotlib 0.99 (or later).

---

[1]http://www.astro.uni-koeln.de/projects/schilke/MAGIX

[2]http://www.astronet-eu.org

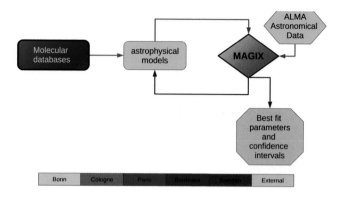

Figure 1.   Schematic view of the CATS framework. It consists of a database part (on the left), provided by French and Swedish partners, and by the MAGIX part (right), provided by Germany.

1.1.   Parallel Computations

Depending on the model, the computational load is heavy. Sometimes, calculating a point of the optimization function takes more than one hour. For some optimization algorithms, hundreds of function calls are necessary; improving the speed will be the next long-term goal. The whole system of MAGIX is being parallelized with parallel calculations in the models and in the optimization algorithms, increasing the computational speed.

1.2.   Pre-Registered Models

At the moment pre-registered models include: myXCLASS (modeling of the star-forming regions with access to the CDMS and the JPL molecular data bases); Sim-Line and RATRAN (computing the profiles of molecular lines); myCloud (computing of 3D data cubes of spectral ranges with arbitrary input geometries and calculating the radiative transfer).

2.   Optimization Algorithms

Most physical models depend on a set of parameters, and finding the best model means finding the parameter set that most closely reproduces the data by some criteria, e.g. a minimum of the χ^2-distribution. Since most models are nonlinear functions of the input parameters, this means finding the global minimum of a multi-dimensional nonlinear function, which is by no means trivial. However, there are algorithms that can achieve the goals in most cases. Examples are the Levenberg-Marquardt (conjugate gradient) method, which is fast, but can get stuck in local minima; Simulated Annealing and Particle Swarm Optimization methods, which are slower, but more robust against local minima. Other, more modern methods, such as Bees, Genetic or Nested Sampling algorithms are included in MAGIX for exploring the solution landscape, checking for the existence of multiple solutions, and giving confidence ranges.

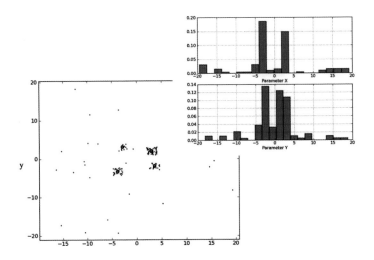

Figure 2.   Bees algorithm results on the Himmelblau function ($f(x, y) = (x^2 + y - 11)^2 + (x + y^2 - 7)^2$; the function has four minima) after 600 function calls. Red points are most probable places for minima. The histograms show the final distributions of parameters values (the probabilities obtained by Nested Sampling algorithm). One can see that the minima are found reasonably well.

2.1.   Bees Algorithm

The Bees Algorithm[3] is an optimization algorithm inspired by the natural foraging behavior of honey bees to find the optimal solution. In its basic version, the algorithm performs a kind of neighborhood search combined with random search and can be used for function optimization. The advantages of the algorithm that it finds multiple minima, converges quickly, lends itself to parallelization, while the main disadvantage is the large computing time, because of many function calls.

2.2.   Genetic Algorithm

The Genetic algorithm is a probabilistic search algorithm that iteratively transforms a set (called a population) of parameter vectors, each with an associated fitness value, into a new population of offspring objects using the Darwinian principle of natural selection and operations that are patterned after naturally occurring genetic operations, such as crossover (recombination) and mutation. The advantages of the algorithm that it investigates the landscape of optimization function, converges quickly, and lends itself to parallelization.

2.3.   Nested Sampling

The Nested Sampling algorithm[4] is a variant of the Monte Carlo technique based on a Bayesian approach, which reduces the dimensionality of the space through integration,

---

[3]http://www.bees-algorithm.com

[4]http://www.inference.phy.cam.ac.uk/bayesys

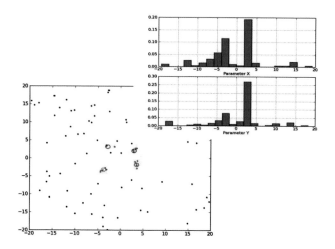

Figure 3.   Genetic algorithm results on the Himmelblau function after 10 iterations (about 300 function calls). Red points are the final population and black points are the initial population. The histograms show the final distributions of parameters values (the probabilities obtained by Nested Sampling algorithm).

allowing not only multiple solutions, but also estimation of the confidence intervals of parameters values and evaluation of Bayesian evidence. In parameter estimation, the evidence factor is usually ignored, since it is independent of the parameters, but the evidence automatically implements Occam's razor: a simpler theory with a compact parameter space will have a larger evidence than more complicated one, unless the latter is significantly better at explaining the data. Thus, the evidence can be used for definition of a number of free parameters in the model. Posterior inferences can be generated using Bayes theorem from the nested sampling process. Then they can be used to calculate inferences of posterior parameters values such as means, standard deviations, covariances, etc., or to construct marginalized posterior distributions. The advantages of the algorithm that it finds multiple minima, gives posterior inferences, typically requires many times fewer samples than standard MCMC methods. The disadvantage is that the algorithm is not easy to parallelize.

3. Outlook

In the final version, it is envisioned that MAGIX will have a Heuristics module that will to be able choose the best combination of algorithms based on user-defined priorities. We are currently working on making registration and instantiation more user-friendly by going to a web-based approach.

   Acknowledgments.   We acknowledge funding from BMBF through the ASTRO-NET Project CATS.

Astronomical Data Analysis Software and Systems XX
ASP Conference Series, Vol. 442
Ian N. Evans, Alberto Accomazzi, Douglas J. Mink, and Arnold H. Rots, eds.
©*2011 Astronomical Society of the Pacific*

The Crates Library: The Redesigned Python Interface for Scripting Languages

Janine Lyn, Doug Burke, Mark Cresitello-Dittmar, Ian Evans, and
Janet DePonte Evans

Smithsonian Astrophysical Observatory,
60 Garden Street, Cambridge, MA 02138, USA

Abstract.   Crates is a Python module produced by the Chandra X-ray Center (CXC) that provides a convenient high-level user interface for accessing and manipulating data stored in a variety of formats. Crates is currently utilized by Chandra's plotting, modeling and fitting tools. This paper will highlight the design changes and improvements made to Crates. This version of Crates has been completely rewritten in Python and has been optimized to conserve time and memory resources through lazy initialization. It provides increased functionality for data and metadata manipulation along with better memory management. In addition, Crates will be able to interface with several different backend modules, allowing the user to effortlessly switch between the CXC Data Model (DM), Virtual Observatory (VO), and pyFITS formats.

1. An Overview of Crates

Crates was developed as a high level I/O interface for general file access. It has been redesigned and implemented entirely in Python to make use of Python's flexibility, extensibility and automatic memory management.

The redesigned Crates library has been separated into two distinct layers. One layer handles all of the file processing needs. It determines the appropriate backend processor to use based on the input file name and reads in the file information as required. The default backend is the Chandra DataModel, and we intend to include backends for PyFITS and the Virtual Observatory.

The second layer is the application interface. It includes several classes to represent file data, header keywords, table columns, and images, along with methods for the user to examine and manipulate the stored input file data.

1.1. The Design Architecture

Crates can be imported as a stand-alone module or imported for use within other applications, such as Sherpa (Chandra's modeling and fitting tool), ChIPS (the Chandra Imaging and Plotting System), and the CIAO (Chandra Interactive Analysis of Observations) contrib scripts.

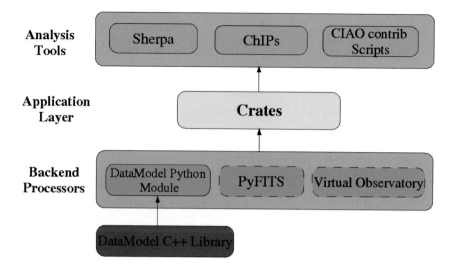

Figure 1.   An illustration showing the relationship between Crates and the backend processors and analysis tools.

2.   Improvements To Crates

The new implementation of Crates incorporates additional file handling capabilities not previously available:

- Access to multiple sections of a dataset,

- Ability to update data files in place,

- Ability to easily create a dataset from scratch,

- Access to the file subspace,

- Dynamically generated named attribute shortcuts to table columns and images.

We have further optimized Crates by using lazy initialization to reduce read-in time and memory requirements. Instantiation of keyword and data objects is deferred until a requesting function call by the user. Once all information has been read in, the backend file pointers are closed to release memory.

3.   Crates Classes

The Crates module provides a set of classes for accessing and modifying data files:

- CrateDataset — Provides access to multiple sections of a dataset.

- Crate — An abstract base class representing one section in a dataset. It stores a collection of keywords, columns and/or image data, and provides access to header keywords.

- TABLECrate — An extension of the Crate class that provides access to tabular column data, which includes regular and virtual columns.

- IMAGECrate — An extension of the Crate class that provides access to the image array and axes.

- CrateData — Stores a column or an image and its associated metadata.

- CrateKey — Stores one header keyword.

Specialized CrateDatasets are also included for data products requiring a unique combination of Crates and specific header keywords.

- PHACrateDataset — Class for spectral file format, Types I and II.

- RMFCrateDataset — Class for the response matrix file.

4. A Scripting Example

The following is a simple example of using the Crates module to read in a file, display that file's information, update data, and write the file in place.

```
>>> from pycrates2 import *              >>> tab.sky
>>> ds = CrateDataset("file.fits")         Name:        sky
>>> print ds                               Shape:       (6117, 2)
   Crate Dataset:                          Datatype:    float32
      File Name:          test.fits        Nsets:       6117
      Read-Write Mode:    rw               Unit:        pixel
      Number of Crates:   2                Desc:        sky coordinates
         1) Crate Type:      <TABLECrate>  Eltype:      Vector
   Crate Name:         EVENTS                 NumCpts:   2
   Crate Number:       1                      Cpts:      ['x', 'y']
   Ncols:              10                  >>> tab.sky.values
   Nrows:              6117               array([[  4311.04150391,    98535.84375   ],
                                                 [  4250.35546875,    98576.4453125 ],
         2) Crate Type:      <TABLECrate>         [  4267.41015625,    98558.9609375 ],
   Crate Name:         GTI7                       ...,
   Crate Number:       2                          [  4512.25976562,    4170.50244141],
   Ncols:              2                          [  4507.57373047,    4162.91894531],
   Nrows:              1                          [  4505.46386719,    4160.11914062]],
>>> tab = ds.get_crate(1)                  dtype=float32)
>>> tab.print_colnames()                   >>> tab.sky.values[0][1] = 99999.000
      Colname                              >>> tab.sky.values
   0)   time                              array([[  4311.04150391,    99999.          ],
   1)   ccd_id                                   [  4250.35546875,    98576.4453125 ],
   2)   node_id                                  [  4267.41015625,    98558.9609375 ],
   3)   expno                                    ...,
   4)   chip(chipx, chipy)                       [  4512.25976562,    4170.50244141],
   5)   tdet(tdetx, tdety)                       [  4507.57373047,    4162.91894531],
   6)   det(detx, dety)                          [  4505.46386719,    4160.11914062]],
   7)   sky(x, y)                         dtype=float32)
   8)   phas                              >>> ds.write()
   9)   pha
```

Based on the input file used in the example script, Figure 2 shows the organizational structure of the Crates classes once the file has been read in. The diagram shows a CrateDataset object representing file.fits which contains two TABLECrate objects named 'EVENTS' and 'GTI7'. Both tables store keyword information in a list of CrateKey objects and column data in a list of CrateData objects.

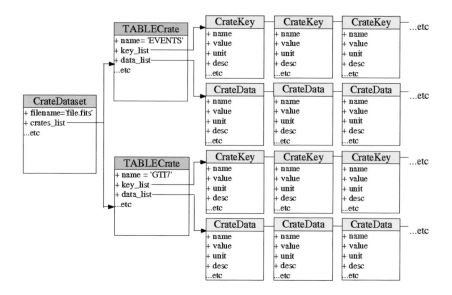

Figure 2.   An illustration showing the relationship between Crates classes using the dataset file.fits from the example script.

5. Conclusion

Both Sherpa and ChIPS analysis tools have Python front ends, and many scientists are writing Python scripts to interact with CIAO. The addition of Crates Python interface will give them further access to perform data analysis using CIAO.

Acknowledgments.   Support for the development of Crates is provided by the National Aeronautics and Space Administration through the Chandra X-ray Center, which is operated by the Smithsonian Astrophysical Observatory for and on behalf of the National Aeronautics and Space Administration contract NAS8-03060.

Astronomical Data Analysis Software and Systems XX
ASP Conference Series, Vol. 442
Ian N. Evans, Alberto Accomazzi, Douglas J. Mink, and Arnold H. Rots, eds.
©*2011 Astronomical Society of the Pacific*

Charming Users into Scripting CIAO with Python

D. J. Burke

Smithsonian Astrophysical Observatory, Cambridge, MA 02138, USA

Abstract. The Science Data Systems group of the Chandra X-ray Center provides a number of scripts and Python modules that extend the capabilities of CIAO.[1] Experience in converting the existing scripts—written in a variety of languages such as bash, csh/tcsh, Perl and S-Lang—to Python, and conversations with users, led to the development of the `ciao_contrib.runtool` module. This allows users to easily run CIAO tools from Python scripts, and utilizes the metadata provided by the parameter-file system to create an API that provides the flexibility and safety guarantees of the command-line. The module is provided to the user community and is being used within our group to create new scripts.

1. Parameter Files in CIAO

The CIAO data analysis system uses parameter files to define the user-interface of tools and scripts that it provides. These parameter files provide the names, types, default values, possible constraints and help text for the parameters, or arguments, that the tools use. Users can query these files to find out what inputs and outputs a tool uses, find what the current values are, or change the values, as shown in Figure 1. We show a typical sequence of commands that a user may write at the command line or in a Shell script in Figure 2 and the equivalent calls using `ciao_contrib.runtool` in Figure 3.

2. Goals of the Python interface

- Prioritize familiarity and simplicity over functionality.

- Map tool names to functions (actually objects).

- Avoid worries about quoting special characters (CIAO filters can contain "awkward" characters such as []!*'").

- Provide names arguments matching those of the tools.

- Type conversion and verification for argument values.

- Provide easy access to screen output of the tool.

- Convert tool failures to Python IOError exceptions.

- Mimic parameter access via object attributes.

- Support running multiple copies of the tool simultaneously.

---

[1]`http://cxc.harvard.edu/ciao/`

```
% plist acis_bkgrnd_lookup

Parameters for /home/ciaouser/cxcds_param4/acis_bkgrnd_lookup.par

      infile =           Event file for which you want background files
      outfile =          ACIS background file(s) to use
     (blname = none)     What block identifier should be added?
    (verbose = 0)        Debug level (0=no debug information)
       (mode = ql)

% pset acis_bkgrnd_lookup verbose=1
% pset acis_bkgrnd_lookup blname=all
pquery: invalid enumerated value : blname
What block identifier should be added?
     (none|name|number|cfitsio) (none): name
```

Figure 1.   An example of the parameter interface being used at the command-line to inspect and change the parameters for a tool.

```
dmcopy "evt2.fits[sky=region(src.reg),energy=500:7000]" src.fits \
       clobber=yes
acis_bkgrnd_lookup src.fits verbose=0
set bfile = 'pget acis_bkgrnd_lookup outfile'
dmcopy "${bfile}[energy=500:7000]" bg.fits
```

Figure 2.   An excerpt from a mythical shell script that filters an event file, finds out the matching "blank-sky" background file for the resulting file, and then filters that file.

```
from ciao_contrib.runtool import dmcopy, acis_bkgrnd_lookup

dmcopy("evt2.fits[sky=region(src.reg),energy=500:7000]",
       "src.fits", clobber=True)
acis_bkgrnd_lookup("src.fits", verbose=0)
bfile = acis_bkgrnd_lookup.outfile
dmcopy(bfile + "[energy=500:7000]", "bg.fits")
```

Figure 3.   Conversion of the shell script from Figure 2 into Python, using the ciao_contrib.runtool module to run CIAO tools and access parameter values.

3. Implementation Details

Four classes are used to provide the required functionality. The base class (CIAO-Parameter) provides the methods to read and write parameter values; it is used to represent the small set of CIAO parameter files which are for configuration and do not have an associated executable. The CIAOTool class is the basis for callable tools, and is a thin layer on top of the Python subprocess module,[2] with most of the code involved with validating the parameter values. The actual implementation is left to its sub-classes: CIAOToolParFile and CIAOToolDirect. Most executables are handled by the former, with the latter class provided to handle the small number of cases which do not support the @@parfile syntax used by the parameter interface (see Section 4).

The module is created by code generation — based on a parsed view of the CIAO parameter files — rather than having the necessary instances created either when the module is loaded or explicitly by the user. The trade off here was ease, and speed, of use for the user versus a more complicated development environment. Since the functionality is encapsulated within a class structure, it would be relatively easy to switch to the run-time approach.

Although CIAO provides a Python module that binds to the parameter library, this interface is low level and does not provide the required functionality — in particular parameter validation without error messages or requiring user interaction — which means that a significant part of the module is essentially replicating the functionality of the parameter library.

Since CIAO tools occasionally use both the stdout and stderr channels to output information, so both channels are combined into one and returned to the caller when the tool finishes. If the tool returns a non-zero exit status the screen output is instead returned to the user as the message payload of an IOError exception. There are several tools which do not set the exit status on certain errors, which will result in the Python routine apparently succeeding.

As the module is intended for use from within a Python, the interactive mode of operation of CIAO tools, where users are prompted for missing or invalid parameter values, is not supported by the module. Attempts to include the tool or parameter information in the Python docstrings for the routines was not successful; once it started to require the use of Python metaclasses the complexities of the interface outweighed the benefits to the user.

4. Running Multiple Copies of a Tool

One of the design goals of the module is to allow the user to easily run multiple copies of a tool simultaneously, whether via the multiprocessing module[3] or by repeated runs of the same script. Simultaneous runs of a tool is likely to cause corruption of the parameter file, since each copy is reading and writing the same file, which can lead to invalid output. To avoid this, each copy of the tool is run with its own unique (temporary) parameter file, supplied using the @@parname functionality of the CIAO parameter library, which is removed once the tool has finished. However a small subset

---

[2]http://docs.python.org/release/2.6.6/library/subprocess.html

[3]http://docs.python.org/release/2.6.6/library/multiprocessing.html

of CIAO tools do not support this syntax (these tools are run with all its parameters set on the command line to reduce the chance of corruption), and some tools, in particular those that are wrappers around other tools, require multiple parameter files.

Whilst the CIAO 4.3 release essentially removes the first problem (only one tool remains in this category), the second problem can only be avoided by explicitly creating separate directories to store the parameter files for each task. This can be achieved using the set_pfiles() routine provided by the module, which changes the user-directory portion of the PFILES environment variable used by the parameter library.

5. Future Work

The main aims of the module have been met,[4] and it is being used by SDS to enhance CIAO (Galle et al. 2011), so future development depends on user feed-back. The main areas that have been identified so far are:

1. Extending it to support other systems with parameter files, notably the FTOOLS package, but other systems, such as that used by TEAL at STScI, are possible,

2. Including support for reading in parameter settings from the history records that CIAO provides (in a similar manner to De Jong et al. 2011),

3. Supporting the piping of tools together to avoid the need to create intermediary files,

4. Completing the parameter interface (there is limited support for some of the more arcane parameter-redirection capabilities provided by the CIAO parameter library).

Acknowledgments.   Support for this work was provided by the National Aeronautics and Space Administration through the Chandra X-ray Center, which is operated by the Smithsonian Astrophysical Observatory for and on behalf of the National Aeronautics and Space Administration contract NAS8-03060. I would like to thank Tom Aldcroft for several stimulating discussions revolving around Python metaclasses.

References

De Jong, J., et al. 2011, in Astronomical Data Analysis Software and Systems XX, edited by I. N. Evans, A. Accomazzi, D. J. Mink, & A. H. Rots (San Francisco, CA: ASP), vol. 442 of ASP Conf. Ser., 371
Galle, E. C., Anderson, C. S., Bonaventura, N. R., Burke, D. J., Fruscione, A., & McDowell, J. C. 2011, in Astronomical Data Analysis Software and Systems XX, edited by I. N. Evans, A. Accomazzi, D. J. Mink, & A. H. Rots (San Francisco, CA: ASP), vol. 442 of ASP Conf. Ser., 131

---

[4]http://cxc.harvard.edy/ciao/scripting/runtool.html

Astronomical Data Analysis Software and Systems XX
ASP Conference Series, Vol. 442
Ian N. Evans, Alberto Accomazzi, Douglas J. Mink, and Arnold H. Rots, eds.
© *2011 Astronomical Society of the Pacific*

The Sherpa Maximum Likelihood Estimator

D. Nguyen, S. Doe, I. Evans, R. Hain, and F. Primini

Harvard-Smithsonian Center for Astrophysics, Cambridge, MA, USA

Abstract.   A primary goal for the second release of the Chandra Source Catalog (CSC) is to include X-ray sources with as few as 5 photon counts detected in stacked observations of the same field, while maintaining acceptable detection efficiency and false source rates. Aggressive source detection methods will result in detection of many false positive source candidates. Candidate detections will then be sent to a new tool, the Maximum Likelihood Estimator (MLE), to evaluate the likelihood that a detection is a real source. MLE uses the Sherpa modeling and fitting engine to fit a model of a background and source to multiple overlapping candidate source regions. A background model is calculated by simultaneously fitting the observed photon flux in multiple background regions. This model is used to determine the quality of the fit statistic for a background-only hypothesis in the potential source region. The statistic for a background-plus-source hypothesis is calculated by adding a Gaussian source model convolved with the appropriate Chandra point spread function (PSF) and simultaneously fitting the observed photon flux in each observation in the stack. Since a candidate source may be located anywhere in the field of view of each stacked observation, a different PSF must be used for each observation because of the strong spatial dependence of the Chandra PSF. The likelihood of a valid source being detected is a function of the two statistics (for background alone, and for background-plus-source). The MLE tool is an extensible Python module with potential for use by the general Chandra user.

1.   Initial Source Candidate Detection in CSC Release 2.0

The first release of the Chandra Source Catalog is the largest catalog of targeted and serendipious X-ray sources oberserved by the Chandra X-ray Observatory, containing ~95,000 sources (Evans et al. 2010). In Release 2, additional, lower-count sources (as few as 5 counts) will be identified. Existing tools in the detect pipeline will be modified in two major ways to detect these fainter source candidates. First, they will be able to process combined stacks of observations of the same sky region at once, leading to better detection of faint sources. Secondly, detect parameters will be tuned to detect source candidates much more aggressively. This will lead to more source detections, but at the price of higher false positive rates.

These changes work in concert with the new Maximum Likelihood Estimator (MLE) step. By detecting fainter and more questionable sources at the detect step, the MLE routine will have a maximum number of potential sources to investigate. MLE's ability to sort true sources from false positives will allow detections of faint sources while eliminating spurious sources.

Figure 1 shows two views of a Chandra Deep Field South (CDFS) region as seen in four observations (obsids 2312, 2313, 2405, 2406). Detections from CSC Release 1.1 processing from a single obsid (2313) are shown to the left, and detections with

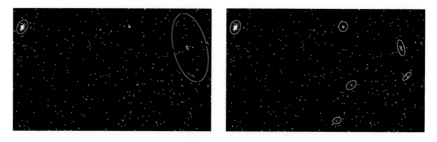

Figure 1.   CSC Rel 1 (left) and prototype Rel 2 detections in a CDFS region.

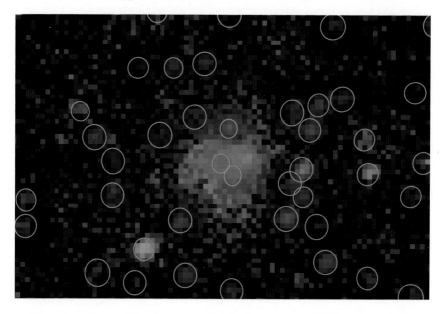

Figure 2.   Detected Sources from multiple Observations of M17, color-coded by source likelihood.

prototype Release 2 code, combining all four obsids and tuning parameters, are shown to the right. Increased source candidate detections from the prototype Release 2 code will need to be classified by the MLE algorithm as either true sources or false positives.

When many sources are grouped together in observed fields the detection requirements are more challenging. In the M17 field shown in Figure 2, the higlighted regions depict the size of the 90% encircled counts fraction of the local Chandra point spread function (PSF), rather than the source regions found in the detect step. Background regions for each source (not shown) typically have similarly orientations to detected source regions, but are approximately two to three times as large, and exclude the source region at the center and all other overlapping source regions to avoid including any source photons as part of the background estimate. However, the PSF is large enough that for the brightest sources in this field it is likely that some photons spill into neighboring source and background regions.

2. Maximum Likelihood Estimator

The Maximum Likelihood Estimator (MLE) suite of tools use the Sherpa modeling and fitting engine to fit a model of a background and source to multiple overlapping candidate source regions by minimizing the Cash statistics. The resulting background and source fit are used to calculate the likelihood for each of the candidate sources identified by the detect pipeline. For an isolated source (see Figure 1), the likelihood is calculated from the statistic for background model alone and the statistic for background plus source model. For crowded fields (see Figure 2), the likelihood calculation is more complicated. The MLE algorithm will need to extract accurate background information and estimate the likelihood for faint sources near bright sources. If MLE does not accurately account for photons which spillover from bright sources into background regions, it can overestimate the background for fainter sources. This could result in too low a likelihood estimate for the faintest sources (false negatives). Distinguishing between source and background photons and spillover photons from bright sources in crowded fields is a current challenge that we are studying. The MLE algorithm is inspired by *emldetect*[1] of the XMM project (Watson et al. 2009; Puccetti et al. 2009).

2.1. Calculating the Likelihood for a Potential Isolated Source

A 2-D constant background model is used to simultaneously fit the observed photon flux in multiple background regions. The fitted 2-D constant model is then used to determine the quality of the fit statistic, C_{bkg}, for a background-only hypothesis in the multiple source regions. The fit statistic, C_{src}, for a background-plus-source hypothesis is done by adding the fitted 2-D constant background model, with its parameter frozen, to a 2-D Gaussian source model convolved with the appropiate Chandra PSF. Since a candidate source may be located anywhere in the field of view of each stacked observation, a different PSF must be used for each observation because of the strong spatial dependence of the Chandra PSF. The fit of the source and background model is simultaneously fitted for each obervation in the stack. The likelihood for the source candidate is given by:

$$Likelihood = -\ln(1 - \Gamma(v/2, \Delta C/2))$$

where v is the number of free parameters in the fit for the source, $\Delta C = C_{bkg} - C_{src}$ and Γ is the incomplete gamma function.

To test MLE, simulations of point sources ranging from zero to thirty counts and from on-axis to thirty arcminutes off-axis were generated. Figure 3 shows the detection of on-axis valid and false sources, shown in green and red, respectively.

2.2. Calculating the Likelihood of Potential Sources in Crowded Fields

An algorithm is in development to calculate the likelihoods of candidate sources in crowded fields. The contributions from nearby bright sources must be taken into account when calculating the likelihood of any potential source in crowded fields. The MLE algorithm of the crowded source case is a generalization of the isolated source case (by adding nearby sources which are brighter then itself but with their parameters kept frozen). The algorithm sorts the potential sources in the field in descending order

---

[1]http://xmm.esac.esa.int/sas/current/doc/emldetect/index.html

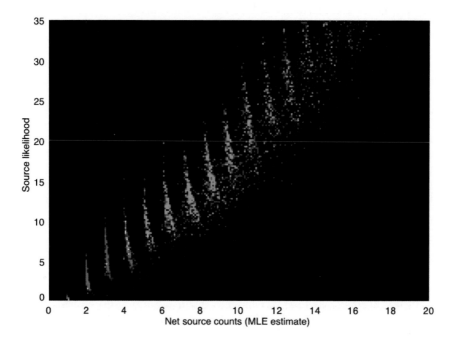

Figure 3.   A 2-D histogram of source likelihood versus net source counts estimated
by MLE for an isolated on-axis source.

of source counts. The likelihood for the brightest source is calculated as though it is an
isolated source. We then iterate down the list of sorted candidate sources calculating
likelihoods similarly but with one minor variation: A 2-D constant background model,
plus the 2-D Gaussians from the nearby brighter sources is used to simultaneously fit
the observed photon flux in multiple background regions. The fitted 2-D constant model
is then used to determine the quality of the fit statistic for a background-only hypothesis
in the multiple source regions. The fit statistic for a background-plus-source hypothesis is done by adding the fitted 2-D constant background model, with its parameters
frozen, to a 2-D Gaussian source model convolved with the Chandra PSF (plus the the
2-D Gaussians from the nearby brighter sources, with their parameters frozen). The
likelihoods for the candidate sources are then calculated using the formula above.

Acknowledgments.   Support of the development of MLE is provided by National
Aeronautics and Space Administration through the Chandra X-ray Center, which is op-
erated by the Smithsonian Astrophysical Observatory for and on behalf of the National
Aeronautics and Space Administration under contract NAS8-03060.

References

Evans, I. N., et al. 2010, ApJS, 189, 37
Puccetti, S., et al. 2009, ApJS, 185, 586
Watson, M. G., et al. 2009, A&A, 493, 339

Astronomical Data Analysis Software and Systems XX
ASP Conference Series, Vol. 442
Ian N. Evans, Alberto Accomazzi, Douglas J. Mink, and Arnold H. Rots, eds.
© 2011 Astronomical Society of the Pacific

Fermi Large Area Space Telescope Offline Software Maintenance Madness

Heather Kelly

For the Fermi LAT Collaboration
SLAC National Accelerator Laboratory, Stanford, CA, USA

Abstract.   The Fermi Observatory, including the Large Area Telescope (LAT), was launched June 11, 2008. The LAT's offline software includes a detailed C++ Monte Carlo simulation built around Geant4 and the Gaudi framework, event by event reconstruction with ROOT output for use within the collaboration, and science tools utilizing FITS which are ultimately distributed to the user community. Within the LAT team, we support Redhat Linux, Windows, and are moving toward Mac OS. We support the use of Visual Studio as a development environment, while we use one build tool for our automated builds across all operating systems. We continue to rely upon nineteen external libraries including: ROOT, Gaudi, Swig, and Xerces. There have been a number of lessons learned, with undoubtedly more to come. This paper will focus on our experiences with our external libraries, maintenance of a large scale offline software project, and support of our developer and user communities.

1. Introduction

The LAT collaboration's offline software includes GlastRelease and the ScienceTools. GlastRelease is our C++ Monte Carlo simulation and data reconstruction software which is utilized as part of the offline flight data processing pipeline. The Science-Tools consist of the software related to scientific analysis of Fermi LAT data, written in C++ with Python interfaces. With eight years ahead of us, we are in the phase of the project where we must move forward to support modern operating systems and compilers to get us through the life of the mission. This means upgrading our external libraries as well as our homegrown software. Meanwhile, it is crucial to our production system that we carefully orchestrate all upgrades to insure stability.

2. Supported Operating Systems

At launch, the LAT team officially supported Redhat Enterprise Linux 3 and Windows Visual Studio 2003. We have since dropped Redhat 3 in favor of Redhat Enterprise 4. We are also expanding to support Redhat Enterprise 5, Mac Snow Leopard, and Visual Studio 2008 on Windows. Some may ask why we support Visual Studio at all, rather than handle Windows via cygwin. We have a handful of proficient Windows developers who are attached to the Visual Studio development environment. Some of Visual Studio's appealing features include:

- Integrated Debugger which allows a developer to go from an error message to setting breakpoints with just a couple of clicks,

- Integrated Editor and Build Properties that allow programmers to set compile and link settings quickly and easily,

- Intellisense editing which provides automated class member completion.

3.   A Tale of Two External Libraries

Of our nineteen external libraries, two are most worthy of mention. ROOT is a data analysis framework that we use in both our GlastRelease and ScienceTools software primarily for its I/O, histogramming, and various math libraries. Gaudi is the Object Oriented framework we chose to base our GlastRelease software implementation.

3.1.   ROOT: A Data Analysis Framework

ROOT is used as our I/O library as well as a data analysis framework by some of our collaborators. GlastRelease primarily makes use of the I/O features, while the Science-Tools also utilizes some of ROOT's math libraries. Among ROOT's many features:

- Machine independent, self-describing file format,

- Object Oriented I/O well suited to our OO design,

- C++ interpreter for command line analysis,

- Python interface (PyROOT).

The ROOT code base has grown substantially since we adopted it as our LAT Offline file format. Fortuantely, ROOT remains modular and we are able to pick and choose what portions of the framework we desire to use. ROOT's documentation and support are very impressive. An online user guide, active mailing lists, and access to the source code provide ample aid to users. The ROOT developers make a point of addressing questions and concerns quickly. ROOT's user support rivals that of commerical software, and in many cases exceeds it. We upgrade ROOT versions about once a year, waiting a month or so after an initial production release to allow bug reports to be addressed. Staying up to date allows us to take advantage of the latest upgrades and bug fixes. ROOT supports a wide variety of operating systems and compilers. Binary distributions are made available for all officially supported platforms. Unfortunately, we have rarely been able to take advantage of these binary distributions due to our adoption of specific versions of Python. There are some complaints about ROOT. There is a steep learning curve to create presentation quality plots. Some users make use of the PyROOT interface which seems to be less daunting.

3.2.   Gaudi: An Object Oriented Framework

We use the Gaudi framework as part of our GlastRelease software which consists of our Monte Carlo simulation and data reconstruction. Gaudi provides a number of basic services:

- Event Data Service (TDS) — a fancy Object Oriented common block,

- Messaging and Logging Services,

- Infrastructure for I/O to support multiple file formats,

- Standard event loop,

- Job parameter handling via ASCII input files,

- Optional Python interface.

While we enjoy the many features Gaudi provides, there are some problems associated with using it. We have been frozen on Gaudi v18r1 since 2006 in anticipation of Fermi's launch. We are now in the process of migrating to Gaudi v21r7. This has been a time consuming process due to the need to jump a number of versions which include some interface changes. We desire to move to support Redhat 5 and gcc 4 as well as Windows with Visual Studio 2008. Officially Gaudi supports Redhat 4 using gcc 3.4.3 and Windows Visual Studio 2003. A full Gaudi installation utilizes twenty-nine external libraries, some of which conflict with the dependencies of some other portions of our software. For example, Geant4 8.0.1 depends on CLHEP 1.9.2.2 while Gaudi v21r7 has moved ahead to CLHEP 1.9.4.4. These versions of CLHEP are incompatible. To alleviate these issues, we use a subset of Gaudi which eliminates most of its twenty-nine dependencies, including CLHEP. This does require some source code modification, but the payoff is worth it. Our other issue with Gaudi is that its documentation is woefully out of date. The user documentation was last updated in 2001. Hence, one must rely on the source code itself in conjunction with the release notes to understand improvements and modfications to the code in subsequent releases. There is a mailing list available where Gaudi users and developers correspond to discuss questions and difficulties.

4. Stability versus Development

Our data processing pipeline has been utilizing a stable version of GlastRelease since launch. Some external upgrades, patches, and bug fixes have been accepted. We use CVS for our code repository and branching to implement required code changes to our stable releases. Unfortunately, there is little confidence and knowledge concerning the use of CVS branches across our development team. We do have a couple of developers willing to tackle the job of maintaining the CVS branches for GlastRelease. In the ScienceTools development, branches are avoided altogether, when possible, in favor of applying patches along the main trunk and rolling out new releases when needed. While we greatly enjoy the stability of our data processing pipeline, this stability is often favored over introducing "unnecessary" patches which can result in improvements being passed over for years at a time.

5. User Support

We have gone to great pains to provide adequate user support. Our online user workbook has been largely written and maintained by a dedicated techinical writer. This workbook has been vital to supporting our distributed team of users and developers across the LAT collaboration. We continue to have weekly offline software meetings, as well as dedicated meetings for special projects. Mailing lists and instant messaging also provide fast communication.

6. Conclusions

While lessons are still being learned, there are some conclusions we can share. External libraries can save a lot of duplicate effort, but they do have a cost. This is code not under our direct control. The ability to quickly upgrade operating systems and compilers may be impacted by the external libraries the software depends on. It is necessary to pay attention to dependencies the external libraries have. Some may be optional, others not. Sometimes multiple external libraries have the same dependencies. This may further complicate your upgrades. Waiting too long to do an upgrade can be very costly. It is much easier to handle incremental upgrades rather than jumping multiple versions of an external library at once. Our experience with ROOT upgrades supports this. It has been fairly painless to upgrade ROOT versions, as we are never too far behind the current production version. Meanwhile, staying with a old version of our Gaudi library has served us well during the first couple of years since Fermi's launch, however, we now have to suffer through the interface modifications necessary to handle a new Gaudi version. Finding and getting to know the experts involved with an external library is helpful. Questions and problems associated with that external library will pop up and having good resources to contact will be vital. Avoid non-standard features at all costs. Interfaces change, and certainly over the long haul of a mission, if you are taking advantage of some quirk in the code of an external library, the rug will likely be pulled out from under you. Finally, beware of your own home grown code. Our choice to adopt an event display built using Fox and Ruby has proven to be a maintenance issue due to the loss of both developers associated with that project. We are now in the situation where we are moving to another event display which does not yet provide all the features of our old one, while the original one lacks any support. Those with Fox and Ruby experience are few and far between.

Acknowledgments.   The *Fermi* LAT Collaboration acknowledges support from a number of agencies and institutes for both development and the operation of the LAT as well as scientific data analysis. These include NASA and DOE in the United States, CEA/Irfu and IN2P3/CNRS in France, ASI and INFN in Italy, MEXT, KEK, and JAXA in Japan, and the K. A. Wallenberg Foundation, the Swedish Research Council and the National Space Board in Sweden. Additional support from INAF in Italy and CNES in France for science analysis during the operations phase is also gratefully acknowledged.

Astronomical Data Analysis Software and Systems XX
ASP Conference Series, Vol. 442
Ian N. Evans, Alberto Accomazzi, Douglas J. Mink, and Arnold H. Rots, eds.
© *2011 Astronomical Society of the Pacific*

Improved Cosmic Ray Rejection for Slitless Spectroscopic Data

M. Kümmel,[1] H. Kuntschner,[1] J. R. Walsh,[1] and H. Bushouse[2]

[1] *STECF/ESO, Karl-Schwarzschild-Str. 2, D-85748 Garching, Germany*

[2] *STScI, 3700 San Martin Drive, Baltimore, MD 21218, USA*

Abstract. We introduce a new method in the aXe software package which allows detection and flagging of deviant pixels when combining dithered slitless spectroscopy data. It follows very closely the techniques that the MultiDrizzle software applies to direct imaging data. We demonstrate the application of the method to WFC3/IR and ACS/WFC slitless data, provide details on the number of previously undetected defects due to cosmic rays and give examples of improved extracted spectra. The limitations of the new method are also discussed.

1. aXedrizzle with Pixel Rejection

A typical slitless spectroscopic data set taken with the Hubble Space Telescope (HST), using the Wide Field Camera 3 (WFC3) IR channel with the G102 or G141 grisms, usually consists of several images with only small shifts or dithers between them. In the aXe slitless spectroscopy extraction software package (Kümmel et al. 2009), we implemented the method *aXedrizzle* that combines individual 2D spectra into a deep 2D slitless spectrum before extracting 1D spectra from the deep image (Kümmel et al. 2005; Kümmel et al. 2009). For the aXe version 2.1, which is part of STSDAS 3.12, we have extended aXedrizzle. Now it is possible to detect and exclude deviant pixel values that are caused by cosmic ray hits or pixels with an unstable response. The primary target of this new method is slitless data taken with the WFC3/IR camera and the grisms G102 and G141.

When combining dithered data, the extended aXedrizzle follows the approach of MultiDrizzle (Koekemoer et al. 2006) for direct imaging. For every object, the following processing steps are executed on its set of individual 2D grism stamp images:

- Drizzle each 2D stamp image individually;

- Median-combine the drizzled, rectified stamps;

- Blot the median image back into the frame of every original 2D stamp;

- Compare the original and the blotted stamps, identify and mask deviating pixel values;

- Drizzle the masked 2D stamps to a co-added, deep 2D grism stamp image;

- Extract 1D spectra from the co-added stamp.

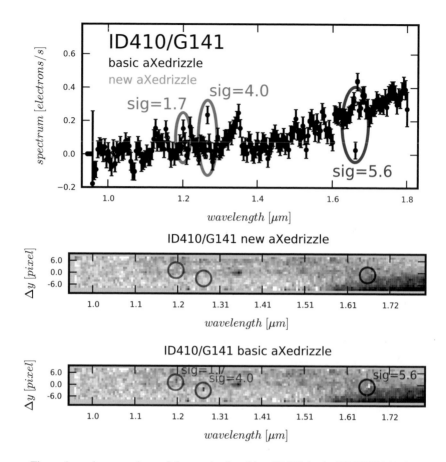

Figure 1.   A comparison of the results for object ID410 in the WFC3/G141 data
set. The upper plot compares the spectra extracted using the basic and the new
aXedrizzle reduction. The lower part shows the corresponding 2D stamp images
from which the spectra were extracted. The three sections where the spectra differ
with *sig* > 1.0 are marked with ellipses, red for higher values in the basic reduction
and blue for lower values. The numbers indicate their significance.

The implementation in aXe uses dedicated STSDAS tasks or customized versions of
the corresponding MultiDrizzle code. The set of new parameters are named as in the
MultiDrizzle task.

2.   The New aXedrizzle Applied to WFC3/IR Slitless Data

As a demonstration, we compare WFC3/G102 and G141 data from the Early Release
Science (ERS) grism field (Straughn et al. 2010) reduced with the basic and the new,
extended aXedrizzle. There are four images per grism (G102 and G141, 4.2 ksec each)
and direct imaging in F098M and F140W (0.8 ksec each). The spectra of ~ 550 sources
are extracted from the G102 and G141 slitless images (see Kümmel et al. 2010, for

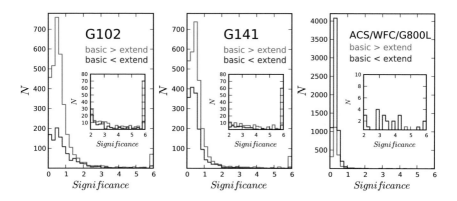

Figure 2.   The histogram of the significance of all differences between the spectra reduced with the basic and the extended aXedrizzle in the WFC3/G102 (left), WFC3/G141 (middle) and ACS/G800L data (right). Differences with larger and smaller values in the basic extraction are marked red and blue, respectively. The insets focus on the region with *sig* > 2.0. The last bins are excess bins.

details). The relevant parameters in the extended aXedrizzle were set to the values recommended for MultiDrizzling WFC3/IR direct images. The average fraction of pixels rejected in aXedrizzle is 0.3% in an image with ∼ 1 ksec exposure time in both the G102 and G141 data. Drastically larger values of 2.0% or more are only achieved in areas with overexposure or persistence from a bright object, which are thus not representative. In order to measure the effect on the extracted spectra, we compute the difference spectrum:

$$spec_{i,diff}(\lambda) = spec_{i,extended}(\lambda) - spec_{i,basic}(\lambda)$$

with $spec_{i,extended}$ and $spec_{i,basic}$ the spectrum of object i extracted with the extended and basic aXedrizzle, respectively. From $spec_{i,diff}(\lambda)$ we collect all the differences, with a difference defined as a set of one or several contiguous spectral elements with $abs(spec_{i,diff}(\lambda)) > 0.01$ e/s. For each set we compute the significance of the difference defined as $sig = MAX(abs(spec_{diff}(\lambda))/error_i(\lambda))$, with $error_i(\lambda)$ the error value for the spectral bin as given by aXe.

Figure 1 shows the spectrum of an object named ID410 in the G141 data, reduced with the basic aXedrizzle (black dots) and with the new, extended aXedrizzle (yellow dots). The lower part of Figure 1 shows the corresponding deep 2D grism stamp images from which the spectra were extracted. The three differences with $sig > 1.0$ are marked with ellipses, with red and blue colors indicating enhancements and depressions in the data reduced with basic aXedrizzle, respectively. From the stamp images it is evident that the blue depression ($sig = 5.6$) in the basic aXedrizzle reduction originates from an individual, unstable pixel and the red enhancements show the typical signatures of cosmic ray hits that had not been detected within the standard calibration pipeline.

Figure 2 shows the histograms of the significance of all differences in G102 and G141 datasets in the left and central panels, respectively. Differences where the spectrum from the basic extraction is enhanced are marked red, the ones with depressions in the basic aXedrizzle are blue. The histograms for G102 and G141 are very similar, and the larger number of differences with enhancements in the basic extraction would be expected if the rejected pixels were affected by cosmic rays.

From Fig. 1 we conservatively set all differences with $sig > 2.0$ as image defects that are corrected with the new aXedrizzle. There are 102 and 185 differences with $sig > 2.0$ in the blue and red diagram for G102 and 34 and 154 for G141. Thus almost every second spectrum from the ~ 550 in the entire data set is significantly improved with the new aXedrizzle.

3. The New aXedrizzle Applied to ACS/WFC/G800L Data

For ACS/G800L data we have compared a dataset extracted with the standard HLA pipeline (Kümmel et al. 2009) that uses `MultiDrizzle` on the slitless data for the identification of cosmics (see Kümmel et al. 2010) and an extraction using the new aXedrizzle. The dataset (association `J8N1ZQH2Q`) chosen for this comparison has four grism images with a total exposure time of $2ksec$ and ~ 900 extracted spectra. The new aXedrizzle yields the best results when using the parameter values for MultiDrizzling ACS direct images (different from the aXedrizzle defaults are `combine_nhigh=1`, `driz_cr_snr=''3.5 3.0''`). An analysis of the differences in the extracted spectra and their significance leads to the histogram in the right panel of Fig. 2. The number of differences with a high significance is rather small. Even for the few $sig > 2.0$ it is difficult to identify on the 2D stamp images an undetected image artifact in one reduction. The new aXedrizzle is thus equivalent to the previously recommended method of using `MultiDrizzle` on the slitless images.

4. Summary

The new aXedrizzle significantly improves the reduction of WFC3/IR slitless data. For ACS/G800L, the new aXedrizzle gives equivalent results to the traditional method of detecting cosmics in `MultiDrizzle`. As in `MultiDrizzle` for direct imaging, the pixel rejection in aXedrizzle improves the results only in case of properly registered data. A careful comparison of the results obtained with and without rejecting pixels in aXedrizzle should always be an important and integral part of the reduction process.

References

Koekemoer, A. M., Fruchter, A. S., Hook, R. N., Hack, W., & Hanley, C. 2006, in The 2005 HST Calibration Workshop: Hubble After the Transition to Two-Gyro Mode, edited by A. M. Koekemoer, P. Goudfrooij, & L. L. Dressel, 423
Kümmel, M., Kuntschner, H., Walsh, J. R., & Bushouse, H. 2010, in The 2010 HST Calibration Workshop, edited by S. Deustua, & C. Oliveira
Kümmel, M., Walsh, J., & Kuntschner, H. 2010, The aXe Manual
Kümmel, M., Walsh, J. R., Pirzkal, N., Kuntschner, H., & Pasquali, A. 2009, PASP, 121, 59
Kümmel, M., et al. 2009, ST-ECF Newsletter, 46, 6
Kümmel, M. W., Walsh, J. R., Larsen, S. S., & Hook, R. N. 2005, in Astronomical Data Analysis Software and Systems XIV, edited by P. Shopbell, M. Britton, & R. Ebert (San Francisco, CA: ASP), vol. 347 of ASP Conf. Ser., 138
Straughn, A. N., Kuntschner, H., Kümmel, M., et al. 2010, ArXiv e-prints. 1005.3071

Astronomical Data Analysis Software and Systems XX
ASP Conference Series, Vol. 442
Ian N. Evans, Alberto Accomazzi, Douglas J. Mink, and Arnold H. Rots, eds.
© *2011 Astronomical Society of the Pacific*

A Database for Data Mining Applications in Astronomy

S. McConnell, G. Henry, R. Sturgeon, and R. Hurley

Department of Computing and Information Systems, Trent University,
1600 West Bank Drive, Peterborough, On, Canada K9J 7B8

Abstract.   Data Mining is defined as the extraction of novel information from large datasets. Astronomy was one of the earliest application areas for data mining, which has now been demonstrated by two decades of successful applications. These applications span a large variety of techniques and problem settings, and resulted in publications beyond astronomical journals. In addition, there exists a wide variety of data-mining techniques, with a large set of possible keywords as a result. This makes the compilation of related work non-trivial. Over the past four years, we have compiled a reference database of data-mining applications in astronomy, which is now available online. The database currently contains over 600 publications and is designed to serve as a tool which enables searches using standard features such as author, year published or keywords contained in the title.

1.   Introduction and Motivation

The application of data-mining techniques in astronomy spans a wide set of techniques and problems and has resulted in publications outside the astronomical community. While most, if not all, of these publications are now available online, the search for related material is rather cumbersome; the phrase *data mining* rarely appears in the title itself, and a multitude of data-mining techniques have been applied in this domain. Example applications include classification of stars, galaxies and planetary nebulae, star/galaxy separation, forecasting of sunspots and of geomagnetic storms, forecasting of seeing, gravitational wave signal detection, antimatter search in cosmic rays, selection of Quasar candidates, detection of expanding HI shells, gamma ray bursts, large-scale matter distribution, binaries, clustering of AGN, cosmic microwave background, source identification, astronomical literature, asteroid taxonomy, and the search for white dwarfs. A typical search using keywords is often ineffective as it is restrictive in the number and types of results it returns, and, at the very least, rather time intensive. The database was designed to overcome these problems and facilitate a more efficient search for background material.

2.   Implementation

The web application uses PHP scripts for submission of the queries to an Oracle database and then displays the results. Currently, the web interface supports various queries, including those by title, type of publication (for example, journal article, proceedings, or technical report), URL, author, keywords, as well as year of publication. The latter also allows for a range search. Access to the database is currently available at `http://cois.trentu.ca/~sabinemcconnell/search.php`. Figure 1 presents an empty search form for the application.

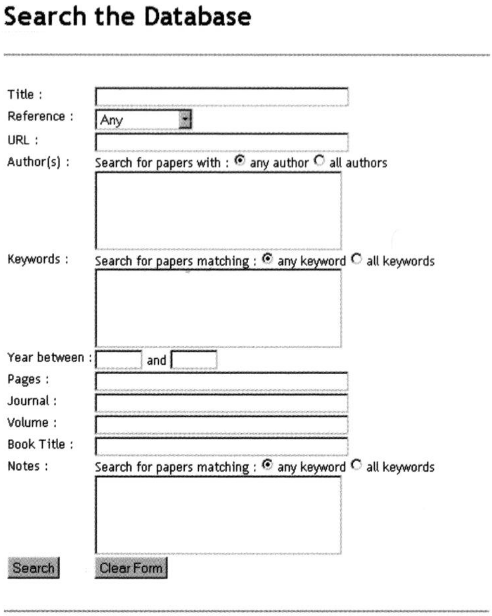

Figure 1.   Empty Search Form.

3. Statistics

The database currently contains over 600 publications that span over two decades of applications. The most commonly used technique is an Artificial Neural Network (ANN), with over 400 publications in the widest collection of application areas. A wide range of predictive and descriptive neural network approaches exist; the most commonly used type in astronomy is that of a feedforward neural network used for prediction. The latter contains a large number of simple processing units (perceptrons) that are interconnected by adaptive weights and organized into layers. The data travels from the input layer through the hidden layers to the output layer, which represents the classes. During training, the weights are adjusted through backwards error propagation to more accurately represent the function to be learned. Typical applications of artificial neural networks in astronomy include the separation of stars and galaxies (Philip et al. 2002) and galaxy classification (Lahav et al. 1996).

Bayesian approaches account for 60 entries in the database; from a data-mining viewpoint, these are divided into Naive Bayesian approaches which assume independence of features, and Bayesian Belief networks. Example applications include the separation of sources from noise (Ritthaler et al. 2007) and classification of variable and transient events (Donalek et al. 2009). Genetic algorithms are evolutionary techniques

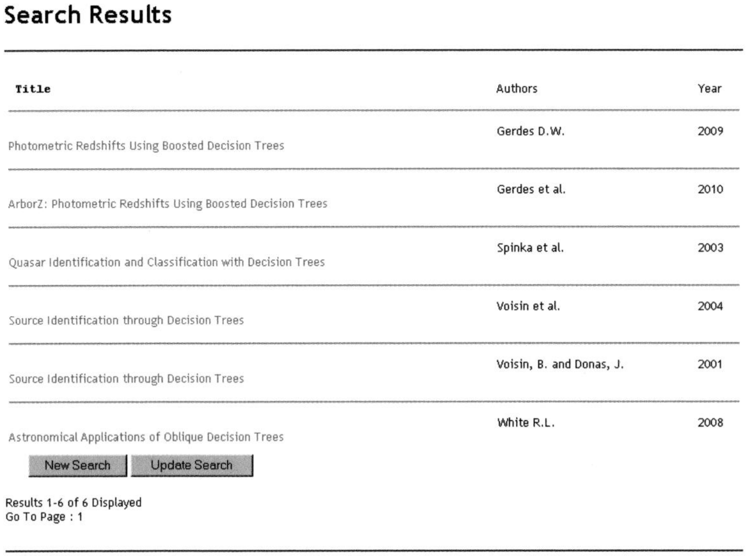

Search Results

Title	Authors	Year
Photometric Redshifts Using Boosted Decision Trees	Gerdes D.W.	2009
ArborZ: Photometric Redshifts Using Boosted Decision Trees	Gerdes et al.	2010
Quasar Identification and Classification with Decision Trees	Spinka et al.	2003
Source Identification through Decision Trees	Voisin et al.	2004
Source Identification through Decision Trees	Voisin, B. and Donas, J.	2001
Astronomical Applications of Oblique Decision Trees	White R.L.	2008

New Search | Update Search

Results 1-6 of 6 Displayed
Go To Page : 1

Figure 2.   Sample Search Results.

that use chromosomes to represent the members of a population which are evaluated through a fitness function. Over a number of iterations, the most suitable members of the current population reproduce through mechanisms such as mutation and crossover. Example applications of the 19 publications in astronomy contained in the reference list include modeling of protoplanetary disks (Hetem & Gregorio-Hetem 2007) and supernovae (Bogdanos & Nesseris 2010) analysis.

A set of techniques which recently gained popularity in this domain are Support Vector Machines (SVMs). This learning technique determines the maximum separation between objects by selecting (small) samples from training data as support vectors. The data is transformed to a higher-dimensional space using non-linear mappings. SVMs are used for regression and classification tasks, and result in typically small models that tend to be less prone to finding local minima and overfitting than ANNs. Example applications of the 11 relevant publications contained in the database include quasar selection (Peng et al. 2010) and solar flare forecasting (Li et al. 2007).

Lastly, applications using Decision Trees, which construct tree-like structures from the training data, are represented in 8 papers. Decision tree algorithms are robust and non-parametric. The data, which is initially associated with the root, is divided recursively into smaller, non-overlapping subsets. Leaves of the tree are associated with classes. Example applications in astronomy include identification of cosmic ray hits (Salzberg et al. 1995) and galaxy classification (Owens et al. 1996). Other entries contained in the database contain references to additional data-mining techniques.

4. Search Results

Figure 2 shows the result returned by an example query referring to decision trees (title) between 2001 and 2010. The title is linked to a reference of the original paper at external sites, or to the paper itself wherever appropriate. Currently, we only display the title, authors and year of publication in the search results.

5. Future Work

Aside from continuous updates of the titles available for search, the project intends to incorporate customization of the search results. This will include the possibility of sorting by the displayed attributes (currently restricted to auther, year, and title) and display of additional information such as keywords and journal. In addition, we plan to add categorizations of the techniques and application areas to facilitate more advanced searches.

References

Bogdanos, C., & Nesseris, S. 2010, in American Institute of Physics Conference Series, edited by J.-M. Alimi & A. Fuözfa, vol. 1241 of AIP Conf. Ser., 200
Donalek, C., Mahabal, A., Djorgovski, S. G., Moghaddam, B., Drake, A., Graham, M., Hensley, B., & Williams, R. 2009, in American Astronomical Society Meeting Abstracts, vol. 214 of American Astronomical Society Meeting Abstracts, 407
Hetem, A., & Gregorio-Hetem, J. 2007, MNRAS, 382, 1707
Lahav, O., Naim, A., Sodré, L., Jr., & Storrie-Lombardi, M. C. 1996, MNRAS, 283, 207
Li, R., Wang, H., He, H., Cui, Y., & Zhan-LeDu 2007, Chinese J. Astron. Astrophys., 7, 441
Owens, E. A., Griffiths, R. E., & Ratnatunga, K. U. 1996, MNRAS, 281, 153
Peng, N., Zhang, Y., & Zhao, Y. 2010, in Software and Cyberinfrastructure for Astronomy, edited by N. M. Radziwill, & A. Bridger (Bellingham, WA: SPIE), vol. 7740 of Proc. SPIE, 77402T
Philip, N. S., Wadadekar, Y., Kembhavi, A., & Joseph, K. B. 2002, A&A, 385, 1119
Ritthaler, M., Luger, G., Young, R. M., J., & Zimmer, P. 2007, in Astronomical Data Analysis Software and Systems XVI, edited by R. A. Shaw, F. Hill, & D. J. Bell (San Francisco, CA: ASP), vol. 376 of ASP Conf. Ser., 413
Salzberg, S., Chandar, R., Ford, H., Murthy, S. K., & White, R. 1995, PASP, 107, 279

Astronomical Data Analysis Software and Systems XX
ASP Conference Series, Vol. 442
Ian N. Evans, Alberto Accomazzi, Douglas J. Mink, and Arnold H. Rots, eds.
© *2011 Astronomical Society of the Pacific*

drPACS: A Simple UNIX Execution Pipeline

Peter Teuben

Astronomy Department, University of Maryland

Abstract.   We describe a very simple yet flexible and effective pipeliner for UNIX commands. It creates a Makefile to define a set of serially dependent commands. The commands in the pipeline share a common set of parameters by which they can communicate. Commands must follow a simple convention to retrieve and store parameters. Pipeline parameters can optionally be made persistent across multiple runs of the pipeline. Tools were added to simplify running a large series of pipelines, which can then also be run in parallel.

1.   Introduction

Pipelines are in common use at large observatory data centers, but are very sophisticated (e.g. Ballester et al. 2011) and often overkill, or simply beyond the patience or skills, of many casual users. Yet, many theoretical or observational projects could very well benefit from this concept of being able to pipeline their dataflow. Often they are an afterthought in small-scale projects. If some effort is taken to pipeline these, the threshold to experiment is lowered, opening the way to more insight. This paper introduces such a concept.

Imagine a series of programs that run a simulation and/or reduce data. If you have made it flexible, it will also have parameters to control this process. But now imagine you have to run many of these pipelines, each on a different set of data (ideally in a different directory) and potentially have to re-run them when something in your pipeline has changed. For example if one of the programs in your pipeline had a bug, or you add or insert steps into the pipeline. You will have to rerun portions of your pipeline for all your directories.

I will describe a generic UNIX solution to this dilemma, to make this process easy. We have used this pipeline in CARMA for PACS data reduction, as well as converted an existing CARMA data reduction pipeline for the STING project. We maintain a website where more information can be found where to download and install the code from.[1]

---

[1]`http://carma.astro.umd.edu/tools/drpacs`

2. Overview

2.1. Concepts

For our purposes a pipeline consists of a serial set of programs controlled by a common pool of parameters that these programs can read and write. These programs can be shell scripts, Python scripts, compiled programs or anything that you would normally run from the command line. Normally a pipeline is run in a project (or run) directory.

2.2. Define and Run the Pipeline

First the pipeline needs to be defined, by providing it the names of the programs that need to run in a certain sequence that we call the pipeline. The simplest way this can be done is by providing them via the command line to the `pipeline` program. Our example starts with 5 program steps, named `step1` through `step5`:

```
pipeline 5 step1 step2 step3 step4 step5 > Pipefile
```

Next, you would probably want to set the values for some of the parameters for the various programs in the pipe. The pipeline programs will have to read them through provided procedures:

```
pipepar -c project=c0184.3B_108PG2130.13 carmaRefant=2
```

and finally you can run the pipeline in one of many ways. For example:

```
pipe all
pipe step3 all
pipe clean all
```

would run the whole pipeline (or formally whatever needed to be done), would only run `step3` and then the remainder of the pipeline, and in the third example clean the pipe from the start and then run the whole pipeline again.

2.3. Advanced Usage

Now imagine for all your runs (directories) you would need to re-run the pipeline with a new parameter in `step3`. Thus `step1` and `step2` do not need to be run again. The `piperun` command can do this in a single line as follows:

```
piperun dirs.txt 'pipepar foo=1.3 ; pipe step3 all'
```

Essentially this will execute the second argument, the text between the singles quotes, in each directory listed in the text file `dirs.txt` provided as the first argument to `piperun`.

Now imagine that each pipeline computed two interesting numbers, `foo` and `bar`, and these were stored in the pipeline parameter database. To produce a simple ascii table of these two variables, which can be used for further analyis, you would issue

```
piperun dirs.txt pipepar -v foo -v bar > foo_bar.tab
```

where the file `foo_bar.tab` contains two columns with the (textual) values as they were stored in the parameter database.

Finally, another useful feature of `drpacs` is the option to save the parameters for a particular "project". This is useful in case the run directory is a scratch directory. The parameter database, which is normally local, will need to be saved, which `drpacs` will do based on it's project name. Assuming this had been done, a new project can be created from scratch as follows:

```
pipesetup a=1 b=2 project=test123
```

and

```
pipe all
```

and

```
pipesave
```

would save the parameter set based on its project name "`test123`".

2.4. Examples

We show two simple examples of scripts that will be `drpacs` compliant, i.e. will read parameters from `drpacs` first, and then the command line, and also save them at the end. A C-shell example:

```
#! /bin/csh -f
#
# (1) define default values in case not given, problem specific
set a=1
set b=2
# (2) pipeline interface to grab old defaults, drpacs compliant
pipepar -s csh > tmp$$.par; source tmp$$.par; rm tmp$$.par
# (3) poor man's command line processor to override parameters
foreach _arg ($*)
  set $_arg
end
# (4) The Actual Code where the work can be done, all problem
#     specific
echo A=$a B=$b
# (5) write pipeline parameters back, drpacs compliant
pipepar a=$a b=$b c=3
```

and a python example:

```
#! /bin/env python
#
import parfile, sys
a=1
b=2
if __name__ == "__main__":
  p = parfile.Parfile('drpacs.def')
```

```
p.argv(sys.argv)
p.set('a',a)
if p.has('b'):
  b = p.get('b')
else:
  p.set('b',b)
p.set('sum',a+b)
p.save()
```

for which you will need the `parfile` module, a simple keyval pair module supplied with the drpacs code. Several more advanced ones are on the market and can be used as effectively. For example `configobj`, the one that STScI's `pytools.teal` uses (C. Sontag, private communication).

2.5. Drawbacks

Simplicity and the usage of the UNIX `make` program under the hood also comes with a price.

First of all, you can only run one pipeline in a given directory. This is not a serious handicap, as the pipelines are easily redefined from a textfile, so a set of `Pipefile`'s could be stored locally. However, the status of each pipeline (e.g. which step was last correctly executed) will be lost. Only the active one will be valid, and switching from one to the other has to be done with care. Parameters between pipelines will have to be shared, there is no namespace.

Secondly, the current version has no dependency on the keywords in the pipeline. For example, if one would change a parameter that is normally defined in step3, the user would need to know this parameter came from step3 and re-execute the pipeline from this step onwards. In principle these dependancies can be coded quite easily via "make" dependancies, in a similar way how the program steps have been coded, but this has not been implemented in the current version.

For each "project" we assume a "directory", although the hierarchy is not specified.

One cannot use the same program twice in the pipeline.

Acknowledgments. The author wishes to thank Ashley Zauderer, Dalton Wu and Roger Curley for their help during the develpment of drpacs.

References

Ballester, P., Bramich, D., Forchi, V., Freudling, W., Garcia-Dabó, C. E., Gebbinck, M. K., Modigliani, A., & Romaniello, M. 2011, in Astronomical Data Analysis Software and Systems XX, edited by I. N. Evans, A. Accomazzi, D. J. Mink, & A. H. Rots (San Francisco, CA: ASP), vol. 442 of ASP Conf. Ser., 261

Astronomical Data Analysis Software and Systems XX
ASP Conference Series, Vol. 442
Ian N. Evans, Alberto Accomazzi, Douglas J. Mink, and Arnold H. Rots, eds.
© *2011 Astronomical Society of the Pacific*

A Method for Measuring Distortion in Wide-Field Imaging with High Order Polynomials

Naoki Yasuda,[1] Yuki Okura,[2] Tadafumi Takata,[2] and Hisanori Furusawa[2]

[1]*Institute for the Physics and Mathematics of the Universe,*
The University of Tokyo, 5-1-5 Kashiwanoha, Kashiwa, 277-8583, Japan

[2]*National Astronomical Observatory of JAPAN, 2-21-1 Osawa,*
Mitaka, Tokyo, 181-8588

Abstract.   In analyzing wide-field images, for example those from the Hyper Suprime-Cam on the Subaru telescope, determining World Coordinate System (WCS) information (for example position and angle) of each CCD is very important. In this paper, we show a method for determining the distortion with high-order polynomials using the TAN-SIP convention, utilizing all of the CCDs in a field of view (FOV) at once.

1. Introduction

For determining a distortion of a FOV, a transformation method is used to fit positions of reference stars in "intermediate world coordinates" (ξ, η) which are projected from "Celestial coordinates" to positions of "Pixel coordinates" (x, y). Unfortunately, offset and rotation are usually not enough to fit these positions, because there is distortion due to the optics and varying pixel scales across the FOV which are made by differences in the paths of light passing through the telescope. This effect can be neglected if we use cameras which observe a narrow field, but we must correct this effect for precise analysis of wide field data. One extended WCS convention proposal, TAN-SIP (http://fits.gsfc.nasa.gov/registry/sip.html) expresses this effect by using a high order polynomial transform. The merits of using all CCD's together are 1) we can measure the higher order part of the distortion which is hard to measure from references in a local region, 2) an increase in precision by using a large number of reference stars, and 3) allowing us to interpolate regions which have a small number of reference stars. Then we obtain WCS information for each CCD from the global WCS.

2. Two Step Determination of CCD Positions and Distortion with TAN-SIP

There are two steps in analysing a distortion of the FOV. The first step is determining relative positions of each CCD from a basis CCD. Global positions of all references (x_G, y_G) which have a local position (x_L, y_L) in ith CCD are determined by using this relative position (Figure 1):

$$x_G = x_L \cos \Delta\theta_i - y_L \sin \Delta\theta_i + \Delta X_i,$$
$$y_G = x_L \sin \Delta\theta_i + y_L \cos \Delta\theta_i + \Delta Y_i,$$

Yasuda et al.

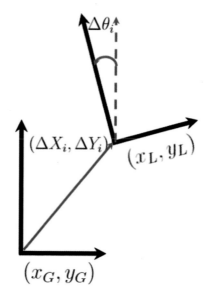

Figure 1.   The relation between Global coordinates and ith CCD coordinates.

where $(\Delta X_i, \Delta Y_i, \Delta \theta_i)$ is local position of ith CCD. In the second step, by fitting this global positions and (ξ, η), high order polynomial distortions are expressed as

$$\begin{pmatrix} \xi \\ \eta \end{pmatrix} = \begin{pmatrix} CD1\_1 & CD2\_1 \\ CD1\_2 & CD2\_2 \end{pmatrix} \begin{pmatrix} x_G + f_G(x_G, y_G) \\ y_G + g_G(x_G, y_G) \end{pmatrix}.$$

3.   After Global Positions Are Determined

3.1.   Determining TANSIP

Next, we determine WCS information for each CCD from the global WCS before finding relative positions. First, we must determine the order of the polynomial for the global fit. Too small an order returns systematic residuals and too large an order forces us to spend unnecessary time calculating. Figure 2 shows systematic residuals from a fit using an insufficient order (in one dimension). We should use a high order so that the systematic residuals become small enough. Then, coefficients of the polynomial are determined by using the least squares method.

3.2.   Determination of TANSIP for Each CCD

WCS information for each CCD is transformed from global WCS using the relative position $(\Delta X_i, \Delta Y_i, \Delta \theta_i)$. The reference pixel for the ith CCD, (CRPIX1, CRPIX2), is obtained from the reference pixel of the global fit (CRPIX1$_G$, CRPIX2$_G$) as follows:

$$\begin{aligned} \text{CRPIX1}_i &= -(\Delta X_i - \text{CRPIX1}_G) \cos \Delta \theta_i - (\Delta Y_i - \text{CRPIX2}_G) \sin \Delta \theta_i, \\ \text{CRPIX2}_i &= -(\Delta Y_i - \text{CRPIX2}_G) \cos \Delta \theta_i + (\Delta X_i - \text{CRPIX1}_G) \sin \Delta \theta_i. \end{aligned}$$

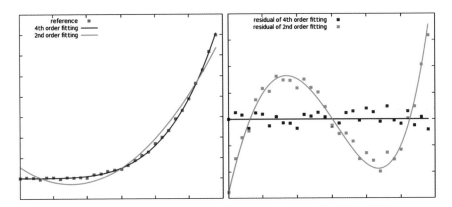

Figure 2.   Simple simulation of polynomial fitting with insufficient order, x and y axes mean arbitrary scale. The left graph shows fitting lines of references with 4th order polynomial distortion by 4th (blue) and 2nd (green) order polynomial. The right graph shows residuals of these fits (4th (blue) and 2nd (green)). Systematic residuals are seen if we use polynomial fitting with insufficient order.

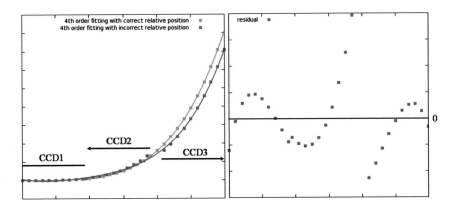

Figure 3.   Simple simulation of polynomial fitting of reference with incorrect relative positions, x and y axes mean arbitrary scale. CCD2 has a slightly negative value and CCD3 has a slightly positive value of incorrect relative position and the references have no random noise. The left graph shows fitting lines of references with correct and incorrect relative position and also shows fitting function (e.g. coefficients) changed by relative positions. The right graph shows residuals of fitting of reference with incorrect relative position. Systematic residuals are seen if incorrect relative positions are used.

Next, fitting functions in global and ith CCD coordinates can be expressed as

$$f_G = \sum_k \sum_l A_G^{kl} x_G^k y_G^l, \qquad f_i = \sum_k \sum_l A_i^{kl} x_i^k y_i^l,$$

respectively. By dividing coefficients into order and spin numbers, coefficients of the polynomial function for the ith CCD are obtained easily from coefficients of the global fit using $\Delta\theta_i$ as follows

$$
\begin{aligned}
A_i^{00} &= A_G^{00}, \\
A_i^{10} &= B\cos(\phi_1 - \Delta\theta_i), \\
A_i^{01} &= B\sin(\phi_1 - \Delta\theta_i),
\end{aligned}
$$

where $B = \sqrt{\left(A_G^{10}\right)^2 + \left(A_G^{01}\right)^2}$ and $\phi_1 = \tan^{-1}\left(A_G^{01}/A_G^{10}\right)$.

$$
\begin{aligned}
A_i^{20} &= B + C\cos(\phi_2 - 2\Delta\theta_i), \\
A_i^{11} &= 2C\sin(\phi_2 - 2\Delta\theta_i), \\
A_i^{02} &= B - C\cos(\phi_2 - 2\Delta\theta_i),
\end{aligned}
$$

where $B = \frac{A_G^{20}+A_G^{02}}{2}$, $C = \sqrt{\left(\frac{A_G^{20}-A_G^{02}}{2}\right)^2 + \left(\frac{A_G^{11}}{2}\right)^2}$, and $\phi_2 = \tan^{-1}\left(A_G^{11}/\left(A_G^{20} - A_G^{02}\right)\right)$.

4. Determination of Relative CCD Positions in a FOV

Because incorrect offset in the global coordinates can not be expressed by smooth polynomials, a systematic residual is made by using an incorrect relative position. Figure 3 shows a simple simulation, making systematic residuals from incorrect relative positions. Therefore we must find relative positions by reducing the systematic residual or minimizing the fit RMS. Figure 3 also shows that the fitting function changes as the relative positions of CCDs (CCD arrangement) change. Therefore, the residuals must be calculated many times for possible combinations of relative positions of the CCDs.

5. Summary

We have determined WCS information for each CCD using global positions of references from all CCDs. This method allows us to measure high order distortion which is hard to measure from only local regions and to also obtain precise fits using a large number of references. We then derived WCS information for each CCD (CRPIX and coefficients of distortion) from that global WCS information. The property of non-polynomial shift which is caused by using incorrect relative positions can be used to determine relative positions of each CCD, but this method still needs an improvement of the algorithm to reduce the processing time.

Part XI

Solar Astronomy

Astronomical Data Analysis Software and Systems XX
ASP Conference Series, Vol. 442
Ian N. Evans, Alberto Accomazzi, Douglas J. Mink, and Arnold H. Rots, eds.
©*2011 Astronomical Society of the Pacific*

Computer Vision for the Solar Dynamics Observatory: First Results and What's Next

P. C. H. Martens,[1,2] and the SDO Feature Finding Team

[1]*Physics Department, Montana State University, P.O. Box 173840, Bozeman, MT 59717-3840*

[2]*Harvard-Smithsonian Center for Astrophysics, MS 58, 60 Garden Street, Cambridge, MA 02138*

Abstract. The Solar Dynamics Observatory (SDO) feature finding team is a large international consortium tasked by NASA to produce a comprehensive system for automated feature recognition for SDO. We are producing robust and very efficient software modules that can keep up with the SDO data stream and detect, trace, and analyze a large number of phenomena, including flares, sigmoids, filaments, and coronal dimmings. Results will be shown for several modules have been inaugurated since the end of SDO commissioning last summer. In addition a description is given of the status of the development of our trainable automated feature finding module.

1. The SDO Computer Vision Project

The SDO Feature Finding Team (FFT) is producing 16 software modules that detect, analyze, and track solar features and events, some in near real time via the SDO data pipeline. Why is there a need for automated feature recognition in SDO data, an element that did not exist in previous solar missions? First, the SDO data stream at about 1.5 Tb per day, nearly 24/7, with a few eclipse interruptions, for the full 10 year duration of the mission, is overwhelming in comparison with all previous solar missions. Figure 1 demonstrates the point by showing archive volume and data rate (radius of balloons) for recent solar missions and ground-based observatories. SDO, when it has completed its 10 year mission, will comprise 90% of all space-mission solar data ever obtained. No amount of solar physicists, even with the help of legions of graduate students, will ever suffice for analyzing all these data.

There is also a second, equally important, argument for automated feature recognition: Solar Physics, as a mature discipline, needs to move from analysis of single events to sets of events and features. Because of the fragmented nature of solar observations (weather, the day/night cycle, eclipses for space missions, observatories at all longitudes, large numbers of wavelength bandpasses, etc.) obtaining uninterrupted observations, good long-term series, and hence reliable statistics for large numbers of events and phenomena, is not a simple matter. With SDO this will become less of an impediment, but at the cost of having huge amounts of data to sift through. Automated feature recognition will produce large catalogs of metadata on solar events and phenomena, metadata that will allow us to find the underlying physical structures and mechanisms for much of solar activity. A good example of such a result in the past is

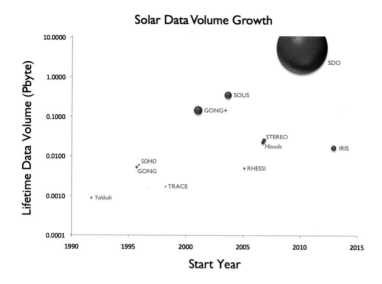

Figure 1.   Life-time data volume and date rate — given by the radius of the balloon — as a function of first light for space-based and ground-based solar observatories. (Figure courtesy of Joe Gurman, NASA-GSFC.)

the detection by Maunder (1904) of the butterfly diagram for sunspots during a solar cycle, a discovery that has greatly enhanced our understanding of the dynamo mechanism in the Sun and in stars.

NASA recognized the need for automated feature recognition for the SDO mission, put out an NRA, and selected my team in the fall of 2008 to carry out this project. We are an international consortium of seven institutions in the US (the Smithsonian Astrophysical Observatory (SAO), Montana State University (MSU), Johns Hopkins Applied Physics Lab (APL), the Southwest Research Institute (SwRI), NASA Marshall Space Flight Center (MSFC), New Mexico State University (NMSU), and Lockheed-Martin Solar and Astrophysics Lab (LMSAL)), and four institutions in Europe (the Royal Observatory of Belgium (ROB), the Academy of Athens (AoA), Trinity College of Dublin (TCD), and the Max Planck Institute at Lindau (MPL)). The team delivers catalogs of metadata for the solar community through separate, robust, and efficient software modules that use standardized interface protocols. Both the codes and the interfaces are publicly available for the development of other community contributed modules. Three of our codes — the ones that detect flares, coronal dimming, and emerging magnetic flux — operate in real-time to produce automated space weather alerts, and immediate event analysis available on-line. The codes operate at LMSAL and at SAO, the latter with a 48 hour latency. Some codes, like the flare code, monitor the data stream constantly, others monitor at a reduced cadence, and some codes are initialized by an event alert from other codes (e.g., the oscillations code is triggered by flares and eruptions). Codes already operating, despite the short time since the start of our grant, are the ones that detect filaments, flares, active regions, coronal dimmings, sigmoids, emerging flux, and polarity inversion lines.

In this paper I will give a brief status report on the results from several of the codes, and then give a more detailed presentation of the general purpose, trainable, feature detection module that we are developing. Because of lack of space I cannot give a description of the flare module, the emerging flux detection module (SWAMIS), the sigmoid module, and the coronal dimming module, all of which are currently in operation. A full description of the Computer Vision project in its entirety is given in Martens et al. (2011).

2. Active Region and Coronal Hole Detection

The *Spatial Possibilistic Clustering Algorithm* (SPoCA), developed by team members at the ROB, produces a segmentation of Extreme Ultra-Violet (EUV) solar images into regions, named here "classes", corresponding to Active Regions (AR), Coronal Holes (CH), and the Quiet Sun (QS); see Barra et al. (2005, 2008, 2009). Other segmentation methods have been proposed; an overview is given in Barra et al. (2009). We have selected SPoCA because of the maturity and flexibility of the program. SPoCA uses a multichannel, fuzzy-logic clustering procedure. It has been applied successfully to a series of EIT image pairs (171 and 195 Å) spanning almost a full solar cycle, see Barra et al. (2009). The classes are determined by minimization of intra-class variance. The method is generic and therefore portable to other instruments, and in particular to SDO/AIA. SPoCA involves a preprocessing by which the limb brightness discontinuity is attenuated.

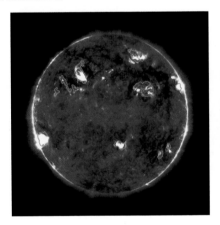 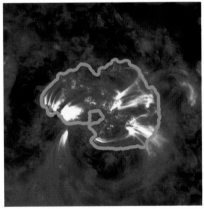

Figure 2.   Left: Full disk AIA 171 Å image from 5 May 2010, together with overlays of segmented Active Regions. A zoomed in cut-out of a single Active Region is shown on the right.

From the maps of e.g., ARs, connected AR pixels are then gathered by means of a region growing technique, see Figure 2 for an example of our results with AIA data. It provides the instantaneous location of the barycenter, the area, the coordinates of the bounding box, and a mask for each AR.

These elements are computed over time and handled as dynamical quantities. The current tracking method uses an optical flow algorithm to locate the barycenter of the AR in the next fuzzy map, where the same region growing technique updates the parameters of the AR being tracked. A starting date for an AR is defined when a new set of connected pixels is identified. An AR 'end-date' is recorded either when the tracking algorithm can no longer find a connected set, or when the AR disappears over the west limb. The algorithm also takes care of merging of ARs.

3. Filament Module

Filaments in the solar chromosphere are well-known, large-scale structures of relatively dense and cool plasma suspended in the hot and thin corona. They are particularly well visible in Hα filtergrams. Filaments and their sources, filament channels, are known to align with photospheric magnetic polarity inversion lines (PILs, Martin 1998). All solar eruptions occur above PILs. In addition, filaments are known to involve helical magnetic fields, twisted beyond their minimum-energy, current-free, magnetic configuration (Martin et al. 1994; Rust & Martin 1994; Pevtsov et al. 2003). Non-potential (i.e., helical) magnetic fields are invariably involved in solar eruptions and give rise to coronal mass ejections (CMEs). Filaments themselves often erupt fully or partially into CMEs, leading to a complete or partial filament disappearance from the solar disk (Gilbert et al. 2000; Asai et al. 2003; Jing et al. 2004). If one knows the sense of twist (chirality) in a filament before its disappearance, then one has additional clues about the magnetic helicity of the CME that might be useful in assessing the CMEs possible geo-effectiveness (e.g., Yurchyshyn et al. 2001; Rust et al. 2005).

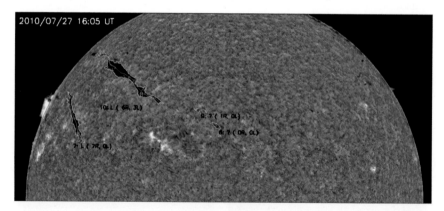

Figure 3.   Results of the filament finder applied to a full disk Hα image from the Global High Resolution Hα Network. Two large and two small filaments are detected and mapped. The bearing (left or right) of the so-called barbs is also determined and indicated.

The "Advanced Automated Filament Detection and Characterization Code" (AFDC) was originally developed by Bernasconi et al. (2005), and is now fully functional, tested, and validated. It has been in continuous operation since March 2010. Images are provided by the Global High-Resolution Hα network, about one to four images per day. The filament code can be used either for statistical study of filament properties over long periods of time spanning a full solar cycle, or if run in near real-time it can provide useful information for space-weather forecasting. When the code detects the disappearance of a large filament it can deliver a CME warning. Where the code has determined the chirality of the disappeared filament it provides information about the orientation of the erupting flux-rope and thus the geo-effectiveness of the associated CME.

Figure 3 shows a snapshot of the automatic tracking of the evolution and eruption of two large filaments in the northern hemisphere of the Sun in July/August 2010.

4. Detection of Coronal Jets

The Coronal Jet (CJ) detection algorithm is being developed at SAO. It is triggered by both a Bright Point (BP) and Coronal Hole (CH) identification by our system. The CH boundary is necessary to assure that the BP is inside the coronal hole. After identifying the boundary of the coronal hole, the BP finder runs on every other image and the pixels within BPs are identified. The CJ detection and parameter determination algorithms work on data cubes covering a box enclosing the BP and extending forward in time. Our methods for determining the CJ parameters are described in detail in Savcheva et al. (2007).

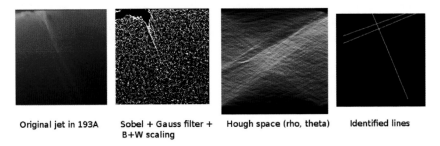

Original jet in 193A   Sobel + Gauss filter +   Hough space (rho, theta)   Identified lines
                      B+W scaling

Figure 4.   Sequence of images illustrating the steps taken by the jet finder algorithm, as described in the text. The AIA 193 Å image on the left of 16 June 2010 00:04:08 UT, is a cutout at a location on the limb, inside the South pole coronal hole, just west of the pole.

The detection algorithm has been tested on XRT images. The algorithm has been proven to work in the general case for low irregularity in the background as is the case with XRT images. Further refinements handle the more dynamic background in corresponding EUV images that are taken with AIA. TRACE images have been used to implement this step. Figure 4 illustrates the sequence of steps in the algorithm behind the jet finder part of the jet module for AIA data. The first plot shows the original jet as seen in AIA 193 Å. The second one shows the same portion of the image after a Sobel and a Gaussian filter ($\sigma = 15$ pixels) and linear scaling have been applied — the resultant image consisting only of pixels with values 0 or 1. The third image shows the part of the ρ–θ space that the scaled image spans after the linear Hough transform has been applied to it. The brightest spot in this image represents the most prominent line in the image in terms of distance from the origin and inclination. The last image shows a reverse Hough transform of the three brightest spots in the Hough space, which is represented by three lines — one for the jet and two for the limb.

5. Polarity Inversion Line Mapping

Identifying the location of *Polarity Inversion Lines* (PILs) — often also called neutral lines — can be of great importance to phenomenological and theoretical studies. Historically, neutral lines in active regions have been useful tools for predicting the locations of flares and coronal mass ejections (Falconer et al. 2002). They also can be used to map out coronal structures (McIntosh 1994) and are associated with filaments and filament channels (e.g., Martin et al. 1994; Chae et al. 2001).

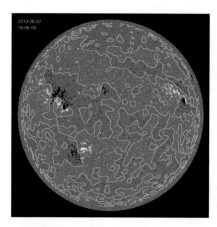 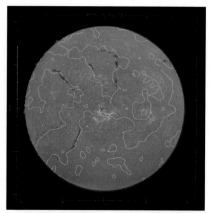

Figure 5.   Left: Polarity inversion contours (in yellow), determined by the PIL module, overlaid on the source SDO/HMI magnetogram. Red dots indicate flare prone regions of high magnetic gradient. Right: Polarity inversion lines in blue from a SOLIS magnetogram, calculated with the same code, overlaid on a nearly simultaneous Hα image from Big Bear Solar Observatory.

We have successfully developed a code that identifies PILs based upon a well-established code previously developed for NSO/KPVT magnetogram data (Jones 2004). See Figure 5 (left) for a recent example of PIL code output overlaid on an HMI magnetogram. Areas of high magnetic gradients are identified and indicated in the figure in red. Projection has been taken into account.

To verify our results early on, we overlaid a PIL map determined from a SOLIS image on an Hα image from BBSO (Figure 5, right). We tested on SOLIS magnetograms because their resolution is closest to HMI.

5.1. Conclusions for Task-Specific Modules

What then is the potential use of the output of all these modules? The answer is that they will allow solar research on datasets that previously would have taken years to compile. With internet access, and a few lines of commands in most programming and visualization packages, one can for example:

- Draw a solar cycle butterfly diagram for active regions.
- Find all filaments that coincide with sigmoids. Correlate sigmoid handedness with filament chirality.
- Correlate EUV jets with small scale flux emergence in coronal holes only.

- Draw Polarity Inversion Line maps with regions of high shear and large magnetic field gradients overlaid, to pinpoint potential flaring regions. Then correlate with actual flare occurrence.

In addition our modules enable the production of automated real time space weather alerts, plus online quicklook data for flares, eruptions, and flux emergence.

We are convinced that the presence of automated feature recognition will facilitate a paradigm shift of solar physics from a discipline mostly focused on the analysis of single events, or very limited sets of events, to a discipline capable of the analysis of very large representative sets of events. This, in turn, will lead to the discovery of statistical patterns leading to the recognition of the underlying physical mechanisms, as well as the prediction of the probability of space weather events.

5.2. The Multi-Purpose Trainable Feature Recognition Module

Humans have an amazing generic feature recognition ability that has been hard to match for computers. For example, a solar scientist can instruct an undergraduate student in less than an hour to recognize sunspots, filaments, loops, and arcades in solar imagery, and the student can then easily produce a catalog of these features from a given set of images. A computer feature-finding algorithm on the other hand takes months to years to develop, and the development needs to be repeated almost from scratch for every new feature.

Motivated by the successful development and implementation of a generic, more human-like, feature detection method for mammography (e.g., Yang et al. 2007) we are in the process of creating an automated solar feature retrieval system that is generic in nature, i.e., the software can detect features of any kind, and rank the returned images based on their similarity to an image provided by the user. Rather than developing a new task-specific application to identify each separate feature, our generic feature recognition software can detect and catalog a wide range of solar features, even involving serendipitous discovery.

There are two steps to our method. The first is implemented at the end of the metadata pipeline, and the second is totally separate from it and can be performed by any user at any time, even on their laptop, without requiring SDO resources. Step one consists of calculating for each image the texture parameters which are then stored in the image texture catalog. Specifically, each SDO image is subdivided in 1024 128×128 pixel sections and for each of those a set of texture parameters, such as entropy, average intensity, standard deviation, kurtosis, etc. is calculated. We have started our project prior to SDO launch, using TRACE images, which are very similar in passbands to AIA/SDO images (Lamb et al. 2008). Assuming that we derive 10 parameters per image section, each 16 bits, the information compression from image to catalog entry is roughly a factor of 1000. Therefore the size of the SDO image-texture catalog will not be prohibitive.

Suppose now that a user wants to build a catalog of, say, sigmoids, by teaching the algorithm how to detect them. The user then downloads a selection of AIA/SDO full-disk images, containing, say, a few dozen sigmoids. In those images the user identifies sigmoids via a simple point-and-click interface. There is no need to identify all sigmoids on the disk. The program then calculates the texture parameters for each of the image segments that contain the identified phenomena. Using these parameters as a feature definition, the entire SDO image-texture catalog can be quickly searched to filter out irrelevant images and then fine tune the search for image segments with similar parameters. These segments will contain sigmoids if the image texture parameters are adequate for the type of image one is analyzing. In a further refinement the user can than teach the module to distinguish between sigmoids with left-handed and right-handed chirality and so on.

Recently our research for this module has focused on the following areas:

- Find the right texture parameters for solar imagery.

- Select the texture parameters that require only modest calculation time, so the module can keep up with the SDO data stream. We have now converged on Mean, Standard Deviation, Third Moment (Skewness), Fourth Moment (Kurtosis), Relative Smoothness, Entropy, Uniformity, Fractal Dimension, Tamura Contrast, and Tamura Directionality.

- Reduce the dimensionality of the image vector obtained. Ten parameters per 128×128 pixels subimage, 1024 such blocks per SDO image, yields a 10,240 element vector per image. Dimensionality is reduced by selecting those vector elements that vary significantly over the imagery. The method is somewhat analogous to calculating the eigenvectors of a matrix and then selecting those with the largest eigenvalues. Lower dimensionality greatly reduces the computing time required to compare image vectors between images.

- Find the most effective distance calculation method, also called dissimilarity measure, e.g., normal Euclidian, angle between image vectors, Spearman, Manhattan metric, etc.

- Find the most effective classifier for this type of imagery. We have tried C4.5 plus AdaBoost, naïve Bayesian, and Support Vector Machines (SVM).

This research is described in full detail in Banda et al. (2011). The net result is condensed in Figure 6. The ordinate denotes the percentage agreement we found for 1200 TRACE images between labeling by human observers (trained solar physicists) and our code for the three classifiers we tested, with reduction in the elements of the image vector as indicated, and using standard Euclidian vector metric, which we found to produce the best results. On the top row we find 95% agreement for the full TRACE image vector (640 elements) between humans and our code for the SVM classifier. SVM also gets very close to that result using the linear dimensionality reduction methods of Principal Component Analysis (PCA) and Locality Preserving Projections (LPP), as well as the non-linear method of Laplacian Eigenmaps (LE), in each casing reducing 640 dimensions to 143, roughly a factor four. Hence, for efficiency's sake, a strong reduction in dimensionality is indicated.

A couple of considerations are worthwhile mentioning. First, human observers frequently make mistakes as well, so a lack of perfect agreement between code and humans is to be expected, and can be entirely due to human error. The patterns of human mistakes are typically very different from those of machines; we are currently conducting further tests with a second set of human observers to classify these errors. Second, our method of image characterization is not invariant for mirror or rotation transformations; yet we find that for the final classification mirroring or rotating a subset of the images makes no difference.

The conclusion is that we are obtaining very encouraging results in developing a trainable feature recognition algorithm for solar imagery. With the right choice of texture parameters, classifiers, dissimilarity measures, and selection of a suitable method

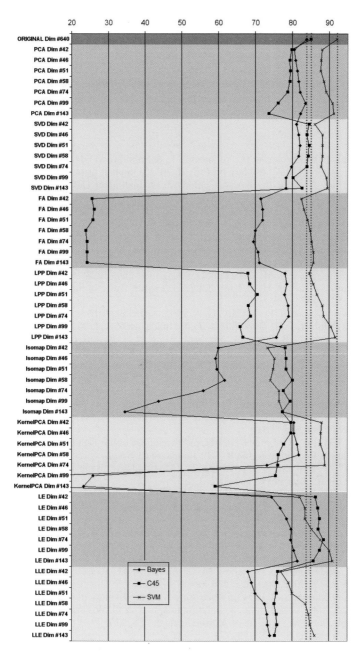

Figure 6.   Comparison of 1200 human labeled TRACE images with labeling by the trainable feature finding module. Percentage agreement is shown for three widely used classifier algorithms, described in the text. The top row shows the results for the full 640 dimensional image vectors, and the remainder of the table shows the results for the dimensionality reduction methods that we tested.

for dimensionality reduction the general trainable feature finding module has potential applicability to any large image data set. A nice example of a rather different application of the same method is given in the paper by Kim in this volume (Kim 2011) describing a method for separating Quasi-Stellar Objects (QSOs) from other types of variable stars, microlensing events, and non-variable stars in the MAssive Compact Halo Objects (MACHO) database by analyzing the spectra using SVM.

References

Asai, A., Ishii, T. T., Kurokawa, H., Yokoyama, T., & Shimojo, M. 2003, ApJ, 586, 624

Banda, J. M., Angryk, R. A., & Martens, P. C. H. 2011, J. Statist. Anal. & Data Mining, Special Issue on: "Best of the NASA Conference on Intelligent Data Understanding 2010", in press

Barra, V., Delouille, V., & Hochedez, J.-F. 2008, Adv. Space Res., 42, 917

Barra, V., Delouille, V., Hochedez, J.-F., & Chainais, P. 2005, in The Dynamic Sun: Challenges for Theory and Observations, edited by D. Danesy, S. Poedts, A. De Groof, & J. Andries, vol. 600 of ESA SP, 71

Barra, V., Delouille, V., Kretzschmar, M., & Hochedez, J. 2009, A&A, 505, 361

Bernasconi, P. N., Rust, D. M., & Hakim, D. 2005, Solar Phys., 228, 97

Chae, J., Martin, S. F., Yun, H. S., Kim, J., Lee, S., Goode, P. R., Spirock, T., & Wang, H. 2001, ApJ, 548, 497

Falconer, D. A., Moore, R. L., & Gary, G. A. 2002, ApJ, 569, 1016

Gilbert, H. R., Holzer, T. E., Burkepile, J. T., & Hundhausen, A. J. 2000, ApJ, 537, 503

Jing, J., Yurchyshyn, V. B., Yang, G., Xu, Y., & Wang, H. 2004, ApJ, 614, 1054

Jones, H. P. 2004, in "Knowledge-Based Intelligent Information and Engineering Systems: 8th International Conference, KES 2004, Wellington, New Zealand", edited by M. G. Negoita, R. J. Howlett, & L. C. Jain, vol. 3215 of "Lecture Notes in Computer Science", 433

Kim, D.-W. 2011, in Astronomical Data Analysis Software and Systems XX, edited by I. N. Evans, A. Accomazzi, D. J. Mink, & A. H. Rots (San Francisco, CA: ASP), vol. TBD of ASP Conf. Ser., TBD

Lamb, R., Angryk, R., & Martens, P. 2008, in Proceedings of the 19th International Conference on Pattern Recognition (ICPR '08), Tampa, FL, USA, 1

Martens, P. C. H., et al. 2011, Solar Phys., SDO Mission Issue, in press

Martin, S. F. 1998, Solar Phys., 182, 107

Martin, S. F., Bilimoria, R., & Tracadas, P. W. 1994, in Solar Surface Magnetism, edited by R. J. Rutten, & C. J. Schrijver (Dordrecht: Kluwer), 303

Maunder, E. W. 1904, MNRAS, 64, 747

McIntosh, P. S. 1994, in X-ray solar physics from Yohkoh, edited by Y. Uchida, T. Watanabe, K. Shibata, & H. S. Hudson (Tokyo: Universal Academy Press), 271

Pevtsov, A. A., Balasubramaniam, K. S., & Rogers, J. W. 2003, ApJ, 595, 500

Rust, D. M., Anderson, B. J., Andrews, M. D., Acuña, M. H., Russell, C. T., Schuck, P. W., & Mulligan, T. 2005, ApJ, 621, 524

Rust, D. M., & Martin, S. F. 1994, in Solar Active Region Evolution: Comparing Models with Observations, edited by K. S. Balasubramaniam, & G. W. Simon (San Francisco, CA: Astron. Soc. Pac.), vol. 68 of Astron. Soc. Pac. CS, 337

Savcheva, A., et al. 2007, PASJ, 59, 771

Yang, L., Jin, R., Sukthankar, R., Zheng, B., Mummert, L., Satyanarayanan, M., Chen, M., & Jukic, D. 2007, in Medical Imaging 2007: Computer-Aided Diagnosis, edited by M. L. Giger, & N. Karssemeijer (Bellingham, WA: SPIE), vol. 6514 of Proc. SPIE

Yurchyshyn, V. B., Wang, H., Goode, P. R., & Deng, Y. 2001, ApJ, 563, 381

Part XII

Virtual Observatory

Astronomical Data Analysis Software and Systems XX
ASP Conference Series, Vol. 442
Ian N. Evans, Alberto Accomazzi, Douglas J. Mink, and Arnold H. Rots, eds.
© *2011 Astronomical Society of the Pacific*

Near-Infrared Variable AGNs Derived by Cross-Identification

Shinjirou Kouzuma[1] and Hitoshi Yamaoka[2]

[1]*School of International Liberal Studies, Chukyo University, Toyota 470-0393, Japan*

[2]*Department of Physics, Kyushu University, Fukuoka 812-8581, Japan*

Abstract. We examined near-infrared variability of Active Galactic Nuclei (AGNs) using cross-identification. We first cross-identified the Two Micron All Sky Survey (2MASS) catalog with two AGN catalogs (namely, Quasars and Active Galactic Nuclei (12th Ed.) and the SDSS-DR5 quasar catalog), and extracted a 2MASS counterpart for a source in the AGN catalogs. Each AGN with a 2MASS counterpart was further cross-identified with DENIS and UKIDSS catalogs. On the basis of standard deviations for magnitude difference between 2MASS/DENIS or 2MASS/UKIDSS, we extracted the sources with magnitude difference over 3σ as near-infrared variable AGNs. We discuss correlations between near-infrared variability and physical parameters (i.e., redshift, luminosity, and a near-infrared color).

1. Introduction

Variability is a characteristic of Active Galactic Nuclei (AGNs). It is believed that the majority of AGNs exhibit rapid and apparently random variability in a wide wavelength range, from X-ray to radio wavelengths. Characteristic timescales of variability range from months to years. Variability is a powerful feature for constraining models and has a potential to probe the structure and dynamics of the central engine.

We extracted near-infrared variable AGNs using cross-identification of catalogs, and studied near-infrared variability of AGNs. We conclude by discussing correlations between near-infrared variability and some physical parameters (redshift, luminosity, and near-infrared color).

2. Catalogs

We used two AGN catalogs and three near-infrared catalogs. The AGN catalogs are the catalog of Quasars and Active Galactic Nuclei (12th Ed., Véron-Cetty & Véron 2006, hereafter QA) and SDSS-DR5 quasar catalog (Schneider et al. 2007, hereafter SQ). The QA catalog includes 85,221 quasars, 1,122 BL Lac objects and 21,737 active galaxies (including 9,628 Seyfert I), with positional accuracy superior to $1''\!.0$. The SQ catalog includes 77,429 quasars in the area covering ~ 5740 deg^2, with positional accuracy superior to $0''\!.2$.

Table 1.　The number of extracted variable AGNs

AGN Catalog	Variability criterion	DENIS J	DENIS K	UKIDSS J	UKIDSS K
QA	$> 3\sigma$	205	41	355	216
	$> 5\sigma$	50	9	147	102
SQ	$> 3\sigma$	43	2	95	47
	$> 5\sigma$	9	1	27	14

The near-infrared catalogs are Two Micron All Sky Survey (2MASS), DEep Near-Infrared Survey of the southern sky (DENIS), and UKIRT Infrared Deep Sky Survey (UKIDSS) catalogs. These catalogs contain magnitudes at both J and K bands.

3.　Extraction of Near-Infrared Variable AGNs

After we cross-identified the 2MASS catalog with QA and SQ catalogs, each 2MASS counterpart was cross-identified with DENIS and UKIDSS catalogs. We used only sources having 2MASS photometric quality flags superior to D ($S/N > 3$) and either DENIS quality flags superior to 90 or UKIDSS photometric error smaller than 0.1 mag at J and K bands.

We next examined near-infrared variability by comparing cataloged magnitudes. Because we compare magnitudes listed in different catalogs (in other words, photometric difference between catalogs may be significant), we first investigated photometric differences of both 2MASS/DENIS and 2MASS/UKIDSS. We picked up normal stars in the 2MASS catalog, and extracted counterparts in DENIS and UKIDSS catalogs. We calculated the magnitude differences (i.e., $J_{2MASS} - J_{DENIS}$ and $J_{2MASS} - J_{UKIDSS}$) for each star. Because photometric accuracy generally depends on magnitude, we divided the magnitude range into bins of 0.2 mag and calculated the standard deviation (σ) in each bin. When an AGN had a magnitude difference greater than 3σ, we considered the source as a variable AGN. The identification of variable AGNs is performed at J and K bands, respectively. The resulting numbers are shown in Table 1.

4.　Properties of Near-Infrared Variability

The majority of the extracted sources have magnitude differences (ΔJ and ΔK) smaller than 1.0 mag, but some exhibit much larger variability. The average values of ΔJ and ΔK are 0.57 and 0.51, respectively. Overall, near-infrared variabilities in our sample have smaller amplitudes than the optical differences shown in Hawkins (2000).

There seems to be linear relations between ΔJ and ΔK (in both J- and K-extractions). Significant phase shifts are unlikely to appear in our sample, and this is consistent with previous studies (Neugebauer et al. 1989; Enya et al. 2002). It should be noted that there are several AGNs having a large magnitude difference at J-band but a small magnitude difference at K-band.

Figure 1 shows the redshift-variability diagrams. A positive correlation between redshift and optical variability has already been presented for quasars in previous stud-

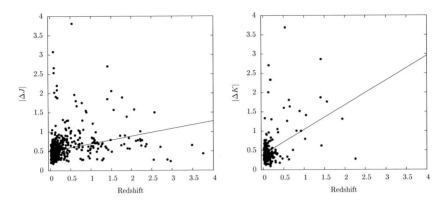

Figure 1.   Redshift versus magnitude difference diagrams. The redshift values are taken from either of the AGN catalogs.

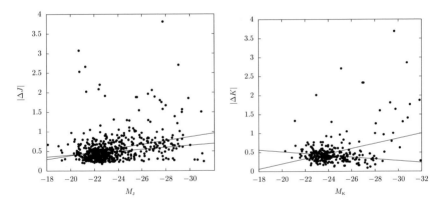

Figure 2.   Absolute magnitude versus magnitude difference diagrams. Positive steeper lines are derived using all sources, and less steeper lines are derived using sources with magnitude difference below 1.0 mag.

ies (Giallongo et al. 1991; Hook et al. 1994; Vanden Berk et al. 2004). The diagrams indicate a correlation with redshift in our sample with correlation coefficients of 0.30 and 0.42, respectively. Therefore, there may possibly be a positive correlation between redshift and variability in near-infrared wavelength.

Figure 2 shows diagrams of absolute magnitude versus magnitude difference. Although optical variability of quasars shows a negative correlation with luminosity (Hook et al. 1994; Cristiani et al. 1996; Giveon et al. 1999; Vanden Berk et al. 2004), our sample appears to indicate a positive correlation. Correlation coefficients are 0.30 (J) and 0.34 (K), respectively. However, we noticed that in the right panel of Fig. 2 most sources with $M_K \lesssim -28$ show large variability though most sources with $M_K \gtrsim -28$ show relatively small variability with a negative correlation. If we calculate correlation coefficients using only sources with the small variability (i.e., sources having magnitude differences below 1.0 mag), correlation coefficients become 0.30 (J) and -0.24

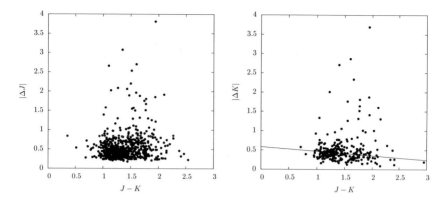

Figure 3.   Color versus magnitude difference diagrams. The $(J - K)$ colors are calculated using 2MASS magnitudes.

(K), respectively. Although sources still show a small positive correlation at J-band, they show a small negative correlation at K-band.

Figure 3 shows $(J - K)$ color versus variability. The near-infrared color is based on 2MASS photometry. There appears to be no correlation between the near-infrared color and variability, although the variability of sources with $|\Delta K| \leq 1.0$ decreases as a function of $(J - K)$ color. Correlation coefficients are 0.12 (J) and 0.05 (K), respectively. Accordingly, near-infrared color has a very weak correlation with near-infrared variability. However, when we calculate correlation coefficients using only sources with magnitude differences below 1.0 mag, they become 0.05 (J) and −0.26 (K), respectively. Therefore, the $(J - K)$ colors of these sources would have no correlation with $|\Delta J|$, but have a negative correlation with $|\Delta K|$.

Acknowledgments.   This publication makes use of data products from the Two Micron All Sky Survey, which is a joint project of the University of Massachusetts and the Infrared Processing and Analysis Center/California Institute of Technology, funded by the National Aeronautics and Space Administration and the National Science Foundation.

References

Cristiani, S., Trentini, S., La Franca, F., Aretxaga, I., Andreani, P., Vio, R., & Gemmo, A. 1996, A&A, 306, 395
Enya, K., Yoshii, Y., Kobayashi, Y., Minezaki, T., Suganuma, M., Tomita, H., & Peterson, B. A. 2002, ApJS, 141, 45
Giallongo, E., Trevese, D., & Vagnetti, F. 1991, ApJ, 377, 345
Giveon, U., Maoz, D., Kaspi, S., Netzer, H., & Smith, P. S. 1999, MNRAS, 306, 637
Hawkins, M. R. S. 2000, A&AS, 143, 465
Hook, I. M., McMahon, R. G., Boyle, B. J., & Irwin, M. J. 1994, MNRAS, 268, 305
Neugebauer, G., Soifer, B. T., Matthews, K., & Elias, J. H. 1989, AJ, 97, 957
Schneider, D. P., et al. 2007, AJ, 134, 102
Vanden Berk, D. E., et al. 2004, ApJ, 601, 692
Véron-Cetty, M., & Véron, P. 2006, A&A, 455, 773

Astronomical Data Analysis Software and Systems XX
ASP Conference Series, Vol. 442
Ian N. Evans, Alberto Accomazzi, Douglas J. Mink, and Arnold H. Rots, eds.
© 2011 Astronomical Society of the Pacific

ETC-42: A VO Compliant Exposure Time Calculator

C. Surace, P.-Y. Chabaud, G. Leleu, N. Apostolakos, and LAM scientists

Laboratoire d'Astrophysique de Marseille, OAMP, CNRS, Université de Provence

Abstract.   We developed at CeSAM (Centre de donnéeS Astrophysiques de Marseille) from LAM a new Virtual Observatory compliant Exposure Time Calculator. This new ETC has been designed to facilitate the integration of new sites, instruments and sources by the user. It is not instrument-specific, but is based on generic XML input data. It is used in several project implementations (EUCLID, EELTs, OPTIMOS) covering a wide wavelength range (from NIR to UV). This paper focuses on the spectroscopic aspects of the tool and defines the structure of the application. It emphasizes the interoperability of the ETC and shows the added value for end users. Equations will be described in another paper in preparation

1.   Yet Another Exposure Time Calculator (ETC)

1.1.   A Global Need

International agencies like ESA, ESO, NASA, as well as other specific Instrument Centers, have developed their own Exposure Time Calculators in order to simulate instrument performance. Most of the time, ETCs are instrument dependent and the output differs from one system to another, even for similar queries. The use of different definitions of noise and flux integration means that these ETCs are not used easily for other projects. After comparing several ETCs and their computation steps, we have defined the independent computation steps needed for any ETC, whatever the project and the nature (earth or space base) of the observational site are. We applied this structure to this new ETC.

1.2.   Why Another ETC

Most ETCs are black boxes and it is not easily possible to include some instrument modification and artifacts or sources (e.g., VISTA,[1] ESO,[2] WPFC2,[3] NOAO[4]). We decided to build up an "open" ETC that will be usable by the astronomical community and by instrument specialists. This ETC is open enough to include any new site, instrument, target, and operation mode without any coding experience. It is generic enough

---

[1]http://www.ast.cam.ac.uk/vdfs/etc

[2]http://www.eso.org/observing/etc/

[3]http://www.stsci.edu/hst/wfpc2/software/wfpc2-etc.html

[4]http://www.noao.edu/gateway/ccdtime/

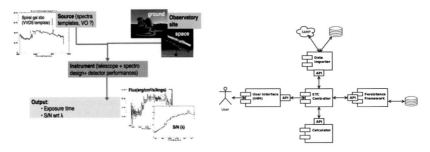

Figure 1.   *Left:* ETC characterization, the ETC has been split in 4 "open" panels. *Right:* The ETC has been defined as multi-tiered application to provide easy access to include new functionality.

to be adaptable to any new project. It is also generic enough to run tests for technical configuration testings.

1.3.   The Goal

As any ETC, the goal is to estimate the exposure time needed depending on source, site, instrument, and observation parameters specifications. Signal to noise ratio, total integration time, observation time specifications, noise components, and signal outputs are the standard outputs of this ETC.

2.   Technical Characteristics

Development:	Java application
Version:	1.0
Dependencies:	JSamp, JFreeChart, XMLBeans
Deployment:	Java webStart, Download
Platform:	Windows, Linux, Mac OS
Availability:	from January 2011
Download:	`http://lamwws.oamp.fr/cesam/votools/etc42/`

3.   User Interfaces

3.1.   Input-Output Panels

The ETC interface is based on four input panels and one output display. These panels are built to provide easy access, modularity and specifics of the instrument, source, and site. The panels are defined to split the input data: on the definition of sources (including spectral and spatial characteristics), site (including atmosphere characteristics), instruments (including telescope, and instrument characteristics) and observing parameters. The outputs include observation Signal-to-Noise Ratio (SNR) with respect to wavelength, noise contributions to the SNR, and the input and output signal. The user can choose pre-calibrated data or can built up his own observational environment. Instruments are defined by the detector, the telescope and the spectroscopic instrument type (Slit, Slitless, IFU). Site parameters depend on the ground or space location of the telescope and include a "seeing limited" option. A Source can be defined using

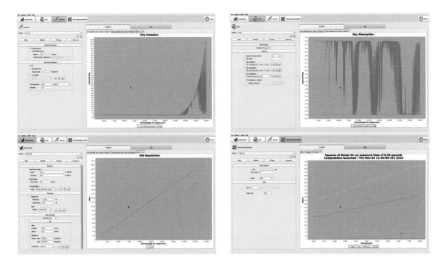

Figure 2.   Panels of ETC: to characterize source, site, instrument, and observing parameters.

its spatial and spectral distribution. Pre-defined standard sources are available. Observing parameters are defined depending on the input/ouputs needed, either with time constraints or with Signal-to-Noise constraints.

3.2.   Computation Process — Display

After launching the computation, results are displayed in the plot window under tabbed panels showing the signal to noise ratio distribution, the output signal and each noise contribution. The plot window is based on the JFreeChart component[5] which has been extended with specific drag and drop functionalities. This extension allows curves to be over-plotted from one tab to another by dragging and dropping the curve legend on the destination tab. Colors, line-styles, interface types, and zoom facilities can be used as included by default in JFreeChart.

3.3.   Input-Output — Interoperability

The ETC is compliant to the Virtual Observatory standards and is able to import a Spectrum into its Source panel, implying a close connection to a spectral tool like VOSpec or Specview for example. Unit Translations are also implemented. The SAMP menu allows the user to connect to any SAMP hub as callable client, enabling input of data from the Virtual Observatory or to broadcast any output signal as VOTable. Each output can also be exported in ASCII format (tabulated wavelength-value data).

Sites, sources and instruments can be exported as XML files which can be passed to another user to be imported in the software. Sessions are handled also to export and save any work in progress. ETC-42 provides also a log of all computations that have been launched during a session. This log can also be exported as an ASCII file.

---

[5]http://www.jfree.org/jfreechart/

Figure 3.   ETC provides an interoperability layer and is VO compliant.

4. CeSAM

The Centre de donnéeS Astrophysiques de Marseille (CeSAM) from Laboratoire Astrophysique de Marseille (LAM) has been set up to provide access to quality controlled data via web based applications, tools, pipelines development and VO compliant applications to astrophysical community (Fenouillet et al. 2011).

5. Conclusion

We are providing to the community a generic ETC. New instruments, sites or sources can be easily imported into this ETC. Compatibility tests with other ETCs and functional improvements are ongoing. New developments of the ETC include:

- Adaptive Optics module, (under development with scientists from LOOM (LAM) and ONERA),
- Direct-image computation,
- Full spectroscopy (IFU, slit-less),
- Fabry-Perot observation mode,
- Images as output (simulation),
- Add new libraries templates (sky, site),
- Include ESO and ESA to participate to the standardization effort, input data, and deployment.

If you wish to participate to the development and want to be 42 beta testers (using real data), please, contact us (delivery by December 2010).

Acknowledgments.   This work has been performed thanks to M H. Aumenier, V. Renault, L. Bouguerra, S. Peze, B. Epinat, B. Meneux and several LAM scientists S. Basa, D. Burgarella, J. G. Cuby, A. Ealet, O. Lefevre, B. Milliard.

References

Fenouillet, T., et al. 2011, in Astronomical Data Analysis Software and Systems XX, edited by I. N. Evans, A. Accomazzi, D. J. Mink, & A. H. Rots (San Francisco, CA: ASP), vol. 442 of ASP Conf. Ser., 17

Astronomical Data Analysis Software and Systems XX
ASP Conference Series, Vol. 442
Ian N. Evans, Alberto Accomazzi, Douglas J. Mink, and Arnold H. Rots, eds.
©*2011 Astronomical Society of the Pacific*

Spectral Analysis in the Virtual Observatory

Thomas Rauch and Ellen Ringat

Institute for Astronomy and Astrophysics,
Kepler Center for Astro and Particle Physics, Eberhard Karls University,
Sand 1, D-72076 Tübingen, Germany

Abstract.   In the German Astronomy Community Grid (GACG), a collaboration of German Astrophysical Virtual Observatory (GAVO) and AstroGrid-D, we provide the VO service *TheoSSA* for the access and the calculation of synthetic stellar Spectral Energy Distributions (SEDs) based on static as well as expanding non-LTE model atmospheres. However, the determination of stellar parameters within a spectral analysis is commonly still done in the "classical way", where the astronomer's experience decides the "best fit". An extension of *TheoSSA* will offer a service to perform an automatical classification based on pre-calculated template SEDs. This will be an option for multi-object spectroscopy. In addition, preliminary spectral analysis based on individually calculated SEDs will be possible. We present our concept and the progress in preparatory work.

1.   Spectral Energy Distributions and Atomic Data

The model-atmosphere code.   In the last years, the VO-service *TheoSSA*[1] was developed to simplify spectral analysis. It is currently based on the Tübingen NLTE Model-Atmosphere Package (*TMAP*, Werner et al. 2003). *TMAP* was developed over the last 25 years and assumes hydrostatic and radiative equilibrium and plane-parallel geometry. It can consider about 1000 NLTE levels and hundreds of millions of lines. With this code, model atmospheres with effective temperatures between $20\,000$ and $200\,000$ K and surface gravities ($\log g$) from 4 to 9 can be calculated as done in various analyses like Werner et al. (2008) for example.

Currently available SEDs.   To enable a first, fast and easy spectral analysis, grids of SEDs already calculated with *TMAP* can be downloaded via *TheoSSA*. This service was built in a project of the German Astrophysical Virtual Observatory,[2] the German contribution to the International Virtual Observatory Alliance (IVOA). It can be controlled with a web-interface where the fundamental parameters can be entered. The result is a list of SEDs within a given parameter range, stated in ASCII format and as metadata to be VO compatible. Therefore it can be used by everybody, non-professionals included.

---

[1]`http://vo.ari.uni-tuebingen.de/ssatr-0.01/TrSpectra.jsp?`

[2]`http://www.g-vo.org/`

In the next step, SEDs from additional codes will be included, beginning with the wind code *HotBlast*. It takes *TMAP* models as an input and calculates expanding stellar atmospheres.

More individual SEDs. For more detailed analysis, the VO-service *TMAW*[3] was developed. The user can calculate more individual SEDs, considering the elements H, He, C, N and O (and Ne and Mg in the near future). The request is generally calculated within two days, the results are sent to the user via email and are automatically ingested into *TheoSSA*, which therefore it is growing in time. Also grids of SEDs can be requested. If the number of requests is larger than 64, they are sent to AstroGrid-D,[4] an institution that provides compute power, storage, and the like. In this way even huge requests can be processed within a short time.

Model atoms. Another service developed together with the model-atmosphere code is the Tübingen Model-Atom Database (*TMAD*[5]). It considers data for the elements H − Ca, including level energies and radiative and collisional transition data for all ionization stages (e.g., Fig. 1). It can be used for every code, although its format was created for *TMAP*, because it was made also accessible in the framework of a GAVO project. Interested users can download the ready-to-use model atoms or tailor their own individual ones.

2.   A Change in Spectral Analysis

As described above, good services for spectral analysis with different accuracies are already available. These analyses are usually done the "classical" way, where the analyzer determines the parameters, according to his experience and personal view. He is doing a "χ^2 fit by eye". Thereby the ionization equilibria are used to determine T_{eff}, the line wings provide information about $\log g$, and the equivalent widths of fine lines are indicative of the abundances.

This is an important way to analyze spectra and perhaps, depending on the experience of the astronomer, cannot be replaced by an automated analysis. But the machine-aided version can provide a first classification of spectra that can simplify the "classical" analysis a lot. Especially in the era of multi-object spectroscopy, where thousands of spectra are obtained within a short time, an automated spectral analysis is important. Therefore we will create a number of tools, which will perform the first steps of an analysis automatically and aid the classical analysis. These tools are described in the next section.

---

[3]http://astro.uni-tuebingen.de/~TMAW/TMAW.shtml

[4]www.gac-grid.de/

[5]http://astro.uni-tuebingen.de/~TMAD/TMAD.html

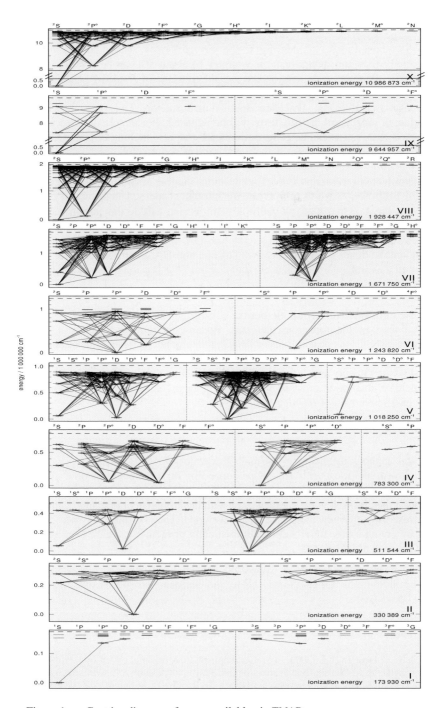

Figure 1.   Grotrian diagram of neon, available via *TMAD*.

3. Extensions of *TheoSSA*

***TheoSSA*-based aids.** In addition to the service *TMAW*, that calculates individually requested SEDs and with which an accuracy of about 5% can be achieved, two Java applets for the direct visualization and simplified usage of *TheoSSA* SEDs will be created. With the first applet uploaded observations can be compared with a selected (*TheoSSA*-) grid of, e.g., H+He models. With scrollbars T_{eff}, $\log g$, the H/He ratio, wavelength range, and the flux level will be adjusted. For a more detailed analysis an uploaded observation can be compared to grids that also include C, N and O. This will allow one to perform analyses with an accuracy of about 20%.

Multi-object spectroscopy. For the enormous number of spectra obtained with multi-object spectroscopy, we want to create an automated spectral analysis that can make a first classification and a coarse analysis. The service will do this in two stages. First, template SEDs are used. Pre-calculated H+He grids available via *TheoSSA* will be employed for a first classification, e.g., DA/non-DA. In the second step more individual fitting will be done. After the lists with DA and non-DA spectra are created, a determination of T_{eff} and $\log g$ with an accuracy of about 20% is possible. A χ^2 fit and a neural network are advantageous for this operation.

Atomic data for iron-group elements. To further develop the atomic data services we will work on the iron-group elements. Because the number of levels of iron-group elements exceeds the possibilities of *TMAP*, they can not be treated like elements with a lower atomic number but need statistical treatment. This is implemented in our program *IrOnIc* (Iron Opacity and Interface, Rauch & Deetjen 2003). At present, it needs days to calculate absorption cross-sections for iron-group elements. We will create a new, fully parallel version of *IrOnIc*.

4. Summary

Spectral analysis by means of Virtual Observatory tools is currently possible. One can perform analyses, with accuracies depending on the amount of time invested, by using the already developed services *TheoSSA*, *TMAW*, and *TMAD*. These analyses are done the classical way. To keep pace with developments in the era of multi-object spectroscopy, some new tools have to be developed. They will simplify spectral analysis a lot by giving a first classification automatically or facilities to ease classical spectral analysis with provided grids.

Acknowledgments. ER is supported by the Deutsche Forschungsgemeinschaft (DFG) under grant WE 1312/41-1, TR by the German Aerospace Center (DLR) under grant 05 OR 0806.

References

Rauch, T., & Deetjen, J. L. 2003, in Stellar Atmosphere Modeling, edited by I. Hubeny, D. Mihalas, & K. Werner (San Francisco, CA: ASP), vol. 288 of ASP Conf. Ser., 103
Werner, K., Deetjen, J. L., Dreizler, S., Nagel, T., Rauch, T., & Schuh, S. L. 2003, in Stellar Atmosphere Modeling, edited by I. Hubeny, D. Mihalas, & K. Werner, vol. 288 of ASP Conf. Ser., 31
Werner, K., Rauch, T., & Kruk, J. W. 2008, A&A, 492, L43

Astronomical Data Analysis Software and Systems XX
ASP Conference Series, Vol. 442
Ian N. Evans, Alberto Accomazzi, Douglas J. Mink, and Arnold H. Rots, eds.
© *2011 Astronomical Society of the Pacific*

ULS: An Upper Limit Server for X-ray and Gamma-ray Astronomy

R. Saxton and C. Diaz-Toledo Gimeno

XMM-Newton SOC, European Space Astronomy Center (ESAC),
28691 Villanueva de la Canada, Madrid, Spain.

Abstract. Catalogs of sources discovered in high-energy, X-ray and Gamma-ray, observations exist for all the major missions and are well supported by the Virtual Observatory (VO) and specific sites such as HEASARC[1] and LEDAS.[2] It is much more difficult for an astronomer to find the upper limit to the flux at a position where no source was detected. Here, we present a web-based tool, which performs an on-the-fly calculation of the upper-limit, using images extracted from archives that support the Simple Image Access Protocol (SIAP). It is designed to work in a mission-independent manner on any photon counts image. Currently, XMM-Newton slew and XMM-Newton pointed data are supported. Extensions to other missions such as ROSAT, Chandra and INTEGRAL are planned.

1. Introduction

Although all-sky X-ray and Gamma-ray surveys have been made from the earliest missions onwards, astronomers currently have to rely on sensitivity maps or home made software to make estimates of the upper limit to the source flux at any given sky position. There is a need for a generally available tool which can return upper limits or flux measurements to compile a long-term light curve, add a point to a spectral energy distribution (SED) or search for variability in a source. The Upper Limit Server (ULS) has been built to satisfy this need and is currently implemented for XMM-Newton pointed (Jansen et al. 2001) and slew (Saxton et al. 2008) data.

2. Design

With the advent of the Virtual Observatory (VO), the SIAP protocol has made it possible to access astronomical images in a predictable manner from distributed archives. The ULS should identify missions with suitable SIAP image servers and offer these as options to the user.

The application requires a client/server architecture where the input and output data are checked and formatted on the user's computer while the calculations are performed on the server machine. The server is required to process several requests si-

---

[1]`http://heasarc.gsfc.nasa.gov`

[2]`http://ledas-www.star.le.ac.uk/`

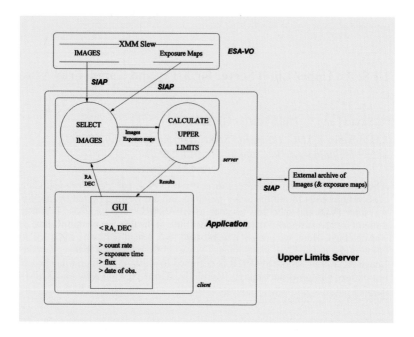

Figure 1.   The top-level design of the ULS application.

multaneously and hence needs to be scaled accordingly. A top level schema of the application is shown in Figure 1.

3.   Implementation

The application, coded in JAVA and javascript, is available from `http://xmm.esac.esa.int/external/xmm_products/slew_survey/upper_limit/uls.shtml`.

In essence the user enters a celestial position (or list of positions) from within a JAVA applet (Fig. 2). The application asks each of the SIAP servers which images they have containing this position and for which energy bands. These images are then requested and a thread launched for each one to calculate the upper limit (or flux measurement) at the source position. C++ code, from the XMM-Newton science analysis system (SAS), is currently used to perform the calculation. This has a start-up overhead which makes the code quite slow. A priority is to recode this as part of the JAVA applet. When the final thread completes, the returned results from each processed image are collated into a results page and displayed in the current browser window.

4.   Calculation

The upper limit to the count rate is calculated by:

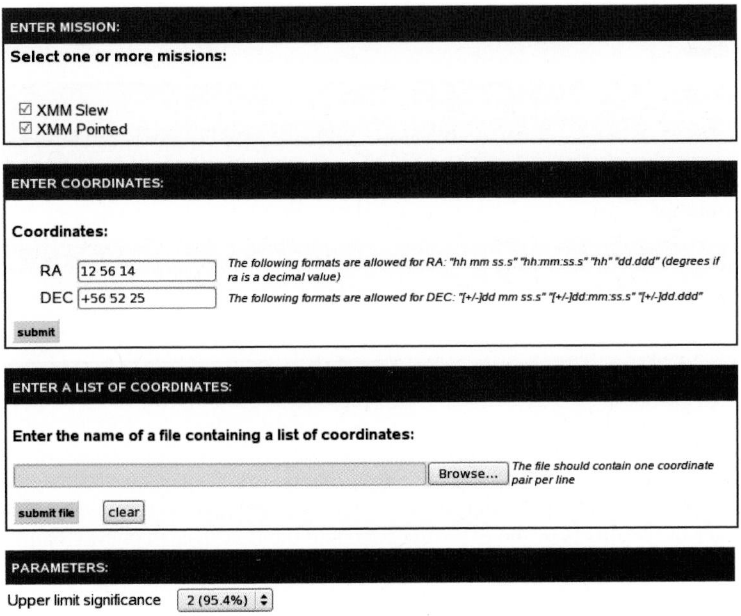

Figure 2.   The front page of the application.

- Extracting the number of counts from within a circle about the source position whose radius is determined by the size of the instrumental point spread function (PSF) and by the estimated accuracy of the image astrometry.

- Extracting the counts from a background area, defined as an annulus about the source whose inner and outer radii are also mission dependent.

- Finding the upper limit to the number of source photons using Bayesian statistics (Kraft et al. 1991) to a user-selectable precision (default is 2-sigma; 95%).

- Calculating the count rate by dividing the number of counts, corrected for the fraction of events which fall outside the extraction region, by the exposure time at the source position, which is obtained from an exposure map also returned by the SIAP server.

- Finally, the count rate is converted into a flux, by a local installation of the PIMMS routine (Mukai 1993). The conversion is made using a typical AGN spectrum of an absorbed power-law of slope 1.7 and NH=3E20. A panel to allow the user to change this model is planned.

5.   Results

The results are presented to the user in a single page ordered by source, then mission and then by observation date within that mission (Fig. 3). The columns show the ob-

[194.05833,56.87361] MKN231

XMM Slew

OBSDATE	COUNT RATE 0.2-12.0	COUNT RATE 0.2-2.0	COUNT RATE 2.0-12.0	EXP TIME (S)	FLUX 0.2-12.0	FLUX 0.2-2.0	FLUX 2.0-12.0	QFLAG
2003-06-30T08:52:01	< 3.578	< 3.578	< 3.578	0.861	< 1.047E-11	< 4.692E-12	< 3.184E-11	Y

XMM Pointed

OBSDATE	COUNT RATE 0.2-12.0	EXP TIME (S)	FLUX 0.2-12.0	QFLAG
2001-06-07T13:59:41	0.149 +/- 0.003	17064.779	4.361E-13 +/- 9.571E-15	Y

Figure 3.   An output page showing the results from XMM-Newton observations of a single object.

servation date, the count rate in each energy band, which may be expanded into the number of photons actually found in the source and background regions by placing the mouse over the results box, the exposure time and the flux in each energy band. Finally a quality flag, giving an indication of any known problems, is displayed.

5.1.   Measurement or Upper Limit

In some cases the source will be detected and a flux measurement with an error can be returned. The simplistic calculation performed within the tool will typically be less ac-curate than fluxes quoted in a source catalog. Ideally, ULS would interrogate available catalogs using VO protocols to check whether a source exists at this position before analysing the image. This is not yet implemented. Given this limitation the ULS re-turns an upper limit by default and will only return a flux, with an error, in cases where the source flux is greater than twice the error on that flux.

6.   Future Plans

The tool should become accessible as a web service so that it can be accessed directly from other VO applications.

ROSAT, Chandra and INTEGRAL archives should be included. We will encounter a problem here in that there is no standard format for the PSF information. The cor-rection for the number of counts falling outside the extraction region will have to be hard-coded in some way. A standardisation of the instrument-specific PSF is an issue which should be addressed by the IACHEC organization (Sembay et al. 2010).

References

Jansen, F., et al. 2001, A&A, 365, 1
Kraft, R., Burrows, D., & Nousek, J. 1991, ApJ, 374, 344
Mukai, K. 1993, Legacy, 3, 21
Saxton, R., Read, A. M., Esquej, P., Freyberg, M. J., Altieri, B., & Bermejo, D. 2008, A&A, 480, 611
Sembay, S., Guainazzi, M., Plucinsky, P., & Nevalainen, J. 2010, AIPC, 1248, 593

Astronomical Data Analysis Software and Systems XX
ASP Conference Series, Vol. 442
Ian N. Evans, Alberto Accomazzi, Douglas J. Mink, and Arnold H. Rots, eds.
©2011 Astronomical Society of the Pacific

Vodka: A Data Keeping-Up Agent for the Virtual Observatory

Omar Laurino[1,2] and Riccardo Smareglia[1]

[1]*INAF — Astronomical Observatory of Trieste*

[2]*Present address: Smithsonian Astrophysical Observatory*

Abstract. Currently, Virtual Observatory (VO) users interact with the VO infrastructure as the main actors. In particular, each time they want to fetch data from the VO they have to choose an application and trigger it. Vodka (VO Data Keeping-up Agent) is a new VO actor which monitors the state of the VO seeking for changes in services and datasets and notifying users for those changes and updates. These snapshots are persistent so that users can manage them and, when new interesting data are found, download them.

1. Synchronous and Asynchronous Access to the Virtual Observatory Data

Users have to trigger their interaction with the Virtual Observatory[1] (hereinafter VO) in order to find and fetch data. In Figure 1 we provide a simple representation of a most general use case: a user wants to fetch data from the VO; in order to do so, users have to start their favourite VO client (e.g. Topcat, Aladin, VAO portal), submit a Data Access Layer (hereinafter DAL) query and retrieve the data.

Figure 1. In the most general VO interaction use case, users have to trigger the data fetching process each time they want to check for data updates.

Whenever the user wants to check for updates, i.e., to retrieve data for the same query, the only way to achieve such a task is to repeat the same operations. Besides, in order to understand what has changed since the first query was submitted, (s)he has to figure out the differences in the files fetched.

We want to stress that this general use case is synchronous in the sense that it requires the user to trigger the DAL clients each time it is needed. This is true noth-

---

[1]http://www.ivoa.net

withstanding the existence of DAL applications that make use of the Universal Worker Service[2] (UWS) IVOA design pattern and are thus asynchronous.

However, the ever growing rate at which data are produced and published suggests the need for a different approach, in which the user is notified whenever new data become available. This need is even more useful for data miners, since the availability of new data can improve the construction of knowledge base datasets used for supervised and unsupervised methods.

2.   Vodka — a New Data Fetching Agent for the Virtual Observatory

We therefore introduced a new agent, named VO Data Keeping-up Agent (Vodka), to which users submit queries to be relayed to the Virtual Observatory and by which they are asynchronously notified of the availability of new data, as depicted in Figure 2.

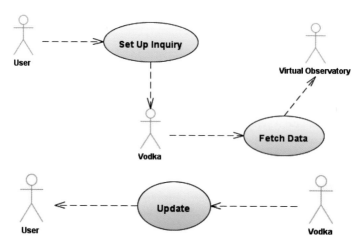

Figure 2.   Vodka is an agent that fetches data asynchronously on behalf of the user.

With Vodka, currently implemented as a web application, we try to achieve the following goals:

- Expose the power of the VO but not its complexity. Queries are created by means of a simple interface, no applications must be downloaded and installed, and there is no need to understand technical VO terms and protocols.

- Make users perceive that the Virtual Observatory is ever growing.

- Make it possible for users to be asynchronously notified when new data are available. This is useful, for example, in order to update knowledge bases used to train machine learning algorithms, and for knowledge discovery in databases in general.

---

[2]http://www.ivoa.net/Documents/UWS/

- Give users a quick glimpse of what data, relevant to their specific research interests, can be found in the Virtual Observatory.
- Make the users' queries and results persistent.

3. Inquiries, Snapshots and Differences

In Vodka, users can create Inquiries. An Inquiry consists of the parameters (see Figure 3) that make up the initial query to the VO DAL services.

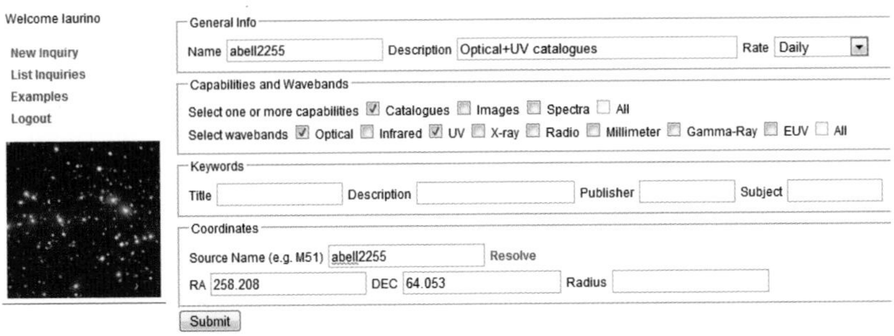
Welcome laurino
New Inquiry
List Inquiries
Examples
Logout
General Info
Name abell2255
Description Optical+UV catalogues
Rate Daily
Capabilities and Wavebands
Select one or more capabilities Catalogues Images Spectra All
Select wavebands Optical Infrared UV X-ray Radio Millimeter Gamma-Ray EUV All
Keywords
Title
Description
Publisher
Subject
Coordinates
Source Name (e.g. M51) abell2255
Resolve
RA 258.208
DEC 64.053
Radius
Submit

Figure 3. The Inquiry setup interface.

With the time rate chosen by the user, Vodka asynchronously submits the same query to the VO, fetching results into snapshots. It also performs differences between snapshots, and notifies the user via e-mail.

For each snapshot, users can navigate through the VO resources that contain data relevant to the inquiry, and download such data.

Figure 4 shows an example of a list of snapshots for an inquiry, and a list of resources for a snapshot.

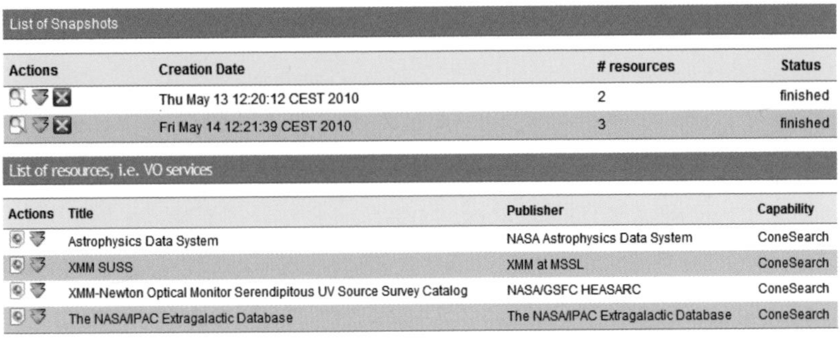
List of Snapshots
Actions
Creation Date
resources
Status
Thu May 13 12:20:12 CEST 2010
2
finished
Fri May 14 12:21:39 CEST 2010
3
finished
List of resources, i.e. VO services
Actions
Title
Publisher
Capability
Astrophysics Data System
NASA Astrophysics Data System
ConeSearch
XMM SUSS
XMM at MSSL
ConeSearch
XMM-Newton Optical Monitor Serendipitous UV Source Survey Catalog
NASA/GSFC HEASARC
ConeSearch
The NASA/IPAC Extragalactic Database
The NASA/IPAC Extragalactic Database
ConeSearch

Figure 4. The list of Snapshots and Resources for an Inquiry.

With Vodka in its current web application implementation, users can:

- Set up inquiries and the updating rate,
- Receive updates directly to their mailboxes,

- View inquiry and snapshot details,

- View the history of incremental differences between snapshots,

- Download a single VOtable for the entire snapshot,

- Download a single VOtable for each resource in a snapshot, as it appeared when the snapshot was taken,

- Download incremental files (new data, old data, missing data).

4. Distribution and Future Updates

Vodka is a web application running at `http://ia2.oats.inaf.it/vodka`.

We are currently working on a new version that will improve query execution speed and provide users with more information. For example, we are working on an algorithm which will guess how much data their query is going to result in and inform them accordingly.

On the other hand, more integration with client VO applications will be added by supporting the Simple Application Messaging Protocol[3] (SAMP).

Astronomical Data Analysis Software and Systems XX
ASP Conference Series, Vol. 442
Ian N. Evans, Alberto Accomazzi, Douglas J. Mink, and Arnold H. Rots, eds.
©*2011 Astronomical Society of the Pacific*

VODance: VO Data Access Layer Service Creation Made Easy

Riccardo Smareglia,* Omar Laurino, and Cristina Knapic

INAF — Astronomical Observatory of Trieste

*smareglia@oats.inaf.it

Abstract.   We present a tool for rapid deployment of Virtual Observatory compliant services. Users who want to publish their datasets to the Virtual Observatory can achieve this goal without having to deal with the technical details of standard services development and without having to move their data. With VODance users just have to provide a database connection to our center that points to their available data and fill out a metadata description form without having to export their data. Data Access Layer services are created on the fly and published, through the Italian Astronomical Archive Center (IA2), to the Virtual Observatory. VODance has been successfully used to publish Cone Search and Simple Image Access Protocol services out of MySQL and Oracle database management systems.

1.   Introduction

The International Virtual Observatory Alliance (IVOA) provides standard protocols and formats for sharing astronomical information. Thus, in order to publish data to the Virtual Observatory community, data providers are to implement web services which are compliant to the Data Access Layer (hereinafter DAL) IVOA standard specifications.

Some of these protocols are widely used by the most popular VO clients, e.g., Topcat, Aladin, and many Virtual Astronomical Observatory web applications.

With VODance we can dynamically create compliant DAL services on the fly at runtime, out of a generic database table or view, either local or remote.

2.   VODance Features

2.1.   Requirements

VODance allows our data center to create on the fly VO compliant DAL services out of a database table or view, either on a local or a remote database.

The basic requirement for VODance was the ability to create an arbitrary number of DAL services at runtime by simply filling in a metadata description of the service and of the individual columns. The data to export are supposed to be present in the form of a database table or view. The only requirement for the database is a JDBC Java driver to be available for the connection.

The basic functional requirement for which VODance has been created is depicted in Figure 1.

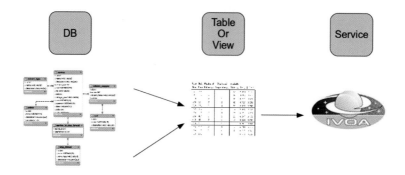

Figure 1.   VODance allows the creation of compliant DAL services out of a database table or view, either on a local or remote database, on the fly at runtime.

2.2.   Service Creation Form

In order to create a new service users can use the administration web application, by means of a simple form. A connection to the database table or view to be exported has to be available on the server on which VODance runs, which is not necessarily the server on which the administration web application runs.

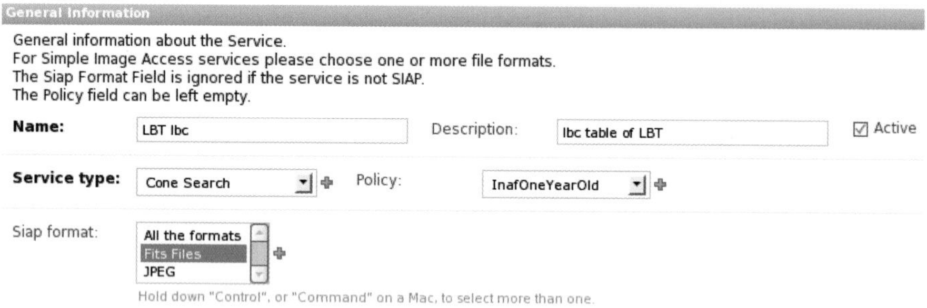

General Information

General information about the Service.
For Simple Image Access services please choose one or more file formats.
The Siap Format Field is ignored if the service is not SIAP.
The Policy field can be left empty.

Name: LBT lbc   Description: lbc table of LBT   ☑ Active

Service type: Cone Search   ▾ ➕   Policy: InafOneYearOld   ▾ ➕

Siap format: All the formats
                 Fits Files   ➕
                 JPEG
                 Hold down "Control", or "Command" on a Mac, to select more than one.

Figure 2.   Service generic information.

The procedure by which a new service is created follows these steps (see Figure 2 and Figure 3):

1. Define general parameters,

2. Define a data access policy, if necessary,

3. Fill in database connection parameters,

4. Define column metadata,

5. Save the newly created service.

As soon as the form is submitted, the new service is up and queryable. The only extra step needed is register the service in a VO resource registry in order for it to be indexed and discoverable.

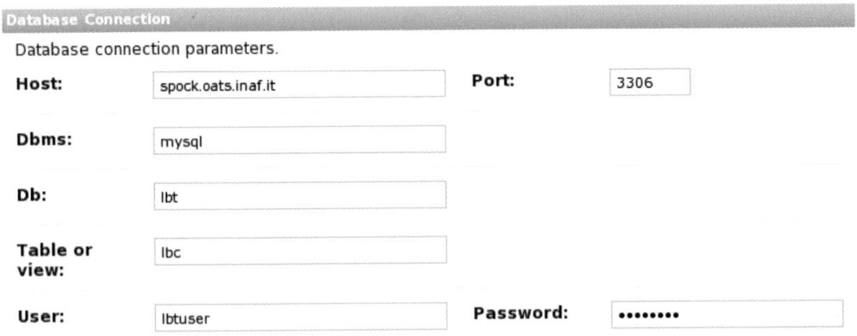

Figure 3.   Database connection information.

3.   Data Access Policy

In some cases, it may be necessary to filter out some rows from the results of the query; for example, if the table contains proprietary data or data from different partners in a collaboration, some records have to be filtered out.

In order to achieve this, one might create a new view on the database. However, VODance offers the capability to create dynamic data access policies, so that no changes are needed in the database.

Policies can be created, combined, added, and removed dynamically at runtime, on the fly, and without touching the database.

4.   Column Metadata

Figure 4.   Column metadata descriptions. Columns can be aggregated into arrays and dynamic type casting can be applied to column data.

When defining column metadata, VODance allows users to set the UCD and units strings for each column. The UCD definition is particularly important for mandatory fields which are used for the query. In particular, by defining the right ascension and declination columns, VODance knows which fields have to be constrained in the query to the specific table. For these columns, moreover, units strings are useful to inform VODance that a conversion between the standard query input (column degrees) and the actual column values (which might be radians, for example) is required.

The column mapper interface also provides two advanced features: aggregate columns and dynamic type casting.

Aggregate columns allow the definition of a new column that contains arrays of values from different columns in the original table. For example, the SIAP protocol dictates that one column must contain an array with the number of pixels of both the x and y image axis. A specific syntax in the form `column1, column2 = column3` informs VODance that values stored in `column1` and `column2` must be inserted in the new `column3` as arrays.

Dynamic type casting is necessary in order to be sure that table columns meet the requirements of the standard DAL service. For example, if a column stores values as floating point numbers while the service specification dictates that the information must be an integer, the casting can be made at runtime. Besides, some JDBC drivers and DBMS specific issues may arise from the query, so that values can be presented to VODance as Java objects that are not easy, or impossible, to serialize to XML.

For example, all the columns created as `NUMBER` in Oracle databases are apparently rendered as `java.math.BigDecimal` objects by the JDBC driver.

Dynamic type casting allows fine grained control over these issues, by allowing the definition of specific and reusable TypeCast entities.

5. Tested Services

We are currently publishing several Cone Search (quasar photometric candidates in the SDSS, photometric redshifts for SDSS galaxies and quasar photometric candidates) and SIAP services (TNG) using two different DBMS's: Oracle and MySQL.

In principle, all DBMS's for which a suitable JDBC driver is available can be used as a backend for retrieving data. However, issues may arise from the specific implementations, in particular for the SQL query syntax.

6. Deployment Environment

VODance is currently deployed in a cluster environment. In particular, a number of shared-nothing instances of application servers host the web application. The internal database is in turn hosted on an high available MySQL cluster, and software load balancing is performed by means of Linux Virtual Server, in a redundant configuration.

A file server web application is deployed in the shared-nothing instances as well, for serving images according to the SIAP services.

The administration user interface is deployed on a different web server that shares the internal database cluster with the web application that actually serves the DAL services.

7. Distribution

Since the architecture depicted in § 6 is quite elaborate, consisting of a stack of different services and servers, VODance will be distributed as a virtual application, i.e. as a disk image intended to be run in a virtual machine.

With this choice we hope to address several issues: first of all, final users will not be required to install package and configurations, except for the unavoidable network configuration; secondly, multiple instances of the application can be combined allowing us to achieve with minimal effort the high level of availability and redundancy that is typical of a production environment.

Astronomical Data Analysis Software and Systems XX
ASP Conference Series, Vol. 442
Ian N. Evans, Alberto Accomazzi, Douglas J. Mink, and Arnold H. Rots, eds.
© *2011 Astronomical Society of the Pacific*

Service Infrastructure for Cross-Matching Distributed Datasets Using OGSA-DAI and TAP

Mark Holliman,[1] Tilaye Alemu,[2] Alastair Hume,[2] Jano van Hemert,[3] Robert G. Mann,[1] Keith Noddle,[1] and Laura Valkonen[3]

[1] *WFAU, Institute for Astronomy, University of Edinburgh*

[2] *EPCC, University of Edinburgh*

[3] *School of Informatics, University of Edinburgh*

Abstract.   One of the most powerful and important goals for VO developers has been to enable cross-match queries between disparate datasets for end users. This has only been achieved within the VO using the early SkyNode infrastructure and has not been reproduced using current IVOA standards. To remedy that situation, the Wide Field Astronomy Unit (WFAU) has worked with the Edinburgh Parallel Computing Center (EPCC) in leveraging the OGSA-DAI grid middleware to enable cross catalog queries on distributed VO services. We have achieved this goal by building a three layer service stack that places the OGSA-DAI software above multiple individual services implementing the IVOA's new Table Access Protocol (TAP), and then a single TAP service is placed above this and presented to the end users. Users can then execute ADQL queries that cross-match between the disparate datasets as though they were in the same database with acceptable performance rates on the resulting data flow. The OGSA-DAI software is able to interrogate any compliant TAP service to acquire the necessary metadata for insertion into the single federated TAP service used for cross-match queries. We are currently testing this distributed infrastructure using the TAP services provided by WFAU for the UKIDSS DR3 and SDSS DR7 datasets in combination with the TAP service available from the Canadian Astronomy Data Center (this last without requiring any action from CADC staff). This forms the basis for a large-scale distributed data mining workflow and similar activities can be readily implemented as more TAP services come online. Future work will involve releasing this infrastructure to the greater astronomical community as an IVOA compliant service for users.

Cross-Matching Distributed Datasets

At present there is no singular tool for cross-matching distributed datasets using current IVOA standards. It is with this aim that we have investigated leveraging the distributed query processing (DQP, Dobrzelecki et al. 2010) functionality of the OGSA-DAI data access and integration middleware (Jackson et al. 2007) to enable cross catalog queries over the IVOA ratified Table Access Protocol (TAP, Dowler et al. 2010) services.

Cross-matching distributed datasets is a non trivial problem due to the difficult nature of query planning and execution over infrastructure that is both disparate and heterogeneous. Astronomy datasets are housed by many individual institutions that often use different backend hardware and software platforms for storing and serving their data. The problems are further compounded by the growing sizes of astronomical datasets which cannot be transported easily or loaded entirely into memory for query execution. The combination of these problems has led to a situation whereby there is no straightforward method for joining datasets offered by different services.

Fortunately there are IVOA standards that address query language and data access which help minimize the problems caused by heterogeneous dataset backend infrastructure. In particular the ADQL standard (Ortiz et al. 2008) unifies the query language necessary for writing queries to astronomical data services. Furthermore, the TAP standard defines a simple yet powerful interface for uniformly accessing astronomical datasets using ADQL queries and producing VOTable output. OGSA-DAI is a data access and integration middleware solution that has been developed to provide a standard interface to distributed underlying databases, and it includes powerful functionality for federating these databases in such a way that they can be accessed as though they were a single database. We have worked on building a service infrastructure that utilizes the powerful dataset federation and query handling capabilities of the OGSA-DAI middleware alongside the uniform TAP service interfaces for obtaining data in order to build a service infrastructure that enables distributed dataset cross-matching.

OGSA-DAI and DQP

The OGSA-DAI server software sits above the database(s) and allows users to build workflows containing queries which it then executes to return query results. OGSA-DAI includes a component called Distributed Query Processing that executes queries (including join operations) across distributed databases. DQP performs the tasks of parsing the query, and builds, optimizes, and executes the query plan. It federates the schemata from the underlying databases into a single schema that acts as though all the tables are within a single database. In order to improve the efficiency of the query execution DQP optimizes the query plan such that as many of the expensive operations as possible are pushed down to the database level for final execution. This ensures that the amount of data that needs to be moved for performing the join operations is minimized.

DQP query plans are executed on one or more OGSA-DAI servers. OGSA-DAI offers a range of possible join algorithms and it is the task of the DQP query plan optimization to choose the most appropriate algorithms for a given query. The performance of these different joins varies significantly depending on the statistics and sizes of the underlying databases. The available join algorithms include:

- In-memory join: One side is transferred and stored in memory, the other side is streamed,

- Partial in-memory join: Gets first results quickly, then all data are stored to disk for joining,

- Ordered merge join: Both inputs are ordered by the join attribute thus supporting an efficient fully streamed join,

- Parallel hash equi-join: Multiple OGSA-DAI servers can be used to implement the join in parallel,

- Batch joins using IN clauses: e.g. SELECT * FROM foo WHERE bar IN (x,y,z).

Table Access Protocol

TAP is a recently approved IVOA protocol for accessing tabular data in a standardized way. Current implementations allow users to submit queries in ADQL through standard HTTP GET and POST methods, and return results in VOTable format. Users can access TAP services using simple clients like web browsers or "wget", or software like VODesktop (AstroGrid 2010) and TAPsh (Demleitner 2010). There are a limited number of TAP services currently published on the VO, but more are expected in the near future. TAP is anticipated to be highly utilized as it offers a more powerful interface to underlying datasets than the current "simple" IVOA protocols like cone-search and image access protocol.

Cross-Matching Infrastructure

We have designed a 3-layer architecture that places the OGSA-DAI middleware above an arbitrary set of TAP services. OSGA-DAI combines the metadata from the underlying datasets, and this is then exposed through a single TAP interface at the top layer, enabling users to perform cross-match queries in ADQL on all of the federated datasets underneath.

We implemented a test infrastructure using an OGSA-DAI deployment (seen in Figure 1) that federates the UKIDSS DR3 TAP service and the Canadian Astronomy Data Center's (CADC) TAP service for CAOM data (CADC 2010). These are independent TAP services, and no coordination or assistance was required from CADC in building the test system. We then submitted several scientifically relevant ADQL queries that cross-matched the different datasets to determine problems and performance for the system. These queries utilized a variety of different join types and expected result set sizes in order to properly stress the service infrastructure. Overall the test system

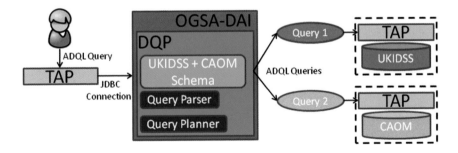

Figure 1.   Cross-match testbed diagram.

successfully demonstrated the viability of the service architecture, returning results for most queries. We have noted those queries that failed and initiated investigation into the causes of their failure to determine at what point in the service infrastructure the failure occurs. The majority of failures were due to query language differences between ADQL and the subset of SQL utilized by the DQP component of OGSA-DAI.

The actual query times varied greatly depending on a number of factors, which included the join algorithm utilized, size of tables to be joined, and result set size. Though not as fast as native systems, the query times were acceptable given the distributed nature of the data and the novelty of the cross-matching capability. For some queries the query plans built by DQP are not as efficient as hand written OGSA-DAI workflows. This is primarily because DQP does not choose the most suitable join algorithm due to the currently limited set of statistics regarding the sizes and attribute value distributions of the datasets. Our main test query selected a portion of the UKIDSS Large Area Survey and then cross-matched those objects against the entire CAOM, and it completes in approximately 70 minutes. Using a hand optimised OGSA-DAI workflow this time can be reduced to around 3 minutes. The main optimisation not currently performed by DQP is the use of a range join algorithm that uses a two sided range index ($x < y$ AND $x > z$) to match values rather than a single-sided range ($x < y$). The other important optimisation is the use of a more compact data transfer format rather than the default XML VOTable representation. It will not be difficult to update DQP to perform these optimisations. This time could be further reduced by placing an OGSA-DAI server geographically close to the CAOM TAP server which would reduce the overall time spent transferring data over the internet.

Future Development

The demonstrated success of our test infrastructure points to a number of possibilities for future development. One ultimate goal is to deploy a registered TAP service utilizing the OGSA-DAI infrastructure to federate data from all the TAP services published on the VO. Such a service would be one of a kind in allowing users to cross-match all the databases and catalogs available. Other possibilities for user services that utilize this system are being explored with input from VO developers and astronomers.

In the meantime we are working on addressing the issues discovered through testing to improve the functionality as well as the robustness of the service infrastructure. In particular we would like to address the issues of error handling through the service layers, and also to better optimize query plans for complex joins of distributed databases. All of this will be investigated; in addition, we plan on implementing more join algorithms, improving join algorithm choice and execution, adding support for ADQL spatial functionality, and improving performance for joins of large (multi-terabyte) databases. A demonstration webpage can be found at: `http://www2.epcc.ed.ac.uk/~ally/aida/Demo1.html`.

References

AstroGrid 2010, Vodesktop. URL `http://www.astrogrid.org/`
CADC 2010, Caom. URL `http://cadcwww.dao.nrc.ca/caom/`
Demleitner, M. 2010, Tapsh. URL `http://vo.ari.uni-heidelberg.de/soft/tapsh`
Dobrzelecki, B., et al. 2010, Phil. Trans. Royal Soc., 368, 4133
Dowler, P., Rixon, G., & Tody, D. 2010, Table Access Protocol, Version 1.0. URL `http://www.ivoa.net/Documents/TAP/`
Jackson, M., et al. 2007, in UK e-Science All Hands, edited by S. J. Cox (Edinburgh: National e-Science Center), vol. 2007 of UK e-Science All Hands
Ortiz, I., et al. 2008, Ivoa astronomical data query language, version 2.0. URL `http://www.ivoa.net/Documents/latest/ADQL.html`

Astronomical Data Analysis Software and Systems XX
ASP Conference Series, Vol. 442
Ian N. Evans, Alberto Accomazzi, Douglas J. Mink, and Arnold H. Rots, eds.
© 2011 Astronomical Society of the Pacific

Cross-Matching Large Photometric Catalogs for Parameterization of Single and Binary Stars

O. Malkov[1] and S. Karpov[2]

[1]*Institute of Astronomy of Rus. Acad. Sci., Russia*

[2]*Special Astrophysical Observatory Rus. Acad. Sci., Russia*

Abstract.   We discuss the methods of reliable cross-matching of large photometric catalogs at various wavelengths (SDSS, GALEX, 2MASS, USNO-B1, etc.) for the purpose of collecting and using multicolor (from UV to IR) photometry of cross-matched objects. We pay special attention to the performance issues and describe several approaches for the speed-up of the match, involving construction of various spatial indices, both planar and spherical. We discuss the means of dealing with ambiguous (multi-candidate) matches using a-priori information. Color-index diagrams, constructed with the obtained photometry, are powerful tools for parameterization of stars. Particularly, detection of a composite flux in photometry can serve as an indication of a photometrically unresolved binarity and can contribute to the parameterization of the components of binary systems. Interstellar extinction values for cross-matched stars can also be calculated from the multicolor photometry.

1.   Introduction

Virtual Observatory facilities allow users to perform fast and correct cross-matching of objects from various surveys. This yields multiphotometry data (color indices) on registered objects and makes it possible to determine stellar parameters. A method of catalogs cross-matching, as well as its application to various areas in the sky and preliminary results of stellar parameterization, are discussed in the present paper.

2.   Cross-Matching Astronomical Catalogs

2.1.   Basics of Cross-Matching

The task of catalog cross-matching is to reliably link the appearances of the same astrophysical source in different catalogs, which may be acquired from observations made at different epochs, with different instruments, in various spectral filters etc. Typical problems that arise here include poor spatial accuracy of one or several catalogs, possible proper motions for near-by objects, and different sensitivities of the surveys, which may lead to significantly variations in object densities. Sole positional information is, therefore, not enough to perform a reliable match; the quoted coordinates of a "true" object may be closer to the truth for some than for others.

A possible solution is to use some prior information when comparing records from different catalogs. If they have measurements in the same photometric filter one may

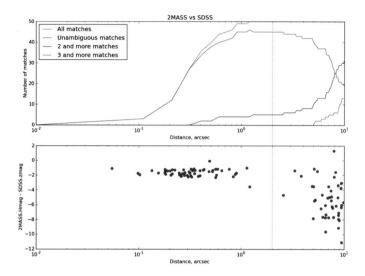

Figure 1.   Upper panel: typical dependence of number of matches on the match radius for a small overlapping region of 2MASS and SDSS. Lower panel: scatter plot of a color; the magnitude difference between measurements in similar spectral bands of two catalogs for matches as a function of separation. The vertical dotted line marks an optimal match radius.

select only objects with roughly equal magnitudes; if they do not, spectral information may be used. Indeed, object colors, or spectral slope, for similar objects are similar too. Reliably matched points will be clustered not only in spatial separation, but also in colors (computed by means of magnitudes from different catalogs). Finding such clusters and performing iterative sigma-clipping filtering, one may choose appropriate limits for match radius and magnitude difference. Fig. 1 illustrates this concept.

Also, the subsequent use of matched objects in spectral modeling and determination of extinction provides one more chance to filter out possibly random coincidences by excluding objects yielding results that are significantly different from ensemble values.

2.2.   Technologies We Use

We used a number of popular and well-established informational technologies in our work.

To locate the electronic versions of catalogs we used the Virtual Observatory registry which provides standard interfaces for searching data services of various types. We got entry points of ConeSearch services for all catalogs we needed and then accessed them to acquire the data for selected fields on the sky in standard VOTable format.

To speed up the procedure of cross-matching (which in straightforward realization requires $\sim N_1 \cdot N_2$ operations to compute and compare all possible pair distances) the spatial indexing may be used. Such an index typically supports fast positional queries to selected points from the neighborhood of a given position, proportional to the logarithm

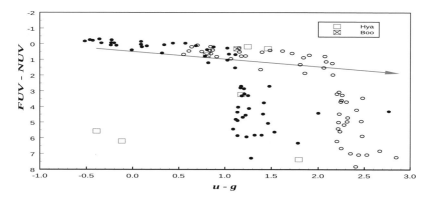

Figure 2.   SDSS/GALEX photometry. Blue squares: SDSS/GALEX objects in Boo (l = 353°, b = +68°, r = 0.1°, crossed squares) and Hya (l = 228°, b = +27°, r = 0.1°, open squares) areas. Filled black circles: Pickles (1998) models recalculated to the AB magnitude system. Open black circles: the same models but reddened by E_{B-V} = 1. Red arrow: reddening curve.

of the number of points, and the total computational complexity of the cross-match may be reduced to ~ $N_1 \cdot \ln N_2$. Widely used sky partitioning schemes, applicable for such indexing, include HEALPix (Górski et al. 2005), Q3C (Koposov & Bartunov 2006) and HTM (Hierarchical Triangular Mesh, Kunszt et al. 2001). In this work we used this last one, developed for SDSS, as a spatial index.

All the code for the actual cross-match and analysis of its results was written in the Python high-level programming language. To access catalog data in VOTable format a modified version of the ATPy package was used, and for the visualization we employed a Matplotlib library and stand-alone TopCat package. HTM partitioning code was taken from EsUtil library, available at http://code.google.com/p/esutil/.

3.   Color Index Diagrams for Observed and Theoretical Points

The cross-matched stars can be put on combined color index diagrams. To determine their stellar intrinsic energy distribution, and hence stellar parameters, one should compare observed and theoretical color indices.

In Fig. 2 we show theoretical and observational points in the (FUV–NUV, u–g) plane, where u and g are SDSS photometric bands, and FUV and NUV are GALEX photometric bands. To simulate stars in the diagram we used the spectral library from Pickles (1998) and the interstellar extinction law from Fluks et al. (1994). SDSS and GALEX response curves were taken from Gunn et al. (1998) and Morrissey et al. (2005), respectively.

There are a number of reasons for the observed disagreement between the empirical and theoretical points in Fig 2: observational photometric uncertainties, catalog misprints, cross-match errors, variability or non-stellar nature of objects, non-standard behavior of the interstellar extinction law in the area, etc. One of possible reasons is unresolved binaries: close binaries, which can be resolved neither photometrically (unless they exhibit mutual eclipses) nor astrometrically.

Unresolved binaries with components of different temperature can exhibit colors different enough from the ones of single stars. Such binaries can be separated from single stars in some color index diagrams. In Malkov et al. (2011a,b) color index diagrams, where the single-binary star separation is possible, were specified.

Obviously, the interstellar reddening complicates the identification of unresolved binaries in color index diagrams. However, for some binaries, color indices can be derived where interstellar reddening does not prevent to discover the pair.

4. Conclusions

A method for reliable cross-matching of catalogs with different object densities is realized in the present study, and a tool for simulation of color index diagrams is constructed. It is shown that combined multicolor photometry can be used for parameterization of the cross-matched stars and for single-binary star separation.

Acknowledgments. This work has been supported by the Russian Foundation for Basic Research (grants No. 08-02-00371, 09-02-00520, 10-02-00426 and 10-07-00342), Federal target-oriented program "Scientific and pedagogical staff for innovation Russia" (contract No. P1195), Federal Science and Innovations Agency (contract No. 02.740.11.0247) and by the program of Presidium of RAS "Leading Scientific Schools Support" (4354.2008.2). We thank Aleksej Mironov and Sergej Sichevskij for collaboration and valuable comments.

References

Fluks, M. A., Plez, B., The, P. S., de Winter, D., Westerlund, B. E., & Steenman, H. C. 1994, A&AS, 105, 311
Górski, K. M., Hivon, E., Banday, A. J., Wandelt, B. D., Hansen, F. K., Reinecke, M., & Bartelmann, M. 2005, ApJ, 622, 759
Gunn, J. E., et al. 1998, AJ, 116, 3040
Koposov, S., & Bartunov, O. 2006, in Astronomical Data Analysis Software and Systems XV, edited by C. Gabriel, C. Arviset, D. Ponz, & E. Solano (San Francisco, CA: ASP), vol. 351 of ASP Conf. Ser., 735
Kunszt, P. Z., Szalay, A. S., & Thakar, A. R. 2001, in Mining the Sky, edited by A. J. Banday, S. Zaroubi, & M. Bartelmann, 631
Malkov, O. Y., Mironov, A. V., & Sichevskij, S. G. 2011a, in Gaia: at the frontiers of astrometry, edited by C. Turon, F. Arenou, & F. Meynadier (EDP Sciences), ELSA 2010 Conference. In press
— 2011b, Ap&SS. In press
Morrissey, P., et al. 2005, ApJ, 619, L7
Pickles, A. J. 1998, PASP, 110, 863

Astronomical Data Analysis Software and Systems XX
ASP Conference Series, Vol. 442
Ian N. Evans, Alberto Accomazzi, Douglas J. Mink, and Arnold H. Rots, eds.
© 2011 Astronomical Society of the Pacific

Development of the "AKARI" Catalog Archive Server: Cross-Identification Using an Open Source RDBMS Without Dividing the Celestial Sphere

Chisato Yamauchi

Institute of Space and Astronautical Science,
Japan Aerospace Exploration Agency,
3-1-1 Yoshinodai, Sagamihara, Kanagawa, 252-5210, Japan

Abstract.   The AKARI Infrared All-Sky Catalogs are important infrared astronomical databases for next generation astronomy. We have developed a Web-based service, AKARI Catalog Archive Server ('AKARI-CAS'), for various types of astronomical research. The service provides useful and attractive search tools, visualization tools and documentation for the AKARI Catalogs. The most remarkable feature of our service is fast dynamic cross-identifications between registered catalogs. In this paper, we present a summary of our service and implementation design of dynamic cross-identifications for the AKARI and 2MASS catalogs. Our performance tests of the cross-identifications included cases that employed the latest low-cost SATA3 (6Gbps) SSDs.

1. "AKARI-CAS" Overview

We show an overview of stored catalogs (Murakami et al. 2007; Yamamura et al. 2010; Ishihara et al. 2010) and web-based service of AKARI-CAS.

Stored Catalogs

Catalog	Objects
AKARI FIS Bright Source Catalog (FIS BSC)	427,071
AKARI IRC Point Source Catalog (IRC PSC)	870,973
2MASS Point Source Catalog (2MASS PSC)	470,992,970
IRAS PSC, IRAS FSC, etc.	

Web-based Service

Search Tool
Radial Search, Rectangular Search, SQL Search,
Cross-id between AKARI Catalogs and SIMBAD/NED list,
Cross-id between AKARI Catalogs and users list

Visualization Tool
Explore Tool, Image List Tool

Shown in above table and Figure 1, AKARI-CAS has powerful search tools and visualization tools. One of the new features of AKARI-CAS is having cached SIM-BAD/NED entries, which can match-up AKARI Catalogs with other catalogs stored

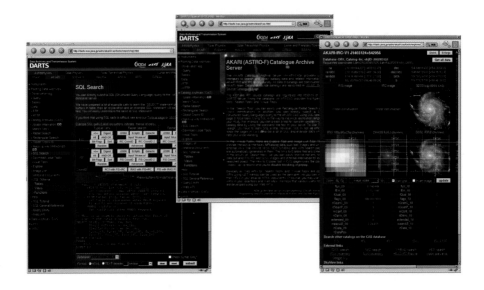

Figure 1.   Top-page (top), 'SQL Search' (left) and 'Explore Tool' (right) of AKARI-CAS. Displaying quick-look images of multi wavelengths and supporting direct input of an SQL statement are distinctive features in our service. Our service covers various demands from general users to power users.

in SIMBAD or NED. Users can perform fast dynamic cross-identifications between registered catalogs directly through SQL queries.

Visit our AKARI-CAS page for more information:
http://darts.isas.jaxa.jp/astro/akari/cas.html

2.   Dynamic Cross-Identification

2.1.   Basic Idea

The object tables in our databases have columns of unit vectors (cx, cy, cz) presenting J2000 source positions. In order to realize a radial search that is fast enough, we created composite indices for (cx, cy, cz), and wrote stored procedures that first catch objects within a cube using index-scan and then drop the objects that fall outside the strict search circle on the celestial sphere. A cross-identification is performed by repeating the radial search.

2.2.   Implementation for AKARI Catalogs and Performance

AKARI Catalogs are relatively small compared with 2MASS PSC (Skrutskie et al. 2006). In such a case, it is possible to store the entire catalog into a database table and create composite index on the FLOAT8 columns cx, cy and cz columns:

```
CREATE INDEX ircobjall_xyz ON IrcObjAll (cx,cy,cz);
```

where IrcObjAll table contains all the objects of IRC PSC.

We created an SQL stored procedure to get the ObjID of the nearest object from a given position in a specified catalog, by using algorithm described in §2.1.

We present our performance test results of matching-up all the FIS objects with all the IRC objects within 0.25 arcmin radius (see §2.4 for hardware configuration):

Condition	Elapsed Time	Count of Radial Search
just after OS rebooting	42.0 sec.	10157 counts/sec.
2nd trial	19.8 sec.	21515 counts/sec.

Results of matching-up FIS BSC with IRC PSC.

The SQL statement for this search:

```
SELECT count(fIrcGetNearestObjIDEq(ra,dec,0.25)) FROM fisobj;
```

This returns 19267 matches. Readers may execute this on our SQL Search page: http://darts.jaxa.jp/astro/akari/cas/tools/search/sql.html

2.3. Implementation for Huge Catalogs and Performance

The next major update of AKARI-CAS will support dynamic cross-identifications between registered catalogs and 2MASS PSC.

2MASS PSC has 470,992,970 objects. The simple method described in §2.2 does not yield satisfactory performance for such huge catalogs due to the big index and the disk I/O bottleneck. Therefore, we tried the following strategy:

(1) Applying the table partitioning technique, we created a special table set for radial search and cross-identification. This table set consisted of objid as primary key, J2000 unit vector (cxi, cyi and czi) columns of type INT4 (from −2e9 to 2e9; converted and scaled from right ascension and declination).

(2) The 2MASS set contains 736 tables (partitions), divided in ranges of declination. Each table has about 0.6 million rows.

(3) We created a composite-expression index on (cxi, cyi, czi) for each table. The indices were created in the range of values from −32400 to 32400 of type INT2, scaled and rounded from (cxi, cyi, czi).

(4) We wrote a stored function in C to return an SQL statement to SELECT appropriate table(s) from above table set for a radial search.

(5) Ww wrote a stored function in PL/pgSQL to execute SQL statement that was returned by the function created in (4).

The sizes of the special tables and indices were about 20GB and 10GB, respectively. They are small enough to be stored in SSD or main memory of current DP (dual-processor) servers.

We attached 'Crucial RealSSD C300', which is a cheap but fast SATA3 SSD, to our hardware described in §2.4, and stored the tables and indices on the SSD. We present our test results of the 2MASS PSC matches from the AKARI IRC PSC and Tycho-2 Catalog (2539913 sources; Høg et al. 2000) using both SATA2 (ICH9R) and SATA3 (ASUS U3S6) interfaces:

Catalog	Radius	Interface	Elap.Time	Cnt of Radial Search
AKARI IRC PSC	0.25 arcmin	SATA2	8.2 min.	1762 counts/sec.
AKARI IRC PSC	0.25 arcmin	SATA3	7.0 min.	2079 counts/sec.
Tycho-2	0.25 arcmin	SATA2	17.8 min.	2379 counts/sec.
Tycho-2	0.25 arcmin	SATA3	15.8 min.	2683 counts/sec.
Tycho-2	1.0 arcsec	SATA3	14.8 min.	2858 counts/sec.

Results of matching-up a catalog with 2MASS PSC.
All tests are measured just after OS rebooting.

In all cases, CPU usage did not reach 80%. We should obtain better results as the I/O performance improves. In the test cases that used Adaptec RAID 2405 with 1TB SATA2 HDD×2 (RAID1), we achieved about half the performance shown in the table above. In other tests using a DP server (Opteron2384) with 32GB memory, we obtained about 3500 counts/sec when we had enough data cached in main memory.

2.4. Configuration of our Performance Tests

- Hardware
  Mainboard: GIGABYTE GA-EX38-DS4 (X38 + ICH9R)
              C1E = Disabled
  CPU: Intel Core2Quad Q9650 (3.0GHz)
  Memory: DDR2-800 8GB
  Storage: Adaptec RAID 2405 + 7200rpm SATA2
              1TB HDD ×2 (RAID1)

- Software
  OS: CentOS-5.5 64-bit
              cpuspeed=off, FS=Ext3, readahead=1024
  RDBMS: PostgreSQL-8.4.5
              shared_buffers=128MB, work_mem=256MB
              effective_cache_size=128MB

References

Høg, E., et al. 2000, A&A, 355, L27
Ishihara, D., et al. 2010, A&A, 514, A1
Murakami, H., et al. 2007, PASJ, 59, 369
Skrutskie, M. F., et al. 2006, AJ, 131, 1163
Yamamura, I., et al. 2010, Akari/fis all-sky survey bright source catalogue version 1.0 release note. URL http://www.ir.isas.jaxa.jp/AKARI/Observation/PSC/Public/RN/AKARI-FIS_BSC_V1_RN.pdf

Astronomical Data Analysis Software and Systems XX
ASP Conference Series, Vol. 442
Ian N. Evans, Alberto Accomazzi, Douglas J. Mink, and Arnold H. Rots, eds.
© *2011 Astronomical Society of the Pacific*

Current Status of the Japanese Virtual Observatory Portal

Yuji Shirasaki,[1] Yutaka Komiya,[1] Masatoshi Ohishi,[1] Yoshihiko Mizumoto,[1] Yasuhide Ishihara,[2] Hiroshi Yanaka,[2] Jumpei Tsutsumi,[2] Takahiro Hiyama,[2] Hiroyuki Nakamoto,[3] and Michito Sakamoto[3]

[1]*NAOJ, 2-21-1 Osawa, Mitaka Tokyo, 181-8588, Japan*

[2]*Fujitsu Ltd., 1-9-3 Nakase, Mihama-ku, Chiba, 261-8588, Japan*

[3]*SEC Co. Ltd., 4-10-1 Youga, Setagaya-ku, Tokyo, 158-0097, Japan*

Abstract. The Japanese Virtual Observatory (JVO) portal is a web portal for accessing astronomical data and analysis system through the Internet. In the years of 2009 and 2010, we have developed two new access interfaces: JVOSky and command-line access interfaces. To enable users to perform all sky searches based on SED properties of celestial objects, we experimentally used the Hadoop for performing cross-match of one billion photometric records in the JVO Digital Universe.

1. JVO Sky

JVO Sky is an on-line data discovery service which displays the coverage of observations made by various instruments on Google Sky (Fig. 1). Using this interface, a user can graphically find sky regions where data of multi-wavelength observations exist. Currently accessible datasets are the data repositories of Subaru Suprime-Cam (image, Shirasaki et al. 2007) and HDS (spectrum), and of Suzaku (X-ray photon event lists). On the map, markers are plotted at the position where data records exist. The number of markers shown in the map is limited to 100, and actual numbers are shown at the bottom of the top panel, such as "spcam:70/3069", which indicates that there are 3069 data records in the displayed area and 70 among them are plotted on the map. An information window for the corresponding data record appears when a user clicks a marker or data record name in the list under the dataset folder. A link to the data download page is provided in the information window.

2. JVO Command

Although a graphical user interface (GUI) is a convenient way for performing a simple query, it is not efficient nor flexible for performing a lot of queries by changing query parameters. This may happen when a user wants to get a large number of data records that may exceed the maximum number that a data service can return. We therefore implemented a command line interface that is accessible through typing commands on the user's computer.

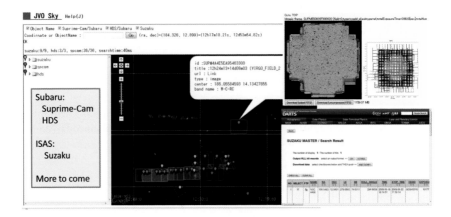

Figure 1.   JVOSky graphical user interface.

Consequently, we provide two kinds of user interfaces on the JVO portal[1] (Fig. 2; Shirasaki et al. 2009); one is a web-based GUI and the other is a CUI which accepts a request sent by the command line tool called *jc*. Fig. 3 shows a syntax of the *jc* command and usage examples. The `search` command is used for data searches based on JVOQL (JVO Query Language, Shirasaki et al. 2006); the `registry` command is used for searching VO data service based on the specified keyword; the `copy21` command is used for copying data on the JVOSpace (a storage service for registered users of JVO portal) to the local machine; the `run` command is used to execute application on the JVO grid system (Shirasaki et al. 2007); and the `join` command is used to combine two VOTables by matching records based on spatial proximity.

3.   Cross-Match Experiment with Hadoop

JVO has a huge astronomical database called Digital Universe (Tanaka et al. 2008), which contains coordinates and photometric information of celestial objects collected from major survey catalogs. Currently we provide a functionality to search for data based on coordinates only. However, a search based on SED properties. would be a legitimatescience use case. In order to provide this kind of search functionality, cross identification among different catalogs should be performed in advance. A search could be conducted against the whole sky, and all the data would be scanned in a reasonable amount of time. To achieve thisfunctionality we developed distributed data search system by means of Hadoop[2] (Fig. 4). This allowed us to finish the cross match of 1G records (1/20 of whole dataset) in 3.7 hours with 70 concurrent tasks. If it was executed with a single task, it would take 9 days.

---

[1]http://jvo.nao.ac.jp/portal/

[2]http://hadoop.apache.org/

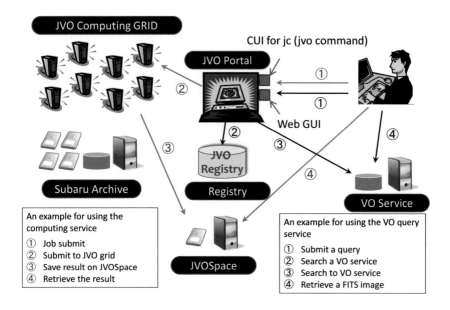

Figure 2.   Diagram of JVO portal system.

```
Syntax of jc (jvo command)
    jc <command>  [<option>] [<argument>] ...

Examples:
    jc search -i <jvoql_file>
    jc registry -k <keyword>
    jc copy2l <source> <destination>
    jc run <program_name> <arguments>

Other commands:
    ls rsync passwd resume suspend abort ps union join
    select
```

Figure 3.   Syntax and examples of the jc command.

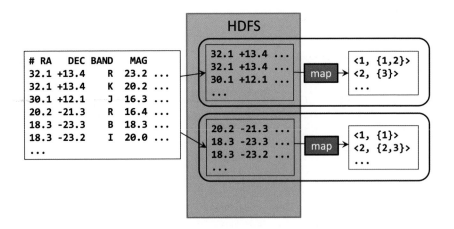

Figure 4.   Application of MapReduce algorithm to the cross match of astronomical catalog. Whole dataset were divided into subsets based on a region of sky, and cross matches were performed in the map function in parallel.

4. Scientific Result Using the JVO

We performed projected cross correlation analysis of AGNs and galaxies at redshifts from 0.3 to 3.0 using the JVO to obtain the Subaru Suprime-Cam images and UKIDSS catalog data around AGNs (Shirasaki et al. 2011). We investigated 1,809 AGNs, which is a sample that is about ten times larger than the one used in previous studies of AGN-galaxy clustering at redshifts larger than 0.6. We found significant excess of galaxies around the AGNs at redshifts from 0.3 to 1.8. We also found that AGNs at higher redshift ranges reside in a denser environment than lower redshift AGNs. It was successfully demonstrated that the use of the archive through the Virtual Observatory system can provide efficient research environment for utilizing a large amount of dataset.

    Acknowledgments.   This work was supported by Grant-in-aid for Information Science (21013048) carried out by the Ministry of Education, Culture, Sports, Science and Technology (MEXT) of Japan. YS is grateful for support under Grant-in-aid for Young Scientists (B) (21740143) carried out by the MEXT of Japan.

References

Shirasaki, Y., Tanaka, M., Ohishi, M., Mizumoto, Y., Yasuda, N., & Takata, T. 2011, PASJ, in press. `0907.5380`
Shirasaki, Y., et al. 2006, in Advanced Software and Control for Astronomy, edited by H. Lewis, & A. Bridger (Bellingham, WA: SPIE), vol. 6274 of Proc. SPIE, 62741D
— 2007, in Astronomical Data Analysis Software and Systems XVI, edited by R. A. Shaw, F. Hill, & D. J. Bell. (San Francisco, CA: ASP), vol. 376 of ASP Conf. Ser., 16
— 2009, in Astronomical Data Analysis Software and Systems XVIII, edited by D. A. Bohlender, D. Durand, & P. Dowler (San Francisco, CA: ASP), vol. 411 of ASP Conf. Ser., 396
Tanaka, M., et al. 2008, in Astronomical Data Analysis Software and Systems XVII, edited by R. W. Argyle, P. S. Bunclark, & J. R. Lewis. (San Francisco, CA: ASP), vol. 394 of ASP Conf. Ser., 261

Astronomical Data Analysis Software and Systems XX
ASP Conference Series, Vol. 442
Ian N. Evans, Alberto Accomazzi, Douglas J. Mink, and Arnold H. Rots, eds.
© *2011 Astronomical Society of the Pacific*

Automated X-ray and Optical Analysis of the Virtual Observatory and Grid Computing

A. Ptak,[1] S. Krughoff,[2] and A. Connolly[2]

[1]*NASA/GSFC Code 662, Greenbelt, MD 20771*

[2]*University of Washington, Physics/Astronomy Bldg Room C319, 3910 15th Ave NE, Seattle, WA 98195-1580*

Abstract. We are developing a system to combine the Web Enabled Source Identification with X-Matching (WESIX) web service, which emphasizes source detection on optical images, with the XAssist program that automates the analysis of X-ray data. XAssist is continuously processing archival X-ray data in several pipelines. We have established a workflow in which FITS images and/or (in the case of X-ray data) an X-ray field can be input to WESIX. Intelligent services return available data (if requested fields have been processed) or submit job requests to a queue to be performed asynchronously. These services will be available via web services (for non-interactive use by Virtual Observatory portals and applications) and through web applications (written in the Django web application framework). We are adding web services for specific XAssist functionality such as determining the exposure and limiting flux for a given position on the sky and extracting spectra and images for a given region. We are improving the queuing system in XAssist to allow for "watch lists" to be specified by users, and when X-ray fields in a user's watch list become publicly available they will be automatically added to the queue. XAssist is being expanded to be used as a survey planning tool when coupled with simulation software, including functionality for NuStar, eRosita, IXO, and the Wide-Field Xray Telescope (WFXT), as part of an end-to-end simulation/analysis system. We are also investigating the possibility of a dedicated iPhone/iPad app for querying pipeline data, requesting processing, and administrative job control. This work was funded by AISRP grant NNG06GE59G.

1. Overview

- WESIX is a web service that runs SExtractor on user-supplied images and cross-matches the source list with catalogs available through SkyPortal. Cross-matched sources can be used to correct the astrometry of the input image.

- XAssist[1] is a program written in Python that automatically processes X-ray data from Chandra, XMM-Newton, and Suzaku using their underlying software systems (CIAO, XMM-SAS and HEASOFT).

- XAssist web services have been developed to query existing pipeline processing of X-ray fields and to request detailed processing of new fields. A quick-look option is under development.

---

[1]http://www.xassist.org

- The final WESIX/XAssist web service will run WESIX on optical images, and run quick-look or query full processing of fields as shown in the flow chart (Figure 1) for corresponding X-ray fields. Since the functionality can be called remotely, the system can be distributed to run on a cluster (and more generally in grid computing).

- Every service will be callable from web applications that we are developing (using the Python framework Django), and possibly also exposed as a RESTful service (which makes it easy for users to execute queries from shell scripts, etc.).

- The combined X-ray and optical source lists will be cross-matched and proper *upper-limits will be computed for unmatched sources.*

- Advanced functionality will include requesting X-ray images, spectra and light curves to be extracted for specified regions.

- We are developing a web application to expand the HEASARC "Data Notifier" concept to also include preliminary analysis, i.e., a "watch list" of sources or regions can be stored, and data that has recently become public with coverage of these regions will be analyzed automatically.

- The automated data analysis capability of this system makes it an ideal back end for a survey planning tool, i.e., to optimize surveying strategy for existing/near-future missions such as Chandra, XMM, and NuStar, and for optimization of proposed missions such as WFXT.[2]

2. XAssist

XAssist reprocesses X-ray data (starting from the "raw" data, in order to apply the most recent calibration), creates detector and exposure maps, and detects sources. It then characterizes the sources by fitting an elliptical Gaussian model to stamp images for each source, which serves to both establish source extent and to define regions for extracting source counts. Sources can also be added manually by listing positions (with optional extents) in a text file for each detector. By default processing occurs over the full energy range but the analysis can be limited to an inputted energy range. Aperture photometry is then used to compute fluxes, and spectra, images and light curves are extracted for each source. "Hard" and "soft" band images (by default in the 0.5–2.0 and 2.0–10.0 keV ranges) are also extracted and hardness ratios are computed for each source. If there are sufficient counts, a simple power-law model is fit to the spectra. K-S tests are performed comparing each source light curve to background light curves. Finally, the sources are optionally cross-matched with several catalogs at HEASARC. See http://www.xassist.org for more details concerning these core operation functions of XAssist. The cross-matching capability will be inherently drastically expanded when XAssist is called via the combined XAssist/WESIX web service sine WESIX more advanced cross-matching capability will be automatically included (see below).

Recently developed features include the ability to limit the processing to a given "region of interest". The intent here is to speed processing when only a given source or region is relevant to the user. This is currently only accessible when running XAssist

---

[2]http://wfxt.pha.jhu.edu

locally but will be added to the web service functionality, particularly when the user watch list functionality is available.

2.1. Queue Processing

XAssist has been used to process Chandra and XMM-Newton fields in pipelines available at http://www.xassist.org. Originally this processing proceeded using simple shell scripts (most notably the "run_all.sh" and "run_xassist.csh" scripts currently still found in the XAssist download). This obviously did not allow for any control of the processing. The new system uses a client/server model with the queue jobs, dataset status, and client status being stored in MySQL tables via sqlalchemy. The server and client are Python scripts which communicate with web services so the clients can be running on the different machines than the server and hence processing can be distributed. Storing the state of the server (where the "state" consists of the queue and the list of active clients) in MySQL allows for a degree of fault tolerance since the system can be easily restarted (e.g., after a power failure) and processing will resume automatically. This approach also allows for jobs to be added to queue (or more generally manipulated) independently of the server application, if necessary (for example, via a Django web application). **Processing can be requested and pipeline data can be queried and downloaded using the xarpc.py XML-RPC client available at http://www.xassist.org**.

3. WESIX

WESIX (Web Enabled Source Identification with XMatching is currently under development at the University of Washington[3] in conjunction with the Virtual Observatory (VO) with funding from the NASA AISR program for the years 2001–2004. WESIX is aimed at enabling users to easily extract source locations and fluxes from imaging data with the optional functionality of cross-matching the resulting catalog with pre-existing data available through the VO. To accomplish this goal, SExtractor is exposed as a callable XML-RPC (originally SOAP) web service enabling a user to specify an image to work on as well as the parameters for source detection and output physical parameters. Cross-matching is accomplished using the VO catalog repository called *SkyPortal*.[4] The SkyPortal web service allows WESIX to upload the SExtractor source catalog as a temporary table in the repository. WESIX then executes an Astronomical Data Query Language (ADQL) call to return matching sources in other catalogs. Any catalog published with full ADQL compatibility may be cross-matched providing a multifrequency catalog of matched sources (together with ancillary catalogs of the sources within each data set that did not match). WESIX is now written in Python which should improve interoperability with XAssist (although web services are by definition intended to be platform-independent).

4. Combined XAssist/WESIX Web Service and Web App

Figure 1 shows a flow chart for our current implementation plan. When an optical image is supplied processing continues with the existing WESIX implementation. When an X-

---

[3]http://nvogre.astro.washington.edu:8080/wesix/

[4]http://openskyquery.org

ray field is requested and the field has already been processed then the existing XAssist source list is returned. In a later implementation user-supplied options will be allowed that will result in custom processing, for example to determine the flux for an extended source. We have been experimenting with using SExtractor for rapid source detection for the case when an X-ray field is requested but no processing has occurred yet. In this case a job request will be submitted to the queue for full processing and the SExtract X-ray source list will be returned in the meantime since full processing can take on the order of hours, depending on the number of sources in the field. So far we have been getting good results, with the XAssist and SExtractor fluxes being well correlated.

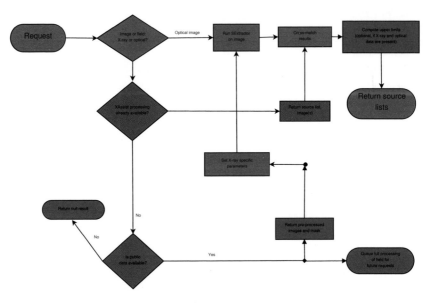

Figure 1.   Flow chart showing the basic stages of execution when the system receives a request to analyze an optical image or X-ray dataset.

An administrative interface to the XAssist queue using the Django web application toolkit has been written and allows the administrator to check which jobs are in the queue and to select fields to be resubmitted for reprocessing (i.e., to apply bug fixes or repeat attempts at processing when network or other errors occurred).

5.   Future

The combined XAssist/WESIX web service should be available in the next six months. We have started to develop a Django front-end web application to allow processing requests for either X-ray or optical data. We have started development of a "watch-list" service to allow users to register and list positions or source names of interest. The X-ray archives at HEASARC will then be periodically queried and new data will be automatically processed when available and users will be notified via email. This functionality will also be incorporated into the web application.

Astronomical Data Analysis Software and Systems XX
ASP Conference Series, Vol. 442
Ian N. Evans, Alberto Accomazzi, Douglas J. Mink, and Arnold H. Rots, eds.
© *2011 Astronomical Society of the Pacific*

Development of a VO Registry Subject Ontology Using Automated Methods

Brian Thomas

National Optical Astronomy Observatory, Tucson, AZ 85719

Abstract.   We report on our initial work to automate the generation of a domain ontology using subject fields of resources held in the Virtual Observatory registry. Preliminary results are comparable to more generalized ontology learning software currently in use. We expect to be able to refine our solution to improve both the depth and breadth of the generated ontology.

1. Introduction

Ontologies promise a rich user interaction with large amounts of data. They may be used to map the heterogeneous semantics which various data repositories use to label their data into a common ontology (or set of ontologies) which describe the aggregate of all available data. This common ontology may in turn then be used to create complex queries which can precisely describe the data of interest using concepts which are familiar to the end user scientist.

The development of such an ontology is a non-trivial matter, however. Problems include the amount of human effort required to both populate and keep up to date individuals (instances) of the ontology as more data may be added after the initial ontology is developed. Furthermore, there are maintenance costs associated with maintaining the ontology itself. The semantics in use at the various data repositories will evolve (e.g., new classes of subjects are added) and the common ontology must evolve to encompass these changes.

The Virtual Observatory (VO) registry presents an interesting test case for developing automated methods to do these tasks. The VO registry contains approximately 30,000 resources, which are simply elements of the VO, such as organizations, data collections or services, that can be described in terms of who curates or maintains it and which can be given a name and a unique identifier. Each resource is allowed to be labeled with one or more subject fields. While its entries each conform to a prescribed data model, the semantics of the VO registry data model (Hanisch et al. 2007) which describe the content (subject) of the data are not constrained and publishers are free to label the subject of the data as they wish.

Our motivation is to create a subject ontology for VO resources which will expand the applicable search results beyond a simple matching against existing terms, taking into account synonyms and hypernyms (parent concepts). For example, a search for resources with the subject of "star" should also turn up all resources not explicitly labeled, such as resources which have sub-classed star subjects like "early-type stars" or "Wolf-Rayet". To enable such a query, we plan to map existing resources into an subject ontology.

2. Methodology

There are many approaches to the generation of ontologies in the literature. Effort is generally directed towards developing generalized solutions, which may handle generation of an ontology from any selected corpus of text regardless of the domain(s) to which they may belong. In our case, we are harvesting information about VO registry resources via their subject fields, and this supplies some advantages not enjoyed by others. First, the subject text typically holds only one or more mostly noun keywords rather than whole sentences or paragraphs (~ 5% of all subject fields contain sentences). We may further assume that all subject text belongs to the same domain. In other words, it is reasonable to assume, for example, that the term "star" is always considered to be an astrophysical object rather than meaning "an asterisk". This assumption, coupled with ignoring the small amount of subject text which are more than keywords, allows us to side-step the use of fancier methods to extract concepts.

Lonsdale et al. (2010) have outlined a general methodology for ontology generation which we have adopted here. They describe a series of primary steps which involve first the selection of concepts and then the retrieval of relationships from a corpus of base text documents and a single source ontology (we don't pursue their last step of constraint discovery here).

We have chosen to use the IVOA Thesaurus, "IVOAT" (Hessman 2008), as the basis for our source ontology. This is an ideal choice as it covers a broad range of concepts in Astronomy similar to the range of subjects in the VO registry. The IVOAT is serialized as a SKOS vocabulary, so to produce the ontology we have used an XSLT stylesheet to transform it, using a simple mapping of transforming SKOS concepts directly into OWL classes, importing SKOS broader relationships to create an "is-a" hierarchy and the `prefLabel` and `altLabel` elements for each concept to record any known synonyms.

In this work, concept selection involves the harvesting of subject text into a corpus of unique instances after filtering out sentence text. This produces a list of about 1100 text instances. We next utilize a simple tokenizer to extract subject concepts from the corpus. Tokenization parses out concepts from text using a small set of regular expressions we have developed. These expressions serve to parse concepts from comma, semi-colon, or space delimited lists, can change casing of concepts from plural, and reformat concept text into standard English from specialized formatting ("star:binary" to "binary_star" for example). We then filter this list to drop any unusual acronyms and/or contractions (such as "cdfsagncxo"). We have adopted the filtering methodology of Yang & Callan (2008), and our filtering is done by first referencing a small local domain dictionary (of Astronomical terms) followed by queries to Wikipedia and then WordNet (we differ from Yang & Callan in that they utilized Google search instead). Any word we fail to identify in any one or more of these sources, is filtered out of candidate subject concepts list. Filtering in this manner results in a list of $\approx$ 450 concepts.

We assemble the list of filtered concepts into an initial, "flat" ontology (i.e., all concepts become classes which inherit from `owl:Thing`) and proceed to merge this ontology with our source ontology, by either making direct lexical matches between

Figure 1.   Extragalactic and stellar object portions of the generated subject ontology. Red underlined items indicate concepts contained within the subject corpus, other subjects are pulled from the IVOAT source ontology during the merge stage.

named classes (or their synonyms), or by indirect matching using hypernyms from the domain dictionary to identify any possible superclasses in the ontology we might use.

3.   Results

Figure 1 shows a portion of our subject ontology generated, with the classes underlined in red representing those classes which have made direct matches to subject concepts. Classes lacking underlining are simply imported from the source IVOAT ontology. This figure shows that for stellar and extragalactic concepts we have achieved some success, but this diagram only represents a small fraction of the corpus of subject terms ($\sim 3\%$). How do we measure our work more meaningfully?

There are a number of more quantitative measures which one might use to gauge the performance of ontology learning software (i.e. software which auto-generates ontologies). Zouaq & Nkambou (2009) give a review of many current measures. Because this is a preliminary work, we have chosen to simply apply a structural evaluation, similar to the class match measure of Alani & Brewster (2006), which evaluates the coverage of an ontology of the "sought terms".

To obtain a scoring result for this measure, we evaluated all subject concepts which were successfully merged into source ontology. For each direct match between a class in the ontology and a subject concept we give a value of 1 (a direct match) and a value of 0.5 for an indirect match which occurs when we match a subject class by using a parent concept of a subject. The overall score is a ratio of the sum of these values divided by the number of subject concepts we made available for ontology generation. This score may thus range from 0 (no matches whatsoever) to 1 (all concepts directly matched). Using this measure we obtained a score of 0.32, which is low. Current ontology learning software averages about 0.3–0.5 (and sometimes even better) when generating an ontology from a corpus of 700 to 1000 sentences (see Zouaq & Nkambou 2009).

Where might problems lie in our approach? A deeper look at the body of subject concepts shows that there are still some failures at parsing the grammar in the subject corpus, and sometimes we have split the text too far as for "solar system" which becomes the separate concepts "solar" and "system". Other problems which lower the score include failed dictionary lookups for synonyms or hypernyms (such as for "globular_cluster" vs. IVOAT class "globularStarCluster") as well as having subject concepts which do not exist in the source ontology, nor have any matching hypernyms and therefore cannot be merged in.

Nevertheless, by this scoring measure, this software has a performance comparable to the lower end of the current average ontology learning software available.

4. Summary

We have shown that some reasonable progress may be made towards the automated generation of a subject ontology for the VO registry. Results show that we have comparable ballpark performance to more generalized solutions which construct ontologies from text corpi.

Because we are operating in a single domain, which contains many specialized concepts, we should be able to outperform these solutions. Possible directions to help increase the depth and breadth of the subject ontology include using WordNet to enhance the dictionary lookups of synonyms and hypernyms during the concept selection and merging stages, improving our local Astronomical dictionary to include more technical terms and hypernyms and using additional source ontologies, such as Ontology of Astronomical Object Types (Derriere et al. 2009).

References

Alani, H., & Brewster, C. 2006, in Proceeding of 4th International EON Workshop, 15th International World Wide Web Conf. (Edinburgh: WWW)
Derriere, S., et al. 2009, Ontology of astronomical object types. URL http://www.ivoa.net/internal/IVOA/IvoaSemantics/OWLDOC-ObjectTypes_tar.gz
Hanisch, R., et al. 2007, Resource Metadata for the Virtual Observatory Version 1.12. URL http://www.ivoa.net/Documents/latest/RM.html
Hessman, F. 2008, A modified version of the 1993 iau thesarus. URL http://volute.googlecode.com/files/ivoa-vocab-1.0.tar.gz
Lonsdale, D., Embley, D. W., Ding, Y., Xu, L., & Hepp, M. 2010, Data and Knowledge Eng., 69, 318
Yang, H., & Callan, J. 2008, in Proceedings of the 2008 international conference on Digital government research, DG.O
Zouaq, A., & Nkambou, R. 2009, Trans. Knowl. Data Eng., 21, 1559

Astronomical Data Analysis Software and Systems XX
ASP Conference Series, Vol. 442
Ian N. Evans, Alberto Accomazzi, Douglas J. Mink, and Arnold H. Rots, eds.
© 2011 *Astronomical Society of the Pacific*

Infrastructure and the Virtual Observatory

Patrick Dowler, Séverin Gaudet, and David Schade

Canadian Astronomy Data Center,
Herzberg Institite of Astrophysics,
National Research Council Canada,
5071 West Saanich Road, Victoria, B.C. V9E 2M7 Canada

Abstract. The modern data center is faced with architectural and software engineering challenges that grow along with the challenges facing observatories: massive data flow, distributed computing environments, and distributed teams collaborating on large and small projects. By using VO standards as key components of the infrastructure, projects can take advantage of a decade of intellectual investment by the IVOA community. By their nature, these standards are proven and tested designs that already exist. Adopting VO standards saves considerable design effort, allows projects to take advantage of open-source software and test suites to speed development, and enables the use of third party tools that understand the VO protocols. The evolving CADC architecture now makes heavy use of VO standards. We show examples of how these standards may be used directly, coupled with non-VO standards, or extended with custom capabilities to solve real problems and provide value to our users. In the end, we use VO services as major parts of the core infrastructure to reduce cost rather than as an extra layer with additional cost and we can deliver more general purpose and robust services to our user community.

1. Introduction

Implementing Virtual Observatory (VO) standards saves time. For all the custom parts of a system architecture we must design and implement. With International Virtual Observatory Alliance (IVOA) standards, the designs are already done by a broad community and the prototyping has proven the designs; we just have to implement and open-source software can help.

Open source software developed at the Canadian Astronomy Data Center (CADC) for implementing IVOA standards is available in the OpenCADC project.[1]

2. Core Services

Credential Delegation Protocol (CDP, Graham et al. 2010b), Group Membership Service (GMS, proposed standard), Table Access Protocol (TAP, Dowler et al. 2010), and VOSpace (Graham et al. 2010a) form the core infrastructure for new CADC user services. In addition, we use the Universal Worker Service (UWS) pattern (Harrison &

---

[1]Currently hosted at `http://code.google.com/p/opencadc/`

Rixon 2010) and common code to implement synchronous and asynchronous web services as well as synchronous web applications.

Users query the archives using TAP — directly, via a web UI, or one of the simpler IVOA services: Simple Image Access Protocol (SIAP, Tody & Plante 2009) or Simple Spectral Access Protocol (SSAP, Tody et al. 2008) for access to calibrated image and spectroscopic data respectively. All astronomical data (files) found via queries and all files stored in VOSpace are delivered by the data web service.

Users store, browse, and retrieve files using VOSpace. Project teams are defined in GMS; teams collaborate by sharing files in VOSpace. The CDP service allows users with X.509 certificates to create and store a short-term proxy certficate so that CADC web services can perform actions on their behalf. For example, a TAP request can persist query results directly to a VOSpace, the SIA service creates and executes a TAP query in authenticated fashion in order to see proprietary content (tables or rows) visible to that user.

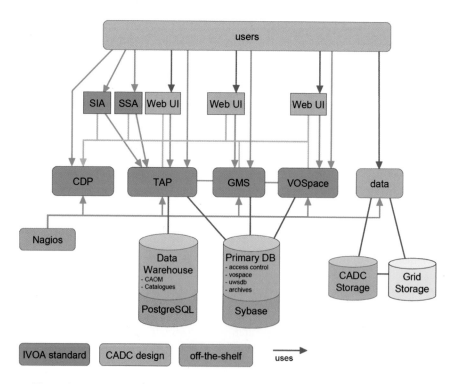

Figure 1.   CADC Web Services Architecture.

The CADC implementation of CDP allows users who login with username and password to use internally generated certificates; this allows the Web UI applications to access web services like TAP, GMS, and VOSpace in authenticated fashion. All authorization checking is delegated to the underlying web services, thuis keeping the user interface applications simple and focussed on presentation and usability. We also realise a large benefit in the implementation of web services since they only accept certficate authentication and are easily deployed on separate systems for operational reasons.

3. Client Software

Using IVOA standards for web services allows users to use many third-party client tools that understand these protocols.[2] Power users can write their own scripts and tools to automate their interactions with web services; using IVOA standards helps these users by letting them use libraries and example code from the community.

Once users become familiar with the system by using the Web UI, they often move on to using the web services directly. By using these web services at the core, we easily ensure a consistent experience for our users. For example, users can put files into VOSpace using a command-line tool and see it through the associated Web UI, or they can perform a simple query through the SIA service, the Web UI, or TAP and get consistent results.

4. User Interface

Web UI applications use web services to perform most of the heavy lifting and concentrate on presentation. Since they interact with web services using standard protocols, they are isolated from implementation details and do not have to worry about protocol changes.

AdvancedSearch (web UI for CAOM tables in TAP) generates an ADQL (Ortiz et al. 2008) query from form input and executes the query using the TAP service; it is isolated from many technical details that are dealth with entirely in TAP. The query result is displayed using an off-the-shelf VOTable (Ochsenbein & Williams 2009) viewer based on XSLT and Javascript. This gives us rich query and result display with low effort by using standards: TAP, ADQL, VOTable. Group Management (web UI for GMS) allows users to create and maintain groups of users.

VOSpaceBrowser (web UI for VOSpace) allows users to graphically browse the VOSpace and download files. Although owners of content in VOSpace must have X.509 certificates, users who browse via the UI can login via certficate or username/password; in either case, the UI uses CDP to get the necessary credentials to access the VOSpace service on their behalf: VOSpace users can safely share their files with team members who do not have X.509 certficates (but do have CADC accounts and are members of the team).

5. Authentication

Web UIs support anonymous access (http), login (username/password authentication on http), and X.509 certificate authentication (https). Web Services support anonymous (http) and X.509 certificate authentication (https) following the IVOA Single-Sign-On profile (SSO, Rixon et al. 2008).

---

[2]See `http://www.ivoa.net/newsletter/` for news about these tools

6. Operational Support

Virtual Observatory Support Interface (VOSI, Graham et al. 2010c) defines a small set of common web service resources to make monitoring and gathering metadata uniform. The *availability* resource reports on the current health and functionality of the service through internal diagnostic tests. We have implemented this resource in all our web services and web applications with a small set of re-usable tests, such as checking that a database is accessible and working, checking that storage space is available, and checking that other web services that are needed are also available. Operations staff have implemented a Nagios plugin that checks the output of the *availability* resource and reports issues immediately.

7. Acronyms

ADQL	Astronomical Data Query Language
CADC	Canadian Astronomy Data Center
CAOM	Common Archive Observation Model
CDP	Credential Delegation Protocol
GMS	Group Membership Service
TAP	Table Access Protocol
VOSI	Virtual Observatory Support Interface
VOSpace	Virtual Observatory Space
VOTable	Virtual Observatory Table
UI	User Interface
XSLT	Extensible Stylesheet Language Transformations

References

Dowler, P., Rixon, G., & Tody, D. 2010, Table Access Protocol, Version 1.0. URL http://www.ivoa.net/Documents/TAP/

Graham, M., Morris, D., Rixon, G., Dowler, P., Schaaf, A., & Tody, D. 2010a, Vospace specification (working draft), version 2.0. URL http://www.ivoa.net/Documents/VOSpace/

Graham, M., Plante, R., Rixon, G., & Taffoni, G. 2010b, Ivoa credential delegation protocol, version 1.0. URL http://www.ivoa.net/Documents/CredentialDelegation/

Graham, M., et al. 2010c, Ivoa support interfaces, version 1.0. URL http://www.ivoa.net/Documents/VOSI/

Harrison, P., & Rixon, G. 2010, Universal worker service pattern, version 1.0. URL http://www.ivoa.net/Documents/UWS/

Ochsenbein, F., & Williams, R. 2009, VOTable Format Definition, Version 1.2. http://www.ivoa.net/Documents/VOTable/

Ortiz, I., et al. 2008, Ivoa astronomical data query language, version 2.0. URL http://www.ivoa.net/Documents/latest/ADQL.html

Rixon, G., et al. 2008, Ivoa single-sign-on profile: Authentication mechanisms, version 1.01. URL http://www.ivoa.net/Documents/latest/SSOAuthMech.html

Tody, D., & Plante, R. 2009, Simple image access specification, version 1.0. URL http://www.ivoa.net/Documents/SIA/

Tody, D., et al. 2008, Simple spectral access specification, version 1.04. URL http://www.ivoa.net/Documents/latest/SSA.html

Part XIII

Visualization

Astronomical Data Analysis Software and Systems XX
ASP Conference Series, Vol. 442
Ian N. Evans, Alberto Accomazzi, Douglas J. Mink, and Arnold H. Rots, eds.
© 2011 Astronomical Society of the Pacific

Build YOUR All-Sky View with Aladin

A. Oberto, P. Fernique, T. Boch, and F. Bonnarel

Centre de données astronomiques de Strasbourg, CDS, France

Abstract. From the need to extend the display outside the boundaries of a single image, the Aladin team recently developed a new feature to visualize wide areas or even all of the sky. This all-sky view is particularly useful for visualization of very large objects and, with coverage of the whole sky, maps from the Planck satellite. To improve on this capability, some catalogs and maps have been built from many surveys (e.g., DSS, IRIS, GLIMPSE, SDSS, 2MASS) in mixed resolutions, allowing progressive display. The maps are constructed by mosaicing individual images. Now, we provide a new tool to build an all-sky view with your own images. From the images you have selected, it will compose a mosaic with several resolutions (HEALPix tessellation), and organize them to allow their progressive display in Aladin. For convenience, you can export it to a HEALPix map, or share it with the community through Aladin from your web site or eventually from the CDS image collection.

1. Introduction

From a set of private images (obtained for example from observations), Aladin will compose an all-sky mosaic to produce a global view (Fernique et al. 2010). To do this, it will resample all of the pixels from the images onto a HEALPix (Hierarchical Equal Area isoLatitude Pixelization, Górski et al. 2005) tessellation. This result is saved on the user's computer together with necessary metadata for a multi-scale visualization through Aladin. HEALPix is a scheme that divides the celestial sphere into twelve pixels (rhombus shape). It produces a geometrically constructed, self-similar and refinable quadrilateral mesh on the sphere.

2. Your Set of Images

With this new feature, the user can ask Aladin to build an All-Sky view from a local set of images. There are several constraints:

- Those images must be in FITS format and must have an astrometric calibration;

- It is assumed that the bitpix value is the same for the whole set of images;

- In order to obtain a better visualization, we recommend having a photometric image calibration, or at least having homogeneous data.

Aladin will process only parts of the sky where images are present. It is not necessary to have all-sky coverage, or even that the images be contiguous. They can be overlapping.

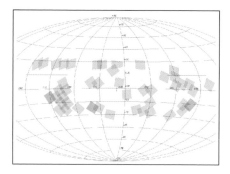

Figure 1.   Example of the coverage of YOUR images in order to build an All-Sky view.

3.   Computation

Once the images are selected, the user may provide output specifications. Aladin suggests a default mosaic building procedure that can be adjusted according to your needs. Depending on the amount of data, the computation may require a very long time and huge disk space. In order to reduce both of these, less consuming parameters can be selected (with consequent loss of data accuracy). First of all, Aladin will browse the data to index the images by the HEALPix diamond number they contribute to in the selected resolution. Then, it will fill each affected diamond with pixel values computed from the images.

All of the computation is done on the user's computer, so the speed will depend on its processor(s). All generated files will be stored on its hard disk.

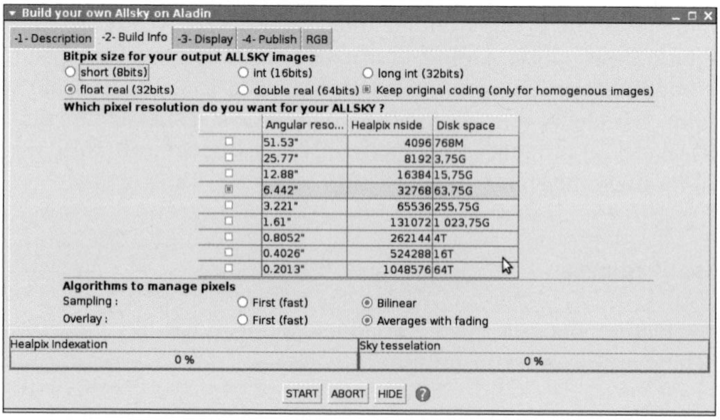

Figure 2.   Panel of parameters of the All-Sky building.

3.1.   Bitpix and Data Values

Aladin will create new FITS files with resampled pixels. Thus the disk space occupied by the data will be duplicated (except for overlaps) for the level with the same resolution as the data. More files are created for the lower resolution levels until we reach the all-sky level. If it is not necessary to keep the original data values, a smaller bitpix value for output can be chosen in order to reduce the disk space used. Conversely, Aladin could keep the original coding.

3.2. Angular Resolution

The default behavior is to produce maps with approximately the same angular resolution as the original images, and another with lower resolution (for progressive display). The resolution is determined by the HEALPix mesh, which divides each pixel into 4 while downsizing the scaling factor. As above, in order to reduce the disk space, a less accurate angular resolution can be selected (in the table shown in Figure 2 we give an estimation of the total size of the created files, for an all-sky coverage).

3.3. Pixel Manipulation

In this version, during the resampling step Aladin suggests two methods: nearest neighbor or bilinear weighting of neighboring pixels. If there is an overlap of several images, it is possible to keep the value of only one of them, or to choose an average. In the future, it will be possible to develop your own Aladin plugin to perform a specific pixel manipulation.

3.4. Output

While the all-sky view is building, a visualization of the current state is shown in Aladin. Eventually, it will be possible to use this all-sky view like a real image in Aladin, and adjust the grey levels for the best visualization. This choice can be stored by converting all the FITS files into JPG files.

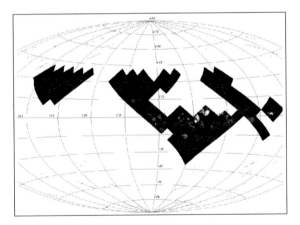

Figure 3.   Example of a preview while processing the diamonds.

4. Public/Private Distribution

All of the generated files are stored on the user's computer. The newly created directory can be opened in Aladin at any time to visualize the result. According to the user's needs, one may want to share this all-sky view. First of all, this dedicated directory could be copied anywhere to share it; and anyone could read it within Aladin. If this directory is placed behind a http access, it could be distributed to a larger audience (and may be published to all Aladin users, if the information is sent to Aladin). To use with other software, it is possible to export the all-sky view in a HEALPix map file, or crop the view in a flat FITS file.

Of course, in Aladin, all capabilities remain available, like superimposing catalogs, displaying additional images (or all-sky) with transparency, RGB composition, etc.

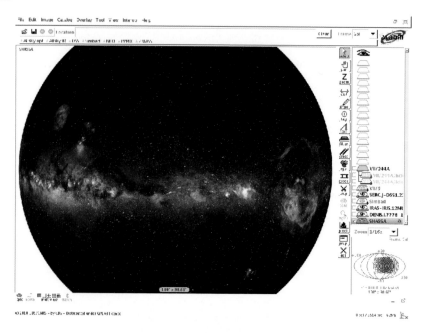

Figure 4.   Aladin display with an All-Sky view.

References

Fernique, P., Boch, T., Oberto, A., & Bonnarel, F. 2010, in Astronomical Data Analysis Software and Systems XIX, edited by Y. Mizumoto, K.-I. Morita, & M. Ohishi (San Francisco, CA: ASP), vol. 434 of ASP Conf. Ser., 163

Górski, K. M., Hivon, E., Banday, A. J., Wandelt, B. D., Hansen, F. K., Reinecke, M., & Bartelmann, M. 2005, ApJ, 622, 759

Astronomical Data Analysis Software and Systems XX
ASP Conference Series, Vol. 442
Ian N. Evans, Alberto Accomazzi, Douglas J. Mink, and Arnold H. Rots, eds.
©*2011 Astronomical Society of the Pacific*

Building a Smart Portal for Astronomy

Sébastien Derriere and Thomas Boch

CDS, Observatoire astronomique de Strasbourg, Université de Strasbourg,
CNRS, UMR 7550, 11 rue de l'Université, F-67000 Strasbourg, France

Abstract. The development of a portal for accessing astronomical resources is not
an easy task. The ever-increasing complexity of the data products can result in very
complex user interfaces, requiring a lot of effort and learning from the user in order to
perform searches. This is often a design choice, where the user must explicitly set many
constraints, while the portal search logic remains simple. We investigated a different
approach, where the query interface is kept as simple as possible (ideally, a simple
text field, like for Google search), and the search logic is made much more complex
to interpret the query in a relevant manner. We will present the implications of this
approach in terms of interpretation and categorization of the query parameters (related
to astronomical vocabularies), translation (mapping) of these concepts into the portal
components metadata, identification of query schemes and use cases matching the input
parameters, and delivery of query results to the user.

1. Introduction: the CDS Portal

The CDS provides three main services: SIMBAD, Aladin and VizieR and each has one
or several dedicated query pages. The CDS portal[1] is an effort to provide simultaneous
access to these services through a unique and simple query page. The input form is a
single text field, where users can type either an object name or sky coordinates. When
the user performs a query, SIMBAD, Aladin and VizieR are queried simultaneously,
and the different results are aggregated into a single result page (see Fig. 1).

If the input string is recognized as a valid object name, the corresponding object
and its coordinates are displayed just below the text input field. This is done on-the-fly
using AJAX and the Sesame name resolver.

The results will be slightly different, depending on the input being an object name
or sky coordinates (Fig. 2). If the query contains an object name, additional data are
retrieved from SIMBAD, and more links are suggested.

2. Towards a Smart Portal

The next step will be to add additional capabilities to the portal, allowing the user to
input parameters other than object names or coordinates, and to use these parameters in
the context of the various services.

---

[1]`http://cdsportal.u-strasbg.fr/`

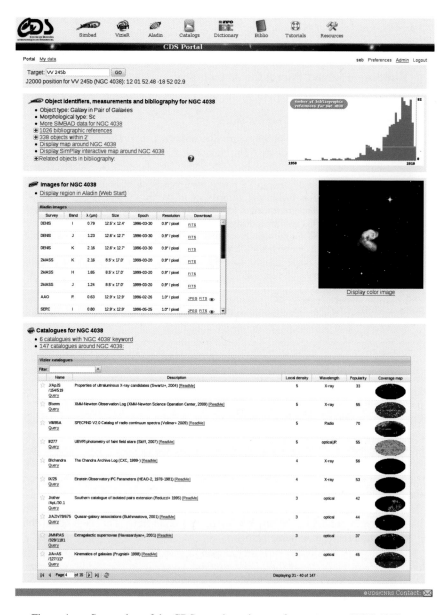

Figure 1.   Screenshot of the CDS portal result page for a query on NGC 4038.

We will try to keep the interface as simple as possible, with a single text input field where users will type their queries in natural language. The portal will dispatch queries to a list of individual services, and aggregate the results in a single page. For this, we need 3 elements: 1) parameters categorization (using vocabularies or services);

SIMBAD **Object identifiers, measurements and bibliography for 12 01 52.48 -18 52 02.9**

⊞ 338 objects within 2'
- Display map around 12 01 52.48 -18 52 02.9
- Display SimPlay interactive map around 12 01 52.48 -18 52 02.9

SIMBAD **Object identifiers, measurements and bibliography for NGC 4038**

- Object type: Galaxy in Pair of Galaxies
- Morphological type: Sc
- More SIMBAD data for NGC 4038
⊞ 1026 bibliographic references
⊞ 338 objects within 2'
- Display map around NGC 4038
- Display SimPlay interactive map around NGC 4038
⊞ Related objects in bibliography:

Figure 2.   CDS portal query output examples in cases where the search was done on sky coordinates (top), or on object name (bottom).

2) mapping from reference vocabularies to service keywords; 3) query templates for each service.

2.1.   Parameters Categorization

We have identified a few categories of terms that can be used in queries to an astronomical portal:

- Astronomical Object Names,
- Sky Coordinates,
- Astronomical Object Types,
- Measurement,
- Author Name,
- Instrument, Mission,
- Dataset Name.

The challenge is to identify in the natural language input text which terms correspond to which categories, and to tag them accordingly. For each category, we will need a standard vocabulary or service allowing us to tag the parameters. For example, the Sesame name resolver can identify astronomical object names in the text query. Vocabularies can be used to identify astronomical object types or measurements, etc.

2.2.   Vocabularies Mapping

A reference vocabulary will be used in the portal. But the individual target services (e.g., SIMBAD, VizieR) do not need to use internally the same vocabulary. In fact, each target service, queried from the portal, already has its own internal keywords or vocabulary. What is needed is a mapping from the portal vocabulary to the services' internal vocabulary.

Example: for astronomical object types, the vocabulary used in the portal can be derived from the Ontology of Astronomical Object types (Cambrésy et al. 2010), while object types recognized by VizieR form a list of keywords, and SIMBAD has its own classification scheme. The mapping will have to translate for example "Globular Cluster" to "Globular_Clusters" (VizieR) or "GlC" (SIMBAD).

2.3. Query Templates

For each target service, we identify several query templates. A query template consists of one or several categorized input parameter(s), and a query link using these parameters.

For example, if we have an object name and a measurement:

- In SIMBAD, we can search the value of this measure for this object;
- In VizieR, we can search for catalogs containing a column corresponding to the measurement and covering the sky area containing our object.

3. Use Case

The user interface remains as simple as it can be: a simple text input. For an example search on *Virgo cluster redshift*, the following scenario unfolds:

(1) The input words are categorized — in this case, the system detects an astronomical object name and a measurement.
(2) The parameters are translated into each service's own keywords. Redshift will be translated to the UCD *src.redshift* in VizieR, and to *redshift* or *RV(Z)* (SIMBAD).
(3) All query templates involving these two categories of parameters are tested.

The user is presented with details on how the query was interpreted, together with the result, for each query template, in each service.

4. Conclusion

This approach is generic enough to be extended to services other than the CDS ones, like ADS or others. The contextual description of service capabilities, together with the definition of standard vocabularies will soon enable smart behaviors in astronomy portals, and aggregation of results from different independent services.

References

Cambrésy, L., Derriere, S., Padovani, P., Preite-Martinez, A., & Richard, A. 2010, Ontology of astronomical object types. URL http://ivoa.net/Documents/Notes/AstrObjectOntology/

Astronomical Data Analysis Software and Systems XX
ASP Conference Series, Vol. 442
Ian N. Evans, Alberto Accomazzi, Douglas J. Mink, and Arnold H. Rots, eds.
©2011 Astronomical Society of the Pacific

VisIVODesktop 3.0: An Interactive Desktop Environment for Astrophysical Visualization

A. Costa,[1] U. Becciani,[1] C. Gheller,[2] P. Massimino,[1] M. Krokos,[3] and A. Grillo[1]

[1]*INAF-Astrophysical Observatory of Catania, Italy*

[2]*Cineca, Italy*

[3]*University of Portsmouth, United Kingdom*

Abstract. The aim of scientific visualization is to create suitable visuals (images and animations) to aid scientists in easily understanding highly-complex datasets. Modern visualization can act as a catalyst in rapidly and intuitively discovering correlations and patterns in large-scale astrophysical datasets without involving numerical algorithms or CPU intensive analysis codes. We introduce VisIVODesktop 3.0, a true multi-platform environment for interactive astrophysical visualization. We outline recently added visualization functionality, namely hybrid and stereoscopic rendering. Hybrid rendering combines volume and point rendering and employs an elliptical, Gaussian distribution function. Opacity transfer functions can be finely tuned to extract interesting details within cosmological structures. We conclude with a summary of our work and future developments.

1. Introduction

Nowadays the technological advances in instrumentation and computing capability impact profoundly on the dramatic growth in the quality and quantity of astrophysical datasets obtained from observational instruments, e.g., sky surveys such as SDSS,[1] or large-scale numerical simulations such as the Millennium II Simulation.[2] The Millennium II simulation models the evolution of a meaningful fraction of the universe by means of 10 billion fluid elements (particles) interacting with each other through gravitational forces; the typical size of a snapshot being 400 Gigabytes approximately. The forthcoming next-generation astrophysical datasets are expected to exhibit massively large sizes (in the order of hundreds of Terabytes). To obtain a meaningful insight into such datasets, astronomers typically employ sophisticated data mining algorithms, often at prohibitively high computational costs. Visual data exploration and discovery tools are then exploited in order to rapidly and intuitively inspect large-scale datasets to identify regions of interest within which to apply time-consuming algorithms. Such tools are based on combining appropriately meaningful data visualizations and user interactions with them. This approach can be a very intuitive way indeed for discovering

---

[1]http://www.sdss.org/

[2]http://www.mpa-garching.mpg.de/galform/millennium-II/index.html

and understanding rapidly new correlations, similarities and data patterns. For on-going processes, e.g., a numerical simulation in progress, visual data exploration and discovery allow constant monitoring and — if anomalies are discovered — prompt correction of the run, thus saving valuable time and resources. To overcome the shortcomings of traditional exploration tools for astrophysical datasets, a new generation of software packages is now emerging, providing robust instruments in the context of large-scale astrophysical datasets (Taylor 2005; Borkin et al. 2005; Comparato et al. 2007; Becciani et al. 2010). The underlying principles are exploitation of high performance architectures — i.e., multicore CPUs and powerful graphics boards, interoperability — different applications can operate simultaneously on shared datasets, and collaborative workflows — permitting several users to work concurrently for exchanging information and sharing visualization experiences. VisIVO is an integrated suite of tools and services designed for the astrophysical community consisting of VisiVO Desktop (Comparato et al. 2007), a desktop environment for interactive visualization on standard PCs, VisIVO Server (Becciani et al. 2010), a grid-enabled platform for high performance visualization, and VisIVO Web,[3] a customized web portal supporting services based on the VisIVO Server functionality. This article introduces VisIVODesktop 3.0, the latest incarnation of VisIVODesktop, which is a true multi-platform environment for interactive astrophysical visualization. We give an overview of VisIVODesktop 3.0, outlining recently added visualization functionality, i.e., hybrid and stereoscopic rendering. We finally summarize our current work and provide pointers to future developments.

2.   VisIVODesktop 3.0

The focus in developing the original VisIVODesktop (Comparato et al. 2007) was to create an open-source environment for interactive visualization of astrophysical datasets (either real-world or numerical simulation) on standard modern PCs. Our software was wrapped around the Multimod Application Framework[4] (MAF). MAF is an open-source framework for rapid development of data analysis and visualisation applications founded on the Visualisation Toolkit.[5] The original development was under Windows XP platforms; subsequently it was ported into Linux systems, more specifically Ubuntu OS. Our intention in developing VisIVODesktop 3.0 was to create a multiplatform suite of tools for astrophysical visualization designed to perform well across a large number of underlying hardware environments, i.e., ranging from standard laptops to high-performance servers. Our development aims at fully exploiting new and emerging technologies such as the MPI paradigm for parallelization of computations and high-performance GPU computing. Another motivation for our development is to provide full compatibility with the Simple Application Messaging Protocol (SAMP, Taylor et al. 2010) which enables astrophysical software tools to interoperate and communicate seamlessly. VisIVODesktop 3.0 is based on the latest stable version of VisIVOServer (Becciani et al. 2010). The defining characteristic of VisIVOServer is that no fixed limits are prescribed for the dimensionality of data tables input for processing, thus

---

[3]http://visivoweb.oact.inaf.it/

[4]http://www.openmaf.org/

[5]http://www.vtk.org/

supporting very large-scale datasets. VisIVOServer is an open-source, multiplatform visualization environment available under Windows XP, Linux, and MacOSX. The current release supports several popular formats, e.g., ASCII, CSV, FITS tables, HDF5, Gadget2 and VOTables. Users can obtain meaningful visualizations while preserving full control of relevant parameters. Next, new visualization functionality is discussed, i.e., hybrid rendering and stereoscopy.

3. Hybrid Rendering

We introduced hybrid rendering for generating meaningful astrophysical visualizations, combining rendering of points and volumes. A point is associated with nearby voxels using an elliptical, Gaussian distribution function. An individual particle's position and velocity influence the value of this function. More specifically, a particle's smoothed effect on neighboring voxels within a given volume is calculated by using the formula (see Fig.1),

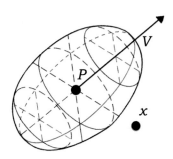

Figure 1. The elliptical Gaussian distribution function depends upon a particle's position P and velocity vector $\vec{V}$.

$$f(x) = S \cdot e^{-\alpha\left((r_{xy}/E)^2 + z^2\right)/R^2}$$

where

- r_{xy} is the distance of x in a direction perpendicular to the velocity vector $\vec{V}$,

- z is the distance of x along the velocity vector $\vec{V}$,

- R represents the radius of a particle's influence,

- E is a user-defined eccentricity factor that controls the elliptical shape of the distribution, it can be a function of $|\vec{V}|$,

- S is a scale factor, and finally,

- α is the exponent factor.

To visualize fine cosmological details within the volume dataset we found useful to employ the following opacity function: $f(x) = \tanh(\alpha \cdot x - \beta) + 1/2$. We offer the user the possibility to tune this function by controlling the parameters α and β in the rendering window. Fig. 2 demonstrates the working scenario with VisIVODesktop 3.0 (16 million particles). The rendering window displays points and volumes simultaneously.

4. Stereoscopy

The VisIVO data exploitation capability can be potentially improved by using 3D stereoscopy. For this reason we are planning to implement and test the following types of stereoscopy.

Figure 2.   A working session using VisIVODesktop 3.0.

- Interlaced stereo using interlaced left/right views: this stereo mode produces a composite image where horizontal lines alternate between left and right views. This technology will be used with polarized glasses.
- Anaglyph stereo visualization. This is to testing purposes and to support a wider set of legacy graphics cards. Anaglyph mode uses two different colors (usually chromatically opposite) to produce the 3D effect.

5. Summary

The development of visual discovery tools for rapid inspection of astrophysical datasets is of paramount importance for future astronomical research, as sizes of datasets are expected to increase dramatically. This article introduced VisIVODesktop 3.0, a true multi-platform environment for interactive astrophysical visualization of large-scale datasets. VisIVODesktop 3.0 is wrapped around Qt4.7 and VTK and ensures full SAMP compatibility. The latest version can be found at sourceforge.net. Initial feedback on VisIVODesktop 3.0 is very encouraging. Once our current development cycle is completed, we are planning a formal evaluation with specific astrophysical communities.[6]

References

Becciani, U., et al. 2010, PASP, 122, 119
Borkin, M. A., Ridge, N. A., Goodman, A. A., & Halle, M. 2005, ArXiv Astrophysics e-prints. arXiv:astro-ph/0506604
Comparato, M., Becciani, U., Costa, A., Larsson, B., Garilli, B., Gheller, C., & Taylor, J. 2007, PASP, 119, 898
Taylor, M., et al. 2010. URL http://www.ivoa.net/Documents/latest/SAMP.html
Taylor, M. B. 2005, in Astronomical Data Analysis Software and Systems XIV, edited by P. Shopbell, M. Britton, & R. Ebert (San Francisco, CA: ASP), vol. 347 of ASP Conf. Ser., 29

---

[6]e.g., the SDSS users, http://www.sdss.org.uk

Astronomical Data Analysis Software and Systems XX
ASP Conference Series, Vol. 442
Ian N. Evans, Alberto Accomazzi, Douglas J. Mink, and Arnold H. Rots, eds.
©*2011 Astronomical Society of the Pacific*

Large-Scale Astrophysical Visualization on Smartphones

U. Becciani,[1] P. Massimino,[1] A. Costa,[1] C. Gheller,[2] A. Grillo,[3] M. Krokos,[4] and C. Petta[5]

[1] *INAF-Astrophysical Observatory of Catania, Italy*

[2] *Cineca, Italy*

[3] *Consorzio Cometa, Italy*

[4] *University of Portsmouth, United Kingdom*

[5] *University of Catania, Italy*

Abstract. Nowadays digital sky surveys and long-duration, high-resolution numerical simulations using high performance computing and grid systems produce multidimensional astrophysical datasets in the order of several Petabytes. Sharing visualizations of such datasets within communities and collaborating research groups is of paramount importance for disseminating results and advancing astrophysical research. Moreover educational and public outreach programs can benefit greatly from novel ways of presenting these datasets by promoting understanding of complex astrophysical processes, e.g., formation of stars and galaxies. We have previously developed VisIVO Server, a grid-enabled platform for high-performance large-scale astrophysical visualization. This article reviews the latest developments on VisIVO Web, a custom designed web portal wrapped around VisIVO Server, then introduces VisIVO Smartphone, a gateway connecting VisIVO Web and data repositories for mobile astrophysical visualization. We discuss current work and summarize future developments.

1. Introduction

An essential part of modern astrophysical research is the necessity to employ computer graphics and scientific visualization tools for displaying appropriately multidimensional data plots and images, either from real-world observations or complex numerical simulations. The availability of large-scale ground-based instruments and space telescopes together with high-performance computational resources can produce astronomical datasets of extremely large sizes, typically requiring storage in a distributed way. This is particularly illustrated by the profound impact of modern digital sky surveys, e.g., the Sloan Digital Sky Survey[1] (SDSS), or on-going efforts for creating archives of complex numerical simulations, e.g., the European Virtual Observatory.[2]

Gaining a comprehensive insight into such datasets typically requires very sophisticated statistical and data analysis algorithms. Often, several data exploration and vi-

---

[1] http://sdss.org.uk/

[2] http://www.euro-vo.org/

sualization tools are employed in the first instance for visual discovery, that is intuitive inspection of datasets to rapidly identify regions of interest to apply time-consuming algorithms. Given current internet downloading speeds, even transferring regions of interest cut out from such large-scale datasets for further processing is prohibitively costly. A typical operational scenario is then to perform full data analysis directly on the server hosting the datasets, so that the overall communication simply consists of user queries and server responses. The variety of possible processing tools that may be needed by scientists makes this scenario sustainable only in the context of highly focused applications, depending upon the support provided by the centers hosting the datasets.

An alternative scenario is to offer advanced exploration and visualization tools to allow astrophysicists to explore their datasets appropriately to identify interesting features, then to restrict downloading for further processing to small-scale manageable datasets characterized by these features. Using appropriate lossless compression algorithms together with statistical decimation schemes, small-scale datasets encapsulating the desired features can be obtained for downloading and further processing at a user's location locally.

This article reviews VisIVO Server (Becciani et al. 2010), a grid-enabled platform for high performance visualization and the latest developments on VisiVO Web,[3] a custom designed web portal supporting services based on the VisIVO Server functionality. We then introduce VisIVO Smartphone,[4] a gateway connecting VisIVO Web and data repositories for mobile astrophysical visualization. We conclude with pointers to future developments.

2.   VisIVO Web

VisIVO Web is a network of custom designed portals for processing and exploration of astrophysical datasets founded on VisIVO Server functionality. The main characteristic of VisIVO Server is the possibility to explore astrophysical datasets with no fixed limits on dimensionality, thus offering support for large-scale datasets. VisIVO Server consists of modules implementing an **Importer**, **Filter** and **Viewer**. To create customized views of 3D renderings from astrophysical datasets, VisIVO Importer is utilized firstly to convert user datasets into an efficient internal format called VisIVO Binary Table (VBT). Then VisIVO Viewer is invoked for display. VisIVO Filters are a collection of several processing modules for constructing customised data tables from VisIVO Importer, supporting a range of operations, e.g., selection of rows/columns, scalar distribution, mathematical operations and randomization.

Currently, VisIVO Web portals are operated in Italy by INAF (Catania and Trieste Observatories) and in the United Kingdom by the University of Portsmouth. Recently, work has focused on custom-designed visual discovery tools suitable for the SDSS community implemented by connecting VisIVO Web and the SDSS server seamlessly. A typical operational scenario involves uploading, managing, processing and exploring datasets through visualizations.

---

[3]http://visivoweb.oact.inaf.it/

[4]http://visivoweb.oact.inaf.it/iphone/

Figure 1.   The VisIVOWeb application (from left to right): (a) uploading datasets, (b) exploring using filters and (c) generating visualizations.

- **Uploading / Managing.** There are two ways for uploading datasets. The first is to upload local user data (i.e., astrophysical tables residing on the PC used to access VisIVO Web). The second is to prompt the user to input the url (with username and password) of the computer system where the actual datasets are located. Once datasets are uploaded users are given full data management functionality.

- **Processing / Exploring.** The users can employ an array of VisIVO Server Filters to process their datasets, e.g., for merging data tables, extracting sub-regions, distributing scalar values throughout regularly spaced grids, decimating, interpolating, or producing new tables from sigma contours. An interactive widget is employed to aid manipulation of datasets.

- **Visualizing.** Users can create a variety of renderings by using fully customized lookup tables and several types of glyphs, e.g., cubes, spheres or cones, scaled by height and/or radius. Furthermore, there is support for generation of movies, e.g., by interpolating appropriately a sequence of visualizations representing snapshots of an evolutionary process obtained as standard VisIVO Web renderings.

The specific installation of VisIVO Web allows the possibility to use a batch queue (LSF or PBS) or gLite grid submission. This feature permits users to exploit high-performance computing facilities for handling heavy computational tasks. To our experience such a task is generation of movies — often lasting several days. Using high performance computing facilities is instrumental for dramatically reducing processing times.

3.   VisIVO Smartphone

VisIVO Smartphone is a Web application that allows a smartphone to exploit VisIVO Web functionality to access large-scale astrophysical datasets residing on a server repository for analysis and visual discovery. Through interactive widgets, customized visualizations (images or movies) can be generated and stored locally. The application notifies users when requested visualizations are available for retrieving (in gif and/or animated gif format) on their smartphones irrespectively of geographical location.

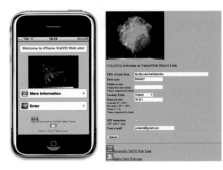

Figure 2.   The VisIVO Smartphone application (from left to right): (a) entry screen, (b) uploading and visualization settings and (c) QR code for initiating application.

VisIVO Smartphone can play a pivotal role within collaborative visualization environments by offering an exciting possibility for research groups and communities to share results and experiences of analysis and exploration of astrophysical datasets, irrespectively of being able to connect to the internet. Moreover this application allows the creation of visualizations on-the-go that can be displayed on smartphones immediately.

Although the current version of VisIVO Smartphone is optimized for the Apple iPhone, it can be easily adapted for other popular smartphones. Currently the prototype application uses the SDSS datasets. The operational scenario requires specifying a url (anonymous HTTP, FTP, or SFTP), a format (e.g., Binary, ASCII, CSV, FITS, VOTables) and finally standard visualization settings. A dataset for testing VisIVO Smartphone containing 2048000 particles with position and velocity properties coming from an N-Body simulation can be found at `ftp://ftp.oact.inaf.it/test.bin`.

4.   Future Work

Our future work will focus on porting VisIVO Server and VisIVO Web to grid environments, specifically on dektop grids and gLite. The exploitation of grid technologies will enhance significantly the performance of VisIVO Smartphone by supplying the required computational capability. A very exciting possibility is deployment of this technology for similar operational scenarios for visual discovery in scientific disciplines apart from astrophysics. Furthermore, a pilot application built upon VisIVO Smartphone is underway for public engagement activities in astrophysics. Our ultimate vision is users (researchers, citizen scientists or even science centers visitors) being able to reserve resources for data analysis and visual discovery as required (possibly involving several production grids) and retrieving results on smartphones on-the-go irrespectively of internet access or geographical location.

References

Becciani, U., et al. 2010, PASP, 122, 119

Astronomical Data Analysis Software and Systems XX
ASP Conference Series, Vol. 442
Ian N. Evans, Alberto Accomazzi, Douglas J. Mink, and Arnold H. Rots, eds.
©2011 Astronomical Society of the Pacific

The GRIDView Visualization Package

Brian R. Kent

Jansky Fellow. National Radio Astronomy Observatory, 520 Edgemont Road, Charlottesville, VA 22903, USA

Abstract. Large three-dimensional data cubes, catalogs, and spectral line archives are increasingly important elements of the data discovery process in astronomy. Visualization of large data volumes is of vital importance for the success of large spectral line surveys. Examples of data reduction utilizing the GRIDView software package are shown. The package allows users to manipulate data cubes, extract spectral profiles, and measure line properties. The package and included graphical user interfaces (GUIs) are designed with pipeline infrastructure in mind. The software has been used with great success analyzing spectral line and continuum data sets obtained from large radio survey collaborations. The tools are also important for multi-wavelength cross-correlation studies and incorporate Virtual Observatory client applications for overlaying database information in real time as cubes are examined by users.

1. GRIDView

GRIDView was developed for the ALFALFA project (Kent et al. 2008) for the purpose of viewing 3D data cubes and extracting spectral profiles and their associated measurements. The software package was written in the Interactive Data Language (IDL) to facilitate compatibility with the Arecibo processing pipeline and reduction library written by Arecibo staff member Phil Perillat. GRIDView includes a comprehensive interface to show a main R.A./Decl. window of data, a secondary window for other wavelength images or Arecibo drift scan spectral weights, controls for data manipulation, and windows for spectral line displays and database catalogs or images relevant to a data reduction session. The graphical interface is shown in Figure 1.

Users are able to slice the cube at specific cz values and view sky maps scaled in linear, logarithmic, and histogram equalization modes. In addition, channels can be averaged with slider controls, and the sky image can be smoothed with boxcar or Gaussian kernels of varying width. These interactive tools allow the user to visually confirm low signal-to-noise detections in data cubes. The interface controls can also switch between polarizations to examine possible radio frequency interference leading to spurious detections. Once sources are identified, users can overlay information from other redshift catalogs. Access to images and catalog information at a rapid rate has greatly increased optical counterpart identification of sources in the ALFALFA survey. Survey participants are heavy end users of the Virtual Observatory and the associated tools and protocols. In addition, catalogs and extracted measurements can be distributed to follow-up websites for review by team members.

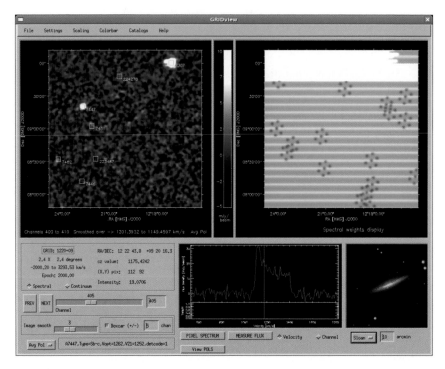

Figure 1.   The GRIDView instance shown is configured for data analysis in the
Arecibo Legacy Fast ALFA survey (Kent et al. 2008). The two sky panels show spectral line data cubes (left) and spectral weights for meridian drift scans (right). The
control panel widgets allow a user full control over data cube manipulation, including spectral and spatial smoothing, continuum views, and dual polarization modes.
Mousing over detections queries online and user-supplied databases for catalog information. In the application depicted, galaxy morphology and redshift information
are shown. The lower middle and right panels show single pixel spectra extracted
from the cube and composite SDSS $g - r - i$ imaging of galaxy NGC 4316.

2.   Spectral Line Measurement

The spectral line measurement tool interfaces with GRIDView by allowing the selection of emission on the sky and the creation of an integrated spectral profile (Figure 2).
Ellipsoidal fits for various isophote levels are created, and the spectra can be interactively baselined. Finally, the user can extract fluxes via the spectral line measurement
algorithm depicted in Springob et al. (2005) and described in Haynes et al. (1999).
Corrections are applied based on comparing continuum sources from ALFALFA to the
NRAO VLA Sky Survey (NVSS) to further improve those from the telescope pointing
models. Comparison of galaxy fluxes can be made via online HI databases (Springob
et al. 2005). Source files can be saved with a complete reduction history of how the
spectra and associated measurements were obtained.

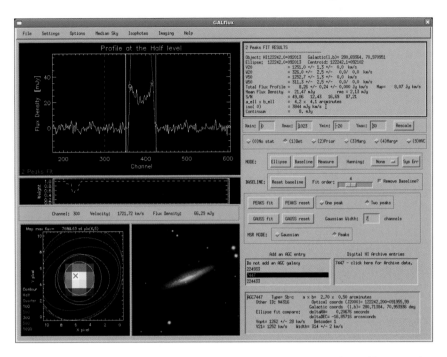

Figure 2.   The spectral line measurement instance shown is configured for neutral hydrogen spectroscopy in the Arecibo Legacy Fast ALFA survey. A weighted integrated spectral profile of galaxy NGC 4316 is created based on ellipsoid isophote fits in the lower left panel. The profile is displayed in the largest panel, allowing the user to zoom and select initial estimates for emission bounds. Interactive baselining and flux measurements can be made with the central control panel. The text display in the upper right displays instructions during the reduction process to guide the user. Centroid positions with telescope pointing corrections, flux measurements, and associated statistical and systematic errors are also shown in the text display. The user can also compare the measurements to previously obtained spectra and available database information.

3.   Data Distribution

The Virtual Observatory has created a suite of tools, standards, and libraries that allow for easy dissemination and statistical analysis of data (Hanisch 2007). ALFALFA catalogs and spectra are exported as XML, ASCII, and binary FITS tables, available from the Digital HI Archive at `http://www.cv.nrao.edu/~bkent/computing/`. The catalogs can be accessed via a ConeSearch web service through any command line or registry service utilized by the Virtual Observatory. As the survey progresses, these services are used by GRIDView itself to plan and determine parameters for multiwavelength followup observations on published ALFALFA catalogs.

4. Science Applications

The GRIDView visualization package has been used with great success to analyze the ALFALFA dataset. Galaxy catalogs with HI measurements have been published (Kent et al. 2008) and are available through the Digital HI Archive. The routines used for spectral line measurements have also been adapted for measurement of aperture synthesis HI studies with the VLA (Kent et al. 2009; Kent 2010). The software implementation used for the ALFALFA survey has resulted in the discovery of a number of intriguing extragalactic HI clouds (Kent et al. 2007).

References

Hanisch, R. J. 2007, in Statistical Challenges in Modern Astronomy IV, edited by G. J. Babu & E. D. Feigelson (San Francisco, CA: ASP), vol. 371 of ASP Conf. Ser., 177

Haynes, M. P., Giovanelli, R., Chamaraux, P., da Costa, L. N., Freudling, W., Salzer, J. J., & Wegner, G. 1999, AJ, 117, 2039

Kent, B. R. 2010, ApJ, 725, 2333

Kent, B. R., Spekkens, K., Giovanelli, R., Haynes, M. P., Momjian, E., Cortés, J. R., Hardy, E., & West, A. A. 2009, ApJ, 691, 1595

Kent, B. R., et al. 2007, ApJ, 665, L15

— 2008, AJ, 136, 713

Springob, C. M., Haynes, M. P., Giovanelli, R., & Kent, B. R. 2005, ApJS, 160, 149

Astronomical Data Analysis Software and Systems XX
ASP Conference Series, Vol. 442
Ian N. Evans, Alberto Accomazzi, Douglas J. Mink, and Arnold H. Rots, eds.
© *2011 Astronomical Society of the Pacific*

dax: DS9 Analysis Extensions in CIAO

Kenny J. Glotfelty, Joseph Miller, and Judy Chen

Smithsonian Astrophysical Observatory, 60 Garden Street, Cambridge, MA 01238, USA

Abstract. *dax* is a suite of scripts that allows various CIAO tools (written to support the Chandra X-ray Observatory) to be run using the analysis framework provided by SAOImage DS9. This allows users to quickly leverage the functionality CIAO provides without having to invest in learning the syntax and semantics of each of the tools. This simplification of the interface benefits all astronomers since many of the CIAO tools are sufficiently generic that they can work with data sets from arbitrary observatories. We present the *dax* design and discuss some of the pitfalls and limitations we encountered while building *dax*.

1. Introduction

CIAO (Fruscione et al. 2006) is the suite of tools and applications developed by the Chandra X-ray Center; it has both Chandra specific and generic analysis tools. SAOImage DS9 (Joye & Mandel 2003) is an independently developed astronomical imager widely used by professionals and amateurs across the electromagnetic spectrum. *dax* is a simple layer in between the two that provides advanced CIAO functionality via the API DS9 provides. By presenting the user with only the most frequently modified parameters the interface is kept minimal as is the knowledge users need to know about any particular CIAO task.

2. Design

An exhaustive description of the DS9 analysis framework is beyond the scope of this paper; however, a few critical elements will be described. External analysis tasks are described in a simple ASCII text file that is loaded by DS9. Each task is described by 4 lines: name, applicable file type (name), launch option (key binding or menu item), and the command to execute along with user parameter options.

The commands and parameters make use of a pseudo Tcl / macro-expansion language where common interface elements are defined with simple `$dollar` macros, e.g., `$image` is used to create or load a new image.

2.1. Initialization

dax is loaded for CIAO users by default by means of a wrapper script around DS9 that uses the `-analysis` command line switch. Alternatively users can setup to have *dax* loaded automatically via DS9 preferences; however, unless symbolic links are used this is fragile as the default path where CIAO is installed changes with each release.

By default, *dax* uses the ChIPS (Germain et al. 2006) plotting package in CIAO to produce plots such as light-curves and spectra. A simple Tcl/Tk start-up script is `-source`'ed that launches a ChIPS server when DS9 starts so that plots are rendered more quickly and without the need for a terminal. A custom call-back is installed when the DS9 window is destroyed to terminate the ChIPS server. If ChIPS is not installed, *dax* will use the BLT plotting package in DS9.

2.2.   Task Selection

CIAO has over 100 tools and applications; *dax* makes only a few available. The following are some considerations that were made when selecting which CIAO tasks *dax* would enable:

Run-Time: GUI users are looking for quick feedback so tasks like the CIAO wavdetect tool which can take a long time to run are not included.

Self contained inputs: If a task needs auxiliary files (e.g., bad pixel files, aspect solution, or dark frames) then its usefulness as a *dax* task is limited. As of DS9 6.1 a new analysis macro is available for selection of additional files; however for multiple files this can stifle the GUI user experience.

Mission independence: Many of the CIAO tasks are mission independent. *dax* only used those that are Chandra-specific when they are well behaved for non-Chandra data.

Output type: One of the *dax* challenges is how to return results in a useful format. If the output of a particular task is more often than not just used as input to another task then it would not be included in *dax*. The output should be a product upon which further analysis can be performed, e.g., by *dax*.

2.3.   Data Sources

The data being displayed in DS9 (most often in the current frame) are easily accessible to *dax*. DS9 provides several methods to access the data depending on the needs of the application. Some pro's and con's of each method are listed below.

$data: FITS image is piped to application. This provides a true WYSIWYG data transfer to the analysis task. So if data have been smoothed or a table (i.e., event file) has been binned then the analysis task will be provided with exactly those pixel values. One must be careful though since binned tables do not retain all the header keywords; only minimal WCS information is provided. Analysis tasks also only have access to image block being displayed; arbitrary FITS extensions are not available (e.g., good-time intervals).

$filename: Path, file-name, and extension are passed to the analysis task. This allows the task to open the appropriate FITS extensions as needed and provides access to the full complement of header keywords. However, this only works when data were loaded from local disk. Data retrieved via image servers, web URLs, SAMPed from another application, or generated by a previous analysis task will have a /tmp path but no such file will exist. One also must be careful as the filtering and binning syntax used by DS9 is the same as funtools[1] which has some incompatibilities with both CIAO and with FTOOLS/CFITSIO.[2]

$xpa: XPA (Mandel & Tody 1995) access point name is passed to the analysis task. XPA provides inter-process communication both via library API and via a suite of command line tools. DS9 provides XPA access points for most of its functionality allowing analysis scripts to GET or SET to retrieve highly customized data elements. This provides an analysis script with the greatest flexibility; however it also is the most costly to implement. Users also need to be aware that multiple, concurrent DS9 GUIs can be running at the same time; and unless users have launched with unique -title arguments they will all respond to any xpaget or xpaset command.

Most *dax* scripts make use of the $data and $filename macros as they are the simplest to use and work most of the time. The $xpa method is primarily reserved for those times when many parameters associated with the data are passed into the script or the task needs to be especially robust.

2.4. Parameterization

dax takes a minimalist approach when gathering input from users. By keeping the tasks small and very specific, users do not need to know all the behind the scenes details; while at the same time similar tasks are grouped into single scripts. For example, the light-curve and spectrum histograms run a single *dax* script that calls the CIAO tool dmextract with the correct settings to create the requested plot. Many *dax* tasks do not require any user parameters to be entered separately.

2.5. Returning Results

DS9 supports 3 different output models: $data, $image, and $text. A fourth, $null, is available when the action of the analysis task is handled via other means. Using XPA, scripts have more fine control over the DS9 GUI. *dax* makes use of all these methods; though even when XPA is used to return results *dax* still uses the $text directive (rather than $null) since it can also capture any processing errors.

dax tries to deliver data in the most useful format for quick evaluation while at the same time trying not to leave the data "trapped" in DS9. See Figure 1.

2.6. Special Consideration

DS9 is capable of displaying much more than simple 2D images from a single FITS file. For example it can now create mosaics from multiple files, 3-color (RGB) images, and display data from cubes up to 10 dimensions. All of these special data types can cause unexpected behavior in *dax* and other analysis tasks that are not expecting such diverse datasets.

Regions are notoriously problematic. Different analysis systems use different syntax for shapes (box expressed as lower-left/upper-right, or center with x and y lengths, or half-lengths); provide different shapes (e.g., DS9's epanda); and imply different intent when multiple shapes are drawn (shapes ANDed or ORed together).

---

[1] https://www.cfa.harvard.edu/~john/funtools/

[2] http://heasarc.gsfc.nasa.gov/ftools/ftools_menu.html

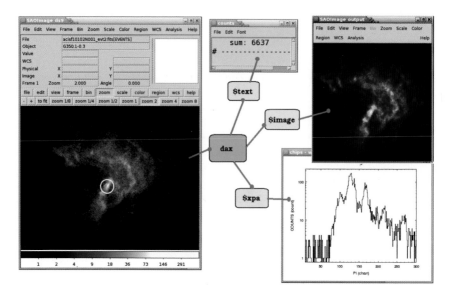

Figure 1.   Examples of *dax* tasks: summing counts, adaptive smoothing, plotting spectrum.

3. Conclusions

DS9 provides easy access to external analysis packages via its feature-rich analysis menu framework. *dax* uses this framework to give users access to many common analysis tasks without having to invest in learning all of CIAO thereby allowing users to quickly get started with their data analysis.

Acknowledgments.   This project is funded by NASA contract NAS8-03060, Chandra X-ray Center.

References

Fruscione, A., et al. 2006, in Observatory Operations: Strategies, Processes, and Systems, edited by D. R. Silva, & R. E. Doxsey (Bellingham, WA: SPIE), vol. 6270 of Proc. SPIE, 62701V

Germain, G., Milaszewski, R., McLaughlin, W., Miller, J., Evans, J. D., Evans, I., & Burke, D. 2006, in Astronomical Data Analysis Software and Systems XV, edited by C. Gabriel, C. Arviset, D. Ponz, & S. Enrique (San Francisco, CA: ASP), vol. 351 of ASP Conf. Ser., 57

Joye, W. A., & Mandel, E. 2003, in Astronomical Data Analysis Software and Systems XII, edited by H. E. Payne, R. I. Jedrzejewski, & R. N. Hook (San Francisco, CA: ASP), vol. 295 of ASP Conf. Ser., 489

Mandel, E., & Tody, D. 1995, in Astronomical Data Analysis Software and Systems IV, edited by R. A. Shaw, H. E. Payne, & J. J. E. Hayes (San Francisco, CA: ASP), vol. 77 of ASP Conf. Ser., 125

Astronomical Data Analysis Software and Systems XX
ASP Conference Series, Vol. 442
Ian N. Evans, Alberto Accomazzi, Douglas J. Mink, and Arnold H. Rots, eds.
©*2011 Astronomical Society of the Pacific*

SAOImage DS9 Frequently Asked Questions

William Joye

*Smithsonian Astrophysical Observatory, 60 Garden Street, Cambridge,
MA 02138, USA*

Abstract. SAOImage DS9 is an astronomical imaging and data visualization application. It supports FITS images and binary tables, multiple frame buffers, region manipulation, and many scale algorithms and colormaps. It provides for easy communication with external analysis tasks and is highly configurable and extensible via XPA and SAMP. This article will highlight some of the more popular questions which have been submitted by users.

1. How SAOImage DS9 got its Name

In 1990, Mike Van Hilst, at the Smithsonian Astrophysical Observatory, Center for Astrophysics, developed SAOImage. SAOImage was first implemented in X10, then reimplemented in X11. In fact, it was one of the first X11 based applications publicly made available. SAOImage was a brilliant program, implementing techniques in scientific visualization 20 years ago that are still being used by today's applications. Since Mike's departure from SAO, SAOImage has been maintained by Doug Mink.

In the mid 1990's, Eric Mandel developed SAOImage: The Next Generation, or SAOtng (Mandel 1997), named after the Star Trek series. SAOtng was based on IRAF's XIMTOOL graphics libraries and Tcl. It explored new GUI interfaces and supported a new external analysis interface, X11 Public Access or XPA (Mandel & Swick 1994; Mandel & Tody 1995; Mandel 1996), which allowed SAOtng to be scripted via a shell, or from other application.

In 1998, while working with Eric, William Joye (Joye & Mandel 2000, 2005) began a complete rewrite of SAOtng, based on the experience developed while supporting SAOtng. For lack of a name, the new project was referred to as DS9, the logical extension of the Star Trek series. The name continues to be in use.

The first versions of SAOImage DS9 were made available in 1999. Since then, the popularity of SAOImage DS9 has grown far beyond expectations.

2. How SAOImage DS9 is Constructed

SAOImage DS9 is a Tcl/Tk application. The GUI is implemented as a very thin layer of Tcl/Tk. A number of Tk canvas widgets written in C++ were developed (Joye & Mandel 1999) to support all the functionality needed. DS9 inherited SAOtng's support of regions, XPA, external analysis support, and the general GUI. However, all the visualization techniques come directly from SAOImage.

The current version of SAOImage DS9 is composed of the SAOTk canvas widgets along with a number of other open source products: Tcl/Tk, AST, BLT , HCompress, HTMLWidget, PLIO, Rice, Tcllib, TclXML, TkCon, TkImg, TkTable, WCSLIB, WCSTools, XMLRPC, XPA, ZIP, zlib, and zvfs. The distributed binaries consist of a self-contained self-extracting archive and application, which provides an independent Tcl/Tk environment without installation.

3.   How SAOImage DS9 Works

To render an image, SAOImage DS9 requires the user to specify a color scale, a contrast/bias pair for the color scale, clip values for the data, and a scale distribution function.

- Step 1. Select a color scale. A color scale is defined as a number of colors (RGB triplets). The number of RGB triplets can vary from just a few to over 200. DS9 supports a number of predefined color scales (e.g., Gray, A, B, I8) or the user may load his own color scale.

- Step 2. Apply a contrast/bias pair. This step takes the result of step 1 and creates a new array with the contrast/bias applied. The length of the new array will be between 200 (for pseudo-color) and 4096 (for true-color).

- Step 3. Calculate the data clip values (low/high data values). The min/max data values may be used or an algorithm may be used to determine the clip data values.

- Step 4. Apply the scale distribution function. This involves taking the result of step 2, and creating yet another array, this time of size 16384, redistributing the colors, based on the scale function selected (See § 3.1).

- Step 5. Based on the data clip values and the value of the data point, index into the result of step 4, which yields either an index into a lookup table (for pseudo-color) and an RGB triplet (for true-color and postscript).

3.1.   Scales

SAOImage DS9 supports 6 scale distribution functions: linear, log, pow, sqrt, square, and histogram equalization.

3.1.1.   Linear

The *linear* scale function is a straight linear distribution from the low to high clip values.

3.1.2.   Log

The *log* function is defined as the following:

$$y = \log(ax + 1)/\log a, \text{ for } 0 < x < 1.$$

The user may specify an exponent a to change the distribution of colors within the colormap. The default value of a is 1000. Typically, optical images respond well at 1000, IR images as low as 100, and high energy bin tables up to 10000. A value of 10000 closely matches the original log function of SAOImage as defined as the following:

$$y = (e^{-10x} - 1)/(e^{-10} - 1).$$

3.1.3. Power

The *pow* scale function is defined as the following:

$$y = (a^x - 1)/a, \text{ for } 0 < x < 1.$$

The user may specify an exponent a to change the distribution of colors within the colormap. The default value of a is 1000.

3.1.4. Square Root

The *sqrt* scale function is defined as the following:

$$y = \sqrt{x}, \text{ for } 0 < x < 1.$$

3.1.5. Square

The *square* scale function is defined as the following:

$$y = x^2, \text{ for } 0 < x < 1.$$

3.1.6. Histogram Equalization

The *histequ* scale function distributes colors based on the frequency of each data value.

3.2. Smoothing

As part of the image rendering process, the user may select one of three types of smoothing kernels, Boxcar, Tophat, and Gaussian. For each, the kernel diameter or width is defined as $2r + 1$.

3.2.1. Gaussian

The Gaussian function is defined as *mean* $= 0$ and $\sigma = r/2$

$$z = \frac{1}{\sqrt{2\pi}\sigma} e^{-\frac{1}{2}(x^2+y^2)/\sigma^2}.$$

3.3. Large Data Files

There are several factors that determine if SAOImage DS9 will be able to load a large file (Joye & Mandel 2004).

3.3.1. 32 bit vs. 64 bit Application

SAOImage DS9 32 bit applications can address up to 4Gb of address space. However, to address very large files, the user may require a 64 bit application. The actual 32 bit application address limit depends on the operating system. Under Linux, for example, the limit appears to be $\sim$ 3Gb. Under Solaris 10, DS9 has a full 4Gb of address space. MacOSX Aqua is only available as 32 bit application and has a limit of $\sim$ 3Gb while MacOSX X11 is a full 64 bit application. Windows is only available as a 32 bit application and currently limited to $\sim$ 2Gb.

3.3.2. Memory Management

There are a number of memory management techniques supported in SAOImage DS9 that will greatly affect the ability and speed of loading large data files.

```
%
$ ds9 foo.fits # uses mmap
$ cat foo.fits | ds9 - # allocates memory
$ xpaset -p ds9 file foo.fits # uses mmap
$ xpaset -p ds9 fits foo.fits # allocates memory
```

Memory Map (mmap) is very fast with a limit of the available physical memory. Allocate is much slower with a limit of the physical memory + swap partition.

3.3.3. Scanning Data

SAOImage DS9 needs to determine the minimum and maximum data values to render an image. For large data files, this can take some time. The user has several options available: scan all data, sample every x value, or use the FITS keyword DATAMIN/MAX or IRAFMIN/MAX. For the best performance, sample the data and set the sample interval between 10 and 100.

Acknowledgments. SAOImage DS9 development has been made possible by funding from the Chandra X-ray Science Center (NAS8-03060) and the High Energy Astrophysics Science Archive Center (NCC5-568).

References

Joye, W., & Mandel, E. 1999, in Astronomical Data Analysis Software and Systems VIII, edited by D. M. Mehringer, R. L. Plante, & D. A. Roberts (San Francisco, CA: ASP), vol. 172 of ASP Conf. Ser., 429
— 2000, in Astronomical Data Analysis Software and Systems IX, edited by N. Manset, C. Veillet, & D. Crabtree (San Francisco, CA: ASP), vol. 216 of ASP Conf. Ser., 91
— 2004, in Astronomical Data Analysis Software and Systems XIII, edited by F. Ochsenbein, M. G. Allen, & D. Egret (San Francisco, CA: ASP), vol. 314 of ASP Conf. Ser., 505
— 2005, in Astronomical Data Analysis Software and Systems XIV, edited by P. Shopbell, M. Britton, & R. Ebert (San Francisco, CA: ASP), vol. 347 of ASP Conf. Ser., 110
Mandel, E. 1996, in Visual Data Exploration and Analysis III, edited by G. G. Grinstein, & R. F. Erbacher (Bellingham, WA: SPIE), vol. 2656 of Proc. SPIE, 214
— 1997, in Astronomical Data Analysis Software and Systems VI, edited by G. Hunt, & H. E. Payne (San Francisco, CA: ASP), vol. 125 of ASP Conf. Ser., 253
Mandel, E., & Swick, R. 1994, in Astronomical Data Analysis Software and Systems III, edited by D. R. Crabtree, R. J. Hanisch, & J. Barnes (San Francisco, CA: ASP), vol. 61 of ASP Conf. Ser., 507
Mandel, E., & Tody, D. 1995, in Astronomical Data Analysis Software and Systems IV, edited by R. A. Shaw, H. E. Payne, & J. J. E. Hayes (San Francisco, CA: ASP), vol. 77 of ASP Conf. Ser., 125

Astronomical Data Analysis Software and Systems XX
ASP Conference Series, Vol. 442
Ian N. Evans, Alberto Accomazzi, Douglas J. Mink, and Arnold H. Rots, eds.
©*2011 Astronomical Society of the Pacific*

Imaging in ChIPS

Joseph Miller,[1] Douglas J. Burke,[1] Ian Evans,[1] Janet D. Evans,[1] and
Warren McLaughlin[2]

[1]*Smithsonian Astrophysical Observatory, Cambridge, MA, USA*

[2]*Northrop Grumman Information Systems, Cambridge, MA, USA*

Abstract.   The Chandra Interactive Plotting System (ChIPS) included in CIAO now allows users to incorporate and manipulate images in their plots. ChIPS uses the Visualization Toolkit (VTK) as a back end to provide basic imaging support, which includes displaying images in pseudo color or RGBA true color, adjusting the translucency of images, and several ways to threshold images. Users also have the ability to enhance them with annotations and place curves and contours directly onto the image. ChIPS imaging support provides a mechanism to adjust the image display resolution as necessary to provide high quality publication ready output. Beyond basic imaging, ChIPS includes the ability to recognize and incorporate WCS metadata into plots. ChIPS accurately calculates the intersections of world coordinate grids and plot axes, ensuring that these elements distort correctly with a tangent plane projection. Multiple image overlays are handled by reprojecting the overlaid images onto the reference image's coordinate system. New zooming and panning functions, and existing limits commands, use the WCS information from the image overlays to update the axes to reflect the new field of view being displayed. Although ChIPS already provides a number of user interactive commands, additional interactive capabilities are being considered for future releases. Enhanced interactive interfaces alongside the ability to script ChIPS in Python provide a more capable and user-friendly system.

Introduction

Since its inception, ChIPS (Germain et al. 2006) has been designed with the intention of allowing users to incorporate images with their visualizations. Taking advantage of the image processing capabilities available in VTK,[1] the graphical backend to ChIPS, ChIPS integrates images into plots alongside other plot objects and provides a number of standard functions to display and manipulate images. Because ChIPS has been written to work with astrophysical data sets, it is necessary that ChIPS understands and displays astrophysical metadata such as WCS information for tangent plane projections (Calabretta & Greisen 2002). ChIPS axes have been modified so that they allow for multiple images from different sources to be overlaid in the same plot, as well as depicting distortions due to the projections.

With the introduction of images into ChIPS, new interactive tools became essential for users to setup their visuals. Functions like zoom and pan were created so a user

---

[1]http://www.vtk.org/

could easily adjust the plot to focus on a desired area. The future of ChIPS will bring more interactive components to its users by providing a graphical user interface (GUI) which allows users to easily produce and format their plots.

1. Imaging

1.1. Basic Imaging

ChIPS offers two modes for displaying images, pseudo color and RGBA true color. Users can fine tune pseudo color images by associating the image with a preloaded (or user defined) color map, adjusting the upper and lower scalar bounds or applying translucency to the image. Furthermore, users can set the number of colors used by the map or invert the mapping.

In the true color mode ChIPS users can combine up to four images to create a single true color image. The first three images are the red, green and blue components of the resultant image. The fourth image is the alpha component to provide translucency to the image.

A number of operations can be done irrespective of the mode. Users can apply different interpolation schemes between pixels, including nearest neighbor, bilinear and bicubic interpolation. They also can adjust the dots per inch (dpi) of the image to set the quality of the hard copy output. Lastly, users can integrate images with other ChIPS objects such as overlaying curves, contours, annotation or other plots.

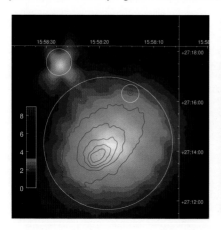 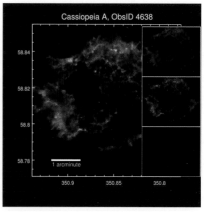

Figure 1.   (Left) Here a pseudo color image is displayed with the color map size reduced to 25 colors. The image is overlaid with a contour and annotated with regions and a color bar. (Right) This plot contains a true color image with the red, green and blue components displayed on the side.

1.2. Beyond the Basics

New algorithms have been added to ChIPS to display WCS information associated with the data accurately and quickly. To do this, ChIPS separated axes into two axis types, Cartesian and a WCS tangent plane projection (WCSTAN). The WCSTAN axis accepts WCS transform information and augments the displayed axis to show the distortion associated with the projection. The data is simply drawn in image coordinates. To place tick marks, ChIPS uses the range of the data in world coordinates to determine an appropriate spacing. ChIPS then places the tick marks in image coordinates by calculating the intersection between the RA/Dec coordinates and the image plane at the axis. Using the tick marks as guides for spacing, the grid lines are drawn by going along the plot area and calculating where each line intersects the plot edge. Each line is subdivided into slices that curve with the projection.

In addition to augmenting axes and grids, ChIPS also allows users to overlay images with data (images, contours or curves) that have different WCS information. ChIPS uses the first image as the reference image coordinates and inverts all other data points from WCS to the reference image coordinates (for images only the extents are transformed and not the individual pixels). To handle rotations, ChIPS renders all images with north up.

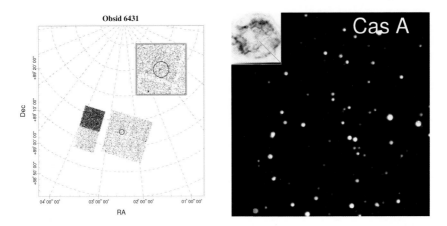

Figure 2.   (Left) This figure shows an image close to a pole. The grid lines are distorted with the projection. In the upper right corner is a 2nd plot with the same image zoomed in on the region. (Right) This plot depicts a true color image taken by the 2MASS survey overlaid with a contour of a Chandra image taken of the same area. The full Chandra image is displayed in the upper left.

1.3.   Manipulating Limits

With the addition of images and the handling of WCS, ChIPS needed new functionality to allow users to change the data range displayed while maintaining the aspect ratio. The limits commands were modified to enforce the aspect ratio when the plot range changes. A new zoom function was added to allow users to programmatically or interactively zoom in or out of the visualization. The zoom function adjusts the range in images coordinates. After which, the extents are converted to world coordinates and updated in the axes. A pan function has also been added, where users can pan to a specified coordinate or interactively drag the data to a new area. Like zoom, pan does all its calculations in image coordinates, converting the input coordinates from world coordinates and then back after the pan is complete.

2. Adding a GUI

Being able to produce high quality visualizations from basic to complex scenes, ChIPS has amassed an extensive command set. This command set is most commonly accessed via the Python scripting language or a limited number of interactive interfaces. The developers of ChIPS are looking to simplify the process of creating plots by expanding the interactive interfaces and adding a more complete GUI. There are many challenges to overcome in designing the GUI, not the least of which is falling into a trap of producing a tool which is more complicated then the one it is trying to improve. One approach being explored is to present ChIPS objects and object hierarchy in a dialog which would give users a graphical approach to updating the object properties. This allows users to make large numbers of modifications to a visualization and fine tune their final product. Functionality like saving and loading state files and scripts, or handling undo and redo would also be added to a general ChIPS dialog. Furthermore, the developers are also looking at ways for users to interact directly with the plots. Selecting, highlighting, moving and resizing objects may be a few of the ways users can interact directly with the objects. These additions will provide users with complete control over how to arrange their visuals.

With GTK[2] as its front end, ChIPS can extend its windows with widgets or add dialogs to the system seamlessly. The back end, VTK, provides methods of ray tracing into a scene to detect objects that have been selected. By capturing the selection event, ChIPS can extend the interactive capabilities in any number of ways.

3. Conclusion

With the addition of images and world coordinate capabilities, ChIPS completes many of the tasks it set out to do when it was first designed. By offering an array of utilities to fine tune images, ChIPS provides the tools necessary to produce publication ready visuals that capture a user's data and intent. With the addition of world coordinates, ChIPS is able to provide visuals which accurately represent the data through algorithms that are fast and precise. Adding a GUI is the next logical step for ChIPS. Designing a system where people can interactively adjust their plots will give users an intuitive and efficient way of creating their scientific visualizations.

Acknowledgments.   Support of the development of ChIPS is provided by the National Aeronautics and Space Administration through the Chandra X-ray Center, which is operated by the Smithsonian Astrophysical Observatory for and on behalf of the National Aeronautics and Space Administration under contract NAS8-03060.

References

Calabretta, M. R., & Greisen, E. W. 2002, A&A, 395, 1077
Germain, G., Milaszewski, R., McLaughlin, W., Miller, J., Evans, J. D., Evans, I., & Burke, D.
    2006, in Astronomical Data Analysis Software and Systems XV, edited by C. Gabriel,
    C. Arviset, D. Ponz, & S. Enrique (San Francisco, CA: ASP), vol. 351 of ASP Conf.
    Ser., 57

---

[2]http://www.gtk.org/

Astronomical Data Analysis Software and Systems XX
ASP Conference Series, Vol. 442
Ian N. Evans, Alberto Accomazzi, Douglas J. Mink, and Arnold H. Rots, eds.
© *2011 Astronomical Society of the Pacific*

Chandra Footprint Service: Visualizing Chandra's Sky Coverage

Aaron C. Watry and Arnold H. Rots

Smithsonian Astrophysical Observatory, 60 Garden St, Cambridge, MA 02138

Abstract.   The Chandra Footprint Service[1] (CFS) provides a visual interface to data in the public Chandra Data Archive and the Chandra Source Catalog. Users of the service can directly see which areas of the sky have been observed by the Chandra X-Ray Observatory near any target of interest. The browser based interface provides visualization of the sky coverage of Chandra instruments for any region of the sky along with tabular data for observations. The service provides an interactive client-based overlay of the instrument region data on top of a Digitized Sky Survey (DSS) background image, previews of observation images, access to data products, and access to the VOTable data used to create the interface.

1. Introduction

The Chandra Footprint Service (CFS) is provided by the Chandra Data Archive (CDA) by combining a front-end and several back-end services into an interactive web site that allows scientists and other users to easily visualize which regions of the sky have been observed by the Chandra X-Ray Observatory. The service uses several Virtual Observatory (VO) standards and existing software and web services in order to create a high-performance, primarily client-based experience.

2. Prior Work

2.1. Virtual Observatory Standards

The CFS uses several existing VO standards, including the VOTable standard[2] which allows encapsulating astronomical data within an XML file; the Simple Image Access Protocol[3] (SIAP) which simplifies searching for images; and the Linear String Serialization[4] (STC-S) of the Space-Time Coordinate (STC) Metadata standard for describing instrument regions in human-readable text.

---

[1] http://cxc.harvard.edu/cda/footprint/cdaview.html

[2] http://www.ivoa.net/Documents/VOTable/

[3] http://www.ivoa.net/Documents/SIA/

[4] http://www.ivoa.net/Documents/Notes/STC-S/

2.2. HLA Footprint Service

The CFS uses C# code from Space Telescope Science Institute (STScI) to create VOTable formatted data which is transmitted to the user. CFS also uses a spherical math library from Johns Hopkins University (JHU) which does high-performance spherical math and region manipulations directly on the database server.

2.3. Existing CXC Services

The CFS utilizes several existing CDA services to look up existing data and to forward users to result sets based upon their searches and selections. These services include WebChaSeR,[5] the Chandra SIAP VO service, and the Chandra public FTP server.

3. Architecture

The CFS uses a split front-end and back-end architecture to help users in finding observations of interest. The back-end is built from several web servers while all data are viewed in the user's web browser. The raw VOTable XML source is available for download to any user who wishes to use it.

3.1. Back End Services

The primary visible server is the public CXC web server. This server:

- Acts as a proxy to the server which generates VOTables containing observation and footprint descriptions,

- Provides a target name resolver,

- Provides a DSS image loading proxy,

- Acts as a proxy to the public CDA SIAP VO service.

The proxy service which forwards the VOTable region data to the web browser can be queried by external services.

3.2. Web Browser Front End

A large amount of development effort has been put into the web-based CFS front end. This front end uses a combination of Asynchronous Javascript and XML (AJAX), XSL Transforms, and the HTML5 <canvas> element. When users load the main page of the web service, they are presented with a search form which the user then enters a target name or coordinates into. The user's browser sends the target name/coordinates to the target resolver, which returns resolved coordinates for the user's search. The service then performs simultaneous queries against the footprint server and the Chandra SIAP service. The service waits for the footprint response and when it is received it uses a series of XSL Transforms to convert the received VOTable into an HTML page.

The generated HTML page consists of a DSS background image, a control form to toggle display of certain region types, a 2-D overlay of the regions for the matching

---

[5]http://cda.harvard.edu/chaser

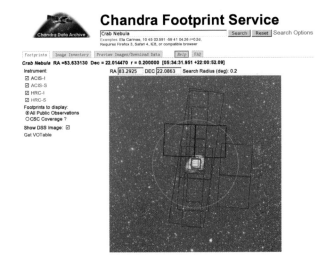

Figure 1.   Overlay with control and search forms.

observations (color-coded by the instrument used) which is rendered on top of the DSS image (Figure 1), and a table-based view of the observations that matched the user's query (Figure 2). A separate tab on the CFS page allows the user to inspect the Chandra images that correspond to selected observations.

3.2.1. Footprint Overlay

The interface allows the user to select the rendered polygons for any observation(s) of interest. The selected observations are toggled in color in the overlay, and the related row in the observation table is selected/deselected as well. This interaction is performed entirely in the user's web browser.

The front end of the CFS uses the <canvas> element defined in the HTML5 specification to perform all rendering operations in the web browser. This eliminates the need to pre-render the overlay on the server as is done in other footprint services which currently exist. The user interface is very responsive as a result of the client-side rendering. This also has the added benefit of reducing server load and network traffic.

3.2.2. Observation Table

The observation table is fully interactive and is connected to the footprint overlay. Objects selected in the overlay are also selected in the table. The reverse is true as well. There are controls to forward the user to the WebChaSeR tool to download selected data. Other controls which instruct the interface to only show Selected, Non-selected, or to show all observations. The table is also configurable. The columns can be sorted by clicking on the column headers and most columns can be filtered for either specific values, ranges of values, or by wildcard matches by entering text into the filter boxes. The list of columns is configurable through the use of a secondary table (not pictured).

Results 1- 14 of 14　　　　　　　　　　　　　　Show 20 ▾ results per page

Click column heading to sort list - Click rows to select
Download Selected ObsIDs in WebChaser
Show selected rows: First | **Mixed** | Only | Not | Reset selection
Text boxes under column headings allow specifying a filter to be applied to columns　Apply Filter　Clear Filter

Observation Date	RA	DEC	Proposal ID	PI Last Name	ObsID	Instrument	Exposure	Grating	JPEG Preview	FITS	«	»
							>20					
Jan 31 2000 1:14AM	05:34:32.0	22:00:52.0	1500806	Weisskopf	758	HRC-S	100.22	LETG	JPEG	FITS		
Feb 2 2000 10:31AM	05:34:32.0	22:00:52.0	1500806	Weisskopf	759	HRC-S	45.43	LETG	JPEG	FITS		
Nov 3 2000 11:42AM	05:34:31.61	22:00:56.5	2500880	HESTER	1994	ACIS-S	23.99	NONE	JPEG	FITS		
Feb 21 2001 3:30AM	05:34:31.61	22:00:56.5	2500880	HESTER	1995	ACIS-S	23.4	NONE	JPEG	FITS		
Nov 25 2000 8:10PM	05:34:31.61	22:00:56.5	2500880	HESTER	1996	ACIS-S	25.7	NONE	JPEG	FITS		
Mar 14 2001 7:34AM	05:34:31.61	22:00:56.5	2500880	HESTER	1997	ACIS-S	22.73	NONE	JPEG	FITS		
Dec 18 2000 8:08PM	05:34:31.61	22:00:56.5	2500880	HESTER	1998	ACIS-S	23.8	NONE	JPEG	FITS		
Jan 9 2001 1:29PM	05:34:31.61	22:00:56.5	2500880	HESTER	1999	ACIS-S	23.57	NONE	JPEG	FITS		
Apr 6 2001 5:35AM	05:34:31.61	22:00:56.5	2500880	HESTER	2000	ACIS-S	23.35	NONE	JPEG	FITS		
Jan 30 2001 11:57PM	05:34:31.61	22:00:56.5	2500880	HESTER	2001	ACIS-S	22.6	NONE	JPEG	FITS		
Apr 14 2002 6:59PM	05:34:32.0	22:11:22.0	3500419	Seward	2796	ACIS-I	20.25	NONE	JPEG	FITS		
Jan 27 2004 1:11AM	05:34:32.0	22:00:52.0	5500128	Seward	4607	ACIS-S	37.74	NONE	JPEG	FITS		
Feb 3 2007 9:58AM	05:34:31.91	22:00:52.1	8500060	Canizares	7587	ACIS-S	46.17	HETG	JPEG	FITS		
Jan 22 2008 3:57PM	05:34:32.0	22:00:52.0	1500806	Weisskopf	9765	HRC-S	96.17	LETG	JPEG	FITS		
Observation Date	RA	DEC	Proposal ID	PI Last Name	ObsID	Instrument	Exposure	Grating	JPEG Preview	FITS	«	»

Results 1- 14 of 14　　　　　　　　　　　　　　Show 20 ▾ results per page

Figure 2.　Table of observations with controls.

4. Performance

The performance of the CFS has been optimized to provide a pleasant experience to as many users as possible. To help achieve this goal, all data loads are done using AJAX asynchronously. As soon as data that are needed are loaded, they are rendered on screen. Once data are loaded, all table and overlay manipulations are performed entirely within the web browser without server interaction. Cutting out return trips to the server has sped up the interface considerably.

The use of the <canvas> element for composition has a large impact on the speed of rendering the ovelay. Benchmarks that compare <canvas> to a pure Javascript implementation show that <canvas> is several orders of magnitude faster while also requiring less memory. Javascript fallbacks are present for users without <canvas>, but they induce a noticeable performance penalty. Internet Explorer 8 is the only current browser which lacks <canvas> support. Internet Explorer 9, Firefox, Safari, Chrome, and Opera all include support for the <canvas> element.

5. Summary

The Chandra Data Archive has developed a new web-based footprint service that uses current web standards to provide high-performance visualization of the regions of the sky that the Chandra X-Ray Observatory has observed. This interface helps the user to select observations of interest through the use of an overlay which renders instrument footprint regions on top of a DSS image. The footprint regions are linked to a data table which lists details for displayed footprints. The interface also helps the user to download data products for selected observations.

Acknowledgments. The authors would like to thank Gretchen Greene and Kim Gillies at Space Telescope Science Institute and Tamás Budavári at Johns Hopkins University for installation assistance during a site visit and for the use of their Spherical Library, initial database schema, and C# APT/SIA code. This work is supported by NASA contract NAS8-03060 (CXC).

Astronomical Data Analysis Software and Systems XX
ASP Conference Series, Vol. 442
Ian N. Evans, Alberto Accomazzi, Douglas J. Mink, and Arnold H. Rots, eds.
© *2011 Astronomical Society of the Pacific*

CSC Sky in Google Earth

Warren McLaughlin,[1] Kenny Glotfelty,[2] and Ian Evans[2]

[1]*Northrop Grumman Information Systems, Cambridge, MA, USA*

[2]*Smithsonian Astrophysical Observatory, Cambridge, MA, USA*

Abstract. CSC Sky provides a visual presentation of Chandra Source Catalog (CSC) data to users through Google Earth. Professional and amateur astronomers can quickly scan the sky or use Google Earth's search features to navigate to a specific location by name or by position. Photometrically correct tri-color images detailing detected X-ray emissions contained in the catalog are displayed within outlined instrument fields of view. Layers and transparency adjustment facilitate overlays with other publicly available information such as optical or infrared data. Real time queries via the catalog public interface provide source markers for the viewable area of the sky. Clicking on any displayed source marker invokes a balloon of principal source properties as well as a link to the source catalog for data product retrieval. The current release of CSC Sky combines real time database access to the Chandra Source Catalog with access to pipeline processed data that were extracted from the catalog archive. This paper will discuss some of the design choices that were made to provide aesthetically pleasing yet scientifically correct images while maintaining an easy-to-use interface. Constraints such as time, data volume, and processing power have their roles in shaping projects. We will note how these factors impacted the development of CSC Sky. Finally, we will conclude by looking at enhancements in features, interoperability, and data presentation.

Overview

CSC Sky[1] is a Keyhole Markup Language (KML) file providing access to observation interval (ObI) data from the Chandra X-ray Observatory. The file works in Google Earth[2] in Sky Mode to present information in an intuitive, easy to navigate geocentric manner. Users are free to zoom and pan through the heavens. Right ascension, declination, and altitude are displayed across the bottom of the window. Numerous precanned layers exist to overlay constellations, real time locations of celestial bodies, and images taken by various missions including the Hubble Space Telescope.

Mosaics from the Sloan Digital Sky Survey (SDSS) and from the Digitized Sky Survey (DSS) provide the cosmic canvas used to present information. For CSC Sky, the data encompass photometrically correct X-ray images from more than 5,000 observations contained in the Chandra Source Catalog[3] (Evans et al. 2010), overlays of

---

[1]http://cxc.harvard.edu/csc/googlecat

[2]http://earth.google.com

[3]http://cxc.harvard.edu/csc

detector fields of view, and ObI details retrieved through WebChaSeR.[4] In addition, CSC Sky also queries the Chandra Source Catalog to provide source properties and location markers for more than 106,000 distinct X-ray sources.

Image Processing

CSC Sky attempts to present X-ray data in Google Earth in an aesthetically pleasing manner without compromising the data's scientific integrity. The image data displayed in CSC Sky were extracted from the Chandra Source Catalog. We started by retrieving exposure maps and images in multiple energy bands for all catalog observations. The data were thresholded to account for any background over-subtraction and to remove artificially bright pixels. The component energy bands were then combined using the Lupton et al. (2004) algorithm to generate tri-color images. For data from the ACIS CCD detector, the soft energy band (0.5–1.2 keV) was mapped to the red channel, the medium energy band (1.2–2 keV) was mapped to the green channel, and the high energy band (2–7 keV) was mapped to the blue channel. Each channel was normalized by the total intensity of all channels and scaled by a nonlinear function to increase the dynamic range. Gaussian smoothing was applied to the resultant images to remove the Moiré pattern visible when Google Earth regrids the images. Since all ObIs are processed independently rather than as a single mosaic, we need to handle overlaps of ObIs. CSC Sky utilizes KML drawOrder tags based on ObI exposure length to determine rendering order.

Instrument Fields of View

CSC Sky displays instrument fields of view (FoVs) of all ObIs that have been taken by Chandra. The FoVs are color coded to specify instrument and gratings configurations and are listed in the navigation panel sequentially by observation identifier (ObsId). Each entry contains the obsid, target name, detector icon, and visibility toggle. Clicking on a FoV or an entry in the list invokes a balloon containing observation information and a link to a WebChaSeR page that provides observation details.

The FoVs are preprocessed into KML region files. The FoV FITS files are retrieved from the archive and header keywords are extracted to obtain observation information. The detector regions are remapped to have RA run between −180° and 180° for display in Google Earth. The data are inserted into template files to generate placemark KML files. To allow independent updates, two sets of placemarks are maintained — one containing all FoVs, and one containing FoVs of ObIs included in the Chandra Source Catalog. The catalog FoVs only need to be updated when new versions of the catalog are released while the list of all FOVs can be updated via a periodic cron job to remain current.

Source Pushpins

Real time queries to the Chandra Source Catalog provide the user with a list of Master sources (X-ray sources on the Sky) when the area of the sky displayed in Google Earth spans an area of less than approximately five arc degrees. The queries are performed via a network link which executes a CGI script to access the catalog's public commandline interface. Using cURL, a query is performed and the resultant VOTable

---

[4]http://cda.harvard.edu/chaser

is piped to a Python script where it is parsed into a KML template to create placemarks for each source and corresponding balloons containing detector specific source properties. Clicking on any of the source markers invokes a balloon which displays a subset of catalog information for the selected source. Our network links are configured to dynamically update two seconds after the end of mouse motion.

Figure 1.   CSC Sky provides color coded instrument fields of view, X-ray images (overlaid here on DSS/STScI optical images), and X-ray source markers and property balloons retrieved through real time queries to the Chandra Source Catalog.

Startup Performance

The initial prototype of CSC Sky utilized a single KML file which provided overlay information at three blocking levels for each observation in the Chandra Source Catalog. At startup, the entire 10 megabyte file was read.

To improve startup time, we converted the file to use network links. The original file was split into over 11,000 component files. A base file, which is loaded at startup, identifies each ObI's position in the sky and contains links to load the appropriate component subfile when the ObI is viewable on the screen. The component files are only accessed when needed and, once accessed, set up links to further load subcomponents as necessary. Since only the base file (2.5 MB) is loaded at startup, the initial load volume was reduced by 75%.

Using zip to compress our base KML file also reduced download volume. The base file was deflated 92% (from 2,487,321 bytes in the KML to 206,334 bytes in the

KMZ). We considered the trade off of bandwidth vs. CPU time (for uncompressing the KMZ on the client side) and opted for the improved download speed.

Data Volume

By taking advantage of KML's level of detail (LOD) functionality and viewRefresh-Mode settings, we reduced the loading of images to only occur when the display is zoomed in sufficiently to require the detail available in the blocked images. This allowed us to rescale our initial images (all 2k × 2k) without significant loss of detail. The top level images were trimmed down to 512 × 512 pixels. The other images were rescaled to be 1k × 1k. By converting our zoomed-in images from 24 bit true color visuals to 8 bit pseudo color we were able to even further reduce our image volume. We determined that the slight degradation in image quality was an acceptable trade off for an approximate 80% savings in image data volume.

Conclusion

Google Earth in Sky Mode's ease of use, support of overlays, and other features make it a powerful tool for data presentation. CSC Sky offers everyone access to the wealth of information available in the Chandra Source Catalog. As an educational tool, it provides students a visual introduction to astrophysics. Educators can add custom tours incorporating various multimedia features such as movie clips and images. As a proposal planning aid, it provides details as to which areas of the sky yearn to be explored. As a professional tool, it serves as an entry point to the Chandra Source Catalog.

While CSC Sky provides much, there is still room for improvement. Due to scheduling constraints, processed mission data including blocked images were used to display X-ray sources. Reprocessing the data to allow super overlays to be created would allow all Chandra ObIs, including those not in the catalog, to be accessed in CSC Sky. A single layer all-sky mosaic of X-ray emissions could potentially be added which could be toggled on or off to allow comparisons with mosaics from other missions.

As the Google Earth interface supports html and javascript within source balloon descriptions, more complex information presentations are conceivable. The balloons could be expanded to add tabs to organize the data by category. Additional content such as observation details, quick look images and plots, and links to data from other missions are possibilities. Finally, access to the data through direct links or through interprocess communication such as Simple Application Messaging Protocol (SAMP) would considerably extend functionality.

Acknowledgments.   Support of the development of CSC Sky is provided by the National Aeronautics and Space Administration through the Chandra X-ray Center, which is operated by the Smithsonian Astrophysical Observatory for and on behalf of the National Aeronautics and Space Administration under contract NAS8-03060. The optical images in Figure 1 appear courtesy of DSS and STScI through Google Earth.

References

Evans, I. N., et al. 2010, ApJS, 189, 37
Lupton, R., Blanton, M. R., Fekete, G., Hogg, D. W., O'Mullane, W., Szalay, A., & Wherry, N. 2004, PASP, 116, 133

Astronomical Data Analysis Software and Systems XX
ASP Conference Series, Vol. 442
Ian N. Evans, Alberto Accomazzi, Douglas J. Mink, and Arnold H. Rots, eds.
©*2011 Astronomical Society of the Pacific*

CSCview: A Graphical User Interface to the Chandra Source Catalog

D. Van Stone, P. Harbo, M. Tibbetts, and P. Zografou

Smithsonian Astrophysical Observatory, 60 Garden St., Cambridge, MA 02138

Abstract.   CSCview is a graphical user interface to the Chandra Source Catalog. The interface uses a Java applet which accesses the catalog through a middle-tier Java EE application server. A tabbed interface allows the user to navigate through the main functions of the application: selecting from a list of catalog versions, creating a query, displaying the results, and downloading products. The application includes other features, such as a persistent set of user preferences and a guided help screen. CSCview had its initial release in 2009 and is being updated to include new features and to accommodate evolving requirements.

1.   Introduction

CSCview[1] is a graphical user interface designed to allow access to the Chandra Source Catalog (CSC). The first release of CSCview in 2009 included a query builder for catalog searches and the ability to retrieve data products from the catalog. More recent releases in 2010 added new features such as image viewing, SAMP[2] messaging, and crossmatching. What follows is a brief overview of the application architecture and a description of the tabbed interface and other features of the application.

2.   CSCview Architecture Overview

CSCview was designed as a Java applet running in a user's web browser, accessing the CSC through a middle-tier Java EE application server. The CSC contains Master Sources (distinct X-ray sources on the sky), Per-ObI Sources (source observation intervals which contributed to a Master Source), and associated data products (Evans et al. 2010). The Chandra Data Archive provides access to the CSC using both an archive server (for product files) and a database server (for properties and product file metadata). See Fig. 1, left.

In CSCview, the available properties are presented as a set of four interconnected tables: the *master_source* table containing the Master Source properties, the *obi_source* table containing the Per-ObI Source properties, the *master_obi_assoc* table specifying

---

[1]`http://cda.harvard.edu/cscview`

[2]Simple Application Messaging Protocol. See `http://www.ivoa.net/Documents/latest/SAMP.html`.

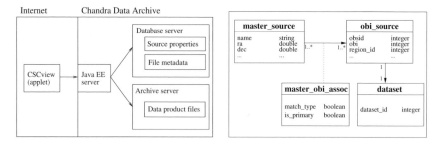

Figure 1.   Left: CSCview architecture layers. Right: The user view of the properties.

which Per-ObI Sources contribute to which Master Sources, and the *dataset* table providing links from the Per-ObI Sources to the data products. See Fig. 1, right.

3.   User Interface Elements

CSCview uses a tabbed interface for navigation through the four pages: selecting a catalog, building a query, displaying results, and downloading products.

On the Catalog Page, users can select which version of the catalog they wish to use: the current catalog release, any previous catalog release, or the live database view which includes any new data processed since the last release.

On the Query Page (Fig. 2), users can choose, combine, and modify any of the example queries, or they can create their own custom query. In a custom query, users can choose the properties they want displayed and in which order, specify their search criteria, and restrict their selection using a cone search or a crossmatch. A crossmatch will match sources in the user's table against the catalog using a specified radius and calculating a basic probability based on position errors. The input table of point sources can be typed in, loaded from disk, or received from other applications with SAMP. There is a Show Language option in the View menu which allows users to see their query expressed in ADQL[3] and make modifications before searching. Lastly, the user can save their query to disk and run it using a command-line interface.

On the Results Page (Fig. 3), users can view the properties and save them to disk in multiple formats, such as tab-separated value or VOTable,[4] or they can send the results to any SAMP-enabled application for further analysis. Users can also preview their sources with the Source Preview (Fig. 4), or they can specify data products to browse for the selected sources.

On the Products Page, users can select products to download to disk, either as individual files or as a single package. The users can also send FITS images and tables to any SAMP-enabled application for further analysis.

When CSCview launches, a help screen ("Getting Started") pops up to guide the new user through using the application. CSCview also provides preferences (such as

---

[3]Astronomical Data Query Language. See http://www.ivoa.net/Documents/latest/ADQL.html.

[4]Virtual Observatory Table. See http://www.ivoa.net/Documents/VOTable.

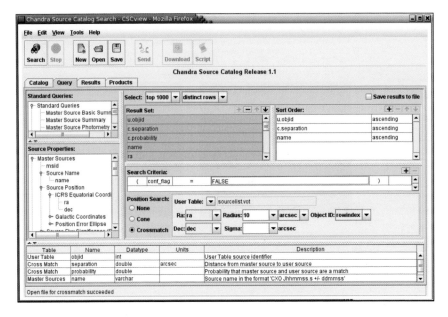

Figure 2.   CSCview: Query Page.

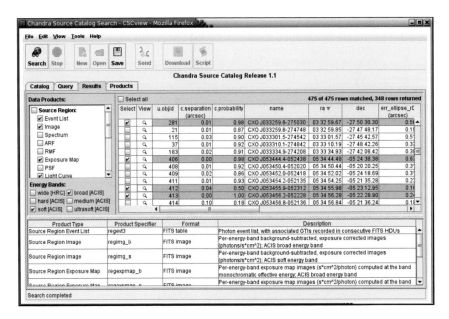

Figure 3.   CSCview: Results Page.

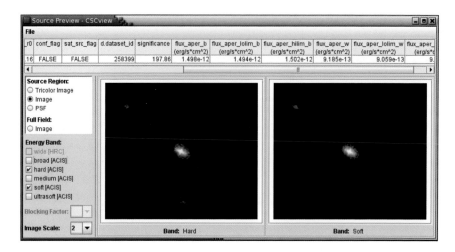

Figure 4.   CSCview: Source Preview.

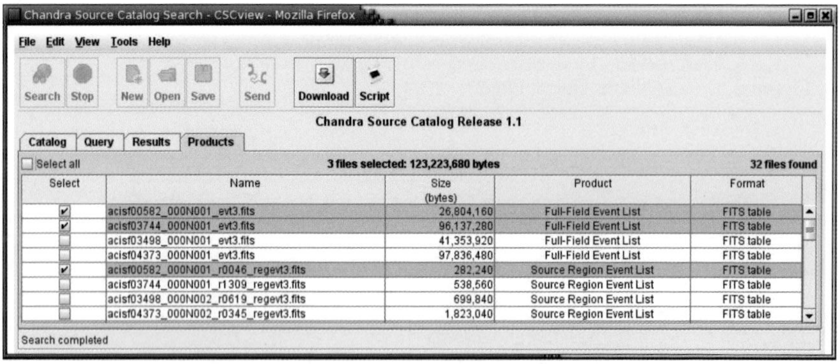

Figure 5.   CSCview: Products Page.

display formats, toolbar appearance, etc.) which are persistent between sessions. New features are expected to be added in the future to meet evolving requirements.

Acknowledgments.   Support of the development of the Chandra Source Catalog is provided by National Aeronautics and Space Administration through the Chandra X-ray Center, which is operated by the Smithsonian Astrophysical Observatory for and on behalf of the National Aeronautics and Space Administration contract NAS8-03060.

References

Evans, I. N., et al. 2010, ApJS, 189, 37

Part XIV

Birds of a Feather Sessions

Astronomical Data Analysis Software and Systems XX
ASP Conference Series, Vol. 442
Ian N. Evans, Alberto Accomazzi, Douglas J. Mink, and Arnold H. Rots, eds.
©*2011 Astronomical Society of the Pacific*

A Journal for the Astronomical Computing Community?

Norman Gray[1] and Robert G. Mann[2]

[1]*School of Physics and Astronomy, University of Glasgow, UK*
[2]*Institute for Astronomy, University of Edinburgh, Edinburgh UK*

Abstract.   One of the Birds of a Feather (BoF) discussion sessions at ADASS XX considered whether a new journal is needed to serve the astronomical computing community. In this paper we discuss the nature and requirements of that community, outline the analysis that led us to propose this as a topic for a BoF, and review the discussion from the BoF session itself. We also present the results from a survey designed to assess the suitability of astronomical computing papers of different kinds for publication in a range of existing astronomical and scientific computing journals. The discussion in the BoF session was somewhat inconclusive, and it seems likely that this topic will be debated again at a future ADASS or in a similar forum.

1.   Introduction

The ADASS conference series is approaching the dawn of its third decade in robust health. The past twenty years have seen a marked increase in the importance of computation in support of astronomical research, and this trend seems set to continue, as the data volumes emerging from detectors and simulation codes increase exponentially. The ADASS proceedings volumes provide a very valuable record of each year's conference, but imperfectly record the activities of the ADASS community, for a number of reasons: (i) appearing up to a year after the conference, they often present out-of-date snapshots of rapidly-developing projects; (ii) being unrefereed, there is no quality threshold, nor are authors pushed to justify and elaborate where needed to provide the best account of their material; (iii) being tied to the annual conference cycle, projects are reported upon when the opportunity arises, not when they have reached appropriate milestones; and (iv) having restricted page lengths, topics receive only brief coverage.

This matters for at least two reasons. Firstly, material from most ADASS conference papers will be published nowhere else, so valuable technical lessons risk being lost. Secondly, as more people pursue a career in astronomical computing, it becomes more important that they have a means of recording their attainment and a track record of refereed journal papers, with associated citation statistics, is what is most readily understood by potential employers and assessors of promotion applications.

Open source publishing systems make the establishment of a community-driven astronomical computing journal possible, but is it necessary? Few papers on computational topics appear in mainstream astronomy journals, but is that a reflection of the journals' editorial policies or a lack of interest or confidence on the part of the community? Would mainstream astronomy journals publish more technical papers if they were submitted? Are there (scientific) computing journals that would welcome these papers? What are the benefits of refereeing to this community and would it devote the time needed to referee? Does the lack of refereed publications hinder the career progress of its members?

2. The Community

To the extent that there is an astronomical software community, it is represented by its attendance at, and support for, ADASS. Indeed, the primary publication outlet for many, or perhaps even most, of the ADASS attendees appears to be the ADASS proceedings, rather than a cluster of specialised or conventional journals. This is an odd situation for an academic discipline, and so the non-appearance of (the presumed non-null set of) software articles worthy of journal publication may have several possible explanations.

(1) The primary output from the discipline is software, not articles: are these a better (or indeed usable) metric for recognition?

(2) The members of the community tend to be in service roles — from those performing routine software development, to the managers of important parts of the astronomical community's infrastructure — and so their career advancement may depend on publication to a lesser extent than conventional astronomers.

(3) There may be no suitable publication outlets, since the existing astronomical journals are uninterested in publishing what they regard as computing science, and computing science journals are uninterested in publishing such applied work.

(4) The community is perhaps such that there is not, or not yet, any *expectation* that software results will be published in journal form, and the community has therefore not developed any shared intuitions about what work is sufficiently valuable, or sufficiently interesting, for formal dissemination and careful preservation.

Problem 1 is too different a question to be considered here. Problem 2 may have been true in the past, but it is surely becoming less true, partly because there is more career crossover between software and observational astronomy now, than there has been in the past, and because with astronomy's accelerating move towards HEP-scale experimentation, and the repeated warnings of the forthcoming 'data deluge', a broader range of software technologies have become integral to present and future astronomical practice. What this means in turn is that there is a growing number of individuals whose principal intellectual excitement, and whose principal contribution to astronomy, is *via* innovative software and system development. These people are not observers or theorists, nor are they computer scientists, but are instead something in between. The term *astroinformatics* seems convenient.

3. The Survey

To test hypothesis 3, we mailed the editors of MNRAS, A&A, ApJ, Earth Science Informatics, Experimental Astronomy, CODATA Data Science Journal, Astronomische Nachrichten and PASP, with five titles and abstracts from last year's ADASS proceedings. This set of five articles was chosen because each seemed typical of one or other class of publication commonly presented at ADASS, and we asked the editors to assess whether the subject matter of the article, independently of its body, would be deemed sufficiently in scope for it to be passed on to a referee.

The article topics were: **alg** — software implementation of scientific algorithms (Bayesian techniques for classification); **app** — application progress report (WWT update); **pipe** — pipeline features and recent developments (detailed report of new IDL pipeline features); **gen** — application of general computing technologies to astronomy (application of Java and HPC techniques to a specific mission); **inf** — development and use of astronomy-specific 'infrastructure' (benefits of HEALPix in a particular application).

Table 1.   Summary of journal survey results

	alg	app	pipe	gen	inf
A&A[1]	yes?	no?[2]	?	no	?
MNRAS[3]	no	no	no	no	no
ApJ[4]	yes?	no?[5]	yes??	no	yes??
ESIn[6]	yes	yes	no?	yes	yes
DSJ[7]	yes	yes	no?	yes	yes
PASP[8]	yes	yes	no?	yes	yes

The responses are summarised in Table 1, and are published in full at the following website: `http://www.roe.ac.uk/~rgm/bof.html`.

Notes to Table 1: 1. Could appear in 'Astronomy Instrumentation' section; must be "of interest to a sizable fraction of the A&A audience"; 2. Issue of VO tools to be discussed soon by editors; 3. "Descriptions of new software appear only if accompanied by new science derived using it"; 4. "[O]ur enthusiasm for techniques papers tend to fall off as they become less concerned with direct results and more divorced from ongoing science projects", but OK if "the paper will be interesting even if the particular instrument never actually gets built"; 5. Might be considered for a WWT special issue; 6. "The topics ... easily fall under the focus of this journal. [...] We consider astronomy informatics a sister domain, related to Earth Science Informatics"; 7. "[D]efinitely in the scope" of DSJ; 8. "We have indeed accepted articles such as the 5 that you sent". No replies were received from Experimental Astronomy or from Astronomische Nachrichten.

4.   Summary of BoF Discussion

Opinion varied widely as to the necessity of a new journal, but there was general enthusiasm at the idea of participating — as readers, authors, referees, and, in some cases, as editors — in such a new journal, if one were to exist.

Some people felt that the increasing importance of computational techniques in astronomy today necessitates the creation of a dedicated journal, and the analogy was made with particle physics, which has long since made a definite split between journals for science results and journals for technical material (experimental details, as well as analysis software, etc). The latter stream is highly valued, both for providing a means of sharing and recording technical knowledge, and for rewarding the efforts of more technical staff, prompting the suggestion that astronomy would benefit from the same system. The case was made that previous attempts to provide such outlets have not met with great enthusiasm from the ADASS community. The AIP's "Computers in Physics" journal had a similar intention to that proposed here (although covering all of physics, rather than just astronomy), but few astronomy-related papers were published there, and the same goes now for Experimental Astronomy.

There was general agreement that the ADASS community would benefit from producing more refereed papers, but most people felt that the opportunities provided by existing journals should be exhausted before serious consideration is made of starting a new one. This might even by coordinated — whether through the dedication of special issues, or more informally amongst authors — in an attempt to produce a journal with a critical mass of material about astronomy computing. It is not clear whether the increasing importance of this domain is better highlighted by the creation of a dedicated journal, or by making a significant presence within an existing journal.

Many people see PASP as the most appropriate outlet for papers on topics requiring a fuller treatment than allowed by the ADASS proceedings, and others mentioned that papers on algorithms can find a home in mainstream astronomy journals, so the problem of *excluded* material is largely restricted to descriptions of pipeline software and the like, whose details should be recorded and made available to their users, but which may lack the conceptual novelty required by most journals (that is, problem 3 may apply only to articles of type **pipe** in Table 1). This both records and advertises the authoring software group's contributions to the astronomical enterprise.

A requirement was identified for an outlet for publishing lessons learnt of the "we did this, but it didn't work because of these reasons" sort. That could be provided by a non-refereed section of a new journal, or, equally, by postings to astro-ph. There is already an "instruments and methods" chapter there, that is currently poorly used by the ADASS community, but it could be transformed by greater use into a suitable vehicle for knowledge exchange within the community.

Another advantage of publishing in astro-ph is that research astronomers are used to looking there, which would be an advantage for some papers in this domain, although, equally, some others may benefit more from publication in a journal (such as Earth Sciences Informatics or CODATA Data Science Journal) that is read by people working on analogous topics in related domains.

5. Conclusions

Our survey reveals that, although the high-impact 'big three' journals do not see ADASS material as naturally in their scope, there are other journals which would be perfectly willing to consider articles, and that conclusion was shared by the BoF participants. There seems little present need for a new journal.

The key question then becomes number 4: why does the community not publish in the journals that are available to it? We look forward to a spike in astroinformatics journal articles in 2011.

Acknowledgments. We are grateful to the journal editors who were generous in examining the sample papers we sent them, and in clarifying the goals they have for their journals. We are grateful to the many people who attended the BoF in Boston, and for their thoughtful comments. We are particularly grateful to Rob Seaman for his persistence in unearthing more journals than we believed possible.

Astronomical Data Analysis Software and Systems XX
ASP Conference Series, Vol. 442
Ian N. Evans, Alberto Accomazzi, Douglas J. Mink, and Arnold H. Rots, eds.
©*2011 Astronomical Society of the Pacific*

Astronomy Visualization for Education and Outreach

Alyssa A. Goodman,[1] Patricia S. Udomprasert,[1] Brian Kent,[2]
Hanna Sathiapal,[3] and Riccardo Smareglia[4]

[1]*Harvard-Smithsonian Center for Astrophysics, 60 Garden St, MS 42,
Cambridge, MA 02138*

[2]*National Radio Astronomy Observatory, 520 Edgemont Road,
Charlottesville, VA, 22903*

[3]*Fingertip Hands-On Exhibits, Birchlenstrasse 10,
8600 Dübendorf, Switzerland*

[4]*Istituto Nazionale Di Astrofisica, Osservatorio Astronomico di Trieste,
Via Tiepolo 11, 34143 Trieste, Italy*

Abstract.   About 50 participants came to a discussion on the benefits and potential obstacles of using astronomy visualization tools for education and public outreach (EPO). Representatives of five different EPO organizations shared information on their project goals and outcomes. Public users need support to learn how to use these programs effectively for education, but the efforts are worthwhile because the thrill that comes from working with real data and the natural beauty of astronomical imagery are great attractors for new science enthusiasts.

1.   Introduction

Scientists and educators have an opportunity to engage the public and improve science education through the use of freely available professional data visualization tools. These tools, coupled with vast archives of data, give educators the potential to reach any demographic, anywhere in the world, so long as the users have a computer and an internet connection. Klahr et al. (2007) have shown that on several different measures, children were able to learn as well with virtual as with physical materials, and the inherent pragmatic advantages of virtual materials in science may make them the preferred instructional medium in many hands-on contexts.

The main obstacle to widespread use of these tools by the public and in classrooms is the learning curve associated with them. Primary and secondary school teachers sometimes require support in astronomy content as well. Appropriate training for teachers and lay-users, whether by the scientists themselves, or by designated instructors trained by the scientists is essential to success of these programs.

2.   Education and Outreach Programs Using Astronomy Visualization Programs

During the BoF, we heard about five organizations around the world that use astronomy visualization tools in education and public outreach. Here we summarize the contributions from each program presented during the BoF.[1]

2.1.   WorldWide Telescope Ambassadors Program (Alyssa Goodman)

The WorldWide Telescope Ambassadors (Goodman et al. 2010) are astronomy experts who use WorldWide Telescope (WWT) to educate the public about astronomy, space, and physics. Ambassadors and learners alike use WWT to create dynamic Tours of the Universe, and they share them in schools, public venues, and online.

In spring 2010, we began a pilot with 6th graders at a middle school near Boston, MA, where students used WWT to supplement their traditional textbook materials while studying curricular subjects like Earth/Sun/Moon relationships. Students also used WWT during a 6-week independent research project, creating Tours to share what they learned with classmates. The students we worked with showed measurable gains in interest in astronomy and science, and in their ability to visualize complex 3-dimensional relationships. The gains found were in comparison to another group of 6th graders at the same school who did not use WWT. In a survey, 71 out of 72 students were highly positive about WWT, telling us that "learning about our Universe by actually seeing and exploring it makes it easier to contemplate and more fun," and "it gave me a better mental map of the universe." One particularly enthusiastic student described working with WWT as "awesome, amazing, cool, incredible (repeat 30 times.)" We are now expanding the program to additional schools in Boston and beyond.

We are currently building a collection of WWT Tours that spans across a broad spectrum of astrophysics topics. The Tour collection will be integrated with WGBH Teachers' Domain, a freely available online service including 500,000 registered users.[2]

2.2.   The European Virtual Observatory for Students and Teachers (Riccardo Smareglia)

The EuroVO-AIDA project (Ramella et al. 2010) aims to extend the benefits of the Virtual Observatory for professional astronomers to students and teachers. The core of our products are a) professional software tools adapted to the special needs of schools and young users (Aladin, Stellarium, SimPlay) and b) a library of use cases presenting simple astronomical problems (such as celestial coordinates, distance determinations, and stellar evolution) and a step-by-step guide to their solution with VO data and tools.

Our products have been developed in two one-year cycles. During the first cycle we used an initial adaption of Aladin and Stellarium and a first set of use cases for a total of more than 200 hours in classes of selected middle- and high-schools. During the second cycle we tested the ability of teachers to use our products autonomously. At the end of the school year we implemented as many suggestions of our testers as possible and released our products to the public. In the end, more than 1000 students and 150 teachers participated to the development of our software and use cases.

---

[1]Speakers at the BoF are listed in parentheses after each group; contributors to the proceedings text are in the author list; program websites and key members of the organizations are listed in the References.

[2]www.teachersdomain.org

At the Trieste Astronomical Observatory we complement VO tools and use cases with observations at our remotely controlled telescope.[3] The combination of the two activities gives an involving glimpse of the "real" astrophysical research.

2.3. HelioLab (André Csillaghy)

HelioLab aims to create awareness among the public that the Sun is actually an object of current research, and that many scientific products are freely available to all (Sathiapal et al. 2010). HelioLab consists of three interlinked modules: the weblab, the schoollab and the expolab. These are linked to online resources such as near real-time images from the Solar Dynamics Observatory (SDO) or the Solar & Heliospheric Observatory (SOHO).

The schoollab is a kit with observation instruments and activities that a scientist brings into the classroom. It supports scientists working with young children, encourages elementary school teachers without a background in science to teach science subjects, and provides an opportunity for K-6 students to get to know a scientist as an active partner in a school project over an extended period of time. The expolab is a similar collection of ready to use educational material for public events. The weblab is a webpage that introduces heliospheric science to the general public. It contains instructions for building a simple observatory at home and provides access to a variety of information related to research in heliophysics. The weblab will be translated into several languages in order to increase the bridging effect to the online resources. HelioLab is an EPO service by HELIO — The Heliophysics Integrated Observatory funded by the European Commission's 7th framework program.

2.4. ALFALFA and Google Sky (Brian Kent)

The Arecibo Legacy FAST ALFA Survey (ALFALFA) has utilized Google Sky as an environment to disseminate radio astronomy data for education and public outreach. Upon processing by members of the spectroscopic survey collaboration, galaxy catalogs and spectra are exported to a database that can be called via the project website, web services, or external applications including Google Sky (Kent et al. 2008). The size and transparency of data symbols can be manipulated in Google Sky KML (Keyhole Markup Language) files to represent quantities such as luminosity and mass. Users can also click on a galaxy to retrieve database information, images, and spectra. The radio data are an excellent complement to the underlying imaging as they show the nearby gas-rich universe within 250 Mpc.

Undergraduate students participating in the ALFALFA program use Google Sky in this manner to study HI mass functions of spiral galaxies. The interactive program environment allows students to engage immediately with the data.

2.5. Network for Astronomy School Education (Carlos Gabriel)

The International Astronomy Union's Network for Astronomy School Education (NASE) provides astronomy training to primary and secondary school teachers in developing countries (Ros et al. 2010). The NASE group organizes courses to train teachers in different regions of the globe, giving them an opportunity to learn the astronomy content and perform hands-on observations at their schools. During a typical course, 40–

---

[3]http://scuole.oats.inaf.it

50 school teachers are taught by 6 NASE and IAU members during 4 days of classes. The teachers who attended then receive support from NASE to reproduce the training they received in their own region or country. Courses have recently been completed in Argentina, Peru, Nicaragua, and Colombia. The educational resources and course materials are currently available in English and Spanish, and translations into Arabic, Portuguese, and French are planned.

3. Conclusions

Several noteworthy points were raised during the discussion. Participants observed that publicly funded astronomical collaborations around the world are required to allocate a certain percentage of their budget to EPO work. However, the reality is that EPO work frequently remains understaffed, and we can make more progress if scientists dedicate time and resources to helping the public understand what we do. Some members of the audience pointed out that there are language and other international barriers to accessing educational content from different countries. We note that several of the organizations here already have translated or have plans to translate their materials into multiple languages. Finally, we discussed the fact that astronomy is frequently left out of "essential" curriculum topics, making it a challenge to engage teachers willing to devote time to it. We argue that astronomy's universal ability to excite, amaze, and inspire, makes it an essential gateway to the study of science, and crucial scientific skills such as observation, and hypothesis making and testing, can be learned through astronomy as well as any other science.

Acknowledgments. The WWTA team would like to thank Microsoft Research for funding the WWTA pilot project discussed in Section 2.1, and we thank all the BoF participants who shared information on their projects and everyone who participated in the discussion.

References

Goodman, A. A., et al. 2010. URL http://www.cfa.harvard.edu/WWTAmbassadors/
Kent, B. R., et al. 2008, AJ, 136, 713
Klahr, D., Triona, L. M., & Williams, C. 2007, J. Res. Sci. Teach., 44, 183
Ramella, M., Iafrate, G., Smareglia, R., Boch, T., Bonnarel, F., Chéreau, F., Fernique, P., & Freistetter, F. 2010, http://wwwas.oats.inaf.it/aidawp5/
Ros, R. M., et al. 2010, http://www.iaucomm46.org/web_nase/index3.html
Sathiapal, H., et al. 2010, http://www.heliolab.eu/

Astronomical Data Analysis Software and Systems XX
ASP Conference Series, Vol. 442
Ian N. Evans, Alberto Accomazzi, Douglas J. Mink, and Arnold H. Rots, eds.
© *2011 Astronomical Society of the Pacific*

Towards HDF5: Encapsulation of Large and/or Complex Astronomical Data

M. Wise,[1] A. Alexov,[2] M. Folk,[3] F. Pierfederici,[4] K. Anderson,[2] and
L. Bähren[2]

[1]*Netherlands Institute for Radio Astronomy (ASTRON),*
Postbus 2, 7990 AA Dwingeloo, The Netherlands

[2]*Astronomical Institute "Anton Pannekoek", University of Amsterdam,*
Postbus 94249, 1090 GE Amsterdam, The Netherlands

[3]*The HDF Group, 1800 South Oak Street, Suite 203, Champaign, IL 61820*

[4]*Space Telescope Science Institute (STScI),*
3700 San Martin Drive, Baltimore, MD 21218

Abstract.　The size and complexity of astronomical data are growing at relentless rates. This increase is especially apparent in the radio community as evidenced by the data challenges faced by many of the SKA pathfinders and other major new radio telescopes such as LOFAR, EVLA, ALMA, ASKAP, MeerKAT, MWA, LWA, and eMER-LIN. Enormous data rates are also becoming a challenge for large optical projects that are currently ramping up including Pan-Starrs and LSST. As progress towards meeting these challenges, ASTRON and the LOFAR project are currently exploring the use of the Hierarchical Data Format, version 5 (HDF5) for LOFAR radio data encapsulation. In this session, we brought together scientists and developers struggling with large and complex datasets as well as those groups currently exploring HDF5 implementations. This paper briefly summarizes the BoF presentations, material discussed and future goals.

1.　LOFAR Overview

A description of The LOw Frequency ARray (LOFAR), as well as additional complimentary information on LOFAR and HDF5, can be found in Anderson et al. (2011), in this volume.

　　LOFAR can be used in many different observing modes: examples of more frequently used modes are Imaging/Visibility, Beam-Forming (BF)/Time-Series and Transient Buffer Board (TBB) dumps. All these observing modes create data diversity and variety. Starting in 2008 the LOFAR project has been concentrating on writing the Interface Control Documents (ICDs) for 6 basic LOFAR data format types. These documents are based on scientific input on how to best store and package the data for maximum scientific clarity and usability. ICDs are discussed in more detail in Anderson et al. (2011). In some of these LOFAR observing modes data are combined from all the stations while in other modes data are stored per individual station, and if needed, at the highest resolution. Because of this flexibility in observing, data rates can be as extreme as 23 TB/hour [~ 6GB/s].

2. Astronomical Data Containers: the Move to HDF5

In order to be able to store LOFAR's large and complex data, we first had to evaluate whether the formats frequently used in astronomy would suffice our needs. The FITS data format is widely used across most wavelengths in astronomy while Casacore Tables are predominantly used for radio data. MBFITS is a FITS format specific to multi-beam radio observations. In addition there are also dozens of different binary formats for various telescopes and instruments, mostly to store time-series data.

  We decided against our own proprietary, binary format solution because software complexity becomes disorderly for different data I/O for every LOFAR format. Additionally all in-house and off-the-shelf tools would then need to be adapted to read these binary data containers. We wanted to find one single format for our primary needs, which were: data diversity, extreme data sizes, parallel I/O, distributed file systems, while at the same time be complementary to our C++/Python in-house software suite. Neither a FITS-based nor CASA-based solution alone could meet all the requirements for our complex and large formats. Our conclusion was that the Hierarchical Data Format, version 5 (HDF5) is the best solution for our data format needs.

  HDF5 is a data model, library, and file format for storing and managing large and complex scientific data (images, N-D arrays, tables, metadata). It supports an unlimited variety of datatypes, and is designed for flexible and efficient I/O and for high volume and complex data. Additionally HDF5 has the following advantages, most of which are all vital to our needs: (1) Self-describing and portable to a diversity of computational environments; (2) No inherent file size limitations; (3) C, C++, Java, Fortran 90, Python interfaces; (4) Can be run on single node or massively parallel/distributed systems; (5) Built-in compression (GNU zlib, but can be replaced with others); (6) Parallel reading and writing (via MPI-I/O); (7) "Chunked" (tiled) data for faster access.

  HDF5 is freely available and has been in use for over 20 years by NASA and other organizations. The HDF Group wrote HDF4/5 and continues to maintain and advance this software tool suite. They are a non-profit company with a staff base of 30 FTEs, are available for consultancy, have an active helpdesk mailing list, and are interested in helping the astronomical community make the transition to using HDF5.

3. LOFAR & HDF5 Current Status

We have integrated the HDF5 format into the LOFAR ICD documents for each type of LOFAR data. Following the hierarchical nature of HDF5, we structured our data formats accordingly, taking into account scientific input (see Fig. 1 & 2 in Anderson et al. 2011). The format is flexible enough to be able to describe all our complex format needs.

  The LOFAR User Software (LUS) repository links against the HDF5 v1.8.x package. LUS is publicly available and contains LOFAR tools, pipelines and the pipeline framework. It contains a C++ Data Access Library (DAL) which is an intermediate layer on top of HDF5. The DAL can be thought of as the software equivalent of the LOFAR ICDs for the purpose of reading, writing and accessing the LOFAR data. Additionally, the LUS contains a DAL Python wrapper, called pyDAL, which gives users access to these same data using a Python interface.

  The LUS is in Beta release and is currently being actively used and further developed by the LOFAR science pipeline teams. We are at the cusp of writing LOFAR data in HDF5. We will spend the next year working out the "kinks" in the system, and evolving the LOFAR HDF5 files based on what we learn about I/O performance and data structure containers. It is likely we will iterate over the LOFAR ICD data formats as we gain experience with HDF5.

4.   The Future Is Bright for HDF5, LOFAR, and Other Projects

LOFAR is in the process of meeting its data size and complexity challenges by investing time and resources in the HDF5-learning curve. Future telescopes have similar challenges. "Big Data" has traditionally been the prerogative of radio astronomy and particle physics, among others. The new optical ground based surveys which have just started operations (e.g. Pann-STARRS) and the ones about to come online (e.g. LSST) are changing the rules of the game.

LSST, an 8 m class survey telescope with a 3.2 giga-pixel camera encompassing a full 9.7 square degrees field of view is going to produce an avalanche of data: 15 TB (16 bit, science and calibration) worth of raw images every 24 hour. This translates into a yearly data volume of 10–11 PB and a final archive of ~ 100 PB after ten years. Traditional astronomical file formats are clearly inadequate to handle this unprecedented amount of data. New formats have to be identified, likely in those scientific fields where high data rates and huge data volumes are routinely handled already. HDF5, being one of the most used formats in data-heavy scientific fields, is extremely promising for the LSST project.

JWST is researching file formats to meet their data complexity needs; the Meer-KAT project is already writing radio visibility data in HDF5 using Python and evaluating the viability of using the LOFAR ICD data formats. The simulation community uses HDF5 (GADGET, ENZO, FLASH). The HDF Group regularly works with the user community to enhance HDF5 functionality as well as add project-specific features; they recently worked closely with the iRODS Grid project to provide an HDF5 virtual file layer on top of an iRODS system which resulted in a major gain in performance.

We hope that upcoming projects facing similar data challenges are not fooled into thinking that project-specific binary is the only solution; the issues with long-term maintenance and lack of astronomy tool-sets are bound to hurt the project and isolate their data users. We encourage the use of HDF5 in the astronomical community. We feel that the time is ripe to solve this issue across wavelengths and projects. HDF5 is mature and is being used extensively in science; Figure 1 conveys the growth of data within NASA's earth-orbiting missions, almost all of which generate data products in some flavor of HDF5 (*e.g.* netCDF-4, HDF-EOS 5).

5.   The Next Generation Astronomical Data Effort

We have set up a moderated email list, called **nextgen-astrodata@astron.nl**. Interested parties are encouraged to sign up via **majordomo@astron.nl** with "**subscribe nextgen-astrodata**" as the only text in the body of the message. We will use this mailing list to keep everyone informed of our progress and hope this helps expand the conversation regarding the future of astronomical data formats. Our near-term task is to set up a wiki and forum for long-term discussion on HDF5 in the astronomical community.

The HDF Group has noted our concerns from the ADASS BoF and would like to work through these issues with our community. One issue raised was a need for tools to validate the data and its organization within a file (akin to an XML validator). The HDF

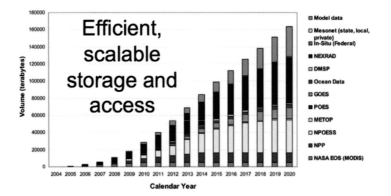

Figure 1.   Shows the increase in NASA's earth-orbiting missions scientific data storage since 2004 in TB, projected up to year 2020. The acronyms are: **NASA EOS (MODIS)**: NASA Earth Observing System (Moderate Resolution Imaging Spectroratiometer); **NPP**: NPOESS Preparatory Project; **NPOESS**: National Polar-orbiting Operational Environmental Satellite System; **METOP**: Operational Meteorology satellite; **POES**: Polar Operational Environmental Satellite; **GOES**: Geostationary Operational Environmental Satellite; **DMSP**: Defense Meteorological Satellite Program; **NEXRAD**: Next-generation Radar

Group acknowledges this as a valuable addition, and is currently developing an XML Infoset description of the HDF5 data model, which may be able to form a basis for validation tools. Another issue raised was the long-term archival risks of storing data in HDF5, given the complexity of the HDF5 format and library, and given the need for both to evolve with changing technologies. This was acknowledged to be a legitimate concern, but also one that the HDF Group continues to focus on. Other HDF5 users, such as NASA's Earth Observing System, provide resources to address the preservation challenge.

LOFAR is a small software project, especially considering the size, scope and complexity of its hardware. We would like to stress that we will need collaborations and additional funding to expand our LOFAR-oriented HDF5 work to a more generic and global astronomy, non-radio-specific, HDF5 data format. We hope to acquire grants with The HDF Group, and possible partners from any of current and future astronomy projects. We intend to approach the IVOA to inquire whether they would be the governing body on HDF5-astronomy standards. This is NOT just a "radio-problem", it's an astronomical problem. Please help us solve it and move into the next generation of data formats for future telescopes and missions.

Acknowledgments.   We would like to thank the LOFAR ICD team for all their hard work on the file formats and assistance with this paper, John Swinbank and Joe Masters for their assistance with the BoF, and the ADASS XX POC for accepting this BoF.

References

Anderson, K. R., Alexov, A., Bähren, L., Grießmeier, J.-M., Wise, M., & Renting, G. A. 2011, in Astronomical Data Analysis Software and Systems XX, edited by I. N. Evans, A. Accomazzi, D. J. Mink, & A. H. Rots (San Francisco, CA: ASP), vol. 442 of ASP Conf. Ser., 53

Part XV

Floor Demonstrations

Astronomical Data Analysis Software and Systems XX
ASP Conference Series, Vol. 442
Ian N. Evans, Alberto Accomazzi, Douglas J. Mink, and Arnold H. Rots, eds.
© *2011 Astronomical Society of the Pacific*

Looking Towards the Future of Radio Astronomy with the CyberSKA Collaborative Portal

Cameron Kiddle,[1] Mircea Andrecut,[1] Adam Brazier,[2] Shami Chatterjee,[2]
Eric Chen,[2] Jim Cordes,[2] Roger Curry,[1] Robert A. Este,[1] Olivier Eymere,[3]
Pavol Federl,[1] Bryan Fong,[4] Arne Grimstrup,[1] Sukhpreet Guram,[1]
Vicky Kaspi,[5] Rafal Klodzinski,[6] Patrick Lazarus,[5] Venkat Mahadevan,[7]
Atallah Mourad,[6] Shibl Mourad,[6] Paolo Pragides,[1] Erik Rosolowsky,[7]
Dina Said,[1] Alexander Samoilov,[6] Christian Smith,[1] Ingrid Stairs,[4] Mark Tan,[4]
Tingxi Tan,[1] A. R. Taylor,[1] and A. G. Willis[8]

[1]*University of Calgary, 2500 University Drive NW,
Calgary, AB, Canada T2N 1N4*

[2]*Cornell University, Ithaca, NY, USA 14853*

[3]*IBM Canada Ltd., 3600 Steeles Ave East, Markham, ON, Canada L3R 9Z7*

[4]*Dept. of Physics and Astronomy, University of British Columbia,
6224 Agricultural Road, Vancouver, BC, Canada V6T 1Z1*

[5]*McGill University, 3600 University Street, Montreal, QC, Canada H3A 2T8*

[6]*Sequence Factory, 6250 Hutchison, Suite 406,
Montreal, QC, Canada H2V 4C5*

[7]*University of British Columbia Okanagan, 3333 University Way,
Kelowna, BC, Canada V1V 1V7*

[8]*Dominion Radio Astrophysical Observatory,
National Research Council Canada, PO Box 248, Penticton, BC, Canada
V2A 6J9*

Abstract.   Advances in radio and digital processing technologies are enabling the construction of radio telescopes that will be able to probe the sky to unprecedented depths at radio wavelengths. With the vast amounts of data that will be produced by such telescopes comes a greater need for a cyberinfrastructure framework to connect the communities of astronomers with the data, processing and visualization tools, and each other. This paper introduces CyberSKA, an on-line, collaborative portal that is aimed at addressing the cyberinfrastructure needs of future radio telescopes such as the Square Kilometer Array (SKA).

1.   Introduction

The Square Kilometer Array (SKA) will be the world's largest radio telescope when completed, and will produce data at unprecedented rates. Most of the key science goals for the SKA will be achieved via large scale survey type observing programs. These will involve very high data rates and volumes, complex and multi-purpose processing and analysis, and will be executed by globally distributed teams of researchers. This drives a need for cyberinfrastructure solutions for data management, processing, visualization and collaboration.

To explore and meet these needs we are developing the CyberSKA on-line, collaborative portal (`http://www.cyberska.org`). As the portal develops, it will provide an environment where astronomers, engineers, technical and administrative staff, educators and students and even the general public can access and share data and computing resources and collaborate more effectively with each other. The portal will enable such functionality without users needing to be aware of the underlying technical details.

As a starting point we are working on supporting the cyberinfrastructure needs of various SKA Pathfinder projects that are making use of the world's current largest radio telescope, the Arecibo Observatory in Puerto Rico. In particular, we are working on supporting the needs of two large scale surveys, GALFACTS (Galactic ALFA Continuum Transit Survey) and PALFA (Pulsar Arecibo L-band Feed Array).

This paper proceeds by providing an overview of the CyberSKA portal in Section 2. Current usage of the portal is then discussed in Section 3. Finally, a discussion of the next steps we plan on taking is given in Section 4.

2.   CyberSKA Portal

CyberSKA is a distributed platform aimed at providing users with access to various data, high performance computing and cloud computing resources, and services at different participating sites via a common collaboration portal. This is depicted in Figure 1. Users are able to get access to all of the services in a transparent manner, without having to worry about the underlying technical details.

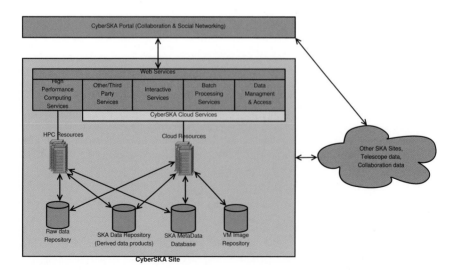

Figure 1.   High level architecture of CyberSKA.

On-line social networking sites such as Facebook (`http://www.facebook.com`) have become increasingly popular over the past several years. Given the success and popularity of such social networking sites, building the CyberSKA portal on top of a social networking platform seemed ideal for enabling collaboration between astronomers and in providing them transparent access to data and computing resource/services. The CyberSKA portal is built on top of Elgg (`http://elgg.org`), an open source social networking platform. Elgg provides many Facebook like features such as contacts, groups, messaging, discussions, activity feeds, blogs, bookmarking, tags, wikis, media/document sharing and so forth. The CyberSKA portal was initially adapted from GeoChronos (`http://geochronos.org/`), another Elgg-based portal targeting the Earth observation science community, and has been significantly customized and extended since. A sample screenshot from the CyberSKA portal depicting the collaborative space made available to a group, in this case a group for members working on the CyberSKA project itself, can be seen in Figure 2.

Figure 2.   Screenshot of example group page on CyberSKA portal.

Various data management, processing, visualization and other services are being integrated with the CyberSKA portal. On the data front, a distributed data management system has been developed (Mahadevan & Rosolowsky 2011). This system is built on top of (Integrated Rule-Oriented Data System) (`http://www.irods.org/`), a tool designed to efficiently manage, move and replicate data between different sites. Associated with the data management system is a query service with a RESTful API that supports spatial, temporal, and spectral queries of FITS images. On the processing side of things, a workflow service allows users to segment, mosaic, extract and compress FITS images retrieved from the query service. There are plans to extend the workflow service to also support image convolution, object identification, spectral analysis, and more. In the area of visualization an on-line tool that enables users to visualize multi-dimensional FITS images directly from the portal has been developed (Federl et al. 2011). It supports interactive panning and zooming, histogram correction, color map adjustments, display of pixel value, region statistics, multiple coordinate systems, grids and selection of frame for multi-dimensional images. A similar visual tool for selecting a region of data to download, with associated parameters, is also available. Furthermore, a third party application API has been established that currently supports single sign-on to remotely hosted applications from the portal using OAuth.

3.   CyberSKA Portal Usage

The portal is currently being used by over 100 members from around the world that are part of communities associated with existing SKA Pathfinder projects such as GAL-FACTS and PALFA. Members are using the portal to establish contact and communicate with each other, establish groups around projects and software, manage personal and project related information/documents/discussions, and visualize FITS files. GAL-FACTS members are also able to select and download GALFACTS survey data, and PALFA members are able to access a variety of useful third party applications.

4.   Next Steps

While much has been accomplished since the CyberSKA project got started earlier this year, there is still much more to do. We plan to enhance and expand the distributed data management system which is currently running at two sites. We will also be establishing cloud computing environments at participating sites to dynamically launch different services and processing tasks based on user demand. A focus on the visualization side of things will be to improve scalability, by taking the currently client-based viewer tool, and providing more server side support. To allow other groups to more easily integrate tools and services with CyberSKA, we will be extending the third party application API to better support communications between applications and the portal. In all of these aspects, transparency to the user and collaboration capabilities will be a high priority.

Acknowledgments.   Funding for CyberSKA is provided by CANARIE, Canada's Advanced Research and Innovation Network, as part of their Network Enabled Platform program. Additional funding for CyberSKA is provided by Cybera, a non-profit organization with a mandate to integrate, leverage and sustain investments in cyberinfrastructure technologies in Alberta, Canada.

References

Federl, P., Grimstrup, A., Kiddle, C., & Taylor, A. R. 2011, in Astronomical Data Analysis Software and Systems XX, edited by I. N. Evans, A. Accomazzi, D. J. Mink, & A. H. Rots (San Francisco, CA: ASP), vol. 442 of ASP Conf. Ser., 467
Mahadevan, V., & Rosolowsky, E. 2011, in Astronomical Data Analysis Software and Systems XX, edited by I. N. Evans, A. Accomazzi, D. J. Mink, & A. H. Rots (San Francisco, CA: ASP), vol. 442 of ASP Conf. Ser., 219

Astronomical Data Analysis Software and Systems XX
ASP Conference Series, Vol. 442
Ian N. Evans, Alberto Accomazzi, Douglas J. Mink, and Arnold H. Rots, eds.
©*2011 Astronomical Society of the Pacific*

Visualization and Logical Binding of Hyperspectral Data Using QuickViz and SAADA

Matthieu Petremand,[1] Laurent Michel,[2] and Mireille Louys[1]

[1]*LSIIT, UMR CNRS 7005, Bd Sébastien Brant, BP 10413, 67412 Illkirch CEDEX, FRANCE*

[2]*Observatoire de Strasbourg, UMR CNRS 7550, 11 rue de l'université, 67000 Strasbourg, FRANCE*

Abstract. Next generation integral-field spectrographs (IFS) such as MUSE will record large hyperspectral cubes and thus generate a huge amount of observational and interpretable data. Their visualization remains a problematic and crucial task before any processing. In addition, observation parameters such as variance values may also be associated to each cube and be involved in the scientific interpretation of the data. In this paper, we propose a new software named QuickViz that provides a set of basic and advanced features enabling the exploration of such hyperspectral images with their related observation parameters. This new software is designed in Java as an Aladin plug-in and thus extends its hyperspectral functionalities while managing all the interactions between cubes and extracted spectra. Moreover, the specifications of the SAADA application allowing to automatically generate astronomical relational databases give the ability to easily explore such large data sets with respect to their complex relationships.

1. Introduction

Future deep-field surveys carried out by next generation integral-field spectrographs will produce a high number of heterogeneous astronomical data sets: hyperspectral cubes, observation parameters (e.g., variance values, PSF widths, data quality flags) and metadata. For instance, the size of a reconstructed MUSE (Laurent et al. 2006) observation with its associated variance cube is about 2.4 GB. Since the amount of data to be analyzed grows, the development of dedicated exploration and visualization tools becomes a mandatory goal. Indeed, new developed software must be designed to ease the joint analysis of such large data sets or the benefit of hyperspectral imagery will be lost. Moreover, additional parameters associated with an observation should no longer be ignored throughout the analysis process as they can give relevant clues for further investigations. For example, variance values and PSF reflect the data quality and can thus be taken into account to check local data quality and select regions of interest. We then propose a new software tool, QuickViz (Petremand et al. 2010), especially dedicated to the visualization of large hyperspectral images together with their associated parameters (see § 2). This innovative tool can be jointly used with a SAADA repository (Michel et al. 2005), automatically designing and populating an astronomical relational database with a set of FITS or VOTable files containing either tables or images. Queries can thus be sent from a web interface to the database so as to select data sets on specified constraints. SAADA's facilities are illustrated in § 3 with a repository housing simulated hyperspectral MUSE data sets. The conclusion and perspectives for further development are given in § 4.

2.   Hyperspectral Visualization with QuickViz

Currently existing software such as Aladin (Bonnarel et al. 2000) or GAIA3D[1] offer well-designed hyperspectral visualization features but don't totally provide a framework for IFS data exploration. For instance, GAIA3D cannot handle more than one cube at the same time whereas Aladin is mainly dedicated to 2D visualization and thus lacks spectral functionalities. Instead of starting the development of a new tool from scratch, we have rather chosen to extend Aladin as it already provides plug-in capabilities, a cross-platform architecture (written in Java) and optimized memory management (up to 3 cubes of 1.2 GB each can be loaded with 1.2 GB of memory). Figure 1 presents the main features of the Aladin/QuickViz duo[2] (circled numbers on Fig. 1 are refered to those in italic font in the following paragraph).

Data extraction from loaded hyperspectral cubes (*1*) can be carried out at each spatial position (*2*) or by computing an averaged spectrum over a circle shaped area (*3*). All newly extracted spectra are stacked in QuickViz (*4*) and available for display, superposition and comparison on spectrum panels (*5*). Users can easily interact with these panels to: browse selected cube's frames thanks to a calibrated cursor (*6*), define selections over spectral ranges (*7*) and use intuitive zoom features (*8*). Several customizable visualization modes can be enabled for data display (*9*) and spectra can be shared between panels thanks to multiview (synchronized or not) capability (*10*). Variance values can be displayed together with data along the spectral axis with the help of two visualization modes: error covering and error bars (*11*). Simple visualization algorithms can also be executed either on frames or spectra and produced outputs such as mean, weighted mean or RGB colored composition (*12*) be used to refine the data analysis or to act as guide images. New algorithms or visualization modes can be easily developed (in Java) and added to QuickViz to fit users' needs. Mainly focused on quick view and advanced visualization, QuickViz can pass selected data on to dedicated IVOA tools (such as VOSpec or SPLAT) for further data analysis via the IVOA SAMP protocol (still in progress).

3.   Logical Binding with SAADA

To illustrate the ability of SAADA[3] to deal with heterogeneous astronomical collections, an online example database has been created (Fig. 2) and filled with 9 simulated MUSE data sets each composed of: a raw image Y, PSF related to each pixel p of Y, a pixtable mapping sensor locations to sky positions and also containing data quality flags and noise variances for each p, a hyperspectral data cube reconstructed from Y with its associated variance cube. A web interface[4] allows a user to send queries on the

---

[1]http://star-www.dur.ac.uk/~pdraper/gaia/gaia3d/index.html

[2]http://lsiit-miv.u-strasbg.fr/paseo/cubevisualization.php

[3]http://saada.u-strasbg.fr

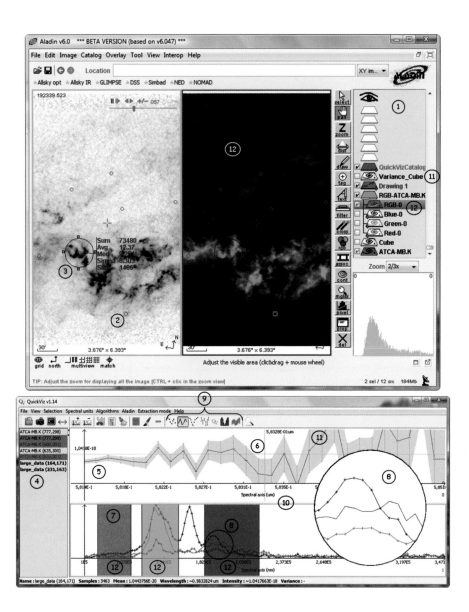

Figure 1.   Aladin/QuickViz duo in action (both in multiview mode). QuickViz (bottom window) takes care both of spectral side and interactions with the spatial domain (e.g., 2D visualization, catalogs, database queries, calibrations) whose management is totally delegated to Aladin (top window). QuickViz, developed in Java, can be easily installed on different operating systems (32-bit or 64-bit architectures) and used to visualize any kind of hyperspectral images if they follow the standard FITS format.

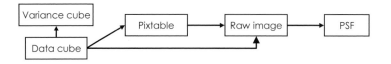

Figure 2.   Database with qualified links generated by SAADA from simulated MUSE data sets. The data cube is notably linked to its pixtable and the raw image from which it has been computed through a reconstruction (resampling) step.

database so as to highlight relevant information. For instance, data cubes can be constrained with data quality flags (*e.g.* number of dead or hot pixels) as well as variance values and output results can be exported to Aladin and QuickViz for visualization. The use of such a SAADA database can greatly improve data exploration and selection, especially in the case of large deep-field surveys where GB of data are nightly recorded.

4.   Conclusion and Perspectives

The size and the complexity of hyperspectral images has greatly increased during the last few years and new analysis tools must be designed to address the management of such a large amount of data. Besides a new hyperspectral visualization tool, we proposed, in this paper, an application of the SAADA software to simulated MUSE data sets. The visualization of observation parameters has been partially solved and current work concern displaying PSF and variance values across the field of view. The latter is performed through a video where the intensity of pixels varies with corresponding variance values, *i.e.* higher variance values induce higher and quicker intensity variations (Petremand et al. 2010). Advanced interactions between Aladin and SAADA repositories are also under discussion so as to directly interrogate SAADA databases from Aladin by providing server definitions in additional GLU mark files.

Acknowledgments.   This work was partially funded by the French Research Agency (ANR) as part of the DAHLIA project (*grant #ANR-08-BLAN-0253*). We thank Roland Bacon and the MUSE project team at CRAL in LYON, for providing the simulated data sets used to develop and train the tools.

References

Bonnarel, F., et al. 2000, A&AS, 143, 33
Laurent, F., Henault, F., Renault, E., Bacon, R., & Dubois, J. P. 2006, PASP, 118, 1564
Michel, L., Nguyen, H. N., & Motch, C. 2005, in Astronomical Data Analysis Software and Systems XIV, edited by P. Shopbell, M. Britton, & R. Ebert (San Francisco, CA: ASP), vol. 347 of ASP Conf. Ser., 71
Petremand, M., Louys, M., Collet, C., Mazet, V., Jalobeanu, A., & Salzenstein, F. 2010, in Proceedings of Astronomical Data Analysis VI, to appear (Monastir, Tunisia)

---

[4]http://saada.u-strasbg.fr/MUSE

Part XVI

Focus Demonstrations

Astronomical Data Analysis Software and Systems XX
ASP Conference Series, Vol. 442
Ian N. Evans, Alberto Accomazzi, Douglas J. Mink, and Arnold H. Rots, eds.
© *2011 Astronomical Society of the Pacific*

Software Package for Solar System Objects Astrometry

A. López García[1,2]

[1]*Astronomical Observatory, University of Valencia, Spain*

[2]*Astronomy and Astrophysics Department, University of Valencia, Spain*

Abstract.   Algorithms for analyzing CCD images in asteroid and satellite astrometric observations have been developed based on the software developed and applied previously to photographic plates. Specific software packages presented are: ephemeris of minor planets and satellites; stellar maps presentation; frame data files; measuring and reduction of CCD fields, residual of observations; and detection of moving objects. The software can be adapted to specific needs of other users.

1.   Ephemerides and Orbit Calculations

Minor Planets to be observed one night can be obtained now from the Minor Planet Center (MPC). As our work goes back 25 years, we have developed our own algorithms. Preliminary, improved and fitted orbits are considered and applied to our observations. Circular and elliptical orbits for a few observations allow for their analysis in "real time." Fitted orbits for all observations provide information about observing quality.

2.   Stellar Map

A stellar map of the observed are of the sky is shown side by side with a CCD frame, allowing for pointing corrections prior to acquiring the final CCD frame. The star catalogs used for this task are the USNO SA2.0 and the GSC++ (a modified version of the GSC developed at the Valencia Observatory).

3.   Observation Frame

For each CCD frame we generate a "frame file" (López García 2005) with observation data and stars coordinates and magnitudes around the center of the frame field. The frame file is used to perform measurements on the image. We make use of the USNO SA2.0 catalog and support both BMP and JPG image formats. The CCD frame name encodes information about the object and epoch of observation. For example, the N00238091210U233703+20F41.bmp frame name encodes the following observational data:

```
N U P: numbered / unnumbered asteroid / planet
00238: asteroid number
091210: yymmdd, date of observation
```

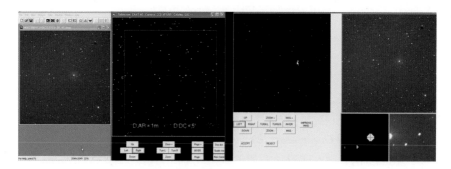

Figure 1.   Map and file of CCD frame.

```
U S W: universal / summer / winter time
233703: hhmmss, initial local time of observation
+20: observing interval, in seconds
F: used in case of filter
4: telescope (observatory) used. 40 cm telescope
1 2: order of observation for one object
```

The initial time of observation is obtained from the FITS header of the CCD frame.

4.   Detection of Moving Objects

We have developed algorithms to distinguish asteroids in sidereal and differential track-ing frames. For two frames with a time interval of a few minutes, it is possible to detect or identify the moving object by comparing both frames. Meteor traces are also de-tected with special algorithms in single and double frames.

5.   Tracking and Asteroid Detection

Depending on the tracking system, we get different stellar and objects images. From that, object identification can be obtained automatically. In sidereal tracking, two meth-ods for identification are available: Object trace versus stellar images and comparison of two frames. In this case the following steps are performed:

(1) Detection by field sweeping.
(2) When an object trace is visible, it is compared with round stellar images.
(3) In frames without traces, two fields are compared.

For differential tracking, we have similarly:

(1) Object image versus stellar traces.
(2) Comparison of two frames.

The steps involved in this process are:

(1) Object detection by sweeping and elimination by model application.

(2) Object discrimination.

(3) Object centering.

In both cases, the main details of detection process are:

(1) Contour detection of a group of images.

(2) Detection and elimination of one object by model application.

(3) Steps (1) and (2) are applied until all objects in the group are detected.

(4) Previous steps are applied for each new group of objects.

(5) Images not fitting the model are eliminated.

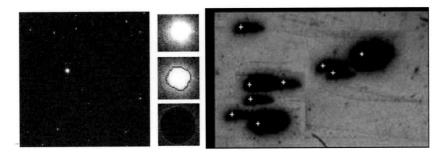

Figure 2.   Object detection and differential motion.

6.   Measurement and Reduction of CCD Fields

Measurement is performed automatically and can be complemented by a manual step. After image detection, the corresponding catalog stars are fitted. The process for satellite detection is similar and the main satellites can be selected for each planet. The steps involved are:

(1) CCD frame and stellar map are shown.

(2) Measuring windows are selected one after another.

(3) Automatic or manual detection of objects in the frame.

(4) Identification of objects in both frame and map.

(5) Fitting of coordinates and magnitude of asteroid.

7.   Residual of Observations

Minor planet observations are checked after fitting. By integrating the orbit we get precise ephemeris for the epoch of observation. Observations with residuals less than a given value are accepted and sent to the MPC.

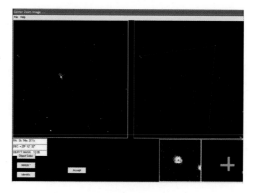

Figure 3.   Final steps of CCD field measurement.

8. Two Field Comparison

When two frames of one field within some time interval are available, we can compare both and detect minor changes between them (meteor traces, moving asteroid; López García 2008a,b). The main steps of this process are:

(1) Object detection and centering by sweeping.
(2) Identification of equivalent triangles.
(3) Reconstruction of working field.
(4) Blinking of fields.
(5) Field subtraction.

9. Final Remarks

The software described here and in López García & Moraño (2004) has been developed in Fortran Power Station and Visual Basic 6.0 under Windows XP, and can easily be adapted. The author is happy to provide source code and assistance to interested parties.

Acknowledgments.   The author thanks the Valencia Astronomical Observatory staff for financial help covering travel expenses and meeting registration.

References

López García, A. 2005, in Astrometry in the Age of the Next Generation of Large Telescopes, edited by P. K. Seidelmann & A. K. B. Monet (San Francisco, CA: ASP), vol. 338 of ASP Conf. Ser., 111
— 2008a, in Mutual Events of the Urania satellites in 2007-2008 and further observations in network, edited by J.-E. Arlot, N. Emelianov, & W. Thuillot (Paris: IMCCE), vol. 1 of IMCCE Publ., 67
— 2008b, in Astronomical Data Analysis Software and Systems XVII, edited by R. W. Argyle, P. S. Bunclark, & J. R. Lewis (San Francisco, CA: ASP), vol. 394 of ASP Conf. Ser., 535
López García, A., & Moraño, J. A. 2004, in Astronomical Data Analysis Software and Systems XIII, edited by F. Ochsenbein, M. G. Allen, & D. Egret (San Francisco, CA: ASP), vol. 314 of ASP Conf. Ser., 547

Astronomical Data Analysis Software and Systems XX
ASP Conference Series, Vol. 442
Ian N. Evans, Alberto Accomazzi, Douglas J. Mink, and Arnold H. Rots, eds.
© *2011 Astronomical Society of the Pacific*

Aladin: An Open Source All-Sky Browser

Thomas Boch, Anaïs Oberto, Pierre Fernique, and François Bonnarel

CDS, Observatoire astronomique de Strasbourg, Université de Strasbourg, CNRS, 11 rue de l'Université, 67000 Strasbourg, France

Abstract. Aladin, developed over 10 years at CDS, has evolved from a simple sky atlas to become a rich and powerful portal able to access, visualize and manipulate images and catalog data. Aladin is widely used in the Virtual Observatory community and beyond. Version 7 of Aladin is open source, and recent developments have been focused on enabling all-sky browsing targeted towards real scientific usage. Hierarchical, multi-resolution image surveys, density maps and catalogs have been created for popular datasets (DSS and Sloan images; Simbad and 2MASS catalogs data, for instance). Users can also easily build their own all-sky sphere from a set of local FITS images and share it through a simple URL link. Aladin features can be extended by external plugins. An SED (Spectral Energy Distribution) plugin, combining fluxes extracted from calibrated images and fluxes in VizieR photometric catalogs has been developed and is available.

Introduction

The first version of Aladin was released in 1999 as a basic FITS image visualizer, able to superimpose catalog data. It has since undergone several major enhancements, including the capability to search and access to Virtual Observatory services (2003), the multi-view mode (2005), and the interoperability with other desktop tools through the SAMP[1] protocol (2008).

Aladin version 7 marks another important milestone. This new release allows access to all-sky data and enables seamless all-sky navigation, following the trend started by NASA World Wind, Google Sky and World Wide Telescope.

Strong emphasis has been on providing astronomers with a flexible tool allowing them to easily browse datasets of interest in an all-sky mode, as shown in the next section.

1. All-Sky Features

We have developed and implemented a technique, based on the HEALPix tesselation and fully described in Fernique et al. (2010), allowing one to build a multi-resolution hierarchical view for different kinds of datasets: images, HEALPix files, coverage maps, and catalogs.

---

[1]Simple Application Messaging Protocol: `http://ivoa.net/samp`

Once this view has been built, the user can easily navigate it by using an intuitive browsing based on zooming and panning.

1.1. Image Surveys

We have built some all-sky views of several popular image surveys, including the Digitized Sky Survey, 2MASS, GLIMPSE in multiple bands, and IRAS.

In order to get some fast network access time, image surveys are loaded by default in JPEG, restricting the range dynamic to 8 bits. Users interested in getting the real pixel values can switch to a (slower) mode which will let them retrieve the full dynamic from FITS files. Combined with a crop tool, this mode can be used to perform photometry measurements.

Users are not limited to image surveys published by the Aladin team. We provide a tool (Oberto et al. 2011) so that they can also create their own multi-resolution views from a set of local FITS images. Generated views can then be easily shared, by making them accessible through a Web server.

1.2. HEALPix FITS Data

Data generated by the Planck mission will be published in HEALPix FITS format. This will also be the case for other upcoming surveys.

Aladin now supports this format natively. Figure 1 shows how a WMAP file including polarization data is displayed in Aladin.

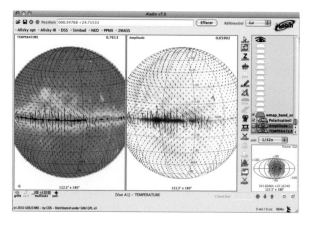

Figure 1.   Temperature and polarization amplitude, as displayed in Aladin. The polarization segments are overlaid in black.

1.3. Coverage Maps

Taking advantage of our all-sky mode, we generated some HEALPix coverage maps for large catalogs, such as 2MASS, DENIS, SDSS DR7 or USNO-B1 (see Figure 2). In such a map, the pixel value represents the actual number of sources in the given HEALPix cell.

Those maps are very useful to compare catalogs footprints and infer regions of common coverage.

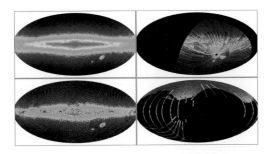

Figure 2.   From top left to bottom right: coverage maps of 2MASS, DENIS, USNO-B1 and SDSS DR7 catalogs, projected in AITOFF.

1.4.   Progressive Catalogs

Progressive display of catalogs are a new feature allowing one to browse very large catalogs without having to load all the sources beforehand. At the lower resolution, when the whole celestial sphere is displayed, we only load and display the more pertinent sources. Pertinence could be the flux of the sources (as for the 2MASS catalog) or the number of bibliographic references (for Simbad sources). As we zoom in, we load fainter and fainter sources, until all sources of the current field of view are displayed on the screen.

We applied this method to reprocess a number of large catalogs, including 2MASS, Tycho2, AKARI or PPMX, all available from the Aladin user interface.

2.   Other New Features

2.1.   Scatter Plots

2D scatter plots can be generated from any catalog loaded in Aladin. Figure 3 shows an example of linked views (Goodman 2010) in Aladin. The left panel displays the position of GLIESE sources overlaid on the SHASSA H-alpha survey, whereas the right panel shows a color-magnitude diagram for the same set of sources. Selecting a set of points in one of the views will be automatically reflected in the other one, allowing easy comparison of the spatial and parameter dimensions.

2.2.   Bookmarks

Common tasks can be easily stored in bookmarks and reused at will. Bookmarks are defined using Aladin script commands. Below is an example of a bookmark which will load an IRAS image for the current field of view and display the corresponding isocontours.

```
function IRASContour($TARGET, $RADIUS) {
   IRAS = get aladin(IRAS,60) $TARGET
   contour 6 nosmooth ; rm IRAS ; select @1
}
```

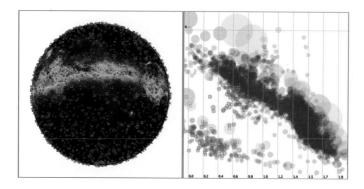

Figure 3.   Left panel: GLIESE sources overlaid on an H-alpha survey. Right panel: color-magnitude 2D plot for GLIESE sources.

3. SED Plugin

A plugin mechanism allows any software developer to take advantage of the Aladin framework in order to add new features.

An SED plugin has been developed internally at CDS.[2] It combines fluxes extracted from images and magnitudes or fluxes coming from catalogs, taking into account proper characterization of those data, to generate and display a SED plot.

4. Licensing and Download

Aladin is open source as version 7 and is released under the GPL 3 license.[3] We firmly believe that open sourcing Aladin will facilitate collaboration and enhance trust in our software as well as its long-term sustainability.

Aladin release 7 is available for download at http://aladin.u-strasbg.fr/java/nph-aladin.pl?frame=downloading.

References

Fernique, P., Boch, T., Oberto, A., & Bonnarel, F. 2010, in Astronomical Data Analysis Software and Systems XIX, edited by Y. Mizumoto, K.-I. Morita, & M. Ohishi (San Francisco, CA: ASP), vol. 434 of ASP Conf. Ser., 163

Goodman, A. A. 2010, in American Astronomical Society Meeting Abstracts #215, vol. 42 of BAAS, #230.03

Oberto, A., Fernique, P., Boch, T., & Bonnarel, F. 2011, in Astronomical Data Analysis Software and Systems XX, edited by I. N. Evans, A. Accomazzi, D. J. Mink, & A. H. Rots (San Francisco, CA: ASP), vol. 442 of ASP Conf. Ser., 609

---

[2]http://aladin.u-strasbg.fr/java/plugins/SedPlugin.jar

[3]http://www.gnu.org/licenses/gpl-3.0.html

Astronomical Data Analysis Software and Systems XX
ASP Conference Series, Vol. 442
Ian N. Evans, Alberto Accomazzi, Douglas J. Mink, and Arnold H. Rots, eds.
©*2011 Astronomical Society of the Pacific*

Advanced Python Scripting Using Sherpa

B. Refsdal, S. Doe, D. Nguyen, A. Siemiginowska, D. Burke, J. Evans, and
I. Evans

Smithsonian Astrophysical Observatory,
60 Garden Street, Cambridge, MA 02138, USA

Abstract. *Sherpa* is a general purpose modeling and fitting application written in
Python. The dynamism of Python allows *Sherpa* to be a powerful and extensible software package ready for the modern challenges of data analysis. Primarily developed
for the Chandra Interactive Analysis of Observations (CIAO) package, it provides a
flexible environment for resolving spectral and image properties, analyzing time series,
and modeling generic types of data. Complex model expressions are supported using *Sherpa's* general purpose definition syntax. *Sherpa's* parameterized data modeling
is achieved using robust optimization methods implementing the forward fitting technique. *Sherpa* includes functions to calculate goodness-of-fit and parameter confidence
limits. CPU intensive routines are written in C++/FORTRAN. But since all other data
structures are contained in Python modules, users can easily add their own data structures, models, statistics or optimization methods to *Sherpa*. We will introduce a scripted
example that highlights *Sherpa's* ability to estimate energy and photon flux errors using
simulations. The draws from these simulations, accessible as NumPy ndarrays, can be
sampled from uni-variate and multi-variate normal distributions and can be binned and
visualized with simple high level functions. We will demonstrate how *Sherpa* can be
extended with user-defined model and statistic classes written in Python. *Sherpa's* open
design even allows users to incorporate prior statistics derived from the source model.

1. Introduction

Sherpa[1] is CIAO's general purpose modeling and fitting package. It provides users with
an object-oriented Python API and a high-level user interface for powerful interactive
use and complex scripting. Users familiar with Python can easily import any of the
Sherpa procedural functions or API objects into an interactive session or access them
in a script. The focus demos described below show *Sherpa's* extensibility with complex
user scripts, including user-defined Python classes.

2. Estimate Flux Errors

The first example estimates the uncertainties on the integrated energy flux of an X-ray
spectral fit using simulations.[2] The spectral fit begins by reading in an X-ray spectrum

---

[1] http://cxc.harvard.edu/sherpa/

[2] http://pysherpa.blogspot.com/2010/11/estimate-flux-errors.html

in FITS format. (*Sherpa* supports multiple I/O packages with a generic I/O interface. By default, *Sherpa* uses the CIAO[3] Datamodel to read FITS and ASCII tables, but users are free to use *pyFITS*,[4] a common Python FITS reader.) Next, the spectrum is filtered and background subtracted to determine the net X-ray counts from 0.5 to 7.0 keV. The fit statistic is chosen to be χ^2 with data variance and the model is defined as an absorbed power-law. Lastly, the fit is performed with `fit()`.

```
>>> load_pha("3c273.pi")
>>> notice(0.5,7.0)
>>> subtract()
>>> set_stat("chi2datavar")
>>> set_model(xsphabs.abs1*xspowerlaw.p1)
>>> fit()
```

Sherpa includes functions to simulate the energy and photon flux given a defined spectrum and a model expression with parameters that are normally distributed. The energy flux simulation runs 100 samples from 0.5 to 7.0 keV and produces a table of integrated fluxes along with the set of parameter values used to calculate the given flux. Simple statistics can be run on the column of fluxes using the Python array package NumPy.[5] (By utilizing common third-party Python modules, users have access to *Sherpa* data for further analysis.)

```
>>> fluxtbl=sample_energy_flux(0.5,7.,num=100)
>>> fluxes=fluxtbl[:,0]
>>> sf=numpy.sort(fluxes)
>>> print "Mean:",numpy.mean(fluxes)
>>> print "Median:",numpy.median(fluxes)
>>> print "95\% quantile:",sf[0.95*(len(fluxes)-1)]
```

Using 7500 simulations, the energy flux distribution can be binned up into a histogram and fit with a simple un-normalized Gaussian model. The fitted parameter pos determines the centroid of the Gaussian while the `fwhm` parameter is converted to σ using $FWHM = \sigma * \sqrt{8 * ln(2)}$ to represent the spread. See Figure 1 for a plot of the energy flux distribution binned as a histogram and fit with a simple Gaussian.

```
>>> hist=get_energy_flux_hist(0.5,7.,num=7500)
>>> load_arrays(hist.xlo,hist.xhi,hist.y,Data1DInt)
>>> set_model(gauss1d.g0)
>>> g0.integrate=False
>>> set_stat("leastsq")
>>> fit()
>>> sigma=g0.fwhm.val/numpy.sqrt(8*numpy.log(2))
>>> print "Sigma: ",sigma,"Position: ",g0.pos.val
```

---

[3]http://cxc.harvard.edu/ciao/

[4]http://www.stsci.edu/resources/software_hardware/pyfits

[5]http://numpy.scipy.org/

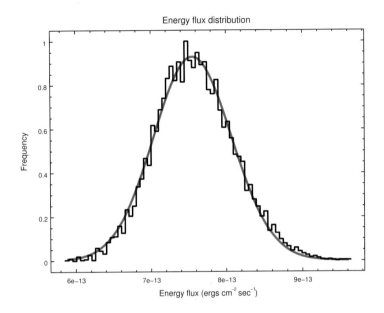

Figure 1.   Fitted Histogram of Energy Flux Distribution.

3.   X-ray Spectral Fit with a User-defined Model and Fit Statistic

The second example highlights *Sherpa's* extensible design which can support user-defined models and a user-defined fit statistic.[6] This X-ray spectral fit, includes a power-law model implemented as a user-defined Python class, `MyPowerLaw`. This power-law class inherits *Sherpa's* model interface from an arithmetic model class in the API and defines parameters in its constructor that represent the power-law index, reference, and normalization. The power-law class also implements class methods that calculate the power-law function over a grid of points and that integrate the function over a grid of bin edges. Each class method uses NumPy uFuncs to calculate the analytical function efficiently with loops that are compiled in C.

The fit statistic in this example is defined as a log-likelihood statistic with prior distributions for each of the model parameters. The log-likelihood, $\mathcal{L}$, is defined as $\mathcal{L} = \sum_i -model_i + data_i * ln(model_i) + P$ where P represents the associated prior distributions, *model* represents the predicted data, and *data* represents the observed data. *Sherpa* natively provides a user statistic class, that can be extended with a user-defined statistic function, however, this example highlights a high-level approach. The user can define a Python function, `calc_stat()`, that does include priors, and attach a reference to that function to the *Sherpa* user statistic class. See Figure 2 for a class diagram of the user-defined statistic class.

---

[6]http://pysherpa.blogspot.com/2010/11/user-defined-model-and-fit-statistic.html

User Statistics Class Diagram

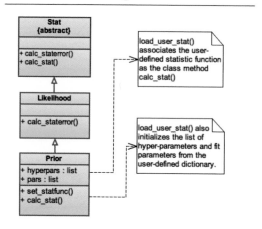

Figure 2.   Class Diagram of User-defined Fit Statistic with a Prior.

The calc_stat() function computes the log-likelihood with a prior given a reference to the statistic instance, the arrays of observed data values, predicted data values, statistical error, systematic error, and associated weights. The function then returns the statistic value and an array of statistic contributions per bin. Again, NumPy uFuncs are used to calculate vectorized mathematical expression efficiently. In the model, there are three thawed parameters for an absorbed power-law model that are assumed each to be normally distributed. The prior distribution P is calculated as the sum of the log of the parameter value sampled from a normal distribution.

The MyPowerLaw class is imported into a *Sherpa* session using the *Sherpa* function add_model(). And the user-defined function calc_stat() is attached to a *Sherpa* statistic class using the *Sherpa* function load_user_stat().

```
>>> add_model(MyPowerLaw)
>>> set_model(xsphabs.abs1*mypowerlaw.p1)
>>> pars=dict(mugamma=0.017, ..., ,alpha=p1.ampl)
>>> load_user_stat("stat",calc_stat,priors=pars)
>>> set_stat(stat)
```

Acknowledgments.   Support of the development of *Sherpa* is provided by National Aeronautics and Space Administration through the Chandra X-ray Center, which is operated by the Smithsonian Astrophysical Observatory for and on behalf of the National Aeronautics and Space Administration contract NAS8-03060.

Astronomical Data Analysis Software and Systems XX
ASP Conference Series, Vol. 442
Ian N. Evans, Alberto Accomazzi, Douglas J. Mink, and Arnold H. Rots, eds.
©*2011 Astronomical Society of the Pacific*

The SPIRE Photometer Interactive Analysis Package SPIA

Bernhard Schulz

California Institute of Technology,
MC100-22, 770 South Wilson Ave., Pasadena, CA 91125, USA

Abstract. The Herschel Common Science System (HCSS) is a substantial Java software package, accompanying the development of the Herschel Mission, supporting all of its phases. In particular, the reduction of data from the scientific instruments for instrument checkout, calibration, and astronomical analysis is one of its major applications. The data reduction software is split up into modules, called "tasks". Agreed-upon sequences of tasks form pipelines that deliver well defined standard products for storage in a web-accessible Herschel Science Archive (HSA). However, as astronomers and instrument scientists continue to characterize instrumental effects, astronomers already need to publish scientific results and may not have the time to acquire a sufficiently deep understanding of the system to apply necessary fixes. There is a need for intermediate level analysis tools that offer more flexibility than rigid pipelines. The task framework within the HCSS and the highly versatile Herschel Interactive Processing Environment (HIPE), together with the rich set of libraries provide the necessary tools to develop GUI-based interactive analysis packages for the Herschel instruments. The SPIRE Photometer Interactive Analysis (SPIA) package, described in this paper, proves the validity of the concept for the SPIRE instrument, breaking up the pipeline reduction into logical components, making all relevant processing parameters available in GUIs, and providing a more controlled and user-friendly access to the complexities of the system.

1. Pipeline Processing versus Interactive Analysis

Ideally, the raw telemetry data that is downlinked from an instrument aboard the Herschel spacecraft is processed in some automatic sequence of processing steps, inverting the transformation function of the instrument. The resulting fluxes for a given filter band and position on the sky are finally made available to the astronomer through the HSA (Leon et al. 2009). This is an easy and straightforward procedure. However, this approach offers no flexibility.

The other extreme is to edit and modify the pipeline script according to the needs of the particular data sets. This offers a maximum of flexibility and is usually needed by instrument experts. The need arises particularly during the early phases of a space observatory mission, when the modules that make up a data reduction pipeline aren't mature yet. Science quality results can also be derived in this way, but they require a substantial investment of time and effort on the side of the astronomer to gather the necessary expertise, learn the intricacies of the scripting language, and the contents of the available software libraries.

This is not an economic way for the general scientist who needs to limit the depth of his instrument involvement to a reasonable level. In this case, an interactive modular approach, providing guidance via GUIs, retains a limited amount of flexibility while avoiding the need to learn about the scripting language in depth.

The typical work pattern consists of loading data, inspecting it, and then reprocessing it with the newest calibration products and algorithms. After that, the data is inspected again and possibly processing parameters are changed before reprocessing another time. When the quality of the intermediate processing level is satisfactory, the analyst advances to the next major step in a similar iterative way. At the end, there are many results, some of which are to be saved for later.

This scenario results in certain requirements, such as simple data retrieval from the HSA, ability of easy data inspection, and the need to split the general workflow into smaller pipeline blocks. It is also necessary to have interactive access to the processing parameters via a GUI. Finally, the I/O of observational data must be simple and reformatting of output products for further processing by external astronomical applications should be provided.

2. Implementation

An early example of an interactive analysis for astronomical space missions, designed to provide guidance through GUIs to the astronomer, was shown by Gabriel et al. (1997) as an IDL implementation for the ISOPHOT instrument (Lemke et al. 1996). We describe here the implementation of the SPIRE Photometer Interactive Analysis (SPIA), which is based on the task framework of the Herschel Common Science System (Ott et al. 2006; Ott & Science Ground Segment Consortium 2010) and its interactive processing environment HIPE for the photometer part of the SPIRE instrument (Griffin et al. 2010) on board of Herschel (Pilbratt et al. 2010). The instrument was chosen because of the author's close familiarity with it, but the concept, as such, is applicable to the other Herschel instruments and sub-instruments as well.

The package resides in a Jython file including several classes, each representing a separate task. The task framework provides automatic GUIs for all input and output parameters that are defined in an initialization section in the code, thus avoiding time consuming SWING programming. The definitions include default values, variable type definitions, as well as text for GUI tooltips, contributing to user-friendliness.

The general structure is shown in Figure 1. The tasks are drawn in blue, data repositories are shown in orange, and products within the HIPE session are shown in yellow. Many viewer and data manipulation tools were already available in HIPE and can be used along with the new tasks. All tasks use the observation context as a handle, which is an object with pointers to all other data products the observation consists of. Tasks perform I/O of observations and calibration products, as well as the interactive processing to the levels 0.5, 1, and 2, representing i) data in engineering units, ii) flux-calibrated time streams with sky-positions, and iii) reconstructed sky maps respectively.

Figure 2 shows a screenshot of a HIPE session with a typical arrangement of certain views, showing the task GUI, the command line, variables, tasks, and the internal outline of an observation context. A number of viewers for signal timelines, flags, and map data are available that can not be shown here for lack of space.

An important improvement, as compared to prior implementations of the Interactive Analysis concept, is the implicit command line support of the task framework. Executing a task by hitting the "Accept" button in the GUI, will also create a command line that can be included into a Jython script to repeat the same reduction procedure on

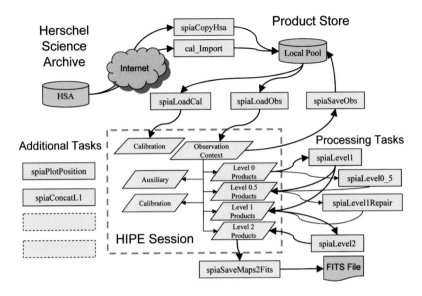

Figure 1.   Block diagram of the different components of the SPIA. Processing tasks that are controlled via GUIs are shown in blue, data repositories are in orange, and data products held within the HIPE session in memory are shown in yellow.

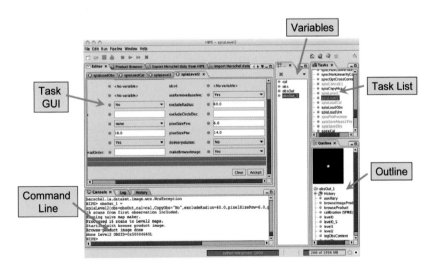

Figure 2.   Screenshot of a HIPE session in a typical arrangement of views (HIPE perspective) for the use with SPIA, showing the task GUI, command line view, variables view, task list view, and outline view. The outline view shows an observation context that includes a thumbnail of a Level 2 map.

other datasets. Thus the interactive analysis can be used as a pathfinder to optimize the data reduction, producing template scripts for later automatic bulk processing of larger datasets without obligating the astronomer to learn a lot about scripting.

A challenge will be to keep the package in sync with the still quite rapid development of the SPIRE pipeline and HIPE. A possible solution is the integration of the package with HIPE to benefit from the test harnesses that could point out any inconsistencies already during the build process.

The SPIA package is currently still distributed separately via the website of the NASA Herschel Science Center (NHSC) at IPAC/Caltech at `https://nhscsci.ipac.caltech.edu/sc/index.php/Spire/SPIA`. The package comes in two forms: 1) a Jython script that needs to be executed in HIPE first, in order to have all tasks of the SPIA available, 2) a HIPE plugin that needs to be installed only once, but is only compatible with version 5 of HIPE and above. HIPE can be downloaded from the website of the the the ESA Herschel Science Center (HSC) in Spain at `http://herschel.esac.esa.int/HIPE_download.shtml`. A user manual for SPIA is available as well at the NHSC site. The package is already being used successfully within the SPIRE instrument team and by some of the general users. An implementation of this concept for the spectrometer part of SPIRE is anticipated.

Acknowledgments. The *Herschel* Interactive Processing Environment (HIPE) is a joint development by the *Herschel* Science Ground Segment Consortium, consisting of ESA, the NASA Herschel Science Center, and the HIFI, PACS and SPIRE consortia. The author thanks the colleagues from the HSC, the NHSC, and the SPIRE ICC for valuable comments and suggestions. Special thanks go to Lijun Zhang, Dave Shupe, Annie Hoac, Paul Balm, Jaime Saiz, Javier Diaz, Jorgo Bakker, and Stephan Ott.

References

Gabriel, C., Acosta-Pulido, J., Heinrichsen, I., Morris, H., & Tai, W. 1997, in Astronomical Data Analysis Software and Systems VI, edited by G. Hunt & H. Payne (San Francisco, CA: ASP), vol. 125 of ASP Conf. Ser., 108
Griffin, M. J., et al. 2010, A&A, 518, L3
Lemke, D., et al. 1996, A&A, 315, L64
Leon, I., et al. 2009, in Astronomical Data Analysis Software and Systems XVIII, edited by D. A. Bohlender, D. Durand, & P. Dowler (San Francisco, CA: ASP), vol. 411 of ASP Conf. Ser., 438
Ott, S., & Science Ground Segment Consortium 2010, in American Astronomical Society Meeting Abstracts, vol. 216 of American Astronomical Society Meeting Abstracts, 413.10
Ott, S., et al. 2006, in Astronomical Data Analysis Software and Systems XV, edited by C. Gabriel, C. Arviset, D. Ponz, & S. Enrique (San Francisco, CA: ASP), vol. 351 of ASP Conf. Ser., 516
Pilbratt, G. L., et al. 2010, A&A, 518, L1

Author Index

Subject Index

ASTRONOMICAL SOCIETY OF THE PACIFIC

THE ASTRONOMICAL SOCIETY OF THE PACIFIC is an international, nonprofit, scientific, and educational organization. Some 120 years ago, on a chilly February evening in San Francisco, astronomers from Lick Observatory and members of the Pacific Coast Amateur Photographic Association—fresh from viewing the New Year's Day total solar eclipse of 1889 a little to the north of the city—met to share pictures and experiences. Edward Holden, Lick's first director, complimented the amateurs on their service to science and proposed to continue the good fellowship through the founding of a Society "to advance the Science of Astronomy, and to diffuse information concerning it." The Astronomical Society of the Pacific (ASP) was born.

The ASP's purpose is to increase the understanding and appreciation of astronomy by engaging scientists, educators, enthusiasts, and the public to advance science and science literacy. The ASP has become the largest general astronomy society in the world, with members from over 70 nations.

The ASP's professional astronomer members are a key component of the Society. Their desire to share with the public the rich rewards of their work permits the ASP to act as a bridge, explaining the mysteries of the universe. For these members, the ASP publishes the Publications of the Astronomical Society of the Pacific (PASP), a well-respected monthly scientific journal. In 1988, Dr. Harold McNamara, the PASP editor at the time, founded the ASP Conference Series at Brigham Young University. The ASP Conference Series shares recent developments in astronomy and astrophysics with the professional astronomy community.

To learn how to join the ASP or to make a donation, please visit http://www.astrosociety.org.

ASTRONOMICAL SOCIETY OF THE PACIFIC
MONOGRAPH SERIES
Published by the Astronomical Society of the Pacific

The ASP Monograph series was established in 1995 to publish select reference titles.
For electronic versions of ASP Monographs, please see
http://www.aspmonographs.org.

INFRARED ATLAS OF THE ARCTURUS SPECTRUM, 0.9-5.3μm
eds. Kenneth Hinkle, Lloyd Wallace, and William Livingston (1995)
ISBN: 1-886733-04-X, e-book ISBN: 978-1-58381-687-5

**VISIBLE AND NEAR INFRARED ATLAS
OF THE ARCTURUS SPECTRUM 3727-9300Å**
eds. Kenneth Hinkle, Lloyd Wallace, Jeff Valenti, and Dianne Harmer (2000)
ISBN: 1-58381-037-4, e-book ISBN: 978-1-58381-688-2

ULTRAVIOLET ATLAS OF THE ARCTURUS SPECTRUM 1150-3800Å
eds. Kenneth Hinkle, Lloyd Wallace, Jeff Valenti, and Thomas Ayres (2005)
ISBN: 1-58381-204-0, e-book ISBN: 978-1-58381-689-9

**HANDBOOK OF STAR FORMING REGIONS: VOLUME I
THE NORTHERN SKY**
ed. Bo Reipurth (2008)
ISBN: 978-1-58381-670-7, e-book ISBN: 978-1-58381-677-6

**HANDBOOK OF STAR FORMING REGIONS: VOLUME II
THE SOUTHERN SKY**
ed. Bo Reipurth (2008)
ISBN: 978-1-58381-671-4, e-book ISBN: 978-1-58381-678-3

A complete list and electronic versions of ASPCS volumes may be found at
http://www.aspbooks.org.

All book orders or inquiries concerning the ASP Conference Series, ASP
Monographs, or International Astronomical Union Volumes published by the ASP
should be directed to:

Astronomical Society of the Pacific
390 Ashton Avenue
San Francisco, CA 94112-1722 USA
Phone: 800-335-2624 (within the USA)
Phone: 415-337-2126
Fax: 415-337-5205
Email: service@astrosociety.org

For a complete list of ASP publications, please visit
http://www.astrosociety.org.